Small elastic deformations of thin shells

Monographs and textbooks on mechanics of solids and fluids

editor-in-chief: G. Æ. Oravas

Mechanics of surface structures

editor: W. A. Nash

1. P. Seide
   Small elastic deformations of thin shells
2. V. Panc
   Theories of elastic plates

Small elastic deformations of thin shells

Paul Seide

Professor of Civil Engineering
University of Southern California
Los Angeles, California
USA

NOORDHOFF INTERNATIONAL PUBLISHING LEYDEN

ISBN 90 286 0064 7
Library of Congress Catalog Card Number 73–94268

Printed in Great Britain by
William Clowes & Sons, Limited, London, Beccles and Colchester

To Joan Cecilia, Richard, and Wendy

Contents

Preface

In the last decade or so the theory of shells has undergone a tremendous increase in development. Formerly a subject of interest only to a few specialists and for which the literature was relatively small, the needs of structures for aerospace missions instigated a torrent of papers on all facets of the theory which also found application in the less glamorous earthbound shell structures important in everyday life. Some idea of the rapidity of the development can be gained from the fact that a bibliography* completed in 1953 listed some 1455 books and papers as the sum total of the literature on shell theory to that date. Three years later, however, a supplement† added another 884 papers to the list, an increase of 60 per cent in that short period of time. The number of papers published since these listings has increased to an extent that does not bear contemplation.

Obviously no single volume could contain all that constitutes the theory of shells and so this book is restricted to that portion of the theory associated with small deformations of elastic shells. Plastic deformations of shells, which is hardly developed, and nonlinear deformations and stability, which would require at least a separate volume, are thus excluded. Even with this restriction, however, the present volume represents a long overdue compromise between completeness and finiteness. In making this compromise I have undoubtedly omitted discussions of many topics and references to many excellent papers which should have been included.

The presentation of the material has evolved out of my own attempts to master the intricacies of shell theory while employed at Space Technology Laboratories (now TRW Systems Group) and, a little later, at Aerospace Corporation. In that context it represents my own view of shell theory and

* W. A. Nash, 'Bibliography on Shells and Shell-Like Structures', David W. Taylor Model Basin Report No. 863, 1954.

† W. A. Nash, 'Bibliography on Shells and Shell-Like Structures (1954–1956)', Univ. of Florida, Eng. and Ind. Exp. Station, Gainsville, Fla., June 1957.

what I would have liked to have been able to read when starting my own studies. For the most part the goal has been a consistent and logical presentation of shell theory with a minimal number of assumptions, as well as thorough coverage of the many available or possible solutions of the theory. The form of the book has been influenced by the pioneering volumes of S. Timoshenko on which the author was 'raised'. Experience in teaching some, though obviously not all, of the material in a one-semester course on the Theory of Shells at the University of Southern California has led to modifications of the text due to the questioning of my students and attempts to clarify various topics. Some of the topics discussed have not appeared elsewhere in the literature. Many that have been were revised to make them consistent with the approach and notation used herein. In some cases minor errors have been corrected. Most emphasis has been placed on the bending theory of shells and less on the membrane theory except as exact or approximate solutions of bending theory, since recent investigations tend to cast doubt on the general applicability of membrane theory. Descriptions of some modern numerical techniques for use in obtaining approximate solutions of the equations of the theory of shells are included at various points where it seemed appropriate to mention them. These numerical techniques require the use of digital computers having large memory capacity and detailed knowledge of numerical analysis. Since the lengthy programs involved are already available for the asking, they are not likely to be redone for economic reasons, so that all that is really required is a knowledge of the basis for them.

Some explanation is necessary in regard to the shell theory that is covered in the present volume. This is what is known as classical shell theory based on the Kirchhoff–Love assumptions of inextensible normals to an undeformed shell reference surface remaining normal to the deformed surface, as well as certain simplifications of the constitutive equations for the shell material. These assumptions are introduced here in the form of postulates about the shell material which relate the classical theory of shells to the theory of elasticity for curvilinearly orthotropic materials. The resultant theory is the most widely accepted and is of a form which corresponds to most intuitive notions about shell theory, namely that it should consist of only the six rigid body equations of equilibrium of a shell element involving the shell thickness and constitutive relations for force and moment stress resultants in terms of the six measures of deformation of the shell reference surface. These ideas are traditional and, of course, find their origins in Newtonian rigid body mechanics and the Bernoulli–Euler theory of beams. Where possible, exact solutions of the equations have been sought on the presumption that an exact solution of approximate equations is at least as good as an approximate solution of approximate equations.

There are other currents in shell theory, however, which are not discussed here. One of these is the question of how the classical theory of shells can be derived consistently from the theory of elasticity for *isotropic* materials. The answer to this question has beckoned countless investigators onward, as did the Holy Grail of the medieval Arthurian legends to the Knights of the Round Table. And just as the Grail was never seen very clearly, so does the answer to the question appear to be nebulous. The best that can be done appears to be the development of equations which are similar to, but not identical with, those of the classical theory of shells as the result of various asymptotic or perturbation expansions of the equations of the theory of isotropic elasticity. While very little has been done with these alternate shell equations, they do serve to lend credence to the results of the classical theory and can be used as the basis for more accurate solutions of various problems, if desired, as an answer to the other question of how to improve the classical theory of shells. Such improvements have, for the most part, been found to be unnecessary for thin shells under most loadings of interest.

Part I is devoted to a general discussion of the theory of shells. Vector notation rather than tensor notation is used to develop the theory. While tensor notation would have been more compact, much of the physical flavor of shell theory appears to be lost when it is used. The elements of the theory of surfaces which are needed are derived in Chapter 1. Some general results are given for arbitrary surfaces coordinates and are then reduced to the simpler forms for lines of curvature coordinates used in the remainder of the development of the theory of shells. Chapter 2 contains the fundamental equations of the theory of elasticity for shell coordinates, the basic assumptions of shell theory, and the consequences of these assumptions for isotropic shells of constant thickness. Approximations to the theory are discussed in Chapter 3.

Applications to axisymmetric deformations of shells of revolution are discussed in Part II. Detailed expositions of results for cylindrical, conical, and spherical shells are given respectively in Chapters 4 to 6. Chapter 7 is devoted to shells of revolution of arbitrary meridian. Finally, Chapter 8 include discussions of torsion and circumferential bending. Part III contains detailed analyses of asymmetric deformations of spherical and cylindrical shells in Chapters 9 and 10 and results for other shells in Chapter 11. Extensions of the theory to include thermal effects, anisotropy, variable thickness, and inhomogeneity such as multiple layers and closely spaced stiffening elements are covered in Chapter 12. Certain aspects of shell dynamics, including free vibrations and modal analysis, are treated in Chapters 13 and 14 which comprise Part IV. In my own course on shell theory, the topics treated in much, though not all, of Chapters 1–5, 7, 13, and 14 appear to be about as much as can be covered in one semester at the rate of three hours per week.

A book does not spring full blown from the author's mind but requires many hours, days, and even years of contemplation. Obviously the author's previous experiences, even in a technical book, influence its form. In looking back over the years I can discern the influences of many people to whom I owe a debt of gratitude I can never repay. Among these are my parents Julius and Sylvia Seide who instilled in me a love of knowledge, my wife Joan who provided the urging and encouragement to complete a volume so long in progress, and finally Eugene E. Lundquist, former Chief of the Structures Research Division of NACA (now NASA) at Langley Air Force Base, Va., who taught me and so many others that the questions which an engineering document should answer are

(a) What is the question?
(b) What was done about it?
(c) What was learned?

PAUL SEIDE

Palos Verdes Estates, California

Part I

The linear theory of shells

1

Geometry of surfaces

1.1 Geometric relations for surfaces

The general theory of thin shells draws so heavily on quantities associated with the geometry of surfaces that some prior knowledge of differential geometry is necessary. In this section, therefore, we shall consider definitions and derivations of relationships pertinent to an arbitrary surface. These will be utilized in the subsequent development of the theory.

A surface is defined by a relationship of the form, for a cartesian system of coordinates,

$$f(x, y, z) = 0, \tag{1.1.1a}$$

or, in terms of arbitrary curvilinear coordinates, λ and μ, by the relations

$$x = x(\lambda, \mu), \tag{1.1.1b}$$

$$y = y(\lambda, \mu), \tag{1.1.1c}$$

$$z = z(\lambda, \mu). \tag{1.1.1d}$$

Eq. (1.1.1b) can be used to define two families of curves on the surface (Fig. 1.1). Along each of the curves of one family, hereafter called λ curves, μ remains constant and λ varies, while along each curve of the other family, hereafter called μ curves, λ remains constant and μ varies. Let the radius vector to the surface be denoted by

$$\mathbf{r} = \mathbf{r}(\lambda, \mu) \tag{1.1.2}$$

and let us define the following scalar quantities

$$g_\lambda = \left(\frac{\partial \mathbf{r}}{\partial \lambda} \cdot \frac{\partial \mathbf{r}}{\partial \lambda}\right)^{1/2} = \left[\left(\frac{\partial x}{\partial \lambda}\right)^2 + \left(\frac{\partial y}{\partial \lambda}\right)^2 + \left(\frac{\partial z}{\partial \lambda}\right)^2\right]^{1/2}, \tag{1.1.3a}$$

$$g_\mu = \left(\frac{\partial \mathbf{r}}{\partial \mu} \cdot \frac{\partial \mathbf{r}}{\partial \mu}\right)^{1/2} = \left[\left(\frac{\partial x}{\partial \mu}\right)^2 + \left(\frac{\partial y}{\partial \mu}\right)^2 + \left(\frac{\partial z}{\partial \mu}\right)^2\right]^{1/2}, \tag{1.1.3b}$$

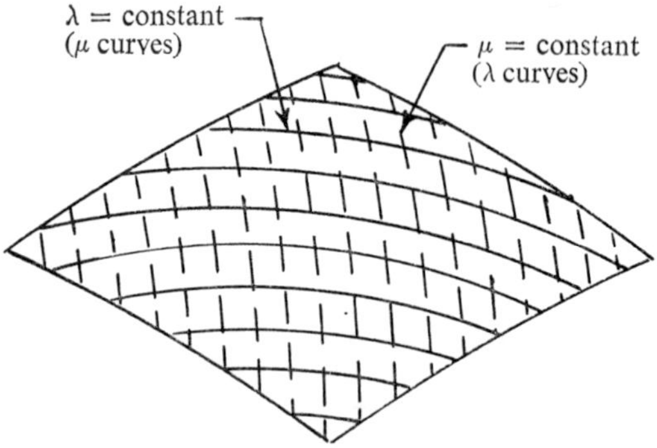

Fig. 1.1. Curvilinear coordinate system for surface

$$g_{\lambda\mu} = \frac{\partial \mathbf{r}}{\partial \lambda}\cdot\frac{\partial \mathbf{r}}{\partial \mu} = \frac{\partial x}{\partial \lambda}\frac{\partial x}{\partial \mu} + \frac{\partial y}{\partial \lambda}\frac{\partial y}{\partial \mu} + \frac{\partial z}{\partial \lambda}\frac{\partial z}{\partial \mu}. \tag{1.1.3c}$$

Then the unit vectors tangent to the surface in the direction of the λ and μ curves, respectively, are given by

$$\mathbf{e}_\lambda = \frac{1}{g_\lambda}\frac{\partial \mathbf{r}}{\partial \lambda}, \tag{1.1.4a}$$

$$\mathbf{e}_\mu = \frac{1}{g_\mu}\frac{\partial \mathbf{r}}{\partial \mu}. \tag{1.1.4b}$$

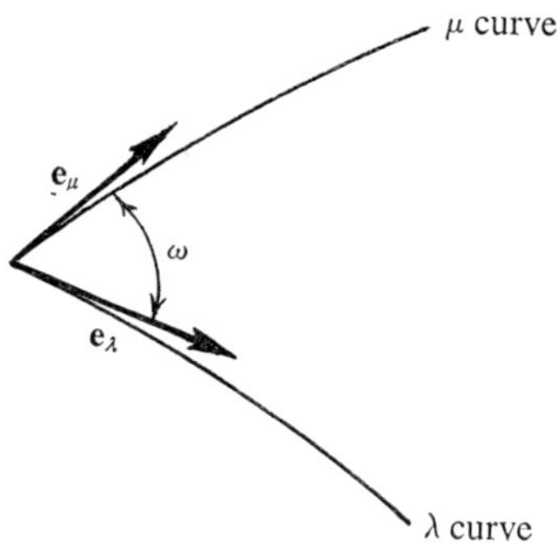

Fig. 1.2. Unit tangent vectors

The angle included between the two tangent vectors (Fig. 1.2), or the angle at which the λ and μ curves intersect, is determined from

$$\cos\omega = \mathbf{e}_\lambda\cdot\mathbf{e}_\mu = \frac{g_{\lambda\mu}}{g_\lambda g_\mu}. \tag{1.1.5}$$

Eq. (1.1.5) indicates that the magnitude of $g_{\lambda\mu}$ is always less than $g_\lambda g_\mu$ since the magnitude of cos ω is less than or equal to unity. The unit vector that is normal to the surface is given by

$$\mathbf{e}_\zeta = \frac{\mathbf{e}_\lambda \times \mathbf{e}_\mu}{\sin \omega}, \tag{1.1.6}$$

so that the unit vectors $\mathbf{e}_\lambda$, $\mathbf{e}_\mu$, and $\mathbf{e}_\zeta$ form a right-handed trihedron (Fig. 1.3).

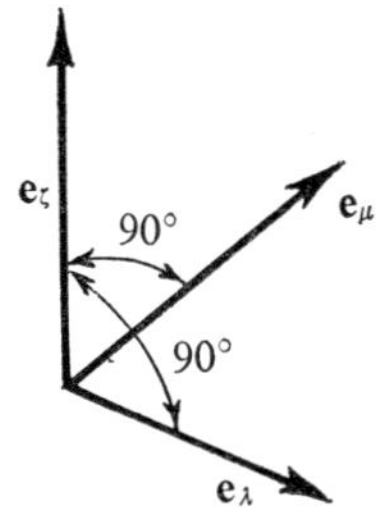

Fig. 1.3. Unit normal vector

The unit tangent and normal vectors satisfy the following vector relationships:

$$\mathbf{e}_\lambda \cdot \mathbf{e}_\lambda = \mathbf{e}_\mu \cdot \mathbf{e}_\mu = \mathbf{e}_\zeta \cdot \mathbf{e}_\zeta = 1, \tag{1.1.7a}$$

$$\mathbf{e}_\lambda \cdot \mathbf{e}_\zeta = \mathbf{e}_\zeta \cdot \mathbf{e}_\lambda = \mathbf{e}_\mu \cdot \mathbf{e}_\zeta = \mathbf{e}_\zeta \cdot \mathbf{e}_\lambda = 0, \tag{1.1.7b}$$

$$\mathbf{e}_\lambda \cdot \mathbf{e}_\mu = \mathbf{e}_\mu \cdot \mathbf{e}_\lambda = \cos \omega, \tag{1.1.7c}$$

$$\mathbf{e}_\lambda \times \mathbf{e}_\lambda = \mathbf{e}_\mu \times \mathbf{e}_\mu = \mathbf{e}_\zeta \times \mathbf{e}_\zeta = 0, \tag{1.1.7d}$$

$$\mathbf{e}_\lambda \times \mathbf{e}_\zeta = -\mathbf{e}_\zeta \times \mathbf{e}_\lambda = -\frac{\mathbf{e}_\mu - \mathbf{e}_\lambda \cos \omega}{\sin \omega}, \tag{1.1.7e}$$

$$\mathbf{e}_\mu \times \mathbf{e}_\zeta = -\mathbf{e}_\zeta \times \mathbf{e}_\mu = \frac{\mathbf{e}_\lambda - \mathbf{e}_\mu \cos \omega}{\sin \omega}. \tag{1.1.7f}$$

The scalar product relations follow from the definition of the scalar product and from the fact that $\mathbf{e}_\zeta$ is perpendicular to $\mathbf{e}_\lambda$ and $\mathbf{e}_\mu$. The first two lines of the vector product relations follow from the definition of the vector product and from Eq. (1.1.6). The last two lines of the vector product relations can be derived by noting that the vector product of two vectors is a vector perpendicular to their plane. Then, for example, the vector $\mathbf{e}_\lambda \times \mathbf{e}_\zeta$ is perpendicular to $\mathbf{e}_\lambda$ and $\mathbf{e}_\zeta$ and thus is a vector in the plane of $\mathbf{e}_\lambda$ and $\mathbf{e}_\mu$ perpendicular to $\mathbf{e}_\lambda$. It may then be expressed as a linear combination of $\mathbf{e}_\lambda$ and $\mathbf{e}_\mu$ as

$$\mathbf{e}_\lambda \times \mathbf{e}_\zeta = A\mathbf{e}_\lambda + B\mathbf{e}_\mu. \tag{1.1.8a}$$

Then the perpendicularity relationship yields

$$\mathbf{e}_\lambda \cdot (\mathbf{e}_\lambda \times \mathbf{e}_\zeta) = 0 = A + B \cos \omega. \tag{1.1.8b}$$

Also

$$\mathbf{e}_\mu \cdot (\mathbf{e}_\lambda \times \mathbf{e}_\zeta) = A \cos \omega + B = -\sin \omega, \tag{1.1.8c}$$

since the triple scalar product $\mathbf{e}_\lambda \times \mathbf{e}_\mu \cdot \mathbf{e}_\zeta$ is the volume of the parallelepiped defined by the three vectors. The solution of Eqs. (1.1.8b) and (1.1.8c) leads to Eq. (1.1.7e). A similar demonstration leads to Eq. (1.1.7f).

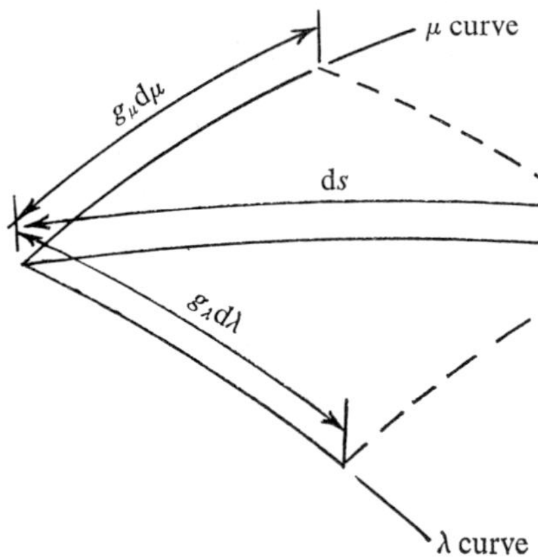

Fig. 1.4. Element of length

A differential element of length on the surface (Fig. 1.4) is given by

$$\mathrm{d}\mathbf{s} = \frac{\partial \mathbf{r}}{\partial \lambda} \mathrm{d}\lambda + \frac{\partial \mathbf{r}}{\partial \mu} \mathrm{d}\mu = g_\lambda \mathbf{e}_\lambda \, \mathrm{d}\lambda + g_\mu \mathbf{e}_\mu \, \mathrm{d}\mu, \tag{1.1.9a}$$

which has the scalar magnitude ds given by

$$\mathrm{d}s^2 = \mathrm{d}\mathbf{s} \cdot \mathrm{d}\mathbf{s} = (g_\lambda \, \mathrm{d}\lambda)^2 + 2(g_\lambda \, \mathrm{d}\lambda)(g_\mu \, \mathrm{d}\mu) \cos \omega + (g_\mu \, \mathrm{d}\mu)^2. \tag{1.1.9b}$$

Eq. (1.1.9b) is called the first fundamental form of the surface. We define the unit tangent vector in the direction of **ds** as

$$\mathbf{e}_s = \frac{\mathrm{d}\mathbf{s}}{\mathrm{d}s} = \frac{\partial \mathbf{r}}{\partial \lambda} \frac{\mathrm{d}\lambda}{\partial s} + \frac{\partial \mathbf{r}}{\partial \mu} \frac{\mathrm{d}\mu}{\mathrm{d}s} = g_\lambda \mathbf{e}_\lambda \frac{\mathrm{d}\lambda}{\mathrm{d}s} + g_\mu \mathbf{e}_\mu \frac{\mathrm{d}\mu}{\mathrm{d}s}. \tag{1.1.9c}$$

The surface area vector is defined by

$$\begin{aligned} \mathrm{d}\mathbf{A} &= (g_\lambda \mathbf{e}_\lambda \, \mathrm{d}\lambda) \times (g_\mu \mathbf{e}_\mu \, \mathrm{d}\mu) \\ &= (g_\lambda \, \mathrm{d}\lambda)(g_\mu \, \mathrm{d}\mu) \mathbf{e}_\zeta \sin \omega, \end{aligned} \tag{1.1.10a}$$

and, since $\mathbf{e}_\zeta$ is a unit vector, has a scalar magnitude equal to the area of a surface element:

$$dA = (g_\lambda \, d\lambda)(g_\mu \, d\mu) \sin \omega. \tag{1.1.10b}$$

The unit tangent vector, $\mathbf{e}_s$, and the unit normal vector, $\mathbf{e}_\zeta$, define a plane which is normal to the surface and which intersects the surface in a plane curve. The curvature vector of this plane curve at any point is defined as the rate of change of $\mathbf{e}_s$ along the curve and lies in the direction of the normal vector $\mathbf{e}_\zeta$; i.e.,

$$\mathbf{e}_s \cdot \frac{d\mathbf{e}_s}{ds} = \frac{1}{2} \frac{d}{ds} (\mathbf{e}_s \cdot \mathbf{e}_s) = 0, \tag{1.1.11}$$

implies that $d\mathbf{e}_s/ds$ is perpendicular to $\mathbf{e}_s$ and, since it lies in the plane of $\mathbf{e}_s$ and $\mathbf{e}_\zeta$, it must lie in the direction of $\mathbf{e}_\zeta$. The scalar magnitude of the curvature vector is called the normal curvature of the surface in the direction of $\mathbf{e}_s$ for, from Fig. 1.5, since $\mathbf{e}_s$ is a unit vector we have

$$d\phi = ds/R_s = |d\mathbf{e}_s|. \tag{1.1.12a}$$

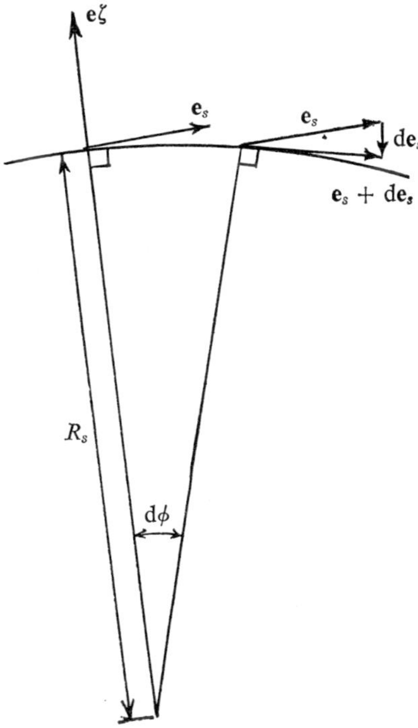

Fig. 1.5. Normal curvature vector

Then

$$\frac{1}{R_s} = -\mathbf{e}_\zeta \cdot \frac{\mathrm{d}\mathbf{e}_s}{\mathrm{d}s} = -\mathbf{e}_\zeta \cdot \frac{\mathrm{d}}{\mathrm{d}s}\left(\frac{\partial \mathbf{r}}{\partial \lambda}\frac{\mathrm{d}\lambda}{\mathrm{d}s} + \frac{\partial \mathbf{r}}{\partial \mu}\frac{\mathrm{d}\mu}{\mathrm{d}s}\right) = -\left(\mathbf{e}_\zeta \cdot \frac{\partial \mathbf{r}}{\partial \lambda}\frac{\mathrm{d}^2\lambda}{\mathrm{d}s^2} + \mathbf{e}_\zeta \cdot \frac{\partial \mathbf{r}}{\partial \mu}\frac{\mathrm{d}^2\mu}{\mathrm{d}s^2}\right)$$

$$-\left[\mathbf{e}_\zeta \cdot \frac{\partial^2 \mathbf{r}}{\partial \lambda^2}\left(\frac{\mathrm{d}\lambda}{\mathrm{d}s}\right)^2 + 2\mathbf{e}_\zeta \cdot \frac{\partial^2 \mathbf{r}}{\partial \lambda\, \partial \mu}\frac{\mathrm{d}\lambda}{\mathrm{d}s}\frac{\mathrm{d}\mu}{\mathrm{d}s} + \mathbf{e}_\zeta \cdot \frac{\partial^2 \mathbf{r}}{\partial \mu^2}\left(\frac{\mathrm{d}\mu}{\mathrm{d}s}\right)\right]^2. \qquad (1.1.12b)$$

The negative sign is included so that a convex surface with an outwardly directed normal has positive normal curvature (Fig. 1.5). The terms in the first pair of parentheses on the extreme right side of Eq. (1.1.12b) vanish since $\mathbf{e}_\zeta$ is perpendicular to $\partial\mathbf{r}/\partial\lambda$ and $\partial\mathbf{r}/\partial\mu$ (or $\mathbf{e}_\lambda$ and $\mathbf{e}_\mu$). If, further, we introduce the following definitions

$$\frac{1}{R_\lambda} = -\frac{1}{g_\lambda^2}\,\mathbf{e}_\zeta \cdot \frac{\partial^2 \mathbf{r}}{\partial \lambda^2}, \qquad (1.1.13a)$$

$$\frac{1}{R_\mu} = -\frac{1}{g_\mu^2}\,\mathbf{e}_\zeta \cdot \frac{\partial^2 \mathbf{r}}{\partial \mu^2}, \qquad (1.1.13b)$$

$$\frac{1}{R_{\lambda\mu}} = -\frac{1}{g_\lambda g_\mu}\,\mathbf{e}_\zeta \cdot \frac{\partial^2 \mathbf{r}}{\partial \lambda\, \partial \mu}, \qquad (1.1.13c)$$

Eq. (1.1.12b) may be written as

$$\frac{1}{R_s} = \frac{1}{R_\lambda}\left(\frac{g_\lambda\,\mathrm{d}\lambda}{\mathrm{d}s}\right)^2 + \frac{2}{R_{\lambda\mu}}\frac{g_\lambda\,\mathrm{d}\lambda}{\mathrm{d}s}\frac{g_\mu\,\mathrm{d}\mu}{\mathrm{d}s} + \frac{1}{R_\mu}\left(\frac{g_\mu\,\mathrm{d}\mu}{\mathrm{d}s}\right)^2 \qquad (1.1.13d)$$

which is called the second fundamental form of the surface. We note that the normal curvatures in the direction of the coordinate curves are $1/R_\lambda$ and $1/R_\mu$, respectively. The quantity $1/R_{\lambda\mu}$ is called the twist of the surface, which can be interpreted as the measure of the warping of an element of area. By considering the sum of the normal curvatures for any two perpendicular directions we can show that the quantity

$$H = \frac{1}{2\sin^2\omega}\left(\frac{1}{R_\lambda} + \frac{1}{R_\mu} - \frac{2\cos\omega}{R_{\lambda\mu}}\right), \qquad (1.1.14a)$$

is independent of the coordinate system. The invariant H is called the mean curvature. The Gaussian curvature of the surface at a point, the product of the maximum and minimum normal curvatures at a point, is given by

$$L = \frac{1}{\sin^2\omega}\left(\frac{1}{R_\lambda R_\mu} - \frac{1}{R_{\lambda\mu}^2}\right), \qquad (1.1.14b)$$

and can be shown to be an invariant as well.

With the foregoing definitions we can obtain expressions for the derivatives of the unit tangential and normal vectors in terms of the unit vectors themselves. These relations are especially useful in the development of the equations of the theory of shells of general shape. We introduce the following notation for a vector in terms of its three components in the directions of $\mathbf{e}_\zeta$ and the vectors perpendicular to $\mathbf{e}_\lambda$ and $\mathbf{e}_\mu$

$$\mathbf{a} = a^\lambda(\mathbf{e}_\lambda - \mathbf{e}_\mu \cos \omega) + a^\mu(\mathbf{e}_\mu - \mathbf{e}_\lambda \cos \omega) + a^\zeta \mathbf{e}_\zeta, \tag{1.1.15}$$

where, by virtue of Eqs (1.1.7), we have

$$a^\lambda = \mathbf{a}\cdot\mathbf{e}_\lambda \csc^2 \omega, \tag{1.1.16a}$$

$$a^\mu = \mathbf{a}\cdot\mathbf{e}_\mu \csc^2 \omega, \tag{1.1.16b}$$

$$a^\zeta = \mathbf{a}\cdot\mathbf{e}_\zeta. \tag{1.1.16c}$$

To obtain some of the components of $\mathbf{a}$ when $\mathbf{a}$ takes on the values

$$\frac{\partial \mathbf{e}_\lambda}{\partial \lambda}, \quad \frac{\partial \mathbf{e}_\lambda}{\partial \mu}, \quad \frac{\partial \mathbf{e}_\mu}{\partial \lambda}, \quad \frac{\partial \mathbf{e}_\mu}{\partial \mu}, \quad \frac{\partial \mathbf{e}_\zeta}{\partial \lambda}, \quad \frac{\partial \mathbf{e}_\zeta}{\partial \mu},$$

we make use of the identity

$$\frac{\partial}{\partial \lambda}\left(\frac{\partial \mathbf{r}}{\partial \mu}\right) = \frac{\partial}{\partial \mu}\left(\frac{\partial \mathbf{r}}{\partial \lambda}\right), \tag{1.1.17a}$$

which becomes, with the aid of Eqs (1.1.4),

$$g_\mu \frac{\partial \mathbf{e}_\mu}{\partial \lambda} + \mathbf{e}_\mu \frac{\partial g_\mu}{\partial \lambda} = g_\lambda \frac{\partial \mathbf{e}_\lambda}{\partial \mu} + \mathbf{e}_\lambda \frac{\partial g_\lambda}{\partial \mu}. \tag{1.1.17b}$$

The scalar product of Eq. (1.1.17b) with $\mathbf{e}_\lambda$ and $\mathbf{e}_\mu$ respectively then yields

$$\mathbf{e}_\lambda \cdot \frac{\partial \mathbf{e}_\mu}{\partial \lambda} = \frac{1}{g_\mu}\left(\frac{\partial g_\lambda}{\partial \mu} - \frac{\partial g_\mu}{\partial \lambda}\cos \omega\right), \tag{1.1.18a}$$

$$\mathbf{e}_\mu \cdot \frac{\partial \mathbf{e}_\lambda}{\partial \mu} = \frac{1}{g_\lambda}\left(\frac{\partial g_\mu}{\partial \lambda} - \frac{\partial g_\lambda}{\partial \mu}\cos \omega\right), \tag{1.1.18b}$$

when use is made of the relation for unit vectors

$$\mathbf{e}_\gamma \cdot \frac{\partial \mathbf{e}_\gamma}{\partial \delta} = \frac{1}{2}\frac{\partial}{\partial \delta}(\mathbf{e}_\gamma\cdot\mathbf{e}_\gamma) = \frac{1}{2}\frac{\partial(1)}{\partial \delta} = 0. \tag{1.1.19}$$

From Eqs. (1.1.7) and (1.1.18) we can obtain the following relations,

$$\begin{aligned}\mathbf{e}_\mu \cdot \frac{\partial \mathbf{e}_\lambda}{\partial \lambda} &= \frac{\partial}{\partial \lambda}(\mathbf{e}_\mu\cdot\mathbf{e}_\lambda) - \mathbf{e}_\lambda\cdot\frac{\partial \mathbf{e}_\mu}{\partial \lambda} \\ &= \frac{1}{g_\mu}\left[\frac{\partial}{\partial \lambda}(g_\mu \cos \omega) - \frac{\partial g_\lambda}{\partial \mu}\right],\end{aligned} \tag{1.1.20a}$$

$$\mathbf{e}_\lambda \cdot \frac{\partial \mathbf{e}_\mu}{\partial \mu} = \frac{\partial}{\partial \mu}(\mathbf{e}_\lambda \cdot \mathbf{e}_\mu) - \mathbf{e}_\mu \cdot \frac{\partial \mathbf{e}_\lambda}{\partial \mu}$$

$$= \frac{1}{g_\lambda}\left[\frac{\partial}{\partial \mu}(g_\lambda \cos \omega) - \frac{\partial g_\mu}{\partial \lambda}\right]. \qquad (1.1.20b)$$

Finally we derive with the aid of Eqs. (1.1.4), (1.1.7), and (1.1.13)

$$\mathbf{e}_\zeta \cdot \frac{\partial \mathbf{e}_\lambda}{\partial \lambda} = \mathbf{e}_\zeta \cdot \frac{\partial}{\partial \lambda}\left(\frac{1}{g_\lambda}\frac{\partial \mathbf{r}}{\partial \lambda}\right)$$

$$= \frac{1}{g_\lambda}\mathbf{e}_\zeta \cdot \frac{\partial^2 \mathbf{r}}{\partial \lambda^2} - \frac{1}{g_\lambda}\frac{\partial g_\lambda}{\partial \lambda}\mathbf{e}_\zeta \cdot \mathbf{e}_\lambda$$

$$= -\frac{g_\lambda}{R_\lambda} = -\mathbf{e}_\lambda \cdot \frac{\partial \mathbf{e}_\zeta}{\partial \lambda}, \qquad (1.1.21a)$$

$$\mathbf{e}_\zeta \cdot \frac{\partial \mathbf{e}_\lambda}{\partial \mu} = \mathbf{e}_\zeta \cdot \frac{\partial}{\partial \mu}\left(\frac{1}{g_\lambda}\frac{\partial \mathbf{r}}{\partial \lambda}\right)$$

$$= \frac{1}{g_\lambda}\mathbf{e}_\zeta \cdot \frac{\partial^2 \mathbf{r}}{\partial \lambda\, \partial \mu} - \frac{1}{g_\lambda}\frac{\partial g_\lambda}{\partial \mu}\mathbf{e}_\zeta \cdot \mathbf{e}_\lambda$$

$$= -\frac{g_\mu}{R_{\lambda\mu}} = -\mathbf{e}_\lambda \cdot \frac{\partial \mathbf{e}_\zeta}{\partial \mu}, \qquad (1.1.21b)$$

and similarly

$$\mathbf{e}_\zeta \cdot \frac{\partial \mathbf{e}_\mu}{\partial \lambda} = -\frac{g_\lambda}{R_{\lambda\mu}} = -\mathbf{e}_\mu \cdot \frac{\partial \mathbf{e}_\zeta}{\partial \lambda}, \qquad (1.1.21c)$$

$$\mathbf{e}_\zeta \cdot \frac{\partial \mathbf{e}_\mu}{\partial \mu} = -\frac{g_\mu}{R_\mu} = -\mathbf{e}_\mu \cdot \frac{\partial \mathbf{e}_\zeta}{\partial \mu}. \qquad (1.1.21d)$$

Then from Eqs. (1.1.15), (1.1.16), (1.1.18), (1.1.20), and (1.1.21) we find that the derivatives of the unit vectors are given by

$$\frac{\partial \mathbf{e}_\lambda}{\partial \lambda} = \frac{1}{g_\mu \sin^2 \omega}\left[\frac{\partial}{\partial \lambda}(g_\mu \cos \omega) - \frac{\partial g_\lambda}{\partial \mu}\right](\mathbf{e}_\mu - \mathbf{e}_\lambda \cos \omega) - \frac{g_\lambda}{R_\lambda}\mathbf{e}_\zeta, \quad (1.1.22a)$$

$$\frac{\partial \mathbf{e}_\lambda}{\partial \mu} = \frac{1}{g_\lambda \sin^2 \omega}\left(\frac{\partial g_\mu}{\partial \lambda} - \frac{\partial g_\lambda}{\partial \mu}\cos \omega\right)(\mathbf{e}_\mu - \mathbf{e}_\lambda \cos \omega) - \frac{g_\mu}{R_{\lambda\mu}}\mathbf{e}_\zeta, \quad (1.1.22b)$$

$$\frac{\partial \mathbf{e}_\mu}{\partial \lambda} = \frac{1}{g_\mu \sin^2 \omega}\left(\frac{\partial g_\lambda}{\partial \mu} - \frac{\partial g_\mu}{\partial \lambda}\cos \omega\right)(\mathbf{e}_\lambda - \mathbf{e}_\mu \cos \omega) - \frac{g_\lambda}{R_{\lambda\mu}}\mathbf{e}_\zeta, \quad (1.1.22c)$$

$$\frac{\partial \mathbf{e}_\mu}{\partial \mu} = \frac{1}{g_\lambda \sin^2 \omega}\left[\frac{\partial}{\partial \mu}(g_\lambda \cos \omega) - \frac{\partial g_\mu}{\partial \lambda}\right](\mathbf{e}_\lambda - \mathbf{e}_\mu \cos \omega) - \frac{g_\mu}{R_\mu}\mathbf{e}_\zeta, \quad (1.1.22d)$$

$$\frac{\partial \mathbf{e}_\zeta}{\partial \lambda} = \frac{g_\lambda}{\sin^2 \omega}\left[\frac{1}{R_\lambda}(\mathbf{e}_\lambda - \mathbf{e}_\mu \cos \omega) + \frac{1}{R_{\lambda\mu}}(\mathbf{e}_\mu - \mathbf{e}_\lambda \cos \omega)\right], \tag{1.1.22e}$$

$$\frac{\partial \mathbf{e}_\zeta}{\partial \mu} = \frac{g_\mu}{\sin^2 \omega}\left[\frac{1}{R_{\lambda\mu}}(\mathbf{e}_\lambda - \mathbf{e}_\mu \cos \omega) + \frac{1}{R_\mu}(\mathbf{e}_\mu - \mathbf{e}_\lambda \cos \omega)\right]. \tag{1.1.22f}$$

Eqs. (1.1.23) are called the derivative equations of Weingarten and Gauss.

A final group of equations which can be derived are the conditions of Codazzi and Gauss, which can be considered to be compatibility conditions for the first and second fundamental quantities. These can be obtained from the condition of continuity of the derivatives of the unit vectors $\mathbf{e}_\lambda$ and $\mathbf{e}_\mu$; i.e.,

$$\frac{\partial}{\partial \mu}\left(\frac{\partial \mathbf{e}_\lambda}{\partial \lambda}\right) = \frac{\partial}{\partial \lambda}\left(\frac{\partial \mathbf{e}_\lambda}{\partial \mu}\right), \tag{1.1.23a}$$

$$\frac{\partial}{\partial \mu}\left(\frac{\partial \mathbf{e}_\mu}{\partial \lambda}\right) = \frac{\partial}{\partial \lambda}\left(\frac{\partial \mathbf{e}_\mu}{\partial \mu}\right). \tag{1.1.23b}$$

With the aid of the derivative equations of Weingarten and Gauss, Eqs. (1.1.24) can be expressed in terms of vector components in the direction of $\mathbf{e}_\lambda$, $\mathbf{e}_\mu$, and $\mathbf{e}_\zeta$. These are of the form

$$A(\mathbf{e}_\mu - \mathbf{e}_\lambda \cos \omega) + B_1 \mathbf{e}_\zeta = 0, \tag{1.1.24a}$$

$$A(\mathbf{e}_\lambda - \mathbf{e}_\mu \cos \omega) + B_2 \mathbf{e}_\zeta = 0. \tag{1.1.24b}$$

Since each of the coefficients of the unit vectors must vanish for Eqs. (1.1.24) to be satisfied, the Codazzi–Gauss conditions are

$$A = B_1 = B_2 = 0, \tag{1.1.25}$$

or, when A, B_1, and B_2 are written out in full and simplified

$$\begin{aligned} &\frac{g_\lambda g_\mu}{\sin^2 \omega}\left(\frac{1}{R_\lambda R_\mu} - \frac{1}{R_{\lambda\mu}^2}\right) + \frac{1}{\sin \omega}\left\{\frac{\partial}{\partial \lambda}\left[\frac{1}{g_\lambda \sin \omega}\left(\frac{\partial g_\mu}{\partial \lambda} - \frac{\partial g_\lambda}{\partial \mu}\cos \omega\right)\right]\right. \\ &\quad \left. + \frac{\partial}{\partial \mu}\left[\frac{1}{g_\mu \sin \omega}\left(\frac{\partial g_\lambda}{\partial \mu} - \frac{\partial g_\mu}{\partial \lambda}\cos \omega\right)\right]\right\} - \frac{\cos \omega}{\sin^4 \omega}\frac{\partial \cos \omega}{\partial \lambda}\frac{\partial \cos \omega}{\partial \mu} \\ &\quad - \frac{1}{\sin^2 \omega}\frac{\partial^2 \cos \omega}{\partial \lambda\, \partial \mu} = 0, \end{aligned} \tag{1.1.26a}$$

$$\begin{aligned} &\frac{\partial}{\partial \mu}\left(\frac{g_\lambda}{R_\lambda}\right) - \frac{\partial}{\partial \lambda}\left(\frac{g_\mu}{R_{\lambda\mu}}\right) - \frac{1}{\sin^2 \omega}\left(\frac{1}{R_\mu} - \frac{\cos \omega}{R_{\lambda\mu}}\right)\left[\frac{\partial g_\lambda}{\partial \mu} - \frac{\partial (g_\mu \cos \omega)}{\partial \lambda}\right] \\ &\quad - \frac{1}{\sin^2 \omega}\left(\frac{1}{R_{\lambda\mu}} - \frac{\cos \omega}{R_\lambda}\right)\left(\frac{\partial g_\mu}{\partial \lambda} - \frac{\partial g_\lambda}{\partial \mu}\cos \omega\right) = 0, \end{aligned} \tag{1.1.26b}$$

$$\frac{\partial}{\partial\lambda}\left(\frac{g_\mu}{R_\mu}\right) - \frac{\partial}{\partial\mu}\left(\frac{g_\lambda}{R_{\lambda\mu}}\right) - \frac{1}{\sin^2\omega}\left(\frac{1}{R_\lambda} - \frac{\cos\omega}{R_{\lambda\mu}}\right)\left[\frac{\partial g_\mu}{\partial\lambda} - \frac{\partial(g_\lambda\cos\omega)}{\partial\mu}\right]$$

$$- \frac{1}{\sin^2\omega}\left(\frac{1}{R_{\lambda\mu}} - \frac{\cos\omega}{R_\mu}\right)\left(\frac{\partial g_\lambda}{\partial\mu} - \frac{\partial g_\mu}{\partial\lambda}\cos\omega\right) = 0. \quad (1.1.26c)$$

The satisfaction of Eqs. (1.1.26b) and (1.1.26c) implies that the derivatives of the unit normal vector are continuous since these equations can also be obtained from the relation

$$\frac{\partial}{\partial\mu}\left(\frac{\partial\mathbf{e}_\zeta}{\partial\lambda}\right) = \frac{\partial}{\partial\lambda}\left(\frac{\partial\mathbf{e}_\zeta}{\partial\mu}\right). \quad (1.1.27)$$

1.2 Lines of curvature

The relations of Section 1.1 are significantly reduced when the λ and μ coordinate curves are chosen to be a particular set of curves for the surface. These curves are called lines of curvature, the coordinates of which will be denoted by α and β, and are defined by the property that along the curves adjacent normals to the surface intersect, obviously at the normal center of curvature. A necessary and sufficient condition for this property to hold is that the unit normal vector, the change in the unit normal vector, and the tangent vector be coplanar (Fig. 1.6) which is represented by the equation

$$\mathbf{e}_\zeta \times \frac{d\mathbf{r}}{ds}\cdot\frac{d\mathbf{e}_\zeta}{ds} = 0. \quad (1.2.1)\dagger$$

Thus, along the α and β lines of curvature we must have, respectively,

$$\mathbf{e}_\zeta \times \frac{\partial\mathbf{r}}{\partial\alpha}\cdot\frac{\partial\mathbf{e}_\zeta}{\partial\alpha} = 0, \quad (1.2.2a)$$

$$\mathbf{e}_\zeta \times \frac{\partial\mathbf{r}}{\partial\beta}\cdot\frac{\partial\mathbf{e}_\zeta}{\partial\beta} = 0. \quad (1.2.2b)$$

Since the relations of Section 1.1 are valid for lines of curvature, the use of Eqns. (1.1.4), (1.1.7), and (1.1.23) in Eqns. (1.2.2) implies the following relationships at any point,

$$\frac{1}{R_{\alpha\beta}} - \frac{\cos\omega}{R_\alpha} = 0, \quad (1.2.3a)$$

† The vector product $\mathbf{e}_\zeta \times d\mathbf{r}/ds$ is a vector perpendicular to both $\mathbf{e}_\zeta$ and $d\mathbf{r}/ds$ by definition. The scalar product of this vector and $d\mathbf{e}_\zeta/ds$ being equal to zero indicates that $d\mathbf{e}_\zeta/ds$ is perpendicular to the vector product and hence lies in the plane of $\mathbf{e}_\zeta$ and $d\mathbf{r}/ds$.

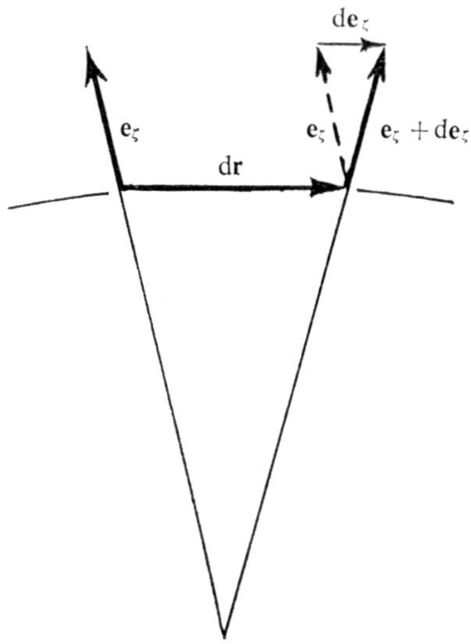

Fig. 1.6. Coplanar vectors

$$\frac{1}{R_{\alpha\beta}} - \frac{\cos\omega}{R_\beta} = 0. \tag{1.2.3b}$$

These can be considered to be a set of simultaneous homogeneous equations for $1/R_{\alpha\beta}$ and $\cos\omega$ which are then satisfied if, and only if,

$$\frac{1}{R_{\alpha\beta}} = \cos\omega = 0. \tag{1.2.4}$$†

Eq. (1.2.4) indicates that lines of curvature are orthogonal trajectories on the undeformed surface and, in addition, that the surface twist of the coordinate curves vanishes so that an element of area can be considered to be almost plane.

For the system of lines of curvature coordinates, the most important relationships for surfaces for the theory of shells, the Weingarten–Gauss derivative relations and the conditions of Codazzi and Gauss, reduce to

$$\frac{\partial \mathbf{e}_\zeta}{\partial \alpha} = \frac{g_\alpha}{R_\alpha}\,\mathbf{e}_\alpha, \tag{1.2.5a}$$

$$\frac{\partial \mathbf{e}_\zeta}{\partial \beta} = \frac{g_\beta}{R_\beta}\,\mathbf{e}_\beta, \tag{1.2.5b}$$

† An exception is the sphere for which any plane curve passing through a point on the sphere is a line of curvature. We shall restrict ourselves to coordinate systems on the sphere which satisfy Eq. (1.2.4).

$$\frac{\partial \mathbf{e}_\alpha}{\partial \alpha} = -\frac{1}{g_\beta}\frac{\partial g_\alpha}{\partial \beta}\mathbf{e}_\beta - \frac{g_\alpha}{R_\alpha}\mathbf{e}_\zeta, \tag{1.2.5c}$$

$$\frac{\partial \mathbf{e}_\alpha}{\partial \beta} = \frac{1}{g_\alpha}\frac{\partial g_\beta}{\partial \alpha}\mathbf{e}_\beta, \tag{1.2.5d}$$

$$\frac{\partial \mathbf{e}_\beta}{\partial \alpha} = \frac{1}{g_\beta}\frac{\partial g_\alpha}{\partial \beta}\mathbf{e}_\alpha, \tag{1.2.5e}$$

$$\frac{\partial \mathbf{e}_\beta}{\partial \beta} = -\frac{1}{g_\alpha}\frac{\partial g_\beta}{\partial \alpha}\mathbf{e}_\alpha - \frac{g_\beta}{R_\beta}e_\zeta, \tag{1.2.5f}$$

$$\frac{\partial}{\partial \alpha}\left(\frac{1}{g_\alpha}\frac{\partial g_\beta}{\partial \alpha}\right) + \frac{\partial}{\partial \beta}\left(\frac{1}{g_\beta}\frac{\partial g_\alpha}{\partial \beta}\right) + \frac{g_\alpha g_\beta}{R_\alpha R_\beta} = 0, \tag{1.2.6a}$$

$$\frac{\partial}{\partial \beta}\left(\frac{g_\alpha}{R_\alpha}\right) = \frac{1}{R_\beta}\frac{\partial g_\alpha}{\partial \beta}, \tag{1.2.6b}$$

$$\frac{\partial}{\partial \alpha}\left(\frac{g_\beta}{R_\beta}\right) = \frac{1}{R_\alpha}\frac{\partial g_\beta}{\partial \alpha}. \tag{1.2.6c}$$

The relationship between the lines of curvature and an arbitrary system of coordinates can be obtained from Eq. (1.2.1) which may be expanded to read

$$\mathbf{e}_\zeta \times \left(\frac{\partial \mathbf{r}}{\partial \lambda}\frac{\mathrm{d}\lambda}{\mathrm{d}s} + \frac{\partial \mathbf{r}}{\partial \mu}\frac{\mathrm{d}\mu}{\mathrm{d}s}\right)\cdot\left(\frac{\partial \mathbf{e}_\zeta}{\partial \lambda}\frac{\mathrm{d}\lambda}{\mathrm{d}s} + \frac{\partial \mathbf{e}_\zeta}{\partial \mu}\frac{\mathrm{d}\mu}{\mathrm{d}s}\right) = 0. \tag{1.2.7a}$$

Then, with the use of Eqs. (1.1.4), (1.1.7), and (1.1.23), we have

$$A(g_\lambda\,\mathrm{d}\lambda)^2 + B(g_\lambda\,\mathrm{d}\lambda)(g_\mu\,\mathrm{d}\mu) - C(g_\mu\,\mathrm{d}\mu)^2 = 0, \tag{1.2.7b}$$

where

$$A = \frac{1}{R_{\lambda\mu}} - \frac{\cos\omega}{R_\lambda}, \tag{1.2.8a}$$

$$B = \frac{1}{R_\mu} - \frac{1}{R_\lambda}, \tag{1.2.8b}$$

$$C = \frac{1}{R_{\lambda\mu}} - \frac{\cos\omega}{R_\mu}. \tag{1.2.8c}$$

The solutions of Eq. (1.2.7b), given by

$$\left(\frac{g_\lambda\,\mathrm{d}\lambda}{g\mu\,\mathrm{d}\mu}\right)_{\alpha,\beta} = \frac{\dfrac{1}{R_\lambda} - \dfrac{1}{R_\mu} \pm 2(H^2-L)^{1/2}\sin^2\omega}{2\left(\dfrac{1}{R_{\lambda\mu}} - \dfrac{\cos\omega}{R_\lambda}\right)}, \tag{1.2.9a}$$

define the perpendicular directions (Fig. 1.7) of the lines of curvature with respect to the arbitrary system of coordinates. We may write

$$\tan\phi = \frac{\sin\omega}{\cos\omega + (g_\lambda\,\mathrm{d}\lambda/g_\mu\,\mathrm{d}\mu)_\alpha}, \tag{1.2.9b}$$

where ϕ is the angle between a line of curvature and the λ axis (Fig. 1.7). The corresponding normal curvatures can be shown to be given by

$$\left.\begin{matrix}1/R_\alpha\\1/R_\beta\end{matrix}\right\} = H \pm (H^2 - L)^{1/2}. \tag{1.2.10}$$

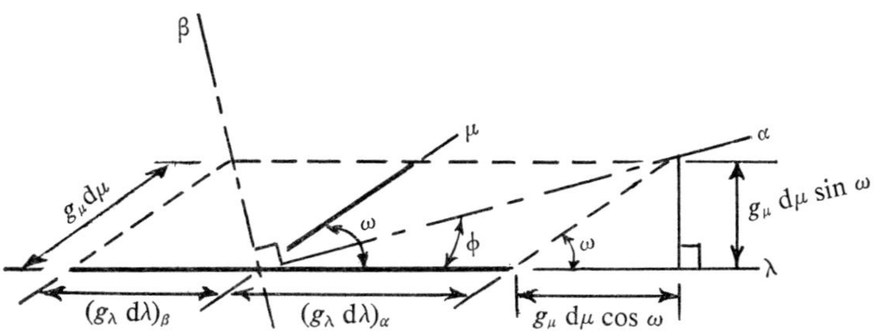

Fig. 1.7. Location of lines of curvature

It can be shown that the normal curvatures in the direction of the lines of curvature are extremal values. From Eqs. (1.1.9b) and (1.1.13) for the first and second fundamental forms of the surface we have

$$\frac{1}{R_s} = \frac{\dfrac{1}{R_\lambda}\left(\dfrac{g_\lambda\,\mathrm{d}\lambda}{g_\mu\,\mathrm{d}\mu}\right)^2 + \dfrac{2}{R_{\lambda\mu}}\left(\dfrac{g_\lambda\,\mathrm{d}\lambda}{g_\mu\,\mathrm{d}\mu}\right) + \dfrac{1}{R_\mu}}{\left(\dfrac{g_\lambda\,\mathrm{d}\lambda}{g_\mu\,\mathrm{d}\mu}\right)^2 + 2\left(\dfrac{g_\lambda\,\mathrm{d}\lambda}{g_\mu\,\mathrm{d}\mu}\right)\cos\omega + 1}, \tag{1.2.11}$$

which is an equation for the normal curvature in any direction as a function of $(g_\lambda\,\mathrm{d}\lambda/g_\mu\,\mathrm{d}\mu)$ since $1/R_\lambda$, $1/R_\mu$, $1/R_{\lambda\mu}$ and $\cos\omega$ are known for a particular coordinate system. The extremal values of $1/R_s$ can be obtained by differentiating Eq. (1.2.11) with respect to $(g_\lambda\,\mathrm{d}\lambda/g_\mu\,\mathrm{d}\mu)$ and equating the result to zero. Then

$$\frac{\partial\left(\dfrac{1}{R_s}\right)}{\partial\left(\dfrac{g_\lambda\,\mathrm{d}\lambda}{g_\mu\,\mathrm{d}\mu}\right)} = 0 = A\left(\frac{g_\lambda\,\mathrm{d}\lambda}{g_\mu\,\mathrm{d}\mu}\right)^2 + B\left(\frac{g_\lambda\,\mathrm{d}\lambda}{g_\mu\,\mathrm{d}\mu}\right) - C, \tag{1.2.12}$$

where A, B, and C are given by Eqs. (1.2.8). But Eq. (1.2.12) is identical with Eq. (1.2.7b) for the direction of the lines of curvature. Therefore, $1/R_s$ takes on its maximum and minimum values in these directions.

1.3 Surfaces of revolution

A very common type of shell structure is one in the shape of a shell of revolution, which can be described as one whose middle surface is generated by the rotation of a plane curve, called the meridian, about a straight line, called the axis of revolution. The position of a point P on the middle surface can be described by the coordinates x, y, z. We may also describe the position of the point P by the angle θ between the x–y plane and the meridian on

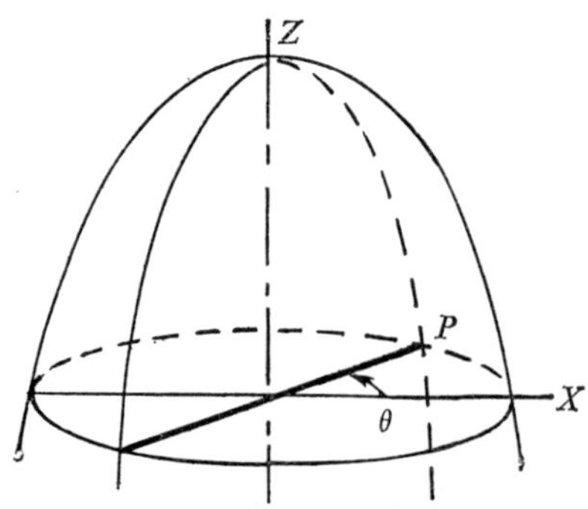

Fig. 1.8. Location of meridian of surface of revolution

which the point lies (Fig. 1.8) and by some measure of location of the point on that particular meridian. The measure of location is usually taken as the angle ϕ at the point in question between the normal to the meridian and the z axis (Fig. 1.9). Sometimes the distance s along the meridian from a reference point to the point P, or the distance r of the point P from the axis of revolution is used. The measurement of location on a meridian through the point is independent of the angle θ and defines a circle in a plane normal to the axis of revolution, called a parallel circle. Thus, the coordinate curves we have chosen are meridians and parallel circles.

We define the equation of the meridian by

$$r = r(z), \tag{1.3.1a}$$

or, in terms of the coordinate ϕ, by the parametric equations

$$\left.\begin{aligned} r &= r(\phi) \\ z &= z(\phi) \end{aligned}\right\}. \tag{1.3.1b}$$

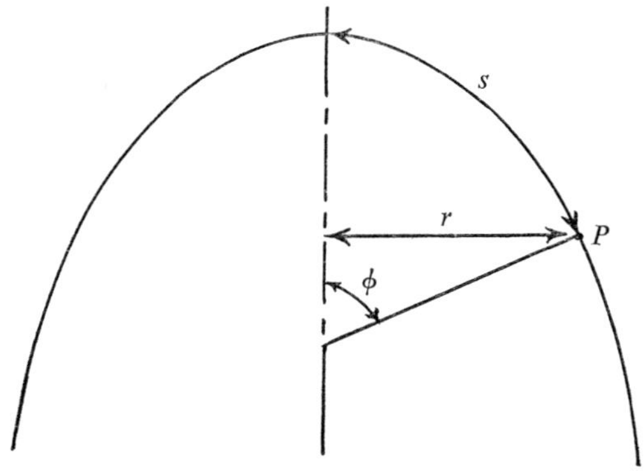

Fig. 1.9. Various coordinates for locating a point on the meridian of a surface of revolution

The element of length of the meridian is given by

$$ds_\phi = [(dr)^2+(dz)^2]^{1/2} = \left[\left(\frac{dr}{d\phi}\right)^2 + \left(\frac{dz}{d\phi}\right)^2\right]^{1/2} d\phi. \tag{1.3.2a}$$

Since the element of length of the meridian is seen to be also given by

$$ds_\phi = \bar{R}_\phi\, d\phi, \tag{1.3.2b}$$

where $\bar{R}_\phi$ is the radius of curvature of the meridian, we have

$$\bar{R}_\phi = \left[\left(\frac{dr}{d\phi}\right)^2 + \left(\frac{dz}{d\phi}\right)^2\right]^{1/2}. \tag{1.3.2c}$$

The radius vector to a point on the surface may be expressed as

$$\mathbf{r} = r(\phi)(\mathbf{i}\cos\theta + \mathbf{j}\sin\theta) + z(\phi)\mathbf{k}. \tag{1.3.3}$$

Then the unit tangent vectors are given by

$$\mathbf{e}_\phi = \frac{\partial\mathbf{r}/\partial\phi}{|\partial\mathbf{r}/\partial\phi|} = \frac{1}{\bar{R}_\phi}\left[\frac{dr}{d\phi}(\mathbf{i}\cos\theta + \mathbf{j}\sin\theta) + \frac{dz}{d\phi}\mathbf{k}\right], \tag{1.3.4a}$$

$$\mathbf{e}_\theta = \frac{\partial\mathbf{r}/\partial\theta}{|\partial\mathbf{r}/\partial\theta|} = \mathbf{j}\cos\theta - \mathbf{i}\sin\theta, \tag{1.3.4b}$$

and the quantities of the first fundamental form are given by

$$g_\phi = \left|\frac{\partial\mathbf{r}}{\partial\phi}\right| = \bar{R}_\phi, \tag{1.3.5a}$$

$$g_\theta = \left|\frac{\partial\mathbf{r}}{\partial\theta}\right| = r, \tag{1.3.5b}$$

$$g_{\phi\theta} = \frac{\partial \mathbf{r}}{\partial \phi} \cdot \frac{\partial \mathbf{r}}{\partial \theta} = 0. \tag{1.3.5c}$$

The last of these equations indicates that the coordinate curves are orthogonal. The unit normal vector is obtained as

$$\mathbf{e}_\zeta = \mathbf{e}_\phi \times \mathbf{e}_\theta = -\frac{1}{\bar{R}_\phi}\left[\frac{\mathrm{d}z}{\mathrm{d}\phi}(\mathbf{i}\cos\theta + \mathbf{j}\sin\theta) - \mathbf{k}\frac{\mathrm{d}r}{\mathrm{d}\phi}\right]. \tag{1.3.6}$$

Since the coordinate curve for constant θ is the meridian which is a plane curve, the radius of curvature of the surface in the direction of $\mathbf{e}_\phi$ is equal to the radius of curvature of the meridian. Thus

$$R_\phi = \bar{R}_\phi. \tag{1.3.7a}$$

The remaining normal curvature and twist are given by

$$\frac{1}{R_\theta} = -\frac{1}{r^2}\mathbf{e}_\zeta \cdot \frac{\partial^2 \mathbf{r}}{\partial \theta^2} = \frac{\sin\phi}{r}, \tag{1.3.7b}$$

$$\frac{1}{R_{\phi\theta}} = -\frac{1}{r\bar{R}_\phi}\mathbf{e}_\zeta \cdot \frac{\partial^2 \mathbf{r}}{\partial \phi\, \partial \theta} = 0. \tag{1.3.7c}$$

The radius of curvature R_θ is the distance along the normal to the shell meridian from the point P to the axis of revolution (Fig. 1.9). Eqs. (1.3.5c) and (1.3.7c) indicate that the coordinate curves (the meridians and parallel circles) are lines of curvature for the surface of revolution. Values of R_ϕ and R_θ for some typical shells of revolution are given in Table 1.1.

The first and second fundamental magnitudes are independent of the angle θ so that the second of the Codazzi–Gauss conditions, given by Eq. (1.2.6b) for lines of curvature coordinates, is automatically satisfied. The first and third of these conditions are equivalent and yield the relationship

$$\frac{1}{R_\phi}\frac{\mathrm{d}}{\mathrm{d}\phi}(R_\theta \sin\phi) = \cos\phi, \tag{1.3.8a}$$

or

$$\frac{\mathrm{d}R_\theta}{\mathrm{d}\phi} = (R_\phi - R_\theta)\cot\phi. \tag{1.3.8b}$$

Note that Eq. (1.3.8a) is equivalent to the identity

$$\frac{\mathrm{d}r}{\mathrm{d}s} = \cos\phi, \tag{1.3.8c}$$

which can be obtained from Fig. 1.9.

Table 1.1 Principal radii of curvature for some shells of revolution

Type of shell	Longitudinal cross-section	Equation of cross-section	R_ϕ	R_θ
(1) Skew elliptical torus	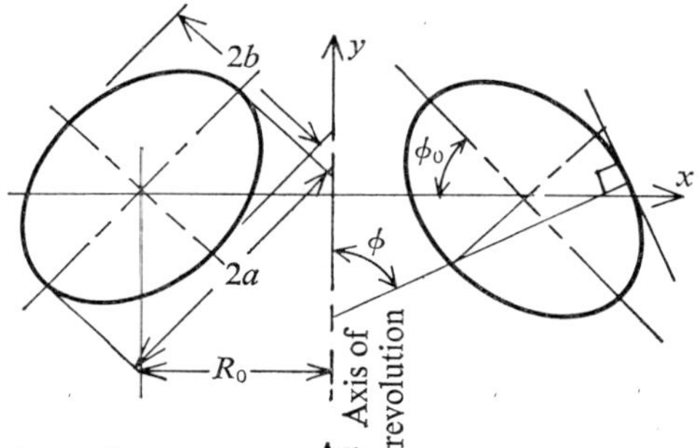	$\left(\frac{\lvert x\rvert \cos\phi_0 - y\sin\phi_0 - R_0}{a}\right)^2 + \left(\frac{\lvert x\rvert \sin\phi_0 + y\cos\phi_0}{b}\right)^2 = 1$	$\frac{a^2}{b}\left[1 + \left(\frac{a^2}{b^2} - 1\right) \times \sin^2(\phi - \phi_0)\right]^{-3/2}$	$\left\{R_0 + \left[1 + \left(\frac{a^2}{b^2} - 1\right) \times \sin^2(\phi - \phi_0)\right]^{-1/2} \times \left[\frac{a^2}{b}\sin(\phi - \phi_0)\cos\phi_0 + b\cos(\phi - \phi_0) \times \sin\phi_0\right]\right\}\csc\phi$
(2) Elliptical torus [(1) with $\phi_0 = 0$]	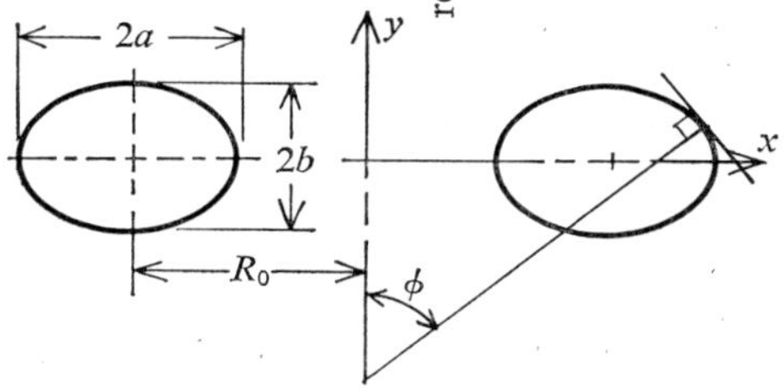	$\left(\frac{\lvert x\rvert - R_0}{a}\right)^2 + \left(\frac{y}{b}\right)^2 = 1$	$\frac{a^2}{b}\left[1 + \left(\frac{a^2}{b^2} - 1\right) \times \sin^2\phi\right]^{-3/2}$	$R_0\csc\phi + \frac{a^2}{b}\left[1 + \left(\frac{a^2}{b^2} - 1\right)\sin^2\phi\right]^{-1/2}$
(3) Circular torus [(1) with $\phi_0 = 0$, $a = b$]	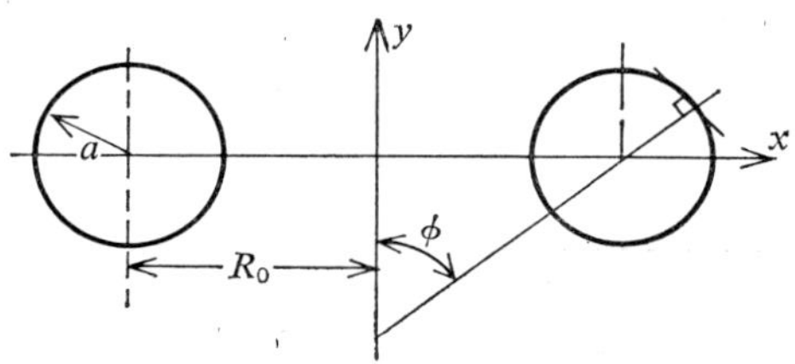	$(\lvert x\rvert - R_0)^2 + y^2 = a^2$	a	$R_0\csc\phi + a$

Table 1.1—*contd.*

Type of shell	Longitudinal cross-section	Equation of cross-section	R_ϕ	R_θ
(4) Ellipsoid [(1) with $\phi_0 = R_0 = 0$]	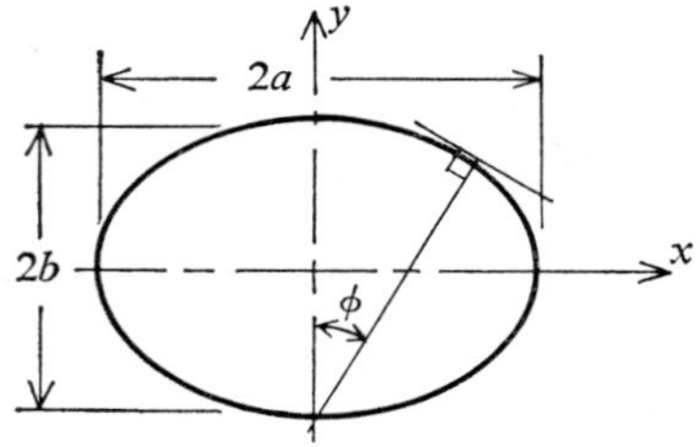	$\left(\frac{x}{a}\right)^2 + \left(\frac{y}{b}\right)^2 = 1$	$\frac{a^2}{b}\left[1 + \left(\frac{a^2}{b^2} - 1\right) \times \sin^2\phi\right]^{-3/2}$	$\frac{a^2}{b}\left[1 + \left(\frac{a^2}{b^2} - 1\right) \times \sin^2\phi\right]^{-1/2}$
(5) Sphere [(1) with $\phi_0 = R_0 = 0$, $a = b$]	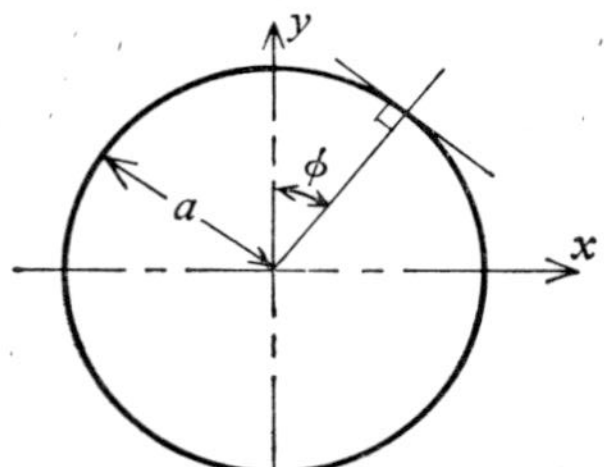	$x^2 + y^2 = a^2$	a	a
(6) Cone	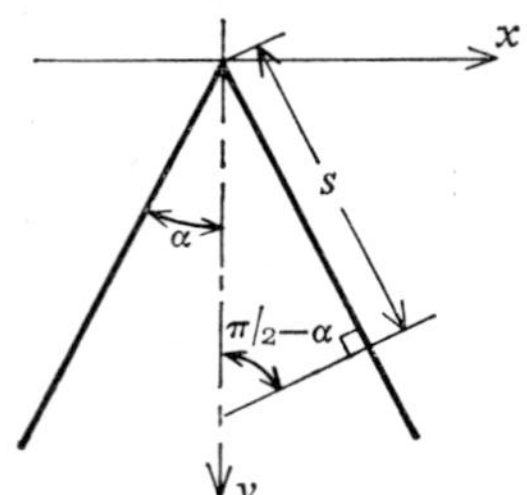	$y = \|x\| \cot\alpha$	∞	$s \tan\alpha$

(7) Cylinder	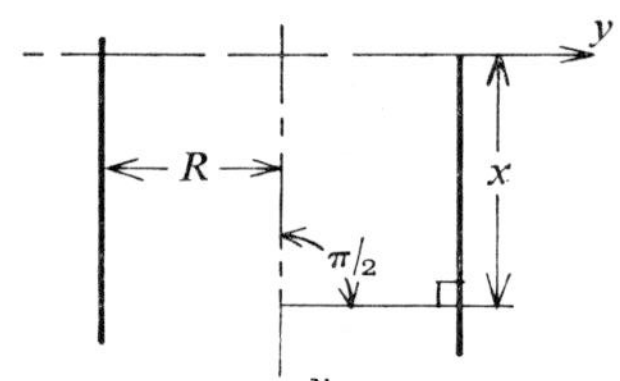	$\lvert y \rvert = R$	∞	R
(8) Hyper-boloid	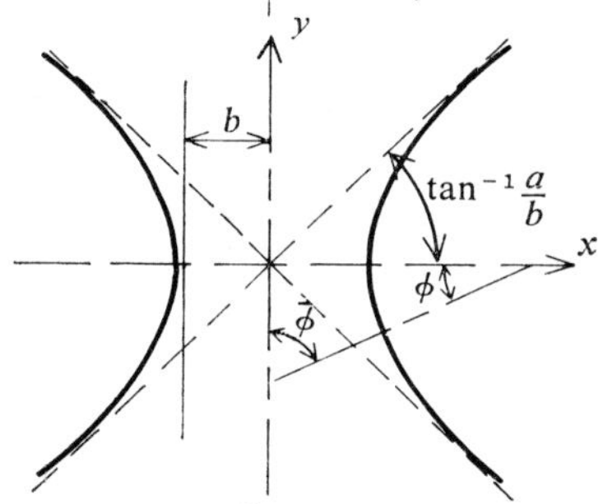	$\left(\frac{x}{a}\right)^2 - \left(\frac{y}{b}\right)^2 = 1$	Rotated about y axis: $-\frac{b^2}{a}\left[\left(\frac{b^2}{a^2}+1\right) \times \sin^2\phi - 1\right]^{-3/2}$ Rotated about x axis: $\frac{a^2}{b}\left[1-\left(1+\frac{a^2}{b}\right)\sin^2\phi\right]^{3/2}$	Rotated about y axis: $\frac{b^2}{a}\left[\left(\frac{b^2}{a^2}+1\right) \times \sin^2\phi - 1\right]^{-1/2}$ Rotated about x axis: $\frac{a^2}{b}\left[1-\left(1+\frac{a^2}{b^2}\right) \times \sin^2\phi\right]^{-1/2}$
(9) Paraboloid	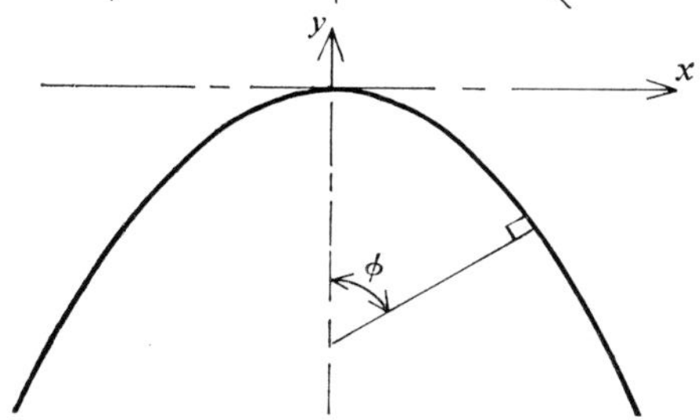	$y = -kx^2$	$\frac{1}{2k\cos^3\phi}$	$\frac{1}{2k\cos\phi}$

1.4 Parallel surfaces

The location of a point in space can be described in terms of the radius vector $\mathbf{r}$ to a point on some surface, hereafter called the generating surface, and the distance ζ along the normal to that surface as (Fig. 1.10)

$$\mathbf{r}^{(\zeta)} = \mathbf{r} + \zeta \mathbf{e}_\zeta, \tag{1.4.1}$$

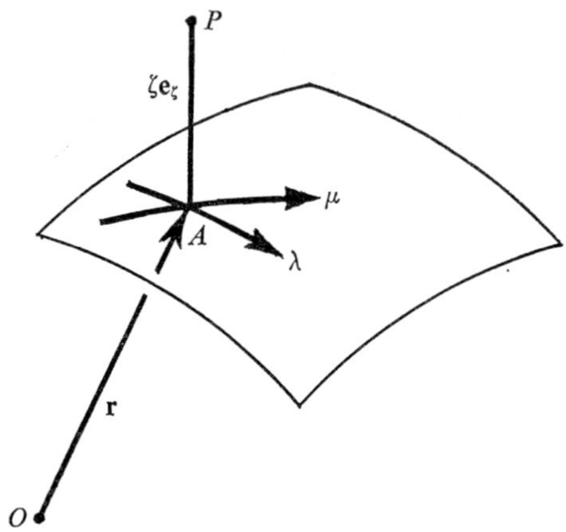

Fig. 1.10. Location of point in space

where both $\mathbf{r}$ and $\mathbf{e}_\zeta$ are functions of curvilinear coordinates λ and μ on the generating surface. If ζ is held constant and we allow λ and μ, or the position of the surface point, to vary, the end of the constant length normal vector ($\zeta\mathbf{e}_\zeta$) describes a surface in space which is said to be parallel to the generating surface. Since such surfaces enter into the theory of shells, we shall consider the description of their geometric properties in terms of those of the generating surface.

The most interesting of these properties come to light when the coordinate curves on the generating surface are lines of curvature. Let us consider the curves traced on a parallel surface as the point on the generating surface moves along these coordinate curves (Fig. 1.11). Since ζ is constant, these traced curves are a function only of α and β and can be considered to be coordinate curves for the parallel surface, described by the same curvilinear coordinates used for the generating surface. Tangent vectors to the coordinate curves on the parallel surface are given, with the aid of the Weingarten–Gauss derivative relations and the Codazzi–Gauss conditions for the generat-

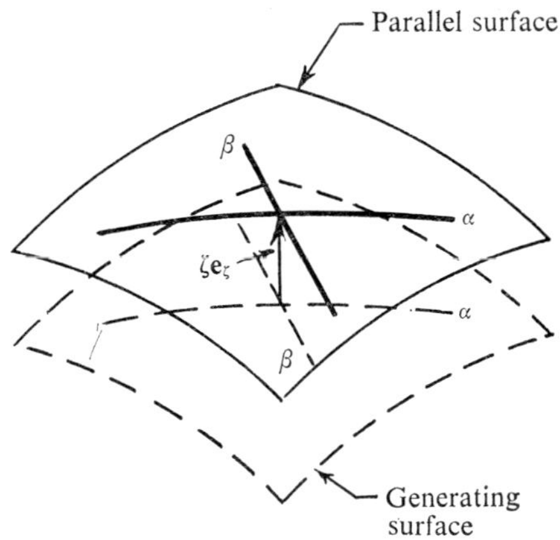

Fig. 1.11. Lines of curvature coordinate curves for parallel surfaces

ing surface, Eqs. (1.2.5) and (1.2.6), by

$$\frac{\partial \mathbf{r}^{(\zeta)}}{\partial \alpha} = \left(1 + \frac{\zeta}{R_\alpha}\right)\frac{\partial \mathbf{r}}{\partial \alpha}, \tag{1.4.2a}$$

$$\frac{\partial \mathbf{r}^{(\zeta)}}{\partial \beta} = \left(1 + \frac{\zeta}{R_\beta}\right)\frac{\partial \mathbf{r}}{\partial \beta}, \tag{1.4.2b}$$

and are thus seen to be parallel to the vectors tangent to the generating surface. It follows that the normal to the generating surface is also normal to the parallel surface and, from the relations of Section 1.2, that the coordinate curves on the parallel surface are lines of curvature.

The small area element enclosed by two adjacent normals and the coordinate curves on the generating and parallel surfaces is, from the definition of lines of curvature, a plane surface. It is interesting to note that for any generating surface coordinate system other than lines of curvature this surface would be warped since, from Eqs. (1.1.23), the trace of the normal vector on the parallel surface is then not parallel to the curve on the generating surface. Thus lines of curvature are the only coordinate curves that will include the normals to the middle surface as part of an orthogonal set for the space between the bounding surfaces of the shell.

Since the point of intersection of adjacent normals is the center of curvature for both the generating and parallel surfaces, it is readily seen that the normal radii of curvature of parallel surfaces are given by

$$R_\alpha^{(\zeta)} = R_\alpha + \zeta, \tag{1.4.3a}$$

$$R_\beta^{(\zeta)} = R_\beta + \zeta, \tag{1.4.3b}$$

and that the first fundamental quantities are given by

$$g_\alpha^{(\zeta)} = \left(1 + \frac{\zeta}{R_\alpha}\right) g_\alpha, \tag{1.4.4a}$$

$$g_\beta^{(\zeta)} = \left(1 + \frac{\zeta}{R_\beta}\right) g_\beta. \tag{1.4.4b}$$

Some other relations that will prove helpful later on are

$$\frac{\partial g_\alpha^{(\zeta)}}{\partial \beta} = \left(1 + \frac{\zeta}{R_\beta}\right) \frac{\partial g_\alpha}{\partial \beta}, \tag{1.4.5a}$$

$$\frac{\partial g_\beta^{(\zeta)}}{\partial \alpha} = \left(1 + \frac{\zeta}{R_\alpha}\right) \frac{\partial g_\beta}{\partial \alpha}, \tag{1.4.5b}$$

which follow from Eqs. (1.4.5) and (1.2.6).

1.5 Small deformations of a surface

Suppose that the surface considered in Section 1.1 is deformed by some small arbitrary displacement of each of its points. The λ and μ coordinate curves will be transformed to a network of curves that can be considered to be a system of coordinate curves for the deformed surface. Since points on these distorted curves can still be described by pairs of values of λ and μ, we take these to be curvilinear coordinates for the deformed surface as well as for the undeformed surface (Fig. 1.12). The geometry of the deformed surface is then described by the system of equations derived in Section 1.1 with the radius vector taken as the radius vector $\mathbf{r}'$ to the deformed surface. Our task then is to relate the distortions of the deformed surface to geometrical quantities for the undeformed surface and to displacements relative to the undeformed surface. The lines of curvature of the surface will be taken as coordinate curves since the relationships have their simplest form for this system of coordinates.

We denote the radius vector to a point on the deformed surface by

$$\mathbf{r}' = \mathbf{r} + \mathbf{u}, \tag{1.5.1a}$$

where $\mathbf{r}$ is the radius vector to the point in its original position and $\mathbf{u}$ is the displacement vector of the point relative to its original position. In terms of displacements in the direction of the unit tangential and normal vectors of the undeformed surface, the displacement vector is given by

$$\mathbf{u} = u_\alpha \mathbf{e}_\alpha + u_\beta \mathbf{e}_\beta + w\mathbf{e}_\zeta. \tag{1.5.1b}$$

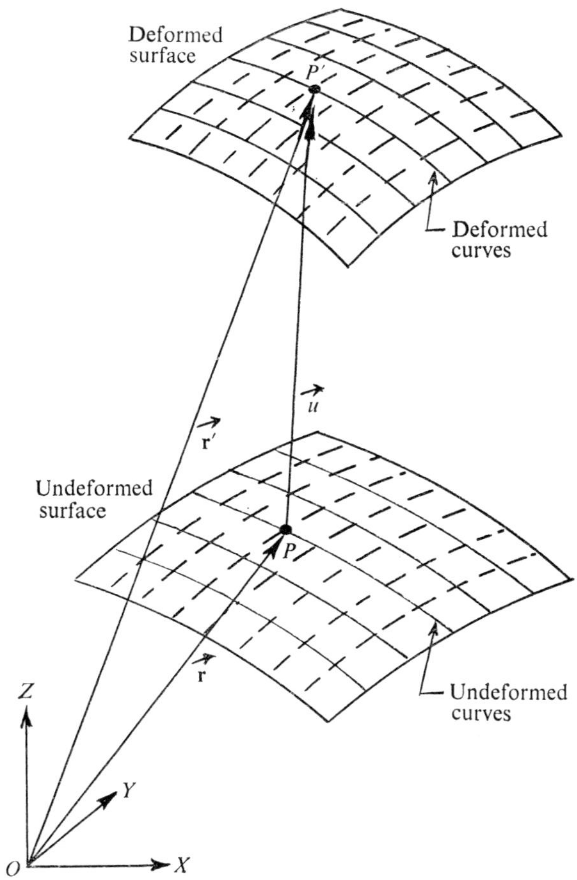

Fig. 1.12. Relationship between undeformed and deformed surfaces

Tangent vectors to the deformed surface in the direction of the distorted α and β coordinate curves (Fig. 1.13) are obtained with the aid of the Weingarten–Gauss relations for the undeformed surface, Eqs. (1.2.5), as

$$\frac{\partial \mathbf{r}'}{\partial \alpha} = g_\alpha[(1 + \epsilon_\alpha)\mathbf{e}_\alpha + \gamma_\alpha \mathbf{e}_\beta + \Theta_\alpha \mathbf{e}_\zeta], \tag{1.5.2a}$$

$$\frac{\partial \mathbf{r}'}{\partial \beta} = g_\beta[\gamma_\beta \mathbf{e}_\alpha + (1 + \epsilon_\beta)\mathbf{e}_\beta + \Theta_\beta \mathbf{e}_\zeta], \tag{1.5.2b}$$

where

$$\epsilon_\alpha = \frac{1}{g_\alpha}\left(\frac{\partial u_\alpha}{\partial \alpha} + \frac{1}{g_\beta}\frac{\partial g_\alpha}{\partial \beta} u_\beta + \frac{g_\alpha}{R_\alpha} w\right), \tag{1.5.3a}$$

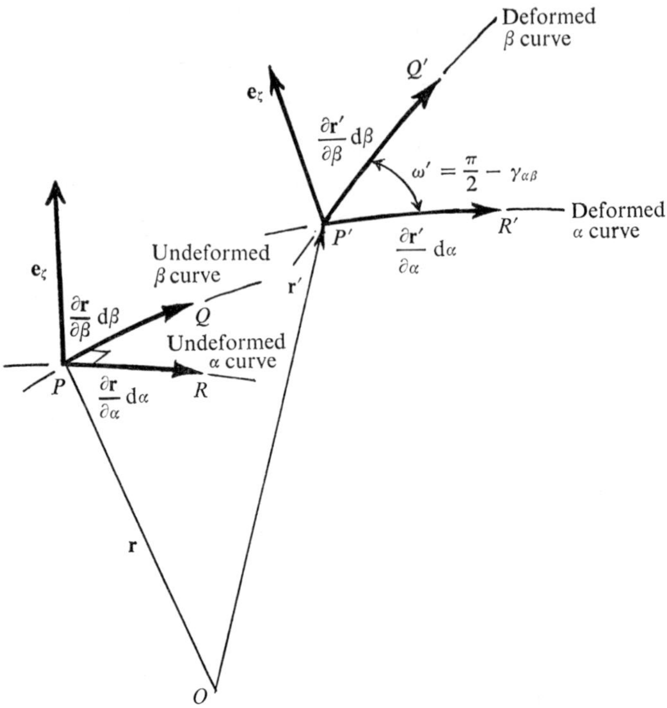

Fig. 1.13. Coordinate curves on undeformed and deformed surface

$$\gamma_\alpha = \frac{1}{g_\alpha}\left(\frac{\partial u_\beta}{\partial \alpha} - \frac{1}{g_\beta}\frac{\partial g_\alpha}{\partial \beta}u_\alpha\right), \tag{1.5.3b}$$

$$\Theta_\alpha = \frac{1}{g_\alpha}\left(\frac{\partial w}{\partial \alpha} - \frac{g_\alpha}{R_\alpha}u_\alpha\right), \tag{1.5.3c}$$

$$\epsilon_\beta = \frac{1}{g_\beta}\left(\frac{\partial u_\beta}{\partial \beta} + \frac{1}{g_\alpha}\frac{\partial g_\beta}{\partial \alpha}u_\alpha + \frac{g_\beta}{R_\beta}w\right), \tag{1.5.3d}$$

$$\gamma_\beta = \frac{1}{g_\beta}\left(\frac{\partial u_\alpha}{\partial \beta} - \frac{1}{g_\alpha}\frac{\partial g_\beta}{\partial \alpha}u_\beta\right), \tag{1.5.3e}$$

$$\Theta_\beta = \frac{1}{g_\beta}\left(\frac{\partial w}{\partial \beta} - \frac{g_\beta}{R_\beta}u_\beta\right). \tag{1.5.3f}$$

Eqs. (1.5.2) define the new position and length of two line elements originally in the direction of the lines of curvature on the undeformed surface. The lengths of these distorted line elements are given by

$$g'_\alpha\,\mathrm{d}\alpha = \left[\frac{\partial \mathbf{r}'}{\partial \alpha}\cdot\frac{\partial \mathbf{r}'}{\partial \alpha}\right]^{1/2}\mathrm{d}\alpha \approx g_\alpha(1+\epsilon_\alpha)\,\mathrm{d}\alpha, \tag{1.5.4a}$$

$$g'_\beta \, \mathrm{d}\beta = \left[\frac{\partial \mathbf{r}'}{\partial \beta} \cdot \frac{\partial \mathbf{r}'}{\partial \beta}\right]^{1/2} \mathrm{d}\beta \approx g_\beta(1 + \epsilon_\beta) \, \mathrm{d}\beta, \tag{1.5.4b}$$

where Eqs. (1.1.7) have been used and products of displacements and their derivatives have been neglected. The engineering strain of the line elements is seen to be given by

$$\bar{\epsilon}_\alpha = \frac{g'_\alpha \, \mathrm{d}\alpha - g_\alpha \, \mathrm{d}\alpha}{g_\alpha \, \mathrm{d}\alpha} \approx \epsilon_\alpha, \tag{1.5.5a}$$

$$\bar{\epsilon}_\beta = \frac{g'_\beta \, \mathrm{d}\beta - g_\beta \, \mathrm{d}\beta}{g_\beta \, \mathrm{d}\beta} \approx \epsilon_\beta. \tag{1.5.5b}$$

The change of the angle between the distorted line elements is given by

$$\cos \omega' = \left(\frac{1}{g'_\alpha} \frac{\partial \mathbf{r}'}{\partial \alpha}\right) \cdot \left(\frac{1}{g'_\beta} \frac{\partial \mathbf{r}'}{\partial \beta}\right) \approx \gamma_\alpha + \gamma_\beta. \tag{1.5.6a}$$

The angle ω' can be expressed as $\pi/2 - \gamma_{\alpha\beta}$ where $\gamma_{\alpha\beta}$, the change in the angle between the originally perpendicular line element, is a small quantity. From Eq. (1.5.6a) we then obtain

$$\gamma_{\alpha\beta} \approx \gamma_\alpha + \gamma_\beta = \frac{g_\alpha}{g_\beta} \frac{\partial}{\partial \beta}\left(\frac{u_\alpha}{g_\alpha}\right) + \frac{g_\beta}{g_\alpha} \frac{\partial}{\partial \alpha}\left(\frac{u_\beta}{g_\beta}\right), \tag{1.5.6b}$$

which is the shear strain of a distorted area element.

The normal curvatures and twist of the deformed surface are obtained from the relations

$$\frac{1}{R'_\alpha} = -\frac{1}{(g'_\alpha)^2} \mathbf{e}'_\zeta \cdot \frac{\partial^2 \mathbf{r}'}{\partial \alpha^2}, \tag{1.5.7a}$$

$$\frac{1}{R'_\beta} = -\frac{1}{(g'_\beta)^2} \mathbf{e}'_\zeta \cdot \frac{\partial^2 \mathbf{r}'}{\partial \beta^2}, \tag{1.5.7b}$$

$$\frac{1}{R'_{\alpha\beta}} = -\frac{1}{g'_\alpha g'_\beta} \mathbf{e}'_\zeta \cdot \frac{\partial^2 \mathbf{r}'}{\partial \alpha \, \partial \beta}. \tag{1.5.7c}$$

where the orientation of the unit normal to the distorted surface is defined by

$$\mathbf{e}'_\zeta = \frac{1}{g'_\alpha g'_\beta \sin \omega'} \frac{\partial \mathbf{r}'}{\partial \alpha} \times \frac{\partial \mathbf{r}'}{\partial \beta} \approx \mathbf{e}_\zeta - \Theta_\alpha \mathbf{e}_\alpha - \Theta_\beta \mathbf{e}_\beta. \tag{1.5.8}$$

On using Eqs. (1.1.7), (1.5.2), (1.5.4), and (1.5.8), and the Weingarten–Gauss relations for the undeformed surface, we have

$$\frac{1}{R_\alpha} \approx \frac{1}{R_\alpha} + \kappa_\alpha - \frac{\epsilon_\alpha}{R_\alpha}, \tag{1.5.9a}$$

$$\frac{1}{R'_\beta} \approx \frac{1}{R_\beta} + \kappa_\beta - \frac{\epsilon_\beta}{R_\beta}, \tag{1.5.9b}$$

$$\frac{1}{R'_{\alpha\beta}} \approx \kappa_{\alpha\beta}, \tag{1.5.9c}$$

where

$$\kappa_\alpha = -\frac{1}{g_\alpha}\left(\frac{\partial \Theta_\alpha}{\partial \alpha} + \frac{1}{g_\beta}\frac{\partial g_\alpha}{\partial \beta}\Theta_\beta\right) = -\frac{1}{g_\alpha}\left[\frac{\partial}{\partial \alpha}\left(\frac{1}{g_\alpha}\frac{\partial w}{\partial \alpha}\right) + \frac{1}{g_\beta^2}\frac{\partial g_\alpha}{\partial \beta}\frac{\partial w}{\partial \beta} - \frac{\partial}{\partial \alpha}\left(\frac{u_\alpha}{R_\alpha}\right) - \frac{u_\beta}{g_\beta}\frac{\partial}{\partial \beta}\left(\frac{g_\alpha}{R_\alpha}\right)\right], \tag{1.5.10a}$$

$$\kappa_\beta = -\frac{1}{g_\beta}\left(\frac{\partial \Theta_\beta}{\partial \beta} + \frac{1}{g_\alpha}\frac{\partial g_\beta}{\partial \alpha}\Theta_\alpha\right) = -\frac{1}{g_\beta}\left[\frac{\partial}{\partial \beta}\left(\frac{1}{g_\beta}\frac{\partial w}{\partial \beta}\right) + \frac{1}{g_\alpha^2}\frac{\partial g_\beta}{\partial \alpha}\frac{\partial w}{\partial \alpha} - \frac{\partial}{\partial \beta}\left(\frac{u_\beta}{R_\beta}\right) - \frac{u_\alpha}{g_\alpha}\frac{\partial}{\partial \alpha}\left(\frac{g_\beta}{R_\beta}\right)\right], \tag{1.5.10b}$$

$$\kappa_{\alpha\beta} = \frac{\gamma_\alpha}{R_\beta} - \frac{1}{g_\beta}\left(\frac{\partial \Theta_\alpha}{\partial \beta} - \frac{1}{g_\alpha}\frac{\partial g_\beta}{\partial \alpha}\Theta_\beta\right) = \frac{\gamma_\beta}{R_\alpha} - \frac{1}{g_\alpha}\left(\frac{\partial \Theta_\beta}{\partial \alpha} - \frac{1}{g_\beta}\frac{\partial g_\alpha}{\partial \beta}\Theta_\alpha\right)$$
$$= -\frac{1}{g_\alpha g_\beta}\left[\frac{\partial^2 w}{\partial \alpha\, \partial \beta} - \frac{1}{g_\alpha}\frac{\partial g_\alpha}{\partial \beta}\frac{\partial w}{\partial \alpha} - \frac{1}{g_\beta}\frac{\partial g_\beta}{\partial \alpha}\frac{\partial w}{\partial \beta} - \frac{g_\alpha^2}{R_\alpha}\frac{\partial}{\partial \beta}\left(\frac{u_\alpha}{g_\alpha}\right) - \frac{g_\beta^2}{R_\beta}\frac{\partial}{\partial \alpha}\left(\frac{u_\beta}{g_\beta}\right)\right]. \tag{1.5.10c}$$

The two forms $\kappa_{\alpha\beta}$ in terms of rotations of line elements are a result of the fact that we can obtain an expression for $\partial^2 r'/\partial\alpha\,\partial\beta$ by differentiating $\partial \mathbf{r}'/\partial\alpha$ with respect to β or $\partial\mathbf{r}'/\partial\beta$ with respect to α.

The quantities g'_α, g'_β, $\cos\omega'$, $1/R'_\alpha$, $1/R'_\beta$, and $1/R'_{\alpha\beta}$ are fundamental quantities for the deformed surface and as such must satisfy three equations of the form of Eq. (1.1.27), with the unprimed quantities replaced by primed quantities. If we introduce Eqs. (1.5.4), (1.5.6), and (1.5.9) into Eqs. (1.1.27), neglect products of displacements and their derivatives, and take into account the fact that the fundamental quantities for the undeformed surface also satisfy Eq. (1.1.27), we find that the strain and change of normal curvature components must satisfy the following relationships:

$$\frac{\partial}{\partial \beta}\left[g_\alpha\left(\kappa_\alpha - \frac{\epsilon_\alpha}{R_\alpha}\right)\right] - \left(\kappa_\beta - \frac{\epsilon_\beta}{R_\beta}\right)\frac{\partial g_\alpha}{\partial \beta} - \frac{1}{g_\beta}\frac{\partial}{\partial \alpha}\left[g_\beta^2\left(\kappa_{\alpha\beta} - \frac{\gamma_{\alpha\beta}}{R_\beta}\right)\right] + g_\alpha\left(\frac{1}{R_\alpha} - \frac{1}{R_\beta}\right)\frac{\partial \epsilon_\alpha}{\partial \beta} = 0, \tag{1.5.11a}$$

$$\frac{\partial}{\partial\alpha}\left[g_\beta\left(\kappa_\beta - \frac{\epsilon_\beta}{R_\beta}\right)\right] - \left(\kappa_\alpha - \frac{\epsilon_\alpha}{R_\alpha}\right)\frac{\partial g_\beta}{\partial\alpha} - \frac{1}{g_\alpha}\frac{\partial}{\partial\beta}\left[g_\alpha^2\left(\kappa_{\alpha\beta} - \frac{\gamma_{\alpha\beta}}{R_\alpha}\right)\right]$$

$$+ g_\beta\left(\frac{1}{R_\beta} - \frac{1}{R_\alpha}\right)\frac{\partial\epsilon_\beta}{\partial\alpha} = 0 \tag{1.5.11b}$$

$$\frac{\partial}{\partial\alpha}\left\{\frac{1}{g_\alpha}\left[\frac{\partial}{\partial\alpha}(g_\beta\epsilon_\beta) - \epsilon_\alpha\frac{\partial g_\beta}{\partial\alpha} - \frac{1}{2g_\alpha}\frac{\partial}{\partial\beta}(g_\alpha^2\gamma_{\alpha\beta})\right]\right\}$$

$$+ \frac{\partial}{\partial\beta}\left\{\frac{1}{g_\beta}\left[\frac{\partial}{\partial\beta}(g_\alpha\epsilon_\alpha) - \epsilon_\beta\frac{\partial g_\alpha}{\partial\beta} - \frac{1}{2g_\beta}\frac{\partial}{\partial\alpha}(g_\beta^2\gamma_{\alpha\beta})\right]\right\}$$

$$+ g_\alpha g_\beta\left(\frac{\kappa_\alpha}{R_\beta} + \frac{\kappa_\beta}{R_\alpha}\right) = 0. \tag{1.5.11c}$$

The foregoing equations are the three compatibility conditions that must be satisfied when distortions of the surface are described by six strains and curvature changes rather than by three displacements. The extensional and shearing strains determine the change of shape of the projection of an element of surface area on the tangent plane, while the changes of normal curvature and twist determine the change of bending and twist of the element out of the tangent plane. The equations are identically satisfied when the strains and curvature changes are replaced by their expressions in terms of displacements.

Supplementary references

[1] Brand, L., *Vector and Tensor Analysis*, John Wiley and Sons Inc., New York (1947).

[2] Coxeter, H. S. M., *Introduction to Geometry*, John Wiley and Sons Inc., New York (1961).

[3] Struik, D. J., *Lectures on Classical Differential Geometry*, Second Edition, Addison Wesley Publishing Co. Inc., New York (1961).

[4] Green, A. E., and Zerna, W., *Theoretical Elasticity*, Second Edition, Oxford University Press, London (1968).

2

The linear theory of transversely rigid shells

2.1 Fundamental definitions

We define a shell as a three-dimensional structure bounded primarily by two arbitrary curved surfaces a relatively small distance apart. The middle surface of the shell is the locus of points equidistant from the bounding surfaces and the local thickness of the shell is the distance between the bounding surfaces measured perpendicular to the middle surface. The theory of shells is usually concerned with shells of constant thickness. In this case, the bounding surfaces of the shell are said to be parallel. In some problems with which we will be concerned, the bounding surfaces are closed surfaces. In most problems, however, the bounding surfaces are assumed to be interrupted by the ruled surfaces or edges defined by curves on the middle surface and the normals to the middle surface along these curves.

The problem for which shell theory attempts to provide a solution can be stated as follows. If

a. The middle surface of the shell is arbitrary,
b. The shell material is elastic and in most cases isotropic,
c. The bounding surfaces are subjected to arbitrary load distribution,
d. The interior of the shell wall is subjected to arbitrary body forces,
e. Stresses and/or displacements consistent with the applied bounding surface loading and body forces are known at each point of the ruled edges,

what distribution of stresses and displacements within the shell wall is predicted by the three-dimensional theory of elasticity? In the following exposition we shall confine our discussions to shells of constant thickness and to loadings for which the displacements, rotations, and strains are small. We shall also tacitly assume that the thickness of the shell is small compared to the least radius of curvature of the middle surface ($h/R < \frac{1}{10}$ to $\frac{1}{20}$). While there is nothing in the approximations of the theory which pinpoints the exact bounds of the limitation on thickness/radius ratio, comparisons with exact shell solutions and asymptotic expansions of the equations for isotropic shells indicate that the 'approximate' theory to be developed is usually adequate in the range given above. Similarly we must also assume that the surface loading does not change rapidly over distances comparable to the value $(Rh)^{1/2}$.

2.2 Equations of three-dimensional elasticity

A shell of constant thickness is defined by the equation of its middle surface and the magnitude of the thickness. The inner and outer bounding surfaces of the shell are surfaces parallel to the middle surface. A system of orthogonal curvilinear coordinates for points within the shell wall is defined by the coordinates α and β corresponding to the lines of curvature on the middle surface and the distance ζ along the normal to the middle surface. A particular value of the coordinate ζ defines a surface parallel to the middle surface and the coordinates α and β define ruled surfaces intersecting each of the parallel surfaces along its lines of curvature.

We will first derive the differential equations of the three-dimensional theory of elasticity for the region between the bounding surfaces of the shell. To obtain equilibrium equations for stresses in the shell wall, we consider a small element in the shell wall bounded by two surfaces parallel to the middle surface of the shell, a distance $\mathrm{d}\zeta$ apart, and four adjacent normals to the middle surface along the lines of curvature, which define four planes intersecting the parallel surfaces along lines of curvature. As is usual in small deformation theory, we assume that the effects of geometry changes of the element on equilibrium of forces are negligible, i.e., essentially, we consider the element to be part of a rigid body so far as equilibrium of forces is concerned. On three perpendicular faces of the element, as shown in Fig. 2.1, forces defined by the vectors

$$-\mathbf{F}_\alpha^{(\zeta)} = -[\sigma_\alpha^{(\zeta)}\mathbf{e}_\alpha + \tau_{\alpha\beta}^{(\zeta)}\mathbf{e}_\beta + \tau_{\alpha\zeta}^{(\zeta)}\mathbf{e}_\zeta]g_\beta^{(\zeta)}\,\mathrm{d}\beta\,\mathrm{d}\zeta, \tag{2.2.1a}$$

$$-\mathbf{F}_\beta^{(\zeta)} = -[\tau_{\alpha\beta}^{(\zeta)}\mathbf{e}_\alpha + \sigma_\beta^{(\zeta)}\mathbf{e}_\beta^{(\zeta)} + \tau_{\beta\zeta}^{(\zeta)}\mathbf{e}_\zeta]g_\alpha^{(\zeta)}\,\mathrm{d}\alpha\,\mathrm{d}\zeta, \tag{2.2.1b}$$

$$-\mathbf{F}_\zeta^{(\zeta)} = -[\tau_{\alpha\zeta}^{(\zeta)}\mathbf{e}_\alpha + \tau_{\beta\zeta}^{(\zeta)}\mathbf{e}_\beta + \sigma_\zeta^{(\zeta)}\mathbf{e}_\zeta]g_\alpha^{(\zeta)}g_\beta^{(\zeta)}\,\mathrm{d}\alpha\,\mathrm{d}\beta, \tag{2.2.1c}$$

act. On the three adjacent perpendicular faces, the forces are given by

$$\mathbf{F}_\alpha^{(\zeta)} + \frac{\partial\mathbf{F}_\alpha^{(\zeta)}}{\partial\alpha}\,\mathrm{d}\alpha, \qquad \mathbf{F}_\beta^{(\zeta)} + \frac{\partial\mathbf{F}_\beta^{(\zeta)}}{\partial\beta}\,\mathrm{d}\beta, \qquad \mathbf{F}_\zeta^{(\zeta)} + \frac{\partial\mathbf{F}_\zeta^{(\zeta)}}{\partial\zeta}\,\mathrm{d}\zeta. \tag{2.2.2}$$

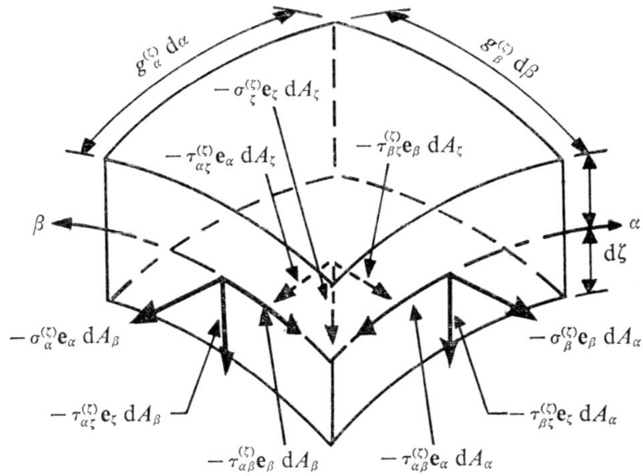

$$dA_\alpha = g_\alpha^{(\zeta)}\,d\alpha\,d\zeta$$
$$dA_\beta = g_\beta^{(\zeta)}\,d\beta\,d\zeta$$
$$dA_\zeta = g_\alpha^{(\zeta)}g_\beta^{(\zeta)}\,d\alpha\,d\beta$$

Fig. 2.1. Forces on faces of element of shell wall

In addition, the element may be subjected to body forces given by

$$\mathbf{F}_b^{(\zeta)} = [\rho_\alpha^{(\zeta)}\mathbf{e}_\alpha + \rho_\beta^{(\zeta)}\mathbf{e}_\beta + \rho_\zeta^{(\zeta)}\mathbf{e}_\zeta]\,g_\alpha^{(\zeta)}g_\beta^{(\zeta)}\,d\alpha\,d\beta\,d\zeta. \tag{2.2.3}$$

The vector equation of equilibrium is obtained by setting the sum of all the forces on the element equal to zero to obtain

$$\mathbf{F}_b^{(\zeta)} + \frac{\partial \mathbf{F}_\alpha^{(\zeta)}}{\partial \alpha}\,d\alpha + \frac{\partial \mathbf{F}_\beta^{(\zeta)}}{\partial \beta}\,d\beta + \frac{\partial \mathbf{F}_\zeta^{(\zeta)}}{\partial \zeta}\,d\zeta = 0, \tag{2.2.4a}$$

or, when Eqs. (2.2.1) and (2.2.3) are used,

$$\begin{aligned}
&\frac{\partial}{\partial\alpha}\{g_\beta^{(\zeta)}[\sigma_\alpha^{(\zeta)}\mathbf{e}_\alpha + \tau_{\alpha\beta}^{(\zeta)}\mathbf{e}_\beta + \tau_{\alpha\zeta}^{(\zeta)}\mathbf{e}_\zeta]\}\\
&\quad + \frac{\partial}{\partial\beta}\{g_\alpha^{(\zeta)}[\tau_{\alpha\beta}^{(\zeta)}\mathbf{e}_\alpha + \sigma_\beta^{(\zeta)}\mathbf{e}_\beta + \tau_{\beta\zeta}^{(\zeta)}\mathbf{e}_\zeta]\}\\
&\quad + \frac{\partial}{\partial\zeta}\{g_\alpha^{(\zeta)}g_\beta^{(\zeta)}[\tau_{\alpha\zeta}^{(\zeta)}\mathbf{e}_\alpha + \tau_{\beta\zeta}^{(\zeta)}\mathbf{e}_\beta + \sigma_\zeta^{(\zeta)}\mathbf{e}_\zeta]\}\\
&\quad + g_\alpha^{(\zeta)}g_\beta^{(\zeta)}[\rho_\alpha^{(\zeta)}\mathbf{e}_\alpha + \rho_\beta^{(\zeta)}\mathbf{e}_\beta + \rho_\zeta^{(\zeta)}\mathbf{e}_\zeta] = 0.
\end{aligned} \tag{2.2.4b}$$

On performing the indicated operations and using the expressions for derivatives of the unit vectors given by Eqs. (1.2.5) (noting that $\mathbf{e}_\alpha$, $\mathbf{e}_\beta$, and $\mathbf{e}_\zeta$ are independent of ζ), expressions for the first fundamental quantities of parallel surfaces [Eqs. (1.4.5)], and the Codazzi–Gauss equations for the middle surface given by Eqs. (1.2.6), we can express the left side of Eq. (2.2.4b) as the sum of three force vectors in the directions of the unit tangential and normal vectors $\mathbf{e}_\alpha$, $\mathbf{e}_\beta$, and $\mathbf{e}_\zeta$. Each of these vectors must vanish for

Eq. (2.2.4b) to be identically satisfied so that the equations of equilibrium of the element can be written as

$$\frac{\partial}{\partial\alpha}\left[g_\beta\left(1+\frac{\zeta}{R_\beta}\right)\sigma_\alpha^{(\zeta)}\right]+\frac{\partial}{\partial\beta}\left[g_\alpha\left(1+\frac{\zeta}{R_\alpha}\right)\tau_{\alpha\beta}^{(\zeta)}\right]$$
$$+\frac{\partial}{\partial\zeta}\left[g_\alpha g_\beta\left(1+\frac{\zeta}{R_\alpha}\right)\left(1+\frac{\zeta}{R_\beta}\right)\tau_{\alpha\zeta}^{(\zeta)}\right]$$
$$+\frac{\partial g_\alpha}{\partial\beta}\left(1+\frac{\zeta}{R_\beta}\right)\tau_{\alpha\beta}^{(\zeta)}-\frac{\partial g_\beta}{\partial\alpha}\left(1+\frac{\zeta}{R_\alpha}\right)\sigma_\beta^{(\zeta)}+\frac{g_\alpha g_\beta}{R_\alpha}\left(1+\frac{\zeta}{R_\beta}\right)\tau_{\alpha\zeta}^{(\zeta)}$$
$$+g_\alpha g_\beta\left(1+\frac{\zeta}{R_\alpha}\right)\left(1+\frac{\zeta}{R_\beta}\right)p_\alpha^{(\zeta)}=0, \tag{2.2.5a}$$

$$\frac{\partial}{\partial\alpha}\left[g_\beta\left(1+\frac{\zeta}{R_\beta}\right)\tau_{\alpha\beta}^{(\zeta)}\right]+\frac{\partial}{\partial\beta}\left[g_\alpha\left(1+\frac{\zeta}{R_\alpha}\right)\sigma_\beta^{(\zeta)}\right]$$
$$+\frac{\partial}{\partial\zeta}\left[g_\alpha g_\beta\left(1+\frac{\zeta}{R_\alpha}\right)\left(1+\frac{\zeta}{R_\beta}\right)\tau_{\beta\zeta}^{(\zeta)}\right]$$
$$+\frac{\partial g_\beta}{\partial\alpha}\left(1+\frac{\zeta}{R_\alpha}\right)\tau_{\alpha\beta}^{(\zeta)}-\frac{\partial g_\alpha}{\partial\beta}\left(1+\frac{\zeta}{R_\beta}\right)\sigma_\alpha^{(\zeta)}+\frac{g_\alpha g_\beta}{R_\beta}\left(1+\frac{\zeta}{R_\alpha}\right)\tau_{\beta\zeta}^{(\zeta)}$$
$$+g_\alpha g_\beta\left(1+\frac{\zeta}{R_\alpha}\right)\left(1+\frac{\zeta}{R_\beta}\right)p_\beta^{(\zeta)}=0, \tag{2.2.5b}$$

$$\frac{\partial}{\partial\alpha}\left[g_\beta\left(1+\frac{\zeta}{R_\beta}\right)\tau_{\alpha\zeta}^{(\zeta)}\right]+\frac{\partial}{\partial\beta}\left[g_\alpha\left(1+\frac{\zeta}{R_\alpha}\right)\tau_{\beta\zeta}^{(\zeta)}\right]$$
$$+\frac{\partial}{\partial\zeta}\left[g_\alpha g_\beta\left(1+\frac{\zeta}{R_\alpha}\right)\left(1+\frac{\zeta}{R_\beta}\right)\sigma_\zeta^{(\zeta)}\right]$$
$$-\frac{g_\alpha g_\beta}{R_\alpha}\left(1+\frac{\zeta}{R_\beta}\right)\sigma_\alpha^{(\zeta)}-\frac{g_\alpha g_\beta}{R_\beta}\left(1+\frac{\zeta}{R_\alpha}\right)\sigma_\beta^{(\zeta)}$$
$$+g_\alpha g_\beta\left(1+\frac{\zeta}{R_\alpha}\right)\left(1+\frac{\zeta}{R_\beta}\right)p_\zeta^{(\zeta)}=0. \tag{2.2.5c}$$

Equilibrium of moments about three orthogonal axes has been satisfied automatically by taking the magnitudes of shear stresses on adjacent faces to be equal, i.e.,

$$\tau_{\alpha\beta}=\tau_{\beta\alpha},\qquad \tau_{\alpha\zeta}=\tau_{\zeta\alpha},\qquad \tau_{\beta\zeta}=\tau_{\zeta\beta}. \tag{2.2.6}$$

The strain components corresponding to the stress at each point in the shell wall are six in number, defined by the extensions of three originally orthogonal line elements passing through the point and the change in the right angle between each pair of line elements. These line elements, in our coordinate system, lie in the direction of the lines of curvature of a surface parallel to the middle surface and the normal to the parallel surface. Three of the strain components, those associated with line elements on the parallel surface, can be obtained from the relations of Section 1.5 and are given by

$$\begin{aligned}\epsilon_\alpha^{(\zeta)} &= \frac{1}{g_\alpha^{(\zeta)}}\left[\frac{\partial u_\alpha^{(\zeta)}}{\partial \alpha} + \frac{1}{g_\beta^{(\zeta)}}\frac{\partial g_\alpha^{(\zeta)}}{\partial \beta} u_\beta^{(\zeta)} + \frac{g_\alpha^{(\zeta)}}{R_\alpha^{(\zeta)}} w^{(\zeta)}\right] \\ &= \frac{1}{g_\alpha\left(1 + \dfrac{\zeta}{R_\alpha}\right)}\left[\frac{\partial u_\alpha^{(\zeta)}}{\partial \alpha} + \frac{1}{g_\beta}\frac{\partial g_\alpha}{\partial \beta} u_\beta^{(\zeta)} + \frac{g_\alpha}{R_\alpha} w^{(\zeta)}\right],\end{aligned} \tag{2.2.7a}$$

$$\begin{aligned}\epsilon^{(\zeta)} &= \frac{1}{g_\beta^{(\zeta)}}\left[\frac{\partial u_\beta^{(\zeta)}}{\partial \beta} + \frac{1}{g_\alpha^{(\zeta)}}\frac{\partial g_\beta^{(\zeta)}}{\partial \alpha} u_\alpha^{(\zeta)} + \frac{g_\beta^{(\zeta)}}{R_\beta^{(\zeta)}} w^{(\zeta)}\right] \\ &= \frac{1}{g_\beta\left(1 + \dfrac{\zeta}{R_\beta}\right)}\left[\frac{\partial u_\beta^{(\zeta)}}{\partial \beta} + \frac{1}{g_\alpha}\frac{\partial g_\beta}{\partial \alpha} u_\alpha^{(\zeta)} + \frac{g_\beta}{R_\beta} w^{(\zeta)}\right],\end{aligned} \tag{2.2.7b}$$

$$\begin{aligned}\gamma_{\alpha\beta}^{(\zeta)} &= \frac{1}{g_\alpha^{(\zeta)}}\left[\frac{\partial u_\beta^{(\zeta)}}{\partial \alpha} - \frac{1}{g_\beta^{(\zeta)}}\frac{\partial g_\alpha^{(\zeta)}}{\partial \beta} u_\alpha^{(\zeta)}\right] + \frac{1}{g_\beta^{(\zeta)}}\left[\frac{\partial u_\alpha^{(\zeta)}}{\partial \beta} - \frac{1}{g_\alpha^{(\zeta)}}\frac{\partial g_\beta^{(\zeta)}}{\partial \alpha} u_\beta^{(\zeta)}\right] \\ &= \frac{1}{g_\alpha\left(1 + \dfrac{\zeta}{R_\alpha}\right)}\left[\frac{\partial u_\beta^{(\zeta)}}{\partial \alpha} - \frac{1}{g_\beta}\frac{\partial g_\alpha}{\partial \beta} u_\alpha^{(\zeta)}\right] \\ &\quad + \frac{1}{g_\beta\left(1 + \dfrac{\zeta}{R_\beta}\right)}\left[\frac{\partial u_\alpha^{(\zeta)}}{\partial \beta} - \frac{1}{g_\alpha}\frac{\partial g_\beta}{\partial \alpha} u_\beta^{(\zeta)}\right].\end{aligned} \tag{2.2.7c}$$

Implied in these equations are the three remaining strain components, for the middle surface normals and the parallel surface lines of curvature are themselves lines of curvature of the ruled surfaces defined by constant α or constant β. Eqs. (2.2.7) then apply to these surfaces if we make the appropriate changes of displacement components and fundamental quantities. Thus the remaining strain components are given by

$$\epsilon_\zeta^{(\zeta)} = \frac{1}{g_\zeta^{(\zeta)}}\left[\frac{\partial w^{(\zeta)}}{\partial \zeta} + \frac{1}{g_\beta^{(\zeta)}}\frac{\partial g_\zeta^{(\zeta)}}{\partial \beta} u^{(\zeta)} - \frac{g_\zeta^{(\zeta)}}{R_\zeta^{(\zeta)}} u_\alpha^{(\zeta)}\right], \tag{2.2.8a}$$

or

$$\epsilon_\zeta^{(\zeta)} = \frac{1}{g_\zeta^{(\zeta)}}\left[\frac{\partial w^{(\zeta)}}{\partial \zeta} + \frac{1}{g_\alpha^{(\zeta)}}\frac{\partial g_\zeta^{(\zeta)}}{\partial \alpha} u_\alpha^{(\zeta)} - \frac{g_\zeta^{(\zeta)}}{R_\zeta^{(\zeta)}} u_\beta^{(\zeta)}\right], \tag{2.2.8b}$$

and

$$\gamma_{\alpha\zeta}^{(\zeta)} = \frac{1}{g_\alpha^{(\zeta)}}\left[\frac{\partial w^{(\zeta)}}{\partial \alpha} - \frac{1}{g_\zeta^{(\zeta)}}\frac{\partial g_\alpha^{(\zeta)}}{\partial \zeta} u_\alpha^{(\zeta)}\right] + \frac{1}{g_\zeta^{(\zeta)}}\left[\frac{\partial u_\alpha^{(\zeta)}}{\partial \zeta} - \frac{1}{g_\alpha^{(\zeta)}}\frac{\partial g_\zeta^{(\zeta)}}{\partial \alpha} w^{(\zeta)}\right], \tag{2.2.8c}$$

$$\gamma_{\beta\zeta}^{(\zeta)} = \frac{1}{g_\beta^{(\zeta)}}\left[\frac{\partial w^{(\zeta)}}{\partial \beta} - \frac{1}{g_\zeta^{(\zeta)}}\frac{\partial g_\beta^{(\zeta)}}{\partial \zeta} u_\beta^{(\zeta)}\right] + \frac{1}{g_\zeta^{(\zeta)}}\left[\frac{\partial u_\beta^{(\zeta)}}{\partial \zeta} - \frac{1}{g_\beta^{(\zeta)}}\frac{\partial g_\zeta^{(\zeta)}}{\partial \alpha} w^{(\zeta)}\right]. \tag{2.2.8d}$$

Since the ζ coordinate curves are straight lines and $d\zeta$ is the length of a line element on the ruled surfaces, we have

$$g_\zeta^{(\zeta)} = 1, \tag{2.2.9a}$$

$$R_\zeta^{(\zeta)} = \infty. \tag{2.2.9b}$$

Thus Eqs. (2.2.8) reduce to

$$\epsilon_\zeta^{(\zeta)} = \frac{\partial w^{(\zeta)}}{\partial \zeta}, \tag{2.2.10a}$$

$$\gamma_{\alpha\zeta}^{(\zeta)} = \frac{\partial u_\alpha^{(\zeta)}}{\partial \zeta} + \frac{1}{g_\alpha(1 + \zeta/R_\alpha)}\left[\frac{\partial w^{(\zeta)}}{\partial \alpha} - \frac{R_\alpha}{g_\alpha} u_\alpha^{(\zeta)}\right], \tag{2.2.10b}$$

$$\gamma_{\beta\zeta}^{(\zeta)} = \frac{\partial y_\beta^{(\zeta)}}{\partial \zeta} + \frac{1}{g_\beta(1 + \zeta/R_\beta)}\left[\frac{\partial w^{(\zeta)}}{\partial \alpha} - \frac{g_\beta}{R_\beta} u_\beta^{(\zeta)}\right]. \tag{2.2.10c}$$

These strains can be interpreted as the unit change of length of a straight line element originally normal to a parallel surface and the inclinations of the straight line from the normal to the deformed parallel surface.

The last required elements of the theory of elasticity are relations between stress and strain. For the purposes of our discussion of the theory of shells, it is convenient to consider the shell material to be isotropic in layers parallel to the middle surface but to have different material properties in the direction normal to the middle surface.† Then the stress–strain relations are given by

$$\epsilon_\alpha^{(\zeta)} = \frac{1}{E}[\sigma_\alpha^{(\zeta)} - \nu\sigma_\beta^{(\zeta)}] - \frac{\nu_\zeta}{E_\zeta}\sigma_\zeta^{(\zeta)}, \tag{2.2.11a}$$

$$\epsilon_\beta^{(\zeta)} = \frac{1}{E}[\sigma_\beta^{(\zeta)} - \nu\sigma_\alpha^{(\zeta)}] - \frac{\nu_\zeta}{E_\zeta}\sigma_\zeta^{(\zeta)}, \tag{2.2.11b}$$

$$\epsilon_\zeta^{(\zeta)} = \frac{1}{E_\zeta}\{\sigma_\zeta^{(\zeta)} - \nu_\zeta[\sigma_\alpha^{(\zeta)} + \sigma_\beta^{(\zeta)}]\}, \tag{2.2.11c}$$

$$\gamma_{\alpha\beta}^{(\zeta)} = \frac{2(1 + \nu)}{E}\tau_{\alpha\beta}^{(\zeta)}, \tag{2.2.11d}$$

† Such a material is said to be transversely isotropic, where the term 'transverse' pertains to directions in tangent planes which are perpendicular to the normal to the shell middle surface. We shall use the term 'transverse' later on, however, to indicate the direction of the normal. Both uses of the term are extant in the literature, the only difference being in the reference direction.

$$\gamma_{\alpha\zeta}^{(\zeta)} = \frac{1}{G_\zeta}\tau_{\alpha\zeta}^{(\zeta)}, \tag{2.2.11e}$$

$$\gamma_{\beta\zeta}^{(\zeta)} = \frac{1}{G_\zeta}\tau_{\beta\zeta}^{(\zeta)}. \tag{2.2.11f}$$

For a completely isotropic shell we have

$$E_\zeta = E \tag{2.2.12a}$$

$$\nu_\zeta = \nu, \tag{2.2.12b}$$

$$G_\zeta = \frac{E}{2(1+\nu)}. \tag{2.2.12c}$$

The equilibrium equations, strain–displacement relations, and stress–strain relations given by Eqs. (2.2.5), (2.2.7), (2.2.10), and (2.2.11), together with appropriate boundary conditions, completely determine the distribution of stress and displacement within the shell wall.

2.3 Assumptions of shell theory

In most cases, solutions of the equations of three-dimensional elasticity theory are extremely difficult to obtain so that approximations to the theory are desirable. The so-called theory of shells is an approximation which replaces the three-dimensional problem by a two-dimensional problem which is somewhat easier to consider. The assumptions underlying the theory of shells have been variously stated and, indeed, the theory can be developed in a number of different ways. However, the hypothesis that leads most consistently to the usually accepted equations of the theory is as follows:

> The distributions of stresses and displacements in a shell composed of a hypothetical material which has infinite resistance to extension in the direction normal to the middle surface and infinite resistance to shearing in planes normal to the middle surface are an adequate representation of the corresponding distributions in a shell composed of an isotropic material.

For a transversely† rigid material we must take

$$\nu_\zeta = 0, \tag{2.3.1a}$$

$$E_\zeta = G_\zeta = \infty \tag{2.3.1b}$$

† The term 'transverse' pertains to the direction of the normal as used here.

in Eqs. (2.2.10), which reduce to

$$\epsilon_{\alpha}^{(\zeta)} = \frac{1}{E}[\sigma_{\alpha}^{(\zeta)} - \nu\sigma_{\beta}^{(\zeta)}], \tag{2.3.2a}$$

$$\epsilon_{\beta}^{(\zeta)} = \frac{1}{E}[\sigma_{\beta}^{(\zeta)} - \nu\sigma_{\alpha}^{(\zeta)}], \tag{2.3.2b}$$

$$\gamma_{\alpha\beta}^{(\zeta)} = \frac{2(1+\nu)}{E}\tau_{\alpha\beta}^{(\zeta)}, \tag{2.3.2c}$$

$$\epsilon_{\zeta}^{(\zeta)} = \gamma_{\alpha\zeta}^{(\zeta)} = \gamma_{\beta\zeta}^{(\zeta)} = 0. \tag{2.3.2d}$$

The stress–strain relations in the form of Eqs. (2.3.2) correspond to those implied by the usual assumptions of shell theory attributed to Love and Kirchhoff, which can be stated as follows:

a. Straight lines normal to the undeformed middle surface remain straight and normal to the deformed middle surface and do not change length.
b. The normal stresses acting on surfaces parallel to the middle surface may be neglected in comparison with the other stresses.

To show this we recall the definitions of $\epsilon_{\zeta}^{(\zeta)}$, $\gamma_{\alpha\zeta}^{(\zeta)}$, and $\gamma_{\beta\zeta}^{(\zeta)}$ given in Section 2.2. Thus, $\epsilon_{\zeta}^{(\zeta)}$ is the extensional strain of a line element originally normal to a surface parallel to the middle surface and $\gamma_{\alpha\zeta}^{(\zeta)}$ and $\gamma_{\beta\zeta}^{(\zeta)}$ are the inclinations of the line element from the normal to the deformed parallel surface. If these three strains are zero, it follows that a line element normal to an undeformed parallel surface remains normal to the deformed parallel surface and does not change length, or as another way of expressing the same facts: surfaces that were parallel before deformation remain parallel after deformation and the same distance apart. Since, from Section 1.4, the normal to a parallel surface is also the normal to the middle surface, it is readily seen that assumption (a) is satisfied. That assumption (b) is satisfied is obvious since the stress–strain relations do not contain $\sigma_{\zeta}^{(\zeta)}$, which is the only place the assumption is actually used in the usual development of the theory. However, whereas the assumptions in the form of the above statements are contradictory, since we assume both the transverse direct stress and extensional strain to vanish, the same assumptions in the form of statements about the material properties are entirely consistent and indicate that the usual version of the theory of shells should be considered as a special case of the theory of elasticity for orthotropic materials rather than an approximation to the theory of elasticity for isotropic materials. It should be added that numerous attempts to derive a theory of shells directly from the theory of elasticity for isotropic materials have shown that it is impossible to obtain so physically plausible a set of equations with so few assumptions and in so natural a manner.

2.4 Displacements and strains

In the section above, we have shown that the transversely rigid material hypothesis of shell theory is equivalent to the Love-Kirchhoff deformation assumptions. It is convenient to use the latter to determine the state of deformation in the shell wall. The radius vector to a point in the undeformed shell wall can be written as

$$\mathbf{r}^{(\zeta)} = \mathbf{r} + \zeta\mathbf{e}_\zeta, \tag{2.4.1a}$$

where $\mathbf{r}$ is the radius vector to a point on the middle surface, $\mathbf{e}_\zeta$ is the unit vector normal to the middle surface at that point, and ζ is the distance along the normal from the middle surface point to the point in the shell wall. The radius vector to the same point in the deformed shell is defined by Kirchhoff's assumption (a) as

$$\mathbf{r}'^{(\zeta)} = \mathbf{r}' + \zeta\mathbf{e}'_\zeta, \tag{2.4.1b}$$

where $\mathbf{r}'$ is the radius vector to the displaced middle surface point and $\mathbf{e}'_\zeta$ is the unit vector normal to the deformed middle surface. The displacement vector of the point in the shell wall relative to its original position is then

$$\begin{aligned}\mathbf{u}^{(\zeta)} &= \mathbf{r}'^{(\zeta)} - \mathbf{r}^{(\zeta)} \\ &= \mathbf{r}' - \mathbf{r} + \zeta(\mathbf{e}'_\zeta - \mathbf{e}_\zeta).\end{aligned} \tag{2.4.2}$$

The displacement vector of the middle surface point relative to its original position can be expressed in terms of displacements in the middle directions of the orthogonal unit vectors tangent, and normal to the undeformed middle surface as

$$\mathbf{u} = \mathbf{r}' - \mathbf{r} = u_\alpha\mathbf{e}_\alpha + u_\beta\mathbf{e}_\beta + w\mathbf{e}_\zeta. \tag{2.4.3}$$

From Eq. (1.5.7c), which gives the unit vector normal to deformed surface in terms of these same unit vectors and the displacements of the middle surface, we have

$$\mathbf{e}'_\zeta = \mathbf{e}_\zeta - \Theta_\alpha\mathbf{e}_\alpha - \Theta_\beta\mathbf{e}_\beta, \tag{2.4.4a}$$

where, from Eqs. (1.5.3c) and (1.5.3f)

$$\Theta_\alpha = \frac{1}{g_\alpha}\left(\frac{\partial w}{\partial \alpha} - \frac{g_\alpha}{R_\alpha}u_\alpha\right), \tag{2.4.4b}$$

$$\Theta_\beta = \frac{1}{g_\beta}\left(\frac{\partial w}{\partial \beta} - \frac{g_\beta}{R_\beta} u_\beta\right). \tag{2.4.4c}$$

Thus, the displacement vector of a point in the shell wall can be expressed as

$$\mathbf{u}^{(\zeta)} = u_\alpha^{(\zeta)}\mathbf{e}_\alpha + u_\beta^{(\zeta)}\mathbf{e}_\beta + w^{(\zeta)}\mathbf{e}_\zeta, \tag{2.4.5a}$$

$$u_\alpha^{(\zeta)} = u_\alpha - \zeta\Theta_\alpha, \tag{2.4.5b}$$

$$u_\beta^{(\zeta)} = u_\beta - \zeta\Theta_\beta, \tag{2.4.5c}$$

$$w^{(\zeta)} = w. \tag{2.4.5d}$$

The tangential displacements are thus seen to be linear across the shell wall and the normal displacement to be constant.

With the displacements given by Eqs. (2.4.5), we can readily find by substitution in Eqs. (2.2.7) and (2.2.10), that the strain components are given by

$$\epsilon_\alpha^{(\zeta)} = \frac{1}{g_\alpha(1 + \zeta/R_\alpha)}\left[\frac{\partial u_\alpha}{\partial \alpha} + \frac{1}{g_\beta}\frac{\partial g_\alpha}{\partial \beta} u_\beta + \frac{g_\alpha}{R_\alpha} w - \zeta\left(\frac{\partial \Theta_\alpha}{\partial \alpha} + \frac{1}{g_\beta}\frac{\partial g_\alpha}{\partial \beta}\Theta_\beta\right)\right]. \tag{2.4.6a}$$

$$\epsilon_\beta^{(\zeta)} = \frac{1}{g_\beta(1 + \zeta/R_\beta)}\left[\frac{\partial u_\beta}{\partial \beta} + \frac{1}{g_\beta}\frac{\partial g_\beta}{\partial \alpha} u_\alpha + \frac{g_\beta}{R_\beta} w - \zeta\left(\frac{\partial \Theta_\beta}{\partial \beta} + \frac{1}{g_\alpha}\frac{\partial g_\beta}{\partial \alpha}\Theta_\alpha\right)\right], \tag{2.4.6b}$$

$$\begin{aligned}\gamma_{\alpha\beta}^{(\zeta)} = {} & \frac{1}{g_\alpha(1 + \zeta/R_\alpha)}\left[\frac{\partial u_\beta}{\partial \alpha} - \frac{1}{g_\beta}\frac{\partial g_\alpha}{\partial \beta} u_\alpha - \zeta\left(\frac{\partial \Theta_\beta}{\partial \alpha} - \frac{1}{g_\beta}\frac{\partial g_\alpha}{\partial \beta}\Theta_\alpha\right)\right] \\ & + \frac{1}{g_\beta(1 + \zeta/R_\beta)}\left[\frac{\partial u_\alpha}{\partial \alpha} - \frac{1}{g_\alpha}\frac{\partial g_\beta}{\partial \alpha} u_\beta - \zeta\left(\frac{\partial \Theta_\alpha}{\partial \beta} - \frac{1}{g_\alpha}\frac{\partial g_\beta}{\partial \alpha}\Theta_\beta\right)\right],\end{aligned} \tag{2.4.6c}$$

$$\epsilon_\zeta^{(\zeta)} = \gamma_{\alpha\zeta}^{(\zeta)} = \gamma_{\beta\zeta}^{(\zeta)} = 0. \tag{2.4.6d}$$

We note that although displacements are linear through the shell wall, strain components are not, since the width of the shell element varies with distance from the middle surface.

With the definitions of the strains and changes of curvature of the middle surface given by Eq. (1.5.3) and (1.5.10), the expressions for the two extensional strain components can be reduced to

$$\epsilon_\alpha^{(\zeta)} = \frac{\epsilon_\alpha + \zeta\kappa_\alpha}{1 + \zeta/R_\alpha}, \tag{2.4.7a}$$

$$\epsilon_\beta^{(\zeta)} = \frac{\epsilon_\beta + \zeta\kappa_\beta}{1 + \zeta/R_\beta}, \tag{2.4.7b}$$

which relate the extensional strains at any point in the shell wall to extensional strains and curvature changes of the middle surface. Eq. (2.4.6c) for the shear strain at any point requires some manipulation. With the use of Eqs. (1.5.3b), (1.5.3e), and (1.5.10c), the shear strain can be written as

$$\gamma_{\alpha\beta}^{(\zeta)} = \frac{\gamma_\alpha + \zeta\left(\kappa_{\alpha\beta} - \dfrac{\gamma_\beta}{R_\alpha}\right)}{1 + \dfrac{\zeta}{R_\alpha}} + \frac{\gamma_\beta + \zeta\left(\kappa_{\alpha\beta} - \dfrac{\gamma_\alpha}{R_\beta}\right)}{1 + \dfrac{\zeta}{R_\beta}}. \tag{2.4.8a}$$

By reducing the expression to one having a common denominator and using the relationship

$$\gamma_{\alpha\beta} = \gamma_\alpha + \gamma_\beta, \tag{2.4.8b}$$

we finally have

$$\gamma_{\alpha\beta}^{(\zeta)} = \frac{\gamma_{\alpha\beta}\left(1 - \dfrac{\zeta^2}{R_\alpha R_\beta}\right) + 2\zeta\kappa_{\alpha\beta}\left[1 + \tfrac{1}{2}\zeta\left(\dfrac{1}{R_\alpha} + \dfrac{1}{R_\beta}\right)\right]}{\left(1 + \dfrac{\zeta}{R_\alpha}\right)\left(1 + \dfrac{\zeta}{R_\beta}\right)}, \tag{2.4.8c}$$

which relates the shear strain at any point to the shear strain and the twist of the middle surface.

Eq. (2.4.8c) is the most common formulation used in the literature. A recent investigation† shows that if we use κ_α and κ_β as primary quantities rather than the actual curvature changes $\kappa_\alpha - \epsilon_\alpha/R_\alpha$ and $\kappa_\beta - \epsilon_\beta/R_\beta$, then we should also use the modified twist

$$\bar{\kappa}_{\alpha\beta} = \kappa_{\alpha\beta} - \frac{1}{4}\left(\frac{1}{R_\alpha} + \frac{1}{R_\beta}\right)\gamma_{\alpha\beta}, \tag{2.4.9}$$

if the equations to be derived are to be expressed as the lines of curvature reduction of a formulation of the theory of shells which is of the same form for all coordinate systems. Then Eq. (2.4.8c) becomes

$$\gamma_{\alpha\beta}^{(\zeta)} = \frac{\left[1 + \tfrac{1}{2}\zeta\left(\dfrac{1}{R_\alpha} + \dfrac{1}{R_\beta}\right)\right](\gamma_{\alpha\beta} + 2\zeta\bar{\kappa}_{\alpha\beta}) + \left[\dfrac{\zeta}{2}\left(\dfrac{1}{R_\alpha} - \dfrac{1}{R_\beta}\right)\right]^2 \gamma_{\alpha\beta}}{\left(1 + \dfrac{\zeta}{R_\alpha}\right)\left(1 + \dfrac{\zeta}{R_\beta}\right)}, \tag{2.4.10}$$

† B. Budiansky and J. L. Sanders, Jr.: On the Best First-Order Linear Shell Theory, *Progress in Applied Mechanics*, The Prager Anniversary Volume, The Macmillan Co., New York, 1963, pp. 129–140.

which has the virtue of reducing to a form similar to that of a flat plate when the shell surface is spherical. The question of form invariance of the equations is immaterial so long as we make no approximations in the theory other than the basic Love-Kirchhoff assumptions. The question gains relevance, however, when other approximations are introduced, as in the following chapter. For this reason the shear strain expressed as in Eq. (2.4.10) shall be used in all subsequent discussions.

2.5 Stress resultants and equilibrium equations

At this point in our development of the theory of shells, it is possible to proceed along several paths. If we were to take an entirely straightforward (but brute force) approach, we could express the stresses $\sigma_\alpha^{(\zeta)}$, $\sigma_\beta^{(\zeta)}$, and $\tau_{\alpha\beta}^{(\zeta)}$ in terms of middle surface displacements by means of Eqs. (2.3.2), (2.4.4), and (2.4.6). Then, from Eqs. (2.2.5), we could solve for $\sigma_\zeta^{(\zeta)}$, $\tau_{\alpha\zeta}^{(\zeta)}$, and $\tau_{\beta\zeta}^{(\zeta)}$ in terms of the middle surface displacements and three arbitrary functions of α and β. Finally, six differential equations for the middle surface displacements and the arbitrary functions could be obtained by equating these expressions for $\sigma_\zeta^{(\zeta)}$, $\tau_\alpha^{(\zeta)}$, and $\tau_{\beta\zeta}^{(\zeta)}$ to their known values on the inner and outer parallel bounding surfaces defined by $\zeta = \pm h/2$. It is possible, however, to avoid having to consider the detailed distributions of transverse normal and shear stresses within the shell wall and to deal only with the known values of these stresses on the surfaces bounding the shell wall by integrating the equations of equilibrium over the wall thickness in a particular fashion.

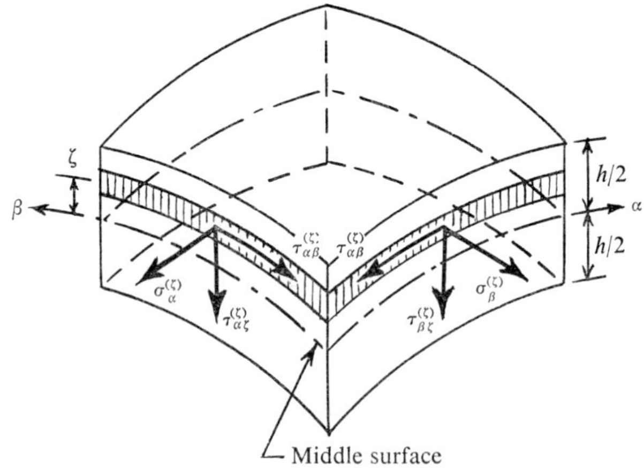

Fig. 2.2. Over-all shell element

Stress resultants are customarily employed, for physical clarity, in the derivation of displacement equations. These quantities, expressed as moments and forces per unit length of middle surface, act on the sides of a shell element bounded by the parallel surfaces $\zeta = \pm h/2$ and planes normal to the middle surface. The sides of this element will be taken in the directions of adjacent lines of curvature (Fig. 2.2). The element treated in Section 2.2 to establish equations of equilibrium for stresses is a small portion of this overall shell element. We can define force components per unit length of the middle surfaces (see Fig. 2.3(a)) by integrating the appropriate stresses over the cross-section on which they act and dividing by the differential length of the middle surface line of curvature. In this way we obtain

$$N_\alpha = \int_{-h/2}^{h/2} \sigma_\alpha^{(\zeta)}\left(1 + \frac{\zeta}{R_\beta}\right) \mathrm{d}\zeta, \tag{2.5.1a}$$

$$N_\beta = \int_{-h/2}^{h/2} \sigma_\beta^{(\zeta)}\left(1 + \frac{\zeta}{R_\alpha}\right) \mathrm{d}\zeta, \tag{2.5.1b}$$

$$N_{\alpha\beta} = \int_{-h/2}^{h/2} \tau_{\alpha\beta}^{(\zeta)}\left(1 + \frac{\zeta}{R_\beta}\right) \mathrm{d}\zeta, \tag{2.5.1c}$$

$$N_{\beta\alpha} = \int_{-h/2}^{h/2} \tau_{\alpha\beta}^{(\zeta)}\left(1 + \frac{\zeta}{R_\alpha}\right) \mathrm{d}\zeta, \tag{2.5.1d}$$

$$Q_\alpha = \int_{-h/2}^{h/2} \tau_{\alpha\zeta}^{(\zeta)}\left(1 + \frac{\zeta}{R_\beta}\right) \mathrm{d}\zeta, \tag{2.5.1e}$$

$$Q_\beta = \int_{-h/2}^{h/2} \tau_{\beta\zeta}^{(\zeta)}\left(1 + \frac{\zeta}{R_\alpha}\right) \mathrm{d}\zeta. \tag{2.5.1f}$$

Bending and twisting moments per unit middle surface length (Fig. 2.3(b)) are similarly defined by

$$M_\alpha = \int_{-h/2}^{h/2} \zeta\sigma_\alpha^{(\zeta)}\left(1 + \frac{\zeta}{R_\beta}\right) \mathrm{d}\zeta, \tag{2.5.2a}$$

$$M_\beta = \int_{-h/2}^{h/2} \zeta\sigma_\beta^{(\zeta)}\left(1 + \frac{\zeta}{R_\alpha}\right) \mathrm{d}\zeta, \tag{2.5.2b}$$

$$M_{\alpha\beta} = \int_{-h/2}^{h/2} \zeta\tau_{\alpha\beta}^{(\zeta)}\left(1 + \frac{\zeta}{R_\beta}\right) \mathrm{d}\zeta, \tag{2.5.2c}$$

$$M_{\beta\alpha} = \int_{-h/2}^{h/2} \zeta\tau_{\alpha\beta}^{(\zeta)}\left(1 + \frac{\zeta}{R_\alpha}\right) \mathrm{d}\zeta. \tag{2.5.2d}$$

The forces and moments defined by Eqs. (2.5.1) and (2.5.2) are in equilibrium with external and body forces acting on the entire shell element.

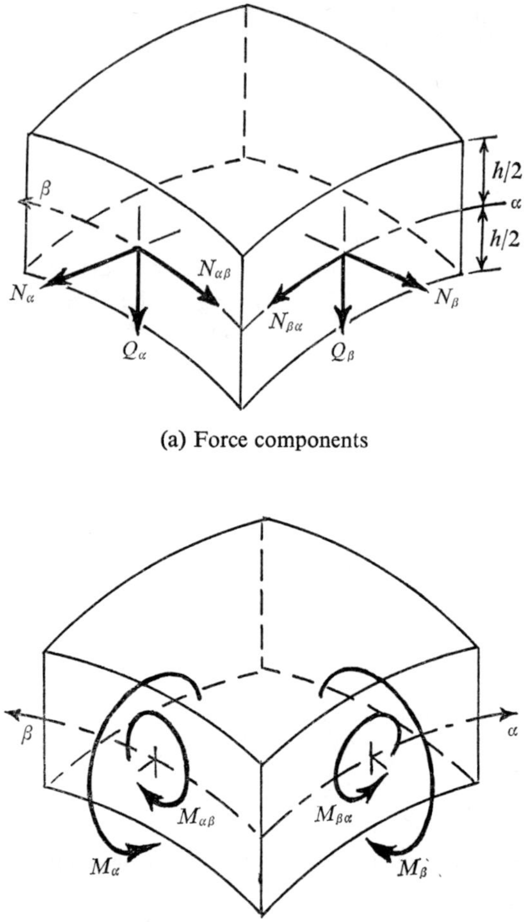

(a) Force components

(b) Moment components

Fig. 2.3. Stress resultants acting on shell element

Since the shell element considered as a rigid body has six degrees of freedom, these forces and moments must satisfy six overall equilibrium conditions which can be obtained by equating the sum of the forces in the directions of the tangent and normal vectors to zero, and by equating the sum of moments about three axes in these same directions to zero. Since the individual stresses making up these force and moment resultants satisfy Eqs. (2.2.5), it is possible to utilize these equations to obtain the overall equations of equilibrium.

We multiply each of Eqs. (2.2.5) by $d\zeta$ and integrate over the shell thickness to obtain, with the use of Eqs. (2.5.1), the following equations of force equilibrium:

$$\frac{\partial}{\partial\alpha}(g_\beta N_\alpha) + \frac{\partial}{\partial\beta}(g_\alpha N_{\beta\alpha}) + \frac{\partial g_\alpha}{\partial\beta} N_{\alpha\beta} - \frac{\partial g_\beta}{\partial\alpha} N_\beta + g_\alpha g_\beta\left(\frac{Q_\alpha}{R_\alpha} + q_\alpha\right) = 0, \tag{2.5.3a}$$

$$\frac{\partial}{\partial\beta}(g_\alpha N_\beta) + \frac{\partial}{\partial\alpha}(g_\beta N_{\alpha\beta}) + \frac{\partial g_\beta}{\partial\alpha} N_{\beta\alpha} - \frac{\partial g_\alpha}{\partial\beta} N_\alpha + g_\alpha g_\beta\left(\frac{Q_\beta}{R_\beta} + q_\beta\right) = 0, \tag{2.5.3b}$$

$$\frac{\partial}{\partial\alpha}(g_\beta Q_\alpha) + \frac{\partial}{\partial\beta}(g_\alpha Q_\beta) - g_\alpha g_\beta\left(\frac{N_\alpha}{R_\alpha} + \frac{N_\beta}{R_\beta} - q_\zeta\right) = 0, \tag{2.5.3c}$$

where q_α, q_β, and q_ζ are resultants per unit middle surface area of surface and body forces acting on the shell element (Fig. 2.4(a)), given by

$$q_\alpha = \left(1 + \frac{h}{2R_\alpha}\right)\left(1 + \frac{h}{2R_\beta}\right)\tau_{\alpha\zeta}^{(h/2)} - \left(1 - \frac{h}{2R_\alpha}\right)\left(1 - \frac{h}{2R_\beta}\right)\tau_{\alpha\zeta}^{(-h/2)} + \int_{-h/2}^{h/2}\left(1 + \frac{\zeta}{R_\alpha}\right)\left(1 + \frac{\zeta}{R_\beta}\right)\rho_\alpha^{(\zeta)}\,\mathrm{d}\zeta, \tag{2.5.4a}$$

$$q_\beta = \left(1 + \frac{h}{2R_\alpha}\right)\left(1 + \frac{h}{2R_\beta}\right)\tau_{\beta\zeta}^{(h/2)} - \left(1 - \frac{h}{2R_\alpha}\right)\left(1 - \frac{h}{2R_\beta}\right)\tau_{\beta\zeta}^{(-h/2)} + \int_{-h/2}^{h/2}\left(1 + \frac{\zeta}{R_\alpha}\right)\left(1 + \frac{\zeta}{R_\beta}\right)\rho_\beta^{(\zeta)}\,\mathrm{d}\zeta, \tag{2.5.4b}$$

$$q_\zeta = \left(1 + \frac{h}{2R_\alpha}\right)\left(1 + \frac{h}{2R_\beta}\right)\sigma_\zeta^{(h/2)} - \left(1 - \frac{h}{2R_\alpha}\right)\left(1 - \frac{h}{2R_\beta}\right)\sigma_\zeta^{(-h/2)} + \int_{-h/2}^{h/2}\left(1 + \frac{\zeta}{R_\alpha}\right)\left(1 + \frac{\zeta}{R_\beta}\right)\rho_\zeta^{(\zeta)}\,\mathrm{d}\zeta. \tag{2.5.4c}$$

Two equations of equilibrium of moments about middle surface axes in the direction of the tangent vectors are obtained by multiplying each of Eqs. (2.2.5a) and (2.2.5b) by $\zeta\,\mathrm{d}\zeta$ and integrating over the shell thickness to obtain, with the aid of Eqs. (2.5.1) and (2.5.2),

$$\frac{\partial}{\partial\alpha}(g_\beta M_\alpha) + \frac{\partial}{\partial\beta}(g_\alpha M_{\beta\alpha}) + \frac{\partial g_\alpha}{\partial\beta} M_{\alpha\beta} - \frac{\partial g_\beta}{\partial\alpha} M_\beta - g_\alpha g_\beta(Q_\alpha - m_\alpha) = 0, \tag{2.5.5a}$$

$$\frac{\partial}{\partial\beta}(g_\alpha M_\beta) + \frac{\partial}{\partial\alpha}(g_\beta M_{\alpha\beta}) + \frac{\partial g_\beta}{\partial\alpha} M_{\beta\alpha} - \frac{\partial g_\alpha}{\partial\beta} M_\alpha - g_\alpha g_\beta(Q_\beta - m_\beta) = 0, \tag{2.5.5b}$$

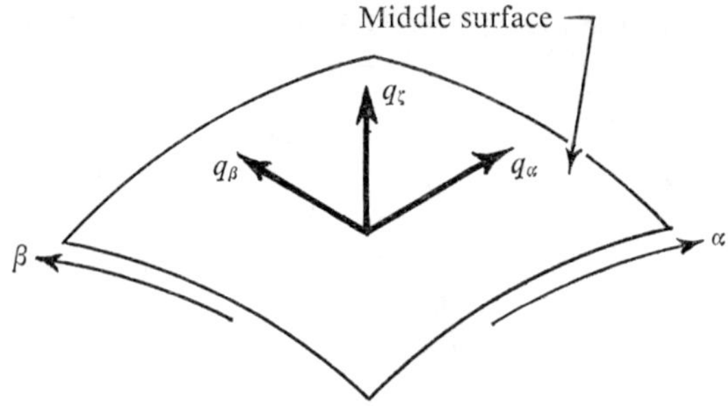

(a) External loading force components

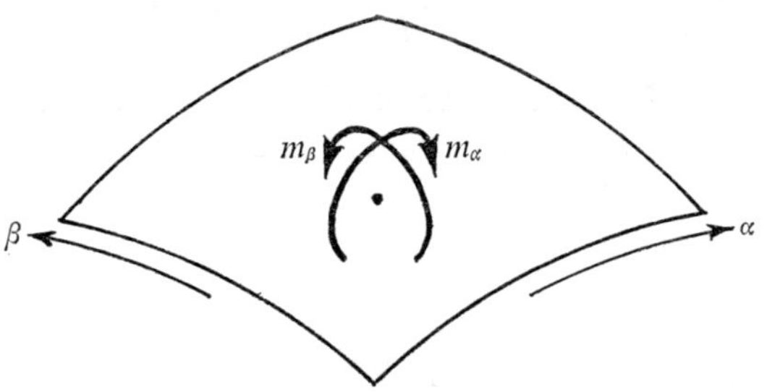

(b) External loading moment components

Fig. 2.4. Surface and body stress resultants

where the resultant moments per unit middle surface area of surface and body forces (Fig. 2.4b) are given by

$$m_\alpha = \left[\left(1 + \frac{h}{2R_\alpha}\right)\left(1 + \frac{h}{2R_\beta}\right)\tau_{\alpha\zeta}^{(h/2)} + \left(1 - \frac{h}{2R_\alpha}\right)\left(1 - \frac{h}{2R_\beta}\right)\tau_{\alpha\zeta}^{(-h/2)}\right]\frac{h}{2} + \int_{-h/2}^{h/2} \zeta\left(1 + \frac{\zeta}{R_\alpha}\right)\left(1 + \frac{\zeta}{R_\beta}\right)\rho_\alpha^{(\zeta)}\,d\zeta, \tag{2.5.6a}$$

$$m_\beta = \left[\left(1 + \frac{h}{2R_\alpha}\right)\left(1 + \frac{h}{2R_\beta}\right)\tau_{\beta\zeta}^{(h/2)} + \left(1 - \frac{h}{2R_\alpha}\right)\left(1 - \frac{h}{2R_\beta}\right)\tau_{\beta\zeta}^{(-h/2)}\right]\frac{h}{2} + \int_{-h/2}^{h/2} \zeta\left(1 + \frac{\zeta}{R_\alpha}\right)\left(1 + \frac{\zeta}{R_\beta}\right)\rho_\beta^{(\zeta)}\,d\zeta. \tag{2.5.6b}$$

The final equation for equilibrium of moments about the normal to the middle surface is automatically satisfied for the microscopic element of the shell wall by the condition that shear stresses on adjoining faces are equal. For the entire shell element this condition becomes merely an algebraic relationship between shear force and twisting moment resultants; viz.,

$$N_{\alpha\beta} - N_{\beta\alpha} = \int_{-h/2}^{h/2} \tau_{\alpha\beta}\left(\frac{1}{R_\beta} - \frac{1}{R_\alpha}\right)\zeta\,d\zeta = \frac{M_{\beta\alpha}}{R_\beta} - \frac{M_{\alpha\beta}}{R_\alpha}. \tag{2.5.7}$$

2.6 Statically equivalent force systems

We shall, at this point, anticipate some of the results that are implied by the principle of virtual work for a shell element. The six equations in ten unknowns derived above can be reduced to five equations in eight unknowns by introducing the following quantities

$$\overline{M}_{\alpha\beta} = \tfrac{1}{2}(M_{\alpha\beta} + M_{\beta\alpha}) = \int_{-h/2}^{h/2} \zeta\tau_{\alpha\beta}^{(\zeta)}\left[1 + \frac{\zeta}{2}\left(\frac{1}{R_\alpha} + \frac{1}{R_\beta}\right)\right] \mathrm{d}\zeta, \qquad (2.6.1a)$$

$$\begin{aligned}\overline{N}_{\alpha\beta} &= N_{\alpha\beta} - \frac{M_{\beta\alpha}}{R_\beta} + \frac{1}{2}\left(\frac{1}{R_\alpha} + \frac{1}{R_\beta}\right)\overline{M}_{\alpha\beta}\\ &= N_{\beta\alpha} - \frac{M_{\alpha\beta}}{R_\alpha} + \frac{1}{2}\left(\frac{1}{R_\alpha} + \frac{1}{R_\beta}\right)\overline{M}_{\alpha\beta}\\ &= \int_{-h/2}^{h/2} \tau_{\alpha\beta}^{(\zeta)}\left[1 + \frac{\zeta}{2}\left(\frac{1}{R_\alpha} + \frac{1}{R_\beta}\right) + \frac{\zeta^2}{4}\left(\frac{1}{R_\alpha} - \frac{1}{R_\beta}\right)^2\right] \mathrm{d}\zeta, \qquad (2.6.1b)\dagger\end{aligned}$$

$$\overline{Q}_\alpha = Q_\alpha + \frac{1}{g_\beta}\frac{\partial M_{\alpha\beta}}{\partial\beta}, \qquad (2.6.1c)$$

$$\overline{Q}_\beta = Q_\beta + \frac{1}{g_\alpha}\frac{\partial M_{\beta\alpha}}{\partial\alpha}, \qquad (2.6.1d)$$

which will be shown later to be more appropriate for the determination of the stress state in a transversely rigid shell since the boundary conditions can be expressed in terms of these quantities. Then, Eq. (2.5.7) is identically satisfied and Eqs. (2.5.3) and (2.5.5) become

$$\begin{aligned}&\frac{\partial}{\partial\alpha}(g_\beta N_\alpha) - \frac{\partial g_\beta}{\partial\alpha}N_\beta + \frac{\partial}{\partial\beta}\left\{g_\alpha\left[\overline{N}_{\alpha\beta} + \frac{1}{2}\left(\frac{3}{R_\alpha} - \frac{1}{R_\beta}\right)\overline{M}_{\alpha\beta}\right]\right\}\\ &\qquad - \boxed{2\frac{\partial}{\partial\beta}\left(\frac{g_\alpha}{R_\alpha}\overline{M}_{\alpha\beta}\right)} + \left[\overline{N}_{\alpha\beta} + \frac{1}{2}\left(\frac{3}{R_\beta} - \frac{1}{R_\alpha}\right)\overline{M}_{\alpha\beta}\right]\frac{\partial g_\alpha}{\partial\beta}\\ &\qquad + g_\alpha g_\beta\left(\frac{1}{R_\alpha}\overline{Q}_\alpha + q_\alpha\right) = 0, \qquad (2.6.2a)\end{aligned}$$

† There are a number of ways effecting the reduction, but the modified quantities given here are consistent with the modified twist $\bar{\kappa}_{\alpha\beta}$ introduced by Eq. (2.4.9). The more commonly used modified quantities differ from those given above only in the use of a modified shear force

$$\overline{N}_{\alpha\beta} = N_{\alpha\beta} - \frac{M_{\beta\alpha}}{R_\beta} = N_{\beta\alpha} - \frac{M_{\alpha\beta}}{R_\alpha} = \int_{-h/2}^{h/2} \tau_{\alpha\beta}^{(\zeta)}\left(1 - \frac{\zeta^2}{R_\alpha R_\beta}\right) \mathrm{d}\zeta,$$

which is consistent with the use of the unmodified twist $\kappa_{\alpha\beta}$.

$$\frac{\partial}{\partial \beta}(g_\alpha N_\beta) - \frac{\partial g_\alpha}{\partial \beta} N_\alpha + \frac{\partial}{\partial \alpha}\left\{g_\beta\left[\overline{N}_{\alpha\beta} + \frac{1}{2}\left(\frac{3}{R_\beta} - \frac{1}{R_\alpha}\right)\overline{M}_{\alpha\beta}\right]\right\}$$

$$- \boxed{2\frac{\partial}{\partial \alpha}\left(\frac{g_\beta}{R_\beta}\overline{M}_{\alpha\beta}\right)} + \left[\overline{N}_{\alpha\beta} + \frac{1}{2}\left(\frac{3}{R_\alpha} - \frac{1}{R_\beta}\right)\overline{M}_{\alpha\beta}\right]\frac{\partial g_\beta}{\partial \alpha}$$

$$+ g_\alpha g_\beta\left(\frac{1}{R_\beta}\overline{Q}_\beta + q_\beta\right) = 0, \tag{2.6.2b}$$

$$\frac{\partial}{\partial \alpha}(g_\beta \overline{Q}_\alpha) + \frac{\partial}{\partial \beta}(g_\alpha \overline{Q}_\beta) - \boxed{2\frac{\partial^2 \overline{M}_{\alpha\beta}}{\partial \alpha\, \partial \beta}} - g_\alpha g_\beta\left(\frac{N_\alpha}{R_\alpha} + \frac{N_\beta}{R_\beta} - q_\zeta\right) = 0, \tag{2.6.2c}$$

$$\frac{\partial}{\partial \alpha}(g_\beta M_\alpha) - \frac{\partial g_\beta}{\partial \alpha} M_\beta + \boxed{2\frac{\partial}{\partial \beta}(g_\alpha \overline{M}_{\alpha\beta})} - g_\alpha g_\beta(\overline{Q}_\alpha - m_\alpha) = 0, \tag{2.6.2d}$$

$$\frac{\partial}{\partial \beta}(g_\alpha M_\beta) - \frac{\partial g_\alpha}{\partial \beta} M_\alpha + \boxed{2\frac{\partial}{\partial \alpha}(g_\beta \overline{M}_{\alpha\beta})} - g_\alpha g_\beta(\overline{Q}_\beta - m_\beta) = 0, \tag{2.6.2e}$$

which are similar in form to Eqs. (2.5.3) and (2.5.5) except for the boxed terms.

Although the quantities $\overline{N}_{\alpha\beta}$, $\overline{M}_{\alpha\beta}$, $\overline{Q}_\alpha$, and $\overline{Q}_\beta$ appear strange, combinations of these can be given a physical interpretation that will enable us to understand better what they represent. Let us consider a plane normal surface element in which lies a segment of, say, an α line of curvature of length ds_α. The twisting moment $M_{\beta\alpha}\, ds_\alpha$ acting on this segment (see Fig. 2.5a) is statically equivalent to the system of forces pictured in Fig. 2.5b, that is, two radial forces $M_{\beta\alpha} - \frac{1}{2}(\partial M_{\beta\alpha}/\partial \alpha)\, ds_\alpha$ and $M_{\beta\alpha} + \frac{1}{2}(\partial M_{\beta\alpha}/\partial s_\alpha)\, ds_\alpha$ acting at the corners of the element and a radial force $(\partial M_{\beta\alpha}/\partial s_\alpha)\, ds_\alpha$ and a tangential force $(M_{\beta\alpha}/R_\alpha)\, ds_\alpha$ acting on the plane surface element. We can show this by adding up all the force components and take moments about the center of curvature, remembering that the angle $d\phi$ is small. Now if we add the statically equivalent normal and tangential force components acting on the surface element to the normal and tangential shear forces $Q_\beta\, ds_\alpha$ and $N_{\beta\alpha}\, ds_\alpha$ acting on the surface we have total equivalent normal and tangential shear forces given by

$$\left(Q_\beta + \frac{\partial M_{\beta\alpha}}{\partial s_\alpha}\right) ds_\alpha = \left(Q_\beta + \frac{1}{g_\alpha}\frac{\partial M_{\beta\alpha}}{\partial \alpha}\right) ds_\alpha = \overline{Q}_\beta\, ds_\alpha, \tag{2.6.3a}$$

$$\left(N_{\beta\alpha} + \frac{M_{\beta\alpha}}{R_\alpha}\right) ds_\alpha = \left[\overline{N}_{\alpha\beta} + \frac{1}{2}\left(\frac{3}{R_\alpha} - \frac{1}{R_\beta}\right)\overline{M}_{\alpha\beta}\right] ds_\alpha, \tag{2.6.3b}$$

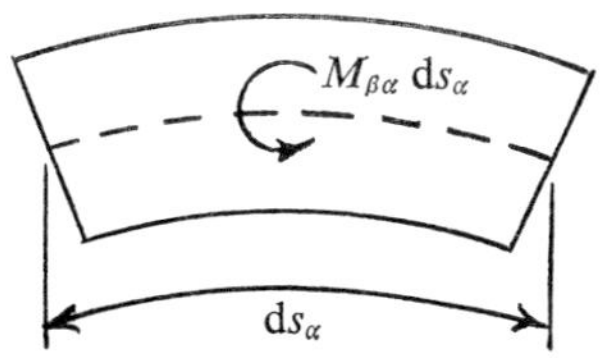

(a) Twisting moment on normal section of shell element

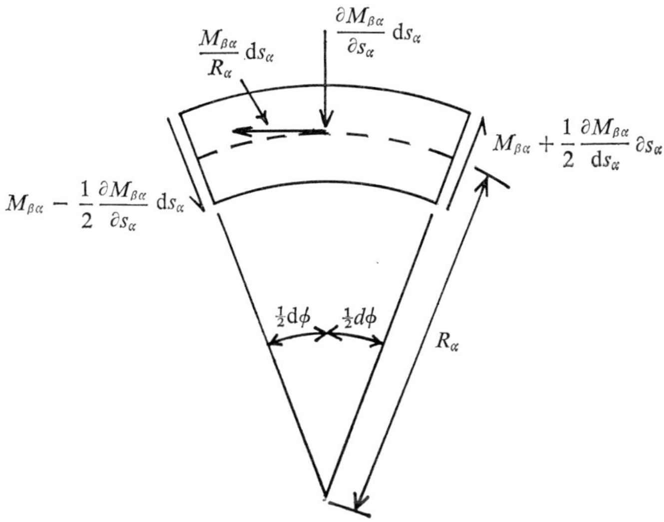

(b) Equivalent forces

Fig. 2.5. Statically equivalent force systems

where the relations

$$ds_\alpha = g_\alpha\,d\alpha = R_\alpha\,d\phi, \tag{2.6.4}$$

have been used. Similarly, along a β line of curvature the equivalent normal and tangential shear forces are

$$\bar{Q}_\alpha\,ds_\beta = \left(Q_\alpha + \frac{1}{g_\beta}\frac{\partial M_{\alpha\beta}}{\partial\beta}\right)ds_\beta, \tag{2.6.5a}$$

$$\left[\bar{N}_{\alpha\beta} + \frac{1}{2}\left(\frac{3}{R_\beta} - \frac{1}{R_\alpha}\right)\bar{M}_{\alpha\beta}\right]ds_\beta = \left(N_{\alpha\beta} + \frac{M_{\alpha\beta}}{R_\beta}\right)ds_\beta. \tag{2.6.5b}$$

If we add all the corner force components of the four normal surface elements making up the sides of the shell element, we obtain the corner forces shown in Fig. 2.6 (the origin of coordinates has been shifted the infinitesimal distance $\frac{1}{2}$ ds_α). Eqs. (2.6.2) then can be shown to represent the overall equations of equilibrium under the equivalent system of forces

$$N_\alpha,\ N_\beta,\ \overline{N}_{\alpha\beta} + \frac{1}{2}\left(\frac{3}{R_\beta} - \frac{1}{R_\alpha}\right)\overline{M}_{\alpha\beta},\ \overline{N}_{\alpha\beta} + \frac{1}{2}\left(\frac{3}{R_\alpha} - \frac{1}{R_\beta}\right)\overline{M}_{\alpha\beta},\ Q_\alpha,\ Q_\beta,$$

the surface moments M_α and M_β, and the concentrated corner forces, with the boxed terms in Eqs. (2.6.2) representing the contribution of the corner forces. Because of the material properties we have assumed for the shell, it will be shown that there is no way of distinguishing between the effects of the physical force system of Section 2.4 or the equivalent force system given above.

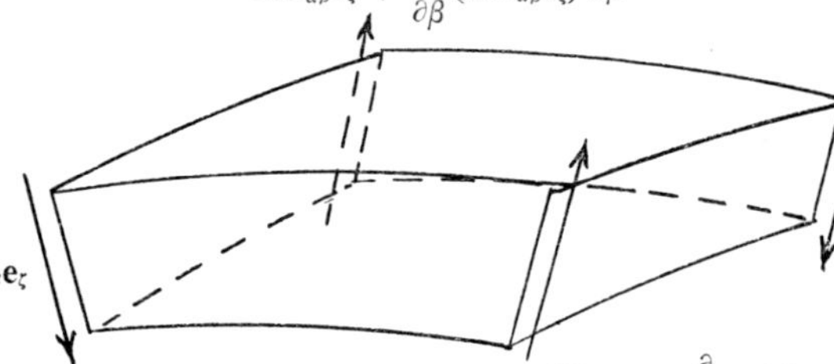

Fig. 2.6. Concentrated forces at element corners

We can further reduce the equations to a set of three equations in six unknowns by eliminating the quantities $\overline{Q}_\alpha$ and $\overline{Q}_\beta$ force Eqs. [Eqs. (3.5.2)] to obtain

$$\frac{\partial}{\partial\alpha}\left[g_\beta\left(N_\alpha + \frac{M_\alpha}{R_\beta}\right)\right] - \left(N_\beta + \frac{M_\beta}{R_\alpha}\right)\frac{\partial g_\beta}{\partial\alpha}$$
$$+ \frac{1}{g_\alpha}\frac{\partial}{\partial\beta}\left\{g_\alpha^2\left[\overline{N}_{\alpha\beta} + \frac{1}{2}\left(\frac{3}{R_\alpha} - \frac{1}{R_\beta}\right)\overline{M}_{\alpha\beta}\right]\right\}$$
$$- g_\beta\left(\frac{1}{R_\beta} - \frac{1}{R_\alpha}\right)\frac{\partial M_\alpha}{\partial\alpha} = -g_\alpha g_\beta\left(\frac{m_\alpha}{R_\alpha} + q_\alpha\right), \tag{2.6.6a}$$

$$\frac{\partial}{\partial\beta}\left[g_\alpha\left(N_\beta + \frac{M_\beta}{R_\alpha}\right)\right] - \left(N_\alpha + \frac{M_\alpha}{R_\beta}\right)\frac{\partial g_\alpha}{\partial\beta}$$
$$+ \frac{1}{g_\beta}\frac{\partial}{\partial\alpha}\left\{g_\beta^2\left[\overline{N}_{\alpha\beta} + \frac{1}{2}\left(\frac{3}{R_\beta} - \frac{1}{R_\alpha}\right)\overline{M}_{\alpha\beta}\right]\right\}$$
$$- g_\alpha\left(\frac{1}{R_\alpha} - \frac{1}{R_\beta}\right)\frac{\partial M_\beta}{\partial\beta} = -g_\alpha g_\beta\left(\frac{m_\beta}{R_\beta} + q_\beta\right), \tag{2.6.6b}$$

$$\frac{\partial}{\partial \alpha}\left\{\frac{1}{g_\alpha}\left[\frac{\partial}{\partial \alpha}(g_\beta M_\alpha) - \frac{\partial g_\beta}{\partial \alpha} M_\beta + \frac{1}{g_\alpha}\frac{\partial}{\partial \beta}(g_\alpha^2 \overline{M}_{\alpha\beta})\right]\right\}$$
$$+ \frac{\partial}{\partial \beta}\left\{\frac{1}{g_\beta}\left[\frac{\partial}{\partial \beta}(g_\alpha M_\beta) - \frac{\partial g_\alpha}{\partial \beta} M_\alpha + \frac{1}{g_\beta}\frac{\partial}{\partial \alpha}(g_\beta^2 \overline{M}_{\alpha\beta})\right]\right\}$$
$$- g_\alpha g_\beta\left(\frac{N_\alpha}{R_\alpha} + \frac{N_\beta}{R_\beta}\right) = -\left[\frac{\partial}{\partial \alpha}(g_\beta m_\alpha) + \frac{\partial}{\partial \beta}(g_\alpha m_\beta) + g_\alpha g_\beta q_\zeta\right], \quad (2.6.6c)$$

where the terms involving m_α and m_β can be interpreted, in a manner similar to that used above for the equivalent forces, as equivalent tangential and normal forces.

2.7 Overall force–strain relations

The final set of relations which we need to complete the development of the differential equations of the theory of transversely rigid shells are expressions for stress resultants in terms of middle surface displacements. These are readily obtained with the use of the stress–strain relations, Eqs. (3.2.2), which can be inverted to yield, with the additional use of Eqs. (3.3.7) and (3.3.9),

$$\sigma_\alpha^{(\zeta)} = \frac{E}{1-\nu^2}\left(\frac{\epsilon_\alpha + \zeta\kappa_\alpha}{1 + \zeta/R_\alpha} + \nu\frac{\epsilon_\beta + \zeta\kappa_\beta}{1 + \zeta/R_\beta}\right), \quad (2.7.1a)$$

$$\sigma_\beta^{(\zeta)} = \frac{E}{1-\nu^2}\left(\frac{\epsilon_\beta + \zeta\kappa_\beta}{1 + \zeta/R_\beta} + \nu\frac{\epsilon_\alpha + \zeta\kappa_\alpha}{1 + \zeta/R_\alpha}\right), \quad (2.7.1b)$$

$$\tau_{\alpha\beta}^{(\zeta)} = \frac{E}{2(1+\nu)}\,\frac{\left[1 + \tfrac{1}{2}\zeta\left(\frac{1}{R_\alpha} + \frac{1}{R_\beta}\right)\right](\gamma_{\alpha\beta} + 2\zeta\bar{\kappa}_{\alpha\beta}) + \left[\frac{2}{\zeta}\left(\frac{1}{R_\alpha} - \frac{1}{R_\beta}\right)\right]^2 \gamma_{\alpha\beta}}{\left(1 + \frac{\zeta}{R_\alpha}\right)\left(1 + \frac{\zeta}{R_\beta}\right)}. \quad (2.7.1c)$$

We can then utilize Eqs. (2.5.1), (2.5.2), (2.6.1), and (2.7.1) to obtain expressions for the overall force–strain relations as

$$N_\alpha = \boxed{K(\epsilon_\alpha + \nu\epsilon_\beta)} + D\left(\frac{1}{R_\beta} - \frac{1}{R_\alpha}\right)\left(\kappa_\alpha - \frac{\epsilon_\alpha}{R_\alpha}\right)(1 + \psi_\alpha), \quad (2.7.2a)$$

$$N_\beta = \boxed{K(\epsilon_\beta + \nu\epsilon_\alpha)} + D\left(\frac{1}{R_\alpha} - \frac{1}{R_\beta}\right)\left(\kappa_\beta - \frac{\epsilon_\beta}{R_\beta}\right)(1 + \psi_\beta), \quad (2.7.2b)$$

$$M_\alpha = \boxed{D\Big\{\kappa_\alpha + \nu\kappa_\beta} + \left(\frac{1}{R_\beta} - \frac{1}{R_\alpha}\right)\left[\epsilon_\alpha - \left(\kappa_\alpha - \frac{\epsilon_\alpha}{R_\alpha}\right)R_\alpha\psi_\alpha\right]\Big\}, \quad (2.7.2c)$$

$$M_\beta = \boxed{D\Big\{\kappa_\beta + \nu\kappa_\alpha} + \left(\frac{1}{R_\alpha} - \frac{1}{R_\beta}\right)\left[\epsilon_\beta - \left(\kappa_\beta - \frac{\epsilon_\beta}{R_\beta}\right)R_\beta\psi_\beta\right]\Big\}, \tag{2.7.2d}$$

$$\begin{aligned}\overline{N}_{\alpha\beta} = \boxed{\frac{1-\nu}{2}K\gamma_{\alpha\beta}} &+ \frac{1-\nu}{8}D\Big\{3\left(\frac{1}{R_\alpha} - \frac{1}{R_\beta}\right)^2\gamma_{\alpha\beta} \\ &+ \left(3 - \frac{R_\alpha}{R_\beta}\right)\left(\frac{1}{R_\beta} - \frac{1}{R_\alpha}\right)\left[\bar\kappa_{\alpha\beta} - \tfrac{1}{4}\gamma_{\alpha\beta}\left(\frac{3}{R_\alpha} - \frac{1}{R_\beta}\right]\right)\psi_\alpha \\ &+ \left(3 - \frac{R_\beta}{R_\alpha}\right)\left(\frac{1}{R_\alpha} - \frac{1}{R_\beta}\right)\left[\bar\kappa_{\alpha\beta} - \tfrac{1}{4}\gamma_{\alpha\beta}\left(\frac{3}{R_\beta} - \frac{1}{R_\alpha}\right)\right]\psi_\beta\Big\},\end{aligned} \tag{2.7.2e}$$

$$\begin{aligned}\overline{M}_{\alpha\beta} = \boxed{(1-\nu)D\Big\{\bar\kappa_{\alpha\beta}} &+ \tfrac{1}{4}\left(1 - \frac{R_\alpha}{R_\beta}\right)\left[\bar\kappa_{\alpha\beta} - \tfrac{1}{4}\gamma_{\alpha\beta}\left(\frac{3}{R_\alpha} - \frac{1}{R_\beta}\right)\right]\psi_\alpha \\ &+ \frac{1}{4}\left(1 - \frac{R_\beta}{R_\alpha}\right)\left[\bar\kappa_{\alpha\beta} - \tfrac{1}{4}\gamma_{\alpha\beta}\left(\frac{3}{R_\beta} - \frac{1}{R_\alpha}\right)\right]\psi_\beta\Big\},\end{aligned} \tag{2.7.2f}$$

with

$$K = \frac{Eh}{1-\nu^2}, \tag{2.7.3a}$$

$$D = \frac{Eh^3}{12(1-\nu^2)}, \tag{2.7.3b}$$

$$\psi_\alpha = \sum_{n=1}^{\infty} \frac{3}{2n+3}\left(\frac{h}{2R_\alpha}\right)^{2n}, \tag{2.7.3c}$$

$$\psi_\beta = \sum_{n=1}^{\infty} \frac{3}{2n+3}\left(\frac{h}{2R_\beta}\right)^{2n}. \tag{2.7.3d}$$

The terms ψ_α and ψ_β are the consequence of using the series expansion

$$\ln\left(\frac{1 + h/2R}{1 - h/2R}\right) = \sum_{n=0}^{\infty} \frac{1}{2n+1}\left(\frac{h}{2R}\right)^{2n}. \tag{2.7.4}$$

The boxed terms are similar in form to those for a flat plate and are exact for a spherical shell. The additional terms arise mainly from the circumstance that the centroids of the normal cross-sections of the shell element are not located by the same value of the ζ coordinate (due to the different normal radii of curvature). As a result there is some coupling in the overall force–strain relations between force components and middle surface curvature changes and between moment components and middle surface extensional and shear strains, except for the spherical shell. This should be no matter for great concern, however, since the force and moment components are also coupled in the equations of equilibrium so that no amount of simplification of the overall force–strain relations will produce force components without attendant moment components, and vice versa, or extensional and shear strains without changes of curvature, and vice versa.

It is also possible to obtain expressions for the individual shear force and moment components ($N_{\alpha\beta}$, $N_{\beta\alpha}$, $M_{\alpha\beta}$, and $M_{\beta\alpha}$) in terms of the shear strain $\gamma_{\alpha\beta}$ and the modified twist $\bar{\kappa}_{\alpha\beta}$. These are unnecessary, however, since they are never needed in the solution of shell problems.

2.8 The principle of virtual work and boundary conditions

Thus far we have considered the derivation of differential equations for displacements which, when used in conjunction with Eqs. (2.7.1), will yield stresses satisfying the prescribed stress conditions on the inner and outer bounding surfaces of the shell wall. If the shell closes on itself, conditions of continuity of displacements and rotations will suffice to determine the constants of integration appearing in the general solution of the problem. In most cases, however, the shell will have edges, assumed normal to the middle surface, on which stresses and/or displacements are prescribed. We shall assume here, for simplicity, that these boundary surfaces intersect the shell middle surface along lines of curvature. Conditions for a general edge boundary are considered in the Appendix.

It is obvious that by restricting ourselves to a material that has infinite transverse stiffness we have imposed certain restrictions on the type of boundary conditions that may be prescribed. For instance, since straight lines normal to the undeformed middle surface remain straight, unchanged in length, and normal to the deformed middle surface, it follows that if the displacements of the middle surface on the boundary curve and the slope of the middle surface normal to the boundary are prescribed, then the displacements of every other point in the boundary surface are automatically prescribed. Thus, possible sets of boundary conditions are that u_α, u_β, w, and Θ_α are prescribed along middle surface β lines of curvature, and that u_α, u_β, w, and Θ_β are prescribed along middle surface α lines of curvature. These are sufficient to completely determine the middle surface displacements and the stress distribution within the shell wall.

If stresses are prescribed, the boundary conditions are not so obvious and some thought must be devoted to their elucidation. Although normal and shearing stresses may be prescribed at every point in the edge surface, these are transmitted to the shell as the distribution of stresses given by Eqs. (2.7.1) and implied by Eqs. (2.2.5), since the normals to the middle surface serve as rigid members for the redistribution of stress. A first impression might be that of the stress resultants of Section 2.5 we can specify the force and moment resultants normal to the boundary surface, the shear force resultants parallel and normal to the middle surface, and the twisting moment, since these would be in equilibrium with the external stress distribution. However, this gives us five conditions on the edge surface, whereas it can be shown that four conditions are sufficient to determine the interior stress–state completely.

By thinking further about the problem, we see that our first conclusion about boundary conditions is based upon the concept of isolating a rigid normal and equilibrating the external forces acting on the normal by force and moment resultants. This simple model does not suffice since normals to the middle surface have four degrees of freedom rather than the five degrees of freedom implied by the model, for if the displacements of the middle surface boundary curve are specified then the rotation of the normal in the direction parallel to the boundary curve is also specified. Thus there are internal constraints implied by the stress–strain law for the transversely rigid material that must be taken into account in obtaining equilibrium of a rigid normal. We can achieve this purpose without considering the exact nature of the constraints by means of the principle of virtual work of the theory of elasticity. This theorem, as will be seen in the development below, also gives us an alternate method of derivation of the equations of equilibrium of shell theory. In essence, therefore, to obtain a complete theory of shells we must abandon the straightforward approach based on the differential equations of elasticity and turn to a variational formulation of the theory of elasticity.

Let us assume that the stresses $\sigma_\zeta^{(\zeta)}$, $\tau_{\alpha\zeta}^{(\zeta)}$, and $\tau_{\beta\zeta}^{(\zeta)}$ are prescribed on the parallel surfaces bounding the shell; the stresses $\sigma_\alpha^{(\zeta)}$, $\tau_{\alpha\beta}^{(\zeta)}$, $\tau_{\beta\zeta}^{(\zeta)}$ are prescribed at every point on the surfaces defined by the normals to the middle surface along the coordinate curves $\alpha = \alpha_1$ and α_2; the stresses $\sigma_\beta^{(\zeta)}$, $\tau_{\alpha\beta}^{(\zeta)}$, and $\tau_{\beta\zeta}^{(\zeta)}$ are prescribed at every point on the corresponding surfaces along the coordinate curves $\beta = \beta_1$ and β_2, and a system of body forces is prescribed at every point within the shell wall. We will also assume that these stresses and body forces are continuous. The theorem of virtual work then states that if a body is in equilibrium under the action of prescribed body and surface forces, the work done by these forces in a small additional displacement is equal to the change in the internal strain energy. From the theory of elasticity† we can

† See, for example, Y. C. Fung, *Foundations of Solid Mechanics*, Prentice-Hall, Inc., 1965, p. 285.

express this theorem as

$$\delta U = \int_{\alpha_1}^{\alpha_2}\int_{\beta_1}^{\beta_2}\int_{-h/2}^{h/2} [\sigma_\alpha^{(\zeta)}\,\delta_\alpha^{(\zeta)} + \sigma_\beta^{(\zeta)}\,\delta\epsilon_\beta^{(\zeta)} + \sigma_\zeta^{(\zeta)}\,\delta\epsilon_\zeta^{(\zeta)}$$

$$+ \tau_{\alpha\beta}^{(\zeta)}\,\delta\gamma_{\alpha\beta}^{(\zeta)} + \tau_{\alpha\zeta}^{(\zeta)}\,\delta\gamma_{\beta\zeta}^{(\zeta)} + \tau_{\beta\zeta}^{(\zeta)}\,\delta\gamma_{\beta\zeta}^{(\zeta)}]g_\alpha^{(\zeta)}\,d\alpha\, g_\beta^{(\zeta)}\,d\beta\, d\zeta$$

$$= \int_{\alpha_1}^{\alpha_2}\int_{\beta_1}^{\beta_2}\Big\{[\sigma_\zeta^{(\zeta)*}\,\delta w^{(\zeta)} + \tau_{\alpha\zeta}^{(\zeta)*}\,\delta u_\alpha^{(\zeta)} + \tau_{\beta\zeta}^{(\zeta)*}\,\delta u_\beta^{(\zeta)}]g_\alpha^{(\zeta)}g_\beta^{(\zeta)}\Big|_{\zeta=-h/2}^{\zeta=h/2}$$

$$+ \int_{-h/2}^{h/2}[\rho_\alpha^{(\zeta)*}\,\delta u_\alpha^{(\zeta)} + \rho_\beta^{(\zeta)*}\,\delta u_\beta^{(\zeta)} + \rho_\zeta^{(\zeta)*}\,\delta w^{(\zeta)}]g_\alpha^{(\zeta)}g_\beta^{(\zeta)}\,d\zeta\Big\}\,d\alpha\,d\beta$$

$$+ \int_{\beta_1}^{\beta_2}\int_{-h/2}^{h/2}\Big\{[\sigma_\alpha^{(\zeta)*}\,\delta u_\alpha^{(\zeta)} + \tau_{\alpha\beta}^{(\zeta)*}\,\delta u_\beta^{(\zeta)} + \tau_{\alpha\zeta}^{(\zeta)*}\,\delta w^{(\zeta)}]g_\beta^{(\zeta)}\Big|_{\alpha=\alpha_1}^{\alpha=\alpha_2}\Big\}\,d\zeta\,d\beta$$

$$+ \int_{\alpha_1}^{\alpha_2}\int_{-h/2}^{h/2}\Big\{[\sigma_\beta^{(\zeta)*}\,\delta u_\beta^{(\zeta)} + \tau_{\alpha\beta}^{(\zeta)*}\,\delta u_\alpha^{(\zeta)} + \tau_{\beta\zeta}^{(\zeta)*}\,\delta w^{(\zeta)}]g_\alpha^{(\zeta)}\Big|_{\beta=\beta_1}^{\beta=\beta_2}\Big\}\,d\zeta\,d\alpha, \tag{2.8.1}$$

where the starred edge stresses and body forces are prescribed quantities. But from the previous discussion of strains and displacements within the shell wall, the virtual displacements must satisfy the following relations,

$$\delta u_\alpha^{(\zeta)} = \delta u_\alpha - \zeta\,\delta\Theta_\alpha, \tag{2.8.2a}$$

$$\delta u_\beta^{(\zeta)} = \delta u_\beta - \zeta\,\delta\Theta_\beta, \tag{2.8.2b}$$

$$\delta w^{(\zeta)} = \delta w, \tag{2.8.2c}$$

$$\delta\epsilon_\alpha^{(\zeta)} = \frac{\delta\epsilon_\alpha + \zeta\,\delta\kappa_\alpha}{1 + \zeta/R_\alpha}, \tag{2.8.2d}$$

$$\delta\epsilon_\beta^{(\zeta)} = \frac{\delta\epsilon_\beta + \zeta\,\delta\kappa_\beta}{1 + \zeta/R_\beta}, \tag{2.8.2e}$$

$$\delta\gamma_{\alpha\beta}^{(\zeta)} = \frac{\left[1 + \tfrac{1}{2}\zeta\left(\dfrac{1}{R_\alpha} + \dfrac{1}{R_\beta}\right)\right](\delta\gamma_{\alpha\beta} + 2\zeta\,\delta\bar{\kappa}_{\alpha\beta}) + \tfrac{1}{4}\zeta^2\left(\dfrac{1}{R_\alpha} - \dfrac{1}{R_\beta}\right)^2\delta\gamma_{\alpha\beta}}{\left(1 + \dfrac{\zeta}{R_\alpha}\right)\left(1 + \dfrac{\zeta}{R_\beta}\right)}, \tag{2.8.2f}$$

$$\delta\epsilon_\zeta^{(\zeta)} = \delta\gamma_{\alpha\zeta}^{(\zeta)} + \delta\gamma_{\beta\zeta}^{(\zeta)} = 0. \tag{2.8.2g}$$

Upon substituting Eqs. (2.8.2) into Eq. (2.8.1), carrying out the integration with respect to ζ and using the notation of Eqs. (2.5.1), (2.5.2), (2.5.4), (2.5.6), and (2.5.8), we obtain

$$\int_{\alpha_1}^{\alpha_2}\int_{\beta_1}^{\beta_2}(N_\alpha\,\delta\epsilon_\alpha + N_\beta\,\delta\epsilon_\beta + \overline{N}_{\alpha\beta}\,\delta\gamma_{\alpha\beta} + M_\alpha\,\delta\kappa_\alpha + M_\beta\,\delta\kappa_\beta + 2\overline{M}_{\alpha\beta}\,\delta\overline{\kappa}_{\alpha\beta})g_\alpha g_\beta\,d\alpha\,d\beta$$
$$= \int_{\alpha_1}^{\alpha_2}\int_{\beta_1}^{\beta_2}(q_\alpha^*\,\delta u_\alpha + q_\beta^*\,\delta u_\beta + q_\zeta^*\,\delta w - m_\alpha^*\,\delta\Theta_\alpha - m_\beta^*\,\delta\Theta_\beta)g_\alpha g_\beta\,d\alpha\,d\beta$$
$$+ \int_{\beta_1}^{\beta_2}(N_\alpha^*\,\delta u_\alpha + N_{\alpha\beta}^*\,\delta u_\beta + Q_\alpha^*\,\delta w - M_\alpha^*\,\delta\Theta_\alpha - M_{\alpha\beta}^*\,\delta\Theta_\beta)g_\beta\Big|_{\alpha=\alpha_1}^{\alpha=\alpha_2}d\beta$$
$$+ \int_{\alpha_1}^{\alpha_2}(N_{\beta\alpha}^*\,\delta u_\alpha + N_\beta^*\,\delta u_\beta + Q_\beta^*\,\delta w - M_{\beta\alpha}^*\,\delta\Theta_\alpha - M_\beta^*\,\delta\Theta_\beta)g_\alpha\Big|_{\beta=\beta_1}^{\beta=\beta_2}d\alpha, \quad (2.8.3)$$

which is the principle of virtual work for the shell in terms of stress resultants and middle surface strain and curvature change components. We note that the change of strain energy is expressed in terms of the stress resultants of Section 2.6.

The principle is not yet suitable for our purpose since $\delta\epsilon_\alpha$, $\delta\epsilon_\beta$, $\delta\gamma_{\alpha\beta}$, $\delta\kappa_\alpha$, $\delta\kappa_\beta$, $\delta\overline{\kappa}_{\alpha\beta}$, $\delta\Theta_\alpha$, and $\delta\Theta_\beta$ are not independent and depend upon the three virtual displacements of points in the middle surface though the relations

$$\delta\epsilon_\alpha = \frac{1}{g_\alpha}\left(\frac{\partial\delta u_\alpha}{\partial\alpha} + \frac{1}{g_\beta}\frac{\partial g_\alpha}{\partial\beta}\,\delta u_\beta + \frac{g_\alpha}{R_\alpha}\,\delta w\right), \quad (2.8.4a)$$

$$\delta\epsilon_\beta = \frac{1}{g_\beta}\left(\frac{\partial\delta u_\beta}{\partial\beta} + \frac{1}{g_\alpha}\frac{\partial g_\beta}{\partial\alpha}\,\delta u_\alpha + \frac{g_\beta}{R_\beta}\,\delta w\right), \quad (2.8.4b)$$

$$\delta\gamma_{\alpha\beta} = \frac{g_\alpha}{g_\beta}\frac{\partial}{\partial\beta}\left(\frac{\delta u_\alpha}{g_\alpha}\right) + \frac{g_\beta}{g_\alpha}\frac{\partial}{\partial\alpha}\left(\frac{\delta u_\beta}{g_\beta}\right), \quad (2.8.4c)$$

$$\delta\Theta_\alpha = \frac{1}{g_\alpha}\left(\frac{\partial\delta w}{\partial\alpha} - \frac{g_\alpha}{R_\alpha}\,\delta u_\alpha\right), \quad (2.8.4d)$$

$$\delta\Theta_\beta = \frac{1}{g_\beta}\left(\frac{\partial\delta w}{g_\beta} - \frac{\partial\beta}{R_\beta}\,\delta u_\beta\right), \quad (2.8.4e)$$

$$\delta\kappa_\alpha = -\frac{1}{g_\alpha}\left[\frac{\partial}{\partial\alpha}\left(\frac{1}{g_\alpha}\frac{\partial\delta w}{\partial\alpha}\right) + \frac{1}{g_\beta^2}\frac{\partial g_\alpha}{\partial\beta}\frac{\partial\delta w}{\partial\beta} - \frac{\partial}{\partial\alpha}\left(\frac{\delta u_\alpha}{R_\alpha}\right) - \frac{1}{g_\beta}\frac{\partial g_\alpha}{\partial\beta}\frac{\delta u_\beta}{R_\beta}\right], \quad (2.8.4f)$$

$$\delta\kappa_\beta = -\frac{1}{g_\beta}\left[\frac{\partial}{\partial\beta}\left(\frac{1}{g_\beta}\frac{\partial\delta w}{\partial\beta}\right) + \frac{1}{g_\alpha^2}\frac{\partial g_\beta}{\partial\alpha}\frac{\partial\delta w}{\partial\alpha} - \frac{\partial}{\partial\beta}\left(\frac{\delta u_\beta}{R_\beta}\right) - \frac{1}{g_\alpha}\frac{\partial g_\beta}{\partial\alpha}\frac{\delta u_\alpha}{R_\alpha}\right],$$

$$\delta\bar{\kappa}_{\alpha\beta} = -\frac{1}{g_\alpha g_\beta}\left[\frac{\partial^2 \delta w}{\partial\alpha\,\partial\beta} - \frac{1}{g_\alpha}\frac{\partial g_\alpha}{\partial\beta}\frac{\partial\delta w}{\partial\alpha} - \frac{1}{g_\beta}\frac{\partial g_\beta}{\partial\alpha}\frac{\partial\delta w}{\partial\beta}\right.$$
$$\left. - \tfrac{1}{4}g_\alpha^2\left(\frac{3}{R_\alpha} - \frac{1}{R_\beta}\right)\frac{\partial}{\partial\beta}\left(\frac{\delta u_\alpha}{g_\alpha}\right) - \tfrac{1}{4}g_\beta^2\left(\frac{3}{R_\beta} - \frac{1}{R_\alpha}\right)\frac{\partial}{\partial\alpha}\left(\frac{\delta u_\beta}{g_\beta}\right)\right]. \quad (2.8.4h)$$

On substituting Eqs. (2.8.4) in Eqs. (2.8.3) and integrating by parts, we find that the principle of virtual work may also be written as

$$\int_{\beta_1}^{\beta_2}\int_{\alpha_1}^{\alpha_2}\Bigg[\!\!\Bigg[\Bigg\langle\frac{\partial}{\partial\alpha}\left[g_\beta\left(N_\alpha + \frac{M_\alpha}{R_\beta}\right)\right] - \frac{\partial g_\beta}{\partial\alpha}\left(N_\beta + \frac{M_\beta}{R_\alpha}\right) + \frac{1}{g_\alpha}\frac{\partial}{\partial\beta}$$
$$\times\left\{g_\alpha^2\left[\bar{N}_{\alpha\beta} + \frac{1}{2}\left(\frac{3}{R_\alpha} - \frac{1}{R_\beta}\right)\bar{M}_{\alpha\beta}\right]\right\}$$
$$- g_\beta\left(\frac{1}{R_\beta} - \frac{1}{R_\alpha}\right)\frac{\partial M_\alpha}{\partial\alpha} + g_\alpha g_\beta\left(\frac{m_\alpha^*}{R_\alpha} + q_\alpha^*\right)\Bigg\rangle\delta u_\alpha$$
$$+\Bigg\langle\frac{\partial}{\partial\beta}\left[g_\alpha\left(N_\beta + \frac{M_\beta}{R_\alpha}\right)\right] - \frac{\partial g_\alpha}{\partial\beta}\left(N_\alpha + \frac{M_\alpha}{R_\beta}\right) + \frac{1}{g_\beta}\frac{\partial}{\partial\alpha}$$
$$\times\left\{g_\beta^2\left[\bar{N}_{\alpha\beta} + \frac{1}{2}\left(\frac{3}{R_\beta} - \frac{1}{R_\alpha}\right)\bar{M}_{\alpha\beta}\right]\right\}$$
$$- g_\alpha\left(\frac{1}{R_\alpha} - \frac{1}{R_\beta}\right)\frac{\partial M_\beta}{\partial\beta} + g_\alpha g_\beta\left(\frac{m_\beta^*}{R_\beta} + q_\beta^*\right)\Bigg\rangle\delta u_\beta$$
$$+\Bigg\langle\frac{\partial}{\partial\alpha}\left\{\frac{1}{g_\alpha}\left[\frac{\partial}{\partial\alpha}(g_\beta M_\alpha) - \frac{\partial g_\beta}{\partial\alpha}M_\beta + \frac{1}{g_\alpha}\frac{\partial}{\partial\beta}(g_\beta^2\bar{M}_{\alpha\beta})\right]\right\}$$
$$+\frac{\partial}{\partial\beta}\left\{\frac{1}{g_\beta}\left[\frac{\partial}{\partial\beta}(g_\alpha M_\beta) - \frac{\partial g_\alpha}{\partial\beta}M_\alpha + \frac{1}{g_\beta}\frac{\partial}{\partial\alpha}(g_\beta^2\bar{M}_{\alpha\beta})\right]\right\}$$
$$- g_\alpha g_\beta\left(\frac{N_\alpha}{R_\alpha} + \frac{N_\beta}{R_\beta} - q_\zeta^*\right) + \frac{\partial}{\partial\alpha}(g_\beta m_\alpha^*)$$
$$+\frac{\partial}{\partial\beta}(g_\alpha m_\beta^*)\Bigg\rangle\delta w\Bigg]\!\!\Bigg]\,\mathrm{d}\alpha\,\mathrm{d}\beta$$
$$+\int_{\beta_1}^{\beta_2}\Bigg\{(N_\alpha - N_\alpha^*)\delta u_\alpha$$
$$+\left[\bar{N}_{\alpha\beta} + \frac{1}{2}\left(\frac{3}{R_\beta} - \frac{1}{R_\alpha}\right)\bar{M}_{\alpha\beta} - N_{\alpha\beta}^* - \frac{M_{\alpha\beta}^*}{R_\beta}\right]\delta u_\beta$$
$$-(M_\alpha - M_\alpha^*)\delta\Theta_\alpha + \left(\bar{Q}_\alpha - Q_\alpha^* - \frac{1}{g_\beta}\frac{\partial M_{\alpha\beta}^*}{\partial\beta}\right)\delta w\Bigg\}g_\beta\,\mathrm{d}\beta\Bigg|_{\alpha=\alpha_1}^{\alpha=\alpha_2}$$

continued over

$$+ \int_{\alpha_1}^{\alpha_2} \Big\{ (N_\beta - N_\beta^*)\delta u_\beta$$

$$+ \left[\overline{N}_{\alpha\beta} + \frac{1}{2}\left(\frac{3}{R_\alpha} - \frac{1}{R_\beta}\right)\overline{M}_{\alpha\beta} - N_{\beta\alpha}^* - \frac{M_{\beta\alpha}^*}{R_\alpha}\right]\delta u_\alpha$$

$$- (M_\beta - M_\beta^*)\delta\Theta_\beta + \left(\overline{Q}_\beta - Q_\beta^* - \frac{1}{g_\alpha}\frac{\partial M_{\beta\alpha}}{\partial \alpha}\right)\delta w \Big\} g_\alpha \, \mathrm{d}\alpha \Big|_{\beta=\beta_1}^{\beta=\beta_2}$$

$$+ (M_{\alpha\beta}^* + M_{\beta\alpha}^* - 2\overline{M}_{\alpha\beta})\delta w \Big|_{\alpha=\alpha_1}^{\alpha=\alpha_2} \Big|_{\beta=\beta_1}^{\beta=\beta_2} = 0, \tag{2.8.5}$$

where $\overline{Q}_\alpha$ and $\overline{Q}_\beta$ are defined respectively by Eqs. (2.6.2d) and (2.6.2e).

Each of the factors of δu_α, δu_β, and δw under the double integral sign must vanish since the equation must be satisfied for arbitrary displacement variations. These expressions are identical with the equilibrium equations [Eqs. (2.6.6)]. The remaining terms represent the equality of work done by the externally applied stresses and the resisting stress resultants and stress couples on the ruled edges of the shell wall. The terms in the line integrals can be identified as the work done by the system of equivalent forces discussed in Section 3.6 which act on the ruled surface. The remaining term represents the work done by the equivalent corner forces which cancel except at the corners of the shell (see Fig. 2.7). Thus we cannot distinguish between the physical stress resultants and the equivalent stress resultants since they do the same amount of work on the structure.

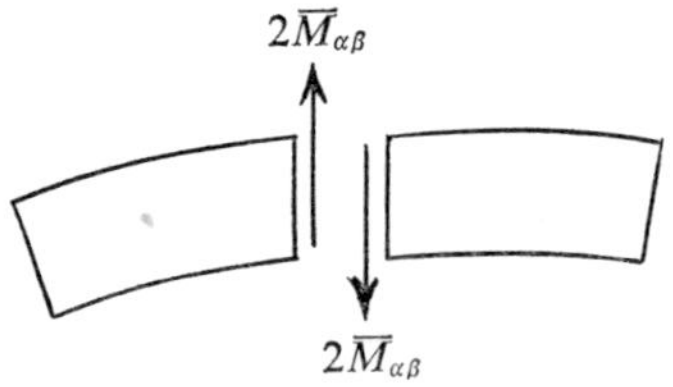

Fig. 2.7. Cancellation of corner forces

Since Eq. (2.8.5) must be satisfied for any virtual displacements, the remaining work expression yields the boundary conditions of the theory of transversely rigid shells. Along α lines of curvature, we must have equilibrium of externally applied and resisting forces of the first equivalent system

$$N_\beta = N_\beta^*, \tag{2.8.6a}$$

$$M_\beta = M_\beta^*, \tag{2.8.6b}$$

$$\overline{N}_{\alpha\beta} + \frac{1}{2}\left(\frac{3}{R_\alpha} - \frac{1}{R_\beta}\right)\overline{M}_{\alpha\beta} = N_{\beta\alpha}^* + \frac{M_{\beta\alpha}^*}{R_\alpha}, \tag{2.8.6c}$$

$$\bar{Q}_\beta = Q_\beta^* + \frac{1}{g_\alpha}\frac{\partial M_{\beta\alpha}^*}{\partial\alpha}, \tag{2.8.6d}$$

and along β lines of curvature we must have

$$N_\alpha = N_\alpha^*, \tag{2.8.7a}$$

$$M_\alpha = M_\alpha^*, \tag{2.8.7b}$$

$$\bar{N}_{\alpha\beta} + \frac{1}{2}\left(\frac{3}{R_\beta} - \frac{1}{R_\alpha}\right)\bar{M}_{\alpha\beta} = N_{\alpha\beta}^* + \frac{M_{\alpha\beta}^*}{R_\beta}, \tag{2.8.7c}$$

$$\bar{Q}_\alpha = Q_\alpha^* + \frac{1}{g_\beta}\frac{\partial M_{\alpha\beta}^*}{\partial\beta}. \tag{2.8.7d}$$

At the corners of the shell we must have

$$M_{\alpha\beta}^* + M_{\beta\alpha}^* = 2\bar{M}_{\alpha\beta}, \tag{2.8.8}$$

These can be expressed in terms of displacements through the use of the previously derived equations of equilibrium, overall force–strain relationships and strain–displacement relationships.

Boundary conditions involving both generalized forces and displacements are readily obtained since Eq. (2.8.5) also indicates what forms these may be permitted to take. Along an α line of curvature we may specify, for example,

$$N_\alpha \quad \text{or} \quad u_\alpha, \tag{2.8.9a}$$

$$\bar{N}_{\alpha\beta} + \frac{1}{2}\left(\frac{3}{R_\beta} - \frac{1}{R_\alpha}\right)\bar{M}_{\alpha\beta} \quad \text{or} \quad u_\beta, \tag{2.8.9b}$$

$$\bar{Q}_\alpha \quad \text{or} \quad w, \tag{2.8.9c}$$

$$M_\alpha \quad \text{or} \quad \Theta_\alpha, \tag{2.8.9d}$$

or appropriate combinations of these, with similar conditions for β lines of curvature. From these we can arrive at some of the more common support conditions, as listed in Table 2.1 for an α line of curvature. Table 2.1 indicates that there is some uncertainty as to what constitutes a simply supported or clamped edge, in contrast to flat plate theory where such edges are uniquely defined. This uncertainty arises from the fact that middle surface stretching and bending are coupled for thin shells so that we have to consider the three displacement components of a point in the shell edge rather than the one displacement component for a flat plate. Thus, there are several imaginable types of support conditions that differ slightly but still qualify for classification under the general headings of simple support or clamping, of which those indicated in Table 2.1 do not exhaust the possibilities.

Table 2.1 Some common boundary conditions

Type of edge support	Possible boundary conditions			
Simply supported	$M_\alpha = 0$	$w = 0$	$\bar{N}_{\alpha\beta} + \frac{1}{2}\left(\frac{3}{R_\beta} - \frac{1}{R_\alpha}\right)\bar{M}_{\alpha\beta} = 0$ or $u_\beta = 0$	$N_\alpha = 0$ or $u_\alpha = 0$
Clamped	$\Theta_\alpha = 0$	$w = 0$	$\bar{N}_{\alpha\beta} + \frac{1}{2}\left(\frac{3}{R_\beta} - \frac{1}{R_\alpha}\right)\bar{M}_{\alpha\beta} = 0$ or $u_\beta = 0$	$N_\alpha = 0$ or $u_\alpha = 0$
Free	$M_\alpha = 0$	$\bar{Q}_\alpha = 0$	$\bar{N}_{\alpha\beta} + \frac{1}{2}\left(\frac{3}{R_\beta} - \frac{1}{R_\alpha}\right)\bar{M}_{\alpha\beta} = 0$	$N_\alpha = 0$

2.9 Strain energy and the principle of minimum potential energy

The fact that the equations of equilibrium and boundary conditions of the theory of shells can be derived from a principle of virtual work suggests that the entire theory can be developed from a variational viewpoint. The principle of virtual work, in itself, is not sufficient to completely specify the state of stress or deformation in the shell wall since it depends on equilibrium and geometric relations, but not on material properties (other than those leading to the geometric relations). We can complete the variational formulation of the theory of shells by starting with the change in strain energy of a shell element for infinitesimal changes in displacements, which can be obtained from Eq. (2.8.3) as

$$\delta U = \int_{\alpha_1}^{\alpha_2} \int_{\beta_1}^{\beta_2} (N_\alpha\, \delta\epsilon_\alpha + N_\beta\, \delta\epsilon_\beta + \bar{N}_{\alpha\beta}\, \delta\gamma_{\alpha\beta} + M_\alpha\, \delta\kappa_\alpha + M_\beta\, \delta\kappa_\beta + 2\bar{M}_{\alpha\beta}\, \delta\bar{\kappa}_{\alpha\beta}) g_\alpha g_\beta\, \mathrm{d}\alpha\, \mathrm{d}\beta. \tag{2.9.1}$$

Since we are dealing with an elastic system, the change in strain energy should depend only on the initial and final equilibrium states and not upon the path described by individual elements in going from one to the other. This condition will be fulfilled only if the integrand in Eq. (2.9.1) is a total differential, in which case there must exist a function W of the strain components† such that

$$N_\alpha = \frac{\partial W}{\partial \epsilon_\alpha}, \tag{2.9.2a}$$

† See, for example, V. V. Novozhilov, *Theory of Elasticity*, Israel Program for Scientific Translations, Jerusalem, 1961, pp. 111–112 (available from The Office of Technical Services, U.S. Dept. of Commerce, Washington, D.C.).

$$N_\beta = \frac{\partial W}{\partial \epsilon_\beta}, \tag{2.9.2b}$$

$$\bar{N}_{\alpha\beta} = \frac{\partial W}{\partial \gamma_{\alpha\beta}}, \tag{2.9.2c}$$

$$M_\alpha = \frac{\partial W}{\partial \kappa_\alpha}, \tag{2.9.2d}$$

$$M_\beta = \frac{\partial W}{\partial \kappa_\beta}, \tag{2.9.2e}$$

$$2\bar{M}_{\alpha\beta} = \frac{W}{\partial \bar{\kappa}_{\alpha\beta}}. \tag{2.9.2f}$$

For a linearly elastic material the function W is given by the expression for the strain energy per unit middle surface area

$$W = \frac{1}{2}\int_{-h/2}^{h/2} [\sigma_\alpha^{(\zeta)}\epsilon_\alpha^{(\zeta)} + \sigma_\beta^{(\zeta)}\epsilon_\beta^{(\zeta)} + \tau_{\alpha\beta}^{(\zeta)}\gamma_{\alpha\beta}^{(\zeta)}]\left(1 + \frac{\zeta}{R_\alpha}\right)\left(1 + \frac{\zeta}{R_\beta}\right) \mathrm{d}\zeta. \tag{2.9.3a}$$

where Eqs. (2.4.6d) have been taken into account. When the stress–strain relations for the shell material [Eqs. (3.5.1)] and the expressions for strains in the shell wall as a function of middle surface deformations Eqs. [(2.4.7) and (2.4.9)] are substituted into Eq. (2.9.3a), we have

$$\begin{aligned} W = \frac{E}{2(1-\nu^2)}\int_{-h/2}^{h/2} \Bigg\langle & (\epsilon_\alpha + \zeta\kappa_\alpha)^2 \frac{1 + \dfrac{\zeta}{R_\beta}}{1 + \dfrac{\zeta}{R_\alpha}} + 2\nu(\epsilon_\alpha + \zeta\kappa_\alpha)(\epsilon_\beta + \zeta\kappa_\beta) \\ & + (\epsilon_\beta + \zeta\kappa_\beta)^2 \frac{1 + \dfrac{\zeta}{R_\alpha}}{1 + \dfrac{\zeta}{R_\beta}} + \frac{1-\nu}{2} \\ & \times \Bigg\{ \gamma_{\alpha\beta}^2 \frac{\left[1 + \frac{1}{2}\zeta\left(\dfrac{1}{R_\alpha} + \dfrac{1}{R_\beta}\right) + \frac{1}{4}\zeta^2\left(\dfrac{1}{R_\alpha} - \dfrac{1}{R_\beta}\right)\right]^2}{\left(1 + \dfrac{\zeta}{R_\alpha}\right)\left(1 + \dfrac{\zeta}{R_\beta}\right)} \\ & + 4\zeta\gamma_{\alpha\beta}\bar{\kappa}_{\alpha\beta}\left[1 + \frac{1}{2}\zeta^2\left(\frac{1}{R_\alpha} - \frac{1}{R_\beta}\right)^2 \frac{1 + \frac{1}{4}\zeta\left(\dfrac{1}{R_\alpha} + \dfrac{1}{R_\beta}\right)}{\left(1 + \dfrac{\zeta}{R_\alpha}\right)\left(1 + \dfrac{\zeta}{R_\beta}\right)}\right] \\ & + 4\bar{\kappa}_{\alpha\beta}^2\zeta^2 \frac{\left[1 + \frac{1}{2}\zeta\left(\dfrac{1}{R_\alpha} + \dfrac{1}{R_\beta}\right)\right]^2}{\left(1 + \dfrac{\zeta}{R_\alpha}\right)\left(1 + \dfrac{\zeta}{R_\beta}\right)} \Bigg\} \Bigg\rangle \mathrm{d}\zeta. \end{aligned} \tag{2.9.3b}$$

By performing the indicated integrations in Eq. (2.9.3b), it can be shown that Eqs. (2.9.2) coincide with Eqs. (2.7.2). From Eqs. (2.9.1), (2.9.2), and (2.9.3) we conclude that

$$\delta U = \int_{\alpha_1}^{\alpha_2}\int_{\beta_1}^{\beta_2} \delta W g_\alpha g_\beta \, d\alpha \, d\beta = \delta\left[\int_{\alpha_1}^{\alpha_2}\int_{\beta_1}^{\beta_2} W g_\alpha g_\beta \, d\alpha \, d\beta\right]. \tag{2.9.4}$$

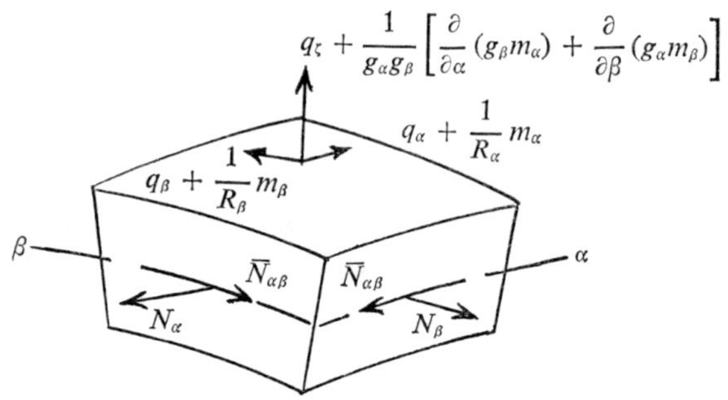

Fig. 2.8. Surface forces and stress resultants in "membrane" force state

Let us now define the following external loading vectors

$$\mathbf{q}^* = q_\alpha^* \mathbf{e}_\alpha + q_\beta^* \mathbf{e}_\beta + q_\zeta^* \mathbf{e}_\zeta, \tag{2.9.5a}$$

$$\mathbf{m}^* = m_\beta^* \mathbf{e}_\alpha + m_\alpha^* \mathbf{e}_\beta, \tag{2.9.5b}$$

$$\mathbf{N}_\alpha^* = N_\alpha^* \mathbf{e}_\alpha + N_{\alpha\beta}^* \mathbf{e}_\beta + Q_\alpha^* \mathbf{e}_\zeta, \tag{2.9.5c}$$

$$\mathbf{N}_\beta^* = N_{\beta\alpha}^* \mathbf{e}_\alpha + N_\beta \mathbf{e}_\beta + Q_\beta^* \mathbf{e}_\zeta, \tag{2.9.5d}$$

$$\mathbf{M}_\alpha^* = M_{\alpha\beta}^* \mathbf{e}_\alpha + M_\alpha^* \mathbf{e}_\beta, \tag{2.9.5e}$$

$$\mathbf{M}_\beta^* = M_\beta^* \mathbf{e}_\alpha + M_{\beta\alpha}^* \mathbf{e}_\beta, \tag{2.9.5f}$$

and the displacement and rotation vectors

$$\mathbf{u} = u_\alpha \mathbf{e}_\alpha + u_\beta \mathbf{e}_\beta + u_\zeta \mathbf{e}_\zeta, \tag{2.9.6a}$$

$$\mathbf{\Theta} = -(\Theta_\beta \mathbf{e}_\alpha + \Theta_\alpha \mathbf{e}_\beta). \tag{2.9.6b}$$

Then the principle of virtual work, Eq. (2.8.3), can be written as

$$\begin{aligned} &\delta \int_{\alpha_1}^{\alpha_2}\int_{\beta_1}^{\beta_2} W g_\alpha g_\beta \, d\alpha \, d\beta - \int_{\alpha_1}^{\alpha_2}\int_{\beta_1}^{\beta_2} (\mathbf{q}^* \cdot \delta\mathbf{u} + \mathbf{m}^* \cdot \delta\mathbf{\Theta}) g_\alpha g_\beta \, d\alpha \, d\beta \\ &\quad - \int_{\beta_1}^{\beta_2} (\mathbf{N}_\alpha^* \cdot \delta\mathbf{u} + \mathbf{M}_\alpha^* \cdot \delta\mathbf{\Theta}) g_\beta \, d\beta \Big|_{\alpha=\alpha_1}^{\alpha=\alpha_2} \\ &\quad - \int_{\alpha_1}^{\alpha_2} (\mathbf{N}_\beta^* \cdot \delta\mathbf{u} + \mathbf{M}_\beta^* \cdot \delta\mathbf{\Theta}) g_\alpha \, d\alpha \Big|_{\beta=\beta_1}^{\beta=\beta_2} = 0, \end{aligned} \tag{2.9.7a}$$

or, since the external loading vectors are known quantities which do not vary,

$$\delta\left[\int_{\alpha_1}^{\alpha_2}\int_{\beta_1}^{\beta_2}(W - \mathbf{q}^* \cdot \mathbf{u} - \mathbf{m}^* \cdot \boldsymbol{\Theta})g_\alpha g_\beta \, d\alpha \, d\beta \right.$$
$$- \int_{\beta_1}^{\beta_2}(\mathbf{N}_\alpha^* \cdot \mathbf{u} + \mathbf{M}_\alpha^* \cdot \boldsymbol{\Theta})g_\beta \, d\beta \Bigg|_{\alpha=\alpha_1}^{\alpha=\alpha_2}$$
$$\left. - \int_{\alpha_1}^{\alpha_2}(\mathbf{N}_\beta^* \cdot \mathbf{u} + \mathbf{M}_\beta^* \cdot \boldsymbol{\Theta})g_\alpha \, d\alpha \Bigg|_{\beta=\beta_1}^{\beta=\beta_2}\right] = 0. \tag{2.9.7b}$$

This formulation of the principle of virtual work is known as the principle of minimum potential energy. The principle can be stated as follows:

> Of all middle surface displacement states which satisfy the boundary conditions on displacements, the correct state makes the potential energy of the shell a minimum.

On carrying out the indicated operations, with W given by Eq. (2.9.3b), we would obtain the equations of equilibrium of the shell [Eqs. (2.6.6)] and the edge boundary conditions [Eq. (2.8.6)] expressed in terms of displacements.

2.10 Stress functions

In the preceding exposition we have discussed the theory of shells from the point of view of obtaining differential equations for middle surface displacements. In some cases, however, boundary conditions are given entirely in terms of stress resultants and middle surface displacements may be of little concern. We will then find it more convenient to solve for the stress distribution directly. We may formulate this problem in two ways, either in terms of middle surface strain components, from which the stress distribution can be found readily by means of Eqs. (2.7.1), or in terms of stress resultants, from which the stress distribution can be found with the use of Eqs. (2.7.5) to (2.7.7) in addition to Eqs. (2.7.1).

The former approach involves the expression of Eqs. (3.5.7) in terms of the six middle surface strain and curvature change components ϵ_α, ϵ_β, $\gamma_{\alpha\beta}$, κ_α, κ_β, and $\bar{\kappa}_{\alpha\beta}$. These, however, are only three differential equations for six quantities. The remaining three equations are the compatibility conditions which the six middle surface strain components must satisfy to assure continuity of the three middle surface displacements. These compatibility conditions are given by Eqs. (1.5.11).

We can carry out the formulation of the problem in terms of stress resultants by using Eqs. (2.6.6) unchanged and by expressing the compatibility conditions in terms of stress resultants by means of Eqs. (2.7.2). This procedure is merely a variation of the scheme outlined for strain components in the paragraph above. We have at our disposal, however, an alternative procedure which is an analog of the middle surface displacement formulation. If we compare the equilibrium equations [Eqs. (2.6.6)] for stress resultants and the compatibility conditions (1.5.1) for strain components, remembering that $\kappa_{\alpha\beta}$ should be replaced by $\bar{\kappa}_{\alpha\beta} + \frac{1}{4}(1/R_\alpha + 1/R_\beta)\gamma_{\alpha\beta}$, it becomes apparent that the form of the equations is similar, with the exception that Eqs. (1.5.1) are homogeneous while Eqs. (2.6.6) are not. Let us then separate the stress resultants into two parts corresponding to a particular solution and the complementary solution of Eqs. (2.6.6). The particular solution can be defined as follows. Assume that the particular solution of the nonhomogeneous equations is given by

$$M_\alpha = M_\beta = \bar{M}_{\alpha\beta} = 0, \tag{2.10.1a}$$

$$N_\alpha = N_\alpha^p, \qquad N_\beta = N_\beta^p, \qquad N_{\alpha\beta} = \bar{N}_{\alpha\beta}^p. \tag{2.10.1b}$$

Then the three defined functions, when substituted into Eqs. (2.6.6), must satisfy the following relationships

$$\frac{1}{g_\alpha g_\beta}\left[\frac{\partial}{\partial\alpha}(g_\beta N_\alpha^p) - N_\beta^p\frac{\partial g_\beta}{\partial\alpha} + \frac{1}{g_\alpha}\frac{\partial}{\partial\beta}(g_\alpha^2\bar{N}_{\alpha\beta}^p)\right] = -\left(q_\alpha + \frac{m_\alpha}{R_\alpha}\right), \tag{2.10.2a}$$

$$\frac{1}{g_\alpha g_\beta}\left[\frac{\partial}{\partial\beta}(g_\alpha N_\beta^p) - N_\alpha^p\frac{\partial g_\alpha}{\partial\beta} + \frac{1}{g_\beta}\frac{\partial}{\partial\alpha}(g_\beta^2\bar{N}_{\alpha\beta}^p)\right] = -\left(q_\beta + \frac{m_\beta}{R_\beta}\right), \tag{2.10.2b}$$

$$\frac{N_\alpha^p}{R_\alpha} + \frac{N_\beta^p}{R_\beta} = q_\zeta + \frac{1}{g_\alpha g_\beta}\left[\frac{\partial}{\partial\alpha}(g_\beta m_\alpha) + \frac{\partial}{\partial\beta}(g_\alpha m_\beta)\right]. \tag{2.10.2c}$$

These equations (more exactly, equations of this form) are known as the membrane equations of the theory of shells, from the resemblance of the stress resultant state to that in a hypothetical membrane. The equations are equivalent to the equations of equilibrium of a shell element subjected to surface loadings given by the right sides of the equations and equilibrated by the system of force resultants N_α, N_β, and $\bar{N}_{\alpha\beta}$ (see Fig. 2.8).

When we turn to the homogeneous set of equations, we note that Eqs. (1.5.11) can be transformed into Eqs. (3.5.7) if κ_β is replaced by N_α, κ_α is replaced by N_β, $\bar{\kappa}_{\alpha\beta}$ replaced by $-\bar{N}_{\alpha\beta}$, ϵ_β is replaced by $-M_\alpha$, ϵ_α is replaced by $-M_\beta$, and $\gamma_{\alpha\beta}$ is replaced by $2\bar{M}_{\alpha\beta}$. But we know that the compatibility equations will be identically satisfied if we express the strain components in terms of the three middle surface displacements by means of Eqs. (1.5.3a), (1.5.3d), (1.5.6b), and (1.5.10). Therefore, the equations of equilibrium will be identically satisfied if we relate the stress resultants to three stress functions χ_α, χ_β, and χ_ζ by means of similar relations. Thus we can write

$$M_\alpha = \frac{1}{g_\beta}\left(\frac{\partial \chi_\beta}{\partial \beta} + \frac{1}{g_\alpha}\frac{\partial g_\beta}{\partial \alpha}\chi_\alpha + \frac{g_\beta}{R_\beta}\chi_\zeta\right), \tag{2.10.3a}$$

$$M_\beta = \frac{1}{g_\alpha}\left(\frac{\partial \chi_\alpha}{\partial \alpha} + \frac{\partial}{g_\beta}\frac{\partial g_\alpha}{\partial \beta}\chi_\beta + \frac{g_\alpha}{R_\alpha}\chi_\zeta\right), \tag{2.10.3b}$$

$$2\overline{M}_{\alpha\beta} = -\left[\frac{g_\alpha}{g_\beta}\frac{\partial}{\partial \beta}\left(\frac{\chi_\alpha}{g_\alpha}\right) + \frac{g_\beta}{g_\alpha}\frac{\partial}{\partial \alpha}\left(\frac{\chi_\beta}{g_\beta}\right)\right], \tag{2.10.3c}$$

$$N_\alpha = N_\alpha^p - \frac{1}{g_\beta}\left[\frac{\chi_\alpha}{g_\alpha}\frac{\partial}{\partial \alpha}\left(\frac{g_\beta}{R_\beta}\right) - \frac{\partial}{\partial \beta}\left(\frac{1}{g_\beta}\frac{\partial \chi_\zeta}{\partial \beta}\right) - \frac{1}{g_\alpha^2}\frac{\partial g_\beta}{\partial \alpha}\frac{\partial \chi_\zeta}{\partial \alpha} + \frac{\partial}{\partial \beta}\left(\frac{\chi_\beta}{R_\beta}\right)\right], \tag{2.10.3d}$$

$$N_\beta = N_\beta^p - \frac{1}{g_\alpha}\left[\frac{\chi_\beta}{g_\beta}\frac{\partial}{\partial \beta}\left(\frac{g_\alpha}{R_\alpha}\right) - \frac{\partial}{\partial \alpha}\left(\frac{1}{g_\alpha}\frac{\partial \chi_\zeta}{\partial \alpha}\right) - \frac{1}{g_\beta^2}\frac{\partial g_\alpha}{\partial \beta}\frac{\partial \chi_\zeta}{\partial \beta} + \frac{\partial}{\partial \alpha}\left(\frac{\chi_\alpha}{R_\alpha}\right)\right], \tag{2.10.3e}$$

$$\overline{N}_{\alpha\beta} = \overline{N}_{\alpha\beta}^p + \frac{1}{g_\alpha g_\beta}\left[\tfrac{1}{4}g_\alpha^2\left(\frac{3}{R_\alpha} - \frac{1}{R_\beta}\right)\frac{\partial}{\partial \beta}\left(\frac{\chi_\alpha}{g_\alpha}\right) + \tfrac{1}{4}g_\beta^2\left(\frac{3}{R_\beta} - \frac{1}{R_\alpha}\right)\frac{\partial}{\partial \alpha}\left(\frac{\chi_\beta}{g_\beta}\right) - \frac{\partial^2 \chi_\zeta}{\partial \alpha\,\partial \beta} + \frac{1}{g_\alpha}\frac{\partial g_\alpha}{\partial \beta}\frac{\partial \chi_\zeta}{\partial \alpha} + \frac{1}{g_\beta}\frac{\partial g_\beta}{\partial \alpha}\frac{\partial \chi_\zeta}{\partial \beta}\right]. \tag{2.10.3f}$$

The three equations that the stress functions must satisfy are then obtained by expressing the strain components in terms of the stress functions and the particular solutions by means of Eqs. (2.7.6), (2.7.7), and (2.10.3) and substituting the result into the compatibility conditions given by Eqs. (1.5.11). We should note that the membrane stress resultants are not an exact particular solution of the equations of the theory of shells since the equations for the stress resultant functions will in general have a particular solution which involves the membrane forces. However, in many problems of practical interest, particularly those of thin shells under surface loads which vary slowly, the error involved in using the membrane stress resultants as a particular solution and neglecting the additional forces and moments due to the stress functions is small except near edges and interior discontinuities or rapid variations of the geometrical parameters of the shell.

2.11 Transverse shear and normal stresses

The last topic of the complete formulation of the theory of transversely rigid shells upon which we shall touch is the calculation of transverse shear and normal stresses in the shell wall. Although these stresses are usually small compared to the stresses in shell wall parallel to the middle surface, they may be required in some instances. Once we have obtained values of the middle surface strain components from the preceding equations we can obtain the transverse shear and normal stresses from the equations of equilibrium of three-dimensional elasticity, Eqs. (3.1.3a), which can be rewritten in the form

$$\frac{\partial}{\partial\zeta}\left[\left(1+\frac{\zeta}{R_\alpha}\right)^2\left(1+\frac{\zeta}{R_\beta}\right)\tau_{\alpha\zeta}^{(\zeta)}\right] = -\frac{1}{g_\alpha g_\beta}\left\{\frac{1}{g_\alpha}\frac{\partial}{\partial\beta}\left[g_\alpha^2\left(1+\frac{\zeta}{R_\alpha}\right)^2\tau_{\alpha\beta}^{(\zeta)}\right]\right.$$
$$+ g_\beta\left(1+\frac{\zeta}{R_\alpha}\right)\left(1+\frac{\zeta}{R_\beta}\right)\frac{\partial\sigma_\alpha^{(\zeta)}}{\partial\alpha} + \frac{\partial g_\beta}{\partial\alpha}\left(1+\frac{\zeta}{R_\alpha}\right)^2[\sigma_\alpha^{(\zeta)} - \sigma_\beta^{(\zeta)}]$$
$$\left. + g_\alpha g_\beta\left(1+\frac{\zeta}{R_\alpha}\right)^2\left(1+\frac{\zeta}{R_\beta}\right)\rho_\alpha^{(\zeta)}\right\}, \tag{2.11.1a}$$

$$\frac{\partial}{\partial\zeta}\left[\left(1+\frac{\zeta}{R_\alpha}\right)\left(1+\frac{\zeta}{R_\beta}\right)^2\tau_{\beta\zeta}^{(\zeta)}\right] = -\frac{1}{g_\alpha g_\beta}\left\{\frac{1}{g_\beta}\frac{\partial}{\partial\alpha}\left[g_\beta^2\left(1+\frac{\zeta}{R_\beta}\right)^2\tau_{\alpha\beta}^{(\zeta)}\right]\right.$$
$$+ g_\alpha\left(1+\frac{\zeta}{R_\alpha}\right)\left(1+\frac{\zeta}{R_\beta}\right)\frac{\partial\sigma_\beta^{(\zeta)}}{\partial\beta} + \frac{\partial g_\alpha}{\partial\beta}\left(1+\frac{\zeta}{R_\beta}\right)^2[\sigma_\beta^{(\zeta)} - \sigma_\alpha^{(\zeta)}]$$
$$\left. + g_\alpha g_\beta\left(1+\frac{\zeta}{R_\alpha}\right)\left(1+\frac{\zeta}{R_\beta}\right)^2\rho_\beta^{(\zeta)}\right\}, \tag{2.11.1b}$$

$$\frac{\partial}{\partial\zeta}\left[\left(1+\frac{\zeta}{R_\alpha}\right)\left(1+\frac{\zeta}{R_\beta}\right)\sigma_\zeta^{(\zeta)}\right] = -\frac{1}{g_\alpha g_\beta}\left\{\frac{\partial}{\partial\alpha}\left[g_\beta\left(1+\frac{\zeta}{R_\beta}\right)\tau_{\alpha\zeta}^{(\zeta)}\right]\right.$$
$$+ \frac{\partial}{\partial\beta}\left[g_\alpha\left(1+\frac{\zeta}{R_\alpha}\right)\tau_{\beta\zeta}^{(\zeta)}\right] - g_\alpha g_\beta\left[\frac{1}{R_\alpha}\left(1+\frac{\zeta}{R_\beta}\right)\sigma_\alpha^{(\zeta)}\right.$$
$$\left.\left. + \frac{1}{R_\beta}\left(1+\frac{\zeta}{R_\alpha}\right)\sigma_\beta^{(\zeta)} - \left(1+\frac{\zeta}{R_\alpha}\right)\left(1+\frac{\zeta}{R_\beta}\right)\rho_\zeta^{(\zeta)}\right]\right\}. \tag{2.11.1c}$$

With the use of the stress–strain relations for the wall material, Eqs. (2.7.1), we can integrate Eqs. (2.11.1) to obtain $\tau_{\alpha\zeta}^{(\zeta)}$, $\tau_{\beta\zeta}^{(\zeta)}$, and $\sigma_\zeta^{(\zeta)}$ in terms of the middle surface strain components and three functions of α and β. These functions can be evaluated by equating the transverse shear and normal stresses to their known values on either bounding surface of the shell. The conditions on the other bounding surface will be automatically satisfied if the middle surface strain components satisfy the equations previously derived.

An alternate method is to express the stresses in terms of stress resultants by means of the stress–strain relations and the overall force–strain relations and to use the overall equations of equilibrium to simplify the resulting expressions. In either case the distribution of stresses will be quite complicated in general but can be simplified by order of magnitude considerations to indicate which terms can be deleted.

As an example of the complexity of the transverse stress distribution in shells, we may integrate Eqs. (2.11.1) with the aid of Eqs. (2.5.4), (2.5.6), (2.6.2), (2.7.1), and (2.7.2) for a spherical shell, with

$$\alpha = \phi, \tag{2.11.2a}$$

$$\beta = \theta, \tag{2.11.2b}$$

$$g_\alpha = R_\alpha = R_\beta = R, \tag{2.11.2c}$$

$$g_\beta = R \sin\phi, \tag{2.11.2d}$$

to obtain the following expressions for the stresses:

$$\begin{aligned}
\left(1+\frac{\zeta}{R}\right)^3 \tau_{\phi\zeta}^{(\zeta)} = {} & \left(1+\frac{h}{2R}\right)^2\left(\frac{1}{2}+\frac{\zeta}{h}\right)\left[3\frac{\zeta}{h}-\frac{1}{2}+\frac{h}{R}\left(2\frac{\zeta^2}{h^2}-\frac{1}{2}\frac{\zeta}{h}+\frac{1}{4}\right)\right]\tau_{\phi\zeta}^{(h/2)} \\
& - \left(1-\frac{h}{2R}\right)^2\left(\frac{1}{2}-\frac{\zeta}{h}\right)\left[3\frac{\zeta}{h}+\frac{1}{2}+\frac{h}{R}\left(2\frac{\zeta^2}{h^2}+\frac{1}{2}\frac{\zeta}{h}+\frac{1}{4}\right)\right]\tau_{\phi\zeta}^{(-h/2)} \\
& + \frac{3}{2}\left(1-4\frac{\zeta^2}{h^2}\right)\left[\left(1+\frac{2}{3}\frac{\zeta}{R}-\frac{h^2}{12R^2}\right)\frac{Q_\phi}{h}\right. \\
& \left. - \int_{-h/2}^{h/2}\left(\frac{\zeta}{h}+\frac{h}{12R}\right)\left(1+\frac{\zeta}{R}\right)^2 \rho_\phi^{(\zeta)}\,\mathrm{d}\zeta\right] + \left(\frac{\zeta}{h}+\frac{1}{2}\right) \\
& \times \int_{\zeta}^{h/2}\left(1+\frac{\zeta}{R}\right)^2 \rho_\phi^{(\zeta)}\,\mathrm{d}\zeta + \left(\frac{\zeta}{h}-\frac{1}{2}\right)\int_{-h/2}^{\zeta}\left(1+\frac{\zeta}{h}\right)^2 \rho_\phi^{(\zeta)}\,\mathrm{d}\zeta \\
& + 4\left(\frac{\zeta^3}{h^3}+\frac{1}{8}\right)\int_{\zeta}^{h/2}\frac{\zeta}{R}\left(1+\frac{\zeta}{R}\right)^2 \rho_\phi^{(\zeta)}\,\mathrm{d}\zeta \\
& + 4\left(\frac{\zeta^3}{h^3}-\frac{1}{8}\right)\int_{-h/2}^{\zeta}\frac{\zeta}{R}\left(1+\frac{\zeta}{R}\right)^2 \rho_\phi^{(\zeta)}\,\mathrm{d}\zeta,
\end{aligned} \tag{2.11.3a}$$

with a similar expression for $\tau_{\theta\zeta}^{(\zeta)}$, and

$$\begin{aligned}
\left(1+\frac{\zeta}{R}\right)^2 \sigma_\zeta^{(\zeta)} = {} & \left(\frac{1}{2}+\frac{\zeta}{h}\right)\left(1+\frac{h}{2R}\right)^2 \sigma_\zeta^{(h/2)} + \left(\frac{1}{2}-\frac{\zeta}{h}\right)\left(1-\frac{h}{2R}\right)^2 \sigma_\zeta^{(-h/2)} \\
& + \left(\frac{1}{2}+\frac{\zeta}{h}\right)\int_{\zeta}^{h/2}\left(1+\frac{\zeta}{R}\right)^2 \rho_\zeta^{(\zeta)}\,\mathrm{d}\zeta - \left(\frac{1}{2}-\frac{\zeta}{h}\right)\int_{-h/2}^{\zeta}\left(1+\frac{\zeta}{R}\right)^2 \rho_\zeta^{(\zeta)}\,\mathrm{d}\zeta \\
& - \frac{h/(2R)}{(1+\zeta/R)\sin\phi}\left\{\left(\frac{1}{2}+\frac{\zeta}{h}\right)^2\int_{\zeta}^{h/2}\left(1+\frac{\zeta}{R}\right)^2\left[\frac{\partial}{\partial\phi}(\rho_\phi^{(\zeta)}\sin\phi)+\frac{\partial\rho_\theta^{(\phi)}}{\partial\theta}\right]\mathrm{d}\zeta\right. \\
& + \left(\frac{1}{2}-\frac{\zeta}{h}\right)^2\int_{-h/2}^{\zeta}\left(1+\frac{\zeta}{R}\right)^2\left[\frac{\partial}{\partial\phi}(\rho_\phi^{(\zeta)}\sin\phi)+\frac{\partial\rho_\theta^{(\zeta)}}{\partial\theta}\right]\mathrm{d}\zeta
\end{aligned}$$

continued over

$$
\begin{aligned}
&+ 4\left(\frac{1}{2} - \frac{\zeta}{h}\right)^2\left(1 + \frac{\zeta}{h}\right) \int_{-h/2}^{\zeta} \frac{\zeta}{h}\left(1 + \frac{\zeta}{R}\right)^2\left[\frac{\partial}{\partial \phi}(\rho_\phi^{(\zeta)} \sin \phi) + \frac{\partial \rho_\theta^{(\zeta)}}{\partial \theta}\right] d\zeta \\
&- 4\left(\frac{1}{2} + \frac{\zeta}{h}\right)^2\left(1 - \frac{\zeta}{h}\right) \int_{\zeta}^{h/2} \frac{\zeta}{h}\left(1 + \frac{\zeta}{R}\right)^2 \left[\frac{\partial}{\partial \phi}(\rho_\phi^{(\zeta)} \sin \phi) + \frac{\partial \rho_\theta^{(\zeta)}}{\partial \theta}\right] d\zeta \\
&+ \frac{1}{2}\left(1 - 4\frac{\zeta^2}{h^2}\right)\Big\langle \left(\frac{1}{2} + \frac{\zeta}{h}\right)\left(1 + \frac{h}{2R}\right)^2\left[\frac{\partial}{\partial \phi}(\tau_{\phi\zeta}^{(h/2)} \sin \phi) + \frac{\partial \tau_{\theta\zeta}^{(h/2)}}{\partial \theta}\right] \\
&- \left(\frac{1}{2} - \frac{\zeta}{h}\right)\left(1 - \frac{h}{2R}\right)^2\left[\frac{\partial}{\partial \phi}(\tau_{\phi\zeta}^{(-h/2)} \sin \phi) + \frac{\partial \tau_{\theta\zeta}^{(-h/2)}}{\partial \theta}\right]\Big\rangle\Big\} \\
&- \left(1 - 4\frac{\zeta^2}{h^2}\right)\left[\frac{1}{Rh}(M_\phi + M_\theta) + \frac{1}{2}\frac{\zeta/h + \frac{1}{4}(h/R)}{1 + \zeta/R}\left(\frac{N_\phi + N_\theta}{R} - q_\zeta\right)\right].
\end{aligned}
\tag{2.11.3b}
$$

These equations are readily seen to yield the correct distribution of stresses on the bounding surfaces $\zeta = \pm h/2$. Corresponding stress distributions for a flat plate are obtained by putting

$$R \to \infty$$

$$R\,\partial\phi \to \partial s$$

$$R \sin \phi \to s$$

yielding equations for transverse stresses in a polar coordinate system.

Supplementary references

[1] Love, A. E. H., *A Treatise on the Mathematical Theory of Elasticity*, Fourth Edition, Dover Publications, New York, 1944, Chapter 24.

[2] Reissner, E., 'A New Derivation of the Equations for the Deformation of Elastic Shells', *Amer. J. Math.*, vol. 63, 1941, pp. 177–184.

[3] Sanders, J. L., Jr., 'An Improved First Approximation for Thin Shells', NASA TR R24, 1959.

[4] Hildebrand, F. B., Reissner, E., and Thomas, G. B., 'Notes on the Foundations of the Theory of Small Displacements of Orthotropic Shells', NACA TN 1833, March 1949.

[5] Novozhilov, V. V., and Finkel'shtein, R. M., 'On the Errors of the Kirchhoff Hypothesis in the Theory of Shells', *Prikl. Mat. Mekh.*, vol. 7, no. 4, 1943, pp. 331–340.

[6] Kil'chevskii, N. A., 'Fundamental Equations of the Theory of Shells and Some Methods for their Integration', *Akad. Nauk URSR Kiev*, no. 4, 1940, pp. 83–147; no. 5, 1940, pp. 73–97; no. 6, 1941, pp. 51–103.

[7] Chien, W. Z., 'The Intrinsic Theory of Thin Plates and Shells', *Quart. Appl. Math.*, vol. 1, no. 4, 1944, pp. 297–327; vol. 2, no. 1, 1944, pp. 43–59; vol. 2, no. 2, 1944, pp. 120–135.

[8] Epstein, P. S., 'On the Theory of Elastic Vibrations of Plates and Shells', *J. Math. Phys.*, vol. 21, no. 3, 1942, pp. 198–209.

[9] Koiter, W. T., 'A Consistent First Approximation in the General Theory of Thin Elastic Shells', *Proc. Symp. Theory of Thin Elastic Shells* (Delft, 1959), North-Holland Publishing Co., Amsterdam, 1960, pp. 12–33.

[10] Green, A. E., 'On the Linear Theory of Thin Elastic Shells', *Proc. Roy. Soc., A*, vol. 266, 1962, pp. 143–160.

[11] Green, A. E., and Naghdi, P. M., 'Some Remarks on the Linear Theory of Shells', *Quart. Jour. Mech. Appl. Math.*, vol. 18, Pt. 3, 1965, pp. 257–276.

[12] Naghdi, P. M., 'Foundations of Elastic Shell Theory', *Progress in Solid Mechanics*, edited by I. N. Snedden and R. Hill, vol. 4, North-Holland Publishing Co., Amsterdam, 1963, pp. 1–90.

[13] Rutten, H. S., 'Asymptotic Approximation in the Three-dimensional Theory of Thin and Thick Elastic Shells', *Proc. 2nd Symp. Theory of Thin Shells* (Copenhagen, 1967), Springer-Verlag, Berlin, 1969, pp. 115–134.

[14] Vekua, I. N., 'On One Version of the Consistent Theory of Elastic Shells', *Proc. 2nd Symp. Theory of Thin Shells* (Copenhagen, 1967), Springer-Verlag, Berlin, 1969, pp. 59–84.

[15] Goldenveiser, A. L., *Theory of Thin Elastic Shells*, Pergamon Press, New York, 1961.

Appendix

Boundary conditions at edges which are not lines of curvature

When the shell edges are not coincident with lines of curvature on the middle surface, we start with the principle of virtual work which we shall state as an equality between the virtual work done by edge stresses and the difference between the change of internal strain energy of the shell wall and the virtual work of surface forces which we denote by δU_E. From Eqs. (2.8.3), (2.8.4), and (2.6.6) we find that δU_E can be expressed as

$$\begin{aligned}\delta U_E = \iint_S & [N_\alpha\,\delta\epsilon_\alpha + N_\beta\,\delta\epsilon_\beta + \overline{N}_{\alpha\beta}\,\delta\gamma_{\alpha\beta} + M_\alpha\,\delta\kappa_\alpha + M_\beta\,\delta\kappa_\beta \\ & + 2\overline{M}_{\alpha\beta}\,\delta\bar{\kappa}_{\alpha\beta} - (q_\alpha\,\delta u_\alpha + q_\beta\,\delta u_\beta + q_\zeta\,\delta w - m_\alpha\,\delta\Theta_\alpha \\ & - m_\beta\,\delta\Theta_\beta)]g_\alpha g_\beta\,d\alpha\,d\beta \\ = \iint_S & \Big\langle \frac{\partial}{\partial\alpha}\Big[g_\beta\Big\{N_\alpha\,\delta u_\alpha + \Big[\overline{N}_{\alpha\beta} + \frac{1}{2}\Big(\frac{3}{R_\beta} - \frac{1}{R_\alpha}\Big)\overline{M}_{\alpha\beta}\Big]\delta u_\beta + \bar{Q}_\alpha\,\delta w \\ & - M_\alpha\,\delta\Theta_\alpha - \frac{1}{g_\beta}\frac{\partial(\overline{M}_{\alpha\beta}\,\delta w)}{\partial\beta}\Big\}\Big] + \frac{\partial}{\partial\beta}\Big[g_\alpha\Big\{N_\beta\,\delta u_\beta \\ & + \Big[\overline{N}_{\alpha\beta} + \frac{1}{2}\Big(\frac{3}{R_\alpha} - \frac{1}{R_\beta}\Big)\overline{M}_{\alpha\beta}\Big]\delta u_\alpha + \bar{Q}_\beta\,\delta w - M_\beta\,\delta\Theta_\beta \\ & - \frac{1}{g_\alpha}\frac{\partial(\overline{M}_{\alpha\beta}\,\delta w)}{\partial\alpha}\Big\}\Big]\Big\rangle\,d\alpha\,d\beta, \end{aligned} \tag{A1}$$

where S is the arbitrarily bounded middle surface of the shell. The surface integral in this last form can be evaluated in terms of edge values by means of Stoke's theorem which, for the orthogonal lines of curvature coordinate system, can be expressed as

$$\iint_S \Big[\frac{\partial}{\partial\alpha}(g_\beta Q) - \frac{\partial}{\partial\beta}(g_\alpha P)\Big]\,d\alpha\,d\beta = \oint_C (P\sin\phi_C + Q\cos\phi_C)\,ds, \tag{A2}$$

where ds is an element of length of the curve C and ϕ_C is the angle at a point between the normal to the curve C at a point on the curve and the α line of curvature passing through that point and the normal to the curve C which

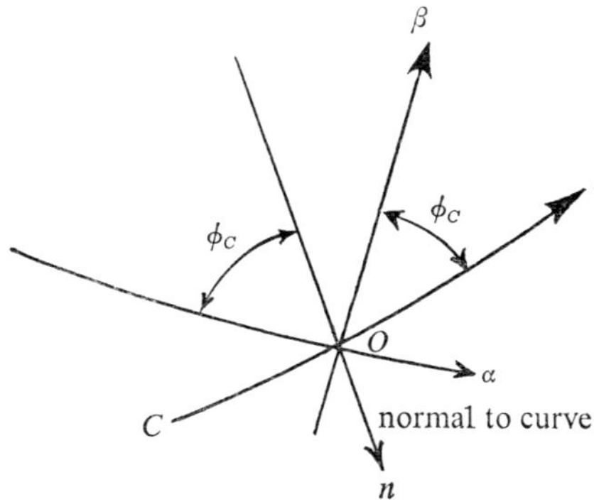

Fig. A.1. Tangent plane of boundary curve at point O

lies in the plane tangent to the middle surface at the point (Fig. A.1). Thus Eq. (A.1) may be written

$$\delta U_E = \oint_C \Bigg\langle \Bigg\{ N_\alpha\,\delta u_\alpha + \left[\bar{N}_{\alpha\beta} + \frac{1}{2}\left(\frac{3}{R_\beta} - \frac{1}{R_\alpha}\right)\bar{M}_{\alpha\beta}\right]\delta u_\beta + \bar{Q}_\alpha\,\delta w$$

$$- M_\alpha\,\delta\Theta_\alpha - \frac{1}{g_\beta}\frac{\partial(\bar{M}_{\alpha\beta}\,\delta w)}{\partial\beta}\Bigg\}\cos\phi_C$$

$$- \Bigg\{N_\beta\,\delta u_\beta + \left[\bar{N}_{\alpha\beta} + \frac{1}{2}\left(\frac{3}{R_\alpha} - \frac{1}{R_\beta}\right)\bar{M}_{\alpha\beta}\right]\delta u_\alpha$$

$$+ \bar{Q}_\beta\,\delta w - M_\beta\,\delta\Theta_\beta - \frac{1}{g_\alpha}\frac{\partial(\bar{M}_{\alpha\beta}\,\delta w)}{\partial\alpha}\Bigg\}\sin\phi_C\Bigg\rangle\,\mathrm{d}s. \tag{A.3}$$

Let us now express the displacements u_α and u_β and the rotations Θ_α and Θ_β in terms of quantities parallel and normal to the edge curve. Then (see Fig. A.2)

$$u_\alpha = u_s \sin\phi_C + u_n \cos\phi_C, \tag{A.4a}$$

$$u_\beta = u_s \cos\phi_C - u_n \sin\phi_C, \tag{A.4b}$$

where u_s and u_n are tangential displacements parallel and normal to the edge curve. Similarly, we have (Fig. A.3)

$$\Theta_\alpha = \Theta_s \sin\phi_C + \Theta_n \cos\phi_C, \tag{A.5a}$$

$$\Theta_\beta = \Theta_s \cos\phi_C - \Theta_n \sin\phi_C. \tag{A.5b}$$

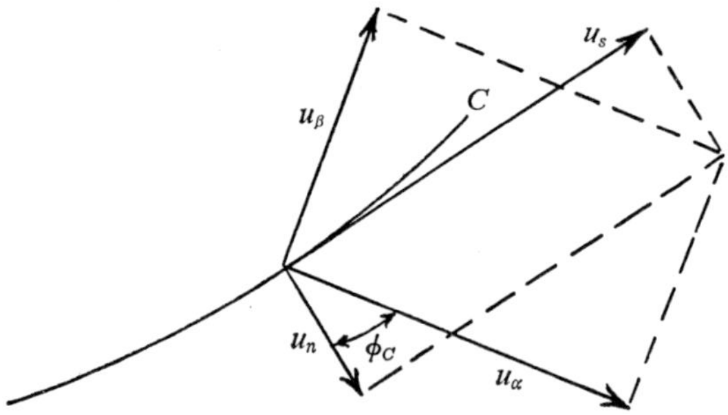

Fig. A.2. Displacement transformation

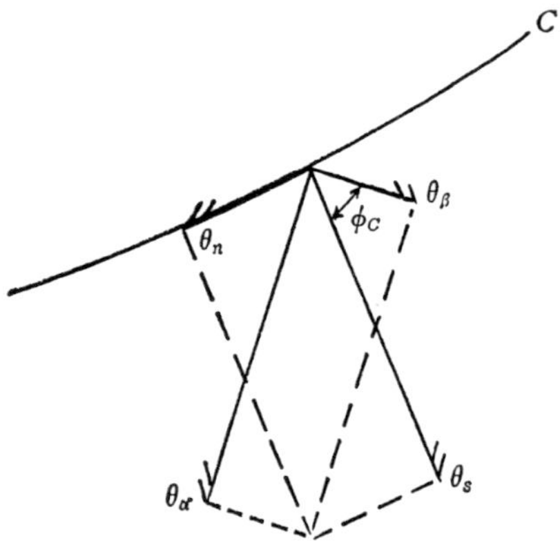

Fig. A.3. Rotation transformation

By inverting Eqs. (A.5) and making use of the definitions of Eqs. (1.5.3), we have

$$\begin{aligned}\Theta_s &= \Theta_\alpha \sin\phi_C + \Theta_\beta \cos\phi_C \\ &= \frac{1}{g_\alpha}\frac{\partial w}{\partial \alpha}\sin\phi_C + \frac{1}{g_\beta}\frac{\partial w}{\partial \beta}\cos\phi_C \\ &\quad - \left(\frac{\sin^2\phi_C}{R_\alpha} + \frac{\cos^2\phi_C}{R_\beta}\right)u_s - \left(\frac{1}{R_\alpha} - \frac{1}{R_\beta}\right)u_n \sin\phi_C \cos\phi_C,\end{aligned} \tag{A.6a}$$

$$\begin{aligned}\Theta_n &= \Theta_\alpha \cos\phi_C - \Theta_\beta \sin\phi_C \\ &= \frac{1}{g_\alpha}\frac{\partial w}{\partial \alpha}\cos\phi_C - \frac{1}{g_\beta}\frac{\partial w}{\partial \beta}\sin\phi_C \\ &\quad - \left(\frac{1}{R_\alpha} - \frac{1}{R_\beta}\right)u_s \sin\phi_C \cos\phi_C - \left(\frac{\cos^2\phi_C}{R_\alpha} + \frac{\sin^2\phi_C}{R_\beta}\right)u_n\end{aligned} \tag{A.6b}$$

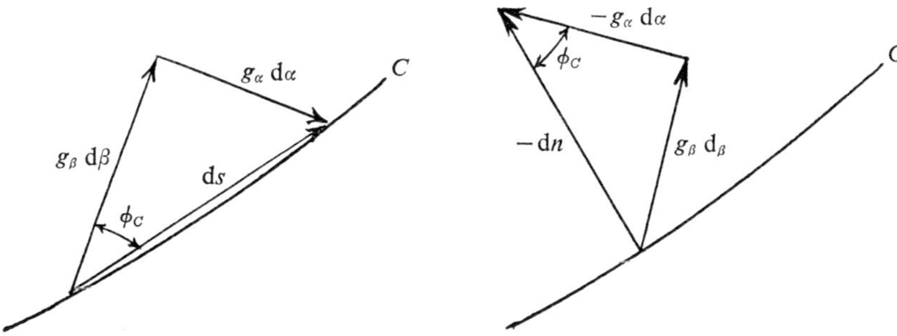

(a) Tangent direction (b) Normal direction

Fig. A.4. Differential lengths

But (see Fig. A.4)

$$\frac{1}{g_\alpha}\frac{\partial(\ldots)}{\partial\alpha}\sin\phi_C + \frac{1}{g_\beta}\frac{\partial(\ldots)}{\partial\beta}\cos\phi_C$$

$$= \frac{\partial(\ldots)}{\partial\alpha}\frac{\partial\alpha}{\partial s} + \frac{\partial(\ldots)}{\partial\beta}\frac{\partial\beta}{\partial s} = \frac{\partial(\ldots)}{\partial s}, \tag{A.7a}$$

$$\frac{1}{g_\alpha}\frac{\partial(\ldots)}{\partial\alpha}\cos\phi_C - \frac{1}{g_\beta}\frac{\partial(\ldots)}{\partial\beta}\sin\phi_C$$

$$= \frac{\partial(\ldots)}{\partial\alpha}\frac{\partial\alpha}{\partial n} + \frac{\partial(\ldots)}{\partial\beta}\frac{\partial\beta}{\partial n} = \frac{\partial(\ldots)}{\partial n} \tag{A.7b}$$

and from Eqs. (1.2.9), (1.2.10), and (1.2.11), with ω equal to $\pi/2$, we can show that

$$\frac{\sin^2\phi_C}{R_\alpha} + \frac{\cos^2\phi_C}{R_\beta} = \frac{1}{R_s}, \tag{A.8a}$$

$$\frac{\cos^2\phi_C}{R_\alpha} + \frac{\sin^2\phi_C}{R_\beta} = \frac{1}{R_n}, \tag{A.8b}$$

$$\left(\frac{1}{R_\alpha} - \frac{1}{R_\beta}\right)\sin\phi_C\cos\phi_C = \frac{1}{R_{sn}}, \tag{A.8c}$$

where $1/R_s$ and $1/R_n$ are the normal curvatures of the middle surface in the directions parallel and normal to the edge curve $1/R_{ns}$ is the twist of the surface for these coordinates. Thus the rotations parallel and perpendicular to the edge can be expressed as

$$\Theta_s = \frac{\partial w}{\partial s} - \frac{u_s}{R_s} - \frac{u_n}{R_{sn}}, \tag{A.9a}$$

$$\Theta_n = \frac{\partial w}{\partial n} - \frac{u_n}{R_n} - \frac{u_s}{R_{sn}}. \tag{A.9b}$$

The substitution of Eqs. (A.4), (A.5), and (A.9) into Eq. (A.3), and integration by parts yields the virtual work of the edge forces, after some manipulation, as

$$\begin{aligned}
\delta U_E = \oint_C \Big\langle &\Big\{\tfrac{1}{2}(N_\alpha - N_\beta)\sin 2\phi_C + \bar{N}_{\alpha\beta}\cos 2\phi_C + \tfrac{1}{2}\Big(\frac{1}{R_\beta} - \frac{1}{R_\alpha}\Big)\bar{M}_{\alpha\beta} \\
&+ \frac{1}{R_s}[\bar{M}_{\alpha\beta}\cos 2\phi_C + \tfrac{1}{2}(M_\alpha - M_\beta)\sin 2\phi_C]\Big\}\delta u_s \\
&+ \Big\{N_\alpha \cos^2\phi_C + N_\beta \sin^2\phi_C - \bar{N}_{\alpha\beta}\sin 2\phi_C \\
&+ \frac{1}{R_{sn}}[\bar{M}_{\alpha\beta}\cos 2\phi_C + \tfrac{1}{2}(M_\alpha - M_\beta)\sin 2\phi_C]\Big\}\delta u_n \\
&+ \Big\{\Big(\bar{Q}_\alpha - \frac{1}{g_\beta}\frac{\partial M_{\alpha\beta}}{\partial\beta}\Big)\cos\phi_C - \Big(\bar{Q}_\beta - \frac{1}{g_\alpha}\frac{\partial \bar{M}_{\alpha\beta}}{\partial\alpha}\Big)\sin\phi_C \\
&+ \frac{\partial}{\partial s}[\bar{M}_{\alpha\beta}\cos 2\phi_C + \tfrac{1}{2}(M_\alpha - M_\beta)\sin 2\phi_C]\Big\}\delta w \\
&- (M_\alpha\cos^2\phi_C + M_\beta\sin^2\phi_C - \bar{M}_{\alpha\beta}\sin 2\phi_C)\delta\Theta_n\Big\rangle\, ds \\
&- [\bar{M}_{\alpha\beta}\cos 2\phi_C + \tfrac{1}{2}(M_\alpha - M_\beta)\sin 2\phi_C]\delta w\Big|_{s_{0+}}^{s_{0-}}.
\end{aligned} \tag{A.10}$$

The last term vanishes if the edge curve is smooth and thus has no discontinuities of direction. If, however, the edge curve has corners so that the angle ϕ_C is discontinuous, the last term can be expressed as

$$\sum_i \delta w_i[\bar{M}_{\alpha\beta_i}(\cos 2\phi_{i+} - \cos 2\phi_{i-}) + \tfrac{1}{2}(M_{\alpha_i} - M_{\beta_i})(\sin 2\phi_{i+} - \sin 2\phi_{i-})], \tag{A.11}$$

where ϕ_{i-} and ϕ_{i+} are the limits of the angle ϕ_C as we approach the corner i from either side (Fig. A.5) and the summation is over the total number of corners.

With the use of the definitions of $\bar{N}_{\alpha\beta}$ and $\bar{M}_{\alpha\beta}$, viz.,

$$\bar{M}_{\alpha\beta} = \tfrac{1}{2}(M_{\alpha\beta} + M_{\beta\alpha}), \tag{A.12a}$$

$$\begin{aligned}
\bar{N}_{\alpha\beta} &= N_{\alpha\beta} - \frac{M_{\beta\alpha}}{R_\beta} + \frac{1}{2}\Big(\frac{1}{R_\alpha} + \frac{1}{R_\beta}\Big)\bar{M}_{\alpha\beta} \\
&= N_{\beta\alpha} - \frac{M_{\alpha\beta}}{R_\alpha} + \frac{1}{2}\Big(\frac{1}{R_\alpha} + \frac{1}{R_\beta}\Big)\bar{M}_{\alpha\beta},
\end{aligned} \tag{A.12b}$$

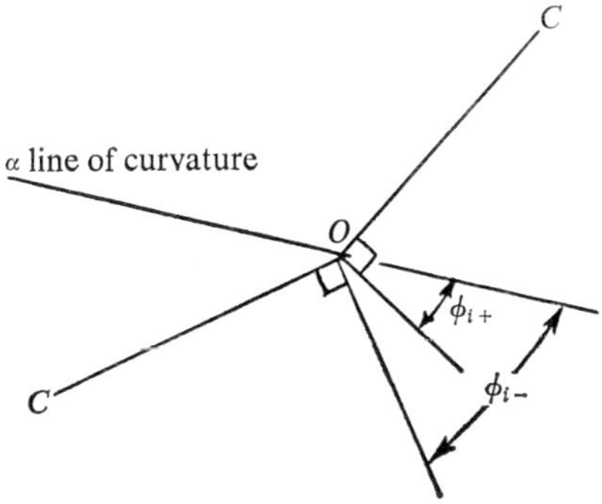

Fig. A.5. Curve C at a corner

and the following relationships between resultant forces and moments acting on the shell edge and physical forces and moments acting on sections along lines of curvature (Fig. A.6)

$$N_n = N_\alpha \cos^2 \phi_C + N_\beta \sin^2 \phi_C - \tfrac{1}{2}(N_{\alpha\beta} + N_{\beta\alpha}) \sin 2\phi_C \tag{A.13a}$$

$$N_{sn} = \tfrac{1}{2}(N_\alpha - N_\beta) \sin 2\phi_C + N_{\alpha\beta} \cos^2 \phi_C - N_{\beta\alpha} \sin^2 \phi_C, \tag{A.13b}$$

$$Q_n = Q_\alpha \cos \phi_C - Q_\beta \sin \phi_C, \tag{A.13c}$$

$$M_n = M_\alpha \cos^2 \phi_C + M_\beta \sin^2 \phi_C - \tfrac{1}{2}(M_{\alpha\beta} + M_{\beta\alpha}) \sin^2 \phi_C, \tag{A.13d}$$

$$M_{sn} = \tfrac{1}{2}(M_\alpha - M_\beta) \sin^2 \phi_C + M_{\alpha\beta} \cos^2 \phi_C - M_{\beta\alpha} \sin^2 \phi_C. \tag{A.13e}$$

Eq. (A.10) may also be expressed as

$$\delta U_E = \oint_C \left[\left(N_{sn} + \frac{M_{sn}}{R_s}\right) \delta u_s + \left(N_n + \frac{M_{sn}}{R_{sn}}\right) \delta u_n + \left(Q_n + \frac{\partial M_{sn}}{\partial s}\right) \delta w - M_n \, \delta \Theta_n\right] \mathrm{d}s + \sum_i (M_{sn_{i+}} - M_{sn_{i-}}) \, \delta w_i. \tag{A.14}$$

Since the virtual work of applied edge forces can be expressed in the same form as that of Eq. (A.14), a possible set of edge conditions is

$$\tfrac{1}{2}(N_\alpha - N_\beta) \sin 2\phi_C + \overline{N}_{\alpha\beta} \cos 2\phi_C + \frac{1}{2}\left(\frac{1}{R_\beta} - \frac{1}{R_\alpha}\right) \overline{M}_{\alpha\beta} + \frac{1}{R_s} [\overline{M}_{\alpha\beta} \cos 2\phi_C + \tfrac{1}{2}(M_\alpha - M_\beta) \sin 2\phi_C] = N^*_{sn} + \frac{1}{R_s} M^*_{sn},$$

or u_s is prescribed along edges (A.15a)

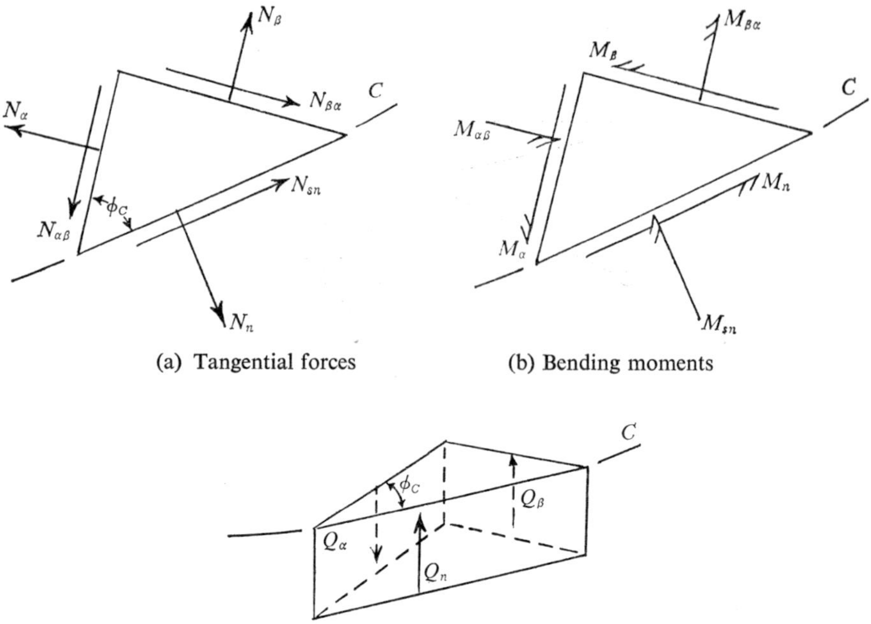

(a) Tangential forces (b) Bending moments

(c) Transverse shear forces

Fig. A.6. Forces and moments on edge element

$$N_\alpha \cos^2 \phi_C + N_\beta \sin^2 \phi_C - \overline{N}_{\alpha\beta} \sin 2\phi_C$$
$$+ \frac{1}{R_{ns}} [\overline{M}_{\alpha\beta} \cos 2\phi_C + \tfrac{1}{2}(M_\alpha - M_\beta) \sin 2\phi_C] = N_n^* + \frac{1}{R_{sn}} M_{sn}^*,$$

or u_n is prescribed along edges (A.15b)

$$\left(\overline{Q}_\alpha - \frac{1}{g_\beta} \frac{\partial \overline{M}_{\alpha\beta}}{\partial \beta}\right) \cos \phi_C - \left(\overline{Q}_\beta - \frac{1}{g_\alpha} \frac{\partial \overline{M}_{\alpha\beta}}{\partial \alpha}\right) \sin \phi_C$$
$$+ \frac{\partial}{\partial s} [\overline{M}_{\alpha\beta} \cos 2\phi_C + \tfrac{1}{2}(M_\alpha - M_\beta) \sin 2\phi_C] = Q_n^* + \frac{\partial M_{sn}^*}{\partial s},$$

or w is prescribed along edges (A.15c)

$$M_\alpha \cos^2 \phi_C + M_\beta \sin^2 \phi_C - \overline{M}_{\alpha\beta} \sin 2\phi_C = M_n^*,$$

or Θ_n is prescribed along edges (A.15d)

$$\overline{M}_{\alpha\beta_i}(\cos 2\phi_{i+} - \cos 2\phi_{i-}) + \tfrac{1}{2}(M_{\alpha_i} - M_{\beta_i})(\sin 2\phi_{i+} - \sin 2\phi_{i-})$$
$$= M_{sn_{i+}}^* - M_{sn_{i-}}^*$$

or w is prescribed at corners. (A.15e)

3

Simplifications of shell theory

3.1 Modification of overall force–strain relationships

The system of equations derived in the previous chapter for the analysis of shell structures is consistent with the primary assumption of infinite transverse stiffness and contains no other approximation. While these equations can be applied easily to spherical and circular cylindrical shells, as will be shown later, their use in the analysis of other types of shells leads to expressions which are discouraging by virtue of their formidable appearance. Various attempts to simplify the theory have been made, therefore, in hopes of achieving esthetic appeal together with a reasonable degree of accuracy. Since no one attempt can be said to be universally accepted or used in practical applications, we shall discuss those simplifications of the general theory of shells which are most likely to be encountered in the literature and which can be related to the previously derived equations most consistently.

As a first step, we shall retain the principle of virtual work of the theory [Eq. (2.8.3)] since it yields the correct equations of overall equilibrium and the boundary conditions for the shell. However, we shall simplify the overall force–strain relations given by the complicated expressions of Section 2.7 by modifying the strain-energy function W given by Eq. (2.9.3b). If we expand the integrals in Eq. (2.9.3b) in a power series in h and retain only terms through h^3, we have

$$W = W_1 + W_2, \tag{3.1.1a}$$

where

$$W_1 = \tfrac{1}{2}K\left(\epsilon_\alpha^2 + 2\nu\epsilon_\alpha\epsilon_\beta + \epsilon_\beta^2 + \frac{1-\nu}{2}\gamma_{\alpha\beta}^2\right) + \tfrac{1}{2}D[\kappa_\alpha^2 + 2\nu\kappa_\alpha\kappa_\beta + \kappa_\beta^2 + 2(1-\nu)\bar{\kappa}_{\alpha\beta}^2] \tag{3.1.1b}$$

and

$$W_2 = \tfrac{1}{2}D\left(\frac{1}{R_\beta} - \frac{1}{R_\alpha}\right)\left[\epsilon_\alpha\left(2\kappa_\alpha - \frac{\epsilon_\alpha}{R_\alpha}\right) - \epsilon_\beta\left(2\kappa_\beta - \frac{\epsilon_\beta}{R_\beta}\right)\right.$$

$$\left. + \tfrac{3}{8}(1 - \nu)\gamma_{\alpha\beta}^2\left(\frac{1}{R_\beta} - \frac{1}{R_\alpha}\right)\right]. \tag{3.1.1c}$$

From Eqs. (2.9.2), the appropriate overall force–strain relations are

$$N_\alpha = K(\epsilon_\alpha + \nu\epsilon_\beta) - D\left(\frac{1}{R_\alpha} - \frac{1}{R_\beta}\right)\left(\kappa_\alpha - \frac{\epsilon_\alpha}{R_\alpha}\right), \tag{3.1.2a}$$

$$N_\beta = K(\epsilon_\beta + \nu\epsilon_\alpha) - D\left(\frac{1}{R_\beta} - \frac{1}{R_\alpha}\right)\left(\kappa_\beta - \frac{\epsilon_\beta}{R_\beta}\right), \tag{3.1.2b}$$

$$\overline{N}_{\alpha\beta} = \frac{1-\nu}{2}\left[K + \frac{3}{4}\left(\frac{1}{R_\alpha} - \frac{1}{R_\beta}\right)^2 D\right]\gamma_{\alpha\beta}, \tag{3.1.2c}$$

$$M_\alpha = D\left[\kappa_\alpha + \nu\kappa_\beta - \left(\frac{1}{R_\alpha} - \frac{1}{R_\beta}\right)\epsilon_\alpha\right], \tag{3.1.2d}$$

$$M_\beta = D\left[\kappa_\beta + \nu\kappa_\alpha - \left(\frac{1}{R_\beta} - \frac{1}{R_\alpha}\right)\epsilon_\beta\right], \tag{3.1.2e}$$

$$\overline{M}_{\alpha\beta} = D(1 - \nu)\bar{\kappa}_{\alpha\beta}, \tag{3.1.2f}$$

which are exactly the same as the equations that we would obtain by deleting terms with ψ_α and ψ_β from Eqs. (2.7.2). For spherical shells, Eqs. (3.1.2) and (2.7.2) are identical.

We may go further and drop the expression W_2 from the strain–energy function, so that

$$W = W_1. \tag{3.1.3}$$

The use of Eqs. (2.9.2) leads to the equations

$$N_\alpha = K(\epsilon_\alpha + \nu\epsilon_\beta), \tag{3.1.4a}$$

$$N_\beta = K(\epsilon_\beta + \nu\epsilon_\alpha), \tag{3.1.4b}$$

$$\overline{N}_{\alpha\beta} = \frac{1-\nu}{2}K\gamma_{\alpha\beta}, \tag{3.1.4c}$$

$$M_\alpha = D(\kappa_\alpha + \nu\kappa_\beta), \tag{3.1.4d}$$

$$M_\beta = D(\kappa_\beta + \nu\kappa_\alpha), \tag{3.1.4e}$$

$$\overline{M}_{\alpha\beta} = D(1 - \nu)\bar{\kappa}_{\alpha\beta}. \tag{3.1.4f}$$

Eqs. (3.1.3) and (3.1.4) are similar to the strain–energy expression and overall force–strain relations for a flat plate and, again, are exact for spherical shells.

The error in the strain–energy function is a measure of the error of stresses and deformations calculated with the use of Eqs. (3.1.2) or (3.1.4) as compared with those obtained with the use of the more complicated expressions of Eq. (2.7.2). To determine the order of magnitude of the error involved in expressing the strain–energy function by means of Eqs. (3.1.1) or (3.1.3) let us introduce the notation

$$\kappa'_\alpha = \tfrac{1}{2}\kappa_\alpha h, \tag{3.1.5a}$$

$$\kappa'_\beta = \tfrac{1}{2}\kappa_\beta h, \tag{3.1.5b}$$

$$\bar{\kappa}'_{\alpha\beta} = \bar{\kappa}_{\alpha\beta} h, \tag{3.1.5c}$$

which are essentially the strains at the bounding surfaces of the shell due to bending and twist. Then Eq. (3.1.1) can be written as

$$W_1 = \frac{1}{2}K\Big\{\epsilon_\alpha^2 + 2\nu\epsilon_\alpha\epsilon_\beta + \epsilon_\beta^2 + \frac{1-\nu}{2}\gamma_{\alpha\beta}^2 + \frac{1}{3}\Big[(\kappa'_\alpha)^2 + 2\nu\kappa'_\alpha\kappa'_\beta + (\kappa'_\beta)^2 + \frac{1-\nu}{2}(\bar{\kappa}'_{\alpha\beta})^2\Big]\Big\}, \tag{3.1.6a}$$

$$W_2 = \frac{1}{6}Kh\left(\frac{1}{R_\beta} - \frac{1}{R_\alpha}\right)\Big[\epsilon_\alpha\kappa'_\alpha - \epsilon_\beta\kappa'_\beta - \frac{h}{12R_\alpha}\epsilon_\alpha^2 + \frac{h}{12R_\beta}\epsilon_\beta^2 + (1-\nu)\frac{3}{32}h\left(\frac{1}{R_\beta} - \frac{1}{R_\alpha}\right)\gamma_{\alpha\beta}^2\Big]. \tag{3.1.6b}$$

By examining the magnitude of W_2 as compared to W_1, which is a positive definite expression so long as $|\nu| < 1$, we can arrive at the conclusion that the maximum order of magnitude of the W_2 expression is h/R as compared to the W_1 expression, where R is the least radius of curvature of the shell middle surface. The critical terms in W_2 are those involving products of the middle surface strains and the outer fiber bending strains. These are of the order of magnitude less than h/R compared to W_1 if the middle surface strains are small compared to the outer fiber bending strains or vice versa. The worst possible condition occurs when both types of strains are of equal order of magnitude, in which case the terms are of order of magnitude h/R compared to W_1. The terms which we have omitted by truncating the strain–energy function at W_2 can be shown to be of the order of magnitude $(h/R)^3$, or less, compared to W_1. Thus, by using Eqs. (3.1.2) we introduce maximum errors of the order of magnitude $(h/R)^3$ into the theory. By using Eqs. (3.1.4), the errors increase to the order of magnitude of h/R.† Since h/R is small in most

† In both cases we should remember that the errors are those compared to the exact theory of transversely rigid shells and not those compared to the exact theory of isotropic shells.

practical applications, we may generally use the simplest overall force–strain relations, Eqs. (3.1.4), rather than the more complicated expressions given by Eqs. (3.1.2) or (2.7.2), for the analysis of thin shells if desired.

3.2 Internal stress distribution

The method of simplification of the strain–energy function while retaining the principle of virtual work has certain implications which we shall examine now. The principle of virtual work in the form of Eq. (2.8.3) is valid if, and only if, the virtual displacements and strains are given by Eq. (2.8.2) and the stress state within the shell wall satisfies the equations of equilibrium, Eqs. (2.2.5), and the appropriate boundary conditions on stresses. Thus when we change the form of the strain–energy function while retaining the principle of virtual work unchanged, we are not making assumptions about the thinness of the shell wall since the form of the principle of virtual work forbids this approximation. The only interpretation of the process available to us, then, is that we are approximating the stress–strain relations in the shell wall, Eqs. (2.7.1), by other expressions such that:

$$N_\alpha = \int_{-h/2}^{h/2} \sigma_\alpha^{(\zeta)}\left(1 + \frac{\zeta}{R_\beta}\right) d\zeta = \frac{\partial W}{\partial \epsilon_\alpha}, \tag{3.2.1a}$$

$$N_\beta = \int_{-h/2}^{h/2} \sigma_\beta^{(\zeta)}\left(1 + \frac{\zeta}{R_\alpha}\right) d\zeta = \frac{\partial W}{\partial \epsilon_\beta}, \tag{3.2.1b}$$

$$\begin{aligned}\bar{N}_{\alpha\beta} = \int_{-h/2}^{h/2} \tau_{\alpha\beta}^{(\zeta)}\left\{1 - \frac{\zeta^2}{R_\alpha R_\beta} + \tfrac{1}{2}\zeta\left(\frac{1}{R_\alpha} + \frac{1}{R_\beta}\right)\right.\\ \left.\times\left[1 + \tfrac{1}{2}\zeta\left(\frac{1}{R_\alpha} + \frac{1}{R_\beta}\right)\right]\right\} d\zeta = \frac{\partial W}{\partial \gamma_{\alpha\beta}},\end{aligned} \tag{3.2.1c}$$

$$M_\alpha = \int_{-h/2}^{h/2} \zeta\sigma_\alpha^{(\zeta)}\left(1 + \frac{\zeta}{R_\beta}\right) d\zeta = \frac{\partial W}{\partial \kappa_\alpha}, \tag{3.2.1d}$$

$$M_\beta = \int_{-h/2}^{h/2} \zeta\sigma_\beta^{(\zeta)}\left(1 + \frac{\zeta}{R_\alpha}\right) d\zeta = \frac{\partial W}{\partial \kappa_\beta}, \tag{3.2.1e}$$

$$\bar{M}_{\alpha\beta} = \int_{-h/2}^{h/2} \zeta\tau_{\alpha\beta}^{(\zeta)}\left[1 + \tfrac{1}{2}\zeta\left(\frac{1}{R_\alpha} + \frac{1}{R_\beta}\right)\right] d\zeta = \frac{\partial W}{\partial \bar{\kappa}_{\alpha\beta}}. \tag{3.2.1f}$$

Guided by the form of the stress–strain relations given by Eq. (2.7.1) and by the form of Eqs. (3.2.1), let us assume that the modified stress–strain relations are given by

$$\left(1 + \frac{\zeta}{R_\beta}\right)\frac{1-\nu^2}{E}\,\sigma_\alpha^{(\zeta)} = A_\alpha^{(\zeta)}\epsilon_\alpha + \zeta B_\alpha^{(\zeta)}\kappa_\alpha + \nu[C_\alpha^{(\zeta)}\epsilon_\beta + \zeta D_\alpha^{(\zeta)}\kappa_\beta], \tag{3.2.2a}$$

$$\left(1 + \frac{\zeta}{R_\alpha}\right) \frac{1 - \nu^2}{E} \sigma_\beta^{(\zeta)} = \nu[C_\beta^{(\zeta)}\epsilon_\alpha + \zeta D_\beta^{(\zeta)}\kappa_\alpha] + A_\beta^{(\zeta)}\epsilon_\beta + \zeta B_\beta^{(\zeta)}\kappa_\beta, \quad (3.2.2b)$$

$$\left(1 - \frac{\zeta^2}{R_\alpha R_\beta}\right) \frac{2(1 + \nu)}{E} \tau_{\alpha\beta}^{(\zeta)} = E^{(\zeta)}\gamma_{\alpha\beta} + 2\zeta F^{(\zeta)}\bar{\kappa}_{\alpha\beta}, \quad (3.2.2c)$$

where the functions multiplying the strain and change of curvature components are functions only of ζ and middle surface curvature components and are independent of the shell thickness. Let us also assume the strain–energy functions to be given by W_1. Then the substitution of Eqs. (3.1.4) and (3.2.2) in Eqs. (3.2.1) yields the following conditions to be satisfied:

$$\int_{-h/2}^{h/2} A_\alpha^{(\zeta)} \, d\zeta = h, \quad (3.2.3a)$$

$$\int_{-h/2}^{h/2} \zeta A_\alpha^{(\zeta)} \, d\zeta = 0, \quad (3.2.3b)$$

$$\int_{-h/2}^{h/2} \zeta B_\alpha^{(\zeta)} \, d\zeta = 0, \quad (3.2.3c)$$

$$\int_{-h/2}^{h/2} \zeta^2 B_\alpha^{(\zeta)} \, d\zeta = \frac{h^3}{12}, \quad (3.2.3d)$$

$$\int_{-h/2}^{h/2} C_\alpha^{(\zeta)} \, d\zeta = h, \quad (3.2.3e)$$

$$\int_{-h/2}^{h/2} \zeta C_\alpha^{(\zeta)} \, d\zeta = 0. \quad (3.2.3f)$$

$$\int_{-h/2}^{h/2} \zeta D_\alpha^{(\zeta)} \, d\zeta = 0, \quad (3.2.3g)$$

$$\int_{-h/2}^{h/2} \zeta^2 D_\alpha^{(\zeta)} \, d\zeta = \frac{h^3}{12}, \quad (3.2.3h)$$

$$\int_{-h/2}^{h/2} E^{(\zeta)} \, d\zeta = h, \quad (3.2.3i)$$

$$\int_{-h/2}^{h/2} \zeta E^{(\zeta)} \frac{\left[1 + \frac{1}{2}\zeta\left(\frac{1}{R_\alpha} + \frac{1}{R_\beta}\right)\right]}{1 - \frac{\zeta^2}{R_\alpha R_\beta}} \, d\zeta = 0 \quad (3.2.3j)$$

$$\int_{-h/2}^{h/2} \zeta F^{(\zeta)} \, d\zeta = -\frac{h^3}{24}\left(\frac{1}{R_\alpha} + \frac{1}{R_\beta}\right), \quad (3.2.3k)$$

$$\int_{-h/2}^{h/2} \frac{\zeta^2 F^{(\zeta)}\left[1 + \frac{1}{2}\zeta\left(\frac{1}{R_\alpha} + \frac{1}{R_\beta}\right)\right]}{1 - \frac{\zeta^2}{R_\alpha R_\beta}} \, d\zeta = \frac{h^3}{12}. \tag{3.2.31}$$

The solution of Eqs. (3.2.3) can be obtained by expanding each of the coefficients of Eqs. (3.2.3) into a power series in ζ, integrating the result, and equating the coefficients of like powers of h on both sides of the equations to obtain

$$A_\alpha^{(\zeta)} = B_\alpha^{(\zeta)} = C_\alpha^{(\zeta)} = D_\alpha^{(\zeta)} = A_\beta^{(\zeta)} = B_\beta^{(\zeta)} = B_\beta^{(\zeta)} = C_\beta^{(\zeta)} = D_\beta^{(\zeta)} = 1, \tag{3.2.4a}$$

$$E^{(\zeta)} = 1 - \tfrac{1}{2}\zeta\left(\frac{1}{R_\alpha} + \frac{1}{R_\beta}\right), \tag{3.2.4b}$$

$$F^{(\zeta)} = 1 - \tfrac{1}{2}\zeta\left(\frac{1}{R_\alpha} + \frac{1}{R_\beta}\right) + \tfrac{1}{4}\zeta^2\left(\frac{1}{R_\alpha} - \frac{1}{R_\beta}\right)^2, \tag{3.2.4c}$$

which can be verified by substitution.

If the strain–energy function is given by the sum of W_1 and W_2 similar reasoning yields

$$A_\alpha^{(\zeta)} = 1 - \left(\frac{1}{R_\alpha} - \frac{1}{R_\beta}\right)\left(1 - \frac{\zeta}{R_\alpha}\right)\zeta, \tag{3.2.5a}$$

$$B_\alpha^{(\zeta)} = 1 - \zeta\left(\frac{1}{R_\alpha} - \frac{1}{R_\beta}\right), \tag{3.2.5b}$$

$$C_\alpha^{(\zeta)} = D_\alpha^{(\zeta)} = C_\beta^{(\zeta)} = D_\beta^{(\zeta)} = 1, \tag{3.2.5c}$$

$$A_\beta^{(\zeta)} = 1 - \left(\frac{1}{R_\beta} - \frac{1}{R_\alpha}\right)\left(1 - \frac{\zeta}{R_\beta}\right)\zeta, \tag{3.2.5d}$$

$$B_\beta^{(\zeta)} = 1 - \zeta\left(\frac{1}{R_\beta} - \frac{1}{R_\alpha}\right), \tag{3.2.5e}$$

$$E^{(\zeta)} = \left[1 - \tfrac{1}{2}\zeta\left(\frac{1}{R_\alpha} + \frac{1}{R_\beta}\right)\right]\left[1 + \tfrac{3}{4}\left(\frac{1}{R_\alpha} - \frac{1}{R_\beta}\right)^2\zeta^2\right], \tag{3.2.5f}$$

$$F^{(\zeta)} = 1 - \tfrac{1}{2}\zeta\left(\frac{1}{R_\alpha} + \frac{1}{R_\beta}\right) + \tfrac{1}{4}\zeta^2\left(\frac{1}{R_\alpha} - \frac{1}{R_\beta}\right)^2. \tag{3.2.5g}$$

Provided the middle surface strains and curvature changes are known, the remaining stresses in the shell wall may be calculated from Eqs. (2.10.1). The use of Eqs. (3.2.4) or (3.2.5) will ensure the identity of derived and prescribed surface stresses.

We see that the simplification of the overall force–strain relation does not necessarily lead to simplified internal stress–strain relations. Indeed, in some instances the stress–strain expressions are more complicated. It is obvious, however, that if the shell wall is thin compared to the least principle radius of curvature the differences in the various expressions will be insignificant and, further, that other terms may be deleted without significant error. Thus, in practice stresses are generally calculated from the relations

$$\sigma_\alpha^{(\zeta)} = \frac{E}{1-\nu^2}[\epsilon_\alpha + \zeta\kappa_\alpha + \nu(\epsilon_\beta + \zeta\kappa_\beta)], \tag{3.2.6a}$$

$$\sigma_\beta^{(\zeta)} = \frac{E}{1-\nu^2}[\epsilon_\beta + \zeta\kappa_\beta + \nu(\epsilon_\alpha + \zeta\kappa_\alpha)], \tag{3.2.6b}$$

$$\tau_{\alpha\beta}^{(\zeta)} = \frac{E}{2(1+\nu)}(\gamma_{\alpha\beta} + 2\zeta\bar{\kappa}_{\alpha\beta}), \tag{3.2.6c}$$

which give a linear variation of the stresses across the wall thickness.

3.3 Approximate reduction of shell theory: Vlasov's equations

Although reducing the overall force–strain relations simplifies the theory of shells in a number of cases, the amount of simplification is insignificant in general. Thus other modifications of the theory have been introduced. Those leading to the most esthetic form of the equations have been motivated in a number of ways in the literature, but appear to be derivable in the simplest manner as follows. In the theory of bending of flat plates the coupling between displacements in the plane of the plate and displacements out of the plane of the plate vanishes. In particular, the tangential displacements do not appear in the expressions for rotations or changes of curvature. Let us then assume that the tangential displacements u_α and u_β can be deleted from the rotation and change of curvature expressions for curved shells as well. Implicit in this assumption is the subsidiary assumption that the middle surface of the shell does not differ radically from a flat plate, for in that case the displacement assumption would be reasonable. We can extend the region of validity of the assumption by noting that if bending effects vary rapidly, the region of interest could be approximated by a series of tangent flat plates for each of which the assumption would be valid. We do not, however, delete the normal displacement terms from the expressions for middle surface strains since these are of an order of magnitude comparable to those due to tangential displacements.

We next derive appropriate equations of equilibrium and boundary conditions by utilizing the principle of virtual work, Eq. (2.8.3), with the substitute relations

$$\kappa_\alpha = -\frac{1}{g_\alpha}\left[\frac{\partial}{\partial\alpha}\left(\frac{1}{g_\alpha}\frac{\partial w}{\partial\alpha}\right) + \frac{1}{g_\beta^2}\frac{\partial g_\alpha}{\partial\beta}\frac{\partial w}{\partial\beta}\right], \tag{3.3.1a}$$

$$\kappa_\beta = -\frac{1}{g_\beta}\left[\frac{\partial}{\partial\beta}\left(\frac{1}{g_\beta}\frac{\partial w}{\partial\beta}\right) + \frac{1}{g_\alpha^2}\frac{\partial g_\beta}{\partial\alpha}\frac{\partial w}{\partial\alpha}\right], \tag{3.3.1b}$$

$$\bar{\kappa}_{\alpha\beta} = -\frac{1}{g_\alpha g_\beta}\left[\frac{\partial^2 w}{\partial_\alpha\partial_\beta} - \frac{1}{g_\alpha}\frac{\partial g_\alpha}{\partial\beta}\frac{\partial w}{\partial\alpha} - \frac{1}{g_\beta}\frac{\partial g_\beta}{\partial\alpha}\frac{\partial w}{\partial\beta}\right], \tag{3.3.1c}$$

and

$$\Theta_\alpha = \frac{1}{g_\alpha}\frac{\partial w}{\partial\alpha}, \tag{3.3.1d}$$

$$\Theta_\beta = \frac{1}{g_\beta}\frac{\partial w}{\partial\beta}, \tag{3.3.1e}$$

to obtain the equilibrium equations

$$\frac{\partial}{\partial\alpha}(g_\beta N_\alpha) - \frac{\partial g_\beta}{\partial\alpha}N_\beta + \frac{1}{g_\alpha}\frac{\partial}{\partial\beta}(g_\alpha^2\bar{N}_{\alpha\beta}) + g_\alpha g_\beta q_\alpha = 0, \tag{3.3.2a}$$

$$\frac{\partial}{\partial\beta}(g_\alpha N_\beta) - \frac{\partial g_\alpha}{\partial\beta}N_\alpha + \frac{1}{g_\beta}\frac{\partial}{\partial\alpha}(g_\beta^2\bar{N}_{\alpha\beta}) + g_\alpha g_\beta q_\beta = 0, \tag{3.3.2b}$$

$$\begin{aligned}\frac{\partial}{\partial\alpha}&\left\{\frac{1}{g_\alpha}\left[\frac{\partial}{\partial\alpha}(g_\beta M_\alpha) - \frac{\partial g_\beta}{\partial\alpha}M_\beta + \frac{1}{g_\alpha}\frac{\partial}{\partial\beta}(g_\alpha^2\bar{M}_{\alpha\beta})\right]\right\}\\ &+ \frac{\partial}{\partial\beta}\left\{\frac{1}{g_\beta}\left[\frac{\partial}{\partial\beta}(g_\alpha M_\beta) - \frac{\partial g_\alpha}{\partial\beta}M_\alpha + \frac{1}{g_\beta}\frac{\partial}{\partial\alpha}(g_\beta^2\bar{M}_{\alpha\beta})\right]\right\}\\ &- g_\alpha g_\beta\left(\frac{N_\alpha}{R_\alpha} + \frac{N_\beta}{R_\beta} - q_\zeta\right) + \frac{\partial}{\partial\alpha}(g_\beta m_\alpha) + \frac{\partial}{\partial\beta}(g_\alpha m_\beta) = 0,\end{aligned} \tag{3.3.2c}$$

the edge conditions

$$N_\alpha = N_\alpha^* \quad \text{or } u_\alpha \text{ is prescribed}, \tag{3.3.3a}$$

$$\bar{N}_{\alpha\beta} = N_{\alpha\beta}^* \quad \text{or } u_\beta \text{ is prescribed}, \tag{3.3.3b}$$

$$M_\alpha = M_\alpha^* \quad \text{or } \frac{1}{g_\alpha}\frac{\partial w}{\partial\alpha} \text{ is prescribed}, \tag{3.3.3c}$$

$$\bar{Q}_\beta = Q_\beta^* + \frac{1}{g_\alpha}\frac{\partial M_{\beta\alpha}^*}{\partial\alpha} \quad \text{or } w \text{ is prescribed}, \tag{3.3.3d}$$

along β lines of curvature, and

$$N_\beta = N_\beta^* \qquad \text{or } u_\beta \text{ is prescribed,} \tag{3.3.4a}$$

$$\bar{N}_{\alpha\beta} = N_{\beta\alpha}^* \qquad \text{or } u_\alpha \text{ is prescribed,} \tag{3.3.4b}$$

$$M_\beta = M_\beta^* \qquad \text{or } \frac{1}{g_\beta}\frac{\partial w}{\partial \beta} \text{ is prescribed,} \tag{3.3.4c}$$

$$\bar{Q}_\alpha = Q_\alpha^* + \frac{1}{g_\beta}\frac{\partial M_{\alpha\beta}^*}{\partial \beta} \qquad \text{or } w \text{ is prescribed,} \tag{3.3.4d}$$

where $\bar{Q}_\alpha$ and $\bar{Q}_\beta$ are defined respectively by Eqs. (2.6.2d) and (2.6.2e). At corners we must have, as before

$$\bar{M}_{\alpha\beta} = (M_{\alpha\beta}^* + M_{\beta\alpha}^*) \quad \text{or } w \text{ is prescribed.} \tag{3.3.5a}$$

From Eqs. (3.3.3b) and (3.3.4b) it appears that we must also assume that at corners

$$N_{\alpha\beta}^* = N_{\beta\alpha}^*. \tag{3.3.5b}$$

Let us now write the stress resultants in the form

$$N_\alpha = N_{\alpha 0} + N_{\alpha 1}, \tag{3.3.6a}$$

$$N_\beta = N_{\beta 0} + N_{\beta 1}, \tag{3.3.6b}$$

$$\bar{N}_{\alpha\beta} = \bar{N}_{\alpha\beta 0} + \bar{N}_{\alpha\beta 1}, \tag{3.3.6c}$$

where $N_{\alpha 0}$, $N_{\beta 0}$, and $\bar{N}_{\alpha\beta 0}$ are particular solutions of the membrane equations

$$\frac{1}{g_\alpha g_\beta}\left[\frac{\partial}{\partial \alpha}(g_\beta N_{\alpha 0}) - N_{\beta 0}\frac{\partial g_\beta}{\partial \alpha} + \frac{1}{g_\alpha}\frac{\partial}{\partial \beta}(g_\alpha^2 \bar{N}_{\alpha\beta 0})\right] = -\, q_\alpha, \tag{3.3.7a}$$

$$\frac{1}{g_\alpha g_\beta}\left[\frac{\partial}{\partial \beta}(g_\alpha N_{\beta 0}) - N_{\alpha 0}\frac{\partial g_\alpha}{\partial \beta} + \frac{1}{g_\beta}\frac{\partial}{\partial \alpha}(g_\beta^2 \bar{N}_{\alpha\beta 0})\right] = -\, q_\beta, \tag{3.3.7b}$$

$$\frac{N_{\alpha 0}}{R_\alpha} + \frac{N_{\beta 0}}{R_\beta} = 0. \tag{3.3.7c}$$

The substitution of Eqs. (3.3.6) into Eqs. (3.3.2a) and (3.3.2b) then yields

$$\frac{\partial}{\partial \alpha}(g_\alpha N_{\alpha 1}) - N_{\beta 1}\frac{\partial g_\beta}{\partial \alpha} + \frac{1}{g_\alpha}\frac{\partial}{\partial \beta}(g_\alpha^2 \bar{N}_{\alpha\beta 1}) = 0, \tag{3.3.8a}$$

$$\frac{\partial}{\partial \beta}(g_\alpha N_{\beta 1}) - N_{\alpha 1}\frac{\partial g_\alpha}{\partial \beta} + \frac{1}{g_\beta}\frac{\partial}{\partial \alpha}(g_\beta^2 \bar{N}_{\alpha\beta 1}) = 0. \tag{3.3.8b}$$

Eqs. (3.3.8) are two equations in three unknowns, a circumstance which suggests that the three stress resultants might be replaced, as in plane elasticity theory, by expressions involving a single stress function which automatically satisfy Eqs. (3.3.8). Eqs. (3.3.8) are in a form similar to the corresponding equations of equilibrium in terms of curvilinear coordinates for the plane stress or plane strain problems of the theory of elasticity. For the plane problem the relationships between the stress resultants and a stress function can be expressed as†

$$N_{\alpha 1} = \frac{1}{g_\beta}\frac{\partial}{\partial \beta}\left(\frac{1}{g_\beta}\frac{\partial \Phi}{\partial \beta}\right) + \frac{1}{g_\alpha^2 g_\beta}\frac{\partial g_\beta}{\partial \alpha}\frac{\partial \Phi}{\partial \alpha}, \tag{3.3.9a}$$

$$N_{\beta 1} = \frac{1}{g_\alpha}\frac{\partial}{\partial \alpha}\left(\frac{1}{g_\alpha}\frac{\partial \Phi}{\partial \alpha}\right) + \frac{1}{g_\beta^2 g_\alpha}\frac{\partial g_\alpha}{\partial \beta}\frac{\partial \Phi}{\partial \beta}, \tag{3.3.9b}$$

$$\overline{N}_{\alpha\beta 1} = -\frac{1}{g_\alpha g_\beta}\left(\frac{\partial^2 \Phi}{\partial \alpha\, \partial \beta} - \frac{1}{g_\alpha}\frac{\partial g_\alpha}{\partial \beta}\frac{\partial \Phi}{\partial \alpha} - \frac{1}{g_\beta}\frac{\partial g_\beta}{\partial \alpha}\frac{\partial \Phi}{\partial \beta}\right). \tag{3.3.9c}$$

However, since the surface with which we are dealing is not flat but curved, Eqs. (3.3.9) do not satisfy Eqs. (3.3.8) exactly, in general. If we substitute Eqs. (3.3.9) into Eqs. (3.3.8) we have

$$\frac{\partial}{\partial \alpha}(g_\alpha N_{\alpha 1}) - N_{\beta 1}\frac{\partial g_\beta}{\partial \alpha} + \frac{1}{g_\alpha}\frac{\partial}{\partial \beta}(g_\alpha^2 \overline{N}_{\alpha\beta 1}) = -\frac{g_\beta}{R_\alpha R_\beta}\frac{\partial \Phi}{\partial \alpha}, \tag{3.3.10a}$$

$$\frac{\partial}{\partial \beta}(g_\alpha N_{\beta 1}) - N_{\alpha 1}\frac{\partial g_\alpha}{\partial \beta} + \frac{1}{g_\beta}\frac{\partial}{\partial \alpha}(g_\beta^2 \overline{N}_{\alpha\beta 1}) = -\frac{g_\alpha}{R_\alpha R_\beta}\frac{\partial \Phi}{\partial \beta}, \tag{3.3.10b}$$

which vanishes identically only when the Gaussian curvature of the shell middle surface vanishes as for flat plates and for cylindrical and conical shells or may be considered to vanish approximately if the Gaussian curvature is small as for shallow shells.‡ If bending effects vary rapidly the stress function Φ is a quantity which increases in magnitude under repeated differentiation. We can then consider Eqs. (3.3.9) to satisfy Eqs. (3.3.8) to a high degree of approximation since derivatives of Φ higher than the first cancel and first-order derivatives are negligible compared to the higher order derivatives. Thus in general the relationships between stress resultants and the stress function Φ are taken as Eqs. (3.3.9).

To obtain the first of what are known as Vlasov's equations we now

† V. V. Novozhilov, *Theory of Elasticity*, Pergamon Press, 1961, p. 378.

‡ We may include the sphere among the surfaces for which the reduced equations of equilibrium are satisfied by adding the term $\Phi/(R_\alpha R_\beta)$ to the expressions for N_α and N_β in Eqs. (3.3.9). The right sides of Eqs. (3.3.10) then become $g_\beta \Phi\, \partial/\partial\alpha\, (1/R_\alpha R_\beta)$ and $g_\alpha \Phi\, \partial/\partial\beta\, (1/R_\alpha R_\beta)$ respectively, which vanish for surfaces of zero or constant Gaussian curvature. The improvement in the equations is negligible, however.

assume the moment–curvature change relationships to be given by Eqs. (3.1.4d), (3.1.4e), and (3.1.4f) as

$$M_\alpha = D(\kappa_\alpha + \nu\kappa_\beta), \tag{3.3.11a}$$

$$M_\beta = D(\kappa_\beta + \nu\kappa_\alpha), \tag{3.3.11b}$$

$$\overline{M}_{\alpha\beta} = D(1 - \nu)\overline{\kappa}_{\alpha\beta}. \tag{3.3.11c}$$

The substitution of Eqs. (3.3.1), (3.3.9), and (3.3.11) into Eq. (3.3.2c) then yields

$$D\nabla^4 w + \nabla_1^2\Phi = q_\zeta + \frac{1}{g_\alpha g_\beta}\left[\frac{\partial}{\partial\alpha}(g_\beta m_\alpha) + \frac{\partial}{\partial\beta}(g_\alpha m_\beta)\right], \tag{3.3.12}$$

where Laplace's operator for the surface is given by

$$\nabla^2(\,) = \frac{1}{g_\alpha g_\beta}\left\{\frac{\partial}{\partial\alpha}\left[\frac{g_\beta}{g_\alpha}\frac{\partial(\,)}{\partial\alpha}\right] + \frac{\partial}{\partial\beta}\left[\frac{g_\alpha}{g_\beta}\frac{\partial(\,)}{\partial\beta}\right]\right\}, \tag{3.3.13a}$$

$$\nabla^4 = \nabla^2\nabla^2, \tag{3.3.13b}$$

$$\nabla_1^2(\,) = \frac{1}{g_\alpha g_\beta}\left\{\frac{\partial}{\partial\alpha}\left[\frac{1}{R_\beta}\frac{g_\beta}{g_\alpha}\frac{\partial(\,)}{\partial\alpha}\right] + \frac{\partial}{\partial\beta}\left[\frac{1}{R_\alpha}\frac{g_\alpha}{g_\beta}\frac{\partial(\,)}{\partial\beta}\right]\right\}. \tag{3.3.13c}$$

Thus far we have replaced the three independent variables u_α, u_β, and w by the two variables w and Φ. Middle surface displacements u_α and u_β are related to the normal displacement w and the stress function Φ by means of the three overall force–strain equations

$$\epsilon_\alpha = \frac{N_\alpha - \nu N_\beta}{Eh}, \tag{3.3.14a}$$

$$\epsilon_\beta = \frac{N_\beta - \nu N_\alpha}{Eh}, \tag{3.3.14b}$$

$$\gamma_{\alpha\beta} = \frac{2(1 + \nu)}{Eh} N_{\alpha\beta}, \tag{3.3.14c}$$

where the middle surface strain components are defined in terms of displacement by Eqs. (1.5.3a), (1.5.3d), and (1.5.6b) and the force components are given by Eqs. (3.3.6), (3.3.7), and (3.3.9). But since we have three relationships for two unknown quantities, it follows that the middle surface strain components must be connected by some relation if the quantities u_α and u_β obtained from any two of the three equations are to be identical. The required compatibility condition or relationship between the strain components ϵ_α, ϵ_β, and $\gamma_{\alpha\beta}$ is one which does not contain the displacements u_α and u_β. An approximation to the required compatibility condition† is provided by Eq. (1.5.10c)

$$\frac{\partial}{\partial\alpha}\left\{\frac{1}{g_\alpha}\left[\frac{\partial}{\partial\alpha}(g_\beta\epsilon_\beta) - \epsilon_\alpha\frac{\partial g_\beta}{\partial\alpha} - \frac{1}{2g_\alpha}\frac{\partial}{\partial\beta}(g_\alpha^2\gamma_{\alpha\beta})\right]\right\}$$
$$+\frac{\partial}{\partial\beta}\left\{\frac{1}{g_\beta}\left[\frac{\partial}{\partial\beta}(g_\alpha\epsilon_\alpha) - \epsilon_\beta\frac{\partial g_\alpha}{\partial\beta} - \frac{1}{2g_\beta}\frac{\partial}{\partial\alpha}(g_\beta^2\gamma_{\alpha\beta})\right]\right\}$$
$$= -g_\alpha g_\beta\left(\frac{\kappa_\alpha}{R_\beta} + \frac{\kappa_\beta}{R_\alpha}\right) \approx g_\alpha g_\beta\nabla_1^2 w, \tag{3.3.15}$$

since we have deleted tangential displacements from the curvature expressions. The substitution of Eqs. (3.3.14) into Eq. (3.3.15) then yields

$$\nabla^2(N_\alpha + N_\beta) - \frac{1+\nu}{g_\alpha g_\beta}\left\{\frac{\partial}{\partial\alpha}\frac{1}{g_\alpha}\left[\frac{\partial}{\partial\alpha}(g_\beta N_\alpha) - N_\beta\frac{\partial g_\beta}{\partial\alpha} + \frac{1}{g_\alpha}\frac{\partial}{\partial\beta}(g_\alpha^2\overline{N}_{\alpha\beta})\right]\right.$$
$$\left.+\frac{\partial}{\partial\beta}\frac{1}{g_\beta}\left[\frac{\partial}{\partial\beta}(g_\alpha N_\beta) - N_\alpha\frac{\partial g_\alpha}{\partial\alpha} + \frac{1}{g_\beta}\frac{\partial}{\partial\alpha}(g^2\overline{N}_{\alpha\beta})\right]\right\}$$
$$= Eh\nabla_1^2(w). \tag{3.3.16a}$$

The terms in braces can be replaced by the quantity

$$-\left[\frac{\partial}{\partial\alpha}(g_\beta q_\alpha) + \frac{\partial}{\partial\beta}(g_\alpha q_\beta)\right] \tag{3.3.16b}$$

by virtue of Eqs. (3.3.2a) and (3.3.2b).‡ We also have, from Eqs. (3.3.6) and (3.3.9),

$$N_\alpha + N_\beta = \nabla^2\Phi + N_{\alpha 0} + N_{\beta 0}. \tag{3.3.16c}$$

Thus the second of Vlasov's equations can be written as

$$\nabla^4\Phi - Eh\nabla_1^2(w) = -\frac{1+\nu}{g_\alpha g_\beta}\left[\frac{\partial}{\partial\alpha}(g_\beta q_\alpha) + \frac{\partial}{\partial\beta}(g_\alpha q_\beta)\right]$$
$$- \nabla^2(N_{\alpha 0} + N_{\beta 0}). \tag{3.3.17}$$

---

† The compatibility equation is exact for cylinders and conical shells. The exact compatibility equation for a sphere, and a slightly better compatibility equation for the general shell, is obtained by adding the quantity $(\epsilon_\alpha + \epsilon_\beta)/R_\alpha R_\beta$ to the left side of Eq. (3.3.15) and the quantity

$$\frac{w}{R_\alpha R_\beta}\left(\frac{1}{R_\alpha} + \frac{1}{R_\beta}\right)$$

to the left side of the equation.

‡ If we had used instead the definition of N_α, N_β, and $\overline{N}_{\alpha\beta}$ given by Eqs. (3.3.9) in the braced-portion of the equation, we would obtain additional terms involving Φ. These, however, can be considered to be negligible in comparison to other terms in the equation involving higher order derivatives of Φ.

We may obtain, finally, a single complex equivalent of Eqs. (3.3.12) and (3.3.18) by multiplying Eq. (3.3.17) by a constant $1/k$ and adding the result to Eq. (3.3.12) to yield

$$D\nabla^4\left(w + \frac{\Phi}{kD}\right) - \frac{Eh}{k}\nabla_1^2\left(w - \frac{k}{Eh}\Phi\right)$$
$$= q_\zeta + \nabla^2(N_{\alpha 0} + N_{\beta 0}) + \frac{1}{g_\alpha g_\beta}$$
$$\times\left\{\frac{\partial}{\partial\alpha}\left[g_\beta\left(m_\alpha - \frac{1+\nu}{k}q_\alpha\right)\right] + \frac{\partial}{\partial\beta}\left[g_\alpha\left(m_\beta - \frac{1+\nu}{k}q_\beta\right)\right]\right\}. \tag{3.3.18a}$$

Now let us choose k such that

$$\frac{1}{kD} = -\frac{k}{Eh}, \tag{3.3.18b}$$

or

$$k = i\frac{[12(1-\nu^2)]^{1/2}}{h}. \tag{3.3.18c}$$

Then Eq. (3.3.18a) becomes

$$D\left[\nabla^4 - \frac{[12(1-\nu^2)]^{1/2}}{h}i\nabla_1^2\right]\left[w + \frac{[12(1-\nu^2)]^{1/2}}{Eh^2}i\Phi\right] = \bar{q}_\zeta, \tag{3.3.19a}$$

where

$$\bar{q}_\zeta = q_\zeta + \nabla^2(N_{\alpha 0} + N_{\beta 0}) + \frac{1}{g_\alpha g_\beta}\left\{\frac{\partial}{\partial\alpha}\left[g_\beta\left(m_\alpha + i\left[\frac{1+\nu}{12(1-\nu)}\right]^{1/2} q_\alpha h\right)\right]\right.$$
$$\left. + \frac{\partial}{\partial\beta}\left[g_\alpha\left(m_\beta + i\left[\frac{1+\nu}{12(1-\nu)}\right]^{1/2} q_\beta h\right)\right]\right\}, \tag{3.3.19b}$$

which is a single complex equation in one complex variable. The equation is simplified, of course, when only normal loading is applied to the shell, for with q_α and q_β equal to zero, $N_{\alpha 0}$ and $N_{\beta 0}$ also vanish. Thus for normal loading alone, Eq. (3.3.19a) reduces to

$$\left[\nabla^4 - \frac{[12(1-\nu^2)]^{1/2}}{Eh^2}i\nabla_1^2\right]\left[w + \frac{[12(1-\nu^2)]^{1/2}}{Eh^2}i\Phi\right] = \frac{q_\zeta}{D}. \tag{3.3.20}$$

Once we have determined w and Φ, the real and imaginary parts of the solution of Eq. (3.3.19a), the force components can be determined from Eqs. (3.3.6), (3.3.7), and (3.3.9). The moment components are given by Eqs. (3.3.1) and (3.3.11). Since we have generally violated the compatibility conditions, it is not possible to obtain unique values for tangential displacements from Eqs. (3.3.14). However, the nonunique terms will contain lower order derivatives of the stress function Φ than the terms common to the various expressions and may therefore be omitted without serious error. We should also note that because we have deleted tangential displacements from change of curvature expressions, rigid body displacements generally will not yield states of zero stress. But for those problems for which rigid body displacements are important, Vlasov's equations should generally not be used.

The final simplified equation of the theory of shells given by Eq. (3.3.19a) represents a remarkable reduction in the theory since not only have we replaced three displacement or six force or strain component equations by a single equation but we have reduced the order of the equations from eight to four. The difference in order, apparently leading to a number of arbitrary constants for Eq. (3.3.19a) which is half the number associated with the original system, is compensated by the fact that the arbitrary constants for Eq. (3.3.19a) are complex, each consisting of two arbitrary quantities. The form of Eq. (3.3.19a) is very similar to that of the equations of the theory of flat plates. Indeed, if we let the principal radii of curvature of the shell middle surface vanish, we have

$$\nabla_1^2(\ldots) \equiv 0, \tag{3.3.21}$$

and Eq. (3.3.19a), or Eqs. (3.3.12) and (3.3.18), can be written as

$$D\nabla^4 w = q_\zeta + \frac{1}{g_\alpha g_\beta}\left[\frac{\partial}{\partial\alpha}(g_\beta m_\alpha) + \frac{\partial}{\partial\beta}(g_\alpha m_\beta)\right], \tag{3.3.22a}$$

$$\nabla^4\Phi = -\left\{\frac{1+\nu}{g_\alpha g_\beta}\left[\frac{\partial}{\partial\alpha}(g_\beta q_\alpha) + \frac{\partial}{\partial\beta}(g_\alpha q_\beta)\right] + \nabla^2(N_{\alpha 0} + N_{\beta 0})\right]\right\}, \tag{3.3.22b}$$

which are, respectively, the deflection equation for transverse bending of flat plates and the stress–function equation for in-plane loading of a plate.

3.4 Shallow shells†

In many applications of shell theory, particularly for shells used as roofs of buildings, the shell middle surface differs little from a plane surface. In this

† An alternate derivation of the same equations directly from the equations of shell theory for arbitrary coordinates is given by A. A. Nazarov, 'On the Theory of Thin Shallow Shells,' NACA Technical Memorandum 1426 (Translation), December 1956.

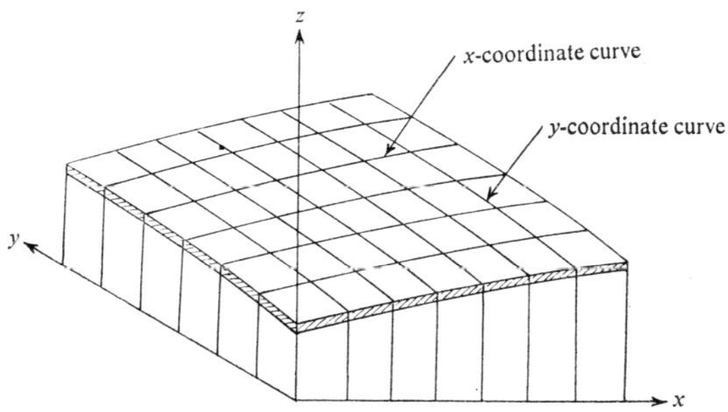

Fig. 3.1. Shallow shell with coordinate curves parallel to Cartesian coordinate planes

case a more convenient set of coordinates for Vlasov's equations are cartesian coordinates x, y in the plane (see Fig. 3.1). With

$$\alpha = \alpha(x, y), \tag{3.4.1a}$$

$$\beta = \beta(x, y), \tag{3.4.1b}$$

$$z = z(x, y), \tag{3.4.1c}$$

we have

$$\frac{\partial z}{\partial \alpha} = \frac{\partial z}{\partial x}\frac{\partial x}{\partial \alpha} + \frac{\partial z}{\partial y}\frac{\partial y}{\partial \alpha}, \tag{3.4.2a}$$

$$\frac{\partial z}{\partial \beta} = \frac{\partial z}{\partial x}\frac{\partial x}{\partial \beta} + \frac{\partial z}{\partial y}\frac{\partial y}{\partial \beta}. \tag{3.4.2b}$$

From the assumptions

$$\frac{\partial z}{\partial x}, \frac{\partial z}{\partial y} \ll 1, \tag{3.4.3}$$

the first fundamental quantities of the shell middle surface, given by Eqs. (1.1.3), become

$$g_\alpha^2 = \left(\frac{\partial x}{\partial \alpha}\right)^2\left[1 + \left(\frac{\partial z}{\partial x}\right)^2\right] + 2\frac{\partial z}{\partial x}\frac{\partial z}{\partial y}\frac{\partial x}{\partial \alpha}\frac{\partial y}{\partial \alpha} + \left(\frac{\partial y}{\partial \alpha}\right)^2\left[1 + \left(\frac{\partial z}{\partial y}\right)^2\right]$$

$$\approx \left(\frac{\partial x}{\partial \alpha}\right)^2 + \left(\frac{\partial y}{\partial \alpha}\right)^2, \tag{3.4.4a}$$

$$g_\beta^2 = \left(\frac{\partial x}{\partial \beta}\right)^2\left[1 + \left(\frac{\partial z}{\partial x}\right)^2\right] + 2\frac{\partial z}{\partial x}\frac{\partial z}{\partial y}\frac{\partial x}{\partial \beta}\frac{\partial y}{\partial \beta} + \left(\frac{\partial y}{\partial \beta}\right)^2\left[1 + \left(\frac{\partial z}{\partial y}\right)^2\right]$$

$$\approx \left(\frac{\partial x}{\partial \beta}\right)^2 + \left(\frac{\partial y}{\partial \beta}\right)^2, \tag{3.4.4b}$$

$$g_{\alpha\beta} = 0 = \frac{\partial x}{\partial \alpha}\frac{\partial x}{\partial \beta}\left[1 + \left(\frac{\partial z}{\partial x}\right)^2\right] + \frac{\partial z}{\partial x}\frac{\partial z}{\partial y} \times \left(\frac{\partial x}{\partial \alpha}\frac{\partial y}{\partial \beta} + \frac{\partial x}{\partial \beta}\frac{\partial y}{\partial \alpha}\right) + \frac{\partial y}{\partial \alpha}\frac{\partial y}{\partial \beta}\left[1 + \left(\frac{\partial z}{\partial y}\right)^2\right]$$

$$\approx \frac{\partial x}{\partial \alpha}\frac{\partial x}{\partial \beta} + \frac{\partial y}{\partial \alpha}\frac{\partial y}{\partial \beta}. \tag{3.4.4c}$$

With the use of the above simplified expressions and Eqs. (A.8) we can show that a transformation of the operators ∇^2 and ∇_1^2 from the lines of curvature coordinate system to the x, y coordinate system yields†

$$\nabla^2(\,) = \frac{1}{g_\alpha g_\beta}\left\{\frac{\partial}{\partial \alpha}\left[\frac{g_\beta}{g_\alpha}\frac{\partial(\,)}{\partial \alpha}\right] + \frac{\partial}{\partial \beta}\left[\frac{g_\alpha}{g_\beta}\frac{\partial(\,)}{\partial \beta}\right]\right\}$$

$$\approx \frac{\partial^2(\,)}{\partial x^2} + \frac{\partial^2(\,)}{\partial y^2}, \tag{3.4.5a}$$

$$\nabla_1^2(\,) = \frac{1}{g_\alpha g_\beta}\left\{\frac{\partial}{\partial \alpha}\left[\frac{1}{R_\beta}\frac{g_\beta}{g_\alpha}\frac{\partial(\,)}{\partial \alpha}\right] + \frac{\partial}{\partial \beta}\left[\frac{1}{R_\alpha}\frac{g_\alpha}{g_\beta}\frac{\partial(\,)}{\partial \beta}\right]\right\}$$

$$\approx \frac{1}{R_y}\frac{\partial^2(\,)}{\partial x^2} - \frac{2}{R_{xy}}\frac{\partial^2(\,)}{\partial x\,\partial y} + \frac{1}{R_x}\frac{\partial^2(\,)}{\partial y^2}, \tag{3.4.5b}$$

where R_x, R_y, and R_{xy} are the normal radii of curvature and twist of the surface for the x, y coordinate curves. Since the slopes are small these may be expressed as

$$\frac{1}{R_x} = -\frac{\partial^2 z/\partial x^2}{[1 + (\partial z/\partial x)^2][1 + (\partial z/\partial x)^2 + (\partial z/\partial y)^2]^{1/2}} \approx -\frac{\partial^2 z}{\partial x^2}, \tag{3.4.6a}$$

$$\frac{1}{R_y} = -\frac{\partial^2 z/\partial y^2}{[1 + (\partial z/\partial y)^2][1 + (\partial z/\partial x)^2 + (\partial z/\partial y)^2]^{1/2}} \approx -\frac{\partial^2 z}{\partial y^2}, \tag{3.4.6b}$$

$$\frac{1}{R_{xy}} = -\frac{\partial^2 z/\partial x\,\partial y}{\{[1 + (\partial z/\partial x)^2][1 + (\partial z/\partial y)^2][1 + (\partial z/\partial x)^2 + (\partial z/\partial y)^2]\}^{1/2}}$$

$$\approx -\frac{\partial^2 z}{\partial x\,\partial y}. \tag{3.4.6c}$$

† Some useful relationships for this process are the following which can be derived from Eqs. (3.4.4)

$$\left(\frac{1}{g_\alpha}\frac{\partial x}{\partial \alpha}\right)^2 \approx \left(\frac{1}{g_\beta}\frac{\partial y}{\partial \beta}\right)^2, \qquad \left(\frac{1}{g_\beta}\frac{\partial x}{\partial \beta}\right)^2 \approx \left(\frac{1}{g_\alpha}\frac{\partial y}{\partial \alpha}\right)^2.$$

Then

$$\nabla_1^2(\,) \approx -\frac{\partial^2 z}{\partial y^2}\frac{\partial^2(\,)}{\partial x^2} + 2\frac{\partial^2 z}{\partial x\,\partial y}\frac{\partial^2(\,)}{\partial x\,\partial y} - \frac{\partial^2 z}{\partial x^2}\frac{\partial^2(\,)}{\partial y^2}. \tag{3.4.7}$$

Thus the complex form of Vlasov's equations may be written for shallow shells as

$$\left[\left(\frac{\partial^2}{\partial x^2} + \frac{\partial^2}{\partial y^2}\right)^2 + \frac{[12(1-\nu^2)]^{1/2}}{h}\, i\left(\frac{\partial^2 z}{\partial y^2}\frac{\partial^2}{\partial x^2} - 2\frac{\partial^2 z}{\partial x\,\partial y}\frac{\partial^2}{\partial x\,\partial y} + \frac{\partial^2 z}{\partial x^2}\frac{\partial^2}{\partial y^2}\right)\right] \times \left[w + \frac{[12(1-\nu^2)]^{1/2}}{Eh^2}\, i\Phi\right] = \frac{\bar{q}_\zeta}{D}. \tag{3.4.8}$$

Edge conditions appropriate to the new coordinate system can be shown by means of Eqs. (A.15), (3.4.4), and a transformation of coordinates to be

$$\frac{\partial^2\phi}{\partial y^2} = N_x^* \qquad \text{or } u_x \text{ is prescribed,} \tag{3.4.9a}$$

$$-\frac{\partial^2\phi}{\partial x\,\partial y} = N_{xy}^* \qquad \text{or } u_y \text{ is prescribed,} \tag{3.4.9b}$$

$$-D\left(\frac{\partial^2 w}{\partial x^2} + \nu\frac{\partial^2 w}{\partial y^2}\right) = M_x^* \qquad \text{or } \frac{\partial w}{\partial x} \text{ is prescribed,} \tag{3.4.9c}$$

$$-D\frac{\partial}{\partial x}\left[\frac{\partial^2 w}{\partial x^2} + (2-\nu)\frac{\partial^2 w}{\partial y^2}\right] = Q_x^* + \frac{\partial M_{xy}^*}{\partial y}, \qquad \text{or } w \text{ is prescribed} \tag{3.4.9d}$$

along y coordinate curves,

$$\frac{\partial^2\phi}{\partial x^2} = N_y^* \qquad \text{or } u_y \text{ is prescribed,} \tag{3.4.10a}$$

$$-\frac{\partial^2\phi}{\partial x\,\partial y} = N_{yx}^* \qquad \text{or } u_x \text{ is prescribed,} \tag{3.4.10b}$$

$$-D\left(\frac{\partial^2 w}{\partial y^2} + \nu\frac{\partial^2 w}{\partial x^2}\right) = M_y^* \qquad \text{or } \frac{\partial w}{\partial y} \text{ is prescribed,} \tag{3.4.10c}$$

$$-D\frac{\partial}{\partial y}\left[\frac{\partial^2 w}{\partial y^2} + (2-\nu)\frac{\partial^2 w}{\partial x^2}\right] = Q_y^* + \frac{\partial M_{yx}^*}{\partial x} \qquad \text{or } w \text{ is prescribed,} \tag{3.4.10d}$$

along x coordinate curves, and

$$-D(1-\nu)\frac{\partial^2 w}{\partial x\,\partial y} = \tfrac{1}{2}(M_{xy}^* + M_{yx}^*), \tag{3.4.11}$$

at corners. In addition, for consistency, we must assume

$$N^*_{xy} = N^*_{yx}, \tag{3.4.12}$$

at corners.

3.5 Exact reduction of shell theory for zero Poisson's ratio

(I) Complex stress resultants

Vlasov's equations are possibly the simplest set of equations which retain the essential characteristics of the general theory of shells. In these equations the normal deflection w and a stress function Φ were taken as variables. It is also possible to reduce the equations of shell theory, although not in so drastic a manner, by other choices of variables. When Poisson's ratio is zero, this reduction can be achieved with no assumptions other than the use of the modified force–strain relations given by Eqs. (3.1.4) which become

$$N_\alpha = Eh\epsilon_\alpha, \tag{3.5.1a}$$
$$N_\beta = Eh\epsilon_\beta, \tag{3.5.1b}$$
$$\overline{N}_{\alpha\beta} = \tfrac{1}{2}Eh\gamma_{\alpha\beta}, \tag{3.5.1c}$$
$$M_\alpha = \tfrac{1}{2}Eh^3\kappa_\alpha, \tag{3.5.1d}$$
$$M_\beta = \tfrac{1}{2}Eh^3\kappa_\beta, \tag{3.5.1e}$$
$$\overline{M}_{\alpha\beta} = \tfrac{1}{2}Eh^3\bar{\kappa}_{\alpha\beta}. \tag{3.5.1f}$$

Let us consider the three equations of equilibrium of stress resultants, Eqs. (2.6.6), and the three equations of compatibility of strains given by Eqs. (1.5.11), with the strain components replaced by the corresponding stress resultants by means of Eqs. (3.5.1). Then we have, on alternating the equations of equilibrium and of compatibility,

$$\frac{\partial}{\partial a}\left[g_\beta\left(N_\alpha + \frac{M_\alpha}{R_\beta}\right)\right]\frac{\partial g_\beta}{\partial \alpha} + \frac{1}{g_\alpha}\frac{\partial}{\partial \beta}\left\{g_\alpha^2\left[\overline{N}_{\alpha\beta} + \frac{1}{2}\left(\frac{3}{R_\alpha} - \frac{1}{R_\beta}\right)\overline{M}_{\alpha\beta}\right]\right\} - g_\beta\left(\frac{1}{R_\beta} - \frac{1}{R_\alpha}\right)\frac{\partial M_\alpha}{\partial \alpha} = -g_\alpha g_\beta\left(q_\alpha + \frac{m_\alpha}{R_\alpha}\right), \tag{3.5.2a}$$

$$\frac{\partial}{\partial \alpha}\left[g_\beta\left(\frac{12M_\beta}{Eh^3} - \frac{N_\beta}{EhR_\beta}\right)\right] - \left(\frac{12M_\alpha}{Eh^3} - \frac{N_\alpha}{EhR_\alpha}\right)\frac{\partial g_\beta}{\partial \alpha} - \frac{1}{g_\alpha}\frac{\partial}{\partial \alpha} \times \left\{g_\alpha^2\left[\frac{12\overline{M}_{\alpha\beta}}{Eh^3} - \frac{1}{2}\left(\frac{3}{R_\alpha} - \frac{1}{R_\beta}\right)\frac{\overline{N}_{\alpha\beta}}{Eh}\right]\right\} + g_\beta\left(\frac{1}{R_\beta} - \frac{1}{R_\alpha}\right)\frac{\partial}{\partial \alpha}\left(\frac{N_\beta}{Eh}\right) = 0, \tag{3.5.2b}$$

$$\frac{\partial}{\partial\beta}\left[g_\alpha\left(N_\beta + \frac{M_\beta}{R_\alpha}\right)\right] - \left(N_\alpha + \frac{M_\alpha}{R_\beta}\right)\frac{\partial g_\alpha}{\partial\beta} + \frac{1}{g_\beta}\frac{\partial}{\partial\alpha}$$
$$\times\left\{g_\beta^2\left[\overline{N}_{\alpha\beta} + \frac{1}{2}\left(\frac{3}{R_\beta} - \frac{1}{R_\alpha}\right)\overline{M}_{\alpha\beta}\right]\right\}$$
$$- g_\alpha\left(\frac{1}{R_\alpha} - \frac{1}{R_\beta}\right)\frac{\partial M_\beta}{\partial\beta} = -g_\alpha g_\beta\left(q_\beta + \frac{m_\beta}{R_\beta}\right), \tag{3.5.2c}$$

$$\frac{\partial}{\partial\beta}\left[g_\alpha\left(\frac{12M_\alpha}{Eh^3} - \frac{N_\alpha}{EhR_\alpha}\right)\right] - \left(\frac{12M_\beta}{Eh^3} - \frac{N_\beta}{EhR_\beta}\right)\frac{\partial g_\alpha}{\partial\beta} = \frac{1}{g_\beta}\frac{\partial}{\partial\alpha}$$
$$\times\left\{g_\beta^2\left[\frac{12\overline{M}_{\alpha\beta}}{Eh^3} - \frac{1}{2}\left(\frac{3}{R_\beta} - \frac{1}{R_\alpha}\right)\frac{\overline{N}_{\alpha\beta}}{Eh}\right]\right\}$$
$$+ g_\alpha\left(\frac{1}{R_\alpha} - \frac{1}{R_\beta}\right)\frac{\partial}{\partial\beta}\left(\frac{N_\alpha}{Eh}\right) = 0, \tag{3.5.2d}$$

$$\frac{\partial}{\partial\alpha}\left\{\frac{1}{g_\alpha}\left[\frac{\partial}{\partial\alpha}(g_\beta M_\alpha) - M_\beta\frac{\partial g_\beta}{\partial\alpha} + \frac{1}{g_\alpha}\frac{\partial}{\partial\beta}(g_\alpha^2\overline{M}_{\alpha\beta})\right]\right\}$$
$$+ \frac{\partial}{\partial\beta}\left\{\frac{1}{g_\beta}\left[\frac{\partial}{\partial\beta}(g_\alpha M_\beta) - M_\alpha\frac{\partial g_\alpha}{\partial\beta} + \frac{1}{g_\beta}\frac{\partial}{\partial\alpha}(g_\beta^2\overline{M}_{\alpha\beta})\right]\right\}$$
$$- g_\alpha g_\beta\left(\frac{N_\alpha}{R_\alpha} + \frac{N_\beta}{R_\beta}\right) = -\left[g_\alpha g_\beta q_\zeta + \frac{\partial}{\partial\alpha}(g_\beta m_\alpha) + \frac{\partial}{\partial\beta}(g_\alpha m_\beta)\right], \tag{3.5.2e}$$

$$\frac{\partial}{\partial\alpha}\left\{\frac{1}{g_\alpha}\left[\frac{\partial}{\partial\alpha}\left(g_\beta\frac{N_\beta}{Eh}\right) - \frac{N_\alpha}{Eh}\frac{\partial g_\beta}{\partial\alpha} - \frac{1}{g_\alpha}\frac{\partial}{\partial\beta}\left(g_\beta^2\frac{\overline{N}_{\alpha\beta}}{Eh}\right)\right]\right\}$$
$$+ \frac{\partial}{\partial\beta}\left\{\frac{1}{g_\beta}\left[\frac{\partial}{\partial\beta}\left(g_\alpha\frac{N_\alpha}{Eh}\right) - \frac{N_\beta}{Eh}\frac{\partial g_\alpha}{\partial\beta} - \frac{1}{g_\beta}\frac{\partial}{\partial\alpha}\left(g_\alpha^2\frac{\overline{N}_{\alpha\beta}}{Eh}\right)\right]\right\}$$
$$+ g_\alpha g_\beta\left(\frac{12M_\beta}{Eh^3R_\alpha} + \frac{12M_\alpha}{Eh^3R_\beta}\right) = 0. \tag{3.5.2f}$$

The six equations [Eqs. (3.5.2)] in six unknowns can be combined to yield three equations in three unknowns. Let us consider, for example, Eqs. (3.5.2a) and (3.5.2b). If we multiply Eq. (3.5.2b) by a constant k and add it to Eq. (3.5.2a), we may express the result as

$$\frac{\partial}{\partial\alpha}\left\{g_\beta\left[\left(N_\alpha + \frac{12k}{Eh^3}M_\beta\right) - \frac{k}{EhR_\beta}\left(N_\beta - \frac{Eh}{k}M_\alpha\right)\right]\right\}$$
$$- \left[\left(N_\beta + \frac{12k}{Eh^3}M_\alpha\right) - \frac{k}{EhR_\alpha}\left(N_\alpha - \frac{Eh}{k}M_\beta\right)\right]\frac{\partial g_\beta}{\partial\alpha} + \frac{1}{g_\alpha}\frac{\partial}{\partial\beta}$$
$$\times\left\{g_\alpha^2\left[\left(\overline{N}_{\alpha\beta} - \frac{12k}{Eh^3}\overline{M}_{\alpha\beta}\right) + \frac{k}{2Eh}\left(\frac{3}{R_\alpha} - \frac{1}{R_\beta}\right)\left(\overline{N}_{\alpha\beta} + \frac{Eh}{k}\overline{M}_{\alpha\beta}\right)\right]\right\}$$
$$+ g_\beta\left(\frac{1}{R_\beta} - \frac{1}{R_\alpha}\right)\frac{k}{Eh}\frac{\partial}{\partial\alpha}\left(N_\beta - \frac{Eh}{k}M_\alpha\right) = -g_\alpha g_\beta\left(q_\alpha + \frac{m_\alpha}{R_\alpha}\right). \tag{3.5.3a}$$

We choose k, now, such that

$$\frac{12k}{Eh^3} = -\frac{Eh}{k}, \tag{3.5.3b}$$

which yields

$$k = i\frac{Eh^2}{(12)^{1/2}}. \tag{3.5.3c}$$

Then, with the notation

$$\tilde{N}_\alpha = N_\alpha + (12)^{1/2} i \frac{M_\beta}{h}, \tag{3.5.4a}$$

$$\tilde{N}_\beta = N_\beta + (12)^{1/2} i \frac{M_\alpha}{h}, \tag{3.5.4b}$$

$$\tilde{N}_{\alpha\beta} = \overline{N}_{\alpha\beta} - (12)^{1/2} i \frac{\overline{M}_{\alpha\beta}}{h}, \tag{3.5.4c}$$

Eq. (3.5.3a) becomes

$$\begin{aligned} &\frac{\partial}{\partial\alpha}\left[g_\beta\left(\tilde{N}_\alpha - \frac{ih\tilde{N}_\beta}{(12)^{1/2}R_\beta}\right)\right] - \left(\tilde{N}_\beta - \frac{ih}{(12)^{1/2}R_\alpha}\tilde{N}_\alpha\right)\frac{\partial g_\beta}{\partial\alpha} \\ &\quad + \frac{1}{g_\alpha}\frac{\partial}{\partial\beta}\left\{g_\alpha^2\tilde{N}_{\alpha\beta}\left[1 + \frac{ih}{2(12)^{1/2}}\left(\frac{3}{R_\alpha} - \frac{1}{R_\beta}\right)\right]\right\} \\ &\quad + g_\beta\frac{ih}{(12)^{1/2}}\left(\frac{1}{R_\beta} - \frac{1}{R_\alpha}\right)\frac{\partial\tilde{N}_\beta}{\partial\alpha} = -g_\alpha g_\beta\left(q_\alpha + \frac{m_\alpha}{R_\alpha}\right). \end{aligned} \tag{3.5.5a}$$

Similar operations on the two remaining pairs of Eqs. (3.4.2) yield

$$\begin{aligned} &\frac{\partial}{\partial\beta}\left[g_\alpha\left(\tilde{N}_\beta - \frac{ih}{(12)^{1/2}R_\alpha}\tilde{N}_\alpha\right)\right] - \left(\tilde{N}_\alpha - \frac{ih}{(12)^{1/2}R_\beta}N_\beta\right)\frac{\partial g_\alpha}{\partial\beta} \\ &\quad + \frac{1}{g_\beta}\frac{\partial}{\partial\alpha}\left\{g_\beta^2\tilde{N}_{\alpha\beta}\left[1 + \frac{ih}{(12)^{1/2}}\left(\frac{3}{R_\beta} - \frac{1}{R_\alpha}\right)\right]\right\} \\ &\quad + g_\alpha\frac{ih}{(12)^{1/2}}\left(\frac{1}{R_\alpha} - \frac{1}{R_\beta}\right)\frac{\partial\tilde{N}_\alpha}{\partial\beta} = -g_\alpha g_\beta\left(q_\beta + \frac{m_\beta}{R_\beta}\right), \end{aligned} \tag{3.5.5b}$$

$$\begin{aligned} &g_\alpha g_\beta\left(\frac{\tilde{N}_\alpha}{R_\alpha} + \frac{\tilde{N}_\beta}{R_\beta}\right) + \frac{ih}{(12)^{1/2}}\frac{\partial}{\partial\alpha} \\ &\quad \times\left\{\frac{1}{g_\alpha}\left[\frac{\partial}{\partial\alpha}(g_\beta\tilde{N}_\beta) - \tilde{N}_\alpha\frac{\partial g_\beta}{\partial\alpha} - \frac{1}{g_\alpha}\frac{\partial}{\partial\beta}(g_\alpha^2\tilde{N}_{\alpha\beta})\right]\right\} \\ &\quad + \frac{ih}{(12)^{1/2}}\frac{\partial}{\partial\beta}\left\{\frac{1}{g_\beta}\left[\frac{\partial}{\partial\beta}(g_\alpha\tilde{N}_\alpha) - \tilde{N}_\beta\frac{\partial g_\alpha}{\partial\beta} - \frac{1}{g_\beta}\frac{\partial}{\partial\alpha}(g_\beta^2\tilde{N}_{\alpha\beta})\right]\right\} \\ &= \left[g_\alpha g_\beta q_\zeta + \frac{\partial}{\partial\alpha}(g_\beta m_\alpha) + \frac{\partial}{\partial\beta}(g_\alpha m_\beta)\right]. \end{aligned} \tag{3.5.5c}$$

Eqs. (3.5.5) are three complex equations for three complex resultants $\tilde{N}_\alpha$, $\tilde{N}_\beta$, and $\tilde{N}_{\alpha\beta}$ from which the six real stress resultants may be obtained by separating the complex quantities into their real and imaginary parts, i.e.,

$$N_\alpha = \operatorname{Re} \tilde{N}_\alpha, \tag{3.5.6a}$$

$$N_\beta = \operatorname{Re} \tilde{N}_\beta, \tag{3.5.6b}$$

$$\bar{N}_{\alpha\beta} = \operatorname{Re} \tilde{N}_{\alpha\beta}, \tag{3.5.6c}$$

$$M_\alpha = \frac{h}{(12)^{1/2}} \operatorname{Im} \tilde{N}_\beta, \tag{3.5.6d}$$

$$M_\beta = \frac{h}{(12)^{1/2}} \operatorname{Im} \tilde{N}_\alpha, \tag{3.5.6e}$$

$$\bar{M}_{\alpha\beta} = -\frac{h}{(12)^{1/2}} \operatorname{Im} \tilde{N}_{\alpha\beta}. \tag{3.5.6f}$$

The system of Eqs. (3.5.5) is of fourth order, or one-half the order of the system of Eqs. (3.5.2).

The displacements of the shell can now be obtained from the solution of the three equations

$$\epsilon_\alpha = \frac{1}{g_\alpha}\left(\frac{\partial u_\alpha}{\partial \alpha} + \frac{1}{g_\beta}\frac{\partial g_\alpha}{\partial \beta} u_\beta + \frac{g_\alpha}{R_\alpha} w\right) = \frac{N_\alpha}{Eh} = \frac{1}{Eh} \operatorname{Re} \tilde{N}_\alpha, \tag{3.5.7a}$$

$$\epsilon_\beta = \frac{1}{g_\beta}\left(\frac{\partial u_\beta}{\partial \beta} + \frac{1}{g_\alpha}\frac{\partial g_\beta}{\partial \alpha} u_\alpha + \frac{g_\beta}{R_\beta} w\right) = \frac{N_\beta}{Eh} = \frac{1}{Eh} \operatorname{Re} \tilde{N}_\beta, \tag{3.5.7b}$$

$$\gamma_{\alpha\beta} = \frac{g_\alpha}{g_\beta}\frac{\partial}{\partial \beta}\left(\frac{u_\alpha}{g_\alpha}\right) + \frac{g_\beta}{g_\alpha}\left(\frac{u_\beta}{g_\beta}\right) = \frac{2\bar{N}_{\alpha\beta}}{Eh} = \frac{2}{Eh} \operatorname{Re} \tilde{N}_{\alpha\beta}, \tag{3.5.7c}$$

subject to the condition that the displacements satisfy the relationships

$$\kappa_\alpha = \frac{12 M_\alpha}{Eh^3} = \frac{(12)^{1/2}}{Eh^2} \operatorname{Im} \tilde{N}_\beta, \tag{3.5.8a}$$

$$\kappa_\beta = \frac{12 M_\beta}{Eh^3} = \frac{(12)^{1/2}}{Eh^2} \operatorname{Im} \tilde{N}_\alpha, \tag{3.5.8b}$$

$$\bar{\kappa}_{\alpha\beta} = \frac{12 \bar{M}_{\alpha\beta}}{Eh^3} = -\frac{(12)^{1/2}}{Eh^2} \operatorname{Im} \tilde{N}_{\alpha\beta}. \tag{3.5.8c}$$

This condition will be satisfied, since the strains obtained from the complex stress resultants satisfy conditions of compatibility, provided the complementary solutions of Eqs. (3.5.7) are restricted to rigid body displacements of the shell.

3.6 Exact reduction of shell theory for zero Poisson's ratio

(II) Complex displacements

An alternate representation of the theory in terms of displacements is possible if we introduce complex displacements $\tilde{u}_\alpha$, $\tilde{u}_\beta$, and $\tilde{w}$ such that

$$\tilde{\epsilon}_\alpha = \frac{1}{g_\alpha}\left(\frac{\partial \tilde{u}_\alpha}{\partial \alpha} + \frac{1}{g_\beta}\frac{\partial g_\alpha}{\partial \beta}\tilde{u}_\beta + \frac{g_\alpha}{R_\alpha}\tilde{w}\right) = \frac{1}{Eh}\tilde{N}_\alpha, \tag{3.6.1a}$$

$$\tilde{\epsilon}_\beta = \frac{1}{g_\beta}\left(\frac{\partial \tilde{u}_\beta}{\partial \beta} + \frac{1}{g_\alpha}\frac{\partial g_\beta}{\partial \alpha}\tilde{u}_\alpha + \frac{g_\beta}{R_\beta}\tilde{w}\right) = \frac{1}{Eh}\tilde{N}_\beta, \tag{3.6.1b}$$

$$\tilde{\gamma}_{\alpha\beta} = \frac{g_\alpha}{g_\beta}\frac{\partial}{\partial \beta}\left(\frac{\tilde{u}_\alpha}{g_\alpha}\right) + \frac{g_\beta}{g_\alpha}\frac{\partial}{\partial \alpha}\left(\frac{\tilde{u}_\beta}{g_\beta}\right) = \frac{2}{Eh}\tilde{N}_{\alpha\beta}. \tag{3.6.1c}$$

We also introduce auxiliary stress resultants N^p_α, N^p_β, $N^p_{\alpha\beta}$, such that

$$N^p_\alpha = \tilde{N}_\alpha - i\frac{Eh^2}{(12)^{1/2}}\tilde{\kappa}_\beta, \tag{3.6.2a}$$

$$N^p_\beta = \tilde{N}_\beta - i\frac{Eh^2}{(12)^{1/2}}\tilde{\kappa}_\alpha, \tag{3.6.2b}$$

$$\overline{N}^p_{\alpha\beta} = \tilde{N}_{\alpha\beta} + i\frac{Eh^2}{(12)^{1/2}}\tilde{\kappa}_{\alpha\beta}, \tag{3.6.2c}$$

where

$$\tilde{\kappa}_\alpha = -\frac{1}{g_\alpha}\left[\frac{\partial}{\partial \alpha}\left(\frac{1}{g_\alpha}\frac{\partial \tilde{w}}{\partial \alpha}\right) + \frac{1}{g_\beta^2}\frac{\partial g_\alpha}{\partial \beta}\frac{\partial \tilde{w}}{\partial \beta} - \frac{\partial}{\partial \alpha}\left(\frac{\tilde{u}_\alpha}{R_\alpha}\right) - \frac{\tilde{u}_\beta}{g_\beta}\frac{\partial}{\partial \beta}\left(\frac{g_\alpha}{R_\alpha}\right)\right], \tag{3.6.3a}$$

$$\tilde{\kappa}_\beta = -\frac{1}{g_\beta}\left[\frac{\partial}{\partial \beta}\left(\frac{1}{g_\beta}\frac{\partial \tilde{w}}{\partial \beta}\right) + \frac{1}{g_\alpha^2}\frac{\partial g_\beta}{\partial \alpha}\frac{\partial \tilde{w}}{\partial \alpha} - \frac{\partial}{\partial \beta}\left(\frac{\tilde{u}_\beta}{R_\beta}\right) - \frac{\tilde{u}_\alpha}{g_\alpha}\frac{\partial}{\partial \alpha}\left(\frac{g_\beta}{R_\beta}\right)\right], \tag{3.6.3b}$$

$$\begin{aligned}\tilde{\kappa}_{\alpha\beta} = -\frac{1}{g_\alpha g_\beta}\Bigg[&\frac{\partial^2 \tilde{w}}{\partial \alpha\,\partial \beta} - \frac{1}{g_\alpha}\frac{\partial g_\alpha}{\partial \beta}\frac{\partial \tilde{w}}{\partial \alpha} - \frac{1}{g_\beta}\frac{\partial g_\beta}{\partial \alpha}\frac{\partial \tilde{w}}{\partial \beta}\\ &- \tfrac{1}{4}g_\alpha^2\left(\frac{3}{R_\alpha} - \frac{1}{R_\beta}\right)\frac{\partial}{\partial \beta}\left(\frac{\tilde{u}_\alpha}{g_\alpha}\right) - \tfrac{1}{4}g_\beta\left(\frac{3}{R_\beta} - \frac{1}{R_\alpha}\right)\frac{\partial}{\partial \alpha}\left(\frac{\tilde{u}_\beta}{g_\beta}\right)\Bigg].\end{aligned} \tag{3.6.3c}$$

Let us now require the complex strains to satisfy compatibility equations of the form of Eqs. (1.5.11). We then obtain

$$\frac{\partial}{\partial\alpha}\left[g_\beta\left(\tilde{N}_\alpha - \frac{ih}{(12)^{1/2}R_\beta}\tilde{N}_\beta\right)\right] - \left(\tilde{N}_\beta - \frac{ih}{(12)^{1/2}R_\alpha}\tilde{N}_\alpha\right)\frac{\partial g_\beta}{\partial\alpha}$$

$$+ \frac{1}{g_\alpha}\frac{\partial}{\partial\beta}\left\{g_\alpha^2\tilde{N}_{\alpha\beta}\left[1 + \frac{ih}{2(12)^{1/2}}\left(\frac{3}{R_\alpha} - \frac{1}{R_\beta}\right)\right]\right\}$$

$$+ g_\beta\frac{ih}{(12)^{1/2}}\left(\frac{1}{R_\beta} - \frac{1}{R_\alpha}\right)\frac{\partial\tilde{N}_\beta}{\partial\alpha}$$

$$= \frac{\partial}{\partial\alpha}(g_\beta N_\alpha^p) - N_\beta^p\frac{\partial g_\beta}{\partial\alpha} + \frac{1}{g_\alpha}\frac{\partial}{\partial\beta}(g_\alpha^2\overline{N}_{\alpha\beta}^p), \tag{3.6.4a}$$

$$\frac{\partial}{\partial\beta}\left[g_\alpha\left(\tilde{N}_\beta - \frac{ih}{(12)^{1/2}R_\alpha}\tilde{N}_\alpha\right)\right] - \left(\tilde{N}_\alpha - \frac{ih}{(12)^{1/2}R_\beta}\tilde{N}_\beta\right)\frac{\partial g_\alpha}{\partial\beta}$$

$$+ \frac{1}{g_\beta}\frac{\partial}{\partial\alpha}\left\{g_\beta^2\tilde{N}_{\alpha\beta}\left[1 + \frac{ih}{2(12)^{1/2}}\left(\frac{3}{R_\beta} - \frac{1}{R_\alpha}\right)\right]\right\}$$

$$+ g_\alpha\frac{ih}{(12)^{1/2}}\left(\frac{1}{R_\alpha} - \frac{1}{R_\beta}\right)\frac{\partial\tilde{N}_\alpha}{\partial\beta}$$

$$= \frac{\partial}{\partial\beta}(g_\alpha N_\beta^p) - N_\alpha^p\frac{\partial g_\alpha}{\partial\beta} + \frac{1}{g_\beta}\frac{\partial}{\partial\alpha}(g_\beta^2\overline{N}_{\alpha\beta}^p), \tag{3.6.4b}$$

$$g_\alpha g_\beta\left(\frac{\tilde{N}_\alpha}{R_\alpha} + \frac{\tilde{N}_\beta}{R_\beta}\right) + \frac{ih}{(12)^{1/2}}\frac{\partial}{\partial\alpha}$$

$$\times\left\{\frac{1}{g_\alpha}\left[\frac{\partial}{\partial\alpha}(g_\beta\tilde{N}_\beta) - \tilde{N}_\alpha\frac{\partial g_\beta}{\partial\alpha} - \frac{1}{g_\alpha}\frac{\partial}{\partial\beta}(g_\alpha^2\tilde{N}_{\alpha\beta}^2)\right]\right\}$$

$$+ \frac{ih}{(12)^{1/2}}\frac{\partial}{\partial\beta}\left\{\frac{1}{g_\beta}\left[\frac{\partial}{\partial\beta}(g_\alpha\tilde{N}_\alpha) - \tilde{N}_\beta\frac{\partial g_\alpha}{\partial\beta} - \frac{1}{g_\beta}\frac{\partial}{\partial\alpha}(g_\beta^2\tilde{N}_{\alpha\beta})\right]\right\}$$

$$= g_\alpha g_\beta\left(\frac{N_\alpha^p}{R_\alpha} + \frac{N_\beta^p}{R_\beta}\right). \tag{3.6.4c}$$

With the use of Eqs. (3.5.5), Eqs. (3.6.4) reduce to the membrane equations

$$\frac{1}{g_\alpha g_\beta}\left[\frac{\partial}{\partial\beta}(g_\beta N_\alpha^p) - N_\beta^p\frac{\partial g_\beta}{\partial\alpha} + \frac{1}{g_\alpha}\frac{\partial}{\partial\beta}(g_\alpha^2\overline{N}_{\alpha\beta}^p)\right] = -\left(q_\alpha + \frac{m_\alpha}{R_\alpha}\right), \tag{3.6.5a}$$

$$\frac{1}{g_\alpha g_\beta}\left[\frac{\partial}{\partial\beta}(g_\alpha N_\beta^p) - N_\alpha^p\frac{\partial g_\alpha}{\partial\beta} + \frac{1}{g_\beta}\frac{\partial}{\partial\alpha}(g_\beta^2\overline{N}_{\alpha\beta}^p)\right] = -\left(q_\beta + \frac{m_\beta}{R_\beta}\right), \tag{3.6.5b}$$

$$\frac{N_\alpha^p}{R_\alpha} + \frac{N_\beta^p}{R_\beta} = q_\zeta + \frac{1}{g_\alpha g_\beta}\left[\frac{\partial}{\partial\alpha}(g_\beta m_\alpha) + \frac{\partial}{\partial\beta}(g_\alpha m_\beta)\right], \tag{3.6.5c}$$

which determine the real auxiliary stress resultants. With N_α^p, N_β^p, and $\bar{N}_{\alpha\beta}^p$ known, equations for the complex displacements may be obtained by eliminating the complex stress resultants from Eqs. (3.6.1) and (3.6.2) to yield

$$\tilde{\epsilon}_\alpha - \frac{ih}{(12)^{1/2}}\tilde{\kappa}_\beta = \frac{1}{Eh}N_\alpha^p, \tag{3.6.6a}$$

$$-\frac{ih}{(12)^{1/2}}\tilde{\kappa}_\alpha = \frac{1}{Eh}N_\beta^p, \tag{3.6.6b}$$

$$\tfrac{1}{2}\tilde{\gamma}_{\alpha\beta} + \frac{ih}{(12)^{1/2}}\tilde{\kappa}_{\alpha\beta} = \frac{1}{Eh}\bar{N}_{\alpha\beta}^p, \tag{3.6.6c}$$

which is a fourth-order system of equations for the displacements $\tilde{u}_\alpha$, $\tilde{u}_\beta$, $\tilde{w}$ when the complex strains are replaced by their expression in terms of complex displacements.

The two systems of equations given by Eqs. (3.6.5) and (3.6.6) are equivalent to Eqs. (3.5.5) and (3.5.7). Complex stress resultants obtained from the complex displacements by means of Eqs. (3.6.1) or (3.6.2) will satisfy Eqs. (3.5.5) by virtue of Eqs. (3.6.4) and (3.6.5). Furthermore, the real part of Eqs. (3.6.1) is identical with Eqs. (3.5.7) so that the real part of each of the complex displacements is identical with the corresponding middle surface displacement of the shell. We may also give an interpretation to the imaginary parts of the complex displacements. Let us consider the imaginary part of Eqs. (3.6.1) and (3.6.2). Then we have, with the use of Eqs. (3.5.4),

$$M_\alpha = \frac{Eh^2}{(12)^{1/2}}\operatorname{Im}\tilde{\epsilon}_\beta, \tag{3.6.7a}$$

$$M_\beta = \frac{Eh^2}{(12)^{1/2}}\operatorname{Im}\tilde{\epsilon}_\alpha, \tag{3.6.7b}$$

$$2\bar{M}_{\alpha\beta} = -\frac{Eh^2}{(12)^{1/2}}\operatorname{Im}\tilde{\gamma}_{\alpha\beta}, \tag{3.6.7c}$$

$$N_\alpha = N_\alpha^p - \frac{Eh^2}{(12)^{1/2}}\operatorname{Im}\tilde{\kappa}_\beta, \tag{3.6.7d}$$

$$N_\beta = N_\beta^p - \frac{Eh^2}{(12)^{1/2}}\operatorname{Im}\tilde{\kappa}_\alpha, \tag{3.6.7e}$$

$$\bar{N}_{\alpha\beta} = \bar{N}_{\alpha\beta}^p + \frac{Eh^2}{(12)^{1/2}}\operatorname{Im}\tilde{\kappa}_{\alpha\beta}. \tag{3.6.7f}$$

Now Eqs. (3.6.7) are of the same form as Eqs. (2.10.3) which express the stress resultants in terms of particular solutions of the equilibrium equations and three stress functions χ_α, χ_β, and χ. On comparing the two sets of equations, we can identify the auxiliary stress resultants of this section as the particular solution of the equations of equilibrium, and the imaginary parts of the complex displacements (which we shall denote by u'_α, u'_β, and w', respectively) as a constant factor times the stress function defined in Section 2.10, the exact relation being

$$u'_\alpha = \frac{Eh^2}{(12)^{1/2}} \chi_\alpha, \tag{3.6.8a}$$

$$u'_\beta = \frac{Eh^2}{(12)^{1/2}} \chi_\beta, \tag{3.6.8b}$$

$$w' = \frac{Eh^2}{(12)^{1/2}} \chi_\zeta. \tag{3.6.8c}$$

3.7 Approximate reduction of shell theory for nonzero Poisson's ratio

When Poisson's ratio is not zero, it is, in general, impossible to reduce the equations of shell theory in a manner similar to the above without violating some of the equations by the deletion of terms. For a shell whose Poisson's ratio is not equal to zero, the equilibrium equations [Eqs. (3.5.2a), (3.5.2c), and (3.5.2e)] remain the same, but the compatibility equations [Eqs. (3.5.2b), (3.5.2d), and (3.5.2f)] are changed by the following replacements

$$N_\alpha \to N_\alpha - \nu N_\beta, \tag{3.7.1a}$$

$$N_\beta \to N_\beta - \nu N_\alpha, \tag{3.7.1b}$$

$$\overline{N}_{\alpha\beta} \to (1 + \nu)\overline{N}_{\alpha\beta}, \tag{3.7.1c}$$

$$M_\alpha \to M_\alpha - \nu M_\beta, \tag{3.7.1d}$$

$$M_\beta \to M_\beta - \nu M_\alpha, \tag{3.7.1e}$$

$$\overline{M}_{\alpha\beta} \to (1 + \nu)\overline{M}_{\alpha\beta}, \tag{3.7.1f}$$

when the overall force–strain relations are assumed to be given by Eq. (3.1.4).

Let us consider the first two equations of the revised set and proceed, as in Section 3.5 by multiplying the second equation by a constant k and adding the result to the first equation to obtain, after some manipulation,

$$
\begin{aligned}
&\frac{\partial}{\partial\alpha}\left\{g_\beta\left[N_\alpha + \frac{12(1-\nu^2)k}{Eh^3}\,\frac{M_\beta - \nu M_\alpha}{1-\nu^2}\right.\right.\\
&\left.\left.-\frac{k}{Eh}\,\frac{N_\beta - \dfrac{Eh}{k}\dfrac{M_\alpha - \nu M_\beta}{1-\nu^2} - \nu\left(N_\alpha - \dfrac{Eh}{k}\dfrac{M_\beta - \nu M_\alpha}{1-\nu^2}\right) \boxed{-\dfrac{2\nu Eh}{k}\dfrac{M_\beta - \nu M_\alpha}{1-\nu^2}}}{R_\beta}\right]\right\}\\
&\qquad -\left[N_\beta + \frac{12(1-\nu^2)k}{Eh^3}\,\frac{M_\alpha - \nu M_\beta}{1-\nu^2}\right.\\
&\left.-\frac{k}{Eh}\,\frac{N_\alpha - \dfrac{Eh}{k}\dfrac{M_\beta - \nu M_\alpha}{1-\nu^2} - \nu\left(N_\beta - \dfrac{Eh}{k}\dfrac{M_\alpha - \nu M_\beta}{1-\nu^2}\right) \boxed{-\dfrac{2\nu Eh}{k}\dfrac{M_\alpha - \nu M_\beta}{1-\nu^2}}}{R_\alpha}\right]\\
&\qquad \times \frac{\partial g_\beta}{\partial\alpha} + \frac{1}{g_\alpha}\frac{\partial}{\partial\beta}\left\langle g_\alpha^2\left\{\overline{N}_{\alpha\beta} - \frac{12(1-\nu^2)k}{Eh^3}\,\frac{\overline{M}_{\alpha\beta}}{1-\nu} + \frac{k}{2Eh}\left(\frac{3}{R_\alpha} - \frac{1}{R_\beta}\right)\right.\right.\\
&\qquad \left.\left.\times\left[(1+\nu)\left(\overline{N}_{\alpha\beta} + \frac{Eh}{k}\,\frac{\overline{M}_{\alpha\beta}}{1-\nu}\right)\boxed{-\frac{2\nu Eh}{k}\,\frac{\overline{M}_{\alpha\beta}}{1-\nu}}\right]\right\}\right\rangle\\
&\qquad + g_\beta\left(\frac{1}{R_\beta} - \frac{1}{R_\alpha}\right)\frac{k}{Eh}\frac{\partial}{\partial\alpha}\\
&\times\left[N_\beta - \frac{Eh}{k}\,\frac{M_\alpha - \nu M_\beta}{1-\nu^2} - \nu\left(N_\alpha - \frac{Eh}{k}\,\frac{M_\beta - \nu M_\alpha}{1-\nu^2}\right)\boxed{-\frac{2\nu Eh}{k}\,\frac{M_\beta - \nu M_\alpha}{1-\nu^2}}\right]\\
&\qquad = -g_\alpha g_\beta\left(q_\alpha + \frac{m_\alpha}{R_\alpha}\right).
\end{aligned}
\tag{3.7.2}
$$

Eq. (3.7.2a) will be in a form similar to Eq. (3.5.3) if we delete the boxed terms, which we may argue are of the order of magnitude of the ratio of the thickness and radius of curvature compared to terms which are retained. Then with

$$k = i\,\frac{Eh^2}{[12(1-\nu^2)]^{1/2}}, \tag{3.7.3a}$$

$$\tilde{N}_\alpha = N_\alpha + i\,\frac{[12(1-\nu^2)]^{1/2}}{h}\,\frac{M_\beta - \nu M_\alpha}{1-\nu^2}, \tag{3.7.3b}$$

$$\tilde{N}_\beta = N_\beta + i\,\frac{[12(1-\nu^2)]^{1/2}}{h}\,\frac{M_\alpha - \nu M_\beta}{1-\nu^2}, \tag{3.7.3c}$$

$$\tilde{N}_{\alpha\beta} = \bar{N}_{\alpha\beta} - i\,\frac{[12(1-\nu^2)]^{1/2}}{h}\,\frac{\bar{M}_{\alpha\beta}}{1-\nu}, \tag{3.7.3d}$$

Eq. (3.7.2) will reduce to

$$\left[1 + \frac{(1+\nu)ih}{[12(1+\nu^2)]^{1/2}R_\alpha}\right]\frac{\partial}{\partial\alpha}(g_\beta\tilde{N}_\alpha) - \tilde{N}_\beta\left[1 + \frac{ih(1+\nu)}{[12(1-\nu^2)]^{1/2}R_\alpha}\right]\frac{\partial g_\beta}{\partial\alpha}$$
$$+ \frac{1}{g_\alpha}\frac{\partial}{\partial\beta}\left\{g_\alpha^2\tilde{N}_{\alpha\beta}\left[1 + \frac{(1+\nu)(3/R_\alpha - 1/R_\beta)ih}{2[12(1-\nu^2)]^{1/2}}\right]\right\}$$
$$- \frac{ih}{[12(1-\nu^2)]^{1/2}}\,\frac{g_\beta}{R_\alpha}\,\frac{\partial(\tilde{N}_\beta + \tilde{N}_\alpha)}{\partial\alpha} = -g_\alpha g_\beta\left(q_\alpha + \frac{m_\alpha}{R_\alpha}\right). \tag{3.7.4a}$$

A similar procedure for the remaining revised equations will yield

$$\left[1 + \frac{(1+\nu)ih}{[12(1-\nu^2)]R_\beta}\right]\frac{\partial}{\partial\beta}(g_\alpha\tilde{N}_\beta) - \tilde{N}_\alpha\left[1 + \frac{ih(1+\nu)}{[12(1-\nu^2)]^{1/2}R_\beta}\right]\frac{\partial g_\alpha}{\partial\beta}$$
$$+ \frac{1}{g_\beta}\frac{\partial}{\partial\alpha}\left\{g_\beta^2\tilde{N}_{\alpha\beta}\left[1 + \frac{ih(1+\nu)(3/R_\beta - 1/R_\alpha)}{[12(1-\nu^2)]^{1/2}}\right]\right\}$$
$$- \frac{ih}{[12(1-\nu^2)]^{1/2}}\,\frac{g_\alpha}{R_\beta}\,\frac{(\tilde{N}_\alpha + \tilde{N}_\beta)}{\partial\beta} = -g_\alpha g_\beta\left(q_\beta + \frac{m_\beta}{R_\beta}\right), \tag{3.7.4b}$$

$$g_\alpha g_\beta\left[\frac{\tilde{N}_\alpha}{R_\alpha} + \frac{\tilde{N}_\beta}{R_\beta} + \frac{ih}{[12(1-\nu^2)]^{1/2}}\nabla^2(\tilde{N}_\alpha + \tilde{N}_\beta)\right]$$
$$- \frac{(1+\nu)ih}{[12(1-\nu^2)]^{1/2}}\left\{\frac{\partial}{\partial\alpha}\frac{1}{g_\alpha}\left[\frac{\partial}{\partial\alpha}(g_\beta\tilde{N}_\alpha) - \tilde{N}_\beta\frac{\partial g_\beta}{\partial\alpha} + \frac{\partial}{\partial\beta}(g_\alpha\tilde{N}_{\alpha\beta})\right]\right.$$
$$\left. + \frac{\partial}{\partial\beta}\frac{1}{g_\beta}\left[\frac{\partial}{\partial\beta}(g_\alpha\tilde{N}_\beta) - \tilde{N}_\alpha\frac{\partial g_\alpha}{\partial\beta} + \frac{\partial}{\partial\alpha}(g_\beta^2\tilde{N}_{\alpha\beta})\right]\right\}$$
$$= \left[g_\alpha g_\beta q_\zeta + \frac{\partial}{\partial\alpha}(g_\beta m_\alpha) + \frac{\partial}{\partial\beta}(g_\alpha m_\beta)\right]. \tag{3.7.4c}$$

Separation of Eqs. (3.7.4) into real and imaginary parts indicates that the terms which have been omitted belong to the equations of equilibrium of shell theory and not to the compatibility conditions. The fact that the compatibility conditions are unchanged indicates that displacements obtained through the use of the overall force–strain relations

$$\epsilon_\alpha = \frac{1}{Eh}\,\mathrm{Re}\,(\tilde{N}_\alpha - \nu\tilde{N}_\beta), \tag{3.7.5a}$$

$$\epsilon_\beta = \frac{1}{Eh}\,\mathrm{Re}\,(\tilde{N}_\beta - \nu\tilde{N}_\alpha), \tag{3.7.5b}$$

$$\gamma_{\alpha\beta} = \frac{2(1+\nu)}{Eh}\,\mathrm{Re}\,\tilde{N}_{\alpha\beta}, \tag{3.7.5c}$$

will yield changes of curvature which also satisfy the overall force–strain relations, again provided that the complementary solutions of Eq. (3.7.5) are restricted to rigid body displacements.

The reduced theory of shells can also be recast in terms of complex displacements in a manner similar to that used in the preceding section. Let us replace Eqs. (3.6.1) and (3.6.2) by

$$\tilde{\epsilon}_\alpha = \frac{1}{Eh}(\tilde{N}_\alpha - \nu\tilde{N}_\beta), \tag{3.7.6a}$$

$$\tilde{\epsilon}_\beta = \frac{1}{Eh}(\tilde{N}_\beta - \nu\tilde{N}_\alpha), \tag{3.7.6b}$$

$$\gamma_{\alpha\beta} = \frac{2(1+\nu)}{Eh}\tilde{N}_{\alpha\beta}, \tag{3.7.6c}$$

$$N_\alpha^p = \tilde{N}_\alpha - \frac{iEh^2}{[12(1-\nu^2)]^{1/2}}\tilde{\kappa}_\beta, \tag{3.7.6d}$$

$$N_\beta^p = \tilde{N}_\beta - \frac{iEh^2}{[12(1-\nu^2)]^{1/2}}\tilde{\kappa}_\alpha, \tag{3.7.6e}$$

$$\bar{N}_{\alpha\beta}^p = \tilde{N}_{\alpha\beta} + \frac{iEh^2}{[12(1-\nu^2)]^{1/2}}\tilde{\kappa}_{\alpha\beta}. \tag{3.7.6f}$$

Then the substitution of Eq. (3.7.6) into the compatibility Eqs. (1.5.11) yields

$$\begin{aligned} &\frac{\partial}{\partial\alpha}(g_\beta\tilde{N}_\alpha) - \left[\tilde{N}_\beta - \frac{ih(1+\nu)}{[12(1-\nu^2)]^{1/2}}\frac{\tilde{N}_\alpha - \tilde{N}_\beta}{R_\alpha}\right]\frac{\partial g_\beta}{\partial\alpha} \\ &\quad + \frac{1}{g_\alpha}\frac{\partial}{\partial\beta}\left\{g_\alpha^2\tilde{N}_{\alpha\beta}\left[1 + \frac{(1+\nu)ih(3/R_\alpha - 1/R_\beta)}{[12(1-\nu^2)]^{1/2}}\right]\right\} \\ &\quad - \frac{g_\beta}{R_\alpha}\frac{ih}{[12(1-\nu^2)]^{1/2}}\frac{\partial}{\partial\alpha}(\tilde{N}_\beta - \nu\tilde{N}_\alpha) \\ &= \frac{\partial}{\partial\alpha}(g_\beta N_\alpha^p) - N_\beta^p\frac{\partial g_\beta}{\partial\alpha} + \frac{1}{g_\alpha}\frac{\partial}{\partial\alpha}(g_\alpha^2\bar{N}_{\alpha\beta}^p), \end{aligned} \tag{3.7.7a}$$

$$\begin{aligned} &\frac{\partial}{\partial\beta}(g_\alpha\tilde{N}_\beta) - \left[\tilde{N}_\alpha - \frac{ih(1-\nu)}{[12(1-\nu^2)]^{1/2}}\frac{\tilde{N}_\beta - \tilde{N}_\alpha}{R_\beta}\right]\frac{\partial g_\alpha}{\partial\beta} \\ &\quad + \frac{1}{g_\beta}\frac{\partial}{\partial\alpha}\left\{g_\beta^2\tilde{N}_{\alpha\beta}\left[1 + \frac{(1+\nu)ih(3/R_\beta - 1/R_\alpha)}{2[12(1-\nu^2)]^{1/2}}\right]\right\} \\ &\quad - \frac{g_\alpha}{R_\beta}\frac{ih}{[12(1-\nu^2)]^{1/2}}\frac{\partial}{\partial\beta}(\tilde{N}_\alpha - \nu\tilde{N}_\beta) \\ &= \frac{\partial}{\partial\beta}(g_\alpha N_\beta^p) - N_\alpha^p\frac{\partial g_\alpha}{\partial\beta} + \frac{1}{g_\beta}\frac{\partial}{\partial\alpha}(g_\beta^2\bar{N}_{\alpha\beta}^p), \end{aligned} \tag{3.7.7b}$$

$$g_\alpha g_\beta\left(\frac{\tilde{N}_\alpha}{R_\alpha}+\frac{\tilde{N}_\beta}{R_\beta}\right)+\frac{ih}{[12(1-\nu^2)]^{1/2}}\frac{\partial}{\partial\alpha}$$

$$\times\left\langle\frac{1}{g_\alpha}\left\{\frac{\partial}{\partial\alpha}\left[g_\alpha(\tilde{N}_\beta-\nu\tilde{N}_\alpha)\right]\right.\right.$$

$$\left.\left.-(\tilde{N}_\alpha-\nu\tilde{N}_\beta)\frac{\partial g_\beta}{\partial\alpha}-\frac{1+\nu}{g_\alpha}\frac{\partial}{\partial\beta}(g_\alpha^2\tilde{N}_{\alpha\beta})\right\}\right\rangle$$

$$+\frac{ih}{[12(1-\nu^2)]^{1/2}}\frac{\partial}{\partial\beta}\left\langle\frac{1}{g_\beta}\left\{\frac{\partial}{\partial\beta}\left[g_\beta(\tilde{N}_\alpha-\nu\tilde{N}_\beta)\right]\right.\right.$$

$$\left.\left.-(\tilde{N}_\beta-\nu\tilde{N}_\alpha)\frac{\partial g_\alpha}{\partial\beta}-\frac{1+\nu}{g_\beta}\frac{\partial}{\partial\alpha}(g_\beta^2\tilde{N}_{\alpha\beta})\right\}\right\rangle$$

$$=g_\alpha g_\beta\left(\frac{N_\alpha^p}{R_\alpha}+\frac{N_\beta^p}{R_\beta}\right). \tag{3.7.7c}$$

But the left side of each of Eqs. (3.7.7) is identical with the corresponding portion of Eqs. (3.7.4). Therefore the complex strain components will be compatible if N_α^p, N_β^p, and $\bar{N}_{\alpha\beta}^p$ are identical with the solutions of Eqs. (3.6.5). By eliminating the complex stress resultants from Eq. (3.7.6), the equations for the determination of the complex displacements becomes

$$\frac{\tilde{\epsilon}_\alpha+\nu\tilde{\epsilon}_\beta}{1-\nu^2}-\frac{ih}{[12(1-\nu^2)]^{1/2}}\tilde{\kappa}_\beta=\frac{N_\alpha^p}{Eh}, \tag{3.7.8a}$$

$$\frac{\tilde{\epsilon}_\beta+\nu\tilde{\epsilon}_\alpha}{1-\nu^2}-\frac{ih}{[12(1-\nu^2)]^{1/2}}\tilde{\kappa}_\alpha=\frac{N_\beta^p}{Eh}, \tag{3.7.8b}$$

$$\frac{1}{2(1-\nu)}\tilde{\gamma}_{\alpha\beta}+\frac{ih}{[12(1-\nu^2)]^{1/2}}\tilde{\kappa}_{\alpha\beta}=\frac{\bar{N}_{\alpha\beta}^p}{Eh}. \tag{3.7.8c}$$

As before, the real parts of the complex displacements correspond to the middle surface displacements. The imaginary parts can, of course, be interpreted as stress functions, but only as ones which satisfy a modified set of equilibrium equations rather than Eqs. (2.6.6).

If we omit terms of the order of h/R_α or h/R_β compared to unity in Eqs. (3.7.4) we have

$$\frac{1}{g_\alpha g_\beta}\left\{\frac{\partial}{\partial\alpha}(g_\beta\tilde{N}_\alpha)-\tilde{N}_\beta\frac{\partial g_\beta}{\partial\alpha}+\frac{1}{g_\alpha}\frac{\partial}{\partial\beta}(g_\alpha^2\tilde{N}_{\alpha\beta})\right.$$

$$\left.-\frac{ih}{[12(1-\nu^2)]^{1/2}}\frac{g_\beta}{R_\alpha}\frac{\partial\tilde{N}}{\partial\alpha}\right\}=-\left(q_\alpha+\frac{m_\alpha}{R_\alpha}\right), \tag{3.7.9a}$$

$$\frac{1}{g_\alpha g_\beta}\left\{\frac{\partial}{\partial\beta}(g_\alpha\tilde{N}_\beta)-\tilde{N}_\alpha\frac{\partial g_\alpha}{\partial\beta}+\frac{1}{g_\beta}\frac{\partial}{\partial\alpha}(g_\beta^2\tilde{N}_{\alpha\beta})\right.$$

$$\left.-\frac{ih}{[12(1-\nu^2)]^{1/2}}\frac{g_\alpha}{R_\beta}\frac{\partial\tilde{N}}{\partial\beta}\right\}=-\left(q_\beta+\frac{m_\beta}{R_\beta}\right), \tag{3.7.9b}$$

$$\frac{\tilde{N}_\alpha}{R_\alpha} + \frac{\tilde{N}_\beta}{R_\beta} + \frac{ih}{[12(1-\nu^2)]^{1/2}} \nabla^2 \tilde{N} = q_\zeta$$

$$+ \frac{1}{g_\alpha g_\beta}\left[\frac{\partial}{\partial\alpha}(g_\beta m_\alpha) + \frac{\partial}{\partial\beta}(g_\alpha m_\beta)\right], \tag{3.7.9c}$$

where

$$\tilde{N} = \tilde{N} + \tilde{N}_\beta,$$

which are the equations of Novozhilov.†

3.8 The membrane theory of shells

The reductions of shell theory discussed thus far retain the essence of the general theory of transversely rigid shells in that we may specify relatively general distributions of forces and moments or of middle surface displacements and slopes on the shell edges. The final simplification which we shall discuss in this chapter abandons the possibility of satisfying all edge conditions and reduces the theory of shells to its simplest possible form.

Let us assume, as is true in many cases, that strains due to changes of curvature of the shell are small compared to the middle surface strain components so that the strain energy function W may be approximated by

$$W = \tfrac{1}{2}K\left(\epsilon_\alpha^2 + 2\nu\epsilon_\alpha\epsilon_\beta + \epsilon_\beta^2 + \frac{1-\nu}{2}\gamma_{\alpha\beta}^2\right). \tag{3.8.1}$$

The use of Eqs. (2.9.2) then leads to the following overall force–strain relations:

$$N_\alpha = K(\epsilon_\alpha + \nu\epsilon_\beta), \tag{3.8.2a}$$

$$N_\beta = K(\epsilon_\beta + \nu\epsilon_\alpha), \tag{3.8.2b}$$

$$\overline{N}_{\alpha\beta} = \frac{1-\nu}{2} K\gamma_{\alpha\beta}, \tag{3.8.2c}$$

$$M_\alpha = M_\beta = \overline{M}_{\alpha\beta} = 0, \tag{3.8.2d}$$

† V. V. Novozhilov, *The Theory of Thin Shells*, P. Noordhoff Ltd, Groningen, The Netherlands, 1959, pp. 75–84. Another formulation of complex shell equations is given by J. Lyell Sanders, Jr, 'On the Shell Equations in Complex Form,' *Proc. 2nd IUTAM Symp. on the Theory of Thin Shells* (Copenhagen, 1967), Springer-Verlag, Berlin, 1969, pp. 135–156.

which, by the methods of Section 3.2, imply that the internal stress–strain relations are given by

$$\sigma_\alpha^{(\zeta)} = \frac{E}{1-\nu^2}\frac{\epsilon_\alpha + \nu\epsilon_\beta}{1+\zeta/R_\beta} = \frac{N_\alpha}{h(1+\zeta/R_\beta)}, \tag{3.8.3a}$$

$$\sigma_\beta^{(\zeta)} = \frac{E}{1-\nu^2}\frac{\epsilon_\beta + \nu\epsilon_\alpha}{1+\zeta/R_\alpha} = \frac{N_\beta}{h(1+\zeta/R_\alpha)}, \tag{3.8.3b}$$

$$\begin{aligned}\tau_{\alpha\beta}^{(\zeta)} &= \frac{E}{2(1+\nu)}\gamma_{\alpha\beta}\frac{1-\frac{1}{2}\zeta(1/R_\alpha + 1/R_\beta)}{1-\zeta^2/R_\alpha R_\beta} \\ &= \frac{\bar{N}_{\alpha\beta}}{h}\frac{1-\frac{1}{2}\zeta(1/R_\alpha + 1/R_\beta)}{1-\zeta^2/R_\alpha R_\beta}.\end{aligned} \tag{3.8.3c}$$

From Eqs. (2.6.2), we then have the reduced equilibrium equations

$$\bar{Q}_\alpha = m_\alpha, \tag{3.8.4a}$$

$$\bar{Q}_\beta = m_\beta, \tag{3.8.4b}$$

and

$$\frac{\partial}{\partial\alpha}(g_\beta N_\alpha) - N_\beta\frac{\partial g_\beta}{\partial\alpha} + \frac{1}{g_\alpha}\frac{\partial}{\partial\beta}(g_\alpha^2\bar{N}_{\alpha\beta}) + g_\alpha g_\beta\left(q_\alpha + \frac{m_\alpha}{R_\alpha}\right) = 0, \tag{3.8.5a}$$

$$\frac{\partial}{\partial\beta}(g_\alpha N_\beta) - N_\alpha\frac{\partial g_\alpha}{\partial\beta} + \frac{1}{g_\beta}\frac{\partial}{\partial\alpha}(g_\beta^2\bar{N}_{\alpha\beta}) + g_\alpha g_\beta\left(q_\beta + \frac{m_\beta}{R_\beta}\right) = 0, \tag{3.8.5b}$$

$$g_\alpha g_\beta\left(\frac{N_\alpha}{R_\alpha} + \frac{N_\beta}{R_\beta} - q_\zeta\right) - \frac{\partial}{\partial\alpha}(g_\beta m_\alpha) - \frac{\partial}{\partial\beta}(g_\alpha m_\beta) = 0. \tag{3.8.5c}$$

The problem of the determination of the stresses and displacements of shell structures is thus reduced to the problem of the solution of Eqs. (3.8.5), a set of three equations in the three unknowns N_α, N_β, and $\bar{N}_{\alpha\beta}$. These equations are identical with Eqs. (2.10.2) and (3.6.5) and are known as the membrane equations of the theory of shells. These equations are a set of three equations in three unknowns so that if appropriate force boundary conditions are prescribed, the state of stress in the shell is completely determined. The existence of the membrane equations characterizes the main difference between plate structures and shell structures, namely that for a shell there is the possibility that applied forces normal to the bounding surfaces can be sustained predominantly by forces tangent to the middle surface whereas for plates bending and twisting effects predominate. The demarcation between the two categories of structures is vague, of course, and depends on the relative dimensions of the middle surface. Thus a structure with relatively small values of radius/thickness ratio and length/radius ratio will experience more bending than one for which the various ratios are relatively large.

Once the stress resultants are known, displacements can be determined from the overall force–strain relationships, written as

$$\epsilon_\alpha = \frac{1}{Eh}(N_\alpha - \nu N_\beta), \tag{3.8.6a}$$

$$\epsilon_\beta = \frac{1}{Eh}(N_\beta - \nu N_\alpha), \tag{3.8.6b}$$

$$\gamma_{\alpha\beta} = \frac{2(1+\nu)}{Eh}\overline{N}_{\alpha\beta}. \tag{3.8.6c}$$

That we cannot satisfy all of the edge boundary conditions of the theory of shells is readily seen from the fact that Eqs. (3.8.5) for the stress resultants are of second order, which indicates that at most one of the stress resultants may be specified at each edge. Eqs. (3.8.6) for the displacements are also a second-order system of differential equations. Since, however, the stress resultants on the right side of these equations contain two undetermined functions, we may be able to satisfy two conditions on displacements on each edge of a shell if the edge stress resultants are not specified. On the other hand, if an edge stress resultant is able to be specified we can only satisfy conditions on one edge displacement.

The question now arises as to exactly what boundary conditions we can impose. This question can be answered, as in the general bending theory, by again turning to the variational principle underlying the theory of shells. We have assumed nothing thus far to invalidate the variational principle given by Eq. (2.8.5). Therefore the principle can be considered to hold for membrane theory as well. We see that the vanishing of each of the line integrals and (3.8.2d) leads to Eqs. (3.8.5). The remainder of the variational principle states that the following line integral must vanish:

$$\int_{\beta_1}^{\beta_2}\left[(N_\alpha - N_\alpha^*)\delta u + \left(\overline{N}_{\alpha\beta} - N_{\alpha\beta}^* - \frac{M_{\alpha\beta}^*}{R_\beta}\right)\delta u_\beta\right] g_\beta \Big|_{\alpha=\alpha_1}^{\alpha=\alpha_2} d\beta$$

$$\times \int_{\beta_1}^{\beta_2}\left[(N_\beta - N_\beta^*)\delta u_\beta + \left(\overline{N}_{\alpha\beta} - N_{\beta\alpha}^* - \frac{M_{\beta\alpha}^*}{R_\alpha}\right)\delta u_\alpha\right] g_\alpha \Big|_{\beta=\beta_1}^{\beta=\beta_2} d\alpha = 0, \tag{3.8.7}$$

where Eqs. (3.8.2d) and (3.8.4) have been taken into account and we have assumed that applied-edge values of M_α, M_β, $\overline{M}_{\alpha\beta}$, $\overline{Q}_\alpha$, and $\overline{Q}_\beta$ are small and can be neglected. However, the generality of the edge conditions implied by Eq. (3.8.7) exceeds the limitations of membrane theory. We may bring the requirements and limitations into agreement by stating the boundary conditions in the following manner. Along β lines of curvature, for example, we may have

$$N_\alpha = N_\alpha^* \quad \text{and} \quad \delta u_\beta = 0, \tag{3.8.8a}$$

or

$$\bar{N}_{\alpha\beta} = N_{\alpha\beta}^* + \frac{M_{\alpha\beta}^*}{R_\beta} \quad \text{and} \quad \delta u = 0, \tag{3.8.8b}$$

or

$$\delta u_\alpha = 0 \quad \text{and} \quad \delta u_\beta = 0, \tag{3.8.8c}$$

which satisfy the condition that one stress resultant and one displacement or no stress resultants and two displacements may be specified along an edge. We see that the displacements which we may specify are the displacements tangent to the middle surface of the shell. It should be emphasized, however, that it is not always possible to satisfy even the reduced number of boundary conditions indicated by Eqs. (3.8.8). For some shell shapes the membrane equations are hyperbolic in nature and permit the satisfaction of boundary conditions along one set of edges but not the other. In such cases it must be assumed that whatever forces or displacements the theory yields at these edges are provided for.

The complementary solutions of Eqs. (3.8.5) and (3.8.6) deserve some attention at this point since they are an explanation of the need to specify displacement conditions on the shell edges. The displacements obtained from these equations yield zero values of the strain components and zero values of the stress resultants. Unlike the results of the general bending theory, these inextensional deformations need not be confined to rigid body displacements. Thus, unless at least one tangential displacement is prescribed along the shell edges, the equilibrium position is not unique even if we eliminate rigid body motions. This freedom of motion is a consequence of the assumption of membrane theory that changes the curvature of the shell middle surface do not introduce any stresses or stress resultants. Therefore, if membrane theory is to be a reasonable approximation to the general bending theory, these unrestrained motions must be small since otherwise they lead to violations of the assumption that stresses due to changes of curvature are small compared to those due to middle surface strain components.

Other conditions exist for membrane theory to be an adequate approximation to the general bending theory. Some of these are obvious, namely the conditions that large moments and transverse shear forces, or normal deflections and changes of slope, should not be applied at the shell edges. In addition neither the shape of the shell nor the surface loading should change abruptly or rapidly since changes of curvature are usually large in the vicinity of these rapid changes and thus violate the assumptions of membrane theory.

In practice the membrane solutions for stresses are used as an estimate for the stress state in the interior of a shell. Conditions at the edges may be either deliberately contrived to conform to membrane theory assumptions or the edges may have to be reinforced in accordance with more accurate analyses, the results of model tests, or other empirical methods to withstand larger stresses due to bending.

3.9 Transformation of the equations of membrane theory

Since the equations of membrane theory correspond to the equations of equilibrium of a shell element subjected only to forces tangent to the middle surface and such that the shear force is symmetric, we can easily obtain equations of equilibrium for coordinate curves other than lines of curvature. A particularly useful set of coordinate curves are those cut by planes parallel to the xz and yz planes of a cartesian coordinate system. We denote the equation of the middle surface by

$$z = z(x, y) \tag{3.9.1}$$

and consider an element of the shell described by the planes of constant x, and $x + \mathrm{d}x$ parallel to the yz plane and the planes of constant y, and $y + \mathrm{d}y$ parallel to the xz plane. The equivalent forces per unit middle surface length acting in the shell middle surface are denoted by $\bar{N}_x$ which is tangent to the middle surface and parallel to the xz plane, $\bar{N}_y$ which is tangent to the middle surface and parallel to the yz plane, and $\bar{N}_{xy}$ which are equivalent shear forces acting on the edges of the element as shown in Fig. 3.2. The projection of these forces on the xy plane are denoted by $\tilde{N}_x^*$, $\tilde{N}_y^*$, and $\tilde{N}_{xy}$ and are defined by

$$\tilde{N}_x^* \, \mathrm{d}y = \bar{N}_x \frac{\mathrm{d}y}{\cos \psi} \cos \chi, \tag{3.9.2a}$$

$$\tilde{N}_y^* \, \mathrm{d}x = \bar{N}_y \frac{\mathrm{d}x}{\cos \chi} \cos \psi, \tag{3.9.2b}$$

$$\tilde{N}_{xy}^* \, \mathrm{d}x = \bar{N}_{xy} \frac{\mathrm{d}x}{\cos \chi} \cos \chi = \bar{N}_{xy} \, \mathrm{d}x, \tag{3.9.2c}$$

or

$$\tilde{N}_{xy}^* \mathrm{d}\, y = \bar{N}_{xy} \frac{\mathrm{d}y}{\cos \psi} \cos \psi = \bar{N}_{xy} \, \mathrm{d}y. \tag{3.9.2d}$$

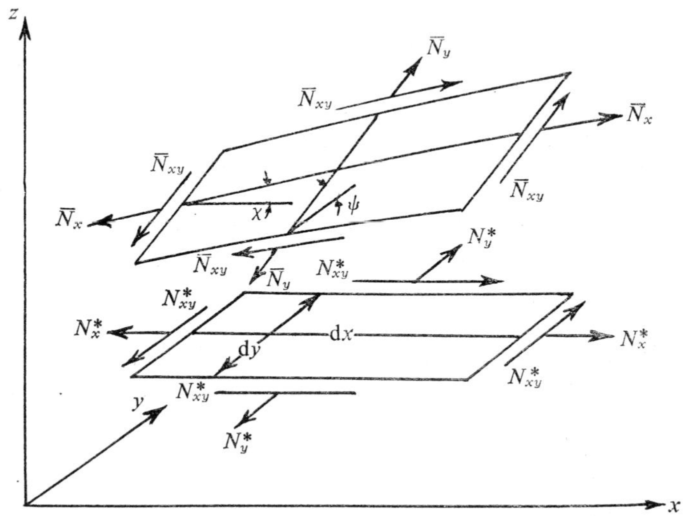

Fig. 3.2. Equivalent membrane forces on shell element and projections on coordinate plane

Thus

$$\tilde{N}_x^* = \bar{N}_x \frac{\cos \chi}{\cos \psi}, \tag{3.9.3a}$$

$$\tilde{N}_y^* = \bar{N}_y \frac{\cos \psi}{\cos \chi}, \tag{3.9.3b}$$

$$\tilde{N}_{xy}^* = \bar{N}_{xy}, \tag{3.9.3c}$$

where

$$\cos \chi = \left[1 + \left(\frac{\partial z}{\partial x}\right)^2\right]^{-1/2}, \tag{3.9.4a}$$

$$\cos \chi = \left[1 + \left(\frac{\partial z}{\partial y}\right)^2\right]^{-1/2}. \tag{3.9.4b}$$

The equations of equilibrium formed for directions parallel to the x and y coordinate axes can be readily obtained from the projection on the xy plane as

$$\frac{\partial \tilde{N}_y^*}{\partial x} + \frac{\partial \tilde{N}_{xy}^*}{\partial y} + \tilde{p}_x^* = 0, \tag{3.9.5a}$$

$$\frac{\partial \tilde{N}_y^*}{\partial y} + \frac{\partial \tilde{N}_{xy}^*}{\partial x} + \tilde{p}_y^* = 0, \tag{3.9.5b}$$

where $\tilde{p}_x^*$ and $\tilde{p}_y^*$ are equivalent surface forces in the x and y axis directions per unit projected area of the middle surface. The equation of equilibrium in the z direction is given by

$$\frac{\partial}{\partial x}\left(\bar{N}_x \frac{\mathrm{d}y}{\cos\psi}\sin\chi\right)\mathrm{d}x + \frac{\partial}{\partial y}\left(\bar{N}_y \frac{\mathrm{d}x}{\cos\chi}\sin\chi\right)\mathrm{d}y$$
$$+ \frac{\partial}{\partial x}\left(\bar{N}_{xy} \frac{\mathrm{d}y}{\cos\psi}\sin\psi\right)\mathrm{d}x + \frac{\partial}{\partial y}\left(\bar{N}_{xy} \frac{\mathrm{d}x}{\cos\chi}\sin\chi\right)\mathrm{d}y$$
$$- \tilde{p}_z^*\,\mathrm{d}x\,\mathrm{d}y = 0, \tag{3.9.6}$$

where $\tilde{p}_z^*$ is the equivalent surface force per unit projection area of the middle surface in the z axis direction. With the use of Eqs. (3.9.3), and the relations

$$\tan\chi = \frac{\partial z}{\partial x}, \tag{3.9.7a}$$

$$\tan\psi = \frac{\partial z}{\partial y}, \tag{3.9.7b}$$

Eq. (3.9.6) becomes

$$\frac{\partial}{\partial x}\left(\tilde{N}_x^* \frac{\partial z}{\partial x}\right) + \frac{\partial}{\partial y}\left(\tilde{N}_y^* \frac{\partial z}{\partial y}\right) + \frac{\partial}{\partial x}\left(\tilde{N}_{xy}^* \frac{\partial z}{\partial y}\right) + \frac{\partial}{\partial y}\left(\tilde{N}_{xy}^* \frac{\partial z}{\partial x}\right) + \tilde{p}_z^* = 0, \tag{3.9.8a}$$

or, upon performing the indicated differentiation and making use of Eqs. (3.9.5)

$$\tilde{N}_x^* \frac{\partial^2 z}{\partial x^2} + \tilde{N}_y^* \frac{\partial^2 z}{\partial y^2} + 2\tilde{N}_{xy}^* \frac{\partial^2 z}{\partial x\,\partial y} = -\tilde{p}_z^* + \tilde{p}_x^* \frac{\partial z}{\partial x} + \tilde{p}_y^* \frac{\partial z}{\partial y}. \tag{3.9.8b}$$

Eqs. (3.9.5) are similar to the equations of equilibrium of elasticity theory for states of general plane stress or of plane strain and may be satisfied by the introduction of a stress function Φ such that

$$\tilde{N}_x^* = \frac{\partial^2\Phi}{\partial y^2} + \int \tilde{p}_x^*\,\mathrm{d}x, \tag{3.9.9a}$$

$$\tilde{N}_y^* = \frac{\partial^2\Phi}{\partial x^2} - \int \tilde{p}_y^*\,\mathrm{d}y, \tag{3.9.9b}$$

$$\tilde{N}_{xy}^* = -\frac{\partial^2\Phi}{\partial x\,\partial y}. \tag{3.9.9c}$$

Then the substitution of Eqs. (3.9.9) into Eqs. (3.9.8b) yields the following relationship which Φ must satisfy:

$$\frac{\partial^2\Phi}{\partial x^2}\frac{\partial^2 z}{\partial y^2} - 2\frac{\partial^2\Phi}{\partial x\,\partial y}\frac{\partial^2 z}{\partial x\,\partial y} + \frac{\partial^2\Phi}{\partial y^2}\frac{\partial^2 z}{\partial x^2}$$

$$= -\tilde{p}_z^* + \frac{\partial}{\partial x}\left[\frac{\partial z}{\partial x}\int \tilde{p}_x^*\,\mathrm{d}x\right] + \frac{\partial}{\partial y}\left[\frac{\partial z}{\partial y}\int \tilde{p}_y^*\,\mathrm{d}y\right]. \tag{3.9.10}$$

It is interesting to note that a solution of Eq. (3.9.10) for a particular shell under a particular system of loading is also a solution for a shell whose coordinates are obtained by stretching the original coordinate system and for which the loads $\tilde{p}_x^*$, $\tilde{p}_y^*$, and p_z^* are increased by similar factors. For if the coordinates of the second shell are related to those of the first by

$$x^1 = \lambda_1 x, \tag{3.9.11a}$$

$$y^1 = \lambda_2 y, \tag{3.9.11b}$$

$$z^1 = \lambda_3 z, \tag{3.9.11c}$$

the surface loads by

$$\tilde{p}_x^{*1}(x, y) = \lambda_1 \tilde{p}_x^*(x, y), \tag{3.9.12a}$$

$$\tilde{p}_y^{*1}(x^1, y^1) = \lambda_2 \tilde{p}_y^*(x, y), \tag{3.9.12b}$$

$$\tilde{p}_z^{*1}(x^1, y^1) = \lambda_3 \tilde{p}_z^*(x, y), \tag{3.9.12c}$$

and the stress functions by

$$\Phi^1(x^1, y^1) = (\lambda_1\,\lambda_2)^2\Phi(x, y), \tag{3.9.13}$$

Equation (3.9.10) can be transformed to

$$\frac{\partial^2\Phi^1}{\partial(x^1)^2}\frac{\partial^2 z^1}{\partial(y^1)^2} - 2\frac{\partial^2\Phi^1}{\partial x^1\,\partial y^1}\frac{\partial^2 z^1}{\partial x^1\,\partial y^1} + \frac{\partial^2\Phi^1}{\partial(y^1)^2}\frac{\partial^2 z^1}{\partial(x^1)^2} + \tilde{p}_z^{*1}$$

$$- \frac{\partial}{\partial x^1}\left[\frac{\partial z^1}{\partial x^1}\int \tilde{p}_x^{*1}\,\mathrm{d}x^1\right] - \frac{\partial}{\partial y^1}\left[\frac{\partial z^1}{\partial y^1}\int \tilde{p}_y^{*1}\,\mathrm{d}y^1\right] = 0, \tag{3.9.14}$$

which is the equation for the stress function of the second shell.

Expressions for the stress distribution corresponding to the equivalent forces $\overline{N}_x$, $\overline{N}_y$, and $\overline{N}_{xy}$ can be obtained by performing appropriate coordinate, stress, and stress resultant transformations upon Eqs. (3.8.3). A more customary procedure is to calculate average stresses by means of the relations

$$\sigma_x = \frac{\overline{N}_x}{h}, \tag{3.9.15a}$$

$$\sigma_y = \frac{\bar{N}_y}{h}, \tag{3.9.15b}$$

$$\tau_{xy} = \frac{\bar{N}_{xy}}{h}, \tag{3.9.15c}$$

remembering, however, that these stresses act upon a skew surface element rather than a rectangular element.

Supplementary references

[1] Flügge, W., *Statik und Dynamik der Schalen*, Springer-Verlag, Berlin, Second Edition, 1957.
[2] Byrne, R., Jr., 'Theory of Small Deformations of a Thin Elastic Shell,' Seminar Reports in Mathematics (UCLA), vol. 2, 1944, pp. 103–152.
[3] Lur'e, A. I., 'General Theory of Thin Elastic Shells,' *Prikl. Mat. Mekh.*, vol. 4, no. 2, 1940, pp. 7–34.
[4] Vlasov, V. Z., *The General Theory of Shells and its Industrial Applications*, Gosstroiizdat, 1949 (Translation Available as NASA TT F-99, April 1964).
[5] Pücher, A., 'Über die Spannungsfunctionen beliebig gekrümmter dünner Schalen,' *Proc. Fifth Int. Cong. Appl. Mech.* (Cambridge, Mass.), 1939, pp. 134–139.
[6] Truesdell, C., 'The Membrane Theory of Shells of Revolution,' *Am. Math. Soc. Trans.*, vol. 58, 1945, pp. 96–146.
[7] Goldenveiser, A. L., *Theory of Thin Elastic Shells*, Pergamon Press, 1961.
[8] Flügge, W., *Stresses in Shells*, Springer-Verlag, Berlin, 1962.
[9] Flügge, W., and Geyling, F., 'A General Theory of Deformations of Membrane Shells,' *Proc. 9th Int. Cong. Appl. Mech.* (Brussels, 1956), vol. 6, 1957, p. 250.

Part II

Axisymmetric deformations of shells of revolution

4

The cylindrical shell

4.1 General relations for cylindrical shells

One of the most commonly encountered problems of shell analysis, and one to which a great deal of effort has been devoted because of its relative simplicity, is the analysis of shells of revolution subjected to loading which is independent of position around the circumference. If the shell closes upon itself in the circumferential direction and the remaining edges are subjected to axisymmetric loading and/or displacement conditions, the stress and displacement fields within the interior of the shell will also be axisymmetric.

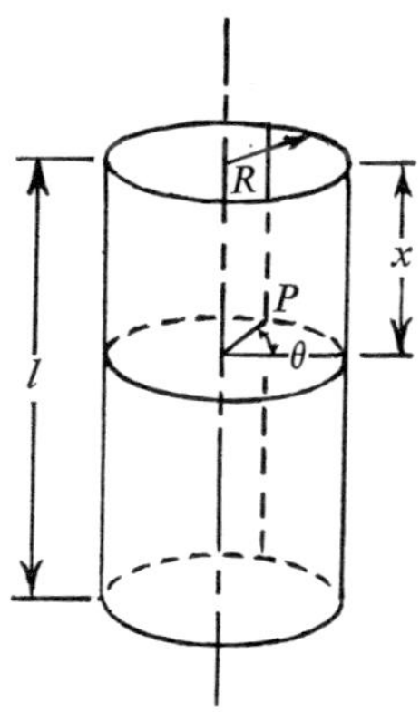

Fig. 4.1. Notation for circular cylindrical shell

Of the types of shells which fall within this class, the circular cylinder will be discussed first since the character of the solutions is obtained in a particularly simple form and is similar to that for other thin shells of revolution.

A cylindrical shell of circular cross-section (Fig. 4.1) has an infinite meridional radius of curvature and a constant circumferential radius of curvature which we shall denote by R. The angle ϕ between the normal to the middle surface and the axis of revolution is constant and equal to a right angle. The lines of curvature coordinate curves are the straight line generators, with the distance from one edge measured along the generator being denoted by x, and the circumferential cross-sectional circles, with the angular measurement around the circumference from a reference generator to a point on another generator being denoted by θ. With this coordinate system, the first fundamental quantities are

$$g_x = 1,$$
$$g_\theta = R. \tag{4.1.1}$$

Since we intend to deal only with axisymmetric deformations, we assume that the meridional and radial displacements u_x and w are independent of θ and that the circumferential displacement u_θ vanishes. From Eqs. (1.5.3) and (1.5.10) we then obtain the following expressions for the strain components and changes of curvature:

$$\gamma_{x\theta} = \bar{\kappa}_{x\theta} = \kappa_\theta = 0, \tag{4.1.2a}$$

$$\epsilon_x = \frac{\mathrm{d}u_x}{\mathrm{d}x}, \tag{4.1.2b}$$

$$\epsilon_\theta = \frac{w}{R}, \tag{4.1.2c}$$

$$\kappa_x = -\frac{\mathrm{d}^2 w}{\mathrm{d}x^2}. \tag{4.1.2d}$$

Because of the axial symmetry of the problem, the stresses $\tau_{x\theta}$ and $\tau_{\theta\zeta}$ and the body forces p_θ vanish so that

$$\bar{Q}_\theta = \bar{N}_{x\theta} = \bar{M}_{x\theta} = q_\theta = m_\theta = 0. \tag{4.1.3}$$

The remaining stress resultants and couples (Fig. 4.2) are independent of θ and, in addition, because of the symmetry of the problem we have

$$\bar{Q}_x = Q_x. \tag{4.1.4}$$

These are related by the equations of equilibrium of the shell element, Eqs. (2.6.2), which can be written as

$$\frac{\mathrm{d}N_x}{\mathrm{d}x} + q_x = 0, \tag{4.1.5a}$$

$$\frac{\mathrm{d}Q_x}{\mathrm{d}x} - \frac{N_\theta}{R} + q_\zeta = 0, \tag{4.1.5b}$$

$$\frac{\mathrm{d}M_x}{\mathrm{d}x} - Q_x + m_x = 0, \tag{4.1.5c}$$

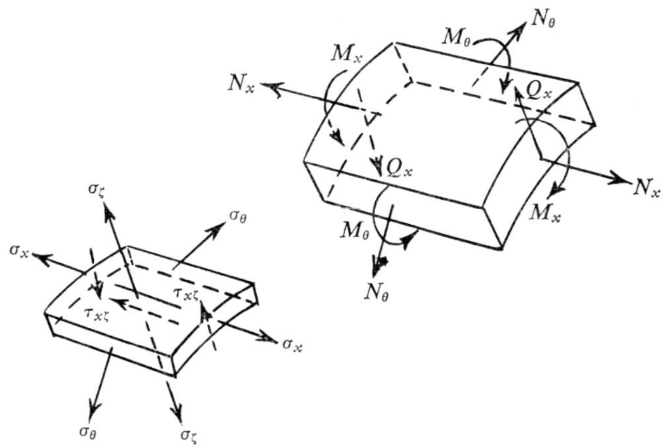

Fig. 4.2. Notation for stresses and stress resultants on cylindrical shell element

with

$$q_x = \left(1 + \frac{h}{2R}\right)\tau_{x\zeta}^{(h/2)} - \left(1 - \frac{h}{2R}\right)\tau_{x\zeta}^{(-h/2)} + \int_{-h/2}^{h/2}\left(1 + \frac{\zeta}{R}\right)\rho_x^{(\zeta)}\,\mathrm{d}\zeta, \tag{4.1.6a}$$

$$q_\zeta = \left(1 + \frac{h}{2R}\right)\sigma_\zeta^{(h/2)} - \left(1 - \frac{h}{2R}\right)\sigma_\zeta^{(-h/2)} + \int_{-h/2}^{h/2}\left(1 + \frac{\zeta}{R}\right)\rho_\zeta^{(\zeta)}\,\mathrm{d}\zeta, \tag{4.1.6b}$$

$$m_x = \frac{h}{2}\left[\left(1 + \frac{h}{2R}\right)\tau_{x\zeta}^{(h/2)} + \left(1 - \frac{h}{2R}\right)\tau_{x\zeta}^{(-h/2)}\right] + \int_{-h/2}^{h/2}\zeta\left(1 + \frac{\zeta}{R}\right)\rho_x^{(\zeta)}\,\mathrm{d}\zeta. \tag{4.1.6c}$$

To complete the mathematical formulation of the problem, we need a set of overall force–strain relations. The simplest set, given by Eqs. (3.1.4), also yields the simplest representation of the theory of axisymmetric deformations of circular cylinders and, as we have seen, is adequate so long as the ratio of the thickness and the least principle radius of curvature, h/R is small. Thus,

$$N_x = K(\epsilon_x + \nu\epsilon_\theta), \tag{4.1.7a}$$

$$N_\theta = K(\epsilon_\theta + \nu\epsilon_x), \tag{4.1.7b}$$

$$M_x = D\kappa_x, \tag{4.1.7c}$$

$$M_\theta = \nu D\kappa_x = \nu M_x. \tag{4.1.7d}$$

Eq. (4.1.5a) can be integrated to yield

$$N_x = \frac{P}{2\pi R} - \int_0^x q_x \, \mathrm{d}x, \tag{4.1.8a}$$

which is independent of the cylinder deformation, with P the total axial load acting on the cross-section defined by x equal to zero. From Eqs. (4.1.7a), (4.1.7b), and (4.1.2c) we have

$$N_\theta = \frac{Eh}{R} w + \nu N_x. \tag{4.1.8b}$$

From Eqs. (4.1.2d), (4.1.5c), and (4.1.7c) the bending moment M_x and the transverse shear force Q_x may be written as

$$M_x = -D \frac{\mathrm{d}^2 w}{\mathrm{d}x^2}, \tag{4.1.9a}$$

$$Q_x - m_x = \frac{\mathrm{d}M_x}{\mathrm{d}x} = -D \frac{\mathrm{d}^3 w}{\mathrm{d}x^3}. \tag{4.1.9b}$$

The equation which w must satisfy is then obtained from Eqs. (4.1.5b), (4.1.8b), and (4.1.9b) as

$$\frac{\mathrm{d}^4 w}{\mathrm{d}x^4} + 4\lambda^4 w = \frac{1}{D}\left(q_\zeta - \frac{\nu}{R} N_x + \frac{\mathrm{d}m_x}{\mathrm{d}x}\right), \tag{4.1.10a}$$

where

$$\lambda = \left[\frac{3(1 - \nu^2)}{R^2 h^2}\right]^{1/4}. \tag{4.1.10b}$$

From Eqs. (4.1.7), (4.1.8), (4.1.9), and (4.1.10) we may obtain an expression for the meridional displacement as

$$u_x = \frac{1}{Eh}\left[\int_0^x (N_x - \nu q_\zeta R) \, \mathrm{d}x - \nu Q_x R\right] + C, \tag{4.1.11}$$

where the constant C is a rigid body translation of the cylinder. Eqs. (4.1.9) and (4.1.10) are similar to those for a uniform beam on an elastic foundation.† Analogous quantities for axisymmetrically deformed cylindrical shells and beams on an elastic foundation are given in Table 4.1.

The general solution of Eq. (4.1.10) can be written as

$$w = C_1 A(\lambda x) + C_2 B(\lambda x) + C_3 A[\lambda(l - x)] + C_4 B[\lambda(l - x)] + w_p(x), \tag{4.1.12}$$

† M. Hetenyi, *Beams on Elastic Foundation*, University of Michigan Press, 1946.

Table 4.1 Analogous quantities for axisymmetrically deformed cylinders and beam on an elastic foundation

Quantity	Beam on elastic foundation	Cylinder
Deflection	w	w
Bending moment	M_x	M_x
Transverse shear force	Q_x	$Q_x - m_x$
Bending stiffness	EI	D
Foundation modulus	k	Eh/R^2
Applied load	q	$q_\zeta - \nu \dfrac{N_x}{R} + \dfrac{dm_x}{dx}$

where

$$A(\xi) = e^{-\xi}\cos\xi = -\frac{1}{2}\frac{dD(\xi)}{d\xi}, \tag{4.1.13a}$$

$$B(\xi) = e^{-\xi}\sin\xi = -\frac{1}{2}\frac{dC(\xi)}{d\xi}, \tag{4.1.13b}$$

$$C(\xi) = e^{-\xi}(\cos\xi + \sin\xi) = A(\xi) + B(\xi) = -\frac{dA(\xi)}{d\xi}, \tag{4.1.13c}$$

$$D(\xi) = e^{-\xi}(\cos\xi - \sin\xi) = A(\xi) - B(\xi) = \frac{dB(\xi)}{d\xi}, \tag{4.1.13d}$$

and $w_p(x)$ is a particular solution of Eq. (4.1.10). The slope, bending moment, and transverse shear force are then given by

$$\frac{1}{\lambda}\frac{dw}{dx} = -C_1C(\lambda x) + C_2D(\lambda x) + C_3C[\lambda(l - x)] - C_4D[\lambda(l - x)] + \frac{1}{\lambda}\frac{dw_p}{dx}, \tag{4.1.14a}$$

$$\frac{M_x}{2\lambda^2 D} = -C_1B(\lambda x) + C_2A(\lambda x) - C_3B[\lambda(l - x)] + C_4A[\lambda(l - x)] - \frac{1}{2\lambda^2}\frac{d^2w_p}{dx^2}, \tag{4.1.14b}$$

$$\frac{Q_x - m_x}{2\lambda^3 D} = -C_1D(\lambda x) - C_2C(\lambda x) + C_3D[\lambda(l - x)] + C_4C[\lambda(l - x)] - \frac{1}{2\lambda^3}\frac{d^3w_p}{dx^3}. \tag{4.1.14c}$$

The functions given by Eqs. (4.1.13) are tabulated in Table 4.2 and are shown

Table 4.2 Value of the functions $A(\xi)$, $B(\xi)$, $C(\xi)$, $D(\xi)$

ξ	$A(\xi)$	$B(\xi)$	$C(\xi)$	$D(\xi)$	ξ	$A(\xi)$	$B(\xi)$	$C(\xi)$	$D(\xi)$
0	1	0	1	1	0.25	0.7546	0.1927	0.9472	0.5619
0.001	0.9990	0.0010	1.0000	0.9980	0.26	0.7451	0.1982	0.9433	0.5469
0.002	0.9980	0.0020	1.0000	0.9960	0.27	0.7357	0.2036	0.9393	0.5321
0.003	0.9970	0.0030	1.0000	0.9940	0.28	0.7264	0.2089	0.9353	0.5175
0.004	0.9960	0.0040	1.0000	0.9920	0.29	0.7171	0.2140	0.9310	0.5030
0.005	0.9950	0.0050	1.0000	0.9900	0.30	0.7078	0.2189	0.9267	0.4888
0.006	0.9940	0.0060	1.0000	0.9880	0.31	0.6985	0.2237	0.9222	0.4748
0.007	0.9930	0.0070	0.9999	0.9861	0.32	0.6893	0.2284	0.9177	0.4609
0.008	0.9920	0.0080	0.9999	0.9841	0.33	0.6801	0.2330	0.9130	0.4472
0.009	0.9910	0.0087	0.9999	0.9821	0.34	0.6710	0.2374	0.9084	0.4337
0.010	0.9900	0.0099	0.9999	0.9801	0.35	0.6620	0.2416	0.9036	0.4204
0.011	0.9890	0.0109	0.9999	0.9781	0.36	0.6530	0.2457	0.8986	0.4072
0.012	0.9880	0.0119	0.9999	0.9761	0.37	0.6440	0.2497	0.8938	0.3943
0.013	0.9870	0.0129	0.9998	0.9742	0.38	0.6351	0.2536	0.8887	0.3815
0.014	0.9860	0.0138	0.9998	0.9722	0.39	0.6262	0.2574	0.8836	0.3688
0.015	0.9850	0.0148	0.9998	0.9702	0.40	0.6174	0.2610	0.8784	0.3564
0.016	0.9840	0.0158	0.9997	0.9683	0.41	0.6087	0.2646	0.8732	0.3441
0.017	0.9830	0.0167	0.9997	0.9663	0.42	0.6000	0.2680	0.8679	0.3320
0.018	0.9820	0.0177	0.9997	0.9643	0.43	0.5913	0.2712	0.8625	0.3201
0.019	0.9810	0.0187	0.9996	0.9624	0.44	0.5827	0.2743	0.8570	0.3084
0.02	0.9800	0.0196	0.9996	0.9604	0.45	0.5742	0.2774	0.8515	0.2968
0.03	0.9700	0.0291	0.9991	0.9409	0.46	0.5657	0.2803	0.8459	0.2853
0.04	0.9600	0.0384	0.9984	0.9216	0.47	0.5573	0.2832	0.8403	0.2742
0.05	0.9501	0.0476	0.9976	0.9025	0.48	0.5489	0.2857	0.8346	0.2632
					0.49	0.5406	0.2883	0.8289	0.2522
0.06	0.9401	0.0565	0.9966	0.8836	0.50	0.5323	0.2908	0.8231	0.2414
0.07	0.9302	0.0653	0.9954	0.8649	0.51	0.5241	0.2932	0.8173	0.2307
0.08	0.9202	0.0738	0.9940	0.8464	0.52	0.5159	0.2954	0.8113	0.2204
0.09	0.9103	0.0822	0.9924	0.8281	0.53	0.5079	0.2976	0.8054	0.2103
					0.54	0.4998	0.2996	0.7994	0.2002
0.10	0.9003	0.0903	0.9906	0.8100	0.55	0.4918	0.3016	0.7934	0.1902
0.11	0.8904	0.0983	0.9887	0.7921	0.56	0.4839	0.3035	0.7873	0.1805
0.12	0.8806	0.1062	0.9867	0.7744	0.57	0.4761	0.3052	0.7813	0.1709
0.13	0.8707	0.1138	0.9844	0.7568	0.58	0.4683	0.3068	0.7752	0.1615
0.14	0.8608	0.1213	0.9821	0.7395	0.59	0.4606	0.3084	0.7690	0.1522
0.15	0.8510	0.1286	0.9796	0.7224	0.60	0.4529	0.3099	0.7628	0.1430
0.16	0.8413	0.1358	0.9770	0.7055	0.61	0.4453	0.3113	0.7566	0.1340
0.17	0.8315	0.1427	0.9742	0.6888	0.62	0.4378	0.3126	0.7503	0.1252
0.18	0.8218	0.1495	0.9713	0.6722	0.63	0.4301	0.3138	0.7442	0.1166
0.19	0.8121	0.1562	0.9683	0.6550	0.64	0.4230	0.3150	0.7379	0.1080
0.20	0.8024	0.1627	0.9651	0.6398	0.65	0.4156	0.3160	0.7315	0.0996
0.21	0.7928	0.1690	0.9618	0.6238	0.66	0.4083	0.3169	0.7252	0.0914
0.22	0.7832	0.1742	0.9583	0.6080	0.67	0.4011	0.3178	0.7189	0.0833
0.23	0.7736	0.1812	0.9547	0.5924	0.68	0.3940	0.3186	0.7126	0.0754
0.24	0.7641	0.1870	0.9511	0.5771	0.69	0.3869	0.3193	0.7062	0.0676

Table 4.2—*contd.*

ξ	$A(\xi)$	$B(\xi)$	$C(\xi)$	$D(\xi)$	ξ	$A(\xi)$	$B(\xi)$	$C(\xi)$	$D(\xi)$
0.70	0.3798	0.3199	0.6997	0.0599	2.5	−0.0658	0.0492	−0.0166	−0.1149
0.71	0.3729	0.3205	0.6933	0.0524	2.6	−0.0636	0.0383	−0.0254	−0.1019
0.72	0.3695	0.3210	0.6869	0.0449	2.7	−0.0608	0.0287	−0.0320	−0.0895
0.73	0.3591	0.3214	0.6805	0.0377	2.8	−0.0573	0.0204	−0.0369	−0.0777
0.74	0.3524	0.3217	0.6741	0.0307	2.9	−0.0534	0.0132	−0.0403	−0.0666
0.75	0.3456	0.3220	0.6676	0.0237	3.0	−0.0493	0.0071	−0.0423	−0.0563
0.76	0.3389	0.3221	0.6611	0.0168	3.1	−0.0450	0.0019	−0.0431	−0.0469
0.77	0.3324	0.3223	0.6547	0.0101	3.2	−0.0407	−0.0024	−0.0431	−0.0383
0.78	0.3259	0.3224	0.6483	0.0035	3.3	−0.0364	−0.0058	−0.0422	−0.0306
0.79	0.3195	0.3224	0.6418	−0.0030	3.4	−0.0323	−0.0085	−0.0408	−0.0237
0.80	0.3131	0.3223	0.6353	−0.0093	3.5	−0.0283	−0.0106	−0.0389	−0.0177
0.81	0.3067	0.3222	0.6289	−0.0155	3.6	−0.0245	−0.0121	−0.0366	−0.0124
0.82	0.3004	0.3221	0.6225	−0.0217	3.7	−0.0210	−0.0131	−0.0341	−0.0079
0.83	0.2943	0.3219	0.6160	−0.0276	3.8	−0.0177	−0.0137	−0.0314	−0.0040
0.84	0.2881	0.3215	0.6096	−0.0334	3.9	−0.0147	−0.0140	−0.0286	−0.0008
0.85	0.2821	0.3212	0.6032	−0.0391	4.0	−0.0120	−0.0139	−0.0258	0.0019
0.86	0.2761	0.3207	0.5968	−0.0446	4.1	−0.0095	−0.0136	−0.0231	0.0040
0.87	0.2702	0.3202	0.5904	−0.0500	4.2	−0.0074	−0.0131	−0.0204	0.0057
0.88	0.2643	0.3197	0.5840	−0.0554	4.3	−0.0054	−0.0125	−0.0179	0.0070
0.89	0.2585	0.3191	0.5776	−0.0606	4.4	−0.0038	−0.0117	−0.0155	0.0079
0.90	0.2527	0.3185	0.5712	−0.0658	4.5	−0.0023	−0.0108	−0.0132	0.0085
0.91	0.2470	0.3178	0.5648	−0.0708	4.6	−0.0011	−0.0100	−0.0111	0.0089
0.92	0.2414	0.3171	0.5584	−0.0757	4.7	0.0001	−0.0091	−0.0092	0.0090
0.93	0.2359	0.3169	0.5521	−0.0805	4.8	0.0007	−0.0082	−0.0075	0.0089
0.94	0.2304	0.3155	0.5459	−0.0851	4.9	0.0014	−0.0073	−0.0059	0.0087
0.95	0.2250	0.3146	0.5396	−0.0896	5.0	0.0019	−0.0065	−0.0046	0.0084
0.96	0.2196	0.3137	0.5333	−0.0941	5.1	0.0023	−0.0057	−0.0033	0.0080
0.97	0.2143	0.3127	0.5270	−0.0984	5.2	0.0026	−0.0049	−0.0023	0.0075
0.98	0.2090	0.3117	0.5207	−0.1027	5.3	0.0028	−0.0042	−0.0014	0.0069
0.99	0.2038	0.3107	0.5145	−0.1069	5.4	0.0029	−0.0035	−0.0006	0.0064
1.0	0.1988	0.3096	0.5083	−0.1108	5.5	0.0029	−0.0029	0.0000	0.0058
1.1	0.1510	0.2967	0.4476	−0.1457	5.6	0.0029	−0.0023	0.0005	0.0052
1.2	0.1091	0.2807	0.3899	−0.1716	5.7	0.0028	−0.0018	0.0010	0.0046
1.3	0.0729	0.2626	0.3355	−0.1897	5.8	0.0027	−0.0014	0.0013	0.0041
1.4	0.0419	0.2430	0.2849	−0.2011	5.9	0.0026	−0.0010	0.0015	0.0036
1.5	0.0158	0.2226	0.2384	−0.2068	6.0	0.0024	−0.0007	0.0017	0.0031
1.6	−0.0059	0.2018	0.1959	−0.2077	6.1	0.0022	−0.0004	0.0018	0.0026
1.7	−0.0235	0.1812	0.1576	−0.2047	6.2	0.0020	−0.0002	0.0019	0.0022
1.8	−0.0376	0.1610	0.1234	−0.1985	6.3	0.0018	0.0001	0.0019	0.0018
1.9	−0.0484	0.1415	0.0932	−0.1899	6.4	0.0017	0.0003	0.0018	0.0015
2.0	−0.0563	0.1230	0.0667	−0.1794	6.5	0.0015	0.0004	0.0018	0.0012
2.1	−0.0618	0.1057	0.0439	−0.1675	6.6	0.0013	0.0005	0.0017	0.0009
2.2	−0.0652	0.0895	0.0244	−0.1548	6.7	0.0011	0.0006	0.0016	0.0006
2.3	−0.0668	0.0748	0.0080	−0.1416	6.8	0.0010	0.0006	0.0015	0.0004
2.4	−0.0669	0.0613	−0.0056	−0.1282	6.9	0.0008	0.0006	0.0014	0.0002

graphically in Fig. 4.3. We see that these functions rapidly decrease in an oscillatory fashion as ξ becomes large. Thus, the functions multiplied by C_1 and C_2 have their maximum values near the end of the cylinder defined by $x = 0$, and become small near the end $x = l$, while the functions multiplied by C_3 and C_4 behave in the opposite manner. The sum of corresponding functions of λx and $\lambda(l - x)$ for the same value of x are functions which are symmetrical about the center of the cylinder ($x = l/2$), while the difference between corresponding functions of λx and $\lambda(l - x)$ for the same value of x are functions which are antisymmetrical about the center.

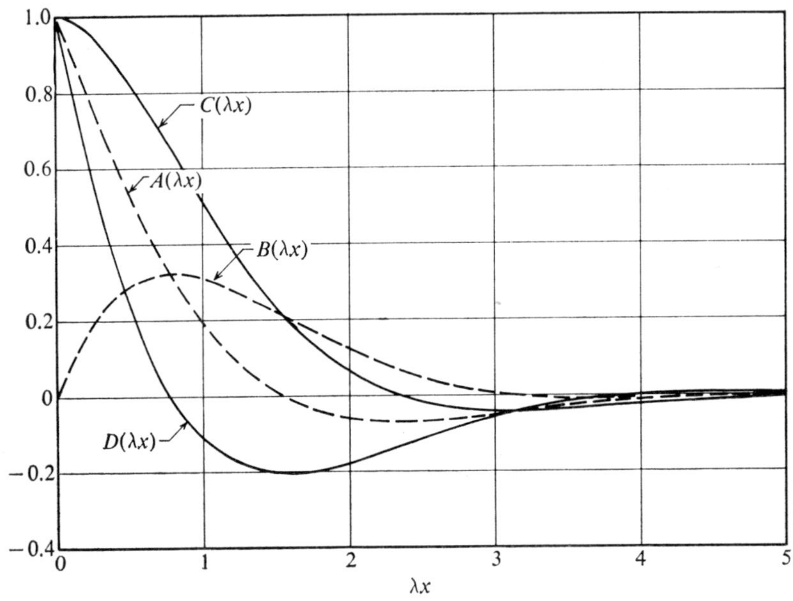

Fig. 4.3. Functions appearing in solution for axisymmetric cylindrical shell deformations

4.2 Edge-loaded cylindrical shells

(a) Semi-infinite cylinder

As an example of the behavior of loaded cylindrical shells, let us consider a semi-infinite cylinder subjected to axisymmetric bending moments M_0 and transverse shear forces Q_0 per unit middle surface circumferential length at the edge $x = 0$ (Fig. 4.4). At $x = \infty$, the deflections and stresses must vanish.

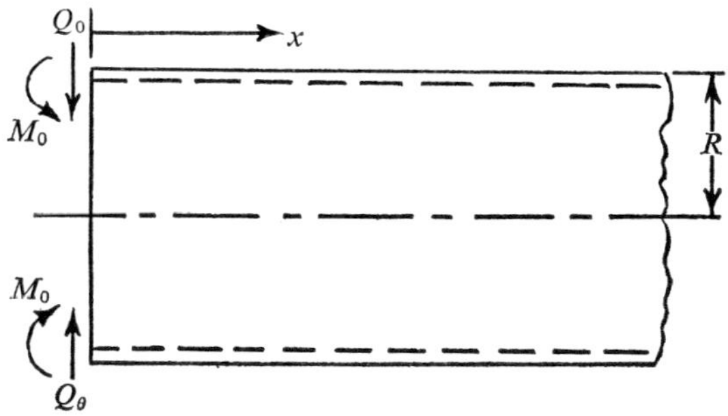

Fig. 4.4. Edge-loaded semi-infinite cylindrical shell

From Eqs. (4.1.14), with w_p equal to zero since we have no surface loading, we must then have

$$C_3 = C_4 = 0. \tag{4.2.1}$$

The remaining coefficients are determined from the loading conditions at $x = 0$,

$$M_x = M_0, \tag{4.2.2a}$$

$$Q_x = Q_0. \tag{4.2.2b}$$

From Eqs. (4.1.14b), (4.1.14c), and (4.2.1) we find the following values for the remaining coefficients

$$C_2 = \frac{M_0}{2\lambda^2 D}, \tag{4.2.3a}$$

$$C_1 = -\frac{1}{2\lambda^2 D}\left(M_0 + \frac{Q_0}{\lambda}\right). \tag{4.2.3b}$$

The radial deflection, slope, bending moment, and shear are thus given by

$$w = -\left[\frac{M_0}{2\lambda^2 D}\,D(\lambda x) + \frac{Q_0}{2\lambda^3 D}\,A(\lambda x)\right], \tag{4.2.4a}$$

$$\frac{dw}{dx} = \frac{M_0}{\lambda D}\,A(\lambda x) + \frac{Q_0}{2\lambda^2 D}\,C(\lambda x), \tag{4.2.4b}$$

$$M_x = M_0 C(\lambda x) + \frac{Q_0}{\lambda}\,B(\lambda x), \tag{4.2.4c}$$

$$Q_x = -\,2M_0\lambda B(\lambda x) + Q_0 D(\lambda x). \tag{4.2.4d}$$

At the edge $x = 0$ at which the loads are applied, the resulting radial deflection and slope are given by

$$w_0 = -\left(\frac{M_0}{2\lambda^2 D} + \frac{Q_0}{2\lambda^3 D}\right), \tag{4.2.5a}$$

$$\frac{dw_0}{dx} = \frac{M_0}{\lambda D} + \frac{Q_0}{2\lambda^2 D}. \tag{4.2.5b}$$

The coefficients of M_0 and Q_0 in Eqs. (4.2.5) are called influence coefficients. These coefficients are the edge deflection and slope per unit axisymmetric applied bending moment or shear force and are multiplied by the total bending moment or shear force to yield total edge deflection and slopes. Relations of the form of Eqs. (4.2.5) are useful in the solution of problems involving combinations of shells of various geometries for which conditions of continuity of deflection, slope, moment, and shear must be satisfied at the junctions of the shells.

(*b*) *Symmetrically loaded finite cylinder*

From Fig. 4.3, where the functions in Eqs. (4.2.4) are plotted, we see that the effects of edge moments or transverse shears are quite localized. At a sufficient distance from the edge, the deformation and stresses become negligible

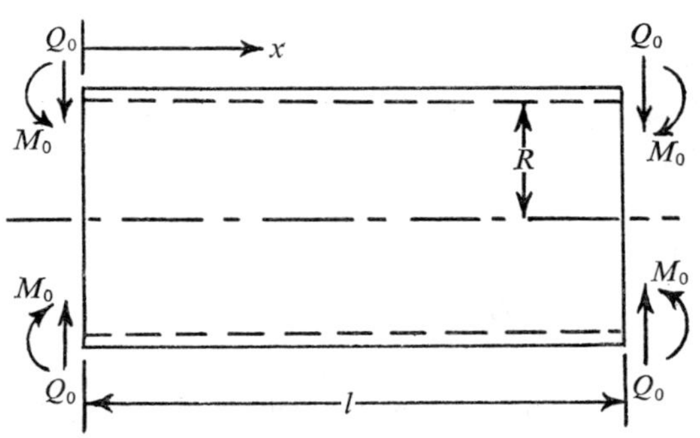

Fig. 4.5. Finite cylinder with edge-loading symmetric about the centre

compared to those near the edge. The value of the distance at which edge effects can be considered negligible can be better determined by considering a finite cylinder. Let us first investigate a cylinder subjected to the edge bending moments and transverse shear forces which are applied symmetrically about the center as shown in Fig. 4.5. For this type of loading the deflection curve is symmetrical about the center of the cylinder so that in Eqs. (4.1.12) and (4.1.14) we must take

$$C_1 = C_3, \tag{4.2.6a}$$

$$C_2 = C_4. \tag{4.2.6b}$$

The coefficients are then determined from the loading conditions at $x = 0$ or at $x = l$ which yield the relations

$$C_1 = -\frac{1}{\Delta(\lambda l)}\left\{\frac{M_0}{2\lambda^2 D}[1 - C(\lambda l)] + \frac{Q_0}{2\lambda^3 D}[1 + A(\lambda l)]\right\} \tag{4.2.8a}$$

$$C_2 = \frac{1}{\Delta(\lambda l)}\left\{\frac{M_0}{2\lambda^2 D}[1 - D(\lambda l)] - \frac{Q_0}{2\lambda^3 D}B(\lambda l)\right\} \tag{4.2.8b}$$

where

$$\begin{aligned}\Delta(\lambda l) &= [1 + A(\lambda l)][1 - D(\lambda l)] + B(\lambda l)[1 - C(\lambda l)]\\ &= 2\,\mathrm{e}^{-\lambda l}(\sinh \lambda l + \sin \lambda l).\end{aligned} \tag{4.2.8c}$$

The edge deflection and slope may now be obtained as

$$w_0 = -\left[\delta_M(\lambda l)\frac{M_0}{2\lambda^2 D} + \delta_H(\lambda l)\frac{Q_0}{2\lambda^3 D}\right], \tag{4.2.9a}$$

$$\frac{\mathrm{d}w_0}{\mathrm{d}x} = \Theta_M(\lambda l)\frac{M_0}{\lambda D} + \Theta_H(\lambda l)\frac{Q_0}{2\lambda^2 D}, \tag{4.2.9b}$$

where

$$\delta_M(\lambda l) = \Theta_H(\lambda l) = \frac{\sinh \lambda l - \sin \lambda l}{\sinh \lambda l + \sin \lambda l}, \tag{4.2.10a}$$

$$\delta_H(\lambda l) = \frac{\cosh \lambda l + \cos \lambda l}{\sinh \lambda l + \sin \lambda l}, \tag{4.2.10b}$$

$$\delta_M(\lambda l) = \frac{\cosh \lambda l - \cos \lambda l}{\sinh \lambda l + \sin \lambda l}. \tag{4.2.10c}$$

The influence coefficients given by Eqs. (4.2.10) satisfy the relationship

$$2\delta_H(\lambda l)\Theta_M(\lambda l) - \delta_M(\lambda l)\Theta_H(\lambda l) = 1. \tag{4.2.10d}$$

The same coefficients describe the edge bending moment and transverse shear force in terms of the edge displacement and slope, for on inverting Eqs. (4.2.9) and using Eqs. (4.2.10d) we have

$$\frac{M_0}{2\lambda^2 D} = \Theta_H(\lambda l)w_0 + \delta_H(\lambda l)\frac{1}{\lambda}\frac{\mathrm{d}w_0}{\mathrm{d}x}, \tag{4.2.11a}$$

$$\frac{Q_0}{2\lambda^3 D} = -\left[2\Theta_M(\lambda l)w_0 + \delta_M(\lambda l)\frac{1}{\lambda}\frac{\mathrm{d}w_0}{\mathrm{d}x}\right]. \tag{4.2.11b}$$

Values of the coefficients δ_H, Θ_M, and $\delta_M = \Theta_H$ are given in the first three columns of Table 4.3. We see that the difference between the influence coefficients for the finite cylinder and the infinite cylinder is less than 1 per cent

Table 4.3 Influence coefficients for edge-loaded cylinders

	Symmetric loading			Antisymmetric loading			Loading at one edge					
λl	$\delta_M(\lambda l)$ or $\Theta_H(\lambda l)$	$\delta_H(\lambda l)$	$\Theta_M(\lambda l)$	$\delta'_M(\lambda l)$ or $\Theta'_H(\lambda l)$	$\delta'_H(\lambda l)$	$\Theta'_M(\lambda l)$	$\delta_{M11}(\lambda l)$ or $\Theta_{H11}(\lambda l)$	$\delta_{H11}(\lambda l)$	$\Theta_{M11}(\lambda l)$	$\delta_{M21}(\lambda l)$ or $\Theta_{H21}(\lambda l)$	$\delta_{H21}(\lambda l)$	$\Theta_{M21}(\lambda l)$
0.5	0.0426	2.004	0.250	24.03	6.006	48.17	12.04	4.005	24.21	−12.00	−2.001	−23.96
1.0	0.165	1.033	0.497	6.044	3.005	6.244	3.105	2.019	3.371	−2.940	−0.986	−2.874
1.5	0.362	0.775	0.730	2.763	2.106	2.141	1.563	1.396	1.435	−1.201	−0.621	−0.705
2.0	0.599	0.738	0.921	1.669	1.537	1.231	1.134	1.137	1.076	−0.535	−0.400	−0.155
2.5	0.820	0.802	1.043	1.220	1.272	0.978	1.020	1.037	1.010	−0.200	−0.235	0.032
3.0	0.972	0.893	1.088	1.029	1.120	0.919	1.000	1.007	1.004	−0.029	−0.113	0.085
3.5	1.043	0.966	1.081	0.958	1.037	0.926	1.001	1.001	1.003	0.043	−0.036	0.078
4.0	1.057	1.005	1.054	0.946	0.997	0.950	1.002	1.001	1.002	0.056	0.004	0.052
4.5	1.044	1.017	1.027	0.958	0.983	0.974	1.001	1.000	1.001	0.043	0.017	0.026
5.0	1.026	1.017	1.009	0.974	0.983	0.991	1.000	1.000	1.000	0.026	0.017	0.009
6.0	1.003	1.006	0.997	0.997	0.994	1.003	1.000	1.000	1.000	0.002	0.006	−0.003
7.0	0.999	1.000	0.997	1.002	1.000	1.003	1.000	1.000	1.000	−0.002	0.000	−0.003
8.0	0.999	0.999	1.000	1.001	1.001	1.001	1.000	1.000	1.000	−0.001	−0.001	−0.001

when λl is greater than about 6, which defines the minimum cylinder length such that the influence of one edge upon the other is negligible.

(c) *Antisymmetrically loaded finite cylinder*

The same is true when the loading at the end $x = l$ is reversed (Fig. 4.6) so that the deformations are antisymmetric about the center of the cylinder. For this case, in Eqs. (4.1.12) and (4.1.14) we take

$$C_1 = -C_3, \tag{4.2.12a}$$

$$C_2 = -C_4, \tag{4.2.12b}$$

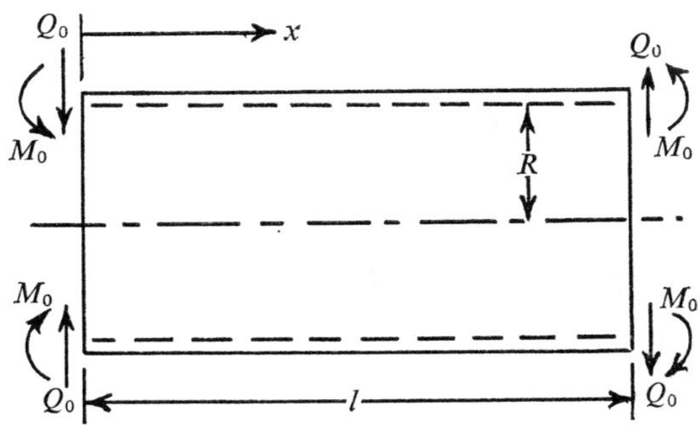

Fig. 4.6. Finite cylinder with edge-loading anti-symmetric about the centre

and again determine the values of the coefficients from the loading conditions at $x = 0$ or $x = l$. Then

$$C_1 = -\frac{1}{\Delta'(\lambda l)}\left\{\frac{M_0}{2\lambda^2 D}[1 + C(\lambda l)] + \frac{Q_0}{2\lambda^3 D}[1 - A(\lambda l)]\right\}, \tag{4.2.13a}$$

$$C_2 = \frac{1}{\Delta'(\lambda l)}\left\{\frac{M_0}{2\lambda^2 D}[1 + D(\lambda l)] + \frac{Q_0}{2\lambda^2 D}B(\lambda l)\right\}, \tag{4.2.13b}$$

where

$$\begin{aligned}\Delta'(\lambda l) &= [1 - A(\lambda l)][1 + D(\lambda l)] - B(\lambda l)[1 + C(\lambda l)] \\ &= 2\,\mathrm{e}^{-\lambda l}(\sinh \lambda l - \sin \lambda l).\end{aligned} \tag{4.2.13c}$$

The deflection and slope at the edge $x = 0$ are now given by

$$w_0 = -\left[\delta'_M(\lambda l)\frac{M_0}{2\lambda^2 D} + \delta'_H(\lambda l)\frac{Q_0}{2\lambda^3 D}\right], \tag{4.3.14a}$$

$$\frac{\mathrm{d}w_0}{\mathrm{d}x} = \Theta'_M(\lambda l)\frac{M_0}{\lambda D} + \Theta'_H(\lambda l)\frac{Q_0}{2\lambda^2 D}, \tag{4.2.14b}$$

where

$$\delta'_M(\lambda l) = \Theta'_H(\lambda l) = \frac{\sinh \lambda l + \sin \lambda l}{\sinh \lambda l - \sin \lambda l}, \tag{4.2.15a}$$

$$\delta'_H(\lambda l) = \frac{\cosh \lambda l - \cos \lambda l}{\sinh \lambda l - \sin \lambda l}, \tag{4.2.15b}$$

$$\Theta'_M(\lambda l) = \frac{\cosh \lambda l + \cos \lambda l}{\sinh \lambda l - \sin \lambda l}. \tag{4.2.15c}$$

These influence coefficients also satisfy a relationship of the form of Equation (4.2.10d) so that the inverse of Equations (4.2.14) is of the form of Eqs. (4.2.11). The coefficients are given in the next three columns of Table 4.3.

(*d*) *Finite cylinder loaded at one edge*

By superimposing the solutions for symmetric and antisymmetric edge loading, we obtain the solution for a finite cylinder loaded only at one end by edge

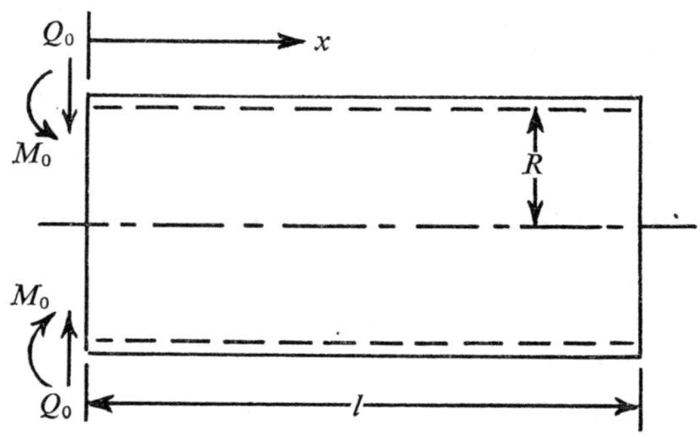

Fig. 4.7. Finite cylinder loaded at one edge

moments and shears (Fig. 4.7). For this case the deflections and slope at the loaded edge are given by

$$w_0 = -\left[\delta_{M_{11}}(\lambda l)\frac{M_0}{2\lambda^2 D} + \delta_{H_{11}}(\lambda l)\frac{Q_0}{2\lambda^3 D}\right], \tag{4.2.16a}$$

$$\frac{dw_0}{dx} = \Theta_{M_{11}}(\lambda l)\frac{M_0}{\lambda D} + \Theta_{H_{11}}(\lambda l)\frac{Q_0}{2\lambda^2 D}, \tag{4.2.16b}$$

where

$$\delta_{M_{11}}(\lambda l) = \Theta_{H_{11}}(\lambda l) = \tfrac{1}{2}[\delta_M(\lambda l) + \delta'_M(\lambda l)] = \frac{\cosh 2\lambda l - \cos 2\lambda l}{\cosh 2\lambda l + \cos 2\lambda l - 2}, \tag{4.2.17a}$$

$$\delta_{H_{11}}(\lambda l) = \tfrac{1}{2}[\delta_H(\lambda l) + \delta'_H(\lambda l)] = \frac{\sinh 2\lambda l - \sin 2\lambda}{\cosh 2\lambda l + \cos 2\lambda l - 2}, \tag{4.2.17b}$$

$$\Theta_{M_{11}}(\lambda l) = \tfrac{1}{2}[\Theta_M(\lambda l) + \Theta'_M(\lambda l)] = \frac{\sinh 2\lambda l + \sin 2\lambda l}{\cosh 2\lambda l + \cos 2\lambda l - 2}. \tag{4.2.17c}$$

These satisfy the relationship

$$2\delta_{H_{11}}(\lambda l)\Theta_{M_{11}}(\lambda l) - \delta_{M_{11}}(\lambda l)\Theta_{H_{11}}(\lambda l) = \frac{\cosh 2\lambda l + \cos 2\lambda l + 2}{\cosh 2\lambda l + \cos 2\lambda l - 2}, \tag{4.2.17d}$$

which simplify the determination of the inverse relationship. At the unloaded edge the radial deflection and slope are given by

$$w_l = -\left[\delta_{M_{21}}(\lambda l)\frac{M_0}{2\lambda^2 D} + \delta_{H_{21}}(\lambda l)\frac{Q_0}{2\lambda^3 D}\right], \tag{4.2.18a}$$

$$\frac{dw_l}{dx} = -\left[\Theta_{M_{21}}(\lambda l)\frac{M_0}{\lambda D} + \Theta_{H_{21}}(\lambda l)\frac{Q_0}{2\lambda^3 D}\right], \tag{4.2.18b}$$

where

$$\delta_{M_{21}}(\lambda l) = \Theta_{H_{21}}(\lambda l) = \tfrac{1}{2}[\delta_M(\lambda l) - \delta'_M(\lambda l)] = -\frac{4\sinh\lambda l\sin\lambda l}{\cosh 2\lambda l + \cos 2\lambda l - 2}, \tag{4.2.19a}$$

$$\delta_{H_{21}}(\lambda l) = \tfrac{1}{2}[\delta_H(\lambda l) - \delta'_H(\lambda l)] = \frac{2(\sinh\lambda l\cos\lambda l - \cosh\lambda l\sin\lambda l)}{\cosh 2\lambda l + \cos 2\lambda l - 2}, \tag{4.2.19b}$$

$$\Theta_{M_{21}}(\lambda l) = \tfrac{1}{2}[\Theta_M(\lambda l) - \Theta'_M(\lambda l)] = -\frac{2(\sinh\lambda l\cos\lambda l + \cosh\lambda l\sin\lambda l)}{\cosh 2\lambda l + \cos 2\lambda l - 2}. \tag{4.2.19c}$$

The influence coefficients for the cylinder loaded at only one edge are given in the last six columns of Table 4.3. We see that now the deflection and slope at the loaded end of the finite cylinder differ by less than 1 percent from the corresponding values for an infinite beam if the length of the cylinder is such that λl is greater than about 3. On the other hand, the deflection and slope of the unloaded end are less than 1 percent of the corresponding values at the loaded end only if λl is greater than about 6, the same value as for the cylinder loaded symmetrically or antisymmetrically at its edges.

From these examples we may conclude that in the analysis of an antisymmetrically end-loaded cylindrical shell, the stresses and deformations at *the more severely loaded end* may be calculated independently of conditions at the other end if the length of the cylinder is such that

$$\lambda l = \left(\frac{3(1-\nu^2)l^4}{R^2h^2}\right)^{1/4} \gtrsim 6. \tag{4.2.20}$$

In most cases the value of λl is so much greater than 6 that the same rule may also be used at the less severely loaded end without significant error. The quantity $1/\lambda$ has the dimensions of a length. From Eq. (4.2.20) we see that $6/\lambda$ or about $5(Rh)^{1/2}$ is the distance from the edge at which edge effects may usually be neglected. This distance is known as the attenuation length of edge effects.

For cylinders which are short ($\lambda l \to 0$), approximate expressions for the various edge influence coefficients may be obtained by expanding the various quantities in power series. Then by using the known expansions for the hyperbolic and trigonometric functions

$$\sinh \xi = \sum_{n=1,3,5\ldots}^{\infty} \frac{1}{n!}\,\xi^n, \tag{4.2.21a}$$

$$\cosh \xi = \sum_{n=0,2,4\ldots}^{\infty} \frac{1}{n!}\,\xi^n, \tag{4.2.21b}$$

$$\sin \xi = \sum_{n=1,3,5\ldots}^{\infty} \frac{(-1)^{n-1}}{n!}\,\xi^n, \tag{4.2.21c}$$

$$\cos \xi = \sum_{n=0,2,4\ldots}^{\infty} \frac{(-1)^n}{n!}\,\xi^n, \tag{4.2.21d}$$

we can obtain

$$\Theta_{H_{11}} = \delta_{M_{11}}(\lambda l) \approx \frac{3}{(\lambda l)^2}, \tag{4.2.22a}$$

$$\delta_{H_{11}}(\lambda l) \approx \frac{2}{\lambda l}, \tag{4.2.22b}$$

$$\Theta_{M_{11}}(\lambda l) \approx \frac{3}{(\lambda l)^3}, \tag{4.2.22c}$$

$$\Theta_{M_{21}} = \delta_{M_{21}}(\lambda l) \approx \frac{3}{(\lambda l)^2}, \tag{4.2.22d}$$

$$\delta_{H_{21}}(\lambda l) \approx \frac{1}{\lambda l}, \tag{4.2.22e}$$

$$\Theta_{M_{21}}(\lambda l) \approx -\frac{3}{(\lambda l)^3}, \tag{4.2.22f}$$

which can be seen to be within 1 percent of the correct values if λl is less than about 0.5. Such a short cylinder is essentially a ring and behaves as if its longitudinal cross-sections (Fig. 4.8) were rigid. To show this, we express

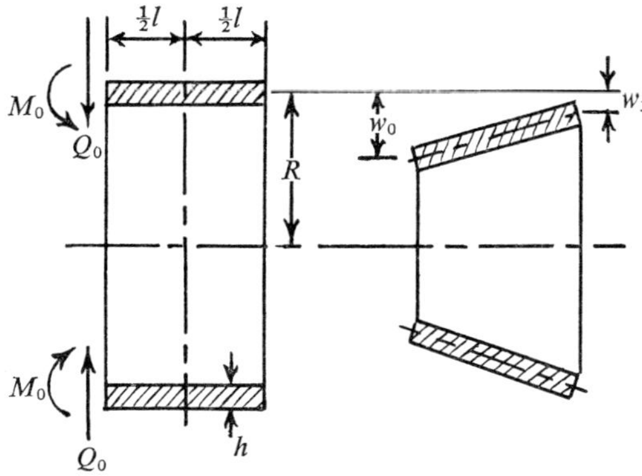

Fig. 4.8. Deformed short cylinder

the deflections and rotations at the loaded and unloaded ends, using the above expressions and the definitions of λ and D, in the form

$$w_0 \approx -\left[\frac{6R^2}{Ehl^2}\left(M_0 + Q_0\frac{l}{2}\right) + \frac{Q_0}{l}\frac{R^2}{Eh}\right], \tag{4.2.23a}$$

$$\frac{dw_0}{dx} \approx \frac{6R^2}{Ehl^2}\left(M_0 + Q_0\frac{l}{2}\right)\Big/\frac{l}{2}, \tag{4.2.23b}$$

$$w_l \approx \frac{6R^2}{Ehl^2}\left(M_0 + Q_0\frac{l}{2}\right) - \frac{Q_0}{l}\frac{R^2}{Eh}, \tag{4.2.23c}$$

$$\frac{dw_l}{dx} \approx \frac{6R^2}{Ehl^2}\left(M_0 + Q_0\frac{l}{2}\right)\Big/\frac{l}{2}. \tag{4.2.23d}$$

The symmetric portion of w_0 and w_l, the term $(Q_0/l)(R^2/Eh)$, can be easily identified as the radial deflection of a ring of radius R and thickness h under the uniform pressure Q_0/l. The remaining terms indicate twisting of the ring under a torque $M_0 + Q_0(l/2)$ about the center of the longitudinal cross-section, since the angle of rotation of the edges is the same and the deflection is given by the product of the angle of rotation and the distance from the center of the cross-section to the edge. The same formulas can be obtained by the more elementary methods of strength of materials.

4.3 Particular integrals for surface-loaded cylindrical shells

Most problems of cylindrical shell analysis involve surface loading in addition to edge loading. The solution of these problems requires the determination of a particular integral of Eq. (4.1.12). We start with consideration of an infinite cylinder subjected to a concentrated uniform ring loading q_0 at its center (Fig. 4.9).† We will assume that the ring loading is applied by means of shear

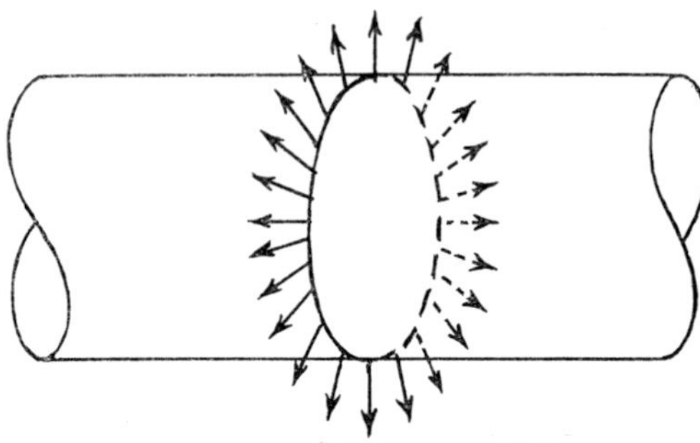

Fig. 4.9. Infinite cylinder with ring shear loading

stresses across the thickness of the cylinder.‡ The solution for this case is very easily determined from Eqs. (4.2.4) and (4.2.5) by considering that portion of the cylinder which extends to the right of the ring loading. We then have a semi-infinite cylinder subjected to edge transverse shear forces Q_0 which, from symmetry, are given by

$$Q_0 = -\tfrac{1}{2}q_0, \tag{4.3.1}$$

† An exact solution of this problem for a long cylindrical shell of an isotropic material is given by J. M. Klosner; 'The Elasticity Solution of a Long Circular Cylindrical Shell Subjected to a Uniform Circumferential Radial Line Load,' *J. Aerospace Sci.*, vol. 29, no. 7, July 1962, pp. 834–841, and is compared to various theories for transversely rigid shells by J. M. Klosner and J. Kempner, 'Comparison of Elasticity and Shell Theory Solutions,' *AIAA J.*, vol. 1, no. 3, March 1963, pp. 627–630. Comparisons for a similar problem involving periodically spaced uniform band loads are given by J. M. Klosner and H. S. Levine, 'Further Comparison of Elasticity and Shell Theory Solutions,' *AIAA J.*, vol. 4, no. 3, March 1966, pp. 467–480. The effect of a flexible elastic core inside the cylinder on deflections and moments has been discussed by J. C. Yao, 'Bending Due to Ring Loading of a Cylindrical Shell With an Elastic Core,' *J. Appl. Mech.*, vol. 32, no. 1, March 1965, pp. 99–103.

‡ We could also assume that the ring loading is applied by means of a concentrated band of loading on one of the bounding surfaces of the cylinder. While the solution of this problem is identical with that for the discontinuous shear problem, the method of solution that would logically be used is considerably more complicated.

and edge bending moments M_0. The value of M_0 is determined from the condition, again from symmetry, that the slope of the cylinder vanishes under the applied load. From the second of Eqs. (4.2.5) we then have

$$\frac{dw_0}{dx} = \frac{M_0}{\lambda D} - \frac{\frac{1}{2}q_0}{2\lambda^2 D} = 0, \tag{4.3.2a}$$

from which

$$M_0 = \frac{q_0}{4\lambda}, \tag{4.3.2b}$$

The solution for the radial deflection to the right of the load is thus

$$w = \frac{q_0}{8\lambda^3 D} C(\lambda x). \tag{4.3.3a}$$

Since the deflections on either side of the load are symmetrical, we may represent the radial deflection equation as

$$w = \frac{q_0}{8\lambda^3 D} C(\lambda |x|), \tag{4.3.3b}$$

where $|x|$ represents the magnitude of x without regard to sign.

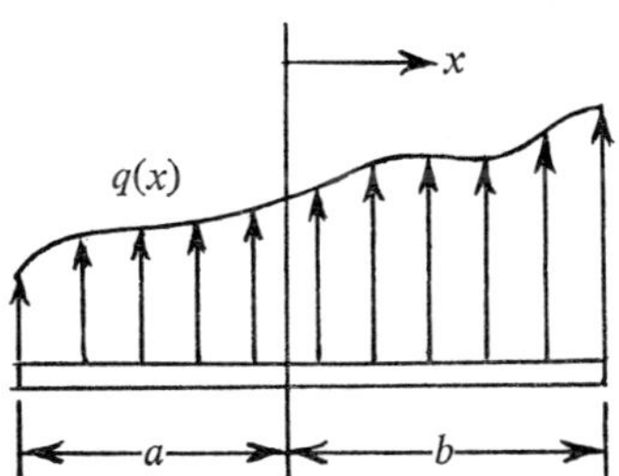

Fig. 4.10. Surface loaded cylinder

Now consider a cylinder loaded by surface stresses and body forces which can be represented, insofar as their effects are concerned, by a distributed equivalent load normal to the middle surface (Fig. 4.10) given by

$$q(x) = q_\zeta - \frac{\nu N_x}{R} + \frac{dm_x}{dx}. \tag{4.3.4}$$

Then a particular integral which satisfies Eq. (4.1.12) can be obtained by considering an infinite cylinder which is loaded by a series of ring shear loads of magnitude $q(x)\,dx$. The increment of deflection at a point x produced by a single ring load located a distance ξ from the origin is given by Eq. (4.3.3b) as

$$\mathrm{d}w(x) = \frac{q(\xi)\,\mathrm{d}\xi}{8\lambda^3 D} C(\lambda\,|x - \xi|). \tag{4.3.5a}$$

The total deflection due to all loads is then

$$w(x) = \frac{1}{8\lambda^3 D} \int_{-a}^{b} q(\xi) C(\lambda\,|x - \xi|)\,\mathrm{d}\xi, \tag{4.3.5b}$$

where $-a$, b are the limits of the extent of the surface loading. Eq. (4.3.5b) may be expressed as one of the following alternative forms

$$w(x) = \frac{1}{8\lambda^3 D}\left[\int_0^{a+x} q(x - \xi) C(\lambda\xi)\,\mathrm{d}\xi + \int_0^{b-x} q(x + \xi) C(\lambda\xi)\,\mathrm{d}\xi\right]$$

$$-a < x < b, \tag{4.3.6a}$$

$$= \frac{1}{8\lambda^3 D}\int_{x-b}^{x+a} q(x - \xi) C(\lambda\xi)\,\mathrm{d}\xi, \quad x > b, \tag{4.3.6b}$$

$$= \frac{1}{8\lambda^3 D}\int_{-x-a}^{-x+b} q(x + \xi) C(\lambda\xi)\,\mathrm{d}\xi, \quad x < -a. \tag{4.3.6c}$$

The evaluation of the integrals of Eqs. (4.3.6a) is facilitated by the use of Eqs. (4.1.13). For example, if the loading function q is uniform over the interval $-L/2 < x < L/2$ and is zero elsewhere, we have

$$w(x) = -\frac{q}{8\lambda^4 D}\left[\int_0^{L/2+x} \frac{\mathrm{d}A(\lambda\xi)}{\mathrm{d}\xi}\,\mathrm{d}\xi + \int_0^{L/2-x} \frac{\mathrm{d}A(\lambda\xi)}{\mathrm{d}\xi}\,\mathrm{d}\xi\right]$$

$$= \frac{q}{8\lambda^4 D}\left\{2 - A\left[\lambda\left(\frac{L}{2} + x\right)\right] - A\left[\lambda\left(\frac{L}{2} - x\right)\right]\right\}, \quad -\frac{L}{2} \le x \le \frac{L}{2}, \tag{4.3.7a}$$

$$w(x) = -\frac{q}{8\lambda^4 D}\int_{x-L/2}^{x+L/2} \frac{\mathrm{d}A(\lambda\xi)}{\mathrm{d}\xi}\,\mathrm{d}\xi$$

$$= \frac{q}{8\lambda^4 D}\left\{A\left[\lambda\left(|x| - \frac{L}{2}\right)\right] - A\left[\lambda\left(|x| + \frac{L}{2}\right)\right]\right\}, \quad |x| > \frac{L}{2}. \tag{4.3.7b}$$

If the uniform loading extends over the entire length of the infinite cylinder, so that L is infinite, Eq. (4.3.7a) reduces to

$$w(x) = \frac{q}{4\lambda^4 D}. \tag{4.3.8}$$

Eq. (4.3.5b) or the alternative forms given by Eqs. (4.3.6), which are solutions of the problem of an infinite cylinder subjected to surface loading, may be used as a particular integral of Eq. (4.1.12) for finite cylinder problems. In many cases involving finite cylinders, however, more convenient particular solutions of Eq. (4.1.12) may be obtained. For instance, if the equivalent load $q(x)$ is given by an expression of the form

$$q(x) = a + bx + cx^2 + dx^3, \tag{4.3.8a}$$

a particular solution of Eq. (4.1.12) is obviously

$$w_p(x) = \frac{a + bx + cx^2 + dx^3}{4\lambda^4 D} = \frac{q(x)}{4\lambda^4 D}. \tag{4.3.8b}$$

For loadings of the form

$$q(x) = a_m x^{4m+i}, \qquad i = 0, 1, 2, 3, \qquad m = 1, 2, 3, \ldots \tag{4.3.9a}$$

it is not difficult to find the particular solution

$$w_p(x) = \frac{q(x)}{4\lambda^4 D} \sum_{k=0}^{m} \frac{(-1)^k (4m + i)!}{[4(m - k) + i]!(2^{1/2}\lambda x)^{4k}}. \tag{4.3.9b}$$

Finally, if $q(x)$ is given by

$$q(x) = a_m \sin \frac{m\pi x}{l} \quad \text{or} \quad b_m \cos \frac{m\pi x}{l}, \tag{4.3.10a}$$

the particular solution

$$w_p(x) = \frac{q(x)}{4\lambda^4 D} \frac{1}{1 + (m\pi/2^{1/2}\lambda l)^4}, \tag{4.3.10b}$$

is readily obtained. Let us note that if λl is large compared to $m\pi/2^{1/2}$ we can neglect the term $(m\pi/2^{1/2}\lambda l)^4$ in the denominator of Eq. (4.3.10b) so that the particular solution for loadings given by Eqs. (4.3.8a) and (4.3.9a) are then of the same form. Note also that Eq. (4.3.10b) is an exact solution for a cylindrical shell which satisfies the simple support edge conditions

$$w = M_x = N_x = 0, \tag{4.3.11a}$$

when $q(x)$ is of the form

$$q(x) = a_m \sin \frac{m\pi x}{l}, \tag{4.3.11b}$$

and the somewhat artificial conditions

$$Q_x = \frac{\partial w}{\partial x} = u_x = 0, \tag{4.3.12a}$$

when $q(x)$ is the form

$$q(x) = b_m \cos \frac{m\pi x}{l} \quad (m \neq 0). \tag{4.3.12b}$$

The approximation

$$w_p(x) \approx \frac{q(x)}{4\lambda^4 D}, \tag{4.3.13}$$

is very often used and is identical with what is obtained from the membrane theory of shells. For axisymmetrically loaded cylinders, the pertinent equations of membrane theory, Eqs. (3.8.5) and (3.8.6), reduce to

$$N_\theta = R\left(q_\zeta + \frac{\mathrm{d}m_x}{\mathrm{d}x}\right), \tag{4.3.14a}$$

$$\frac{\mathrm{d}N_x}{\mathrm{d}x} = -q_x, \tag{4.3.14b}$$

$$\frac{w}{R} = \frac{1}{Eh}(N_\theta - \nu N_x), \tag{4.3.14c}$$

which yield

$$w = \frac{R^2}{Eh}\left(q_\zeta + \frac{\mathrm{d}m_x}{\mathrm{d}x} - \nu\frac{N_x}{R}\right) \equiv \frac{q(x)}{4\lambda^4 D}. \tag{4.3.14d}$$

The previous particular solutions then indicate that the membrane solution is adequate as a particular integral of the equations of bending provided the load varies relatively slowly over the length of the cylinder so that the length over which the load $q(x)$ varies significantly is large compared to the attenuation length of end effects. We may also conclude *for these cases* that bending effects are negligible in the interior of the cylinder since the deformations calculated by bending theory or by membrane theory are practically identical. It is then permissible, as is usually done, to superimpose the membrane solution for the middle surface force resultants and an edge effect solution to obtain the complete stress state. Further, if the load $q(x)$ is constant or linear the membrane equations are an exact particular solution of the equations of bending,† since the associated bending moments vanish and the transverse shear force is equal to m_x.

† We should remember, however, that we are using approximate stress–strain relations in both solutions.

4.4 Some solutions of cylindrical shell problems

(a) Compressed cylinder with rigid end rings

Let us turn to some problems which can be solved easily with the use of the previously derived solutions. Consider a cylinder of length L which is stiffened at its ends by rigid rings and compressed to axisymmetric axial forces N_x (Fig. 4.11). We wish to determine the effect of the end restraint on the axial shortening of the cylinder.

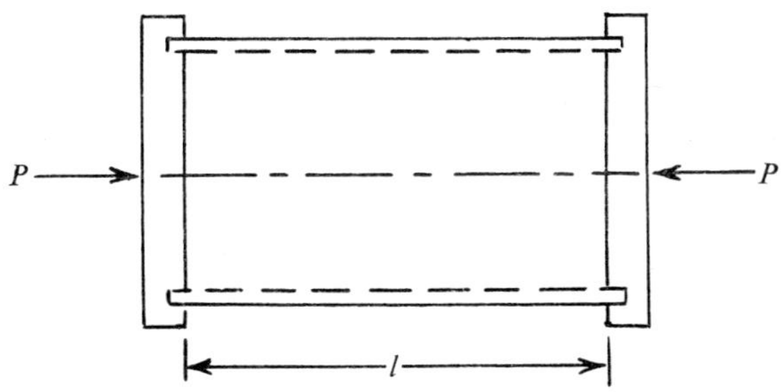

Fig. 4.11. Restrained cylinder under axial compression

To solve this problem we use the method of superposition by assuming first that the end rings are removed. Then the cylinder is subjected only to the axial loads N_x and from Eqs. (4.1.11), (4.3.4), and (4.3.8b) can be found to change length by the amount

$$\Delta u_{x_a} = (u_x)_{x=l} - (u_x)_{x=0} = -\frac{1}{Eh}\int_0^l N_x \,\mathrm{d}x = -\frac{N_x l}{Eh}, \tag{4.4.1a}$$

and to expand radially the uniform amount

$$w_p = \frac{\nu N_x R}{Eh}. \tag{4.4.1b}$$

For the edges to be moved to their original radial position we must superimpose edge bending moments and transverse shear forces which yield deformations symmetric about the center of the cylinder. If the rings have infinite torsional stiffness and thus do not permit rotation of the edges, the conditions which the edge loads must satisfy are

$$\left.\begin{aligned} w_0 &= -w_p \\ \frac{\mathrm{d}w_0}{\mathrm{d}x} &= 0. \end{aligned}\right\} \tag{4.4.2}$$

Then from Eqs. (4.2.11) the edge bending moments and transverse shear forces are

$$M_0 = -\nu\Theta_H(\lambda l)\frac{N_x}{2\lambda^2 R}, \tag{4.4.3a}$$

$$Q_0 = \nu\Theta_M(\lambda l)\frac{N_x}{\lambda R}. \tag{4.4.3b}$$

The change in length of the shell due to the edge loads can be found from Eq. (4.1.11) as

$$\Delta u_{x_b} = \frac{2\nu^2\Theta_M(\lambda l)}{\lambda l}\frac{N_x l}{Eh}, \tag{4.4.4}$$

which represents a decrease in the amount by which the shell shortens and thus a stiffening of the shell. The total amount of shortening is given by the sum of Eqs. (4.4.1a) and (4.4.4) as

$$\Delta u_x = -(\Delta u_{x_a} + \Delta u_{x_b}) = \left[1 - \frac{2\nu^2\Theta_M(\lambda l)}{\lambda l}\right]\frac{N_x l}{Eh}. \tag{4.4.5}$$

The quantity in braces approaches $1 - \nu^2$ as λl becomes small and 1 when λl becomes large, indicating that for short cylinders the outward radial deformation is almost totally restrained so that the shell is nearly in a state of plane strain, whereas for long cylinders the effect of edge restraint is negligibly

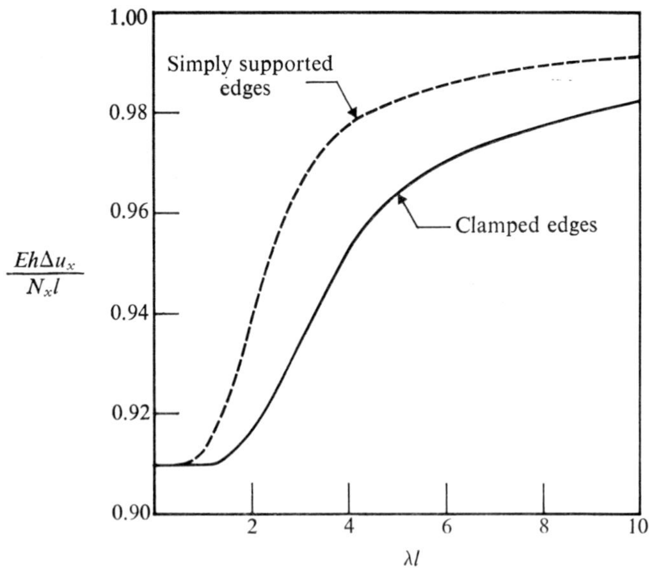

Fig. 4.12. Change of length of compressed cylinder ($\nu = 0.3$)

small. The variation of the quantity in braces with λl is shown in Fig. 4.12 with Poisson's ratio ν taken as 0.3.

If the end rings were torsionally weak, the edge conditions would change to

$$w_0 = -w_p, \tag{4.4.6a}$$

$$M_0 = 0, \tag{4.4.6b}$$

which yields

$$Q_0 = \frac{\nu/2}{\delta_H(\lambda l)} \frac{N_x}{\lambda R}. \tag{4.4.7}$$

The total shortening is then

$$\Delta u_x = \left[1 - \frac{\nu^2}{\lambda l \delta_H(\lambda l)}\right] \frac{N_x l}{Eh}, \tag{4.4.8}$$

which again varies from $(1 - \nu^2)N_x l/Eh$ for very short cylinders to $N_x l/Eh$ for very long cylinders. Values of the bracketed quantity are also shown in Fig. 4.12 and indicate that the effect of edge restraint for clamped edges is about twice as great as that for simply supported edges.

(b) Pressurized cylinder with rigid movable end plates

Let us now consider a cylinder built into rigid end plates (Fig. 4.13). When the cylinder is pressurized internally, it is subjected to an axial tension, obtained by equating the pressure force acting on the internal cross-section of each cylinder (the shaded area in Fig. 4.14) to the resisting force in the cylinder wall to yield

$$N_x = \tfrac{1}{2}pR, \tag{4.4.9}$$

and a uniform radial surface load per unit middle surface area, given by Eq. (4.1.6b) as

$$q_\zeta \approx p, \tag{4.4.10}$$

where terms of the order h/R compared to unity are neglected.

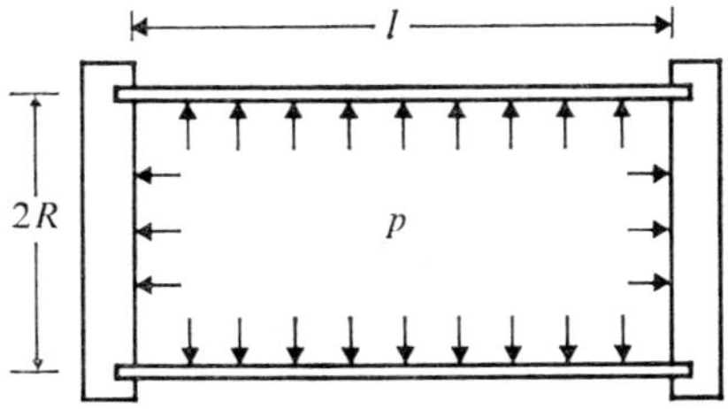

Fig. 4.13. Internally pressurized cylinder with rigid movable end plates

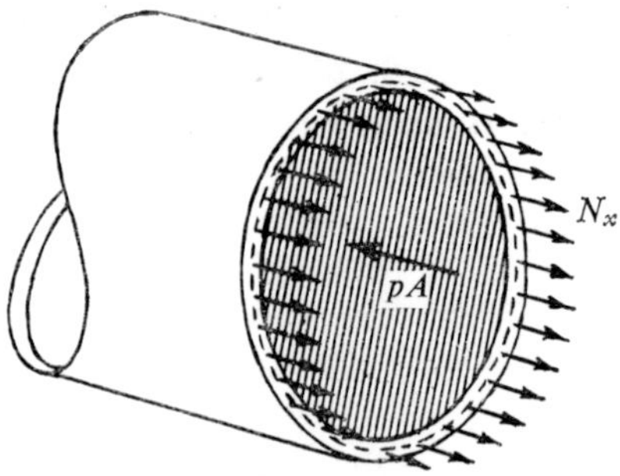

Fig. 4.14. Forces on cross-section of pressurized cylinder

To analyze the structure we first assume that the cylinder is cut at its edges so that it is free to expand uniformly under the applied loading. The expansion is given by Eqs. (4.3.4) and (4.3.8) as

$$w_p \approx \frac{pR^2}{Eh}\left(1 - \frac{\nu}{2}\right), \tag{4.4.11}$$

with attendant uniform circumferential forces given by Eqs. (4.1.8b) as

$$N_{\theta_p} \approx pR. \tag{4.4.12}$$

Our problem now is to determine the symmetric edge transverse shear forces and bending moments required to force the edges back to their original position. The solution of this problem is similar to that for the axially compressed cylinder and is obtained by replacing the value of w_p given by Eq. (4.4.1b) by that given by Eq. (4.4.11) in Eq. (4.4.3) to yield

$$M_0 = -\Theta_H(\lambda l)\frac{p}{2\lambda^2}\left(1 - \frac{\nu}{2}\right), \tag{4.4.13a}$$

$$Q_0 = \theta_M(\lambda l)\frac{p}{\lambda}\left(1 - \frac{\nu}{2}\right). \tag{4.4.13b}$$

The resulting deflection, force, and moment distributions may then be obtained by utilizing Eqs. (4.4.13) in conjunction with Eqs. (4.2.6), (4.2.8), and (4.1.14). We can show that the maximum moment always occurs at the edge.

For a long cylinder the maximum meridional stress, which is the maximum stress in the cylinder, is about twice the nominal circumferential stress in the unrestrained cylinder and is given by

$$\sigma_x \approx \left[\frac{1}{2} + \left(1 - \frac{\nu}{2}\right)\left(\frac{3}{1-\nu^2}\right)^{1/2}\right]\frac{pR}{h}. \tag{4.4.14}$$

If the pressure p acts only on the curved surface of the cylinder, N_x is equal to zero and the quantity $(1 - \nu/2)$ in Eqs. (4.4.11) and (4.4.13) is replaced by unity. The same solutions apply to a long pressurized cylinder stiffened by rigid rings with equal spacing l (Fig. 4.15), the former when the long cylinder has closed ends upon which pressure acts, the latter when the pressure acts only on the curved surfaces.

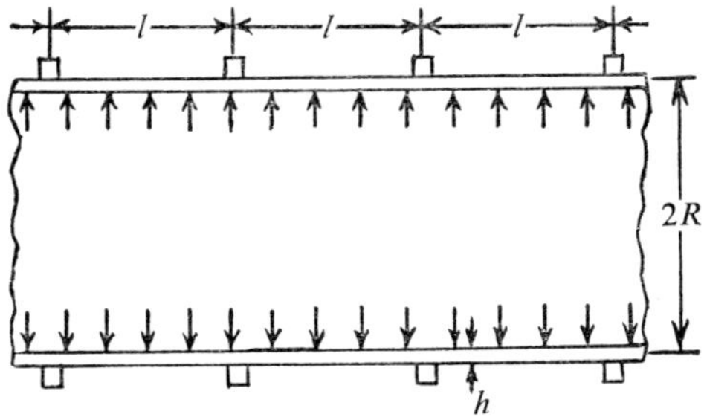

Fig. 4.15. Ring-stiffened pressurized cylinder

(c) *Effect of restraint of longitudinal movement*

If the pressurized cylinder is built into a rigid immovable wall (Fig. 4.16) instead of being closed by rigid end plates, the solutions given above satisfy the condition of zero radial deformation and rotation at the ends, but violate the condition that the ends must remain a fixed distance apart. For a cylinder

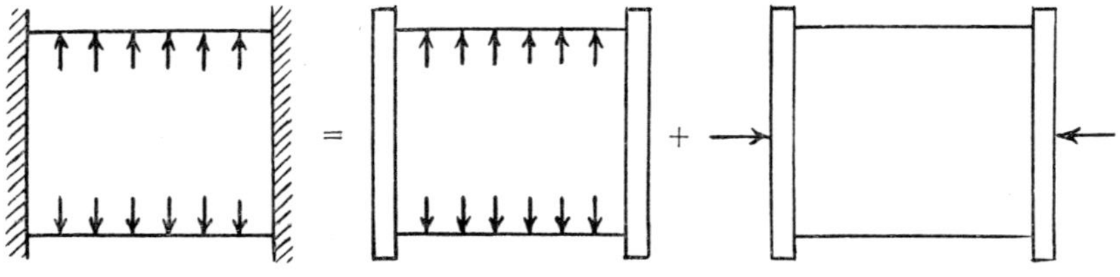

Fig. 4.16. Superposition of problems for pressurized cylinder built into a rigid wall

with pressure loading only on the curved surfaces, the use of Eqs. (4.1.11), (4.4.10), and Eq. (4.4.13b) modified to delete the effect of axial tension, yields the amount the cylinder shortens as

$$\Delta u_x \approx \frac{\nu pRl}{Eh}\left[1 - 2\,\frac{\theta_M(\lambda l)}{\lambda l}\right]. \tag{4.4.15}$$

From Eq. (4.4.5), we know that if we apply a tensile force N_x to a cylinder with edges which are not permitted to expand or rotate the amount by which the cylinder lengthens is

$$\Delta u_x = \left[1 - 2\nu^2 \frac{\theta_M(\lambda l)}{\lambda l}\right] \frac{N_x l}{Eh}. \tag{4.4.16}$$

Therefore, an axial tensile force given by

$$N_x = \nu p R \frac{1 - 2[\theta_M(\lambda l)/\lambda l]}{1 - 2\nu^2[\theta_M(\lambda l)/\lambda l]}, \tag{4.4.17a}$$

must be superimposed on the cylinder, as well as additional symmetric edge bending moments and transverse shear forces given by Eqs. (4.4.3) and (4.4.17a) as

$$M_0' = -\nu^2 \theta_H(\lambda l) \frac{p}{2\lambda^2} \left[\frac{1 - 2[\theta_M(\lambda l)/\lambda l]}{1 - 2\nu^2[\theta_M(\lambda l)/\lambda l]}\right], \tag{4.4.17b}$$

$$Q_0' = -\nu^2 \theta_M(\lambda l) \frac{p}{\lambda} \left[\frac{1 - 2[\theta_M(\lambda l)/\lambda l]}{1 - 2\nu^2[\theta_M(\lambda l)/\lambda l]}\right]. \tag{4.4.17c}$$

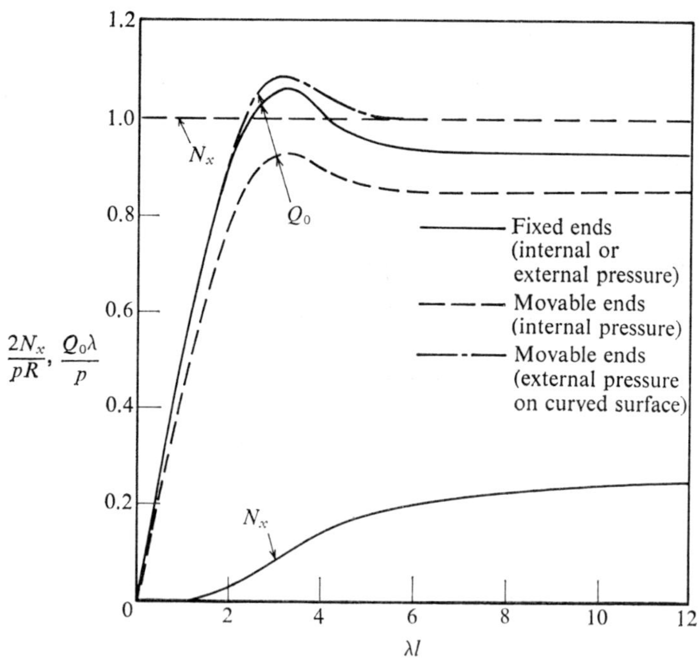

(a) Force resultants

Fig. 4.17. Comparison of stress resultants for pressurized cylinders with movable and fixed ends

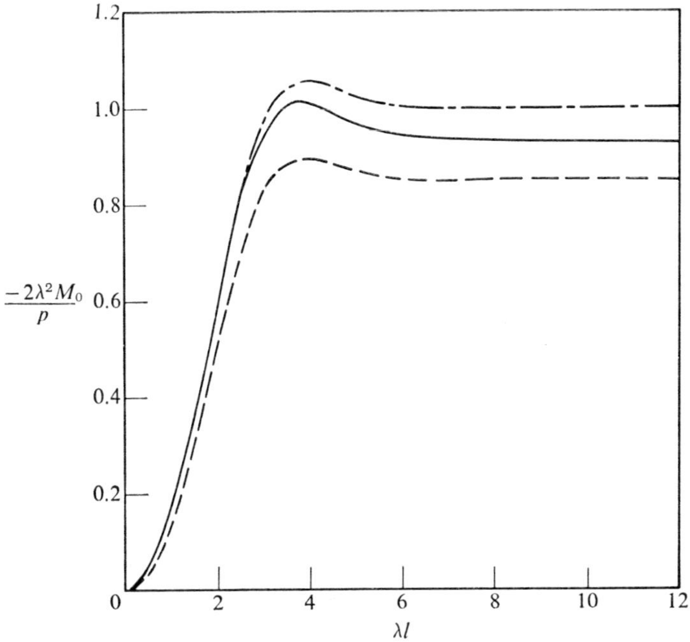

(b) Moment resultant

Fig. 4.17. Concluded

The total edge bending moments and edge transverse shear forces are then

$$M_{0_T} = -\frac{p}{2\lambda^2}\frac{(1-\nu^2)\theta_H(\lambda l)}{1-2\nu^2[\theta_M(\lambda l)/\lambda l]}, \tag{4.4.18a}$$

$$Q_{0_T} = \frac{p}{\lambda}\frac{(1-\nu^2)\theta_M(\lambda l)}{1-2\nu^2[\theta_M(\lambda l)/\lambda l]}. \tag{4.4.18b}$$

A comparison of the results given by Eqs. (4.4.9), (4.4.13), (4.4.17a), (4.4.18) is shown in Fig. 4.17 for cylinders with ν equal to 0.3. The edge bending moment and transverse shear force for the cylinder with ends fixed longitudinally are seen to be larger than those for the cylinder with movable rigid end plates. The axial force is considerably less, however. The net effect for a long cylinder is to reduce the maximum stress in the cylinder to the value

$$\sigma_x \approx \{\nu + [3(1-\nu^2)]^{1/2}\}\frac{pR}{h}, \tag{4.4.19}$$

which is less than twice the nominal stress if ν is less than 0.5.

(d) *Effect of end plate flexibility*

In example (b) the cylinder was restrained by end plates whch were assumed to be rigid against bending or stretching. Let us consider what changes occur in the solution when the pressurized cylinder is closed by flat flexible end

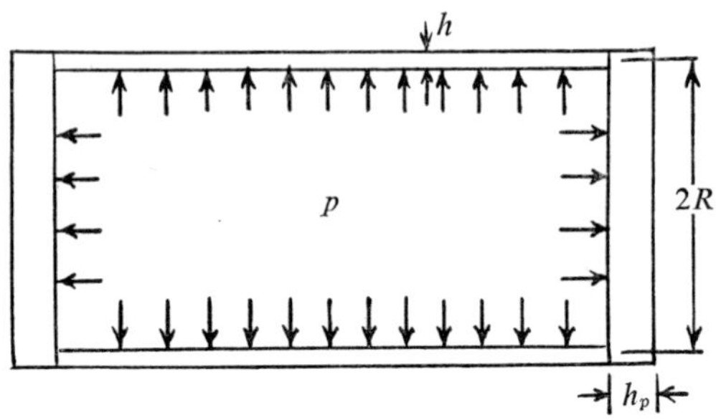

Fig. 4.18. Pressurized tank with flexible end plates

plates (Fig. 4.18). With end plates rigid against bending but perfectly flexible in stretching, the cylinder will expand the uniform amount given by Eq. (4.4.11) as

$$w_1 \approx \frac{pR^2}{Eh}\left(1 - \frac{\nu}{2}\right), \tag{4.4.20}$$

where we have neglected terms of the order of h/R compared to unity because of approximations made later on. To rejoin the cylinder and the end plates, we apply symmetrical end moments M_0 and shear forces Q_0 which produce an end radial deflection and slope, given by Eqs. (4.2.9) as

$$w_2 = -\left[\delta_M(\lambda l)\frac{M_0}{2\lambda^2 D} + \delta_H(\lambda l)\frac{Q_0}{2\lambda^3 D}\right], \tag{4.4.21a}$$

$$\frac{\mathrm{d}w_2}{\mathrm{d}x} = \theta_M(\lambda l)\frac{M_0}{\lambda D} + \theta_H(\lambda l)\frac{Q_0}{2\lambda^2 D}. \tag{4.4.21b}$$

The total radial deformation is then the sum of Eqs. (4.4.20) and (4.4.21) and the total slope is given by Eq. (4.4.21b)

Now we must consider the deformations of the end plate. From a more or less rigorous point of view we should consider the fact that the application of bending moments, transverse shear forces, and axial forces at the cylinder ends implies a certain distribution of direct and shearing stresses across the cylinder thickness, which are reacted by identical distributions of stresses acting on the surface of the end plate [Fig. 4.19(a)]. The deformation of the end plate under the distributed stress system can be obtained with the use of flat plate theory† but would lead to relatively complicated equations for the slope and radial estension at a location corresponding to the juncture of the middle surface of the cylinder and the end plate surface. If we are willing to ignore terms of the order of h/R compared to unity, much simplified equations can be obtained by first replacing the stresses transmitted by the cylinder to the surface of the plate by the equivalent moments M_0 and the normal

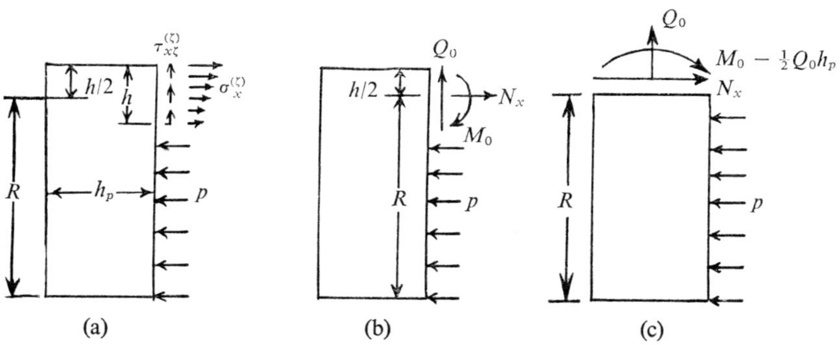

Fig. 4.19. Various approximations for end plate analysis

and tangential forces N_x and Q_0 shown in Fig. 4.19(b). We consider the radius of the flat plate to be R and the pressure to be applied over its entire surface, as shown in Fig. 4.19(c). The reaction forces on the plate surface are then replaced by moments $M_0 - \frac{1}{2}Q_0h_p$ and normal and tangential forces Q_0 and N_x applied to the edge of the flat plate. We can now find the edge rotation of the plate to be

$$\frac{\mathrm{d}w_3}{\mathrm{d}r} = -\frac{(M_0 - \frac{1}{2}Q_0h_p + \frac{1}{8}pR^2)R}{(1 + \nu_p)D_p}, \tag{4.4.22}$$

where the minus sign indicates that the rotation is clockwise as compared to the counter-clockwise rotation which is positive for the cylinder.

The radial deformation of the plate consists of two parts. One is the uniform radial extension due to the edge radial load Q_0 which is given by

$$w_4 = (1 - \nu_p)\frac{Q_0R}{E_ph_p}. \tag{4.4.23a}$$

The other is the radial movement of the surface which is in contact with the

† S. Timoshenko and S. Woinowsky-Krieger, *Theory of Plates and Shells*, 2nd Edition, McGraw Hill Book Co., Inc., New York, 1959, pp. 51–69.

cylinder due to the clockwise rotation of the plate edge about the middle surface of the plate (Fig. 4.20) which is given by

$$w_3 = \tfrac{1}{2}h_p \frac{\mathrm{d}w_3}{\mathrm{d}r} = -\frac{(M_0 - \frac{1}{2}Q_0 h_p + \frac{1}{8}pR^2)Rh_p}{2(1+\nu_p)D_p}. \tag{4.4.23b}$$

The total radial deformation is then the sum of Eqs. (4.4.23a) and (4.4.23b).

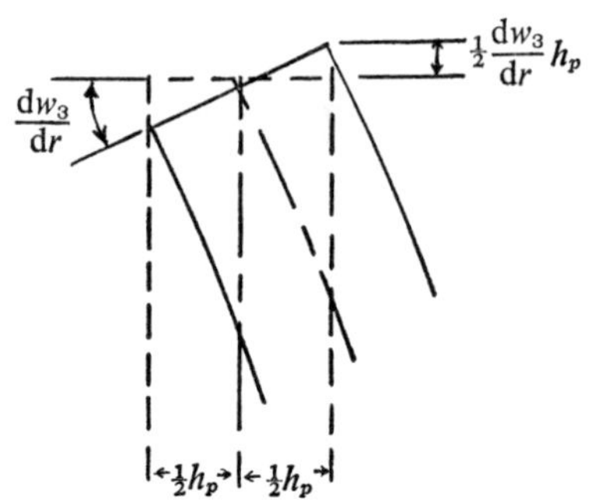

Fig. 4.20. Detail of edge plate deformation

The conditions that the plate and the cylinder must have the same radial deformation and rotation; i.e.

$$w_1 + w_2 = w_3 + w_4, \tag{4.4.24a}$$

$$\frac{\mathrm{d}w_2}{\mathrm{d}x} = \frac{\mathrm{d}w_3}{\mathrm{d}r}, \tag{4.4.24b}$$

leads to the following expressions for M_0 and Q_0:

$$M_0 = -\frac{p}{4\lambda^2\Delta}\left\{(2-\nu+\alpha_3)[\theta_H(\lambda l) - \alpha_1] + \frac{2\alpha_3}{\alpha_2}[\delta_H(\lambda l) + \tfrac{2}{3}\alpha_1\alpha_2]\right\}, \tag{4.4.25a}$$

$$Q_0 = \frac{p}{2\lambda\Delta}\left\{(2-\nu+\alpha_3)\left[\theta_M(\lambda l) + \frac{\alpha_1}{\alpha_2}\right] + \frac{\alpha_3}{\alpha_2}[\delta_M(\lambda l) - \alpha_1]\right\}, \tag{4.4.25b}$$

where

$$\alpha_1 = \frac{3(1-\nu_p)}{[3(1-\nu^2)]^{1/2}}\frac{E}{E_p}\left(\frac{h}{h_p}\right)^2, \tag{4.4.26a}$$

$$\alpha_2 = [3(1-\nu^2)]^{1/4}\frac{h_p}{(Rh)^{1/2}}, \tag{4.4.26b}$$

$$\alpha_3 = \frac{3(1-\nu_p)}{2}\frac{E}{E_p}\left(\frac{h}{h_p}\right)^2\frac{R}{h}, \tag{4.4.26c}$$

$$\Delta = 1 + \alpha_1\left[\delta_M(\lambda L) + \theta_H(\lambda L) + \tfrac{4}{3}\alpha_2\theta_M(\lambda L) + \frac{2}{\alpha_2}\,\delta_H(\lambda L)\right] + \tfrac{1}{3}\alpha_1^2. \tag{4.4.26d}$$

Values of M_0 and Q_0 for various values of R/h_p and h_p/h for an infinite cylinder and end plate of the same material are given in Table 4.4.†

Table 4.4 Values of bending moment and transverse shear at the edge of an infinite pressurized cylinder closed by a flat end plate ($E_p = E$, $\nu_p = \nu = 0.3$)
(a) $M_0/4pR^2$

R/h_p \ h_p/h	0.8	1.0	1.2	1.6	2.0
20	−0.0260	−0.0248	−0.0235	−0.0207	−0.0179
40	−0.0274	−0.0264	−0.0253	−0.0228	−0.0201
50	−0.0278	−0.0268	−0.0258	−0.0234	−0.0209
150	−0.0292	−0.0286	−0.0278	−0.0261	−0.0241
250	−0.0296	−0.0291	−0.0285	−0.0271	−0.0254

(b) $Q_0/2pR$

R/h_p \ h_p/h	0.8	1.0	1.2	1.6	2.0
20	0.3106	0.3349	0.3508	0.3627	0.3555
40	0.4344	0.4740	0.5022	0.5308	0.5305
50	0.4855	0.5313	0.5647	0.6009	0.6048
150	0.8476	0.9371	1.0082	1.1041	1.1477
250	1.1004	1.2201	1.3178	1.4583	1.5357

Further calculations indicate that the stresses in the cylinder are maximum at the edges in the range of parameters considered above and are many times the nominal circumferential stress. The stresses in the end closure are generally a maximum at the edges in this range, but become maximum at the center as the plate thickness increases relative to the cylinder thickness. The magnitude of these maximum stresses decreases as the thickness of the end closure increases with the cylinder dimensions remaining constant.

† G. W. Watts and H. A. Lang, 'The Stresses in a Pressure Vessel with a Flat Head Closure,' *Trans. ASME*, vol. 74, 1952, pp. 1053–1090.

The similar problem of a pressurized cylinder with end flanges has also been investigated.†

4.5 Pressurized cylinder with an abrupt thickness change

Another problem of interest is that of the stress distribution in a closed pressurized tank which is composed of two cylinders of different thickness (Fig. 4.21). Although in practice there is usually a small distance over which the thickness changes gradually to avoid stress concentrations, the thickness change is considered to be abrupt for simplicity. In addition, we assume that each cylinder is long enough so that the joint in question may be considered independently of any others. For generality, the middle surfaces of the two cylinders are displaced from each other by a distance d, as shown in the

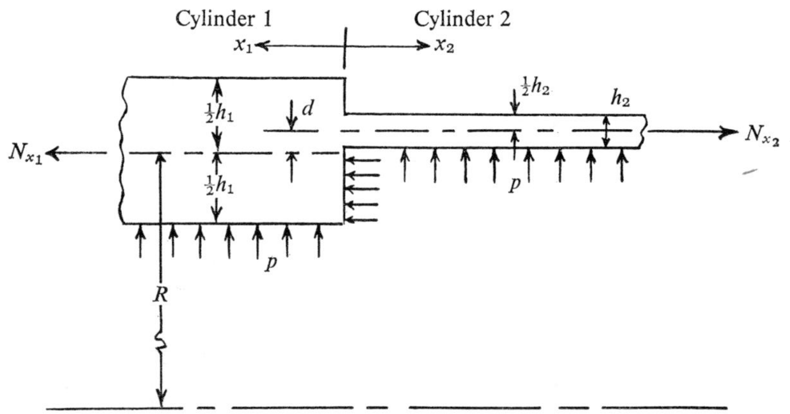

Fig. 4.21. Pressurized cylinder with an abrupt thickness change

figure. The thickness of each cylinder and the middle surface displacement are assumed to be small compared to the radius of each cylinder.

The preliminary analysis of the structure is similar to that of Section 4.4. The radial load per unit middle surface area acting on each cylinder is

$$q_{\zeta 1} \approx q_{\zeta 2} \approx p. \tag{4.5.1}$$

† E. O. Holmberg and K. Axelson, 'Analysis of Stresses in Circular Plates and Rings with Application to Cylinders with Rigidly Attached Flat Heads and Flanges,' *Trans. ASME*, vol. 193, pp. 13–23; see also P. Seide, 'The Effect of Shear Deformations on the Bending of Thick Flanged Cylindrical Shells Under Internal Pressure,' TRW Systems, Report No. GM-TR-283 (EM7-16), November 18, 1957.

In addition, since the tank is closed, each cylinder is subjected to axial tension

$$N_{x1} \approx N_{x2} \approx \tfrac{1}{2}p\bar{R}, \tag{4.5.2}$$

where $\bar{R}$ is an average radius used for both cylinders. Actually there is a small difference between the two axial forces which is counterbalanced by the pressure force acting horizontally on Cylinder 1 in Fig. 4.21. We assume that each cylinder is cut at its joint so that it is free to uniformly expand under the loads given by Eqs. (4.5.1) and (4.5.2). Then the free expansion of each cylinder is given by an amount

$$w_{1p} \approx \left(1 - \frac{\nu}{2}\right)\frac{p\bar{R}^2}{Eh_1}, \tag{4.5.3a}$$

$$w_{2p} \approx \left(1 - \frac{\nu}{2}\right)\frac{p\bar{R}^2}{Eh_2}. \tag{4.5.3b}$$

The corresponding uniform circumferential forces are given by

$$N_{\theta 1} \approx N_{\theta 2} \approx p\bar{R}. \tag{4.5.4}$$

The two cylinders now have a discontinuity at their junction since w_{1p} and w_{2p} are not equal. To bring the two cylinders together at their common

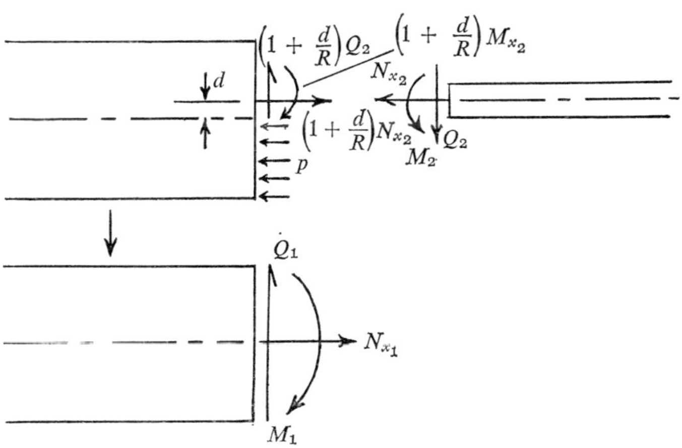

Fig. 4.22. Forces acting on cylinders at the junction

edge we must superimpose edge moments and shear forces as shown in Fig. 4.22. From the figure we see that the net bending moment and transverse shear force per unit middle surface length of Cylinder 1 is given by

$$Q_1 \approx Q_2, \tag{4.5.5a}$$

$$M_1 \approx M_2 + N_{x2}d, \tag{4.5.5b}$$

where the bending moment due to the horizontal pressure has been neglected since it is small compared to the moment of the axial tension N_{x2}. Two additional equations are obtained from the conditions that the deflection and slope must be continuous at the joint. The total deflection of each cylinder is the sum of the free expansion given by Eq. (4.5.3) and the deflection due to edge moments and shears. The latter may be found from Eq. (4.2.5a) with the precaution that care be taken to include the proper signs for the moments and shears on each cylinder. We thus have

$$w_{1p} - \frac{M_1}{2\lambda_1^2 D_1} + \frac{Q_1}{2\lambda_1^3 D_1} = w_{2p} - \frac{M_2}{2\lambda_2^2 D_2} - \frac{Q_2}{2\lambda_2^3 D_2}. \tag{4.5.6a}$$

Since the uniform expansion produces no change of slope, we must equate the slopes due to the applied edge moments and shears, again taking care to use proper signs. From Eq. (4.2.5b), we have

$$-\frac{M_1}{\lambda_1 D_1} + \frac{Q_1}{2\lambda_1^2 D_1} = \frac{M_2}{\lambda_2 D_2} + \frac{Q_2}{2\lambda_2^2 D_2}. \tag{4.5.6b}$$

The solution of Eqs. (4.5.5) and (4.5.6), together with Eqs. (4.5.2) and (4.5.3), can readily be obtained as

$$M_1 = \left[\frac{d}{h_1}\,\delta_1(c) - \frac{2-\nu}{[3(1-\nu^2)]^{1/2}}\,\delta_2(c)\right] p\bar{R}h_1, \tag{4.5.7a}$$

$$M_2 = -\left\{\frac{d}{h_1}\,[\tfrac{1}{2} - \delta_1(c)] + \frac{2-\nu}{[3(1-\nu^2)]^{1/2}}\,\delta_2(c)\right\} p\bar{R}h_1, \tag{4.5.7b}$$

$$Q_1 = Q_2 = \left[\frac{d}{h_1}\,\delta_3(c) + \frac{2-\nu}{[3(1-\nu^2)]^{1/2}}\,\delta_4(c)\right] \lambda_1 p\bar{R}h_1, \tag{4.5.7c}$$

where

$$c = \frac{h_2}{h_1}, \tag{4.5.8a}$$

$$\delta_1(c) = \frac{1 + 2c^{3/2} + c^2}{2[(1+c^2)^2 + 2c^{3/2}(1+c)]}, \tag{4.5.8b}$$

$$\delta_2(c) = \frac{c(1-c)^2(1+c)}{4[(1+c^2)^2 + 2c^{3/2}(1+c)]}, \tag{4.5.8c}$$

$$\delta_3(c) = \frac{c^{3/2}(1+c^{1/2})}{[(1+c^2)^2 + 2c^{3/2}(1+c)]}, \tag{4.5.8d}$$

$$\delta_4(c) = \frac{c^{1/2}(1+c^{3/2})(1-c)}{2[(1+c^2)^2 + 2c^{3/2}(1+c)]}. \tag{4.5.8e}$$

The four functions of c are shown in Fig. 4.23. The longitudinal distribution of the forces and moments can be obtained now by utilizing Eqs. (4.2.4) in conjunction with Eqs. (4.5.7) and (4.5.8).

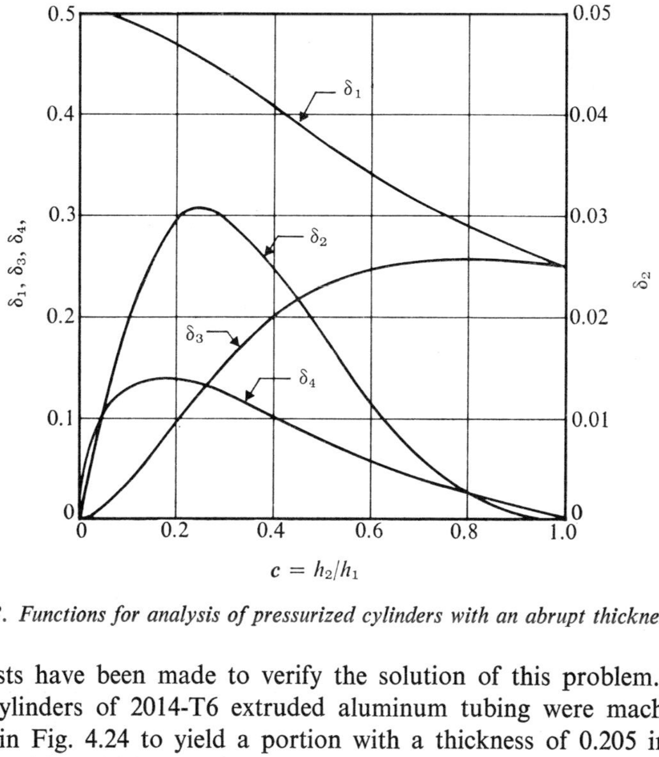

Fig. 4.23. Functions for analysis of pressurized cylinders with an abrupt thickness change

Tests have been made to verify the solution of this problem.† Three small cylinders of 2014-T6 extruded aluminum tubing were machined as shown in Fig. 4.24 to yield a portion with a thickness of 0.205 inch and another with a thickness of 0.082 inch, or a value of c equal to 0.4. One cylinder had a continuous middle surface, the second a continuous inner surface, and the third a continuous outer surface, corresponding to values of d/h_1 of 0, -0.3, and $+0.3$ respectively. Small strain gages were used to measure circumferential and meridional strains in the vicinity of the abrupt thickness discontinuity. These measurements were converted to stress data by means of the stress–strain relations given by Eqs. (2.3.2) and were compared to approximate stress values obtained with the use of the theoretical solution given above and Eqs. (4.1.15).‡

† W. C. Morgan and P. T. Bizon, 'Experimental Investigation of Stress Distributions Near Abrupt Change in Wall Thickness in Thin-Walled Pressurized Cylinders,' NASA TN D-1200, June 1962.

‡ The quantity ζ/R in Eq. (4.1.15) was neglected in comparison with unity.

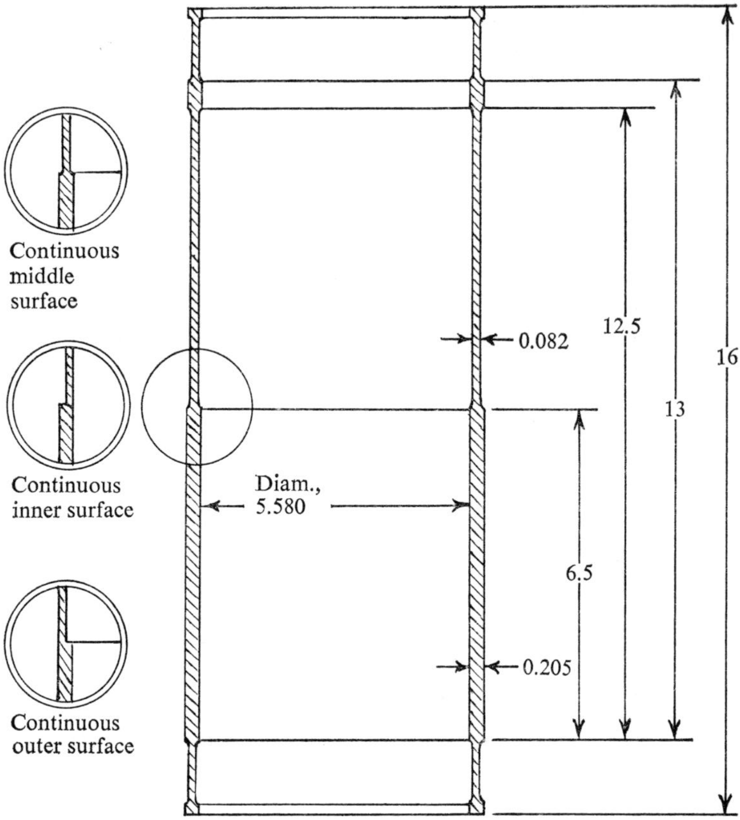

Fig. 4.24. Geometry of three small cylinders. (Dimensions in inches.)

Comparisons of theoretical and experimentally determined meridional and circumferential at the inner and outer surfaces of the cylinders for an internal pressure of 621 psi are shown in Figs. 4.25. We see that despite the approximations made by using the theory of transversely rigid shells and by dropping small terms in comparison with unity the agreement between theory and experiment is quite good. The average deviation of experimental and theoretical stresses for all three cylinders is about 6 percent. The error might be reduced somewhat for these cylinders, for which the radius/thickness ratio is about 14 for the thicker portion and 35 for the thinner portion, by including all the small terms which have been deleted in the approximate theory of transversely rigid shells and in the application of the theory. The difference between the two calculations becomes insignificant, however, if the radius/thickness ratio of each cylinder is greater than about 50.

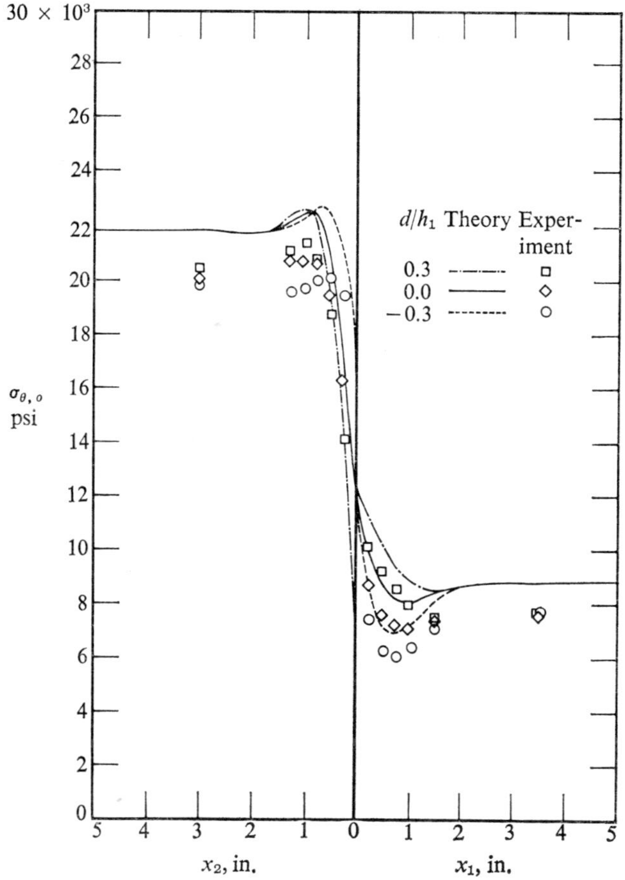

(a) Circumferential stress at the outer surface

Fig. 4.25. Comparison of theoretical and experimental stresses for a pressurized cylinder with an abrupt thickness change

Fig. 4.25. Continued over

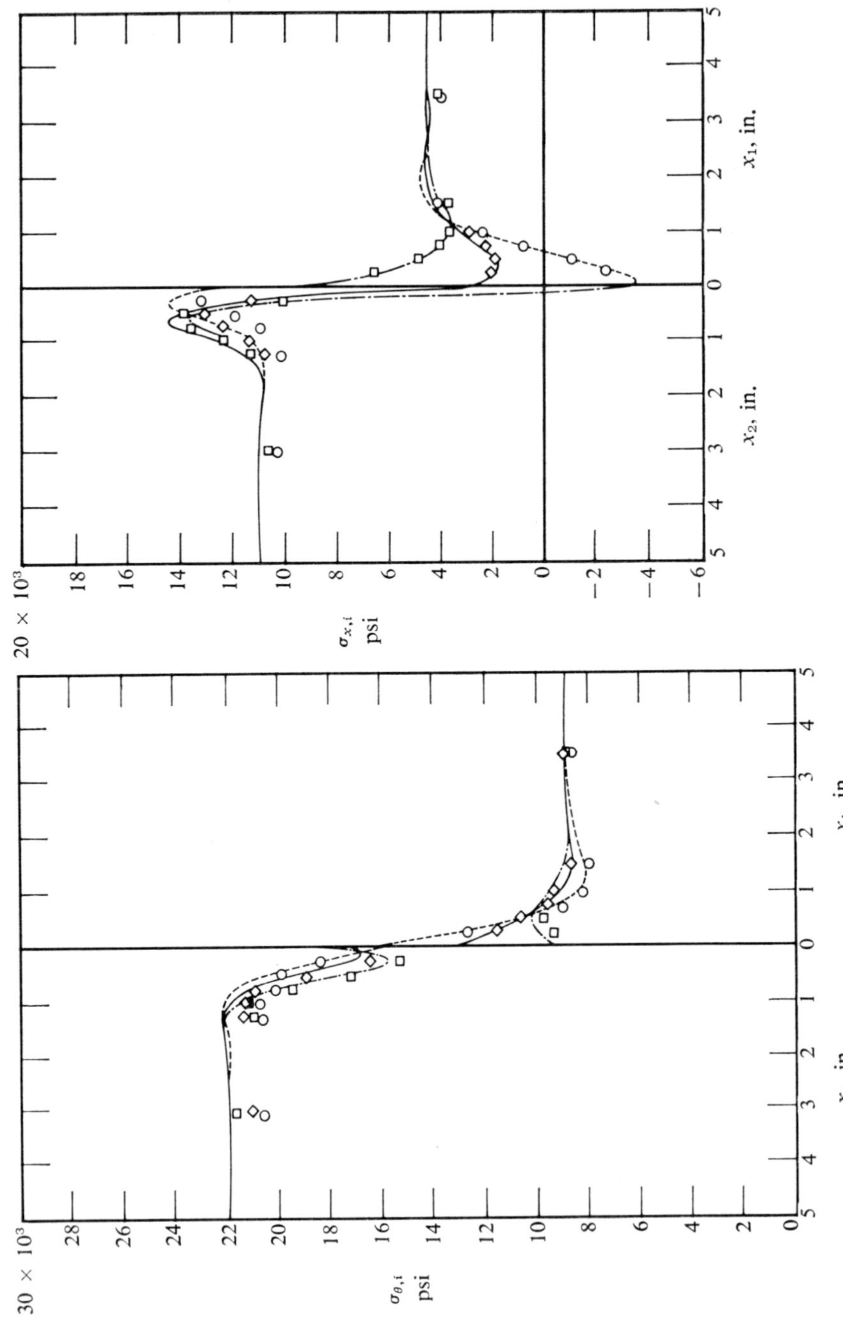

(b) Circumferential stress at the inner surface

(c) Axial stress at the inner surface

Fig. 4.25. Continued

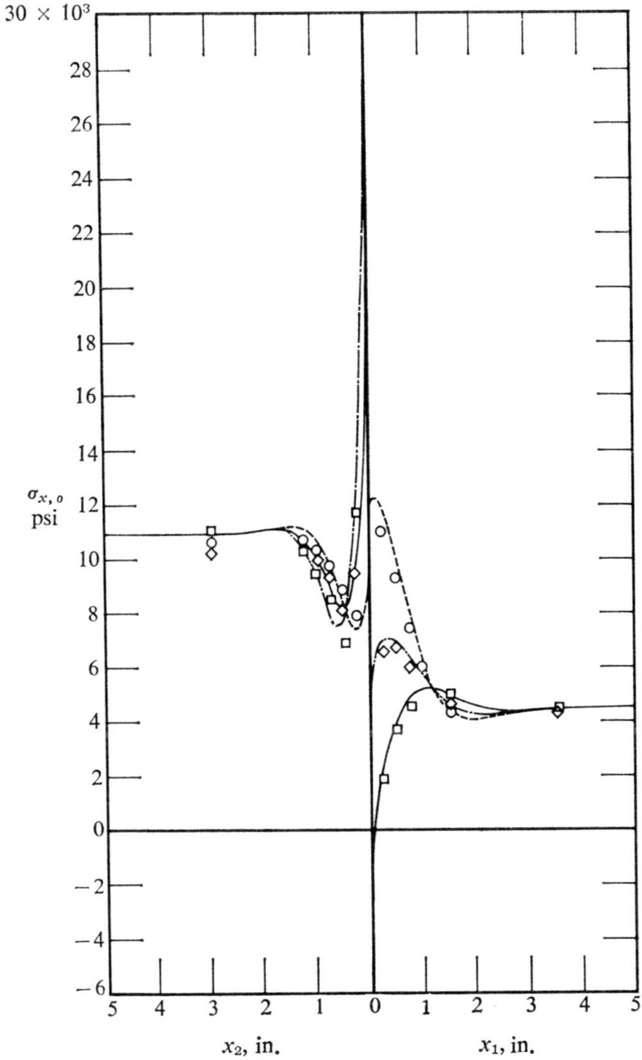

(d) Axial stress at the outer surface

Fig. 4.25. Concluded

4.6 Cylinder in a rigid collar†

A final problem of cylindrical shell analysis which we shall treat is one which is somewhat more difficult to solve than those we have discussed previously since the form of the solution is implicitly related to the geometry of the structure. Consider a long cylinder to which an undersized rigid collar is fitted (Fig. 4.26).‡ We would like to determine the stress state in the cylinder

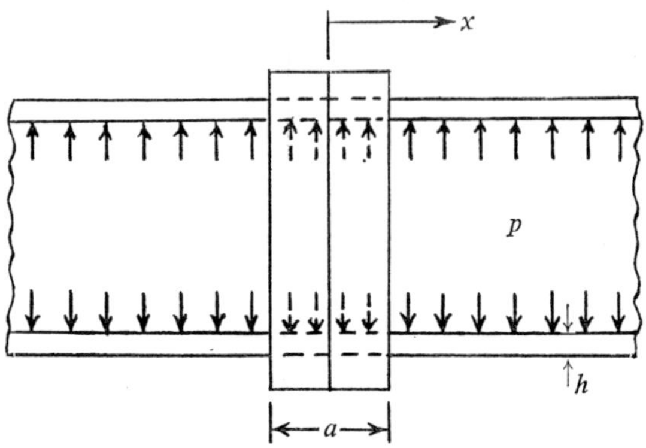

Fig. 4.26. Cylinder in a rigid collar

and the distribution of load between it and the collar under the assumption that there is no bond between the two so that no tangential shear forces or tensile forces are permitted. If the collar is short or the shell is stiff (we will determine how short or how stiff later on) we can assume that the cylinder will touch the collar only at the corners, so that each corner of the collar exerts only a concentrated ring loading on the cylinder (see Fig. 4.27). Thus the problem to be solved is that of finding the value q_0 of the ring loads such that the deflection of the cylinder under the ring loads is equal to the difference δ_0 between the radii of the rigid collar and the cylinder. We may use the solution of Section 4.3 for an infinite cylinder subjected to a ring loading at its

† This problem is discussed by P. Seide, 'Expansion of a Cylindrical Shell Restrained by a Rigid Ring Support,' *Proc. IASS Symp. Shell Structures and Climatic Influences*, Univ. of Calgary, Alberta, Canada, July 3–6, 1972, pp. 347–354, and by F. W. Barton and T. H. Dawson, 'Pressurized Tubes Constrained by Rigid Collars,' *ASCE J. Eng. Mech.*, vol. 99, no. EM3, June 1973, pp. 607–611.

‡ A treatment of the cylinder with a flexible collar is given by R. A. Eubanks, 'Shrink Fit of Arbitrary Length Sleeves on Thin Cylindrical Shells,' *Developments in Mechanics*, edited by S. Ostrach and R. H. Scanlan, vol. 2, pt. 2, Solid Mechanics, Pergamon Press, 1965, pp. 84–101.

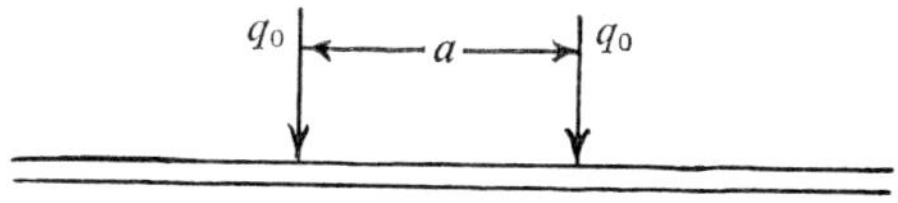

Fig. 4.27. Force distribution for short collars

center. By shifting the origin of coordinates to $x = a/2$ for the ring loading on the right and to $x = -a/2$ for the ring loading to the left in that solution, and superimposing the resulting equations for each of the loadings, we find that the deflections, slopes, moments, and shear forces in the cylinder may be written as

$$w = -\frac{q_0}{8\lambda^3 D}\left[C\left(\lambda\left|x+\frac{a}{2}\right|\right) + C\left(\lambda\left|x-\frac{a}{2}\right|\right)\right], \tag{4.6.1a}$$

$$\frac{dw}{dx} = \frac{q_0}{4\lambda^2 D}\left[B\left(\lambda\left|x+\frac{a}{2}\right|\right)\operatorname{sgn}\left(x+\frac{a}{2}\right) + B\left(\lambda\left|x-\frac{a}{2}\right|\right)\operatorname{sgn}\left(x-\frac{a}{2}\right)\right], \tag{4.6.1b}$$

$$M_x = \frac{q_0}{4\lambda}\left[D\left(\lambda\left|x+\frac{a}{2}\right|\right) + D\left(\lambda\left|x-\frac{a}{2}\right|\right)\right], \tag{4.6.1c}$$

$$Q_x = \frac{q_0}{2}\left[A\left(\lambda\left|x+\frac{a}{2}\right|\right)\operatorname{sgn}\left(x+\frac{a}{2}\right) + A\left(\lambda\left|x-\frac{a}{2}\right|\right)\operatorname{sgn}\left(x-\frac{a}{2}\right)\right]. \tag{4.6.1d}$$

Since the deflection under the ring loads must be equal to $-\delta_0$ we find q_0 to be given by

$$\frac{q_0}{8\lambda^3 D} = \frac{\delta_0}{1 + C(\lambda a)}. \tag{4.6.2}$$

Some deflection curves of the cylinder beneath the collar are shown in Fig. 4.28. For small values of λa the maximum negative deflection occurs at the center and increases with λa. However, when λa is increased beyond a value of 1.193 the numerical value of the center deflection starts to decrease and eventually becomes less than δ_0 (Fig. 4.29). Since the deflection beneath the collar must always be negative and numerically greater than δ_0, the solution given above is valid only until the center of the cylinder touches the rigid collar, or so long as λa is less than the value which satisfies the equation

$$w(0) = -\frac{2\delta_0 C(\lambda a/2)}{1 + c(\lambda a)} = -\delta_0. \tag{4.6.3a}$$

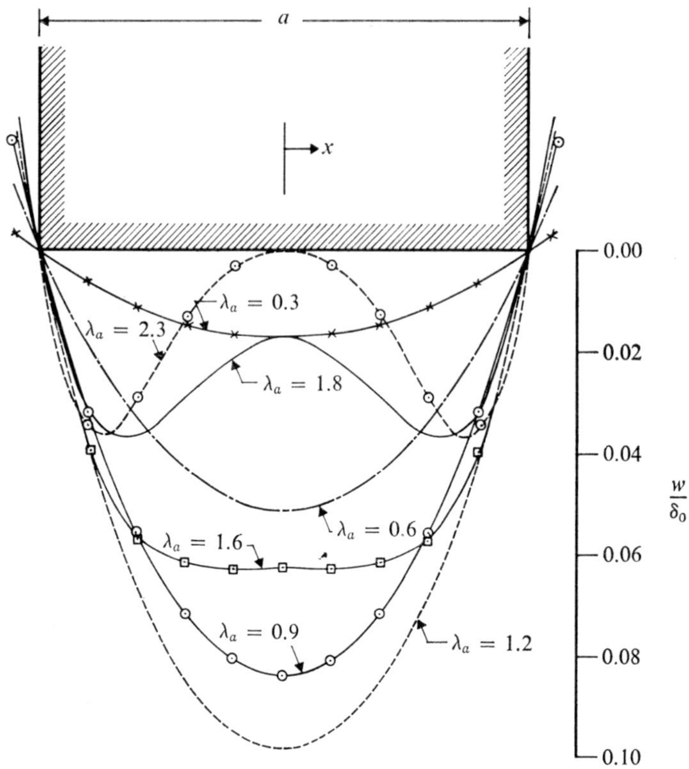

Fig. 4.28. Deflection curves for short collars

Trial and error solution of Eq. (4.6.3a) yields

$$\lambda a \approx 1.858, \tag{4.6.3b}$$

at which point $q_0/8\lambda^3 D$ is equal to $0.905\delta_0$.

When λa is equal to 1.858, the center of the cylinder just touches the rigid collar, with no pressure between the two. For larger values of λa, the collar and the cylinder are in contact at three stations and the cylinder is subjected to three concentrated ring loadings, as shown in Fig. 4.30. By superimposing the solution for a semi-infinite cylinder with a central ring load on the solution given by Eqs. (4.6.1) we obtain the solution for the new problem as

$$w = -\left\{\frac{q_0}{8\lambda^3 D}\left[C\left(\lambda\left|x + \frac{a}{2}\right|\right) + C\left(\lambda\left|x - \frac{a}{2}\right|\right)\right] + \frac{q_1}{8\lambda^3 D}C(\lambda\,|x|)\right\}, \tag{4.6.4a}$$

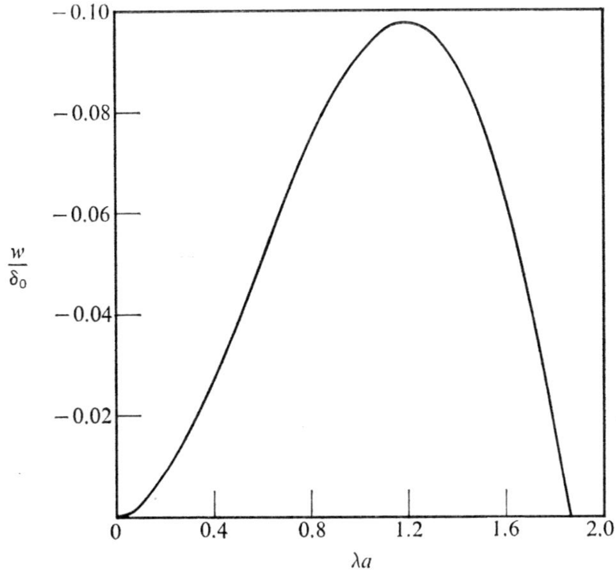

Fig. 4.29. Deflection between cylinder center and collar

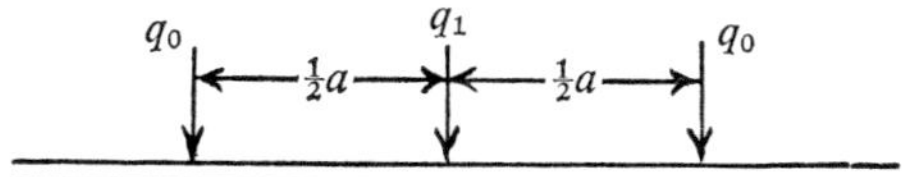

Fig. 4.30. Force distribution for medium-length collars

$$\frac{dw}{dx} = \frac{q_0}{4\lambda^2 D}\left[B\left(\lambda\left|x+\frac{a}{2}\right|\right)\operatorname{sgn}\left(x+\frac{a}{2}\right) + B\left(\lambda\left|x-\frac{a}{2}\right|\right)\operatorname{sgn}\left(x-\frac{a}{2}\right)\right]$$

$$+ \frac{q_1}{4\lambda^2 D} B(\lambda\,|x|)\operatorname{sgn} x, \tag{4.6.4b}$$

$$M_x = \frac{q_0}{4\lambda}\left[D\left(\lambda\left|x+\frac{a}{2}\right|\right) + D\left(\lambda\left|x-\frac{a}{2}\right|\right)\right] + \frac{q_1}{4\lambda} D(\lambda\,|x|), \tag{4.6.4c}$$

$$Q_x = \frac{q_0}{2}\left[A\left(\lambda\left|x+\frac{a}{2}\right|\right)\right]\operatorname{sgn}\left(x+\frac{a}{2}\right)\operatorname{sgn} A\left(\lambda\left|x-\frac{a}{2}\right|\right)\operatorname{sgn}\left(x-\frac{a}{2}\right)\Big]$$

$$+ \frac{q_1}{2} A(\lambda\,|x|)\operatorname{sgn} x. \tag{4.6.4d}$$

The values of q_0 and q_1 are now found by requiring the deflection of the cylinder to be equal to $-\delta_0$ at $x = \pm a/2$ and at $x = 0$. Then

$$\frac{q_0}{8\lambda^8 D} = \frac{1 - C(\lambda a/2)}{1 + C(\lambda a) - 2[C(\lambda a/2)]^2}\,\delta_0, \tag{4.6.5a}$$

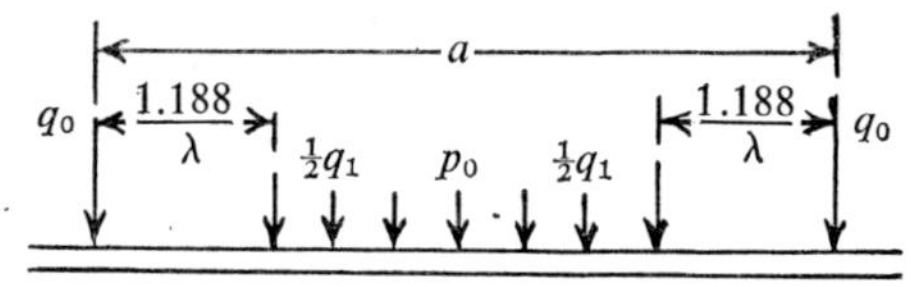

Fig. 4.31. Force distribution for long collars

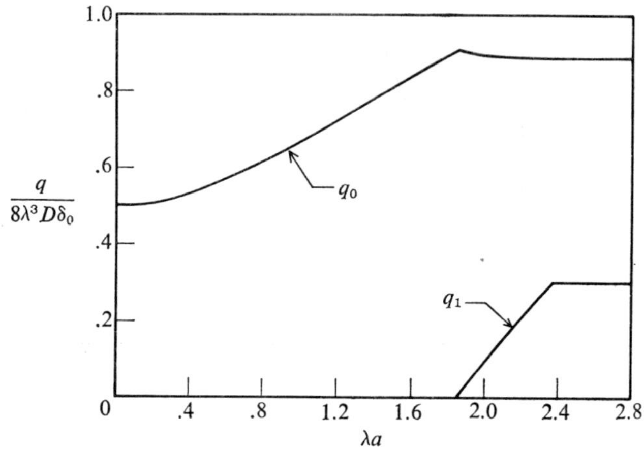

Fig. 4.32. Forces between cylinder and collar

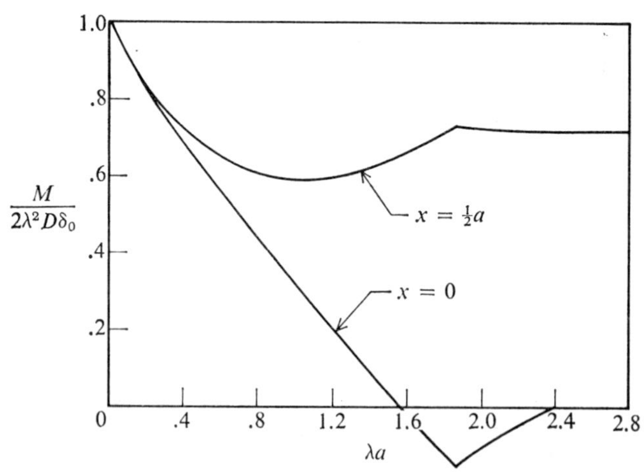

Fig. 4.33. Moments in cylinder

$$\frac{q_1}{8\lambda^8 D} = \frac{1 + C(\lambda a) - 2C(\lambda a/2)}{1 + C(\lambda a) - 2[C(\lambda a/2)]} \delta_0. \tag{4.6.5b}$$

This solution is valid again, so long as the deflections in the region $-a/2 < x < a/2$ are negative and numerically greater than δ_0. However, the condition is violated if λa surpasses a value such that the cylinder moment at the center of the collar vanishes. For, if the bending moment were positive, the deflection curve would bend toward the collar at the center and thus would be numerically less than δ_0 on either side of the center at which the deflection and the slope vanish. The value of λa at which the center moment becomes zero is given by Eqs. (4.6.4c) and (4.6.5) as the solution of the transcendental equation

$$D(\lambda a) + 4B(\lambda a/2) - 1 = 0. \tag{4.6.6a}$$

which has the trial and error solution

$$\lambda a \approx 2.376. \tag{4.6.6b}$$

For values of λa greater than 2.376, the portion of the cylinder defined by

$$-\frac{a}{2}\left(1 - \frac{2.376}{\lambda a}\right) \leq x \leq \frac{a}{2}\left(1 - \frac{2.376}{\lambda a}\right), \tag{4.6.7}$$

remains in contact with the rigid collar. Outside of this region the deflection curve, and hence the slopes, bending moments, and shear forces, is identical with that for a cylinder for which λa is equal to 2.376. The force distribution between the collar and the cylinder then consists of a uniform pressure

$$p_0 = Eh\delta_0/R^2 \tag{4.6.8}$$

in the region defined by Eq. (4.6.7), concentrated ring loads $\frac{1}{2}q_1$ at each edge of this region, and concentrated ring loads q_0 at the ends of the collar as shown in Fig. 4.31. The values of q_0 and q_1 are given by Eq. (4.6.5) with λa equal to 2.376 as

$$\frac{q_0}{8\lambda^3 D} = 0.884\delta_0, \tag{4.6.9a}$$

$$\frac{q_1}{8\lambda^3 D} = 0.299\delta_0. \tag{4.6.9b}$$

The variation of q_0 and q_1 as a function of the length of the collar is shown in Fig. 4.32. The variation of the moment at the ends and center of the collar is shown in Fig. 4.33. Stresses in the cylinder can be obtained with the use of the equations previously derived.

5

The conical shell

5.1 General relations for conical shells

The middle surface of a right circular conical shell is generated by revolving a straight line about an axis, the two intersecting at some angle α. The point of intersection of the generator and the axis is called the vertex of the cone. Since the radius of curvature of the generators of the shell is infinite, we locate a point on the middle surface by the distance s from the vertex measured along the straight-line generator passing through the point and the circumferential angle θ which locates the generator with respect to a reference generator (see Fig. 5.1). With these lines of curvature coordinates, the first and second fundamental quantities are

$$g_s = 1, \tag{5.1.1a}$$

$$g_\theta = s \sin \alpha, \tag{5.1.1b}$$

$$R_s = \infty, \tag{5.1.1c}$$

$$R_\theta = s \tan \alpha. \tag{5.1.1d}$$

For axisymmetric deformations we have

$$u_\theta = 0, \tag{5.1.2a}$$

$$u_s = u_s(s), \tag{5.1.2b}$$

$$w = w(s), \tag{5.1.2c}$$

which lead to the following strain components and curvature changes from Eqs. (1.5.3), (1.5.6b), and (1.5.10):

$$\epsilon_s = \frac{\mathrm{d}u_s}{\mathrm{d}s}, \tag{5.1.3a}$$

$$\epsilon_\theta = \frac{1}{s}(u_s + w \cot \alpha), \tag{5.1.3b}$$

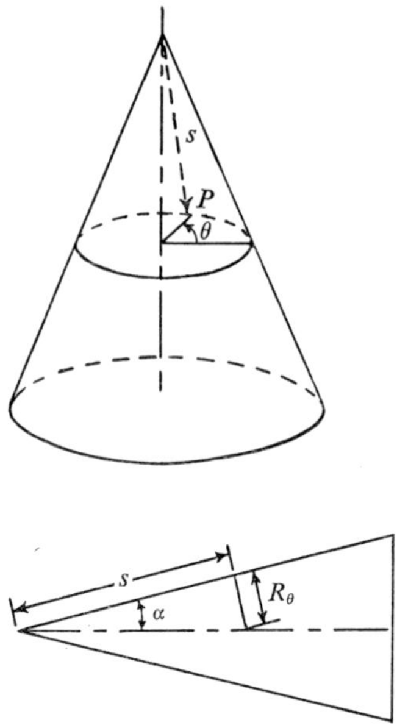

Fig. 5.1. Notation for conical shell

$$\kappa_s = -\frac{d^2w}{ds^2}, \tag{5.1.3c}$$

$$\kappa_\theta = -\frac{1}{s}\frac{dw}{ds}, \tag{5.1.3d}$$

$$\gamma_{s\theta} = \bar{\kappa}_{s\theta} = 0, \tag{5.1.3e}$$

we also have

$$\bar{Q}_\theta = \bar{N}_{s\theta} = \bar{M}_{s\theta} = q_\theta = m_\theta = 0, \tag{5.1.4a}$$

with the remaining stress resultants and couples independent of θ, as well as

$$\bar{Q}_s = Q_s. \tag{5.1.4b}$$

The equations of equilibrium of the conical shell element then become (Fig. 5.2)

$$\frac{d}{ds}(sN_s) - N_\theta + sq_s = 0, \tag{5.1.5a}$$

$$\frac{d}{ds}(sQ_s) - N_\theta \cot \alpha + sq_\zeta = 0, \tag{5.1.5b}$$

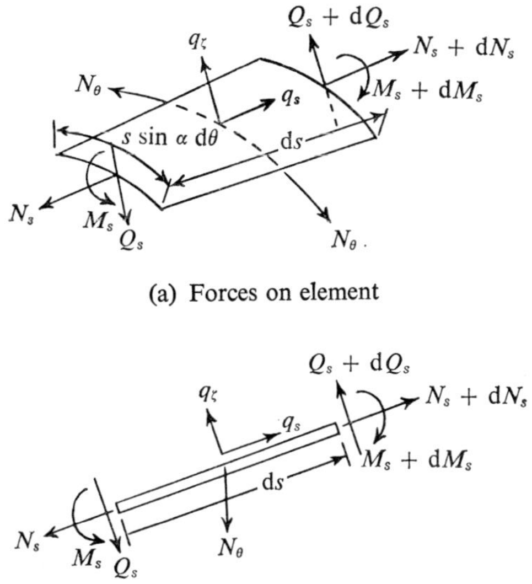

(a) Forces on element

(b) Projected view

Fig. 5.2. Equilibrium of conical shell element

$$\frac{\mathrm{d}}{\mathrm{d}s}(sM_s) - M_\theta - s(Q_s - m_s) = 0, \tag{5.1.5c}$$

with

$$q_s = \left(1 + \frac{h}{2R_\theta}\right)\tau_{s\zeta}^{(h/2)} - \left(- 1 \frac{h}{2R_\theta}\right)\tau_{s\zeta}^{(-h/2)} + \int_{-h/2}^{h/2} \left(1 + \frac{\zeta}{R_\theta}\right)\rho_s^{(\zeta)} \, \mathrm{d}\zeta, \tag{5.1.6a}$$

$$q_\zeta = \left(1 + \frac{h}{2R_\theta}\right)\sigma_\zeta^{(h/2)} - \left(1 - \frac{h}{2R_\theta}\right)\sigma_\zeta^{(-h/2)} + \int_{-h/2}^{h/2} \left(1 + \frac{\zeta}{R_\theta}\right)\rho_\zeta^{(\zeta)} \, \mathrm{d}\zeta, \tag{5.1.6b}$$

$$m_s = \frac{h}{2}\left[\left(1 + \frac{h}{2R_\theta}\right)\tau_{s\zeta}^{(h/2)} + \left(1 - \frac{h}{2R_\theta}\right)\tau_{s\zeta}^{(-h/2)}\right] + \int_{-h/2}^{h/2} \zeta\left(1 + \frac{\zeta}{R_\theta}\right)\rho_s^{(\zeta)} \, \mathrm{d}\zeta. \tag{5.1.6c}$$

The assumed overall force–strain relations [Eqs. (3.1.4)] are given by

$$N_s = K\left[\frac{\mathrm{d}u_s}{\mathrm{d}s} + \frac{\nu}{s}(u_s + w \cot \alpha)\right], \tag{5.1.7a}$$

$$N_\theta = K\left[\frac{1}{s}(u_s + w\cot\alpha) + \nu\frac{\mathrm{d}u_s}{\mathrm{d}s}\right], \tag{5.1.7b}$$

$$M_s = -D\left(\frac{\mathrm{d}^2w}{\mathrm{d}s^2} + \frac{\nu}{s}\frac{\mathrm{d}w}{\mathrm{d}s}\right), \tag{5.1.7c}$$

$$M_\theta = -D\left(\frac{1}{s}\frac{\mathrm{d}w}{\mathrm{d}s} + \nu\frac{\mathrm{d}^2w}{\mathrm{d}s^2}\right). \tag{5.1.7d}$$

As is the case for the cylinder, the theory of axisymmetric bending of conical shells can also be expressed entirely in terms of the normal displacement w. As a start, we note that the bending moments are given only in terms of w. On substituting Eq. (5.1.7c) and (5.1.7d) into Eq. (5.1.5c) we obtain an expression for the transverse shear force Q_s in terms of w as

$$Q_s = m_s - D\frac{\mathrm{d}}{\mathrm{d}s}\left[\frac{1}{s}\frac{\mathrm{d}}{\mathrm{d}s}\left(s\frac{\mathrm{d}w}{\mathrm{d}s}\right)\right]. \tag{5.1.8}$$

We may also express the force components N_s and N_θ in terms of w. We eliminate N_θ from Eqs. (5.1.5a) and (5.1.5b) to obtain the equation of equilibrium of forces parallel to the axis of revolution:

$$\frac{\mathrm{d}}{\mathrm{d}s}[s(N_s\cos\alpha - Q_s\sin\alpha)] + s(q_s\cos\alpha - q_\zeta\sin\alpha) = 0. \tag{5.1.9a}$$

This may be integrated to yield

$$\begin{aligned} N_s &= Q_s\tan\alpha - \frac{1}{s}\int_{s_1}^{s} s(q_s - q_\zeta\tan\alpha)\,\mathrm{d}s + \frac{P}{2\pi s\sin\alpha\cos\alpha} \\ &= m_s\tan\alpha + \frac{P}{2\pi s\sin\alpha\cos\alpha} \\ &\quad - \frac{1}{s}\int_{s_1}^{s} s(q_s - q_\zeta\tan\alpha)\,\mathrm{d}s - D\frac{\mathrm{d}}{\mathrm{d}s}\left[\frac{1}{s}\frac{\mathrm{d}}{\mathrm{d}s}\left(s\frac{\mathrm{d}w}{\mathrm{d}s}\right)\right]\tan\alpha, \end{aligned} \tag{5.1.9b}$$

where P is the total load in the direction of the axis of revolution of the cone at $s = s_1$ (see Fig. 5.3). From Eq. (5.1.5b) we have

$$\begin{aligned} N_\theta &= \left[\frac{\mathrm{d}}{\mathrm{d}s}(sQ_s) + sq_\zeta\right]\tan\alpha \\ &= \left\langle -D\frac{\mathrm{d}}{\mathrm{d}s}\left\{s\frac{\mathrm{d}}{\mathrm{d}s}\left[\frac{1}{s}\frac{\mathrm{d}}{\mathrm{d}s}\left(s\frac{\mathrm{d}w}{\mathrm{d}s}\right)\right]\right\} + \frac{\mathrm{d}}{\mathrm{d}s}(sm_s) + sq_\zeta\right\rangle\tan\alpha \end{aligned} \tag{5.1.10}$$

The equation for the normal displacement w is found by utilizing the fact that the meridional displacement u_s can be determined from Eqs.

(5.1.3a), (5.1.3b), (5.1.7a), and (5.1.7b) as the solution of either of the following equations

$$\frac{\mathrm{d}u_s}{\mathrm{d}s} = \frac{N_s - \nu N_\theta}{Eh}, \tag{5.1.11a}$$

$$u_s = s\left(\frac{N_\theta - \nu N_s}{Eh}\right) - w \cot \alpha. \tag{5.1.11b}$$

In order for the two solutions to be compatible, we must have

$$N_s - \nu N_\theta = \frac{\mathrm{d}}{\mathrm{d}s}\left[s(N_\theta - \nu N_s) - Ehw \cot \alpha\right], \tag{5.1.12a}$$

or, when Eqs. (5.1.9b) and (5.1.10) are used,

$$\left[\frac{\mathrm{d}^4}{\mathrm{d}s^4} + \frac{12(1 - \nu^2)\cot \alpha}{s^2 h^2}\right]\left(s\frac{\mathrm{d}w}{\mathrm{d}s}\right) = \frac{q(s)}{D}, \tag{5.1.12b}$$

where

$$q(s) = \frac{1}{s^2}\left[\frac{\mathrm{d}}{\mathrm{d}s}\left(s^3 \frac{\mathrm{d}m_s}{\mathrm{d}s}\right) + \int_{s_1}^{s} s(q_s \cot \alpha - q_\zeta)\,\mathrm{d}s - \frac{P}{2\pi \sin^2 \alpha}\right]$$
$$+ \frac{1}{s}\frac{\mathrm{d}}{\mathrm{d}s}(s^2 q_\zeta) + \nu q_s \cot \alpha. \tag{5.1.12c}$$

An expression for u_s in terms of w can be obtained from Eqs. (5.1.9b), (5.1.10), and (5.1.11b) as

$$u_s = \frac{s^2 \tan \alpha}{Eh}\left(\frac{\mathrm{d}}{\mathrm{d}s} + \frac{1 - \nu}{s}\right)\left\{m_s - D\frac{\mathrm{d}}{\mathrm{d}s}\left[\frac{1}{s}\frac{\mathrm{d}}{\mathrm{d}s}\left(s\frac{\mathrm{d}w}{\mathrm{d}s}\right)\right] + \frac{1}{s}\int_{s_1}^{s} sq_\zeta\,\mathrm{d}s\right\}$$
$$- \frac{\nu P}{2\pi Eh \sin \alpha \cos \alpha} - w \cot \alpha + \frac{\nu}{Eh}\int_{s_1}^{s} sq_s\,\mathrm{d}s. \tag{5.1.13a}$$

Another expression which is similar in form to that for the cylindrical shell is obtained by using Eq. (5.1.11a) rather than (5.1.11b):

$$u_s = \frac{\tan \alpha}{Eh}\left[-\nu s Q_s + \frac{M_s + M_\theta}{1 + \nu} + \int_{s_1}^{s}(m_s - \nu s q_\zeta)\,\mathrm{d}s\right.$$
$$\left.- \int_{s_1}^{s}\frac{1}{s}\int_{s_1}^{s} s(q_s \cot \alpha - q_\zeta)\,\mathrm{d}s\,\mathrm{d}s + \frac{P}{2\pi \sin^2 \alpha}\ln\frac{s}{s_1}\right] + \text{constant}.$$

5.2 Solution of the homogeneous deflection equation

The solution of the homogeneous equation for w may be obtained by differentiating Eq. (5.1.12b) once with respect to s to obtain

$$s\frac{\mathrm{d}^2}{\mathrm{d}s^2}\left(s\frac{\mathrm{d}^2 f}{\mathrm{d}s^2}\right) + \frac{12(1 - \nu^2)\cot \alpha}{h^2} f = 0, \tag{5.2.1a}$$

$$f = \frac{d}{ds}\left(s\frac{dw}{ds}\right). \tag{5.2.1b}$$

Eq. (5.2.1a) may also be written in the factored form

$$\left[s\frac{d^2}{ds^2} - i\frac{[12(1-\nu^2)]\cot\alpha}{h}\right]\left[s\frac{d^2}{ds^2} + i\frac{[12(1-\nu^2)]\cot\alpha}{h}\right]f = 0.$$

Since the two factors of Eq. (5.2.1c) are commutative operators, the solution of the equation is the sum of the solutions of the independent equations

$$\left[s\frac{d^2}{ds^2} - i\frac{[12(1-\nu^2)]^{1/2}\cot\alpha}{h}\right]f_1 = 0, \tag{5.2.2a}$$

$$\left[s\frac{d^2}{ds^2} + i\frac{[12(1-\nu^2)]^{1/2}\cot\alpha}{h}\right]f_2 = 0. \tag{5.2.2b}$$

Furthermore, we need consider only one of these equations to obtain the complete solution. Since Eq. (5.2.2a) and Eq. (5.2.2b) are complex conjugate equations, their solutions are complex conjugates. Thus the real and imaginary parts of the two independent solutions of either of the equations are solutions of Eq. (5.2.1a).

In Eq. (5.2.2a) then let us make the substitutions

$$u^2 = 4i[12(1-\nu^2)]^{1/2}s\cot\alpha/h, \tag{5.2.3a}$$

$$f_1 = u\frac{dg}{du}. \tag{5.2.3b}$$

Then Eq. (5.2.2a) becomes

$$\frac{d}{du}\left(\frac{d^2g}{du^2} + \frac{1}{u}\frac{dg}{du} - g\right) = 0, \tag{5.2.4a}$$

which can be integrated once to yield

$$\frac{d^2g}{du^2} + \frac{1}{u}\frac{dg}{du} - g = 0. \qquad (5.2.4b)†$$

But this is Bessel's differential equation for the Bessel functions of zeroth order and argument iu. In terms of the argument u, the solution is usually written as

$$g = AI_0(u) + BK_0(u), \qquad (5.2.4c)‡$$

† The constant of integration which could be introduced is immaterial and can be set equal to zero.

‡ See, for example, H. B. Dwight, *Tables of Integrals and Other Mathematical Data*, Third Edition, The Macmillan Co., 1957, pp. 174–191.

where $I_0(u)$ and $K_0(u)$ are zeroth-order modified Bessel functions of the first and second kinds. Since u is itself complex, $I_0(u)$ and $K_0(u)$ are also complex and can be expressed as

$$I_0(u) = I_0(i^{1/2}\xi) = \text{ber}\,\xi + i\,\text{bei}\,\xi, \tag{5.2.5a}$$

$$K_0(u) = K_0(i^{1/2}\xi) = \text{ker}\,\xi + i\,\text{kei}\,\xi, \tag{5.2.5b}$$

where

$$\xi = 2[12(1-\nu^2)]^{1/4}\left(\frac{s\cot\alpha}{h}\right)^{1/2} = 2[12(1-\nu^2)]^{1/4}\left(\frac{R_\theta}{h}\right)^{1/2}\cot\alpha. \tag{5.2.5c}$$

The relationship between ξ, R_θ/h, and α is indicated graphically in Fig. 5.3.

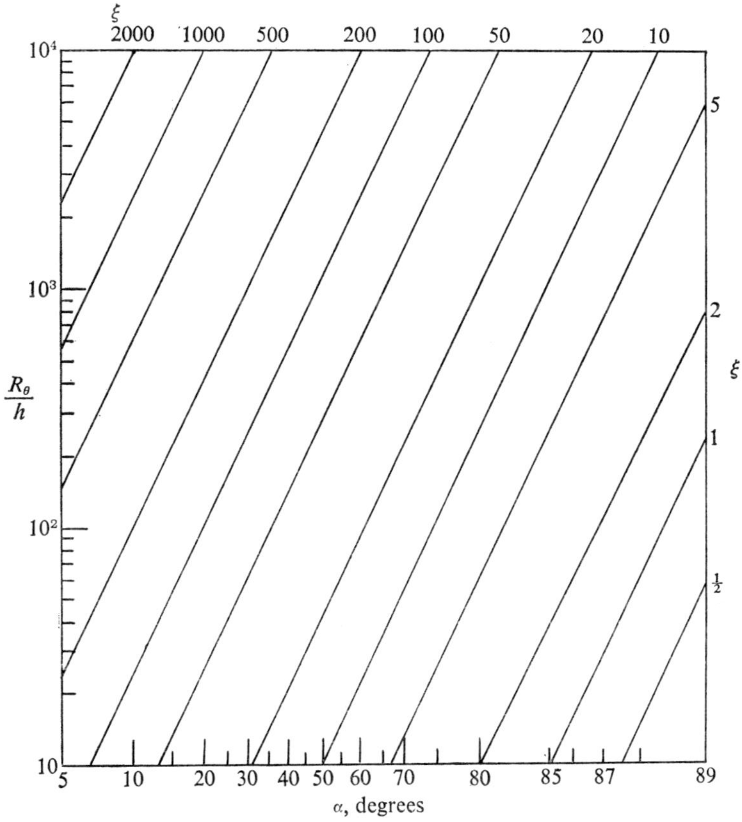

Fig. 5.3. Dependence of ξ *on* R_θ/h *and* α ($\nu = 0.3$)

From Eq. (5.2.3b) we then have

$$f_1 = u\frac{\mathrm{d}g}{\mathrm{d}u} = \xi\frac{\mathrm{d}g}{\mathrm{d}\xi}$$
$$= \xi[A(\mathrm{ber}'\,\xi + i\,\mathrm{bei}'\,\xi) + B(\mathrm{ker}'\,\xi + i\,\mathrm{kei}'\,\xi)], \tag{5.2.6}$$

where the prime indicates differentiation with respect to ξ. Thus the general solution in real form of Eq. (5.2.1a) can be written as

$$f = \frac{\mathrm{d}}{\mathrm{d}s}\left(s\frac{\mathrm{d}w}{\mathrm{d}s}\right) = \frac{[12(1-\nu^2)]^{1/2}\cot\alpha}{h}\frac{1}{\xi}\frac{\mathrm{d}}{\mathrm{d}\xi}\left(\xi\frac{\mathrm{d}w}{\mathrm{d}\xi}\right)$$
$$= \frac{[12(1-\nu^2)]^{1/2}\cot\alpha}{h}\,\xi$$
$$\times\ [A_1\,\mathrm{ber}'\,\xi - A_2\,\mathrm{bei}'\,\xi + A_3\,\mathrm{ker}'\,\xi - A_4\,\mathrm{kei}'\,\xi]. \tag{5.2.7}$$

On performing the required integrations, with the use of the relations

$$\int \xi\,\mathrm{ber}\,\xi\,\mathrm{d}\xi = \xi\,\mathrm{bei}'\,\xi, \tag{5.2.8a}$$

$$\int \xi\,\mathrm{bei}\,\xi\,\mathrm{d}\xi = -\xi\,\mathrm{ber}'\,\xi, \tag{5.2.8b}$$

and with similar relations for the ker and kei functions, we have

$$w = A_1(\xi\,\mathrm{bei}'\,\xi - 2\,\mathrm{bei}\,\xi) + A_2(\xi\,\mathrm{ber}'\,\xi - 2\,\mathrm{ber}\,\xi)$$
$$+ A_3(\xi\,\mathrm{kei}'\,\xi - 2\,\mathrm{kei}\,\xi) + A_4(\xi\,\mathrm{ker}'\,\xi - 2\,\mathrm{ker}\,\xi)$$
$$- A_5\sin\alpha, \tag{5.2.9}$$

where A_5 is a rigid body displacement in the direction of the axis of revolution. We should note that dependence of w on only the parameter ξ is a result of the use of Eqs. (5.1.7) which implies the neglect of terms of the order of h/R_θ. This assumption is violated in the vicinity of the apex of the cone.

With the normal displacement w known in the form of Eq. (5.2.9), we may obtain expressions for the stress resultants, couples, and the meridional displacement as

$$N_s = Q_s\tan\alpha$$
$$= \frac{8[12(1-\nu^2)]^{1/2}E\cot^2\alpha}{\xi^2}\left[A_1\left(\mathrm{bei}\,\xi + \frac{2}{\xi}\mathrm{ber}'\,\xi\right) + A_2\left(\mathrm{ber}\,\xi\right.\right.$$
$$\left.\left. - \frac{2}{\xi}\mathrm{bei}'\,\xi\right) + A_3\left(\mathrm{kei}\,\xi + \frac{2}{\xi}\mathrm{ker}'\,\xi\right) + A_4\left(\mathrm{ker}\,\xi - \frac{2}{\xi}\mathrm{kei}'\,\xi\right)\right] \tag{5.2.10a}$$

$$N_\theta = -N_s + \frac{4[12(1-\nu^2)]^{1/2} E \cot^2 \alpha}{\xi}$$
$$\times [A_1 \operatorname{bei}' \xi + A_2 \operatorname{ber}' \xi + A_3 \operatorname{kei}' \xi + A_4 \operatorname{ker}' \xi], \quad (5.2.10b)$$

$$M_s = -\frac{4Eh \cot^2 \alpha}{\xi} \left\{ A_1 \left[\operatorname{ber}' \xi - \frac{2(1-\nu)}{\xi} \left(\operatorname{ber} \xi - \frac{2}{\xi} \operatorname{bei}' \xi \right) \right] \right.$$
$$- A_2 \left[\operatorname{bei}' \xi - \frac{2(1-\nu)}{\xi} \left(\operatorname{bei} \xi + \frac{2}{\xi} \operatorname{ber}' \xi \right) \right]$$
$$+ A_3 \left[\operatorname{ker}' \xi - \frac{2(1-\nu)}{\xi} \left(\operatorname{ker} \xi - \frac{2 \operatorname{kei}' \xi}{\xi} \right) \right]$$
$$\left. - A_4 \left[\operatorname{kei}' \xi - \frac{2(1-\nu)}{\xi} \left(\operatorname{kei} \xi + \frac{2}{\xi} \operatorname{ker}' \xi \right) \right] \right\} \quad (5.2.10c)$$

$$M_\theta = -\frac{4Eh \cot^2 \alpha}{\xi} \left\{ A_1 \left[\nu \operatorname{ber}' \xi + \frac{2(1-\nu)}{\xi} \left(\operatorname{ber} \xi - \frac{2}{\xi} \operatorname{bei}' \xi \right) \right] \right.$$
$$- A_2 \left[\nu \operatorname{bei}' \xi + \frac{2(1-\nu)}{\xi} \left(\operatorname{bei} \xi + \frac{2}{\xi} \operatorname{ber}' \xi \right) \right]$$
$$+ A_3 \left[\nu \operatorname{ker}' \xi + \frac{2(1-\nu)}{\xi} \left(\operatorname{ker} \xi - \frac{2}{\xi} \operatorname{kei}' \xi \right) \right]$$
$$\left. - A_4 \left[\nu \operatorname{kei}' \xi + \frac{2(1-\nu)}{\xi} \left(\operatorname{kei} \xi + \frac{2}{\xi} \operatorname{ker}' \xi \right) \right] \right\} \quad (5.2.10d)$$

$$u_s = -2 \left\{ A_1 \left[\nu \operatorname{bei} \xi + \frac{2(1+\nu)}{\xi} \operatorname{ber}' \xi \right] \right.$$
$$+ A_2 \left[\nu \operatorname{ber} \xi - \frac{2(1+\nu)}{\xi} \operatorname{bei}' \xi \right]$$
$$+ A_3 \left[\nu \operatorname{kei} \xi + \frac{2(1+\nu)}{\xi} \operatorname{ker}' \xi \right]$$
$$\left. + A_4 \left[\nu \operatorname{ker} \xi - \frac{2(1+\nu)}{\xi} \operatorname{kei}' \xi \right] \right\} \cot \alpha + A_5 \cos \alpha, \quad (5.2.10e)$$

where the relationships

$$\operatorname{ber}'' \xi = -\left(\operatorname{bei} \xi + \frac{1}{\xi} \operatorname{ber}' \xi \right), \quad (5.2.11a)$$

$$\operatorname{bei}'' \xi = \operatorname{ber} \xi - \frac{1}{\xi} \operatorname{bei}' \xi, \quad (5.2.11b)$$

and similar relations for the ker and kei functions have been used. The angle of rotation of the meridian is given by

$$\frac{dw}{ds} = \frac{2[12(1 - \nu^2)]^{1/2} \cot \alpha}{h} \left[A_1 \left(\operatorname{ber} \xi - \frac{2}{\xi} \operatorname{bei}' \xi \right) - A_2 \left(\operatorname{bei} \xi + \frac{2}{\xi} \operatorname{ber}' \xi \right) \right.$$
$$\left. + A_3 \left(\operatorname{ker} \xi - \frac{2}{\xi} \operatorname{kei}' \xi \right) - A_4 \left(\operatorname{kei} \xi + \frac{2}{\xi} \operatorname{ker}' \xi \right) \right]. \qquad (5.2.12)$$

In most problems we are interested in force components and displacements parallel and perpendicular to the axis of revolution. These are given by

$$V = N_s \cos \alpha - Q_s \sin \alpha = 0, \qquad (5.2.13a)$$

$$H = N_s \sin \alpha + Q_s \cos \alpha = \frac{N_s}{\sin \alpha}, \qquad (5.2.13b)$$

$$\frac{u_H}{\cos \alpha} = u_s \tan \alpha + w$$
$$= A_1 \left[\xi \operatorname{bei}' \xi - 2(1 + \nu) \left(\operatorname{bei} \xi + \frac{2}{\xi} \operatorname{ber}' \xi \right) \right]$$
$$+ A_2 \left[\xi \operatorname{ber}' \xi - 2(1 + \nu) \left(\operatorname{ber} \xi - \frac{2}{\xi} \operatorname{bei}' \xi \right) \right]$$
$$+ A_3 \left[\xi \operatorname{kei}' \xi - 2(1 + \nu) \left(\operatorname{kei} \xi + \frac{2}{\xi} \operatorname{ker}' \xi \right) \right]$$
$$+ A_4 \left[\xi \operatorname{ker}' \xi - 2(1 + \nu) \left(\operatorname{ker} \xi - \frac{2}{\xi} \operatorname{kei}' \xi \right) \right], \qquad (5.2.13c)$$

$$u_V = u_s \cos \alpha - w \sin \alpha = u_H \cot \alpha - w \csc \alpha. \qquad (5.2.13d)$$

The ber, bei, ker, and kei functions and their derivatives which appear in the previous equations are qualitatively similar to the more familiar functions of cylindrical shell theory. The ber and bei functions and their derivatives are oscillatory functions with exponentially increasing amplitudes while the ker and kei functions and their derivatives are oscillatory functions with exponentially decreasing amplitudes. A short table of these functions covering the range $0 \leq \xi \leq 10$ in steps of 0.1 is given as Table 5.1. More extensive tables are also available.†

† H. H. Lowell, "Tables of the Bessel–Kelvin Functions Ber, Bei, Ker, Kei, and Their Derivatives for the Argument Range 0(0.01)107.50," NASA TR R32, 1959.

Table 5.1 Values of the Bessel–Kelvin functions and their derivatives

x	ber x	bei x	ber′ x	bei′ x
0	+1	0	0	0
0.1	+0.999 998	+0.002 500	−0.000 063	+0.050 000
0.2	+0.999 975	+0.010 000	−0.000 500	+0.099 999
0.3	+0.999 873	+0.022 500	−0.001 687	+0.149 994
0.4	+0.999 600	+0.039 998	−0.004 000	+0.199 973
0.5	+0.999 023	+0.062 493	−0.007 812	+0.249 919
0.6	+0.997 975	+0.089 980	−0.013 498	+0.299 798
0.7	+0.996 249	+0.122 449	−0.021 433	+0.349 562
0.8	+0.993 601	+0.159 886	−0.031 989	+0.399 147
0.9	+0.989 751	+0.202 269	−0.045 537	+0.448 463
1.0	+0.984 382	+0.249 566	−0.062 446	+0.497 397
1.1	+0.977 138	+0.301 731	−0.083 082	+0.545 808
1.2	+0.967 629	+0.358 704	−0.107 806	+0.593 523
1.3	+0.955 429	+0.420 406	−0.136 972	+0.640 338
1.4	+0.940 075	+0.486 734	−0.170 928	+0.686 008
1.5	+0.921 072	+0.557 560	−0.210 011	+0.730 251
1.6	+0.897 891	+0.632 726	−0.254 545	+0.772 740
1.7	+0.869 971	+0.712 037	−0.304 838	+0.813 105
1.8	+0.836 722	+0.795 262	−0.361 182	+0.850 927
1.9	+0.797 524	+0.882 122	−0.423 845	+0.885 737
2.0	+0.751 734	+0.972 292	−0.493 067	+0.917 014
2.1	+0.698 685	+1.065 388	−0.569 061	+0.944 181
2.2	+0.637 690	+1.160 970	−0.652 000	+0.966 609
2.3	+0.568 049	+1.258 529	−0.742 019	+0.983 607
2.4	+0.489 048	+1.357 485	−0.839 203	+0.994 429
2.5	+0.399 968	+1.457 182	−0.943 583	+0.998 269
2.6	+0.300 092	+1.556 878	−1.055 132	+0.994 263
2.7	+0.188 706	+1.655 742	−1.173 750	+0.981 488
2.8	+0.065 112	+1.752 851	−1.299 264	+0.958 965
2.9	−0.071 368	+1.847 176	−1.431 414	+0.925 659
3.0	−0.221 380	+1.937 587	−1.569 847	+0.880 482
3.1	−0.385 531	+2.022 839	−1.714 104	+0.822 298
3.2	−0.564 376	+2.101 573	−1.863 617	+0.749 924
3.3	−0.758 407	+2.172 310	−2.017 690	+0.662 139
3.4	−0.968 039	+2.233 446	−2.175 495	+0.557 699
3.5	−1.193 598	+2.283 250	−2.336 059	+0.435 296
3.6	−1.435 305	+2.319 864	−2.498 253	+0.293 662
3.7	−1.693 260	+2.341 298	−2.660 779	+0.131 487
3.8	−1.967 423	+2.345 433	−2.822 164	−0.052 527
3.9	−2.257 599	+2.330 022	−2.980 743	−0.259 654
4.0	−2.563 417	+2.292 690	−3.134 654	−0.491 137

Table 5.1—*contd.*

x	ber x	bei x	ber′ x	bei′ x
4.1	−2.884 306	+2.230 943	−3.281 821	−0.748 167
4.2	−3.219 480	+2.142 168	−3.419 951	−1.031 862
4.3	−3.567 911	+2.023 647	−3.546 520	−1.343 252
4.4	−3.928 307	+1.872 564	−3.658 765	−1.683 251
4.5	−4.299 087	+1.686 017	−3.753 681	−2.052 635
4.6	−4.678 357	+1.461 037	−3.828 010	−2.452 013
4.7	−5.063 886	+1.194 601	−3.878 240	−2.881 799
4.8	−5.453 076	+0.883 657	−3.900 599	−3.342 181
4.9	−5.842 942	+0.525 147	−3.891 061	−3.833 085
5.0	−6.230 082	+0.116 034	−3.845 339	−4.354 141
5.1	−6.610 653	−0.346 663	−3.758 901	−4.904 641
5.2	−6.980 346	−0.865 840	−3.626 967	−5.483 505
5.3	−7.334 363	−1.444 260	−3.444 527	−6.089 232
5.4	−7.667 394	−2.084 517	−3.206 356	−6.719 859
5.5	−7.973 596	−2.788 980	−2.907 032	−7.372 913
5.6	−8.246 576	−3.559 747	−2.540 959	−8.045 365
5.7	−8.479 373	−4.398 579	−2.102 401	−8.733 576
5.8	−8.664 445	−5.306 845	−1.585 513	−9.433 252
5.9	−8.793 667	−6.285 446	−0.984 382	−10.139 389
6.0	−8.858 316	−7.334 747	−0.293 080	−10.846 224
6.1	−8.849 080	−8.454 495	+0.494 289	−11.547 179
6.2	−8.756 062	−9.643 739	+1.383 522	−12.234 815
6.3	−8.568 793	−10.900 737	−2.380 248	−12.900 779
6.4	−8.276 250	−12.222 863	+3.489 851	−13.535 755
6.5	−7.866 891	−13.606 512	+4.717 382	−14.129 423
6.6	−7.328 688	−15.046 993	+6.067 462	−14.670 413
6.7	−6.649 176	−16.538 425	+7.544 180	−15.146 266
6.8	−5.815 515	−18.073 624	+9.150 973	−15.543 406
6.9	−4.814 556	−19.643 992	+10.890 504	−15.847 109
7.0	−3.632 930	−21.239 403	+12.764 523	−16.041 489
7.1	−2.257 144	−22.848 079	+14.773 723	−16.109 484
7.2	−0.673 695	−24.456 480	+16.917 585	−16.032 856
7.3	+1.130 800	−26.049 184	+19.194 204	−15.792 207
7.4	+3.169 457	−27.608 771	+21.600 121	−15.367 001
7.5	+5.454 962	−29.115 712	+24.130 125	−14.735 602
7.6	+7.999 382	−30.548 263	+26.777 064	−13.875 334
7.7	+10.813 965	−31.882 362	+29.531 637	−12.762 551
7.8	+13.908 912	−33.091 540	+32.382 176	−11.372 739
7.9	+17.293 128	−34.146 834	+35.314 428	−9.680 623
8.0	+20.973 956	−35.016 725	+38.311 326	−7.660 318

Table 5.1—*contd.*

x	ber x	bei x	ber′ x	bei x'
8.1	+24.956 881	−35.667 081	+41.352 754	−5.285 490
8.2	+29.245 215	−36.061 120	+44.415 316	−2.529 555
8.3	+33.839 755	−36.159 401	+47.472 095	+0.634 098
8.4	+38.738 423	−35.919 830	+50.492 416	+4.231 841
8.5	+43.935 873	−35.297 700	+53.441 618	+8.289 519
8.6	+49.423 085	−34.245 761	+56.280 822	+12.832 116
8.7	+55.186 932	−32.714 319	+58.966 717	+17.883 387
8.8	+61.209 725	−30.651 388	+61.451 355	+23.465 444
8.9	+67.468 741	−28.002 868	+63.681 961	+29.598 302
9.0	+73.935 730	−24.712 783	+65.600 771	+36.299 384
9.1	+80.576 411	−20.723 570	+67.144 889	+43.582 976
9.2	+87.349 953	−15.976 414	+68.246 178	+51.459 634
9.3	+94.208 443	−10.411 662	+68.831 185	+59.935 547
9.4	+101.096 360	−3.969 285	+68.821 114	+69.011 850
9.5	+107.950 032	+3.410 573	+68.131 840	+78.683 888
9.6	+114.697 114	+11.786 984	+66.673 989	+88.940 434
9.7	+121.256 066	+21.217 532	+64.353 071	+99.762 855
9.8	+127.535 652	+31.757 531	+61.069 692	+111.124 240
9.9	+133.434 460	+43.459 153	+56.719 839	+122.988 479
10.0	+138.840 466	+56.370 459	+51.195 258	+135.309 302

x	ker x	kei x	ker′ x	kei′ x
0	+ ∞	−0.785 398	− ∞	0
0.1	+2.420 474	−0.776 851	−9.960 959	+0.145 975
0.2	+1.733 143	−0.758 125	−4.922 949	+0.222 927
0.3	+1.337 219	−0.733 102	−3.219 865	+0.274 292
0.4	+1.062 624	−0.703 800	−2.352 070	+0.309 514
0.5	+0.855 906	−0.671 582	−1.819 800	+0.333 204
0.6	+0.693 121	−0.637 450	−1.456 539	+0.348 164
0.7	+0.561 378	−0.602 176	−1.190 943	+0.356 310
0.8	+0.452 882	−0.566 368	−0.987 335	+0.359 043
0.9	+0.362 515	−0.530 511	−0.825 869	+0.357 443
1.0	+0.286 706	−0.494 995	−0.694 604	+0.352 370
1.1	+0.222 845	−0.460 130	−0.585 905	+0.344 521
1.2	+0.168 946	−0.426 164	−0.494 643	+0.334 474
1.3	+0.123 455	−0.393 292	−0.417 227	+0.322 712
1.4	+0.085 126	−0.361 665	−0.351 055	+0.309 642
1.5	+0.052 935	−0.331 396	−0.294 182	+0.295 608
1.6	+0.026 030	−0.302 566	−0.245 115	+0.280 904
1.7	+0.003 691	−0.275 229	−0.202 682	+0.265 777
1.8	−0.014 696	−0.249 417	−0.165 942	+0.250 439
1.9	−0.029 661	−0.225 142	−0.134 128	+0.235 066
2.0	−0.041 665	−0.202 400	−0.106 601	+0.219 808

Table 5.1—*contd.*

x	ker x	kei x	ker′ x	kei′ x
2.1	−0.051 107	−0.181 173	−0.082 823	+0.204 790
2.2	−0.058 339	−0.161 431	−0.062 337	+0.190 114
2.3	−0.063 671	−0.143 136	−0.044 748	+0.175 864
2.4	−0.067 374	−0.126 242	−0.029 712	+0.162 107
2.5	−0.069 688	−0.110 696	−0.016 930	+0.148 895
2.6	−0.070 826	−0.096 443	−0.006 136	+0.136 269
2.7	−0.070 974	−0.083 422	+0.002 904	+0.124 256
2.8	−0.070 296	−0.071 571	+0.010 399	+0.112 875
2.9	−0.068 939	−0.060 826	+0.016 534	+0.102 136
3.0	−0.067 029	−0.051 122	+0.021 476	+0.092 043
3.1	−0.064 679	−0.042 396	+0.025 374	+0.082 592
3.2	−0.061 985	−0.034 582	+0.028 360	+0.073 775
3.3	−0.059 033	−0.027 620	+0.030 555	+0.065 579
3.4	−0.055 897	−0.021 446	+0.032 066	+0.057 988
3.5	−0.052 639	−0.016 003	+0.032 989	+0.050 982
3.6	−0.049 316	−0.011 231	+0.033 409	+0.044 539
3.7	−0.045 972	−0.007 077	+0.033 403	+0.038 636
3.8	−0.042 647	−0.003 487	+0.033 040	+0.033 248
3.9	−0.039 374	−0.000 411	+0.032 380	+0.028 348
4.0	−0.036 179	+0.002 198	+0.031 478	+0.023 911
4.1	−0.033 084	+0.004 386	+0.030 382	+0.019 908
4.2	−0.030 108	+0.006 194	+0.029 132	+0.016 314
4.3	−0.027 262	+0.007 661	+0.027 767	+0.013 101
4.4	−0.024 557	+0.008 826	+0.026 319	+0.010 243
4.5	−0.022 000	+0.009 721	+0.024 815	+0.007 715
4.6	−0.019 595	+0.010 379	+0.023 279	+0.005 492
4.7	−0.017 344	+0.010 829	+0.021 733	+0.003 550
4.8	−0.015 248	+0.011 097	+0.020 194	+0.001 865
4.9	−0.013 305	+0.011 210	+0.018 677	+0.000 415
5.0	−0.011 512	+0.011 188	+0.017 193	−0.000 820
5.1	−0.009 865	+0.011 052	+0.015 754	−0.001 861
5.2	−0.008 359	+0.010 821	+0.014 368	−0.002 726
5.3	−0.006 989	+0.010 512	+0.013 039	−0.003 433
5.4	−0.005 749	+0.010 139	+0.011 774	−0.004 000
5.5	−0.004 632	+0.009 716	+0.010 576	−0.004 440
5.6	−0.003 632	+0.009 255	+0.009 447	−0.004 769
5.7	−0.002 740	+0.008 766	+0.008 388	−0.005 000
5.8	−0.001 952	+0.008 258	+0.007 400	−0.005 146
5.9	−0.001 258	+0.007 739	+0.006 481	−0.005 217
6.0	−0.000 653	+0.007 216	+0.005 632	−0.005 224

Table 5.1—*contd.*

x	ker x	kei x	ker′ x	kei′ x
6.1	−0.000 130	+0.006 696	+0.004 850	−0.005 176
6.2	+0.000 319	+0.006 183	+0.004 133	−0.005 083
6.3	+0.000 699	+0.005 681	+0.003 479	−0.004 951
6.4	+0.001 017	+0.005 194	+0.002 885	−0.004 788
6.5	+0.001 278	+0.004 724	+0.002 349	−0.004 600
6.6	+0.001 488	+0.004 274	+0.001 867	−0.004 393
6.7	+0.001 653	+0.003 846	+0.001 437	−0.004 171
6.8	+0.001 777	+0.003 440	+0.001 054	−0.003 939
6.9	+0.001 866	+0.003 058	+0.000 716	−0.003 701
7.0	+0.001 922	+0.002 700	+0.000 421	−0.003 460
7.1	+0.001 951	+0.002 367	+0.000 163	−0.003 218
7.2	+0.001 956	+0.002 057	−0.000 058	−0.002 979
7.3	+0.001 940	+0.001 770	−0.000 247	−0.002 745
7.4	+0.001 907	+0.001 507	−0.000 407	−0.002 517
7.5	+0.001 860	+0.001 267	−0.000 539	−0.002 296
7.6	+0.001 800	+0.001 048	−0.000 646	−0.002 084
7.7	+0.001 731	+0.000 850	−0.000 732	−0.001 881
7.8	+0.001 655	+0.000 671	−0.000 798	−0.001 689
7.9	+0.001 572	+0.000 512	−0.000 817	−0.001 507
8.0	+0.001 486	+0.000 370	−0.000 880	−0.001 336
8.1	+0.001 397	+0.000 244	−0.000 899	−0.001 177
8.2	+0.001 306	+0.000 134	−0.000 907	−0.001 028
8.3	+0.001 216	+0.000 038	−0.000 904	−0.000 890
8.4	+0.001 126	−0.000 044	−0.000 893	−0.000 763
8.5	+0.001 037	−0.000 115	−0.000 875	−0.000 647
8.6	+0.000 951	−0.000 174	−0.000 850	−0.000 540
8.7	+0.000 868	−0.000 223	−0.000 820	−0.000 444
8.8	+0.000 787	−0.000 263	−0.000 787	−0.000 357
8.9	+0.000 710	−0.000 295	−0.000 750	−0.000 278
9.0	+0.000 637	−0.000 319	−0.000 711	−0.000 208
9.1	+0.000 568	−0.000 337	−0.000 671	−0.000 146
9.2	+0.000 503	−0.000 349	−0.000 629	−0.000 091
9.3	+0.000 442	−0.000 355	−0.000 588	−0.000 043
9.4	+0.000 386	−0.000 357	−0.000 546	−0.000 002
9.5	+0.000 333	−0.000 356	−0.000 505	+0.000 034
9.6	+0.000 285	−0.000 351	−0.000 464	+0.000 064
9.7	+0.000 240	−0.000 343	−0.000 425	+0.000 090
9.8	+0.000 200	−0.000 333	−0.000 387	+0.000 111
9.9	+0.000 163	−0.000 321	−0.000 350	+0.000 128
10.0	+0.000 129	−0.000 308	−0.000 316	+0.000 141

5.3 Edge-loaded conical shells

(*a*) *Semi-infinite cones*

The homogeneous solution of the deflection equation and the associated deformations, stress resultants, and stress couples derived from this solution are sufficient for the investigation of the behavior of conical shells which are loaded only at their edges. Let us first consider a truncated conical shell which extends to infinity and which is loaded by edge moments and forces normal

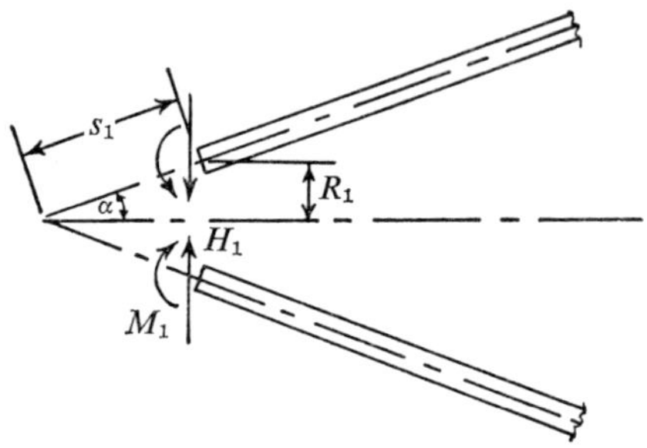

Fig. 5.4. Infinite conical shell with edge loads

to the shell axis of revolution at the truncated edge (Fig. 5.4). Since deformations and forces should vanish at infinity, the ber and bei functions and their derivatives cannot be used in the solution. Accordingly we set A_1 and A_2 equal to zero in Eqs. (5.2.9), (5.2.10), (5.2.12), and (5.2.13), leaving the two undetermined coefficients A_3 and A_4. The edge conditions for the determination of these coefficients are the equality of the edge moment and the applied moment M_1 and of the edge force perpendicular to the axis of revolution and the applied force H_1 (recalling the sign convention for these quantities). With the use of Eqs. (5.2.10a), (5.2.10c), and (5.2.13b), we then obtain

$$A_3 = \frac{\xi_1}{4\Delta_1 E \cot^2 \alpha} \left\{ \frac{M_1}{h} \left(\operatorname{ker} \xi_1 - \frac{2}{\xi_1} \operatorname{kei}' \xi_1 \right) - \frac{\xi_1 H_1 \sin \alpha}{2[12(1-\nu^2)]^{1/2}} \right.$$
$$\left. \times \left[\operatorname{kei}' \xi_1 - \frac{2(1-\nu)}{\xi_1} \left(\operatorname{kei} \xi_1 + \frac{2}{\xi_1} \operatorname{ker}' \xi_1 \right) \right] \right\}, \tag{5.3.1a}$$

$$A_4 = -\frac{\xi_1}{4\Delta_1 E \cot^2 \alpha} \left\{ \frac{M_1}{h} \left(\operatorname{kei} \xi_1 + \frac{2}{\xi_1} \operatorname{ker}' \xi_1 \right) + \frac{\xi_1 H_1 \sin \alpha}{2[12(1-\nu^2)]^{1/2}} \right.$$
$$\left. \times \left[\operatorname{ker}' \xi_1 - \frac{2(1-\nu)}{\xi_1} \left(\operatorname{ker} \xi_1 - \frac{2}{\xi_1} \operatorname{kei}' \xi_1 \right) \right] \right\}, \tag{5.3.1b}$$

with

$$\Delta_1 = -(\text{ker}\ \xi_1\ \text{ker}'\ \xi_1 + \text{kei}\ \xi_1\ \text{kei}'\ \xi_1) + \frac{2(1-\nu)}{\xi_1}$$
$$\times \left[\left(\text{ker}\ \xi_1 - \frac{2}{\xi_1}\text{kei}'\ \xi_1\right)^2 + \left(\text{kei}\ \xi_1 + \frac{2}{\xi_1}\text{ker}'\ \xi_1\right)^2\right]. \qquad (5.3.1c)$$

Quantities which are of particular interest are the edge deformation in the direction of H_1 and the rotation, which are obtained from Eqs. (5.2.12), (5.2.13c), and (5.3.1) as

$$u_{H_1} = -\left[\frac{H_1 \cos\alpha}{2\lambda_1^3 D}\,\delta_{H_{11}}(\xi_1) + \frac{M_1}{2\lambda_1^2 D}\,\delta_{M_{11}}(\xi_1)\right]\cos\alpha, \qquad (5.3.2a)$$

$$\frac{dw_1}{ds} = \frac{H_1\cos\alpha}{2\lambda_1^2 D}\,\Theta_{H_{11}}(\xi_1) + \frac{M_1}{\lambda_1 D}\,\Theta_{M_{11}}(\xi_1), \qquad (5.3.2b)$$

with

$$\lambda_1 = \left[\frac{3(1-\nu^2)}{(R_1/h\cos\alpha)^2}\right]^{1/4}, \qquad (5.3.3a)$$

and

$$\delta_{H_{11}}(\xi_1) = 2^{-1/2}\,[(\text{ker}'\ \xi_1)^2 + (\text{kei}'\ \xi_1)^2]/\Delta_1$$
$$+ \frac{4}{\xi_1}\left[2^{-1/2} - \frac{(1-\nu)^2}{\xi_1}\,\Theta_{M_{11}}(\xi_1)\right], \qquad (5.3.3b)$$

$$\delta_{M_{11}}(\xi_1) = \Theta_{H_{11}}(\xi_1)$$
$$= \frac{1}{\Delta_1}\left\{\text{kei}\ \xi_1\ \text{ker}'\ \xi_1 - \text{ker}\ \xi_1\ \text{kei}'\ \xi_1 + \frac{2}{\xi_1}\,[(\text{ker}'\ \xi')^2\right.$$
$$\left. + (\text{kei}'\ \xi_1)^2]\right\}, \qquad (5.3.3c)$$

$$\Theta_{M_{11}}(\xi_1) = \frac{2^{-1/2}}{\Delta_1}\left[\left(\text{ker}\ \xi_1 - \frac{2}{\xi_1}\text{kei}'\ \xi_1\right)^2 + \left(\text{kei}\ \xi_1 + \frac{2}{\xi_1}\text{ker}'\ \xi_1\right)^2\right]. \qquad (5.3.3d)$$

Finally, the distance by which the edge moves in the direction of the axis of revolution, assuming that the axial displacement vanishes at infinity, is given by Eqs. (5.2.9), (5.2.10e), (5.2.13d), and (5.3.1), after some manipulation, as

$$u_{V_1} = \frac{M_1}{2\lambda_1^2 D}\left\{\Theta_{H_{11}}(\xi_1)\sin\alpha + \frac{4}{\xi_1^2}\left[1 - \frac{2^{3/2}(1-\nu)}{\xi_1}\,\Theta_{M_{11}}(\xi_1)\right]\csc\alpha\right\}$$
$$+ \frac{H_1\cos\alpha}{2\lambda_1^3 D}\left\{\delta_{H_{11}}(\xi_1)\sin\alpha - \frac{2^{1/2}}{\xi_1}\left[\nu + \frac{4(1-\nu)}{\xi_1^2}\,\delta_{M_{11}}(\xi_1)\right]\csc\alpha\right\}. \qquad (5.3.3e)$$

The coefficients given by Eqs. (5.3.3) can be shown to satisfy the relationship

$$2\delta_{H_{11}}(\xi_1)\Theta_{M_{11}}(\xi_1) - \delta_{M_{11}}(\xi_1)\Theta_{H_{11}}(\xi_1) = 1 + \frac{2^{5/2}\nu}{\xi_1}\Theta_{M_{11}}(\xi_1). \tag{5.3.4}$$

Eqs. (5.3.2) are in a form very similar to Eqs. (4.2.5) for the cylindrical shell. For instance, λ_1 for the conical shell contains the edge radius of curvature of the cone, $R_1/\cos\alpha$, and thus corresponds to the quantity λ for the cylindrical shell which contains the cylinder radius of curvature R. The resemblance is further enhanced when we examine the variation of the influence coefficients with ξ, shown in Fig. 5.5,† with ν taken as 0.3, and note

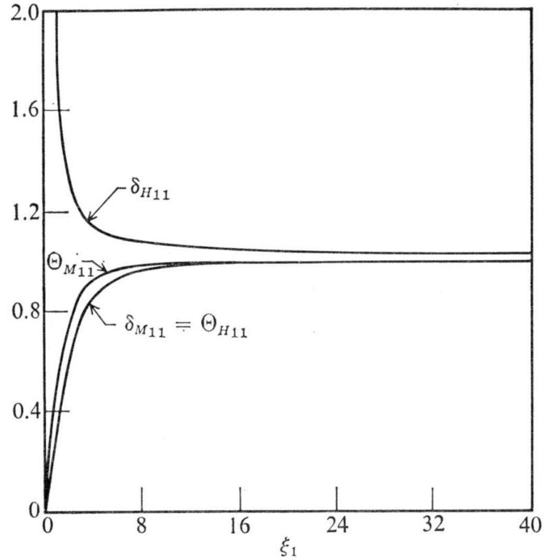

Fig. 5.5. Edge rotation and displacement coefficients for the small end of a long truncated cone

that for large ξ_1 all of the functions are very close to unity, which is expected since a large value of ξ_1 corresponds to either a thin shell or to a nearly cylindrical shell. The similarity of conical and cylindrical shell behavior in the range into which most conical shells fall in practice can be better shown if, instead of deformations perpendicular and parallel to the axis of revolution

† See also C. E. Taylor and E. Wenk, Jr., 'Analysis of Stresses in the Conical Elements of Shell Structures,' *Proc. Second U.S. National Congress of Applied Mechanics*, A.S.M.E., 1955, pp. 323–331.

to the cone, we consider deformations perpendicular and parallel to the cone generators. These can be written as

$$\begin{aligned} w_1 &= u_{H_1} \cos\alpha - u_{V_1} \sin\alpha \\ &= -\frac{M_1}{2\lambda_1^2 D}\left\{\Theta_{H_{11}}(\xi_1) + \frac{4}{\xi_1^2}\left[1 - \frac{2^{3/2}(1-\nu)}{\xi_1}\,\Theta_{M_{11}}(\xi_1)\right]\right\} \\ &\quad - \frac{H_1\cos\alpha}{2\lambda_1^3 D}\left\{\delta_{H_{11}}(\xi_1) - \frac{2^{1/2}}{\xi_1}\left[\nu + \frac{4(1-\nu)}{\xi_1^2}\,\delta_{M_{11}}(\xi_1)\right]\right\}, \end{aligned} \tag{5.3.5a}$$

$$\begin{aligned} u_{s_1} &= u_{H_1}\sin\alpha + u_{V_1}\cos\alpha \\ &= \left\{\frac{M_1}{2\lambda_1^2 D}\,\frac{4}{\xi_1^2}\left[1 - \frac{2^{3/2}(1-\nu)}{\xi_1}\,\Theta_{M_{11}}(\xi_1)\right]\right. \\ &\quad \left. - \frac{H_1\cos\alpha}{2\lambda_1^3 D}\,\frac{2^{1/2}}{\xi_1}\left[\nu + \frac{4(1-\nu)}{\xi_1^2}\,\delta_{M_{11}}(\xi_1)\right]\right\}\cot\alpha \\ &= \frac{M_1\tan\alpha}{Eh}\left[1 - \frac{2^{3/2}(1-\nu)}{\xi_1}\,\Theta_{M_{11}}(\xi_1)\right] - \frac{H_1\cos\alpha}{E}\left(\frac{R_1/\cos\alpha}{h}\right) \\ &\quad \times\left[\nu + \frac{4(1-\nu)}{\xi_1^2}\,\delta_{M_{11}}(\xi_1)\right]. \end{aligned} \tag{5.3.5b}$$

In addition, we have the edge slope, given by

$$\frac{dw_1}{ds} = \frac{M_1}{\lambda_1 D}\,\Theta_{M_{11}}(\xi_1) + \frac{H_1\cos\alpha}{2\lambda_1^2 D}\,\Theta_{H_{11}}(\xi_1). \tag{5.3.5c}$$

For large values of ξ_1 these may be approximated by

$$w_1 \approx -\left(\frac{M_1}{2\lambda_1^2 D} + \frac{H_1\cos\alpha}{2\lambda_1^3 D}\right), \tag{5.3.6a}$$

$$\frac{dw_1}{ds} \approx \frac{M_1}{\lambda_1 D} + \frac{H_1\cos\alpha}{2\lambda_1^2 D}, \tag{5.3.6b}$$

$$u_{s_1} \approx \frac{M_1\tan\alpha}{Eh} - \nu\,\frac{H_1\cos\alpha}{E}\left(\frac{R_1/\cos\alpha}{h}\right). \tag{5.3.6c}$$

If we note that $H_1\cos\alpha$ is the shear force normal to the cone generator (Fig. 5.6), Eqs. (5.3.6a) and (5.3.6b) are of a form identical with Eqs. (4.2.5) for the cylindrical shell. The term containing H_1 in Eq. (5.3.6c) is identical with that of Eq. (4.1.11), but the term containing M_1 has no counterpart in cylindrical shell theory. However, the coefficient of M_1 in Eq. (5.3.6c) is of the order of h/R_θ compared to that in Eq. (5.3.6a)† while the coefficient of

† We can exclude the possibility of $\tan\alpha$ being large since large values of ξ_1 imply, from the definition of ξ, that $(R_\theta/h)^{1/2}$ is much greater than $\tan\alpha$.

H_1 in Eq. (5.3.6c) is of the order of $(h/R_\theta)^{1/2}$ compared to that in Eq. (5.3.6a). Thus the edge deformation of a long edge loaded cone such that ξ_1 is large consists primarily of deflection normal to the cone generator and is equivalent to that of a circular cylinder of thickness h and radius equal to the edge radius of curvature of the cone ($R_1/\cos\alpha$) which is loaded by edge moments M_1 and edge shear forces $H_1 \cos\alpha$.

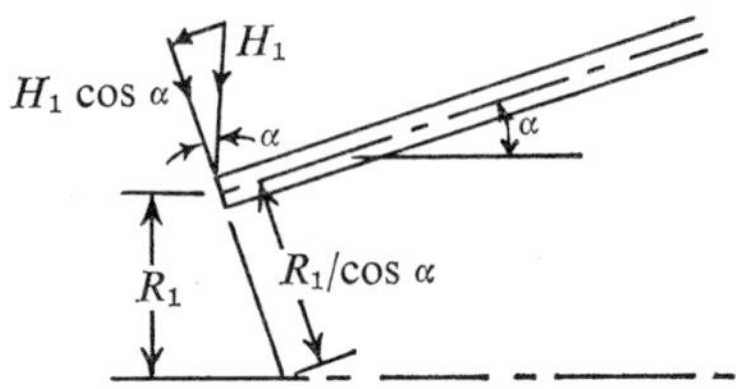

Fig. 5.6. Decomposition of edge load

The coincidence of cone and cylinder edge deformations suggests that the distribution of stress resultants and stress couples along the generator of the cone may be similar to those in a cylinder, provided ξ is large. Let us investigate this in detail. We note that if ξ is sufficiently large the ker, kei, ker′, and kei′ functions may be approximated by the first term of asymptotic expressions† as

$$\mathrm{ker}\,\xi \approx \left(\frac{\pi}{2\xi}\right)^{1/2} e^{-\xi/2^{1/2}} \cos\alpha_2, \tag{5.3.7a}$$

$$\mathrm{kei}\,\xi \approx -\left(\frac{\pi}{2\xi}\right)^{1/2} e^{-\xi/2^{1/2}} \sin\alpha_2, \tag{5.3.7b}$$

$$\mathrm{ker}'\,\xi \approx -\left(\frac{\pi}{2\xi}\right)^{1/2} e^{-\xi/2^{1/2}} \cos\alpha_1, \tag{5.3.7c}$$

$$\mathrm{kei}'\,\xi \approx \left(\frac{\pi}{2\xi}\right)^{1/2} e^{-\xi/2^{1/2}} \sin\alpha_1, \tag{5.3.7d}$$

where

$$\left.\begin{matrix}\alpha_1\\ \alpha_2\end{matrix}\right\} = \frac{\xi_1}{2^{1/2}} \mp \frac{\pi}{8}. \tag{5.3.7e}$$

Then, from Eqs. (5.3.1), we have

$$A_3 \approx \tfrac{1}{2}\left(\frac{\xi_1^3}{\pi}\right)^{1/2} \frac{e^{\xi_1/2^{1/2}}}{E\cot^2\alpha}\left[\frac{M_1}{h}\cos\alpha_2 - \frac{\xi_1 H_1 \sin\alpha}{2[12(1-\nu^2)]^{1/2}}\sin\alpha_1\right], \tag{5.3.8a}$$

† H. H. Lowell, loc. cit.

$$A_4 \approx \frac{1}{2}\left(\frac{\xi_1^3}{\pi}\right)^{1/2} \frac{e^{\xi_1/2^{1/2}}}{E\cot^2\alpha}\left[\frac{M_1}{h}\sin\alpha_2 + \frac{\xi_1 H_1 \sin\alpha}{2[12(1-\nu^2)]^{1/2}}\cos\alpha_1\right]. \quad (5.3.8b)$$

Terms in Eqs. (5.3.1) of the same order of those omitted in the approximations for the ker, kei, ker′, and kei′ functions have been deleted for consistency. Using the same approximations in Eqs. (5.2.13) and (5.2.14) we can obtain

$$w \approx -\left[\frac{M_1}{2\lambda_1^2 D} D\left(\frac{\xi-\xi_1}{2^{1/2}}\right) + \frac{H_1\cos\alpha}{2\lambda_1^3 D} A\left(\frac{\xi-\xi_1}{2^{1/2}}\right)\right] \approx \frac{u_H}{\cos\alpha}, \quad (5.3.9a)$$

$$\frac{dw}{ds} \approx \left(\frac{\xi_1}{\xi}\right)^{1/2}\left[\frac{M_1}{\lambda_1 D} A\left(\frac{\xi-\xi_1}{2^{1/2}}\right) + \frac{H_1\cos\alpha}{2\lambda_1^2 D} C\left(\frac{\xi-\xi_1}{2^{1/2}}\right)\right], \quad (5.3.9b)$$

$$N_s\cot\alpha = Q_s \approx \left(\frac{\xi_1}{\xi}\right)^{5/2}\left[-2\lambda_1 M_1 B\left(\frac{\xi-\xi_1}{2^{1/2}}\right) + H_1\cos\alpha\, D\left(\frac{\xi-\xi_1}{2^{1/2}}\right)\right], \quad (5.3.9c)$$

$$M_s \approx \left(\frac{\xi_1}{\xi}\right)^{3/2}\left[M_1 C\left(\frac{\xi-\xi_1}{2^{1/2}}\right) + \frac{H_1\cos\alpha}{\lambda_1} B\left(\frac{\xi-\xi_1}{2^{1/2}}\right)\right], \quad (5.3.9d)$$

$$M_\theta \approx \nu M_s, \quad (5.3.9e)$$

$$N_\theta \approx \frac{Eh}{R_1/\cos\alpha}\left(\frac{\xi_1}{\xi}\right) w, \quad (5.3.9f)$$

$$u_s \approx -\frac{\nu Q_s}{E}\frac{R_1/\cos\alpha}{h}\left(\frac{\xi}{\xi_1}\right)^2, \quad (5.3.9g)$$

where

$$A\left(\frac{\xi-\xi_1}{2^{1/2}}\right),\quad B\left(\frac{\xi-\xi_1}{2^{1/2}}\right),\quad C\left(\frac{\xi-\xi_1}{2^{1/2}}\right),\quad \text{and}\quad D\left(\frac{\xi-\xi_1}{2^{1/2}}\right),$$

are the cylinder functions defined by Eqs. (4.1.13). We note that the distance along the generator from the loaded edge is given by

$$x = s - s_1 = \frac{\xi_1}{2^{3/2}\lambda_1}\left[\left(\frac{\xi}{\xi_1}\right)^2 - 1\right] \quad (5.3.10a)$$

from which we obtain

$$\frac{\xi}{\xi_1} = \left[1 + \frac{2^{3/2}\lambda_1 x}{\xi_1}\right]^{1/2}, \quad (5.3.10b)$$

$$\frac{\xi-\xi_1}{2^{1/2}} = \frac{\xi_1}{2^{1/2}}\left\{\left[1 + \frac{2^{3/2}\lambda_1 x}{\xi_1}\right]^{1/2} - 1\right\}. \quad (5.3.10c)$$

When $2^{3/2}\lambda_1 x/\xi_1$ is small, Eq. (5.3.10c) may be approximated by

$$\frac{\xi - \xi_1}{2^{1/2}} \approx \frac{\xi_1}{2^{1/2}}\left[1 + \frac{1}{2}\frac{2^{3/2}\lambda_1 x}{\xi_1} + \cdots - 1\right] \approx \lambda_1 x, \tag{5.3.10d}$$

while ξ/ξ_1 is essentially equal to unity. In this case Eqs. (5.3.9) and Eqs. (4.2.4) are identical. The quantity $2^{3/2}\lambda_1 x/\xi_1$ becomes large as x, the meridional distance from the small end, increases so that the above approximations become invalid. However, for large values of x the effects of end loads become negligible so that the difference between the asymptotic solution given by Eqs. (5.3.9) and the cylindrical shell solution is of no consequence.

There still remains the question of how large ξ_1 must be for the various approximations to be valid. A preliminary estimate can be obtained by examining the influence coefficients shown in Fig. 5.5 and noting that $\delta_{H_{11}}$, which approaches unity most slowly, differs from unity by 2 percent or less when ξ_1 is greater than about 35. A better idea of what error is involved in using the cylindrical shell approximation can be obtained, of course, by calculating the actual meridional distributions of the various quantities and comparing them with the approximate results. Calculations made with Eqs. (5.2.9), (5.2.10), and (5.3.1) indicate that for some quantities such as the normal displacement the lower limit of ξ_1 can be reduced to 5 to 10, while for quantities such as M_θ the lower limit should be increased to 50 or above. An average value of ξ_1 of about 20 or so is a reasonable lower limit for which the cylindrical shell approximation is adequate. Calculations indicate that this same average value of ξ_1 applies for the adequacy of Eqs. (5.3.9), the asymptotic solution, which will yield values of the various quantities which are somewhat better than those given by the cylindrical shell approximation. A comparison of exact and asymptotic solutions for normal deflection and circumferential moment is shown in Table 5.2 for a long conical shell subjected to edge moment. In general, it appears that the approximate solutions for conical shells are adequate for quick preliminary analysis or design, but that the more accurate equations are preferable. The availability of extensive tables of Bessel–Kelvin functions, together with the use of manual or high-speed digital computing equipment, makes the use of the accurate equations not too difficult a task.

(*b*) *Complete cones*

The problem of edge loading of a conical shell which tapers to a point (Fig. 5.7) can be considered in a similar manner. For the various deformations and stress resultants to be finite at the origin, the coefficients A_3 and A_4 in Eqs.

(5.2.9), (5.2.10), (5.2.12), and (5.2.13) must vanish since the functions they multiply become singular there. The equations for the problem of the complete cone then differ from those for the edge-loaded infinite cone only in the substitution of the ber and bei functions and their derivatives for the ker and

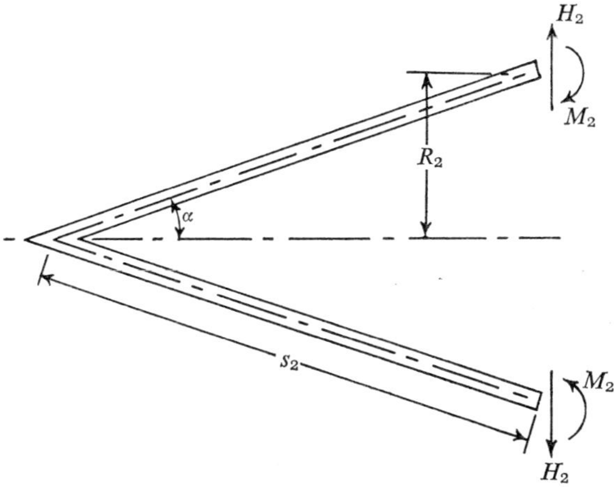

Fig. 5.7. Complete conical shell with edge loading

kei functions and their derivatives in the Eqs. (5.3.1) to (5.3.3), and by replacing the subscript 2 to denote quantities appropriate to the large end of the cone. Thus, the influence coefficients are given by

$$\delta_{H_{22}}(\xi_2) = \frac{2^{-1/2}}{\Delta_2}[(\mathrm{ber}'\,\xi_2)^2 + (\mathrm{bei}'\,\xi_2)^2] + \frac{4}{\xi_2}\left[2^{-1/2} - \frac{(1-\nu)^2}{\xi_2}\,\Theta_{M_{11}}(\xi_2)\right], \tag{5.3.11a}$$

$$\delta_{M_{22}}(\xi_2) = \Theta_{H_{22}}(\xi_2) = \frac{1}{\Delta_2}\left\{\mathrm{bei}\,\xi_2\,\mathrm{ber}'\,\xi_2 - \mathrm{ber}\,\xi_2\,\mathrm{bei}'\,\xi_2 + \frac{2}{\xi_2}[(\mathrm{ber}'\,\xi_2)^2 + (\mathrm{bei}'\,\xi_2)^2]\right\}, \tag{5.3.11b}$$

$$\Theta_{M_{22}}(\xi_2) = \frac{2^{-1/2}}{\Delta_2}\left[\left(\mathrm{ber}\,\xi_2 - \frac{2}{\xi_2}\,\mathrm{bei}'\,\xi_2\right)^2 + \left(\mathrm{bei}\,\xi_2 + \frac{2}{\xi_2}\,\mathrm{ber}'\,\xi_2\right)^2\right], \tag{5.3.11c}$$

Table 5.2 Comparison of exact and approximate values of normal deflection and circumferential bending moment

(a) $-\dfrac{2\lambda_1^2 Dw}{M_1}$

$\lambda_1 x$ \ ξ_1	∞	50		20		10		5		2	
	Cylinder	Exact	Eq. (5.3.9a)	Exact	Eq. (5.3.9a)	Exact	Eq. (5.3.9a)	Exact	Eq. (5.3.9a)	Exact	Eq. (5.3.9a)
0.00	1.0000	1.0029	1.0000	1.0070	1.0000	1.0115	1.0000	1.0032	1.0000	0.8431	1.0000
0.25	0.5619	0.5645	0.5652	0.5696	0.5701	0.5791	0.5782	0.5923	0.5938	0.5369	0.6372
0.50	0.2414	0.2442	0.2469	0.2506	0.2550	0.2650	0.2680	0.2972	0.2931	0.3283	0.3611
0.75	0.0237	0.0258	0.0294	0.0321	0.0379	0.0484	0.0519	0.0916	0.0790	0.1818	0.1544
1.00	−0.1108	−0.1104	−0.1067	−0.1063	−0.0918	−0.0918	−0.0895	−0.0466	−0.0669	0.0773	0.0018
1.25	−0.1815	−0.1840	−0.1806	−0.1838	−0.1785	−0.1746	−0.1738	−0.1347	−0.1608	0.0028	−0.1089
1.50	−0.2068	−0.2127	−0.2098	−0.2176	−0.2132	−0.2159	−0.2164	−0.1863	−0.2160	−0.0502	−0.1872
1.75	−0.2020	−0.2115	−0.2092	−0.2217	−0.2183	−0.2284	−0.2299	−0.2121	−0.2431	−0.0871	−0.2406
2.00	−0.1794	−0.1920	−0.1902	−0.2070	−0.2045	−0.2219	−0.2240	−0.2198	−0.2504	−0.1121	−0.2749
2.25	−0.1482	−0.1631	−0.1618	−0.1820	−0.1820	−0.2038	−0.2061	−0.2152	−0.2443	−0.1282	−0.2945
2.50	−0.1149	−0.1310	−0.1301	−0.1523	−0.1511	−0.1795	−0.1817	−0.2028	−0.2294	−0.1375	−0.3032
2.75	−0.0835	−0.0996	−0.0991	−0.1220	−0.1212	−0.1527	−0.1546	−0.1857	−0.2092	−0.1418	−0.3035
3.00	−0.0563	−0.0715	−0.0712	−0.0936	−0.0931	−0.1260	−0.1274	−0.1661	−0.1864	−0.1424	−0.2976
3.25	−0.0343	−0.0478	−0.0477	−0.0685	−0.0682	−0.1007	−0.1018	−0.1458	−0.1627	−0.1402	−0.2873
3.50	−0.0177	−0.0289	−0.0289	−0.0473	−0.0470	−0.0780	−0.0786	−0.1256	−0.1394	−0.1359	−0.2738
3.75	−0.0079	−0.0146	−0.0146	−0.0301	−0.0299	−0.0582	−0.0584	−0.1065	−0.1172	−0.1303	−0.2582
4.00	0.0019	−0.0043	−0.0044	−0.0166	−0.0165	−0.0415	−0.0413	−0.0887	−0.0967	−0.1236	−0.2413
4.25	0.0064	0.0026	0.0025	−0.0066	−0.0065	−0.0277	−0.0272	−0.0726	−0.0782	−0.1164	−0.2238
4.50	0.0085	0.0068	0.0067	0.0006	0.0007	−0.0166	−0.0159	−0.0582	−0.0617	−0.1087	−0.2060
4.75	0.0090	0.0090	0.0089	0.0054	0.0055	−0.0079	−0.0071	−0.0455	−0.0473	−0.1010	−0.1883
5.00	0.0084	0.0096	0.0096	0.0083	0.0084	−0.0013	−0.0004	−0.0346	−0.0348	−0.0932	−0.1711
5.25	0.0072	0.0093	0.0093	0.0098	0.0099	0.0035	0.0045	−0.0252	−0.0243	−0.0856	−0.1544
5.50	0.0058	0.0084	0.0084	0.0103	0.0104	0.0068	0.0078	−0.0173	−0.0154	−0.0782	−0.1385
5.75	0.0043	0.0072	0.0072	0.0100	0.0101	0.0090	0.0100	−0.0107	−0.0081	−0.0711	−0.1234
6.00	0.0031	0.0059	0.0058	0.0093	0.0093	0.0103	0.0111	−0.0053	−0.0022	−0.0643	−0.1092

Table 5.2—*contd.*

(b) $\dfrac{M_\theta}{\nu M_1}$

$\lambda_1 x$ \ ξ_1	∞	50		20	
	Cylinder	Exact	Eq. (5.3.9e)	Exact	Eq. (5.3.9e)
0.00	1.0000	0.8281	1.0000	0.5700	1.0000
0.25	0.9472	0.8101	0.9377	0.6104	0.9237
0.50	0.8231	0.7192	0.8080	0.5712	0.7867
0.75	0.6676	0.5948	0.6520	0.4918	0.6303
1.00	0.5083	0.4631	0.4960	0.3980	0.4790
1.25	0.3623	0.3399	0.3552	0.3054	0.3455
1.50	0.2384	0.2338	0.2372	0.2224	0.2350
1.75	0.1400	0.1480	0.1439	0.1529	0.1482
2.00	0.0667	0.0824	0.0723	0.0976	0.0831
2.25	0.0157	0.0350	0.0252	0.0556	0.0366
2.50	−0.0166	0.0030	−0.0070	0.0252	0.0050
2.75	−0.0347	−0.0168	−0.0260	0.0044	−0.0148
3.00	−0.0422	−0.0275	−0.0354	−0.0089	−0.0260
3.25	−0.0427	−0.0317	−0.0380	−0.0166	−0.0311
3.50	−0.0388	−0.0315	−0.0364	−0.0202	−0.0320
3.75	−0.0341	−0.0288	−0.0323	−0.0210	−0.0302
4.00	−0.0258	−0.0246	−0.0270	−0.0200	−0.0269
4.25	−0.0191	−0.0199	−0.0215	−0.0180	−0.0230
4.50	−0.0132	−0.0154	−0.0162	−0.0155	−0.0188
4.75	−0.0083	−0.0113	−0.0116	−0.0128	−0.0149
5.00	−0.0046	−0.0078	−0.0078	−0.0102	−0.0114
5.25	−0.0019	−0.0050	−0.0048	−0.0078	−0.0083
5.50	0.0000	−0.0028	−0.0025	−0.0058	−0.0058
5.75	0.0011	−0.0012	−0.0008	−0.0041	−0.0038
6.00	0.0017	−0.0001	−0.0003	−0.0027	−0.0022

with

$$\Delta_2 = -(\mathrm{ber}\,\xi_2\,\mathrm{ber}'\,\xi_2 + \mathrm{bei}\,\xi_2\,\mathrm{bei}'\,\xi_2) + \frac{2(1-\nu)}{\xi_2} \times \left[\left(\mathrm{ber}\,\xi_2 - \frac{2}{\xi_2}\mathrm{bei}'\,\xi_2\right)^2 + \left(\mathrm{bei}\,\xi_2 + \frac{2}{\xi_2}\mathrm{ber}'\,\xi_2\right)^2\right]. \tag{5.3.11d}$$

The variation with ξ_2 of the influence coefficients for the complete cone is shown in Fig. 5.8.† These coefficients are numerically less for the same value of ξ_2, than those shown in Fig. 5.5, a circumstance which might be expected intuitively since it seems apparent that it would be more difficult to bend the edge of a cone which tapers to a point than of one which expands to infinity.

† C. E. Taylor and E. Wenk, Jr., loc. cit.

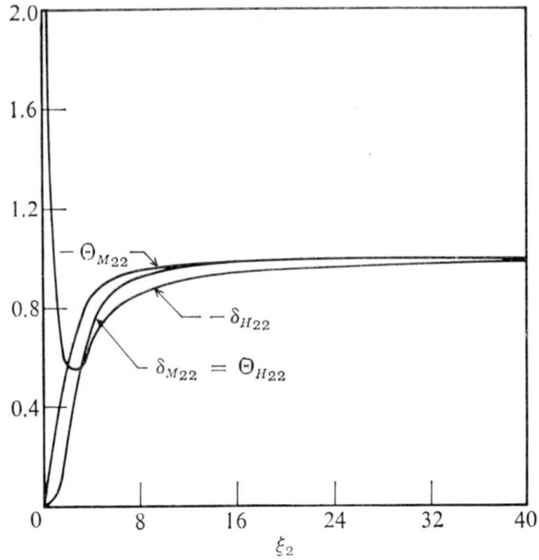

Fig. 5.8. Edge rotation and displacement coefficients for the large end of a complete cone

For large values of ξ_2 the coefficients approach unity and, as before, indicate that the cone behavior is similar to that of a cylinder. When the asymptotic expansions of the ber, bei, ber′, and bei′ functions are used, after much manipulation we obtain

$$w \approx -\left(\frac{\xi}{\xi_2}\right)^{1/2}\left[\frac{M_2}{2\lambda_2^2 D} D\left(\frac{\xi_2-\xi}{2^{1/2}}\right) - \frac{H_2 \cos\alpha}{2\lambda_2^3 D} A\left(\frac{\xi_2-\xi}{2^{1/2}}\right)\right] \approx \frac{u_H}{\cos\alpha}, \tag{5.3.12a}$$

$$\frac{\mathrm{d}w}{\mathrm{d}s} \approx -\left(\frac{\xi_2}{\xi}\right)^{1/2}\left[\frac{M_2}{\lambda_2 D} A\left(\frac{\xi_2-\xi}{2^{1/2}}\right) - \frac{H_2 \cos\alpha}{2\lambda_2^2 D} C\left(\frac{\xi_2-\xi}{2^{1/2}}\right)\right], \tag{5.3.12b}$$

$$N_s \cot\alpha = Q_s \approx \left(\frac{\xi_2}{\xi}\right)^{5/2}\left[2\lambda_2 M_2 B\left(\frac{\xi_2-\xi}{2^{1/2}}\right) + H_2 \cos\alpha\, D\left(\frac{\xi_2-\xi}{2^{1/2}}\right)\right], \tag{5.3.12c}$$

$$M_s \approx \left(\frac{\xi_2}{\xi}\right)^{3/2}\left[M_2 C\left(\frac{\xi_2-\xi}{2^{1/2}}\right) - \frac{H_2 \cos\alpha}{\lambda_2} B\left(\frac{\xi_2-\xi}{2^{1/2}}\right)\right], \tag{5.3.12d}$$

$$M_\theta \approx \nu M_s, \tag{5.3.12e}$$

$$N_\theta \approx \frac{Eh}{R_2/\cos\alpha}\left(\frac{\xi_2}{\xi}\right) w, \tag{5.3.12f}$$

$$u_s \approx -\frac{\nu Q_s}{E}\frac{R_2/\cos\alpha}{h}\left(\frac{\xi}{\xi_2}\right)^2. \tag{5.3.12g}$$

These can be further approximated by the equivalent cylinder equations when the meridional distance from the edge

$$x = s_2 - s, \tag{5.3.13a}$$

or the parameter

$$\frac{2^{3/2}\xi_2 x}{\xi_2} = 1 - \left(\frac{\xi}{\xi_2}\right)^2, \tag{5.3.13b}$$

is used, leading to

$$\frac{\xi}{\xi_2} = \left[1 - \frac{2^{3/2}\lambda_2 x}{\xi_2}\right]^{1/2} \approx 1, \tag{5.3.13c}$$

$$\frac{\xi_2 - \xi}{2^{1/2}} = \frac{\xi_2}{2^{1/2}}\left\{1 - \left[1 - \frac{2^{3/2}\lambda_2 x}{\xi_2}\right]^{1/2}\right\} \approx \lambda_2 x, \tag{5.3.13d}$$

when $2^{3/2}\lambda_2 x/\xi_2$ is small compared to unity. Although the equivalent cylinder expressions might be expected not to be as accurate for a complete cone as for one which extends to infinity, since the parameter ξ_2 and consequently the accuracy of the asymptotic equations, *decreases* away from the edge, calculations indicate that this is not necessarily so. In Figs. 5.9 the meridional distributions of deformations and stress resultants are shown for both a complete cone and an infinite cone having the same edge cross-sectional radius and semi-vertex angle such that ξ_1 and ξ_2 are both equal to 20 and are compared with the equivalent cylinder distributions. We see that for all quantities, both exact distributions are approximated about as well by the cylinder distribution, which is almost the average of the two.

(c) *Finite conical frustums*

The problem of a finite conical frustum subjected to edge loading (Fig. 5.10) can be treated in a similar manner, but all four constants of integration of the general conical shell equations have to be considered, rather than just two. The end result is so complicated and contains so many parameters that few calculated analytical results are available. With high-speed computing equipment and tables of the Bessel–Kelvin functions, an accurate detailed analysis should not be difficult, however. In those cases where the ends of the conical frustum are remote from each other, it is sufficiently accurate, as for the cylindrical shell, to assume that loads applied at one end do not have an effect on the other. For conical shells for which the minimum value of ξ is such that cylinder-like behavior can be expected, ξ greater than 20 or so, we may use the asymptotic equations [Eqs. (5.3.9) or (5.3.12)] and our knowledge

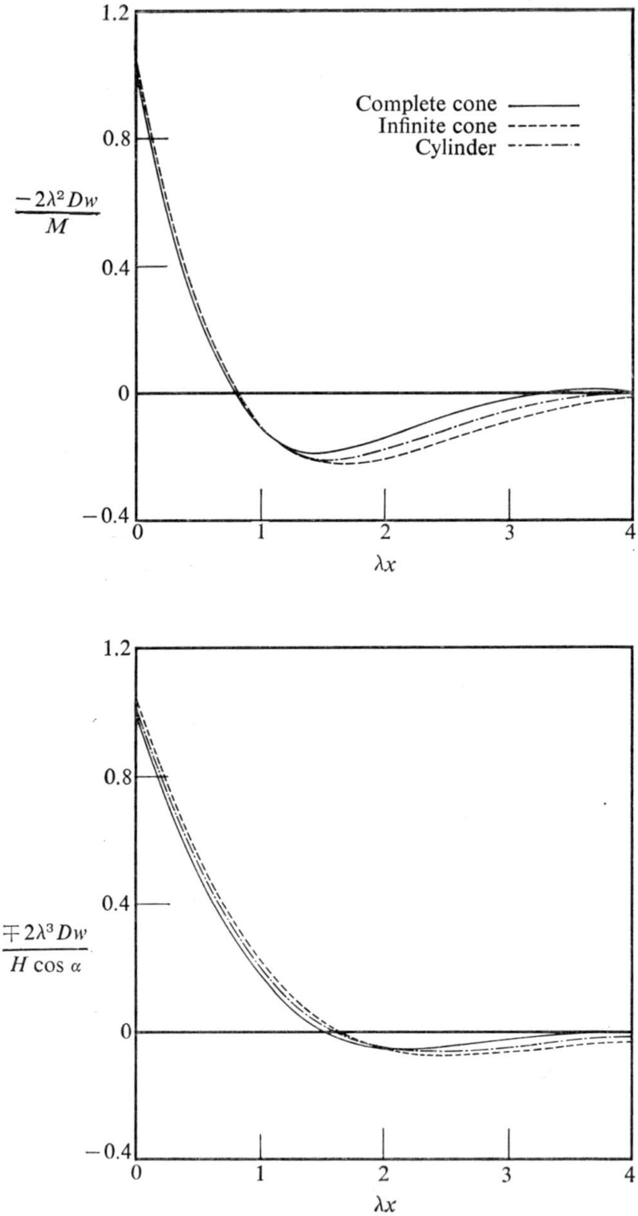

(a) Normal displacement

Fig. 5.9. Variation of deformations and stress resultants for complete and infinite edge-loaded cones ($\xi_1 = \xi_2 = 20$, $\nu = 0.3$)

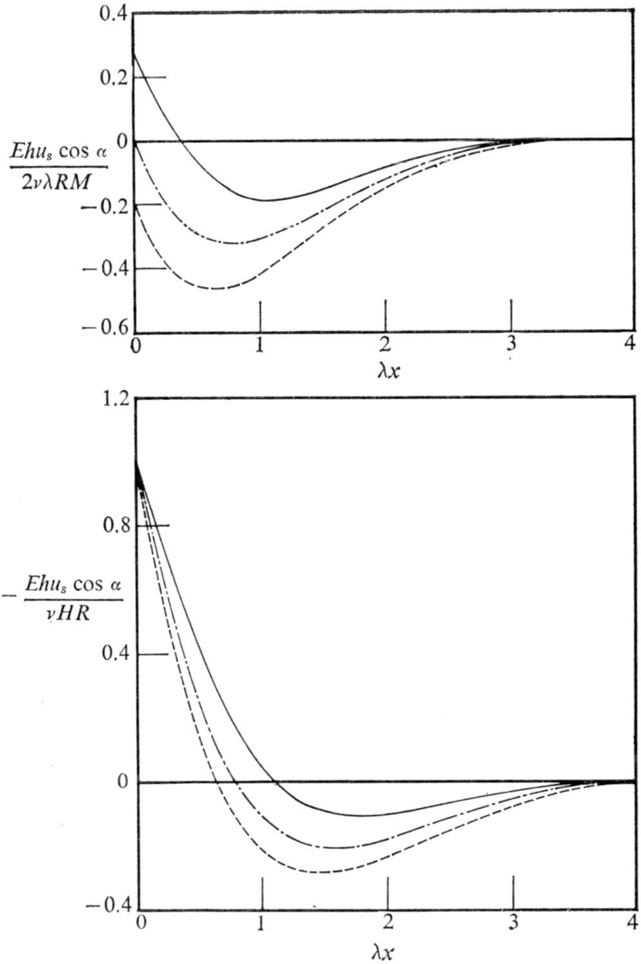

(b) Meridional displacement

Fig. 5.9. Continued

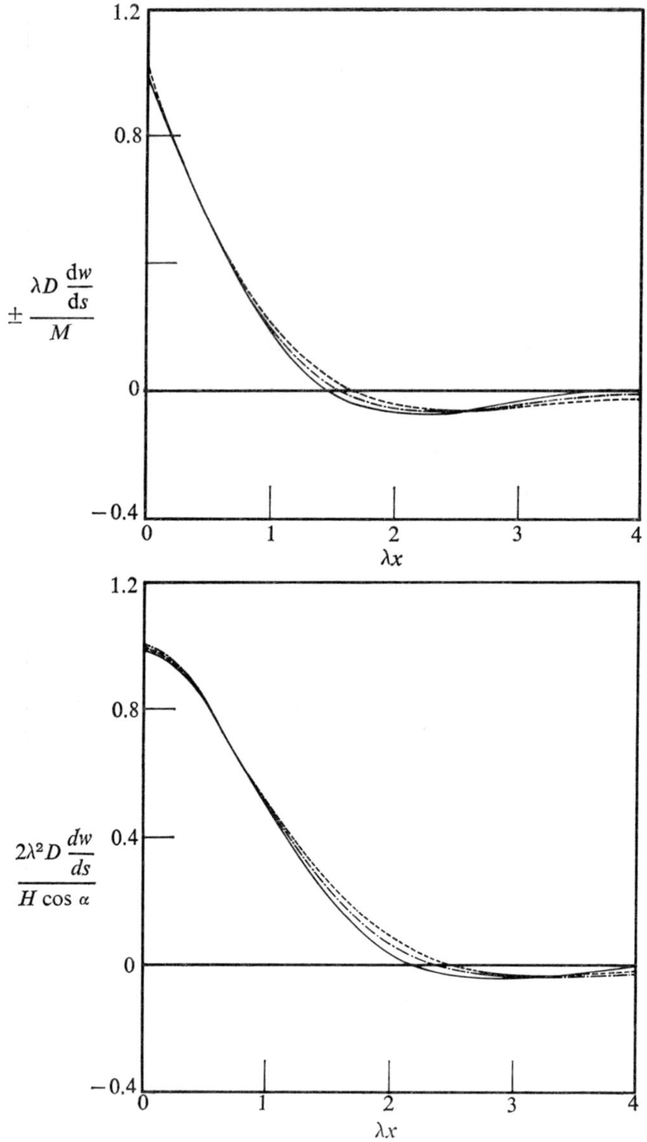

(c) Change of slope

Fig. 5.9. Continued

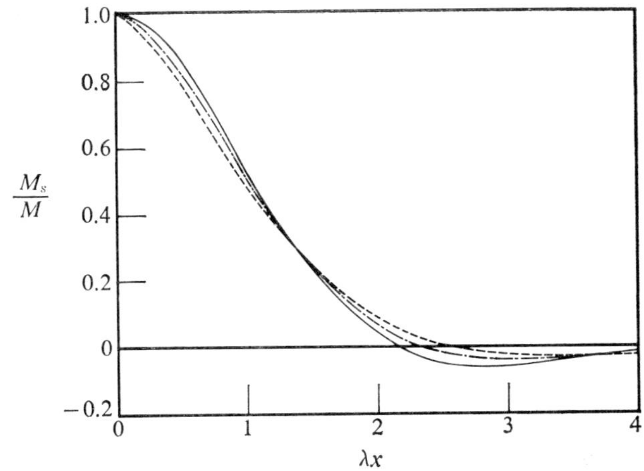

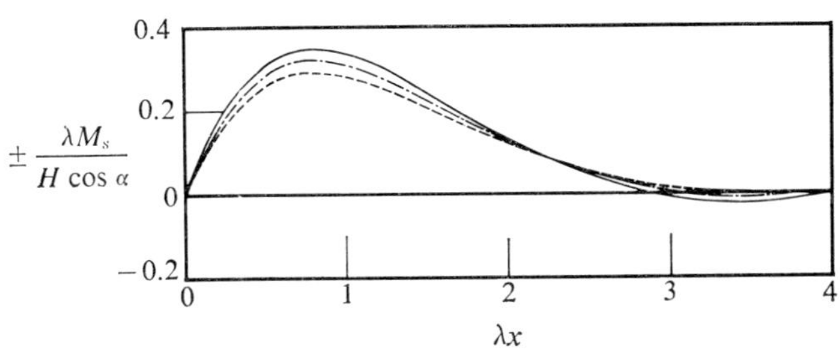

(d) Meridional moment

Fig. 5.9. Continued

of thc behavior of cylindrical shells to obtain immediately the estimate that end interactions may be neglected if the quantity

$$\mu = \frac{\xi_2 - \xi_1}{2^{1/2}} = \frac{2[3(1-\nu^2)]^{1/4}}{[h\cot\alpha]^{1/2}}[(s_2)^{1/2} - (s_1)^{1/2}] \tag{5.3.14}$$

is greater than 6, a result which is also indicated by a more accurate analysis.†

† J. H. Baltrukonis, 'Influence Coefficients for Edge-Loaded Short, Thin Conical Frustums,' *J. Appl. Mech.*, June 1959, vol. 20, no. 2, pp. 241–245. See also F. Dubois, 'Uber die Festigkeit der Kegelschale,' Dissertation, E.T.H. Zurich, 1917.

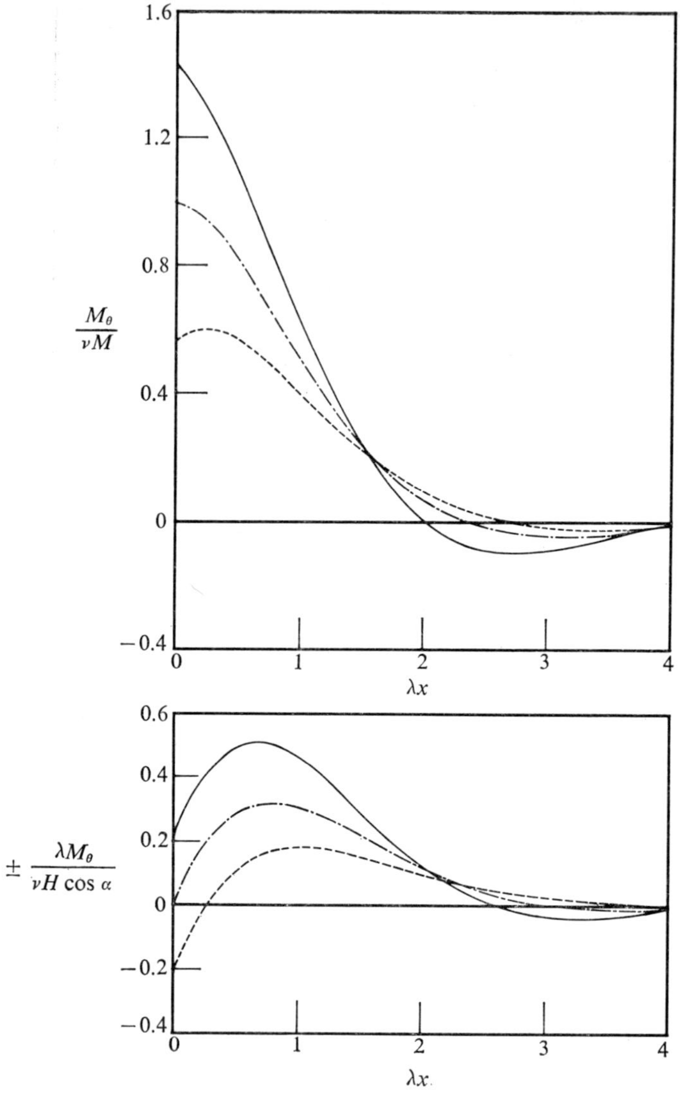

(e) Circumferential moment

Fig. 5.9. Continued

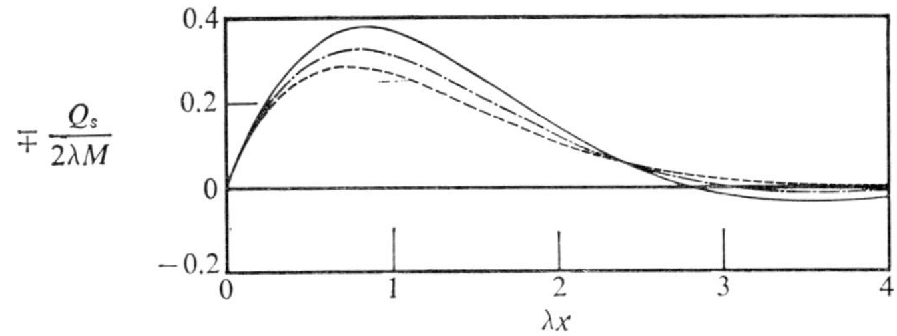

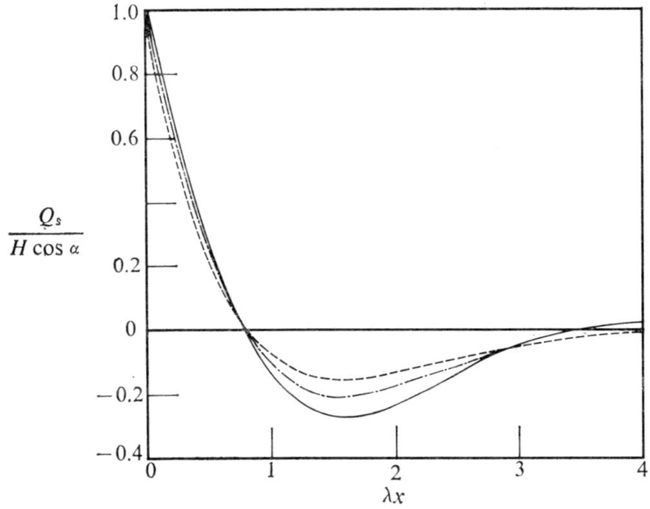

(f) Transverse shear

Fig. 5.9. Continued

We may also use the asymptotic expansions to obtain the result that for sufficiently large values of ξ the edge influence coefficients for a finite cone should be closely approximated by those for a cylindrical shell, which have already been calculated. To show this, let us write the asymptotic expressions for the displacement normal to the axis of revolution, the rotation, the meridional bending moment, and the force normal to the axis of revolution as

$$\frac{u_H}{\xi^{1/2}\cos\alpha} \approx A_1 A\left(\frac{\xi-\xi_1}{2^{1/2}}\right) + A_2 B\left(\frac{\xi-\xi_1}{2^{1/2}}\right) + A_3 A\left(\frac{\xi_2-\xi}{2^{1/2}}\right) + A_4 B\left(\frac{\xi_2-\xi}{2^{1/2}}\right), \tag{5.3.15a}$$

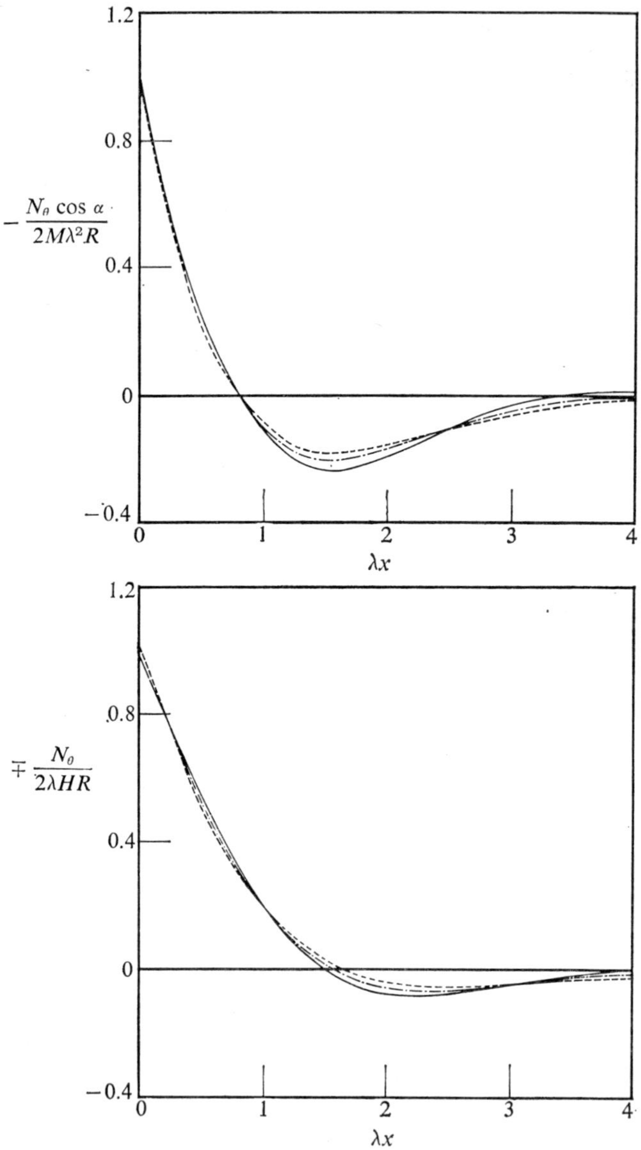

(g) Circumferential force

Fig. 5.9. Concluded

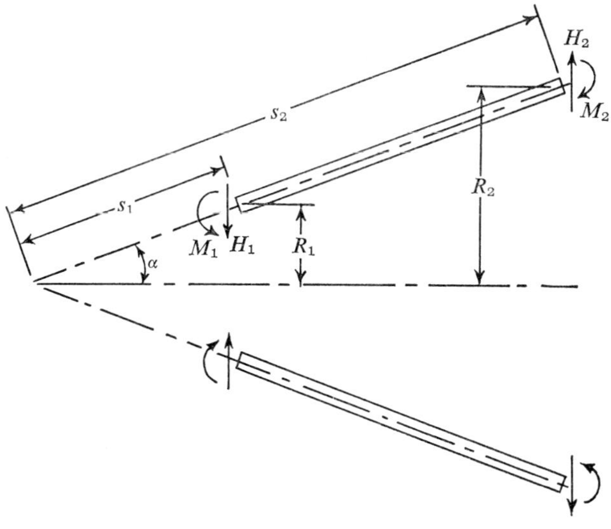

Fig. 5.10. Finite truncated cone with edge loading

$$\frac{\xi^{1/2}}{\xi_1\lambda_1}\frac{\mathrm{d}w}{\mathrm{d}s} \approx -A_1C\left(\frac{\xi-\xi_1}{2^{1/2}}\right) + A_2D\left(\frac{\xi-\xi_1}{2^{1/2}}\right)$$

$$+ A_3C\left(\frac{\xi_2-\xi}{2^{1/2}}\right) - A_4D\left(\frac{\xi_2-\xi}{2^{1/2}}\right), \tag{5.3.15b}$$

$$\frac{\xi^{3/2}}{2(\xi_1\lambda_1)^2D}M_s \approx -A_1B\left(\frac{\xi-\xi_1}{2^{1/2}}\right) + A_2A\left(\frac{\xi-\xi_1}{2^{1/2}}\right)$$

$$- A_3B\left(\frac{\xi_2-\xi}{2^{1/2}}\right) + A_4A\left(\frac{\xi_2-\xi}{2^{1/2}}\right), \tag{5.3.15c}$$

$$\frac{\xi^{5/2}H\cos\alpha}{2(\xi_1\lambda_1)^3D} \approx -A_1D\left(\frac{\xi-\xi_1}{2^{1/2}}\right) - A_2C\left(\frac{\xi-\xi_1}{2^{1/2}}\right)$$

$$+ A_3D\left(\frac{\xi_2-\xi}{2^{1/2}}\right)A_4\left(\frac{\xi_2-\xi}{2^{1/2}}\right), \tag{5.3.15d}$$

which can be seen to be identical in form with Eqs. (4.1.13) and (4.1.14) for the cylinder. In these equations the relationship

$$\xi_1\lambda_1 = \xi_2\lambda_2, \tag{5.3.16}$$

has been used. For a finite conical frustum loaded only at one end, the equations for the determination of the constants of integration are identical with those for the edge-loaded cylindrical shell. Thus, after some manipulation, we may write the edge coefficient equations as

$$\begin{Vmatrix} -\dfrac{\lambda_1^{1/2} u_{H_1}}{\cos\alpha} \\ \lambda_1^{-1/2}\dfrac{dw_1}{dx} \\ -\dfrac{\lambda_2^{1/2} u_{H_2}}{\cos\alpha} \\ -\lambda_2^{-1/2}\dfrac{dw_2}{dx} \end{Vmatrix} = \begin{Vmatrix} \delta_{H_{11}}(\mu) & \delta_{M_{11}}(\mu) & \delta_{H_{12}}(\mu) & \delta_{M_{12}}(\mu) \\ \Theta_{H_{11}}(\mu) & 2\Theta_{M_{11}}(\mu) & \Theta_{H_{12}}(\mu) & 2\Theta_{M_{12}}(\mu) \\ \delta_{H_{21}}(\mu) & \delta_{M_{21}}(\mu) & \delta_{H_{22}}(\mu) & \delta_{M_{22}}(\mu) \\ \Theta_{H_{21}}(\mu) & 2\Theta_{M_{21}}(\mu) & \Theta_{H_{22}}(\mu) & 2\Theta_{M_{22}}(\mu) \end{Vmatrix} \begin{Vmatrix} \dfrac{H_1\cos\alpha}{2\lambda_1^{5/2}D} \\ \dfrac{M_1}{2\lambda_1^{3/2}D} \\ -\dfrac{H_2\cos\alpha}{2\lambda_2^{5/2}D} \\ \dfrac{M_2}{2\lambda_2^{3/2}D} \end{Vmatrix}$$

where

$$\delta_{H_{11}}(\mu) = \delta_{H_{22}}(\mu), \tag{5.3.18a}$$
$$\Theta_{M_{11}}(\mu) = \Theta_{M_{22}}(\mu), \tag{5.3.18b}$$
$$\delta_{M_{11}}(\mu) = \Theta_{H_{11}}(\mu) = \delta_{M_{22}}(\mu) = \Theta_{H_{22}}(\mu), \tag{5.3.18c}$$
$$\delta_{H_{12}}(\mu) = \delta_{H_{21}}(\mu), \tag{5.3.18d}$$
$$\Theta_{M_{12}}(\mu) = \Theta_{M_{21}}(\mu), \tag{5.3.18e}$$
$$\delta_{M_{12}}(\mu) = \Theta_{H_{12}}(\mu) = \delta_{M_{21}}(\mu) = \Theta_{H_{21}}(\mu), \tag{5.3.18f}$$

and are identical with the corresponding coefficients tabulated for the cylindrical shell. The reasonably good accuracy of the approximation is indicated in Fig. 5.11 by a comparison of some of the coefficients with values given by a more accurate analysis.† In the exact solution of the problem, Eqs. (5.3.17) retain their form but the influence coefficients vary with ξ_1 and ξ_2 in addition to μ. The equations which the influence coefficients actually satisfy are the less restrictive relations

$$\delta_{M_{11}} = \Theta_{H_{11}}, \tag{5.3.19a}$$
$$\delta_{H_{12}} = \delta_{H_{21}}, \tag{5.3.19b}$$
$$\delta_{M_{12}} = \Theta_{H_{21}}, \tag{5.3.19c}$$
$$\Theta_{H_{12}} = \delta_{M_{21}}, \tag{5.3.19d}$$
$$\Theta_{M_{12}} = \Theta_{M_{21}}, \tag{5.3.19e}$$
$$\delta_{M_{22}} = \Theta_{H_{22}}, \tag{5.3.19f}$$

which are reciprocity relationships required for conservation of energy. As we shall see later on, Eqs. (5.3.17) and (5.3.18) require only slight modifications to make them applicable as an approximation for shells of revolution with arbitrary meridional shape.

† J. H. Baltrukonis, loc. cit.

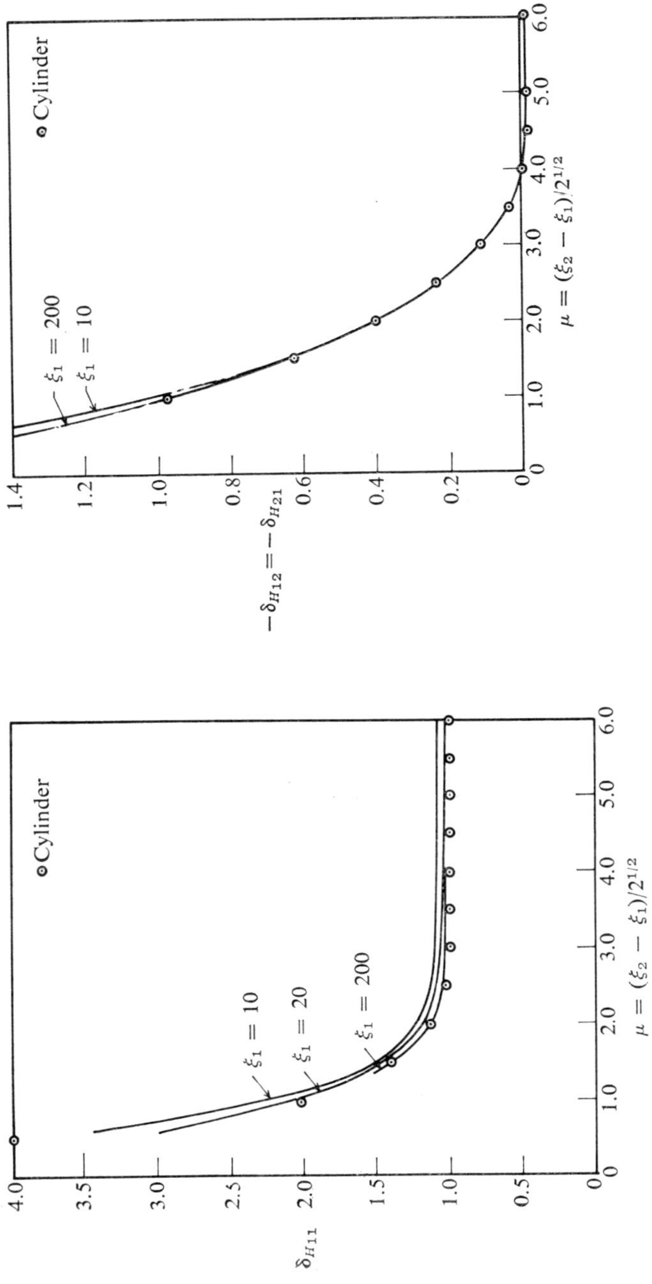

(a) Deflection at small end due to shear force at small end

(b) Deflection at one end due to shear force at the other end

Fig. 5.11. Comparison of exact and approximate influence coefficients for finite conical frustums

Fig. 5.11. Continued over

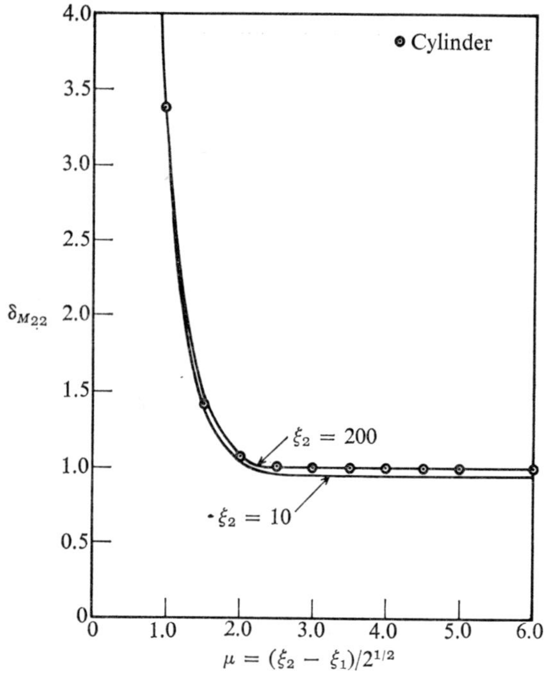

(c) Deflection at large end due to moment at large end

Fig. 5.11. Concluded

5.4 Particular solutions of the conical shell equations

When the conical shell is loaded on its bounding surfaces, or by an axial force P applied to the edges of the shell, we must add a particular solution of Eq. (5.1.12b) to the homogeneous solution given by Eq. (5.2.9). These particular solutions can always be found by the method of variation of parameters with the use of the four solutions of the homogeneous equation, but the resulting expression is quite complicated and involves integrals which are not easily obtained. If $q(s)$ is of the form

$$q(s) = q_n s^n, \tag{5.4.1}$$

some particular solutions can be found readily. For n equal to -2, -1, 0, or 1, a particular solution of Eq. (5.1.12b) is

$$s\,\frac{dw_{pn}}{ds} = \frac{q_n \tan^2 \alpha}{Eh}\, s^{n+2}, \tag{5.4.2}$$

which leads to

$$w_{pn} = \frac{q_n \tan^2 \alpha}{Eh} \ln s, \quad n = -2, \tag{5.4.3a}$$

$$w_{pn} = \frac{q_n \tan^2 \alpha}{Eh} \frac{s^{n+2}}{n+2}, \quad n = -1, 0, 1. \tag{5.4.3b}$$

If n is equal to or greater than 2, the particular solution is given by

$$s \frac{dw_{pn}}{ds} = \frac{q_n s^{n+2} \tan^2 \alpha}{Eh} \sum_{k=0}^{m} (-1)^k \frac{n!(n+2)!}{(n+2)!(n-2k)!} \left(\frac{2}{\xi}\right)^{4k}, \tag{5.4.4a}$$

where

$$\begin{aligned} m &= \frac{n-1}{2} \quad \text{if } n \text{ is odd,} \\ &= \frac{n}{2} \quad \text{if } n \text{ is even.} \end{aligned} \tag{5.4.4b}$$

Then

$$w_{pn} = \frac{q_n s^{n+2} \tan^2 \alpha}{(n+2)Eh} \sum_{k=0}^{m} (-1)^k \frac{n+2}{n+2-2k} \frac{n!(n+2)!}{(n-2k)!(n-2k+2)!} \left(\frac{2}{\xi}\right)^{4k}. \tag{5.4.5}$$

For n equal to $-2, -1, 0$, or 1 the particular solution can be expressed as

$$s \frac{dw_{pn}}{ds} = \frac{q(s) s^2 \tan^2 a}{Eh}. \tag{5.4.6}$$

For other values of n the same solution is approximately correct if the range of values of ξ is such that $(2n/\xi)^4$ is much less than unity. The form of the solution given by Eq. (5.4.6) is identical with that which is obtained from the membrane theory of shells. For, from Eqs. (3.8.5), (3.8.6), and (5.1.3) we have

$$N_\theta = \left[s q_\zeta + \frac{d}{ds}(s m_s) \right] \tan \alpha, \tag{5.4.7a}$$

$$\frac{d}{ds}(s N_s) - N_\theta + s q_s = 0, \tag{5.4.7b}$$

$$\frac{du_s}{ds} = \frac{N_s - \nu N_\theta}{Eh}, \tag{5.4.7c}$$

$$\frac{u_s + w \cot \alpha}{s} = \frac{N_\theta - \nu N_s}{Eh}. \tag{5.4.7d}$$

The elimination of u_s from Eqs. (5.4.7c) and (5.4.7d) yields

$$\frac{\mathrm{d}w}{\mathrm{d}s} = \frac{\tan\alpha}{Eh}\left\{\frac{\mathrm{d}}{\mathrm{d}s}(sN_\theta) - N_s - \nu\left[\frac{\mathrm{d}}{\mathrm{d}s}(sN_s) - N_\theta\right]\right\}, \tag{5.4.8a}$$

or, with the use of Eq. (5.4.7b),

$$s\frac{\mathrm{d}w}{\mathrm{d}s} = \frac{s\tan\alpha}{Eh}\left\{\frac{\mathrm{d}}{\mathrm{d}s}(sN_\theta) - \frac{1}{s}\int_{s_1}^{s} N_\theta\,\mathrm{d}s + \frac{1}{s}\int_{s_1}^{s} sq_s\,\mathrm{d}s + \nu s q_s - \frac{P}{2\pi s\sin^2\alpha}\right\}, \tag{5.4.8b}$$

which reduces to the required expression when N_θ is replaced by the values given by Eq. (5.4.7a). Thus in many cases the displacements given by membrane theory may be used as an adequate particular solution of the bending equations.

Under the same conditions, namely that $(2n/\xi)^4$ is less than unity, the direct stress resultants given by membrane theory are an adequate approximation to the corresponding values given by the particular solution. For example, let us consider the case

$$m_s = q_s = P = 0, \tag{5.4.9a}$$

$$q_\zeta = q_{\zeta n}s^n. \tag{5.4.9b}$$

Then

$$q(s) = \frac{(n+3)(n+1)}{n+2}q_{\zeta n}s^n, \tag{5.4.10}$$

and if n is integral and greater than or equal to -2,

$$s\frac{\mathrm{d}w}{\mathrm{d}s} = \frac{(n+3)(n+1)}{n+2}q_{\zeta n}\frac{s^{n+2}\tan^2\alpha}{Eh} \times \sum_{k=0}^{m}(-1)^k\frac{n!(n+2)!}{(n-2k)!(n-2k+2)!}\left(\frac{2}{\xi}\right)^{4k}. \tag{5.4.11}$$

From Eqs. (5.1.8), (5.1.9b), (5.1.10), (5.1.7c), and (5.1.7d) we then have

$$N_s = \frac{q_{\zeta n}s^{n+1}}{n+2}\tan\alpha\sum_{k=0}^{m+1}(-1)^k\frac{(n+1)!(n+3)!}{(n-2k+1)!(n-3k+3)!}\left(\frac{2}{\xi}\right)^{4k}, \tag{5.4.12a}$$

$$N_\theta = q_{\zeta n}s^{n+1}\tan\alpha \times \sum_{k=0}^{m+1}(-1)^k\frac{n+2-2k}{n+2}\frac{(n+1)!(n+3)!}{(n-2k+1)!(n-2k+3)!}\left(\frac{2}{\xi}\right)^{4k}. \tag{5.4.12b}$$

The first term of which is the membrane solution. Note that in general N_θ is larger than N_s. The remaining stress resultants are given by

$$Q_s = \frac{q_{\zeta n} s^{n+1}}{n+2} \sum_{k=1}^{m+1} (-1)^k \frac{(n+1)!(n+3)!}{(n-2k+1)!(n-2k+3)!} \left(\frac{2}{\xi}\right)^{4k}, \tag{5.4.13a}$$

$$M_s = q_{\zeta n} s^{n+2} \times \sum_{k=1}^{m+1} (-1)^k \frac{n+3+\nu-2k}{n+2} \frac{(n+1)!(n+3)!}{(n-2k+2)!(n-2k+4)!} \left(\frac{2}{\xi}\right)^{4k}, \tag{5.4.13b}$$

$$M_\theta = q_{\zeta n} s^{n+2} \times \sum_{k=1}^{m+1} (-1)^k \frac{1+\nu(n-2k+3)}{n+2} \frac{(n+1)!(n+3)!}{(n-2k+2)!(n-2k+4)!} \left(\frac{2}{\xi}\right)^{4k}. \tag{5.4.13c}$$

The transverse shear force Q_s can be seen to be negligible compared to N_s when $(2n/\xi)^4 \tan \alpha$ is very much less than unity. The stresses due to moments are negligible compared to those due to direct forces when we impose the somewhat more stringent requirement that $(2n/\xi)^2$ is very much less than unity. In this case the maximum stress due to M_s, the larger moment, is small compared to the direct stress due to N_θ, the larger tangential force.

5.5 Complete conical shell subjected to external pressure

We have discussed exact solutions for edge-loaded conical shells and compared them with an approximate solution corresponding to an equivalent cylindrical shell. A comparison of stress resultants obtained by the two methods does not reveal to us everything we want to know since we are usually more interested in stresses than in stress resultants and particularly in the maximum stress rather than in the entire stress distribution. To illustrate the effect of various approximations on the solutions for the stress distributions, let us consider a complete conical shell which is clamped at its base and which is subjected to uniform external pressure p (Fig. 5.12a). For this problem, the surface loading is given by

$$q_\zeta = -p\left(1 + \frac{h}{2R_\theta}\right), \tag{5.5.1a}$$

$$q_s = m_s = P = 0, \tag{5.5.1b}$$

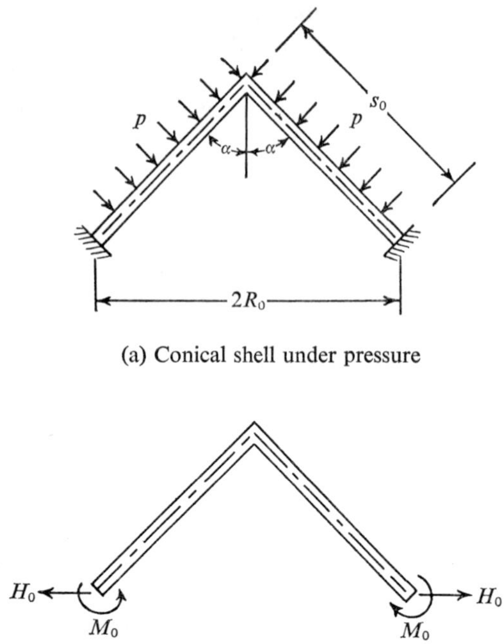

Fig. 5.12. Clamped complete conical shell subjected to uniform external pressure

and, from Eq. (5.1.12c),

$$q(s) = \frac{1}{s}\frac{\mathrm{d}}{\mathrm{d}s}(s^2 q_\zeta) - \frac{1}{s^2}\int_0^s s q_\zeta \,\mathrm{d}s = -\tfrac{3}{2}p. \tag{5.5.1c}$$

The use of Eqs. (5.4.2) and (5.4.3b), then yields

$$\frac{\mathrm{d}w_p}{\mathrm{d}s} = -\frac{3ps\tan^2\alpha}{2Eh}, \tag{5.5.2a}$$

$$w_p = -\frac{3ps^2\tan^2\alpha}{4Eh}. \tag{5.5.2b}$$

The stress resultants corresponding to this particular solution are obtained from Eqs. (5.4.7), (5.4.8), (5.1.9b), and (5.1.10) as

$$M_{sp} = M_{\theta p} = \frac{ph^2\tan^2\alpha}{8(1-\nu)}, \tag{5.5.3a}$$

$$Q_{sp} = 0, \tag{5.5.3b}$$

$$N_{sp} = -\frac{ps \tan\alpha}{2}\left(1 + \frac{h}{s\tan\alpha}\right), \tag{5.5.3c}$$

$$N_{\theta p} = -\,ps\tan\alpha\left(1 + \frac{h}{2s\tan\alpha}\right). \tag{5.5.3d}$$

Finally, the meridional displacement can be obtained from the above equations and Eq. (5.1.13b) as

$$u_{sp} = -\frac{(1-2\nu)ps^2\tan\alpha}{4Eh}\left(1 + 2\,\frac{1-\nu}{1-2\nu}\,\frac{h}{s\tan\alpha}\right). \tag{5.5.4}$$

At the base of the shell the particular solution results in an inward horizontal radial displacement

$$\begin{aligned} u_{Hp} &= (u_{sp}\sin\alpha + w_p\cos\alpha)_{s=s_0} \\ &= -\frac{(2-\nu)ps_0^2\sin^2\alpha}{2Eh\cos\alpha}\left(1 + \frac{1-\nu}{2-\nu}\,\frac{h}{s_0\tan\alpha}\right), \end{aligned} \tag{5.5.5}$$

and a change of slope given by Eq. (5.5.2a). These quantities can be nullified by the application of horizontal loads H_0 and bending moments M_0 at the base (Fig. 5.12b) which produce a base horizontal radial displacement and change of slope given by Eqs. (5.3.2) and (5.3.11) as

$$u_{H_0} = -\left[\frac{H_0\cos\alpha}{2\lambda_0^3 D}\,\delta_{H_{22}}(\xi_0) + \frac{M_0}{2\lambda_0^2 D}\,\delta_{M_{22}}(\xi_0)\right]\cos\alpha, \tag{5.5.6a}$$

$$\frac{dw_0}{ds} = \frac{H_0\cos\alpha}{2\lambda_0^2 D}\,\Theta_{H_{22}}(\xi_0) + \frac{M_0}{\lambda_0 D}\,\Theta_{M_{22}}(\xi_0). \tag{5.5.6b}$$

When the sum of Eqs. (5.5.2a) and (5.5.6b) and of Eqs. (5.5.4) and (5.5.6b) are set equal to zero, we obtain values of the required edge horizontal forces and bending moments as

$$H_0\cos\alpha = -\frac{ps_0}{2^{1/2}\xi_0}\,\frac{2\langle 2-\nu+[\{8(1-\nu)[3(1-\nu^2)]^{1/2}\cot^2\alpha\}/\xi_0^2]\rangle \times \Theta_{M_{22}}(\xi_0) + 3(2^{3/2}/\xi_0)\Theta_{H_{22}}(\xi_0)}{1 + \dfrac{2^{5/2}\nu}{\xi_0}\,\Theta_{M_{11}}(\xi_0)}, \tag{5.5.7a}$$

$$M_0 = \frac{ps_0 h\tan\alpha}{4[3(1-\nu^2)]^{1/2}}\,\frac{\langle 2-\nu+[\{8(1-\nu)[3(1-\nu^2)]^{1/2}\cot^2\alpha\}/\xi_0^2]\rangle \times \delta_{M_{22}}(\xi_0) + 3(2^{3/2}/\xi_0)\delta_{H_{22}}}{1 + \dfrac{2^{5/2}\nu}{\xi_0}\,\Theta_{M_{22}}(\xi_0)}, \tag{5.5.7b}$$

where the relationships

$$\frac{h}{s_0 \tan\alpha} = \frac{8[3(1-\nu^2)]^{1/2}\cot^2\alpha}{\xi_0^2}, \tag{5.5.8a}$$

$$\lambda_0 s_0 = 2^{-3/2}\xi_0, \tag{5.5.8b}$$

have been used. The additional stress resultants due to the end loads given by Eqs. (5.5.7) can now be obtained with the use of the relations of Sections 5.2 and 5.3.

Now let us redo the same problem using (1) the equivalent cylinder approximation and (2) the asymptotic solution for the distribution due to edge forces. From Eqs. (5.3.12) we may obtain the pertinent equations as

$$\frac{u_H}{\cos\alpha} \approx \left[\frac{H_0\cos\alpha}{2\lambda_0^3 D}A(\psi) - \frac{M_0}{2\lambda_0^2 D}D(\psi)\right]\left(\frac{\xi}{\xi_0}\right)^{n/2} \approx w, \tag{5.5.9a}$$

$$\frac{dw}{ds} \approx \left[\frac{H_0\cos\alpha}{2\lambda_0^2 D}C(\psi) - \frac{M_0}{\lambda_0 D}A(\psi)\right]\left(\frac{\xi_0}{\xi}\right)^{n/2}, \tag{5.5.9b}$$

$$N_s\cot\alpha = Q_s \approx [2\lambda_0 M_0 B(\psi) + H_0\cos\alpha D(\psi)]\left(\frac{\xi_0}{\xi}\right)^{5n/2}, \tag{5.5.9c}$$

$$M_s \approx \left[M_0 C(\psi) - \frac{H_0\cos\alpha}{\lambda_0}B(\psi)\right]\left(\frac{\xi_0}{\xi}\right)^{3n/2}, \tag{5.5.9d}$$

$$M_\theta \approx \nu M_s, \tag{5.5.9e}$$

$$N_\theta \approx \frac{Eh}{R_0/\cos\alpha}w\left(\frac{\xi_0}{\xi}\right)^n, \tag{5.5.9f}$$

where

$$n = 0, \quad \psi = \lambda_0 x = \frac{\xi_0}{2^{3/2}}\left[1 - \left(\frac{\xi}{\xi_0}\right)^2\right] \tag{5.5.10a}$$

for the equivalent cylinder approximation, and

$$n = 1, \quad \psi = \frac{\xi_0 - \xi}{2^{1/2}} \tag{5.5.10b}$$

for the asymptotic approximation. In both cases we have

$$-\delta_{H22} \approx \delta_{M22} \approx \Theta_{H22} \approx -\Theta_{M22} \approx 1, \tag{5.5.11}$$

so that the edge force and bending moment are given by

$$H_0\cos\alpha \approx \frac{2^{1/2}ps_0}{\xi_0}\left\{2 - \nu + \frac{8(1-\nu)[3(1-\nu^2)]^{1/2}\cot^2\alpha}{\xi_0^2} - 3\frac{2^{1/2}}{\xi_0}\right\}, \tag{5.5.12a}$$

$$M_0 \approx \frac{ps_0 h \tan\alpha}{4[3(1-\nu^2)]^{1/2}}\left\{2-\nu+\frac{8(1-\nu)[3(1-\nu^2)]^{1/2}\cot\alpha}{\xi_0^2}-3\frac{2^{3/2}}{\xi_0}\right\}. \tag{5.5.12b}$$

Calculations were made for a conical shell with

$\alpha = 60°$,

$\dfrac{R_0/\cos\alpha}{h} = 90.8$,

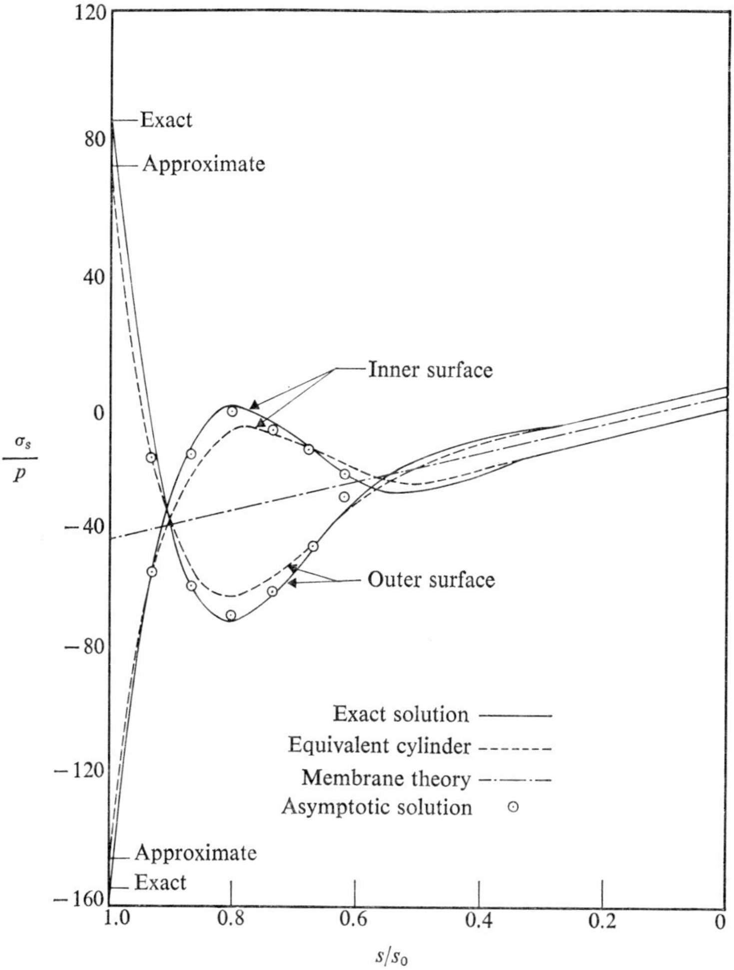

(a) Meridional stress

Fig. 5.13. Comparison of various methods of conical shell analysis

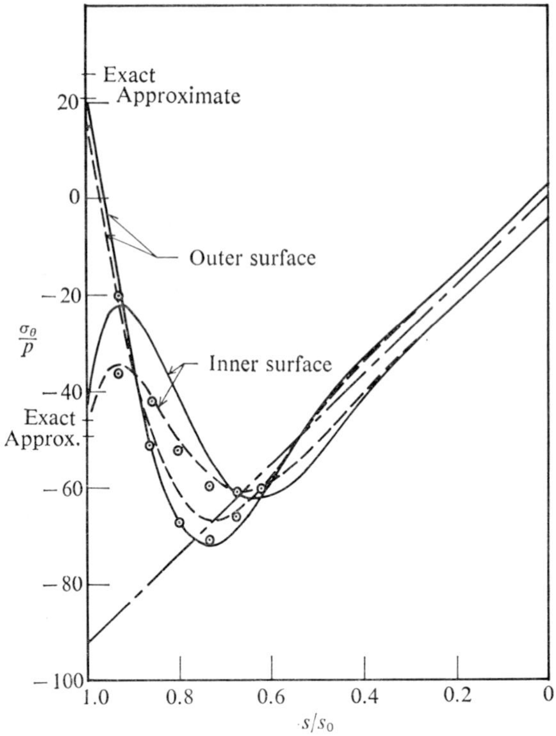

(b) Circumferential stress

Fig. 5.13. Concluded

which yields a value of ξ_0 of 20. The edge forces given by the two solutions differ by about 8 percent and are shown in Table 5.3. Meridional and circumferential stresses calculated by the exact and approximate methods are

Table 5.3 Exact and approximate edge loadings

Quantity	Eq. (5.5.7)	Eq. (5.5.12)
H_0/ph	11.98	11.09
M_0/ph^2	19.12	17.63

shown in Fig. 5.13. The approximate maximum meridional stress, obtained from either the asymptotic or equivalent cylinder solutions, which occurs at the outer surface at the edge, differs from the exact value by about 6 percent.

The distribution of the meridional stresses given by the asymptotic solution is seen to agree somewhat better with the exact solution as we move away from the edge than does the equivalent cylinder solution. The circumferential stress distributions are similarly in reasonably good agreement. The maximum circumferential stress given by the asymptotic solution differs from the exact value by about 2 percent whereas the value given by the equivalent cylinder method differs from the exact value by about 6 percent. Also included for comparison is the stress distribution predicted by the membrane theory of shells which can be seen to be a reasonable approximation away from the clamped edge. For larger values of ξ_0 the agreement between the exact and equivalent cylinder or asymptotic distributions would be improved and the membrane solution would be valid over a greater portion of the shell.

Supplementary references

[1] Flügge, W., and Blythe, W., "Axisymmetric Apex Loadings on Conical Shells," *J. Eng. Mech. Div., ASCE,* vol. 94, no. EM 1, February 1968, pp. 57–77.

[2] Min-Yuan, M. A., "Sur le calcul des pièces coniques de révolution travaillant à la flexion," Dissertation, Univ. of Grenoble, France, 1958. Also abridged in Bulletin S.F.M. No. 23 (Editions Science et Industrie), pp. 1–16.

[3] Watts, G. W., and Lang, H. A., "Stresses in a Pressure Vessel with a Conical Head," *Trans. ASME,* vol. 74, April 1952, pp. 315–326.

[4] Wenk, E., and Taylor, C. E., "Analysis of Stresses at the Reinforced Intersection of Conical and Cylindrical Shells," David Taylor Model Basin Report 826, 1953.

[5] Borg, M. F., "Observations of Stresses and Strains Near Intersections of Conical and Cylindrical Shells," David Taylor Model Basin Report 911, March 1956.

6

The spherical shell

6.1 General relations for spherical shells

The spherical shell is the final shell of revolution for which equations for the normal radial deformation can be solved in terms of known functions. In this case, however, although the functions are defined, they are untabulated and only recently have attempts been made to calculate rigorous results for comparison with those of various approximate approaches. The equations for the spherical shell are more directly analogous to the equations for the flat plate than are those for shells of any other middle surface shape since the curvature of the middle surface is independent of position and of orientation of the coordinate curves. The other distinguishing feature is that the overall force–strain relationships which have been adopted as convenient and adequate approximations for the cylinder and the cone are exact for the spherical shell.

Let us describe the middle surface of the sphere by the spherical coordinates ϕ and θ (Fig. 6.1), the coordinate curves for which are lines of

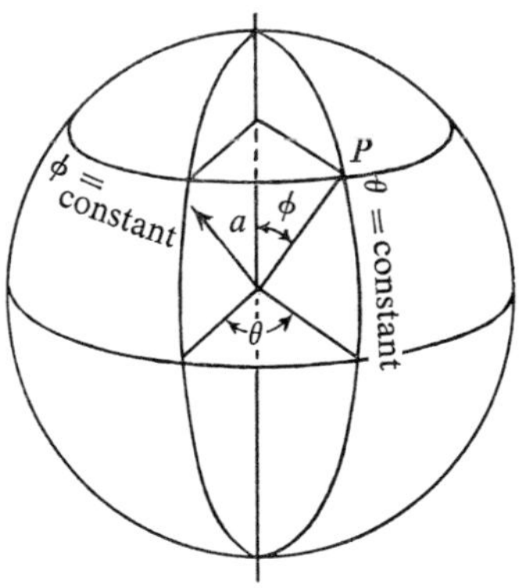

Fig. 6.1. Notation for spherical shell

curvature. The first and second fundamental quantities are then readily obtained as

$$g_\phi = R_\phi = R_\theta = a, \tag{6.1.1a}$$

$$g_\theta = a \sin\phi. \tag{6.1.1b}$$

The axisymmetric equations of equilibrium for the stress resultants and couples (Figs. 6.2) are given by Eqs. (2.5.3) and (2.5.5) as

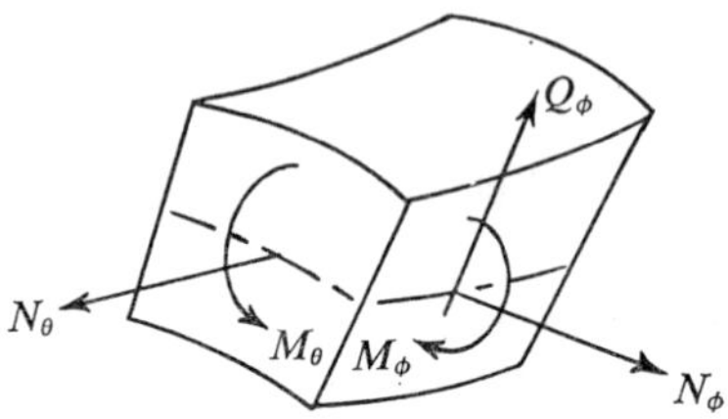

Fig. 6.2. Axisymmetric stress resultants for spherical shells

$$\frac{d}{d\phi}(N_\phi \sin\phi) - N_\theta \cos\phi + (Q_\phi + q_\phi a)\sin\phi = 0, \tag{6.1.2a}$$

$$\frac{d}{d\phi}(Q_\phi \sin\phi) - (N_\phi + N_\theta - q_\zeta a)\sin\phi = 0, \tag{6.1.2b}$$

$$\frac{d}{d\phi}(M_\phi \sin\phi) - M_\theta \cos\phi - (Q_\phi - m_\phi)a\sin\phi = 0, \tag{6.1.2c}$$

with

$$q_\phi = \left(1 + \frac{h}{2a}\right)^2 \tau_{\phi\zeta}^{(h/2)} - \left(1 - \frac{h}{2a}\right)^2 \tau_{\phi\zeta}^{(-h/2)} + \int_{-h/2}^{h/2} \left(1 + \frac{\zeta}{a}\right)^2 \rho_\phi^{(\zeta)}\, d\zeta, \tag{6.1.2d}$$

$$q_\zeta = \left(1 + \frac{h}{2a}\right)^2 \sigma_\zeta^{(h/2)} - \left(1 - \frac{h}{2a}\right)^2 \sigma_\zeta^{(-h/2)} + \int_{-h/2}^{h/2} \left(1 + \frac{\zeta}{a}\right)^2 \rho_\zeta^{(\zeta)}\, d\zeta, \tag{6.1.2e}$$

$$m_\phi = \frac{h}{2}\left[\left(1 + \frac{h}{2a}\right)^2 \tau_{\phi\zeta}^{(h/2)} + \left(1 - \frac{h}{2a}\right)^2 \tau_{\phi\zeta}^{(-h/2)}\right] + \int_{-h/2}^{h/2} \zeta\left(1 + \frac{\zeta}{a}\right)^2 \rho_\phi^{(\zeta)}\, d\zeta. \tag{6.1.2f}$$

From Eqs. (2.5.2b) and (2.5.7c) the axisymmetric strain and curvature change components are

$$\epsilon_\phi = \frac{1}{a}\left(\frac{du_\phi}{d\phi} + w\right), \tag{6.1.3a}$$

$$\epsilon_\theta = \frac{1}{a}(u_\phi \cot\phi + w), \tag{6.1.3b}$$

$$\kappa_\phi = -\frac{1}{a^2}\left(\frac{d^2w}{d\phi^2} - \frac{du}{d\phi}\right) = -\frac{1}{a^2}\left(\frac{d^2w}{d\phi^2} + w\right) + \frac{1}{a}\epsilon_\phi, \tag{6.1.3c}$$

$$\kappa_\theta = -\frac{1}{a^2}\left(\frac{dw}{d\phi} - u_\phi\right)\cot\phi = -\frac{1}{a^2}\left(\frac{dw}{d\phi}\cot\phi + w\right) + \frac{1}{a}\epsilon_\theta. \tag{6.1.3d}$$

The axisymmetric overall force–strain relationships can be obtained from Eqs. (2.7.2) as

$$N_\phi = K(\epsilon_\phi + \nu\epsilon_\theta), \tag{6.1.4a}$$

$$N_\theta = K(\epsilon_\theta + \nu\epsilon_\phi), \tag{6.1.4b}$$

$$\begin{aligned} M_\phi &= D(\kappa_\phi + \nu\kappa_\theta), \\ &= -\frac{D}{a^2}\left[\frac{d^2w}{d\phi^2} + w + \nu\left(\frac{dw}{d\phi}\cot\phi + w\right)\right] + \frac{h^2}{12a}N_\phi, \end{aligned} \tag{6.1.4c}$$

$$\begin{aligned} M_\theta &= D(\kappa_\theta + \nu\kappa_\phi) \\ &= -\frac{D}{a^2}\left[\frac{dw}{d\phi}\cot\phi + w + \nu\left(\frac{d^2w}{d\phi^2} + w\right)\right] + \frac{h^2}{12a}N_\theta. \end{aligned} \tag{6.1.4d}$$

Eqs. (6.1.4a) and (6.1.4b) can be seen to be identical in form with those for flat plates in a state of generalized plane stress. The bending moment equations can be made more recognizable if we rearrange them to read

$$M_\phi - \frac{h^2}{12a}N_\phi = -\frac{D}{a^2}\left[\frac{d^2w}{d\phi^2} + w + \nu\left(\frac{dw}{d\phi}\cot\phi + w\right)\right], \tag{6.1.5a}$$

$$M_\theta - \frac{h^2}{12a}N_\theta = -\frac{D}{a^2}\left[\frac{dw}{d\phi}\cot\phi + w + \nu\left(\frac{dw^2}{d\phi^2} + w\right)\right], \tag{6.1.5b}$$

and note that $h^2/12a$ is the distance from the middle surface to the centroid of an infinitesimal normal cross-section. Then the left sides of Eq. (6.1.5) are the respective moments about the centroidal surface (Fig. 6.3) and it is these moments which can be expressed, as in flat plate theory, entirely in terms of normal displacements.

We shall now proceed to operate on the foregoing equations for spherical shells to reduce them to a more workable form and to express all quantities in terms of the radial displacement w. First, an equation for the transverse shearing force Q_ϕ can be obtained as follows. We multiply Eq. (6.1.2a) by $h^2/12a$ and subtract the result from Eq. (6.1.2c) to yield

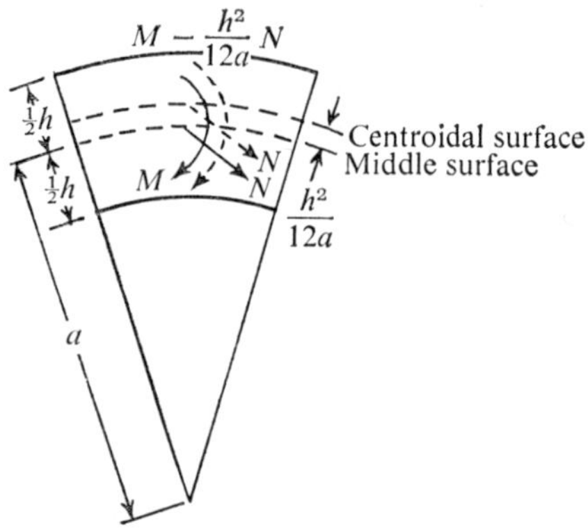

Fig. 6.3. Moments and forces referred to middle and centroidal surfaces

$$\frac{\mathrm{d}}{\mathrm{d}\phi}\left[\left(M_\phi - \frac{h^2}{12a}N_\phi\right)\sin\phi\right] - \left(M_\theta - \frac{h^2}{12a}N_\theta\right) - \left[Q_\phi\left(1 + \frac{h^2}{12a^2}\right) - \left(m_\phi - \frac{h^2}{12a}q_\phi\right)\right]a\sin\phi = 0. \tag{6.1.6a}$$

By utilizing Eqs. (6.1.5) we obtain the desired result

$$Q_\phi = \left(1 + \frac{1}{12}\frac{h^2}{a^2}\right)^{-1}\left[m_\phi - \frac{h^2}{12a}q_\phi - \frac{D}{a}\frac{\mathrm{d}}{\mathrm{d}\phi}\left(\nabla^2 + \frac{2}{a^2}\right)w\right], \tag{6.1.6b}$$

where ∇^2 is the two-dimensional Laplace's operator in spherical coordinates for axisymmetric displacements, given by

$$\nabla^2 = \frac{1}{a^2}\left(\frac{\mathrm{d}^2}{\mathrm{d}\phi^2} + \cot\phi\frac{d}{\mathrm{d}\phi}\right). \tag{6.1.7}$$

Next we may obtain an expression for N_ϕ by multiplying Eq. (6.1.2a) by $\sin\phi$ and Eq. (6.1.2b) by $\cos\phi$, and by subtracting one from the other to obtain

$$\frac{\mathrm{d}}{\mathrm{d}\phi}(N_\phi\sin^2\phi - Q_\phi\sin\phi\cos\phi) + (q_\phi\sin\phi - q_\zeta\cos\phi)a\sin\phi = 0, \tag{6.1.8a}$$

which is the equation of equilibrium of forces in the direction of the axis of revolution of a spherical shell segment. This can be integrated to yield

$$N_\phi = Q_\phi\cot\phi + \frac{1}{\sin^2\phi}\left[\int_{\phi_0}^{\phi}(q_\zeta\cos\phi - q_\phi\sin\phi)a\sin\phi\,\mathrm{d}\phi + \frac{P}{2\pi a}\right], \tag{6.1.8b}$$

or, from Eq. (6.1.6b),

$$N_\phi = \left(1 + \frac{1}{12}\frac{h^2}{a^2}\right)^{-1}\left[m_\phi - \frac{h^2}{12a}q_\phi - \frac{D}{a}\frac{\mathrm{d}}{\mathrm{d}\phi}\left(\nabla^2 + \frac{2}{a^2}\right)w\right]\cot\phi$$
$$+ \frac{1}{\sin^2\phi}\left[\int_{\phi_0}^{\phi}(q_\zeta\cos\phi - q_\phi\sin\phi)a\sin\phi\,\mathrm{d}\phi + \frac{P}{2\pi a}\right], \tag{6.1.8c}$$

where the constant of integration P can be interpreted as the resultant in the direction of the axis of the sphere of edge forces applied at one end of the

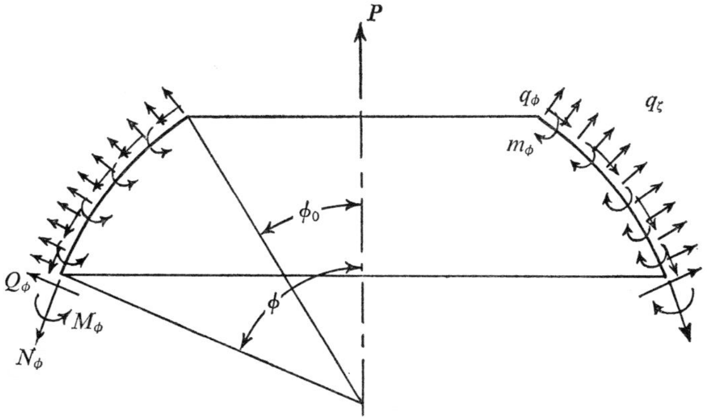

Fig. 6.4. Forces acting on spherical segment

segment (Fig. 6.4). From either Eq. (6.1.2a) or Eq. (6.1.2b), and Eqs. (6.1.6b) and (6.1.8c) we can also obtain

$$N_\theta = \left(1 + \frac{1}{12}\frac{h^2}{a^2}\right)^{-1}\frac{\mathrm{d}}{\mathrm{d}\phi}\left[m_\phi - \frac{h^2}{12a}q_\phi - \frac{D}{a}\frac{\mathrm{d}}{\mathrm{d}\phi}\left(\nabla^2 + \frac{2}{a^2}\right)w\right]$$
$$- \frac{1}{\sin^2\phi}\left[\int_{\phi_0}^{\phi}(q_\zeta\cos\phi - q_\phi\sin\phi)a\sin\phi\,\mathrm{d}\phi + \frac{P}{2\pi a}\right] + q_\zeta a. \tag{6.1.9}$$

The bending moments M_ϕ and M_θ are readily expressed in terms of the radial deflection with the use of Eqs. (6.1.4c), (6.1.4d), (6.1.8c), and (6.1.9).

The equation which the radial deflection w must satisfy is obtained, in a manner similar to that for the cylinder and the cone, by noting that Eqs. (6.1.3a), (6.1.3b), (6.1.5a), (6.1.5b), (6.1.8c), and (6.1.9) yield two equations for the determination of the tangential displacement u_ϕ. For these to be compatible, the following relationship, obtained by eliminating u_ϕ from Eqs. (6.1.3a) and (6.1.3b) must hold:

$$\frac{\mathrm{d}}{\mathrm{d}\phi}\left[\left(\epsilon_\theta - \frac{w}{a}\right)\tan\phi\right] - \left(\epsilon_\phi - \frac{w}{a}\right) = 0, \tag{6.1.10a}$$

or, with the use of Eqs. (6.1.5a) and (6.1.5b),

$$\frac{\mathrm{d}}{\mathrm{d}\phi}[(N_\phi + N_\theta)\tan\phi] - (N_\phi + N_\theta) - (1+\nu) \times \left\{\frac{1}{\cos\phi}\left[\frac{\mathrm{d}}{d\phi}(N_\phi \sin\phi) - N_\theta\cos\phi\right] + N_\phi \tan^2\phi\right\} = \frac{Eh}{a}\sin\phi\frac{\mathrm{d}}{\mathrm{d}\phi}\left(\frac{w}{\cos\phi}\right). \tag{6.1.10b}$$

The substitution of Eqs. (6.1.8c) and (6.1.9) into Eq. (6.1.10b) then yields

$$\sin\phi\frac{\mathrm{d}}{\mathrm{d}\phi}\frac{1}{\cos\phi}\left\{\left[1 + \frac{a^4}{4\chi^4+1}\nabla^2\left(\nabla^2 + \frac{2}{a^2}\right)\right]\frac{Ehw}{a} - q_\zeta a - \left(1+\frac{h^2}{12a^2}\right)^{-1}\frac{1}{\sin\phi}\frac{\mathrm{d}}{\mathrm{d}\phi}\left[\left(m_\phi - \frac{h^2}{12a}q_\phi\right)\sin\phi\right]\right\} - (1+\nu)q_\phi\tan\phi + \frac{1+\nu}{\cos^2\phi}\left[\int_{\phi_0}^{\phi}(q_\zeta\cos\phi - q_\phi\sin\phi)a\sin\phi\,\mathrm{d}\phi + \frac{P}{2\pi a}\right] = 0, \tag{6.1.10c}$$

with

$$\chi = \left\{\frac{1}{4}\left[(1-\nu^2)\left(\frac{12a^2}{h^2}+1\right) - 1\right]\right\}^{1/4}. \tag{6.1.11}$$

Eq. (6.1.10c) may be integrated to yield

$$\left[1 + \frac{a^4}{4\chi^4+1}\nabla^2\left(\nabla^2+\frac{2}{a^2}\right)\right]w = \frac{a^2}{Eh}q(\phi) + C\cos\phi, \tag{6.1.12a}$$

where

$$q(\phi) = q_\zeta + \left(1+\frac{1}{12}\frac{h^2}{a^2}\right)^{-1}\frac{1}{\sin\phi}\frac{\mathrm{d}}{d\phi}\left[\left(\frac{m_\phi}{a} - \frac{h^2}{12a^2}q_\phi\right)\sin\phi\right] + (1+\nu) \times \left\{\cos\phi\int\left[q_\phi - \frac{1}{\cos\phi\sin\phi}\int_{\phi_0}^{\phi}(q_\zeta\cos\phi - q_\phi\sin\phi)\sin\phi\,\mathrm{d}\phi\right] \times \frac{\mathrm{d}\phi}{\cos\phi} + \frac{P}{2\pi a^2}\left(\cos\phi\ln\cot\frac{\phi}{2} - 1\right)\right\}, \tag{6.1.12b}$$

and C is a rigid body translation of the spherical segment in the direction of the shell axis of revolution. Thus, the problem of axisymmetric bending of a spherical segment has been reduced to the problem of finding the solution of a fourth-order ordinary differential equation for the radial displacement, since once the solution for w is known, all of the other quantities are determined.

An explicit expression for u_ϕ can be obtained by combining Eqs. (6.1.2b), (6.1.3b), (6.1.4a), (6.1.4b), (6.1.6b), (6.1.8b), and (6.1.12a) to yield, after much manipulation,

$$u_\phi = -\frac{(1+\nu)a}{Eh}\left\{Q_\phi + \frac{P}{2\pi a}\left((\cot\phi + \sin\phi \ln\cot\frac{\phi}{2}\right) - 2\sin\phi \int \frac{d\phi}{\sin\phi}\left[q_\zeta + \frac{1}{\sin^2\phi}\int_{\phi_0}^{\phi}(q_\zeta\cos\phi - q_\phi\sin\phi)\,d\phi\right]\right\} - C\sin\phi. \quad (6.1.13)$$

In many applications the change of slope of the shell middle surface and the displacement and force components in the directions parallel and perpendicular to the segment axis are needed. These are given by

$$\Theta = \frac{1}{a}\left(\frac{dw}{d\phi} - u_\phi\right), \quad (6.1.14a)$$

$$u_H = u_\phi\cos\phi + w\sin\phi, \quad (6.1.14b)$$

$$u_V = w\cos\phi - u_\phi\sin\phi, \quad (6.1.14c)$$

$$H = N_\phi\cos\phi + Q_\phi\sin\phi = \frac{Q_\phi}{\sin\phi} + \frac{a\cos\phi}{\sin^2\phi}\left[\frac{P}{2\pi a^2} + \int_{\phi_0}^{\phi}(q_\zeta\cos\phi - q_\phi\sin\phi)\sin\phi\,d\phi\right], \quad (6.1.14d)$$

$$V = N_\phi\sin\phi - Q_\phi\cos\phi = \frac{a}{\sin\phi}\left[\frac{P}{2\pi a^2} + \int_{\phi_0}^{\phi}(q_\zeta\cos\phi - q_\phi\sin\phi)\sin\phi\,d\phi\right]. \quad (6.1.14e)$$

6.2 A particular solution of the bending equations

The general solution of Eq. (6.1.12a) is the sum of a particular solution $w^{(p)}$ and the general solution $w^{(h)}$ of the homogeneous equation obtained by setting the right side of Eq. (6.1.12a) equal to zero. The particular solutions for the terms involving C and P are easiest to obtain and are given by

$$w^{(p)} = (1+\nu)\frac{P}{2\pi Eh}\left(\cos\phi\ln\cot\frac{\phi}{2} - 1\right) + C\cos\phi, \quad (6.2.1a)$$

which satisfies the equation

$$\left(\nabla^2 + \frac{2}{a^2}\right) w^{(p)} = 0. \tag{6.2.1b}$$

We then have

$$N_\phi^{(p)} = -N_\theta^{(p)} = \frac{P}{2\pi a \sin^2 \phi}, \tag{6.2.2a}$$

$$M_\phi^{(p)} = M_\theta^{(p)} = Q_\phi^{(p)} = \Theta^{(p)} = 0, \tag{6.2.2b}$$

$$u_\phi = -\frac{(1+\nu)P}{2\pi Eh}\left(\cot\phi + \sin\phi \ln \cot\frac{\phi}{2}\right) - C\cos\phi, \tag{6.2.2c}$$

$$H = \frac{P\cos\phi}{2\pi a \sin^2\phi}, \tag{6.2.2d}$$

$$V = \frac{P}{2\pi a \sin\phi}, \tag{6.2.2e}$$

$$u_H = -\frac{(1+\nu)P}{2\pi Eh \sin\phi}, \tag{6.2.2f}$$

$$u_V = \frac{(1+\nu)P}{2\pi Eh} \ln\cot\frac{\phi}{2} + C. \tag{6.2.2g}$$

Note that in this solution the transverse shear forces Q_ϕ and the bending moments M_ϕ and M_θ vanish throughout the shell, so that the particular solution is exactly a membrane force state. It is also interesting to note that the slope of the shell does not change despite the deformation of the middle surface. The solution given above is exact for a spherical segment subjected to loads applied in the direction of the meridian at both ends (Fig. 6.5). We

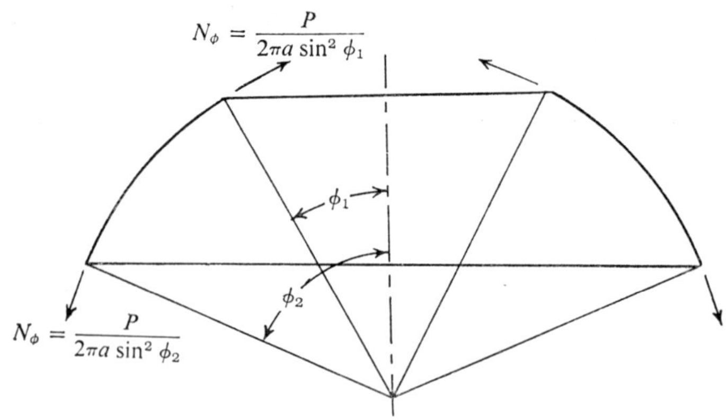

Fig. 6.5. Spherical segment loaded meridionally at the edges

shall see later how this stress and deformation state is modified by the application of the vertical force P in a manner other than as the resultant of a meridional force system.

Other particular solutions depend on the particular variations of the body forces and surface stresses and will be treated later.

6.3 Some relations for the complementary solution

The solution of the homogeneous equation

$$\left[1 + \frac{a^4}{4\chi^4 + 1} \nabla^2\left(\nabla^2 + \frac{2}{a^2}\right)\right] w^{(h)} = 0, \tag{6.3.1a}$$

is facilitated by rewriting the equation in the form

$$(a^2\nabla^2 + 1 + 2i\chi^2)(a^2\nabla^2 + 1 - 2i\chi^2)w^{(h)} = 0. \tag{6.3.1b}$$

Since the factors are commutative, the solution of Eq. (6.3.1b) is given by the sum of the solutions of the second-order complex partial differential equations

$$(a^2\nabla^2 + 1 + 2i\chi^2)w_1^{(h)} = 0, \tag{6.3.2a}$$

$$(a^2\nabla^2 + 1 - 2i\chi^2)w_2^{(h)} = 0. \tag{6.3.2b}$$

We need deal only with, say, Eq. (6.3.2a), since if we denote a complex solution of this equation as

$$w_1^{(h)} = \operatorname{Re} w_1^{(h)} + i \operatorname{Im} w_2^{(h)}, \tag{6.3.3a}$$

the complex conjugate function

$$w_2^{(h)} = \operatorname{Re} w_1^{(h)} - i \operatorname{Im} w_2^{(h)}, \tag{6.3.3b}$$

can be shown to be a solution of Eq. (6.3.2b). We can thus express the solution of Eq. (6.3.1a) in the real form as

$$w^{(h)} = A_1 \operatorname{Re} w_{11}^{(h)} + A_2 \operatorname{Im} w_{11}^{(h)} + A_3 \operatorname{Re} w_{12}^{(h)} + A_4 \operatorname{Im} w_{12}^{(h)}, \tag{6.3.4}$$

where $w_{11}^{(h)}$ and $w_{12}^{(h)}$ are two independent complex solutions of Eq. (6.3.2a). The following useful relationships are obtained by substituting Eq. (6.3.3a) into Eq. (6.3.2a) and equating real and imaginary parts to zero:

$$(a^2\nabla^2 + 1) \operatorname{Re} \begin{Bmatrix} w_{11}^{(h)} \\ \\ w_{12}^{(h)} \end{Bmatrix} = 2\chi^2 \operatorname{Im} \begin{Bmatrix} w_{11}^{(h)} \\ \\ w_{12}^{(h)} \end{Bmatrix}, \tag{6.3.5a}$$

$$(a^2\nabla^2 + 1)\,\mathrm{Im}\begin{Bmatrix} w_{11}^{(h)} \\ \\ w_{12}^{(h)} \end{Bmatrix} = -\,2\chi^2\,\mathrm{Re}\begin{Bmatrix} w_{11}^{(h)} \\ \\ w_{12}^{(h)} \end{Bmatrix}. \tag{6.3.5b}$$

For the time being, let us consider spherical shell segments loaded only at their edges, so that we may put q_ϕ, q_ζ, and m_ϕ equal to zero. Then by means of Eqs. (6.1.5), (6.1.6b), (6.1.8c), (6.1.9), (6.3.4), and (6.3.5) we can express the various stress resultants as

$$Q_\phi = N_\phi \tan\phi = \frac{D}{a^3}\frac{4\chi^4 + \nu^2}{4\chi^4 + 1}\left[A_1\left(2\chi^2\frac{\mathrm{d}\,\mathrm{Im}\,w_{11}^{(h)}}{\mathrm{d}\phi} + \frac{\mathrm{d}\,\mathrm{Re}\,w_{11}^{(h)}}{\mathrm{d}\phi}\right)\right.$$
$$\left. -\,A_2\left(2\chi^2\frac{\mathrm{d}\,\mathrm{Re}\,w_{11}^{(h)}}{\mathrm{d}\phi} - \frac{\mathrm{d}\,\mathrm{Im}\,w_{11}^{(h)}}{\mathrm{d}\phi}\right) + \text{similar terms involving } w_{12}^{(h)}\right], \tag{6.3.6a}$$

$$N_\theta = -N_\phi + \frac{Eh}{a}\,w^{(h)}, \tag{6.3.6b}$$

$$M_\phi = -\frac{D}{a^2}\left\{A_1\left[2\chi^2\,\mathrm{Im}\,w_{11}^{(h)} + \nu\,\mathrm{Re}\,w_{11}^{(h)} - (1-\nu)\left(\frac{4\chi^4-\nu}{4\chi^4+1}\frac{\mathrm{d}\,\mathrm{Re}\,w_{11}^{(h)}}{\mathrm{d}\phi}\right.\right.\right.$$
$$\left.\left. -\,2\chi^2\frac{1+\nu}{4\chi^4+1}\frac{\mathrm{d}\,\mathrm{Im}\,w_{11}^{(h)}}{\mathrm{d}\phi}\right)\cot\phi\right] - A_2\left[2\chi^2\,\mathrm{Re}\,w_{11}^{(h)} - \nu\,\mathrm{Im}\,w_{11}^{(h)}\right.$$
$$\left. +\,(1-\nu)\left(\frac{4\chi^4-\nu}{4\chi^4+1}\frac{\mathrm{d}\,\mathrm{Im}\,w_{11}^{(h)}}{\mathrm{d}\phi} + 2\chi^2\frac{1+\nu}{4\chi^4+1}\frac{\mathrm{d}\,\mathrm{Re}\,w_{11}^{(h)}}{\mathrm{d}\phi}\right)\cot\phi\right]$$
$$\left. +\ \text{similar terms involving } w_{12}^{(h)}\right\}, \tag{6.3.6c}$$

$$M_\theta = -\frac{D}{a^2}\left\{A_1\left[\nu(2\chi^2\,\mathrm{Im}\,w_{11}^{(h)} + \nu\,\mathrm{Re}\,w_{11}^{(h)}) + (1-\nu)\left(\frac{4\chi^4-\nu}{4\chi^4+1}\frac{\mathrm{d}\,\mathrm{Re}\,w_{11}^{(h)}}{\mathrm{d}\phi}\right.\right.\right.$$
$$\left.\left. -\,2\chi^2\frac{1+\nu}{4\chi^4+1}\frac{\mathrm{d}\,\mathrm{Im}\,w_{11}^{(h)}}{\mathrm{d}\phi}\right)\cot\phi\right] - A_2\left[\nu(2\chi^2\,\mathrm{Re}\,w_{11}^{(h)} - \nu\,\mathrm{Im}\,w_{11}^{(h)})\right.$$
$$\left. -\,(1-\nu)\left(\frac{4\chi^4-\nu}{4\chi^4+1}\frac{\mathrm{d}\,\mathrm{Im}\,w_{11}^{(h)}}{\mathrm{d}\phi} + 2\chi^2\frac{1+\nu}{4\chi^4+1}\frac{\mathrm{d}\,\mathrm{Re}\,w_{11}^{(h)}}{\mathrm{d}\phi}\right)\cot\phi\right]$$
$$\left. +\ \text{similar terms involving } w_{12}^{(h)}\right\}. \tag{6.3.6d}$$

The meridional displacement is given by Eq. (6.1.13) as

$$u_\phi = -\frac{(1+\nu)a}{Eh}\,Q_\phi. \tag{6.3.6e}$$

We also have

$$\Theta = \frac{1}{a}\left[\frac{\mathrm{d}w^{(h)}}{\mathrm{d}\phi} + \frac{(1+\nu)a}{Eh} Q_\phi\right], \tag{6.3.7a}$$

$$u_H = w^{(h)} \sin\phi - \frac{(1+\nu)a}{Eh} Q_\phi \cos\phi, \tag{6.3.7b}$$

$$u_V = w^{(h)} \cos\phi + \frac{(1+\nu)a}{Eh} Q_\phi \sin\phi, \tag{6.3.7c}$$

$$H = \frac{Q_\phi}{\sin\phi}, \tag{6.3.7d}$$

$$V = 0. \tag{6.3.7e}$$

6.4 Solutions by means of Legendre functions†

In order to use the relations of the previous section we must obtain solutions of Eq. (6.3.2a). These can be readily defined, for if we define the complex quantity

$$\mu = -\tfrac{1}{2} + \{[\chi^4 + (\tfrac{5}{8})^2]^{1/2} + \tfrac{5}{8}\}^{1/2} + i\{[\chi^4 + (\tfrac{5}{8})^2]^{1/2} - \tfrac{5}{8}\}^{1/2}, \tag{6.4.1a}$$

which is a solution of the equation

$$\mu(\mu + 1) = 1 + 2i\chi^2, \tag{6.4.1b}$$

and introduce the coordinate transformation

$$x = \cos\phi. \tag{6.4.2}$$

Eq. (6.3.2a) becomes

$$\left[(1 - x^2)\frac{\mathrm{d}^2}{\mathrm{d}x^2} - 2x\frac{\mathrm{d}}{\mathrm{d}x} + \mu(\mu + 1)\right] w_1^{(h)} = 0. \tag{6.4.3}$$

But this is Legendre's differential equation, one of the solutions of which is the Legendre function of the first kind of complex order μ, $P_\mu(x)$. Another

† For a more detailed discussion of Legendre functions and their properties, see, for example:

(a) *Higher Transcendental Functions*, vol. I, Bateman Manuscript Project, California Institute of Technology, edited by A. Erdelyi, W. Magnus, F. Oberhettinger, and F. G. Tricomi, McGraw-Hill Book Co., Inc., 1953.

(b) *Formulas and Theorems for the Functions of Mathematical Physics*, by W. Magnus and F. Oberhettinger, Chelsea Pub. Co., 1954.

(c) *The Theory of Spherical and Ellipsoidal Harmonics*, by E. W. Hobson, Cambridge University Press, 1931.

independent solution of the equation, since μ is not an integer, is $P_\mu(-x)$. On expressing the solutions in terms of the coordinate ϕ, we may take the solutions of Eq. (6.3.2a) as†

$$w_{11}^{(h)} = P_\mu(\cos\phi), \tag{6.4.4a}$$

$$w_{11}^{(h)} = P_\mu[\cos(\pi - \phi)]. \tag{6.4.4b}$$

If we know how $P_\mu(\cos\phi)$ varies with ϕ between 0 degrees and 180 degrees, we also know the variation of $P_\mu[\cos(\pi - \phi)]$ for the same range of latitude angle, since the two functions are equal for equal angular distances from the equator of the spherical shell. Alternatively, the solution of the equation is defined for ϕ between 0 degrees and 180 degrees if we know values of both $P_\mu(\cos\phi)$ and $P_\mu[\cos(\pi - \phi)]$ for ϕ between 0 degrees and 90 degrees. The sum of the functions, from the above discussion, is symmetric about the equator ($\phi = 90$ degrees), while the difference is antisymmetric about the equator.

Although the Legendre functions of complex order have not been tabulated, the knowledge that they are solutions of Eq. (6.3.2a) is useful since the behavior of the functions has been studied in some detail. Qualitatively, $P_\mu(\cos\phi)$ is finite when ϕ is equal to 0 degrees and for large values of χ oscillates with increasing amplitude until it becomes singular when ϕ is equal to 180 degrees. The function $P_\mu[\cos(\pi - \phi)]$ varies, of course, in exactly the opposite manner. Quantitatively, a general series expression for $P_\mu(\cos\phi)$ is given by

$$P_\mu(\cos\phi) = \sum_{n=0}^{\infty} \frac{\prod_{p=0}^{p=n}(p^2 - \mu^2)}{\mu(n - \mu)(n!)^2}\left(\sin\frac{\phi}{2}\right)^{2n}, \tag{6.4.5}$$

which is convergent for ϕ less than 180 degrees.

We also required the first derivative of the real and imaginary parts of the Legendre function. While an expression for these could be obtained by

---

† The solutions of Legendre's equation may also be expressed in terms of hypergeometric functions which are symmetric and antisymmetric about the equator of the sphere and which are given by infinite series which converge in the vicinity of the equator about as well as does the Legendre function $P_\mu(\cos\phi)$ in the vicinity of the apex. These series are

$$F\left(-\frac{\mu}{2}, \frac{1}{2} + \frac{1}{2}\mu, \frac{1}{2}, \cos^2\phi\right) = \sum_{n=0}^{\infty} \frac{(-\frac{1}{2}\mu)_n(\frac{1}{2} + \frac{1}{2}\mu)_n}{(\frac{1}{2})_n n!}(\cos\phi)^{2n}$$

$$\cos\phi\, F\left(\frac{1}{2} - \frac{1}{2}\mu, 1 + \frac{1}{2}\mu, \frac{3}{2}, \cos^2\phi\right) = \sum_{n=0}^{\infty} \frac{(\frac{1}{2} - \frac{1}{2}\mu)_n(1 + \frac{1}{2}\mu)_n}{(\frac{3}{2})_n n!}(\cos\phi)^{2n+1},$$

where

$$(a)_0 = 1$$

$$(a)_n = a(a + 1)(a + 2)\dots(a + n - 1), \quad n = 1, 2, 3, \dots.$$

differentiating Eq. (6.4.5) term by term, a better procedure is to utilize one of the properties of Legendre functions which can be written in the form

$$\frac{dP_\mu(\cos\phi)}{d\phi} = \frac{-2\mu(\mu+1)}{\sin\phi}\int_0^{\sin^2(\phi/2)} P_\mu(\cos\phi)\, d\left(\sin^2\frac{\phi}{2}\right). \tag{6.4.6}$$

Although Eqs. (6.4.5) and (6.4.6) can theoretically be used to cover the range of ϕ from 0 degrees to 180 degrccs, the series are satisfactory for practical calculation only so long as χ and/or ϕ are reasonably small. The degree of smallness depends on how many terms we are willing to calculate, which, in turn, will depend on whether we use a slide rule, a manual desk computer, or an electronic digital computer. To help extend the practical range of values of χ and ϕ, there are several other representations of the Legendre functions which may be used. We may, for example, calculate values of $P_\mu(\cos\phi)$ and its first derivative in the vicinity of ϕ equal to 180 degrees by utilizing the equation for $P_\mu[\cos(\pi-\phi)]$ for ϕ in the vicinity of 0 degrees, given by

$$P_\mu[\cos(\pi-\phi)] = \left\{\cos\pi\mu + \frac{2}{\pi}\left[\ln\left(\tan\frac{\phi}{2}\right) + \gamma + \psi(\mu+1)\right]\sin\pi\mu\right\}P_\mu(\cos\phi$$
$$-\frac{2}{\pi}\sin\pi\mu\sum_{n=1}^{\infty}\frac{\prod_{p=0}^{p=n}(p^2-\mu^2)}{\mu(n-\mu)(n!)^2}\left(1+\frac{1}{2}+\frac{1}{3}+\cdots+\frac{1}{n}\right)\left(\sin\frac{\phi}{2}\right)^{2n}. \tag{6.4.7}$$

Here γ is the Euler–Mascheroni constant and $\psi(\mu+1)$ is the psi function.†

For intermediate values of ϕ, a faster converging series representation of the Legendre function may be used. This is given by

$$P_\mu(\cos\phi) = \left(\frac{2}{\pi\sin\phi}\right)^{1/2}\frac{\Gamma(\mu+1)}{\Gamma(\mu+\frac{3}{2})}$$
$$\times\sum_{n=0}^{\infty}\frac{(-1)^n[(\frac{1}{2})_n]^2}{n!(2\sin\phi)^n(\mu+\frac{3}{2})_n}\sin\left[(\mu+n+\tfrac{1}{2})\phi+(n+\tfrac{1}{2})\frac{\pi}{2}\right]. \tag{6.4.8a}$$

The derivative is given by a similar expression

$$\frac{dP_\mu(\cos\phi)}{d\phi} = (\mu+1)\left(\frac{2}{\pi\sin\phi}\right)^{1/2}\frac{\Gamma(\mu+1)}{\Gamma(\mu+\frac{3}{2})}$$
$$\times\sum_{u=0}^{\infty}\frac{(-1)^n(\frac{3}{2})_n(-\frac{1}{2})_n}{n!(2\sin\phi)^n(\mu+\frac{3}{2})_n}\cos\left[(\mu+n+\tfrac{1}{2})\phi+(n+\tfrac{1}{2})\frac{\pi}{2}\right]. \tag{6.4.8b}$$

These series are convergent so long as ϕ is greater than 30 degrees and less than 150 degrees. For values of ϕ outside of this range, the series is asymptotic. The accuracy of the asymptotic expansions depends on both the param-

† See footnote p. 223.

eter μ (or χ) and the angle ϕ. For constant ϕ we obtain results that become more and more accurate as χ increases, whereas for constant χ the accuracy decreases as ϕ approaches 0 degrees or 180 degrees. This can easily be seen when ϕ is near zero degrees since the series diverge ($\sin\phi \to 0$) rather than approaching the known finite values of the Legendre function and its derivative.

If χ and $\chi\phi$ are so large that terms of order $1/\chi$ or less may be neglected compared to unity, the equations for the deformation and stress resultants in the case of spherical segments take on a form similar to that of the cylinder and the cone. In this case we have

$$\begin{aligned} P_\mu(\cos\phi) \approx {} & \frac{(2^{1/2}+1)^{1/2} - i(2^{1/2}-1)^{1/2}}{4(\pi\chi\sin\phi)^{1/2}} \\ & \times \{e^{\chi\phi}[\cos\chi\phi + \sin\chi\phi + i(\cos\chi\phi - \sin\chi\phi)] \\ & + e^{-\chi\phi}[\cos\chi\phi + \sin\chi\phi - i(\cos\chi\phi - \sin\chi\phi)]\}, \end{aligned} \tag{6.4.9a}$$

$$\begin{aligned} P_\mu[\cos(\pi-\phi)] \approx {} & \frac{(2^{1/2}+1)^{1/2} - i(2^{1/2}-1)^{1/2}}{4(\pi\chi\sin\phi)^{1/2}} \\ & \times \langle e^{\chi(\pi-\phi)}\{\cos\chi(\pi-\phi) + \sin\chi(\pi-\phi) \\ & \qquad + i[\cos\chi(\pi-\phi) - \sin\chi(\pi-\phi)]\} \\ & + e^{-\chi(\pi-\phi)}\{\cos\chi(\pi-\phi) + \sin\chi(\pi-\phi) \\ & \qquad - i[\cos\chi(\pi-\phi) - \sin\chi(\pi-\phi)]\}\rangle, \end{aligned} \tag{6.4.9b}$$

and we may write the homogeneous solution for the normal deformation as

$$\begin{aligned} w^{(h)} \approx {} & \left(\frac{\sin\phi_1}{\sin\phi}\right)^{1/2} \{A_1 A[\chi(\phi-\phi_1)] + A_2 B[\chi(\phi-\phi_1)]\} \\ & + \left(\frac{\sin\phi_2}{\sin\phi}\right)^{1/2} \{A_3 A[\chi(\phi_2-\phi)] + A_4 B[\chi(\phi_2-\phi)]\}, \quad \phi_1 \le \phi \le \phi_2, \end{aligned} \tag{6.4.9c}$$

where $A(x)$ and $B(x)$ are the cylinder functions defined by Eqs. (4.1.13) and ϕ_1 and ϕ_2 are angular coordinates denoting the ends of the spherical segment. The use of Eqs. (6.3.6) and (6.3.7) and the deletion of terms of the order of $1/\chi$ or less then yields approximate equations resembling those derived for the cylinder and for the cone

$$\begin{aligned} \frac{2\chi Q_\phi}{Eh/a} = \frac{2\chi N_\phi \tan\phi}{Eh/a} = \frac{2\chi H \sin\phi}{Eh/a} = -\frac{2\chi u_\phi}{1+\nu} \\ \approx -\left(\frac{\sin\phi_1}{\sin\phi}\right)^{1/2} \{A_1 D[\chi(\phi-\phi_1)] + A_2 C[\chi(\phi-\phi_1)]\} \\ + \left(\frac{\sin\phi_2}{\sin\phi}\right)^{1/2} \{A_3 D[\chi(\phi_2-\phi)] + A_4 C[\chi(\phi_2-\phi)]\}, \end{aligned} \tag{6.4.10a}$$

$$N_\theta \approx \frac{Eh}{a} w^{(h)}, \tag{6.4.10b}$$

$$\begin{aligned} -a^2 M_\phi \approx & \left(\frac{\sin \phi_1}{\sin \phi}\right)^{1/2} \{A_1 B[\chi(\phi - \phi_1)] - A_2 A[\chi(\phi - \phi_1)]\} \\ & + \left(\frac{\sin \phi_2}{\sin \phi}\right)^{1/2} \{A_3 B[\chi(\phi_2 - \phi)] - A_4 A[\chi(\phi_2 - \phi)]\} \end{aligned} \tag{6.4.10c}$$

$$M_\theta \approx \nu M_\phi, \tag{6.4.10d}$$

$$\begin{aligned} \Theta \approx & \left(\frac{\sin \phi_1}{\sin \phi}\right)^{1/2} \{-A_1 C[\chi(\phi - \phi_1)] + A_2 D[\chi(\phi - \phi_1)]\} \\ & + \left(\frac{\sin \phi_2}{\sin \phi}\right)^{1/2} \{A_3 C[\chi(\phi_2 - \phi)] - A_4 D[\chi(\phi_2 - \phi)]\}, \end{aligned} \tag{6.4.10e}$$

$$\frac{u_H}{\sin \phi} \approx w^{(h)}. \tag{6.4.10f}$$

The resemblance between cylinder and sphere solutions is strengthened when we note that χ/a is equivalent to λ for the cylinder, i.e.,

$$\frac{\chi}{a} \approx \frac{[3(1 - \nu^2)]^{1/4}}{(ah)^{1/2}}, \tag{6.4.11}$$

and that $a(\phi - \phi_1)$ and $a(\phi_2 - \phi)$ are, respectively, the distances from the ends of the spherical segment measured along the meridian. It is obvious that under the conditions for which the above approximate equations are valid conclusions already reached for the cylinder and the cone can be readily adapted to the spherical shell.

6.5 Shallow spherical shells

When the slope of the spherical shell is small we can introduce approximations which render the solution of Eq. (6.3.2a) much more tractable. Let us make the approximation

$$\cot \phi \approx \frac{1}{\phi}, \tag{6.5.1}$$

which is in error by less than 2 percent if ϕ is less than about 15 degrees, and assume that χ is very large compared to unity. Then Eq. (6.3.2a) may be written as

$$\frac{d^2 w_1^{(h)}}{d\phi^2} + \frac{1}{\phi} \frac{d w_1^{(h)}}{d\phi} + 2i\chi^2 w_1^{(h)} = 0. \tag{6.5.2a}$$

But this is Bessel's equation for which the solution can be expressed in terms of the Bessel–Kelvin functions which appeared in the solution for the conical shell. Then

$$w_1^{(h)} \approx A(\text{ber}\,\xi - i\,\text{bei}\,\xi) + B(\text{ker}\,\xi - \text{kei}\,\xi), \tag{6.5.2b}$$

where

$$\xi = (2)^{1/2}\chi\phi. \tag{6.5.2c}$$

We can then write

$$\text{Re}\, w_{11}^{(h)} \approx \text{ber}\,\xi, \tag{6.5.3a}$$

$$\frac{\text{d Re}\, w_{11}^{(h)}}{\text{d}\phi} \approx 2^{1/2}\chi\,\text{ber}'\,\xi, \tag{6.5.3b}$$

$$\text{Im}\, w_{11}^{(h)} \approx -\,\text{bei}\,\xi, \tag{6.5.3c}$$

$$\frac{\text{d Im}\, w_{11}^{(h)}}{\text{d}\phi} \approx -\,2^{1/2}\,\text{bei}'\,\xi, \tag{6.5.3d}$$

$$\text{Re}\, w_{12}^{(h)} \approx \text{ker}\,\xi, \tag{6.5.3e}$$

$$\frac{\text{d Re}\, w_{12}^{(h)}}{\text{d}\phi} \approx 2^{1/2}\chi\,\text{ker}'\,\xi, \tag{6.5.3f}$$

$$\text{Im}\, w_{12}^{(h)} \approx -\,\text{kei}\,\xi, \tag{6.5.3g}$$

$$\frac{\text{d Im}\, w_{12}^{(h)}}{\text{d}\phi} \approx -\,2^{1/2}\chi\,\text{kei}'\,\xi, \tag{6.5.3h}$$

where primes indicate differentiation with respect to ξ.

With the use of Eqs. (6.3.6) and (6.3.7) the pertinent deformations and stress resultants may now be written as

$$w^{(h)} \approx u_V \approx A_1\,\text{ber}\,\xi - A_2\,\text{bei}\,\xi + A_3\,\text{ker}\,\xi - A_4\,\text{kei}\,\xi, \tag{6.5.4a}$$

$$\begin{aligned} 2^{1/2}\chi u_H \approx \xi\Big[&A_1\Big(\text{ber}\,\xi - \frac{1+\nu}{\xi}\,\text{bei}'\,\xi\Big) - A_2\Big(\text{bei}\,\xi + \frac{1+\nu}{\xi}\,\text{ber}'\,\xi\Big) \\ &+ A_3\Big(\text{ker}\,\xi - \frac{1+\nu}{\xi}\,\text{kei}'\,\xi\Big) - A_4\Big(\text{kei}\,\xi + \frac{1+\nu}{\xi}\,\text{ker}'\,\xi\Big)\Big], \end{aligned} \tag{6.5.4b}$$

$$\frac{a\Theta}{2^{1/2}\chi} \approx A_1\,\text{ber}'\,\xi - A_2\,\text{bei}'\,\xi + A_3\,\text{ker}'\,\xi - A_4\,\text{kei}'\,\xi, \tag{6.5.4c}$$

$$\frac{H}{Eh/a} \approx \frac{N_\phi}{Eh/a} \approx \frac{1}{\xi}\,(A_1\,\text{bei}'\,\xi + A_2\,\text{ber}'\,\xi + A_3\,\text{kei}'\,\xi + A_4\,\text{ker}'\,\xi), \tag{6.5.4d}$$

$$\frac{N_\theta}{Eh/a} \approx A_1\left(\text{ber } \xi - \frac{1}{\xi}\text{ bei}' \xi\right) - A_2\left(\text{bei } \xi + \frac{1}{\xi}\text{ ber}' \xi\right)$$

$$+ A_3\left(\text{ker } \xi - \frac{1}{\xi}\text{ kei}' \xi\right) - A_4\left(\text{kei } \xi + \frac{1}{\xi}\text{ ker}' \xi\right), \tag{6.5.4e}$$

$$\frac{a^2 M_\phi}{2\chi^2 D} \approx A_1\left(\text{bei } \xi + \frac{1-\nu}{\xi}\text{ ber}' \xi\right) + A_2\left(\text{ber } \xi - \frac{1-\nu}{\xi}\text{ bei}' \xi\right)$$

$$+ A_3\left(\text{kei } \xi + \frac{1-\nu}{\xi}\text{ ker}' \xi\right) + A_4\left(\text{ker } \xi - \frac{1-\nu}{\xi}\text{ kei}' \xi\right) \tag{6.5.4f}$$

$$\frac{a^2 M_\theta}{2\chi^2 D} \approx A_1\left(\nu \text{ bei } \xi - \frac{1-\nu}{\xi}\text{ ber}' \xi\right) + A_2\left(\nu \text{ ber } \xi + \frac{1-\nu}{\xi}\text{ bei}' \xi\right)$$

$$+ A_3\left(\nu \text{ kei } \xi - \frac{1-\nu}{\xi}\text{ ker}' \xi\right) + A_4\left(\nu \text{ ker } \xi + \frac{1-\nu}{\xi}\text{ kei } \xi\right), \tag{6.5.4g}$$

where the approximations

$$\cos\phi \approx 1, \tag{6.5.5a}$$

$$\sin\phi \approx \phi, \tag{6.5.5b}$$

have been used and only first-order terms have been retained in the equations to match the accuracy of the differential equation (6.5.2a).

The above expressions are obviously much simpler to calculate than the exact solution. In a later section we will see how a similar derivation can be used to extend the range of ϕ and still retain the relative simplicity of shallow shell theory. Another formulation of shallow shell theory which leads directly to equations containing only first-order terms is also available.† One of the approximations used therein is the reverse of Eq. (6.5.5b), in that we replace ϕ by $\sin\phi$ in the equation rather than the other way round. We shall see, interestingly enough, that the small angle assumption stated as Eq. (6.5.5b) has the wider range of application.

6.6 Closed spherical cap under edge loading

With the nature of the solutions of Eq. (6.3.2a) known, we can now tackle some basic axisymmetric problems for spherical shells. Let us consider a

† E. Reissner, 'Stresses and Small Displacements of Shallow Spherical Shells, I,' *Jour. Math. and Physics*, vol. 25, 1946, pp. 80–85.

spherical cap of included angle $2\phi_0$ which is closed at the apex ($\phi = 0$) and which is subjected to uniformly distributed moments M_0 and horizontal shear forces H_0 per unit width of circumference at the edge (Fig. 6.6). At the

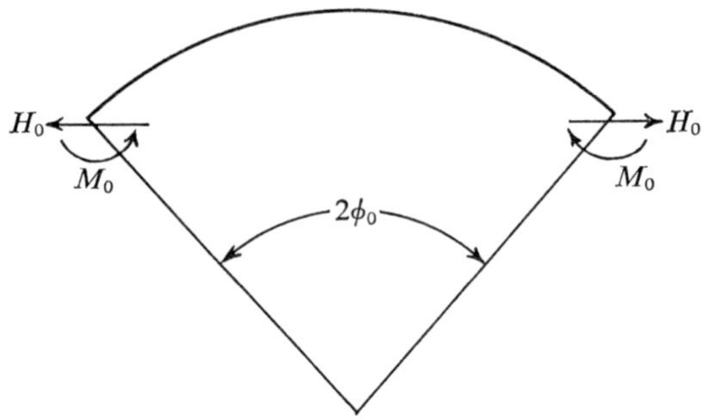

Fig. 6.6. Edge loaded spherical cap

apex, the various displacements and stress resultants should be regular. We can then immediately reduce the complexity of the solution by setting the coefficients A_3 and A_4 in Eqs. (6.3.6) and (6.3.7) equal to zero, since the terms multiplied by these coefficients involve $P_\mu[\cos(\pi - \phi)]$ which is singular at the apex. The constant P is also equal to zero since the applied edge loading has no vertical force resultant, and, for convenience, we set the rigid body translation C equal to zero. We must also investigate the behavior of terms involving the product of the first derivative of the Legendre function $P_\mu(\cos\phi)$ and $\cot\phi$ which appear in some of the equations, since although the derivative vanishes at the apex, the quantity $\cot\phi$ becomes infinite. By means of Eq. (6.4.5) we can obtain the following result

$$\operatorname*{L}_{\phi\to 0} \frac{\mathrm{d}\,\mathrm{Re}\, P_\mu(\cos\phi)}{\mathrm{d}\phi}\cot\phi = \operatorname*{L}_{\phi\to 0} \frac{\mathrm{d}\,\mathrm{Im}\, P_\mu(\cos\phi)}{\mathrm{d}\phi}\cot\phi = \frac{1}{2}. \tag{6.6.1}$$

The boundary conditions which determine the coefficients A_1 and A_2 are that at $\phi = \phi_0$

$$M_\phi = M_0, \tag{6.6.2a}$$

$$H = H_0 \quad \text{or} \quad Q_\phi = H_0 \sin\phi_0. \tag{6.6.2b}$$

Then Eqs. (6.3.6a), (6.3.6c), and (6.4.4) yield

$$A_1 = -\frac{1}{\Delta_0}\left(\frac{4\chi^4 + 1}{4\chi^4 + \nu^2}\beta_2 \frac{a^3 H_0 \sin\phi_0}{D} + \alpha_2 \frac{a^2 M_0}{D}\right), \tag{6.6.3a}$$

$$A_2 = -\frac{1}{\Delta_0}\left(\frac{4\chi^4 + 1}{4\chi^4 + \nu^2}\beta_1 \frac{a^3 H_0 \sin\phi_0}{D} + \alpha_1 \frac{a^2 M_0}{D}\right), \tag{6.6.3b}$$

with

$$\alpha_1 = 2\chi^2 \frac{\mathrm{d}\,\mathrm{Im}\, P_\mu(\cos\phi_0)}{\mathrm{d}\phi} + \frac{\mathrm{d}\,\mathrm{Re}\, P_\mu(\cos\phi_0)}{\mathrm{d}\phi} \tag{6.6.4a}$$

$$\alpha_2 = 2\chi^2 \frac{\mathrm{d}\,\mathrm{Re}\, P_\mu(\cos\phi_0)}{\mathrm{d}\phi} - \frac{\mathrm{d}\,\mathrm{Im}\, P_\mu(\cos\phi_0)}{\mathrm{d}\phi} \tag{6.6.4b}$$

$$\beta_1 = 2\chi^2\,\mathrm{Im}\, P_\mu(\cos\phi_0) + \nu\,\mathrm{Re}\, P_\mu(\cos\phi_0) - \frac{1-\nu}{4\chi^4+1}\Bigg[(4\chi^4-\nu) \times \frac{\mathrm{d}\,\mathrm{Re}\, P_\mu(\cos\phi_0)}{\mathrm{d}\phi} - 2(1+\nu)\,\chi^2 \frac{\mathrm{d}\,\mathrm{Im}\, P_\mu(\cos\phi_0)}{\mathrm{d}\phi}\Bigg]\cot\phi_0 \tag{6.6.4c}$$

$$\beta_2 = 2\chi^2\,\mathrm{Re}\, P_\mu(\cos\phi_0) - \nu\,\mathrm{Im}\, P_\mu(\cos\phi_0) + \frac{1-\nu}{4\chi^4+1}\Bigg[(4\chi^4-\nu) \times \frac{\mathrm{d}\,\mathrm{Im}\, P_\mu(\cos\phi_0)}{\mathrm{d}\phi} + 2(1+\nu)\chi^2 \frac{\mathrm{d}\,\mathrm{Re}\, P_\mu(\cos\phi_0)}{\mathrm{d}\phi}\Bigg]\cot\phi_0 \tag{6.6.4d}$$

$$\begin{aligned}\Delta_0 &= \alpha_2\beta_1 - \alpha_1\beta_2 \\ &= (4\chi^4+\nu)\Bigg[\mathrm{Im}\, P_\mu(\cos\phi_0)\frac{\mathrm{d}\,\mathrm{Re}\, P_\mu(\cos\phi_0)}{\mathrm{d}\phi} - \mathrm{Re}\, P_\mu(\cos\phi_0)\frac{\mathrm{d}\,\mathrm{Im}\, P_\mu(\cos\phi_0)}{\mathrm{d}\phi}\Bigg] \\ &\quad - 2(1-\nu)\chi^2\Bigg\langle \mathrm{Re}\, P_\mu(\cos\phi_0)\frac{\mathrm{d}\,\mathrm{Re}\, P_\mu(\cos\phi_0)}{\mathrm{d}\phi} + \mathrm{Im}\, P_\mu(\cos\phi_0)\frac{\mathrm{d}\,\mathrm{Im}\, P_\mu(\cos\phi_0)}{\mathrm{d}\phi} \\ &\quad + \Bigg\{\Bigg[\frac{\mathrm{d}\,\mathrm{Re}\, P_\mu(\cos\phi_0)}{\mathrm{d}\phi}\Bigg]^2 + \Bigg[\frac{\mathrm{d}\,\mathrm{Im}\, P_\mu(\cos\phi_0)}{\mathrm{d}\phi}\Bigg]^2\Bigg\}\cot\phi_0\Bigg\rangle.\end{aligned} \tag{6.6.4e}$$

The influence coefficients for the spherical cap are then given by Eqs. (6.3.4), (6.3.7a), and (6.3.7b)

$$\left.\frac{u_H}{\sin\phi}\right|_{\phi=\phi_0} = \delta_{H_{11}}\frac{a^3 H_0\sin\phi_0}{2\chi^3 D} - \delta_{M_{11}}\frac{a^2 M_0}{2\chi^2 D}, \tag{6.6.5a}$$

$$\Theta|_{\phi=\phi_0} = \Theta_{H_{11}}\frac{a^2 H_0\sin\phi_0}{2\chi^2 D} - \Theta_{M_{11}}\frac{a M_0}{\chi D}, \tag{6.6.5b}$$

where

$$\Theta_{M11} = \frac{2\chi^3}{\Delta_0}\left\{\left[\frac{\mathrm{d}\,\mathrm{Re}\,P_\mu(\cos\phi_0)}{\mathrm{d}\phi}\right]^2 + \left[\frac{\mathrm{d}\,\mathrm{Im}\,P_\mu(\cos\phi_0)}{\mathrm{d}\phi}\right]^2\right\}, \tag{6.6.6a}$$

$$\begin{aligned}\delta_{M11} = \Theta_{H11} = \frac{4\chi^4}{\Delta_0}\Bigg\{&\mathrm{Re}\,P_\mu(\cos\phi_0)\,\frac{\mathrm{d}\,\mathrm{Re}\,P_\mu(\cos\phi_0)}{\mathrm{d}\phi}\\ &+ \mathrm{Im}\,P_\mu(\cos\phi_0)\,\frac{\mathrm{d}\,\mathrm{Im}\,P_\mu(\cos\phi_0)}{\mathrm{d}\phi}\\ &+ \frac{1}{2\chi^2}\Bigg[\mathrm{Im}\,P_\mu(\cos\phi_0)\,\frac{\mathrm{d}\,\mathrm{Re}\,P_\mu(\cos\phi_0)}{\mathrm{d}\phi}\\ &\qquad - \mathrm{Re}\,P_\mu(\cos\phi_0)\,\frac{\mathrm{d}\,\mathrm{Im}\,P_\mu(\cos\phi_0)}{\mathrm{d}\phi}\Bigg]\Bigg\},\end{aligned} \tag{6.6.6b}$$

$$\begin{aligned}\delta_{H11} = \frac{4\chi^5}{\Delta_0}\,\frac{4\chi^4+1}{4\chi^4+\nu^2}\Bigg\{&[\mathrm{Re}P_\mu(\cos\phi_0)]^2 + [\mathrm{Im}\,P_\mu(\cos\phi_0)]^2\\ &+ \frac{\nu}{\chi^2}\Bigg[\mathrm{Im}\,P_\mu(\cos\phi_0)\,\frac{\mathrm{d}\,\mathrm{Re}\,P_\mu(\cos\phi_0)}{\mathrm{d}\phi}\\ &- \mathrm{Re}\,P_\mu(\cos\phi_0)\,\frac{\mathrm{d}\,\mathrm{Im}\,P_\mu(\cos\phi_0)}{\mathrm{d}\phi}\Bigg]\cot\phi_0\Bigg\}\\ &- \frac{4(1+\nu)\chi^3}{4\chi^4+\nu^2}\left[1 + \frac{1-\nu}{2\chi}\,\Theta_{M11}\cot\phi_0\right]\cot\phi_0.\end{aligned} \tag{6.6.6c}$$

Unlike the infinitely long cylinder or the complete or infinite conical shell, the influence coefficients are not expressed in terms of a single parameter. However, calculations indicate that the influence coefficients are very nearly a function of the parameter $\chi\phi_0$ which we would obtain for shallow shells. In Fig. 6.7 various values of the influence coefficients calculated from the above equations with the use of Eqs. (6.4.5) and (6.4.6), for χ equal to 4, 6, 8, and 10 and for various values of the angle ϕ_0† as well as other available values‡

---

† The values of ϕ were limited to those for which the series converged to four significant figures in sixteen terms or less.

‡ H. E. Williams, 'Some Exact Solutions of the Problem of Axisymmetric Bending of Thin Spherical Shells,' Jet Propulsion Laboratory, California Institute of Technology, Report No. TR32-416, April 1, 1963; E. H. Dill and D. R. Stone, 'Influence Coefficients for Spherical Caps,' *AIAA Launch and Space Vehicle Shell Structures Conference*, Palm Springs, California, April 1–3, 1963.

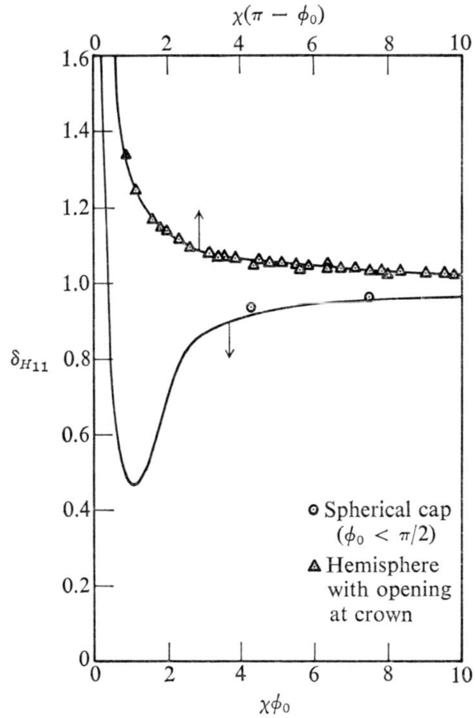

(a) Displacement due to shear force

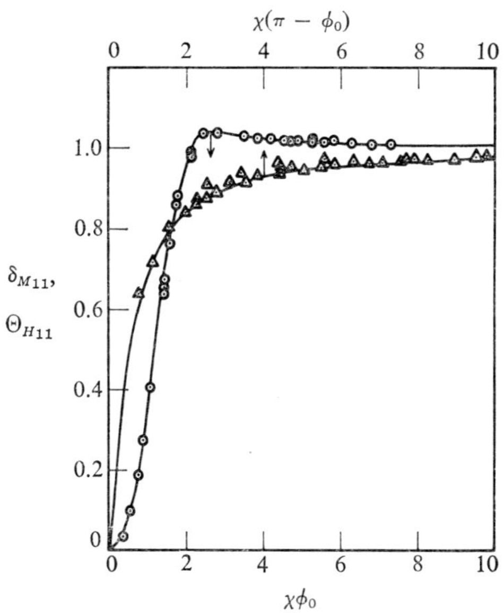

(b) Displacement due to moment or rotation due to shear force

Fig. 6.7. Comparison of exact and approximate influence coefficients for spherical caps

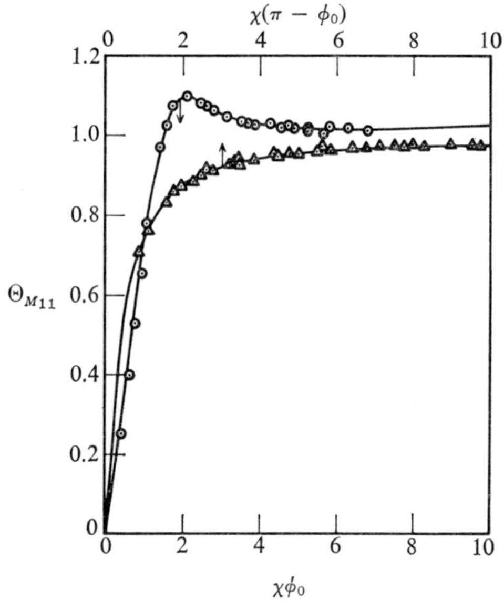

(c) Rotation due to moment
Fig. 6.7 Concluded.

are shown by the circles as a function of $\chi\phi_0$ and can be seen to practically define a single curve in each case. The solid curves which essentially pass through the points are the results of shallow shell theory, which can be obtained merely by substituting the ber and bei functions and their derivatives (the ker and kei functions and their derivatives become infinite at the apex) as indicated by Eqs. (6.5.3) into Eqs. (6.6.5c) and (6.6.7), replacing cot ϕ by $1/\phi$, and deleting higher order terms to obtain

$$\Theta_{M_{11}} \approx \frac{1}{2^{1/2}\bar{\Delta}_0} [(\text{ber}' \, \xi_0)^2 + (\text{bei}' \, \xi_0)^2], \tag{6.6.7a}$$

$$\delta_{M_{11}} = \Theta_{H_{11}} \approx \frac{1}{\bar{\Delta}_0} (\text{ber} \, \xi_0 \, \text{ber}' \, \xi_0 + \text{bei} \, \xi_0 \, \text{bei}' \, \xi_0), \tag{6.6.7b}$$

$$\delta_{H_{11}} \approx \frac{1}{2^{1/2}\bar{\Delta}_0} \left[(\text{ber} \, \xi_0)^2 + (\text{bei} \, \xi_0)^2 + \frac{2\nu}{\xi_0} (\text{ber} \, \xi_0 \, \text{bei}' \, \xi_0 - \text{bei} \, \xi_0 \, \text{ber}' \, \xi_0)\right]$$

$$- \frac{1+\nu}{\xi_0} \left(2^{1/2} + \frac{1-\nu}{\xi_0} \Theta_{M_{11}}\right), \tag{6.6.7c}$$

with

$$\bar{\Delta}_0 \approx \text{ber}\,\xi_0\,\text{bei}'\,\xi_0 - \text{bei}\,\xi_0\,\text{ber}'\,\xi_0 - \frac{1-\nu}{\xi_0}\,[(\text{ber}'\,\xi_0)^2 + (\text{bei}'\,\xi_0)^2], \tag{6.6.7d}$$

and

$$\xi_0 = 2^{1/2}\chi\phi_0. \tag{6.6.7e}$$

It is apparent that the approximate shallow shell solution is adequate for values of ϕ_0 greater than the previously stated limit of 15 degrees so far as the influence coefficients are concerned. As $\chi\phi_0$ becomes large the values of the influence coefficients approach unity, which is to be expected since for a sufficiently large value of this parameter the asymptotic solution given by Eq. (6.4.21c), with the corresponding implication of an equivalent cylinder approach, should apply.

For a given value of χ the value of ϕ_0 is, of course, limited to values less than 180 degrees. When ϕ_0 approaches 180 degrees, both the shallow shell and the asymptotic approximations which predict influence coefficients of unity if $\chi\phi_0$ is sufficiently large become invalid. We may remedy the situation by noting that the problem of finding the influence coefficients for an edge-loaded cap with an included angle near 360 degrees is actually the problem of finding the influence coefficients for a spherical shell containing a small opening, or that part of the sphere complementary to a spherical cap (Fig. 6.8). If we assume that significant effects are confined to a region near the edge, we may again use shallow shell theory. In this case, however, we use that part of the solution containing ker and kei functions and their derivatives

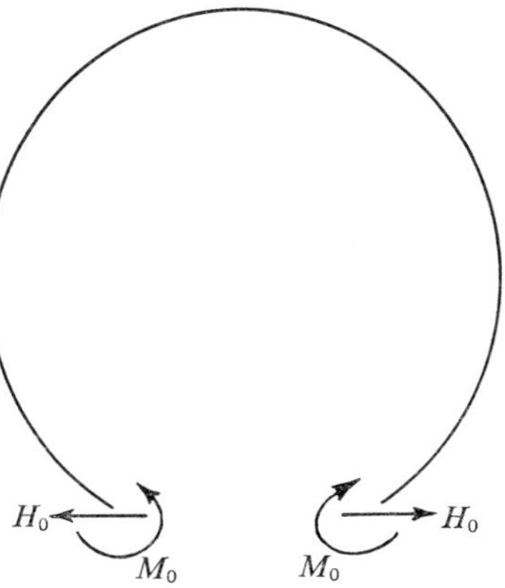

(a) Cap with included angle near 360°

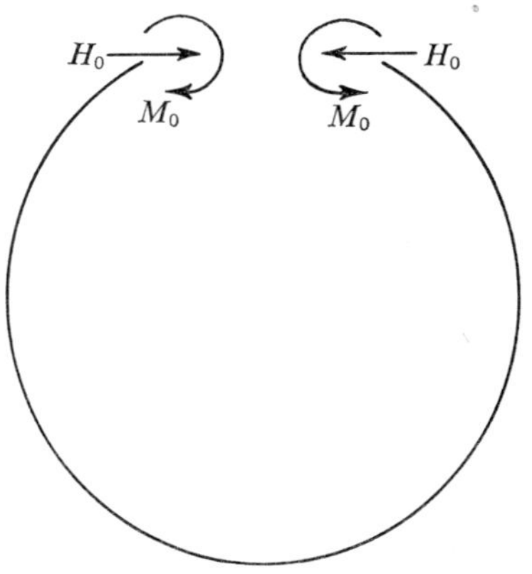

(b) Complement to spherical cap with small included angle

Fig. 6.8. Spherical caps with large included angle

which have the desired property of diminishing with distance from the edge. To obtain the influence coefficients we need only to replace the ber and bei functions and their derivatives in Eqs. (6.6.8) by the ker and kei functions, with due regard for the sign convention for shear forces and slope changes. The influence coefficients thus calculated are shown by the dashed curves in Fig. 6.7. The triangles shown are values calculated by a more accurate method† (for a hemisphere, however) and are seen to be in quite good agreement with the shallow shell approximations.

6.7 Tangent cone approximation

The knowledge that the effects of edge loading are generally restricted to the immediate vicinity of the edge suggests another type of approximation, that of approximating the deformations and stress resultants of the spherical cap by those in a tangent cone (Fig. 6.9). For the tangent cone we have

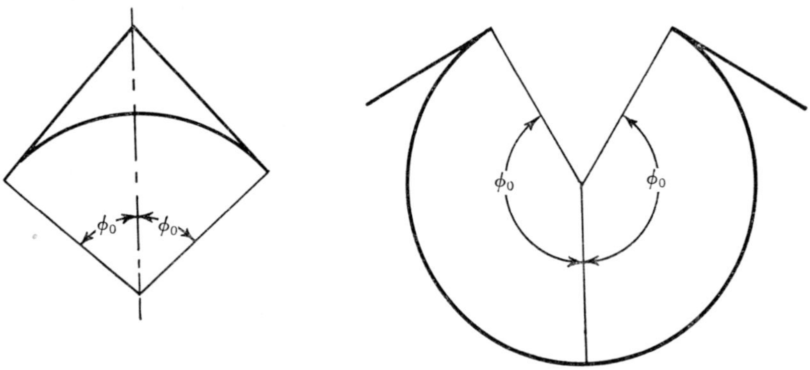

(a) $\phi_0 < 90°$ (b) $\phi_0 > 90°$

Fig. 6.9. Tangent cones

† G. D. Galletly, 'Influence Coefficients for Open Crown Hemispheres,' *J. Engineering for Power*, vol. 82, no. 1, January 1960, pp. 73–81. See also G. D. Galletly, 'Influence Coefficients for Hemispherical Shells with Small Openings at the Vertex,' *J. Applied Mechanics*, vol. 22, no. 1, March 1955, pp. 20–24. The problem of spherical frustums is treated by P. Stern and E. Y. W. Tsui, 'On the Bending of Spherical Shells,' *J. Eng. Mech. Div., ASCE*, vol. 92, no. EM 3, June 1966, pp. 53–66; and by J. C. Gerdeen and F. W. Niedenfuhr, 'Influence Numbers for Shallow Spherical Shells of Circular Ring Planform,' *Developments in Mechanics*, edited by S. Ostrach and R. H. Scanlan, part 2, Solid Mechanics, Pergamon Press, 1965, pp. 278–323.

$$R_\theta = a, \tag{6.7.1a}$$

$$a = \frac{\pi}{2} - \phi_0, \tag{6.7.1b}$$

so that in Eqs. (5.3.3) and (5.3.11) we put

$$\xi_1 = \xi_2 = 2[12(1 - \nu^2)]^{1/4}\left(\frac{a}{h}\right)^{1/2} \tan \phi_0,$$

$$\approx 2^{3/2}\chi \tan \phi_0 \approx 2^{3/2}\chi\phi_0, \tag{6.7.2}$$

to obtain influence coefficients for comparison with those for spherical shells. The comparison shown in Fig. 6.10 indicates that the success of the approximation is mixed. For instance, for spherical caps having a small included angle $2\phi_0$ (Fig. 6.9(a)), the influence coefficients $\delta_{H_{11}}$ for extension due to transverse loading for the cone and the sphere are in generally very good agreement except for a small range of $\chi\phi_0$ where the tangent cone is as much as 20 percent less stiff than the spherical cap. The coefficient $\delta_{M_{11}} = \Theta_{H_{11}}$, the measure of the rotation due to transverse load or the extension due to bending

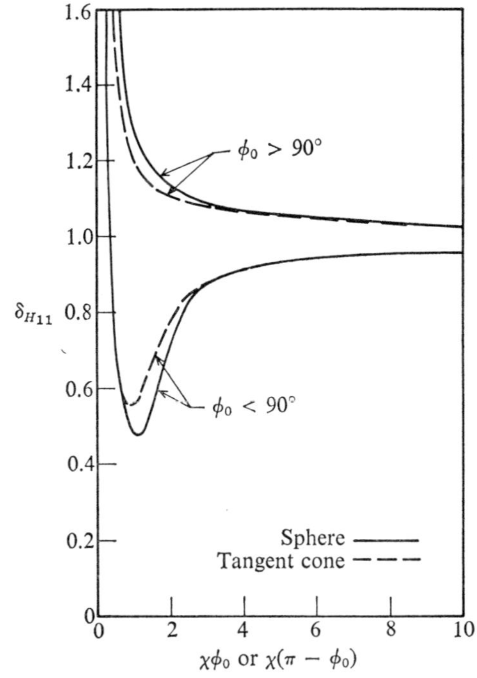

(a) Displacement due to shear force

Fig. 6.10. Comparison of influence coefficients for spherical cap and tangent conical shell

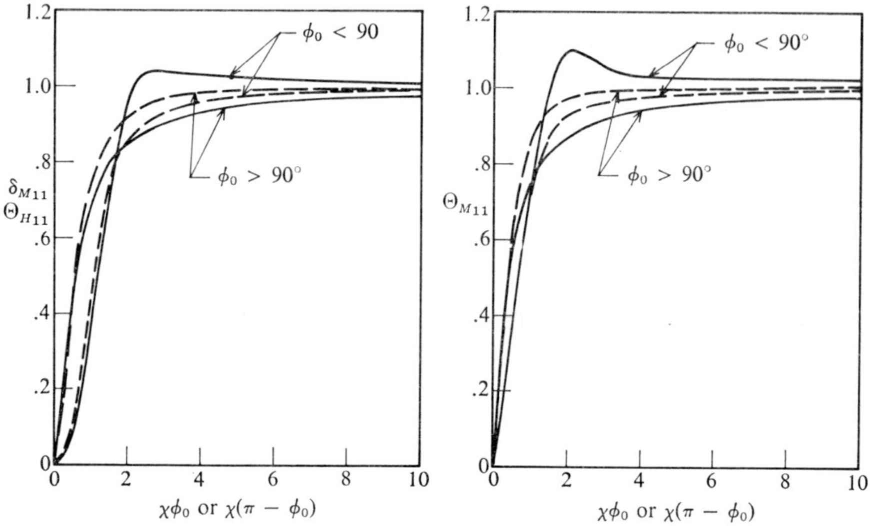

(b) Displacement due to moment or rotation due to shear force

(c) Rotation due to moment

Fig. 6.10. Concluded

moment, for the tangent cone is about 33 percent greater than that for the spherical cap for small values of $\chi\phi_0$ and as much as 14 percent less for intermediate values of $\chi\phi_0$. The coefficient $\Theta_{M_{11}}$, the measure of rotation due to bending moment, for the tangent cone is a good approximation to that for the spherical cap for small values of $\chi\phi_0$ but is as much as 15 percent less for intermediate values of $\chi\phi_0$. In fact, the equivalent cylinder approximation

$$\delta_{M_{11}} = \Theta_{H_{11}} \approx \Theta_{M_{11}} \approx 1, \tag{6.7.3}$$

is somewhat better than the tangent cone approximation when $\chi\phi_0$ is greater than about 2. On the other hand, the tangent cone approximation appears to yield a good representation of the distribution of the stress resultants in the immediate vicinity of the edge. Calculations for meridional bending moments in a shallow spherical cap with a value $2^{1/2}\chi\phi_0$ of 5 and in the tangent cone for which ξ_0 is equal to 10 are shown in Fig. 6.11 as a function of distance from the edge. We see that the two distributions are identical near the edge but diverge as the meridional distance increases. For comparison purposes, the distribution in the equivalent cylinder is also shown and can be seen to be a reasonable, though somewhat less accurate, approximation.

The tangent cone approximation can be improved by introducing a series of tangent conical frustums which approximate the contour of the

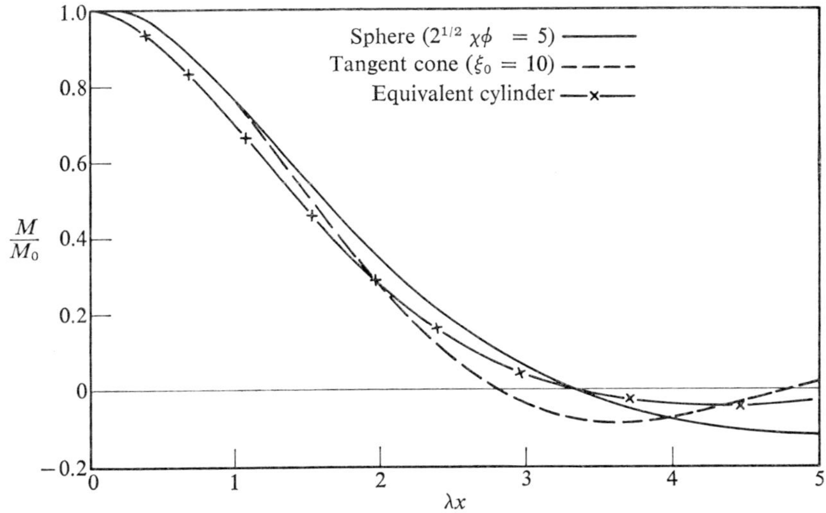

Fig. 6.11. Comparison of various analysis methods for the internal moment distribution in a shallow spherical cap

sphere as in Fig. 6.12. The more conical frustums we use, the better the approximation will be. A method such as this need not be restricted to complete spherical shells but can be used for spherical segments and in general, for shells of revolution of any shape.† Its use depends, however, on the availability of a high speed digital computer to handle the large amount of algebra that is implied.

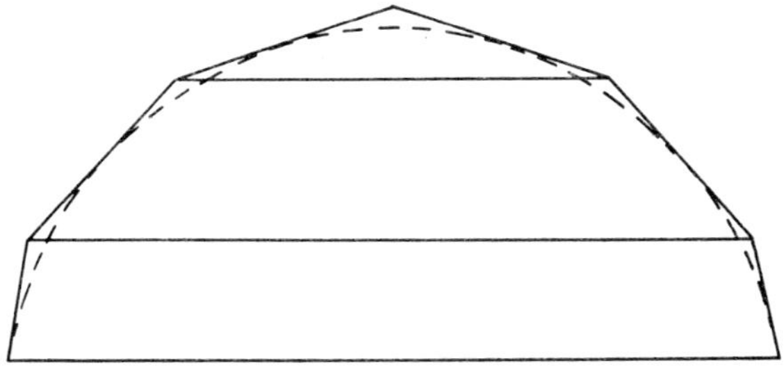

Fig. 6.12. Approximation of a spherical cap by tangent conical frustums

† R. R. Meyer and M. B. Harmon, 'Conical Segment Method for Analyzing Open Crown Shells of Revolution for Edge Loading,' *AIAA J.*, vol. 1, no. 4, April 1963, pp. 886–891.

6.8 Particular solutions for spherical shells

Particular solutions for spherical shells can be found readily if we make use of the fact that any function which is bounded and has a finite number of discontinuities can be expanded in an infinite series of orthogonal functions. For the sphere the appropriate set of orthogonal functions are the Legendre polynomials $P_n(\cos\phi)$ which are defined by Eq. (6.4.5) with μ equal to n, an integer. For each value of n we have a polynomial in $\cos\phi$ with a finite number of terms, i.e.,

$$P_0(\cos\phi) = 1, \tag{6.8.1a}$$

$$P_1(\cos\phi) = \cos\phi, \tag{6.8.1b}$$

$$P_2(\cos\phi) = -\tfrac{1}{2}(1 - 3\cos^2\phi), \tag{6.8.1c}$$

$$P_3(\cos\phi) = -\tfrac{1}{2}(3\cos\phi - 5\cos^3\phi), \tag{6.8.1d}$$

etc.

These satisfy the relations

$$[a^2\nabla^2 + n(n+1)]P_n(\cos\phi) = 0, \tag{6.8.2a}$$

$$\int_0^\pi P_q(\cos\phi)P_r(\cos\phi)\sin\phi\, d\phi = \begin{cases} 0 & q \neq r, \\ 2/(2q+1) & q = r. \end{cases} \tag{6.8.2b}$$

Then if we assume that the particular solution for the radial deformation w can be expressed as

$$w_p = \sum_{n=0}^{\infty} w_{pn}P_n(\cos\phi), \tag{6.8.3a}$$

we have from Eqs. (6.1.12a) and (6.8.2a)

$$\sum_{n=0}^{\infty}\left[1 + \frac{n(n^2-1)(n+2)}{4\chi^4+1}\right]w_{pn}P_n(\cos\phi) = \frac{a^2}{Eh}q(\phi). \tag{6.8.3b}$$

If for a spherical shell segment the variation of $q(\phi)$ with ϕ is assumed on those portions of a complete sphere where it is not defined, the use of Eq. (6.8.2b) then allows us to write an expression for w_{pn} as

$$w_{pn} = \frac{2n+1}{2[1 + \{[n(n^2-1)(n+2)]/(4\chi^4+1)\}}\frac{a^2}{Eh}\int_0^\pi q(\phi)P_n(\cos\phi)\sin\phi\, d\phi \tag{6.8.3c}$$

We note that if $q(\phi)$ is adequately represented by only the first few terms of a series of Legendre polynomials, an approximate expression for w_p may be written as

$$w_p \approx \frac{a^2}{Eh} q(\phi), \tag{6.8.4a}$$

since, if

$$q(\phi) \approx \sum_{n=0}^{M} q_n P_n(\cos \phi) \quad (M \ll 2^{1/2}), \tag{6.8.4b}$$

Eq. (6.8.3b) may be reduced to

$$\sum_{n=0}^{M} \left[1 + \frac{n(n^2 - 1)(n + 2)}{4\chi^4 + 1}\right] w_{pn} P_n(\cos \phi) \approx \sum_{n=0}^{M} w_{pn} P_n(\cos \phi)$$

$$= w_p \approx \frac{a^2}{Eh} q(\phi). \tag{6.8.4c}$$

This approximate expression is practically identical with what we would have obtained had we used the membrane theory of shells. Eq. (6.8.4a) is exact if $q(\phi)$ satisfies the equations

$$a^2 \nabla^2 q(\phi) = 0, \tag{6.8.5a}$$

yielding

$$q(\phi) = A \ln \tan \frac{\phi}{2} + B, \tag{6.8.5b}$$

or

$$(a^2 \nabla + 2) q(\phi) = 0, \tag{6.8.6a}$$

yielding

$$q(\phi) = C\left(\cos \phi \ln \cot \frac{\phi}{2} - 1\right) + D \cos \phi, \tag{6.8.6b}$$

which have been discussed in Section 6.2. We also note the result that if $q(\phi)$ is of the form

$$q(\phi) = q_0 \cos \phi \ln \sin \phi. \tag{6.8.7a}$$

Then a particular solution of Eq. (6.1.12a) is

$$w^{(p)} = \frac{a^2}{Eh} q_0 \cos \phi \left(\ln \sin \phi - \frac{6}{4\chi^4 + 1}\right). \tag{6.8.7b}$$

Since the constant term in the parentheses is generally small compared to $\ln \sin \phi$, we have again

$$w^{(p)} \approx \frac{a^2}{Eh} q(\phi), \tag{6.8.7c}$$

The solutions given by Eqs. (6.8.5) to (6.8.7) include many loading cases of interest. For example, when the bounding surfaces of the sphere are subjected to uniform pressure (Fig. 6.13), we have

$$\sigma_\zeta^{(h/2)} = -p_o, \tag{6.8.8a}$$

$$\sigma_\zeta^{(-h/2)} = -p_i, \tag{6.8.8b}$$

$$\tau_{\phi\zeta}^{(h/2)} = \tau_{\phi\zeta}^{(-h/2)} = \rho_\phi^{(\zeta)} = \rho_\zeta^{(\zeta)} = 0. \tag{6.8.8c}$$

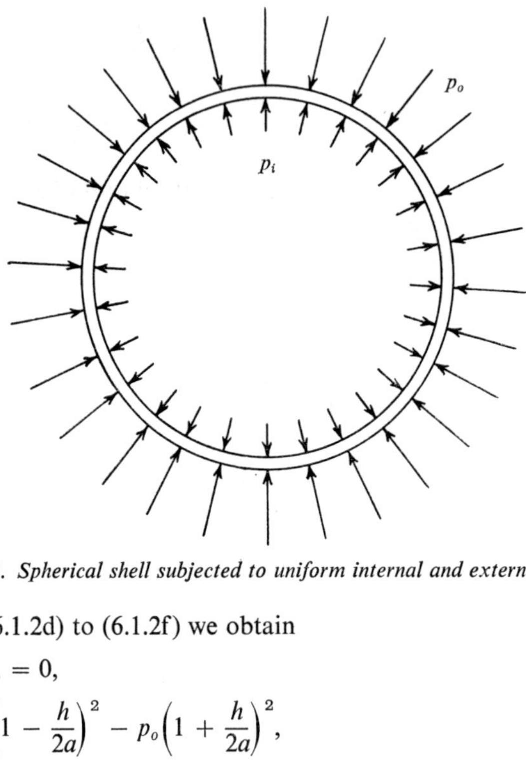

Fig. 6.13. Spherical shell subjected to uniform internal and external pressure

From Eqs. (6.1.2d) to (6.1.2f) we obtain

$$q_\phi = m_\phi = 0, \tag{6.8.9a}$$

$$q_\zeta = p_i\left(1 - \frac{h}{2a}\right)^2 - p_o\left(1 + \frac{h}{2a}\right)^2, \tag{6.8.9b}$$

to yield from Eq. (6.1.12b) (with ϕ_0 equal to zero)

$$q(\phi) = \frac{1-\nu}{2} q_\zeta = \text{constant}, \tag{6.8.9c}$$

which is covered by Eq. (6.8.5b).

For a shell loaded only by its own weight (Fig. 6.14) we have

$$\sigma_\zeta^{(h/2)} = \sigma_\zeta^{(-h/2)} = \tau_{\phi\zeta}^{(h/2)} = \tau_{\phi\zeta}^{(-h/2)} = 0, \tag{6.8.10a}$$

$$\rho_\phi^{(\zeta)} = \gamma \sin\phi, \tag{6.8.10b}$$

$$\rho_\zeta^{(\zeta)} = -\gamma\cos\phi, \tag{6.8.10c}$$

where γ is the weight per unit volume of the shell material. Then

$$q_\phi = \gamma h\left(1 + \frac{1}{12}\frac{h^2}{a^2}\right)\sin\phi, \tag{6.8.11a}$$

$$q_\zeta = -\gamma h\left(1 + \frac{1}{12}\frac{h^2}{a^2}\right)\cos\phi, \tag{6.8.11b}$$

$$m_\phi = \frac{1}{6}\frac{\gamma h^3}{a}\sin\phi, \tag{6.8.11c}$$

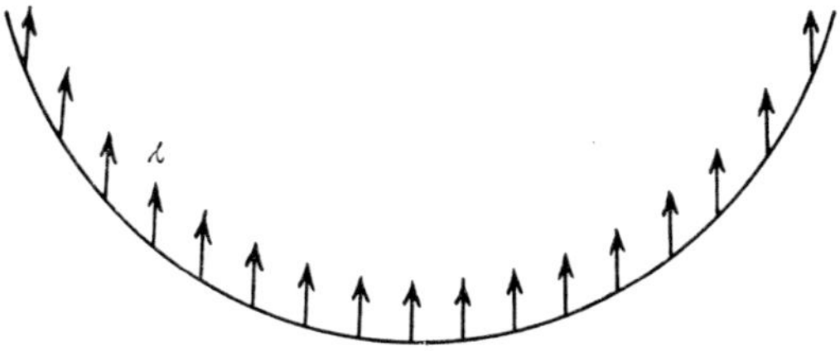

Fig. 6.14. Shell loaded by its own weight

and

$$q(\phi) = -\gamma h\left\{\frac{1 + \dfrac{1}{48}\dfrac{h^4}{a^4}}{1 + \dfrac{1}{12}\dfrac{h^2}{a^2}}\cos\phi + \left(1 + \frac{h^2}{12a^2}\right)[\cos\phi\ln(1 + \cos\phi) - 1]\right\}. \tag{6.8.11d}$$

From Eqs. (6.8.5b), (6.8.6b), and (6.8.7b) the particular solution for the radial deflection can be found to be given by

$$w_p = \frac{a^2}{Eh}\left[q(\phi) + \frac{1}{2(1-\nu)}\frac{h^2}{a^2}\gamma h\cos\phi\right]. \tag{6.8.12}$$

Finally for the case of hydrostatic pressure on the surfaces as shown in Fig. 6.15 the pressure variation yields

$$q_\phi = m_\phi = 0, \tag{6.8.13a}$$

$$q_\zeta = -[p_1 + \gamma a(1 - \cos\phi)], \tag{6.8.13b}$$

from which we obtain

$$q(\phi) = \gamma a\left\{\cos\phi + \frac{1+\nu}{3}[\cos\phi\ln(1 + \cos\phi) - 1]\right\} - \frac{1-\nu}{2}(p_1 + \gamma a). \tag{6.8.14}$$

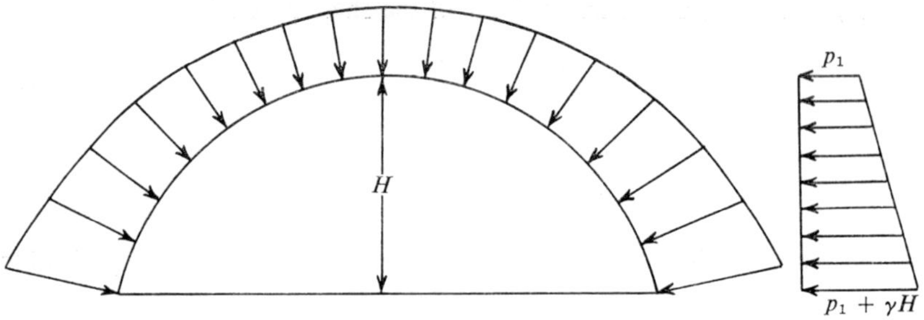

Fig. 6.15. Shell under hydrostatic pressure

The particular solution is thus given by

$$w^{(p)} = \frac{a^2}{Eh}\left[q(\phi) - \frac{2}{1-\nu} \frac{\frac{1}{12}\frac{h^2}{a^2}}{1 + \frac{1}{12}\frac{h^2}{a^2}} \gamma a \cos\phi \right]. \tag{6.8.15}$$

In the cases discussed above we see that the difference between the exact particular solution and the approximate membrane solution is of the order of $(h/a)^2$ compared to unity. We can use these results to show that meridional and circumferential stress resultants differ from those given by membrane theory by quantities of the same order of magnitude, while the contribution of the moments obtained from the exact particular solution to the stresses is of the order of h/a compared to unity. Thus, it would appear that the results of membrane theory are a reasonably good approximation to the exact particular solution for the sphere, with the restriction, as for the cone and cylinder, that the surface loadings should not vary rapidly.

6.9 Complete sphere subjected to internal or external pressure

The particular solution given in the previous section for a sphere subjected to uniform pressure gives us a solution of a problem of transversely rigid shell theory which can be compared to an exact solution of isotropic elasticity theory. For a complete spherical shell subjected to pressure on both the inner and outer surfaces we have the nonvanishing quantities

$$w = \frac{1-\nu}{2}\frac{a^2}{Eh} q_\zeta, \tag{6.9.1a}$$

$$N_\phi = N_\theta = \tfrac{1}{2} q_\zeta a, \tag{6.9.1b}$$

while the remaining stress resultants and displacements vanish. The stress distribution within the shell wall is then given by

$$\sigma_\phi^{(\zeta)} = \sigma_\theta^{(\zeta)} = \frac{\frac{1}{2} q_\zeta a}{1 + \frac{\zeta}{a}}, \tag{6.9.2a}$$

$$\sigma_\zeta^{(\zeta)} = - \frac{p_i \left(\frac{1}{2} - \frac{\zeta}{h}\right)\left(1 - \frac{h}{2a}\right)^2 + p_o \left(\frac{1}{2} + \frac{\zeta}{h}\right)\left(1 + \frac{h}{2a}\right)^2}{\left(1 + \frac{\zeta}{a}\right)^2}. \tag{6.9.2b}$$

From isotropic elasticity theory we can obtain the following results

$$\begin{aligned}\sigma_\phi^{(\zeta)} = \sigma_\theta^{(\zeta)} &= \frac{p_i a}{2h} \left(\frac{1 - \frac{h}{2a}}{1 + \frac{\zeta}{a}}\right)^3 \frac{2\left(1 - \frac{\zeta}{a}\right)^3 + \left(1 + \frac{h}{2a}\right)^3}{3\left(1 + \frac{h^2}{12a^2}\right)} \\ &\quad - \frac{p_o a}{2h} \left(\frac{1 - \frac{h}{2a}}{1 + \frac{\zeta}{a}}\right)^3 \frac{2\left(1 + \frac{\zeta}{a}\right)^3 + \left(1 - \frac{h}{2a}\right)^3}{3\left(1 + \frac{h^2}{12a^2}\right)},\end{aligned} \tag{6.9.3a}$$

$$\begin{aligned}\sigma_\zeta^{(\zeta)} &= - \left\{ p_i \left(\frac{1}{2} - \frac{\zeta}{h}\right) \left(\frac{1 - \frac{h}{2a}}{1 + \frac{\zeta}{a}}\right)^3 \left[1 + \frac{h}{a}\left(\frac{1}{2} + \frac{\zeta}{h}\right) \frac{1 + \frac{1}{3}\frac{\zeta}{a}}{1 + \frac{1}{12}\frac{h^2}{a^2}}\right] \right. \\ &\quad \left. + p_o \left(\frac{1}{2} + \frac{\zeta}{h}\right) \left(\frac{1 + \frac{h}{2a}}{1 + \frac{\zeta}{a}}\right)^3 \left[1 - \frac{h}{a}\left(\frac{1}{2} - \frac{\zeta}{h}\right) \frac{1 + \frac{1}{3}\frac{\zeta}{a}}{1 + \frac{1}{12}\frac{h^2}{a^2}}\right] \right\},\end{aligned} \tag{6.9.3b}$$

$$\begin{aligned}w^{(\zeta)} &= \frac{a^2}{Eh} \left\{ p_i \left(\frac{1 - \frac{h}{2a}}{1 + \frac{\zeta}{a}}\right)^3 \left[\frac{(1 - 2\nu)\left(1 + \frac{\zeta}{a}\right)^3 + \frac{1 + \nu}{2}\left(1 + \frac{h}{2a}\right)^3}{3\left(1 + \frac{1}{12}\frac{h^2}{a^2}\right)}\right] \right. \\ &\quad \left. - p_o \left(\frac{1 + \frac{h}{2a}}{1 + \frac{\zeta}{a}}\right)^3 \left[\frac{(1 - 2\nu)\left(1 + \frac{\zeta}{a}\right)^3 + \frac{1 + \nu}{2}\left(1 - \frac{h}{2a}\right)^3}{3\left(1 + \frac{1}{12}\frac{h^2}{a^2}\right)}\right] \right\}.\end{aligned} \tag{6.9.3c}$$

Table 6.1 Comparison of isotropic and transversely rigid solutions for spherical shell under internal and external pressure

ζ	Quantity	$a/h = 5$		$a/h = 10$		$a/h = 20$	
		Isotropic	Transversely Rigid	Isotropic	Transversely Rigid	Isotropic	Transversely rigid
$-h/2$		0.9266	0.9000	0.9567	0.9500	0.9767	0.9750
0	$\dfrac{2\sigma_\phi^{(\zeta)}}{p_i a/h}$	0.8067	0.8100	0.9017	0.9025	0.9504	0.9506
$h/2$		0.7266	0.7364	0.8567	0.8595	0.9267	0.9274
$-h/2$		1.3266	1.3444	1.1567	1.1605	1.0767	1.0776
0	$-\dfrac{2\sigma_\phi^{(\zeta)}}{p_O a/h}$	1.2067	1.2100	1.1017	1.1025	1.0504	1.0506
$h/2$		1.1266	1.1000	1.0567	1.0500	1.0267	1.0250
$-h/2$		1.0000	1.0000	1.0000	1.0000	1.0000	1.0000
0	$-\dfrac{\sigma_\zeta^{(\zeta)}}{p_i}$	0.4008	0.4050	0.4501	0.4513	0.4750	0.4753
$h/2$		0.0000	0.0000	0.0000	0.0000	0.0000	0.0000
$-h/2$		0.0000	0.0000	0.0000	0.0000	0.0000	0.0000
0	$-\dfrac{\sigma_\zeta^{(\zeta)}}{p_O}$	0.5992	0.6050	0.5499	0.5513	0.5250	0.5253
$h/2$		1.0000	1.0000	1.0000	1.0000	1.0000	1.0000
$-h/2$		0.3843	↑	0.3648	↑	0.3568	↑
0	$\dfrac{Ehw^{(\zeta)}}{a^2 p_i}$	0.3064	0.2835	0.3291	0.3159	0.3398	0.3327
$h/2$		0.2543	↓	0.2998	↓	0.3243	↓
$-h/2$		0.4643	↑	0.4048	↑	0.3768	↑
0	$-\dfrac{Ehw^{(\zeta)}}{a^2 p_O}$	0.3864	0.4235	0.3691	0.3859	0.3598	0.3677
$h/2$		0.3343	↓	0.3398	↓	0.3443	↓

Although the expressions look quite different, the calculations shown in Table 6.1 indicate that the stresses given by Eqs. (6.9.2) for the transversely rigid shell and by Eqs. (6.9.3) for the isotropic shell are almost identical. Even when the middle surface radius is as small as five times the thickness of the shell (the outside radius is 11/9 times the inside radius) the difference in the stresses is less than 3 percent.

The radial deformations of the shell predicted by the two solutions are somewhat different. As shown in Table 6.1 the transversely rigid shell theory predicts a radial deformation which is constant through the thickness, whereas the deformations in a completely elastic shell, of course, vary through the thickness. The deformation given by transversely rigid shell theory appears to coincide with that of a point in the isotropic shell wall that is nearer the unloaded surface than the loaded surface. The ratio of the deflection at the middle surface of the isotropic shell and that of the transversely rigid shell differs from unity by a quantity that is approximately of the order of $h/2a$. The ratio of the deformation at the loaded surface of the isotropic shell to that of the transversely rigid shell differs from unity by an amount which is approximately of the order of h/a. Thus, when the radius of the shell is large compared to the thickness, deformations as well as stresses for the transversely rigid shell are in satisfactory agreement with those for the isotropic shell.

6.10 Spherical shell with concentrated loads at the poles†

Another basic problem which we can study is the problem of a complete spherical shell with concentrated loads at the poles (Fig. 6.16). For this problem, we have to use both the function $P_\mu(\cos\phi)$ and the complementary function $P_\mu[\cos(\pi-\phi)]$. From the symmetry of loading about the shell equator, however, the deformations should also be symmetrical about the equator. Since the sum of $P_\mu(\cos\phi)$ and $P_\mu[\cos(\pi-\phi)]$ meets the required symmetry condition, we have

$$A_1 = A_3, \tag{6.10.1a}$$

$$A_2 = A_4, \tag{6.10.1b}$$

in Eqs. (6.3.4) and (6.3.6). In addition, we use the membrane solution given by Eqs. (6.2.1) and (6.2.2) with P equal to the negative of the applied load P_0.

† A similar analysis is given by W. T. Koiter, 'A Spherical Shell Under Point Loads at its Poles,' *Advances in Applied Mechanics*, Prager Anniversary Volume, 1963, pp. 155–169.

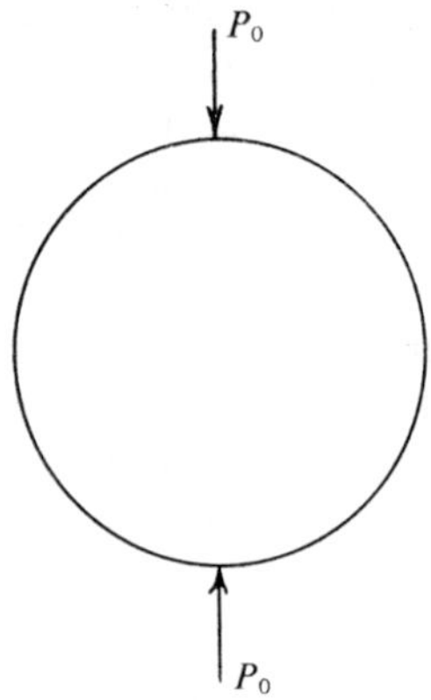

Fig. 6.16. Complete spherical shell loaded at its poles

We have no obvious boundaries at which conditions are specified in this problem and at first glance we might be tempted to delete the bending terms (terms multiplied by A_1 and A_2) since the membrane solution alone is in equilibrium with the applied external loads. However, we note that the membrane deformations become infinite at the poles, a circumstance which cannot be permitted because the shell material has been assumed to be transversely rigid. Since the sum of $P_\mu(\cos\phi)$ and $P_\mu[\cos(\pi - \phi)]$ also becomes infinite at the poles, we can investigate the possibility of cancelling the singularities of the membrane solution. For convenience later on, we shall absorb a constant $(2/\pi)\sin\pi\mu$ into the constants of integration and deal with the real and imaginary parts of the function

$$\bar{P}_\mu(\phi) = \frac{\pi}{2}\csc\pi\mu\{P_\mu(\cos\phi) + P_\mu[\cos(\pi - \phi)]\}, \tag{6.10.2}$$

in the solution.

From Eqs. (6.2.2b), (6.2.2c), (6.3.4), (6.3.6a), (6.3.6e), and (6.3.7a) we can obtain the following relationships:

$$u_\phi + a\Theta = A_1\frac{\mathrm{d}\,\mathrm{Re}\,\bar{P}_\mu(\phi)}{\mathrm{d}\phi} + A_2\frac{\mathrm{d}\,\mathrm{Im}\,\bar{P}_\mu(\phi)}{\mathrm{d}\phi} + \frac{(1+\nu)P_0}{2\pi Eh}\left(\cot\phi - \sin\phi\ln\tan\frac{\phi}{2}\right), \tag{6.10.3a}$$

$$\frac{4\chi^4 - \nu}{2(1+\nu)\chi^2}u_\phi - \frac{a\Theta}{2\chi^2} = A_1\frac{\mathrm{d}\,\mathrm{Im}\,\bar{P}_\mu(\phi)}{\mathrm{d}\phi} - A_2\frac{\mathrm{d}\,\mathrm{Re}\,\bar{P}_\mu(\phi)}{\mathrm{d}\phi} + \frac{(4\chi^4 - \nu)P_0}{4\pi^2 Eh}\left(\cot\phi - \sin\phi\ln\tan\frac{\phi}{2}\right). \tag{6.10.3b}$$

These can be combined into the single complex equation

$$u_\phi + a\Theta + i\left[\frac{4\chi^4 - \nu}{2(1+\nu)\chi^2}u_\phi - \frac{a\Theta}{2\chi^2}\right]$$
$$= \frac{\mathrm{d}}{\mathrm{d}\phi}\left\{(A_1 - iA_2)\bar{P}_\mu(\phi) + \frac{(1+\nu)P_0}{2\pi Eh}\left[1 + i\frac{4\chi^4 - \nu}{2(1+\nu)\chi^2}\right]\left(\cos\phi \ln\tan\frac{\phi}{2} + 1\right)\right\}. \tag{6.10.3c}$$

Now if u_ϕ and Θ are to be finite (in fact, zero) at each pole, it follows that the right side of Eq. (6.10.3c) must contain no singularities and must vanish at the apex. To investigate the conditions which must then be satisfied, we need the expansion of $\bar{P}(\phi)$ and its derivative in the vicinity of either pole. From Eqs. (6.4.5), (6.4.6), and (6.4.7) we have

$$\bar{P}_\mu(\phi) \approx \ln\tan\frac{\phi}{2} + \text{constant} + \text{terms which vanish at the poles.} \tag{6.10.4a}$$

$$\frac{\mathrm{d}\,\bar{P}_\mu(\phi)}{\mathrm{d}\phi} \approx \frac{1}{\sin\phi} + \text{terms which vanish at the poles.} \tag{6.10.4b}$$

Then, for no singularities to exist we must have

$$A_1 - iA_2 = -\frac{(1+\nu)P_0}{2\pi Eh}\left[1 + i\frac{4\chi^4 - \nu}{2(1+\nu)\chi^2}\right], \tag{6.10.5a}$$

or

$$A_1 = -\frac{(1+\nu)P_0}{2\pi Eh}, \tag{6.10.5b}$$

$$A_2 = \frac{(4\chi^2 - \nu)P_0}{4\chi^2\pi Eh}, \tag{6.10.5c}$$

which is the desired solution. A similar technique can be used for other problems containing singularities at one or both poles.

We note from Eqs. (6.1.14a) and (6.10.3c) that the normal displacement is the real part of the term in braces on the right side of Eq. (6.10.3c) and is finite at the origin since the term in braces is finite. From Eqs. (6.10.3c), (6.10.4), and (6.10.5) we can find the values of the displacement at the origin as

$$w_{\phi=0} = -\frac{\chi^2 P_0}{\pi Eh}\left\{\left(1 - \frac{\nu}{4\chi^4}\right)\frac{\pi}{2}\left[\frac{\sinh\pi\alpha_2}{\cosh\pi\alpha_2 - \cos\pi\alpha_1} - \operatorname{Im}\psi(\mu+1)\right] + \frac{1+\nu}{2\chi^2}\left[\frac{\pi}{2}\frac{\sin\pi\alpha_1}{\cosh\pi\alpha_2 - \cos\pi\alpha_1} + \operatorname{Re}\psi(\mu+1) + \gamma - 1\right]\right\}, \tag{6.10.6a}$$

where

$$\alpha_1 = \text{Re}\,\mu = -\tfrac{1}{2} + \{[\chi^4 + (\tfrac{5}{8})^2]^{1/2} + \tfrac{5}{8}\}^{1/2}, \tag{6.10.6b}$$

$$\alpha_2 = \text{Im}\,\mu = \{[\chi^4 + (\tfrac{5}{8})^2]^{1/2} - \tfrac{5}{8}\}^{1/2}. \tag{6.10.6c}$$

When χ and hence α_1 and α_2 are large, we may approximate Eq. (6.10.6a) by

$$w_{\phi=0} \approx -\frac{\chi^2 P_0}{4Eh}\left\{1 + \frac{1}{\pi\chi^2}\,[\tfrac{4}{3} + (1 + \nu)(\ln 2\chi^2 + 2\gamma - 2)]\right\} \tag{6.10.7}$$

The term in braces is shown in Fig. 6.17, with Poisson's ratio taken as 0.3.

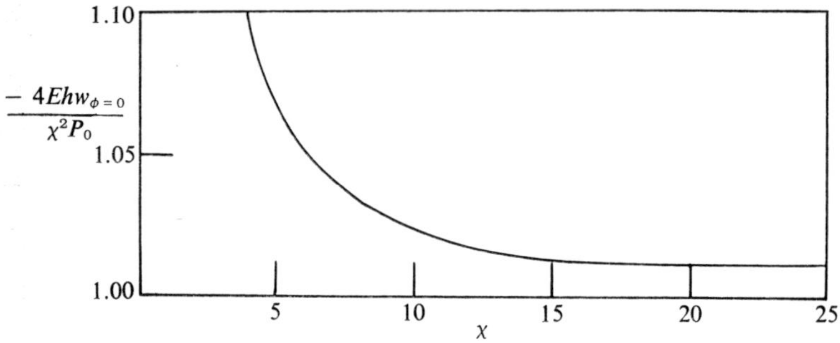

Fig. 6.17. Radial deflection under concentrated loads at the poles

We see that for most practical purposes the term in braces may be taken equal to unity so that the deflection under each concentrated load may be expressed as

$$w_{\phi=0} \approx -\frac{\chi^2 P_0}{4Eh} \approx -\frac{[3(1 - \nu^2)]^{1/2}}{4}\frac{P_0 a}{Eh^2}. \tag{6.10.8}$$

To get some idea of the stiffness of the spherical shell, let us compare the deflection with that of a flat circular simply supported plate having a radius a, a thickness h and subjected to a central concentrated load P_0. This deflection is given by†

$$w_{r=0} = -\frac{3 + \nu}{1 + \nu}\frac{3(1 - \nu^2)}{4\pi}\frac{P_0 a^2}{Eh^3}, \tag{6.10.11a}$$

which differs from that of the sphere by the large factor of

$$\frac{3 + \nu}{1 + \nu}\frac{[3(1 - \nu^2)]^{1/2}}{\pi}\frac{a}{h}.$$

† S. P. Timoshenko and S. Woinowsky-Krieger: *Theory of Plates and Shells*, Second Edition, McGraw-Hill Book Co., Inc. 1959, pp. 67–69.

Another way of expressing the same result is that the circular plate of thickness h would have to have the small radius

$$b = \frac{\{\pi[(1 + \nu)/(3 + \nu)]\}^{1/2}}{[3(1 - \nu^2)]^{1/4}} (ah)^{1/2} \approx 1.11 \frac{a}{\chi}, \tag{6.10.11b}$$

to have the same central deflection as that of the sphere. This radius is of the order of magnitude of the attenuation length of edge effects. We may explain the increase of this effective radius as χ increases by noting that the bending extends over a greater portion of the shell as the shell becomes thicker.

Although the displacements are regular at the apex, the meridional and circumferential force components N_ϕ and N_θ become infinite there as do the shear force Q_ϕ and the moments M_ϕ and M_θ. From Eqs. (6.2.2), (6.3.6), (6.6.1), (6.10.4), and (6.10.5) the singular force resultants can be shown to vary as follows in the vicinity of the apex:

$$Q_\phi \approx \frac{P_0}{2\pi} \frac{1}{(a\phi)}, \tag{6.10.12a}$$

$$M_\phi \approx M_\theta \approx \frac{(1 + \nu)P_0}{4\pi} \ln (a\phi), \tag{6.10.12b}$$

$$N_\phi \approx N_\theta \approx -\frac{(1 + \nu)P_0}{4\pi a} \ln (a\phi). \tag{6.10.12c}$$

The singularities of the shear force Q_ϕ and the moments M_ϕ and M_θ are identical with those in a flat circular plate subjected to a central concentrated load.† The addition of bending terms to the membrane solution for the problem does not get rid of the singular variation of the direct forces N_ϕ and N_θ but does reduce the rapidity of increase to infinity. Further investigation indicates, however, that the increase of the direct forces is more rapid than the increase of bending moments. The direct force singularities generally can be neglected since consideration of stresses in the vicinity of the apex readily shows that the stresses due to N_ϕ and N_0 are of the order of h/a compared to the maximum stresses due to the bending moments M_ϕ and M_θ and are thus small for thin shells. Finally, the singularity can be seen to vanish when the radius becomes infinite, i.e., when the apex can be considered to be part of a plane surface, and thus is not in conflict with the results of flat plate theory.

For practical calculation of effects in the vicinity of the apex the problem can be redone with the use of the shallow shell equations. From Eqs. (6.2.2) and (6.5.4) we have

† S. P. Timoshenko and S. Woinowsky-Krieger: *Theory of Plates and Shells*, Second Edition, McGraw-Hill Book Co., Inc., 1959, pp. 67–69.

$$u_V \approx \frac{1+\nu}{2\pi}\frac{P_0}{Eh}\ln\left(\frac{2^{3/2}\chi}{\xi}\right) + A_1 \operatorname{ber}\xi - A_2 \operatorname{bei}\xi + A_3 \operatorname{ker}\xi - A_4 \operatorname{kei}\xi, \tag{6.10.13a}$$

$$\begin{aligned} 2^{1/2}\chi u_H \approx{}& \frac{(1+\nu)\chi^2 P_0}{\pi Eh}\frac{1}{\xi} \\ &+ \xi\left[A_1\left(\operatorname{ber}\xi - \frac{1+\nu}{\xi}\operatorname{bei}'\xi\right) - A_2\left(\operatorname{bei}\xi + \frac{1+\nu}{\xi}\operatorname{ber}'\xi\right)\right. \\ &\left. + A_3\left(\operatorname{ker}\xi - \frac{1+\nu}{\xi}\operatorname{kei}'\xi\right) - A_4\left(\operatorname{kei}\xi + \frac{1+\nu}{\xi}\operatorname{ker}'\xi\right)\right], \end{aligned} \tag{6.10.13b}$$

$$\frac{a\Theta}{2^{1/2}\chi} \approx A_1 \operatorname{ber}'\xi - A_2 \operatorname{bei}'\xi + A_3 \operatorname{ker}'\xi - A_4 \operatorname{kei}'\xi. \tag{6.10.13c}$$

We note that the terms with ker ξ, and kei ξ, and their derivatives are the only terms which have singularities at the apex so that the coefficients A_3 and A_4 will be determined by the conditions of finiteness at the apex. The condition of zero slope at the apex then requires that A_3 be zero since only ker′ (ξ) becomes infinite at the apex. The condition that u_H be zero at the origin is then satisfied if

$$A_4 = -\frac{\chi^2 P_0}{\pi Eh}. \tag{6.10.14}$$

At this point we run into an apparent inconsistency. For the displacement u_V to be finite at the origin we must have

$$A_3 = -\frac{1+\nu}{2\pi}\frac{P_0}{Eh}, \tag{6.10.15}$$

whereas for finiteness of slope it must vanish. However we note that A_3 given by Eq. (6.10.15) is of the order of $1/\chi^2$ compared to A_4 and that the expressions for displacements and stress resultants contain terms of the order of $1/\chi^2$ and thus of the same order as terms we have already dropped from the shallow shell equations. It is obvious then that to be consistent we must drop the membrane displacement from the expression for the vertical displacement and put A_3 equal to zero.

The coefficients A_1 and A_2 are determined by conditions at a boundary away from the apex. For the complete spherical shell we can assume that the boundary is far enough away so that the displacements and stress resultants at the boundary are negligible, or effectively at infinity. Then A_1 and A_2 can be set equal to zero. The solution for displacements and stress resultants in the vicinity of the apex is then

$$w \approx u_V \approx \frac{\chi^2 P_0}{\pi E h} \operatorname{kei} \xi, \tag{6.10.16a}$$

$$u_H \approx \frac{\chi P_0 \xi}{2^{1/2} \pi E h} \left[\operatorname{kei} \xi + \frac{1+\nu}{\xi} \left(\operatorname{ker}' \xi + \frac{1}{\xi} \right) \right], \tag{6.10.16b}$$

$$\Theta \approx \frac{2^{1/2} \chi^3 P_0}{\pi E h a} \operatorname{kei}' \xi, \tag{6.10.16c}$$

$$N_\phi \approx -\frac{\chi^2 P_0}{\pi \alpha \xi} \left(\operatorname{ker}' \xi + \frac{1}{\xi} \right) \approx H, \tag{6.10.16d}$$

$$N_\theta \approx \frac{\chi^2 P_0}{\pi a} \left[\operatorname{kei} \xi + \frac{1}{\xi} \left(\operatorname{ker}' \xi + \frac{1}{\xi} \right) \right], \tag{6.10.16e}$$

$$M_\phi \approx -\frac{P_0}{2\pi} \left(\operatorname{ker} \xi - \frac{1-\nu}{\xi} \operatorname{ker}' \xi \right), \tag{6.10.16f}$$

$$M_\theta \approx -\frac{P_0}{2\pi} \left(\nu \operatorname{ker} \xi + \frac{1-\nu}{\xi} \operatorname{kei}' \xi \right), \tag{6.10.16g}$$

$$Q_\phi \approx -\frac{\chi P_0}{2\pi a} \operatorname{ker}' \xi. \tag{6.10.16h}$$

At the apex the displacement under the load is

$$w_{\phi=0} = -\frac{\chi^2 P_0}{4Eh}, \tag{6.10.17}$$

which corresponds to the first term of Eq. (6.10.7). Similarly, the bending moments M_ϕ and M_θ and the transverse shear force Q_ϕ have the correct singularities at the apex. The middle surface forces N_ϕ and N_θ do not have the correct singular behavior near the apex but, as noted previously, this error is unimportant.

6.11 Spherical cap loaded at the apex

By superimposing the solution for a spherical cap subjected to edge bending moments and shear forces on the solution for the complete sphere we can find the solution for a spherical cap loaded at the apex and supported at the base. For example, consider the problem of a spherical cap loaded at its apex by a concentrated load and supported at its base by a ring which has infinite torsional stiffness but finite stretching stiffness (Fig. 6.18). One boundary condition at ϕ equal to ϕ_0 is then that the sum of the rotations obtained from the solution for the complete sphere loaded at the poles and from the solution for the edge-loaded spherical cap vanishes. The other relates the radial

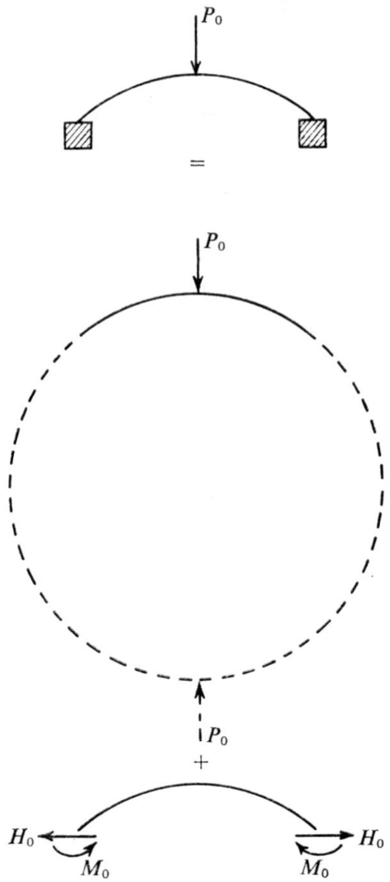

Fig. 6.18. Superposition of problems for spherical cap loaded at the apex

expansion of the ring to the outwardly directed force applied to the ring, along with the continuity condition of equal expansion of the ring and spherical cap and the equilibrium condition that the outward force applied to the ring is the negative of the total horizontal edge force applied to the cap, as

$$u_H = -\frac{Ha^2}{E_R A_R}, \tag{6.11.1}$$

where E_R and A_R are the Young's modulus and area, respectively, of the edge ring.

From Eqs. (6.3.6), (6.3.7), and (6.10.5), the pertinent quantities in the complete sphere loaded at the poles are

$$a\Theta_1|_{\phi=\phi_0} = \frac{\chi^2 P_0}{\pi E h}\left(1 + \frac{\nu^2}{4\chi^4}\right)\frac{\mathrm{d}}{\mathrm{d}\phi}\,\mathrm{Im}\,\bar{P}_\mu(\phi_0), \tag{6.11.2a}$$

$$\left.\frac{u_{H_1}}{\sin\phi}\right|_{\phi=\phi_0} = \left\langle\left(1 - \frac{\nu}{4\chi^4}\right)\mathrm{Im}\,\bar{P}_\mu(\phi_0) - \frac{1+\nu}{2\chi^2}\left\{\mathrm{Re}\,\bar{P}_\mu(\phi_0)\right.\right.$$
$$\left.\left. + \left[\frac{\mathrm{d}}{\mathrm{d}\phi}\,\mathrm{Re}\,\bar{P}_\mu(\phi_0) + \frac{\nu}{2\chi^2}\frac{\mathrm{d}}{\mathrm{d}\phi}\,\mathrm{Im}\,\bar{P}_\mu(\phi_0)\right]\cot\phi_0\right\}\right\rangle\frac{\chi^2 P_0}{\pi E h}, \tag{6.11.2b}$$

$$H_1 \sin\phi\,|_{\phi=\phi_0} = \frac{P_0}{2\pi a}\left[\frac{\mathrm{d}}{\mathrm{d}\phi}\,\mathrm{Re}\,\bar{P}_\mu(\phi_0) + \frac{\nu}{2\chi^2}\frac{\mathrm{d}}{\mathrm{d}\phi}\,\mathrm{Im}\,\bar{P}_\mu(\phi_0)\right], \tag{6.11.2c}$$

while in the edge-loaded spherical cap we have

$$H_2\,|\,\phi = \phi_0 = H_0, \tag{6.11.3a}$$

$$a\Theta_2|_{\phi=\phi_0} = \Theta_{H11}^{(\phi_0)}\frac{a^3 H_0 \sin\phi_0}{2\chi^2 D} - \Theta_{M11}^{(\phi_0)}\frac{a^2 M_0}{\chi D}, \tag{6.11.3b}$$

$$\left.\frac{u_{H_2}}{\sin\phi}\right|_{\phi=\phi_0} = \delta_{H11}^{(\phi_0)}\frac{a^3 H_0 \sin\phi_0}{2\chi^3 D} - \delta_{M11}^{(\phi_0)}\frac{a^2 M_0}{2\chi^2 D}. \tag{6.11.3c}$$

Then H_0 and M_0 are determined by the relations

$$a(\Theta_1 + \Theta_2) = \Theta_{H11}^{(\phi_0)}\frac{a^3 H_0 \sin\phi_0}{2\chi^2 D} - \Theta_{M11}^{(\phi_0)}\frac{a^2 M_0}{\chi D}$$
$$+ \frac{\chi^2 P_0}{\pi E h}\left(1 + \frac{\nu^2}{4\chi^2}\right)\frac{\mathrm{d}}{\mathrm{d}\phi}\,\mathrm{Im}\,\bar{P}_\mu(\phi_0) = 0, \tag{6.11.4a}$$

$$\frac{u_{H_1} + u_{H_2}}{\sin\phi}\,|\,\phi = \phi_0 = \delta_{H11}^{(\phi_0)}\frac{a^3 H_0 \sin\phi_0}{2\chi^3 D} - \delta_{M11}^{(\phi_0)}\frac{a^2 M_0}{2\chi^2 D}$$
$$+ \frac{\chi^2 P_0}{\pi E h}\left\langle\left(1 - \frac{\nu}{4\chi^4}\right)\mathrm{Im}\,\bar{P}_\mu(\phi_0) - \frac{1+\nu}{2\chi^2}\left\{\mathrm{Re}\,\bar{P}_\mu(\phi_0)\right.\right.$$
$$\left.\left. + \left[\frac{\mathrm{d}}{\mathrm{d}\phi}\,\mathrm{Re}\,\bar{P}_\mu(\phi_0) + \frac{\nu}{2\chi^2}\frac{\mathrm{d}}{\mathrm{d}\phi}\,\mathrm{Im}\,\bar{P}_\mu(\phi_0)\right]\cot\phi_0\right\}\right\rangle$$
$$= -\frac{(H_1 + H_2)a^2}{E_R A_R \sin\phi}\,|\,\phi = \phi_0 = -\frac{a^2}{E_R A_R \sin\phi_0}\left\{H_0 + \frac{P_0}{2\pi a \sin\phi_0}\left\{\frac{\mathrm{d}}{\mathrm{d}\phi}\right.\right.$$
$$\left.\left.\mathrm{Re}\,\bar{P}_\mu(\phi_0) + \frac{\nu}{2\chi^2}\frac{\mathrm{d}}{\mathrm{d}\phi}\,\mathrm{Im}\,\bar{P}_\mu(\phi_0)\right]\right\}, \tag{6.11.4b}$$

which are readily solved.

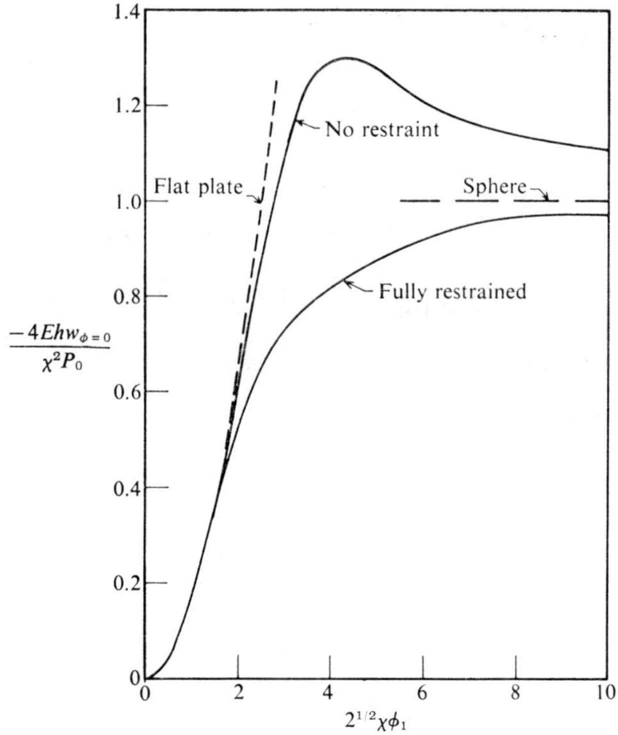

(a) Radial deflection at apex

Fig. 6.19. Results for spherical cap loaded at the apex

The same problem can be solved by shallow shell theory.† The deflection under the load relative to the base and the edge bending moment obtained from this solution are shown in Fig. 6.19 as a function of the parameter $2^{1/2}\chi\phi_0$, denoting the extent of the spherical cap, for the cases when there is no edge ring and when there is an infinitely stiff edge ring. For very small values of $2^{1/2}\chi\phi_0$ the deflections and edge moments for the two edge conditions are identical and equal to the corresponding values for a clamped flat plate of thickness h and radius

$$r = a \sin \phi_0 \approx a\phi_0, \tag{6.11.5a}$$

† E. Reissner, 'Stresses and Small Displacements of Shallow Spherical Shells, II,' *J. Math. and Physics*, vol. 25, 1946, pp. 279–300.

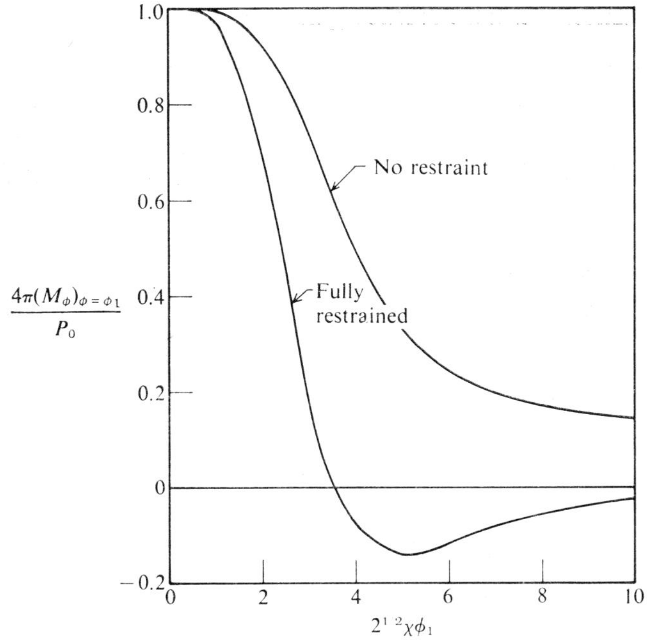

(b) Edge meridional moment

Fig. 6.19. Concluded

which are given by†

$$\frac{4Ehw_0}{\chi^2 P_0} \approx -\frac{1}{2\pi}(2^{1/2}\chi\phi_0)^2, \tag{6.11.5b}$$

$$\frac{4\pi M_\phi}{P_0} = 1. \tag{6.11.5c}$$

As the cap becomes deeper or thinner the deflection under the load very quickly approaches the value for the complete sphere and the edge moment becomes small. It is interesting to note that a fully restrained cap stimulates a complete sphere somewhat sooner than does the cap with no edge restraint.

† S. P. Timoshenko and S. Woinowsky-Krieger: *Theory of Plates and Shells*, Second Edition, McGraw-Hill Book Co., Inc. 1959, pp. 67–69.

6.12 Effect of method of load application

The singularities of bending moment, direct force, and transverse shear which appear in the solution for the spherical shell subjected to concentrated loads at the poles vanish if the load is applied uniformly over a small area (Fig. 6.20), or if it is applied through very stiff inserts (Fig. 6.21). In the first case

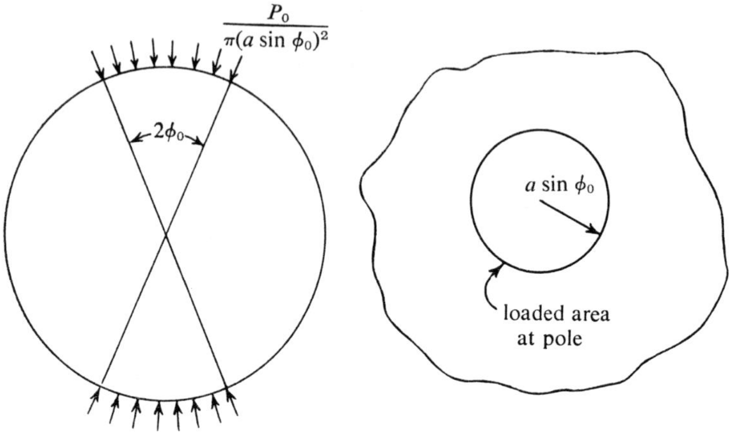

Fig. 6.20. Sphere loaded by uniform pressure at the poles

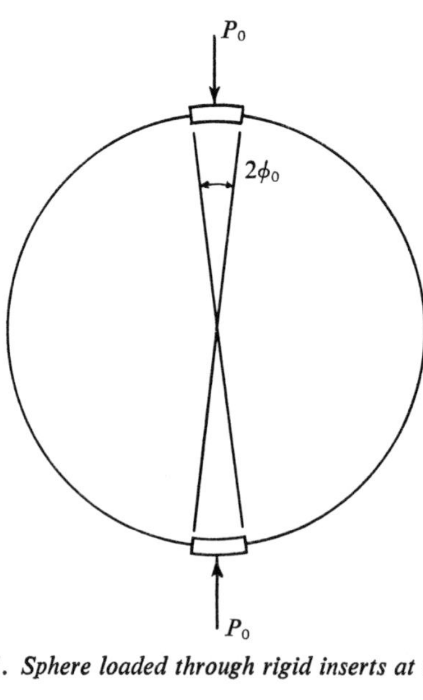

Fig. 6.21. Sphere loaded through rigid inserts at the poles

we must combine the solutions for a spherical cap subjected to uniform external pressure and edge loading with that for an edge-loaded spherical shell with the pole caps removed (Fig. 6.22) so as to ensure continuity of horizontal deflection, change of slope, moment, and horizontal force at the junction of the shell segments. For the second case we superimpose the solu-

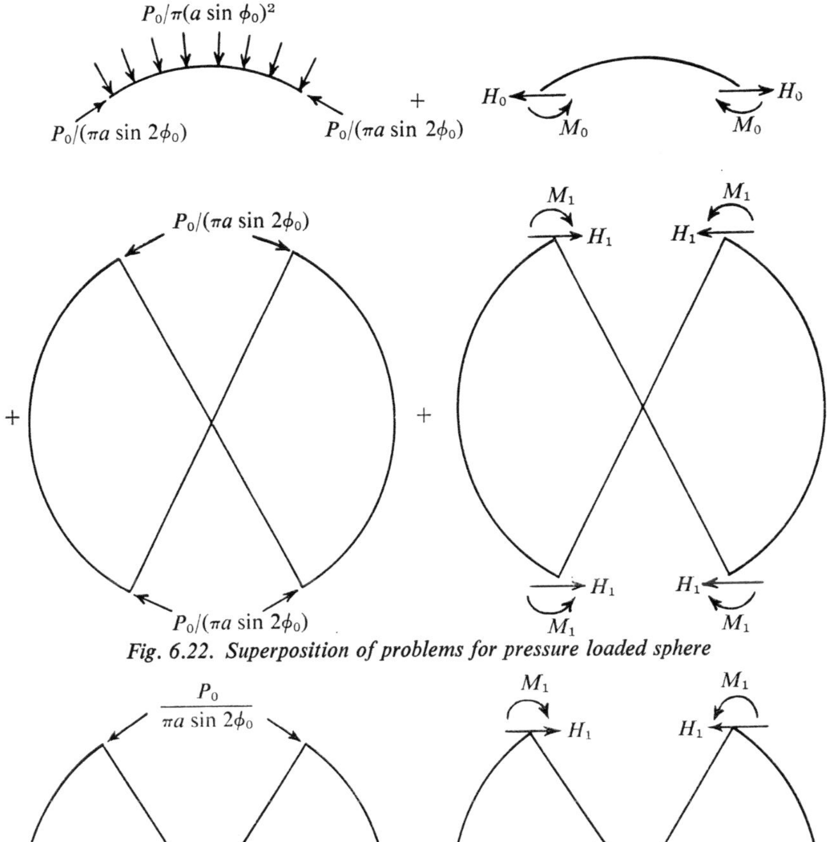

Fig. 6.22. Superposition of problems for pressure loaded sphere

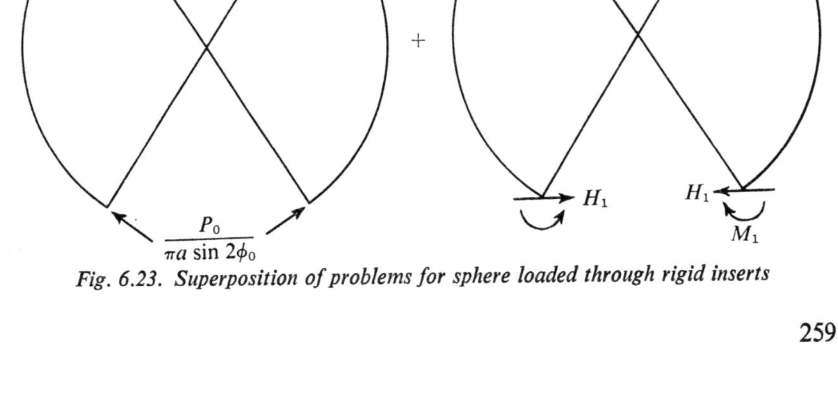

Fig. 6.23. Superposition of problems for sphere loaded through rigid inserts

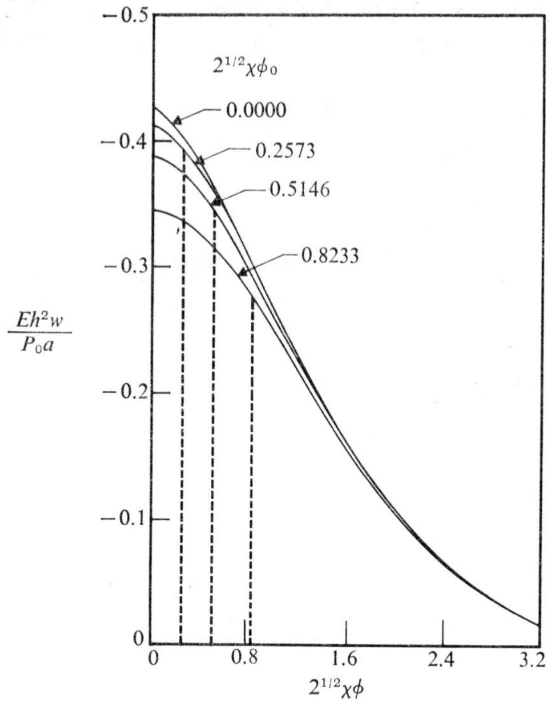

(a) Radial displacement

Fig. 6.24. Stresses and displacements in a sphere loaded by uniform pressure distributed over a small area near the poles

tions for an edge-loaded spherical shell with the pole caps removed (Fig. 6.23) to ensure zero edge rotation and horizontal deflection at the junction with the rigid inserts. These problems may be solved exactly with the use of the Legendre function solutions or approximately with the use of the shallow shell solutions†‡ if the angle ϕ_0 is small and the shell is thin. Some results obtained by means of shallow shell theory are shown in Figs. 6.24 and 6.25.

The former problem has been investigated experimentally as well as theoretically, and results for a one-tenth scale steel model of a nuclear reactor containment building have been obtained.§ The model was a sphere having a

† E. Reissner, op. cit.

‡ P. P. Bijlaard, 'On the Stresses from Loads in Spherical Pressure Vessels,' *Bull. U.S. Weld. Res. Comm.*, 34, 1956.

§ A. S. Tooth and R. M. Kenedi, 'The Influence Line Technique of Shell Analysis,' *Proc. Colloquium in Simplified Calculation Methods of Shell Structures*, Brussels, September 4–6, 1961, North-Holland Publishing Co., Amsterdam, 1962, pp. 44–63.

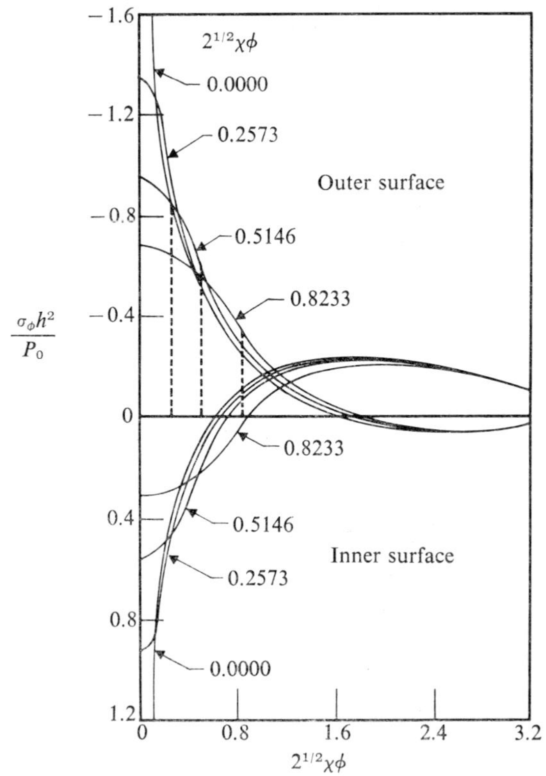

(b) Meridional stress

Fig. 6.24. Continued

nominal radius of 81 inches and a thickness of $\frac{3}{8}$ inch. A radial load was uniformly distributed over a circular area 0.55-inch in diameter and was reacted by a system of forces sufficiently far from the loaded area so that there were no interaction efforts. The value of χ for the sphere is readily calculated, with ν equal to 0.28, as

$$\chi \approx [3(1-\nu^2)]^{1/4}\left(\frac{a}{h}\right)^{1/2} = \{3[1-(0.28)^2]\}^{1/4}\left(\frac{81}{\frac{3}{8}}\right)^{1/2} \approx 19.0.$$

The loaded area subtended an angle

$$\phi_0 \approx \frac{0.275}{81} \text{ radian} \approx 0.0034 \text{ radian},$$

leading to

$$2^{1/2}\,\chi\phi_0 \approx 0.092.$$

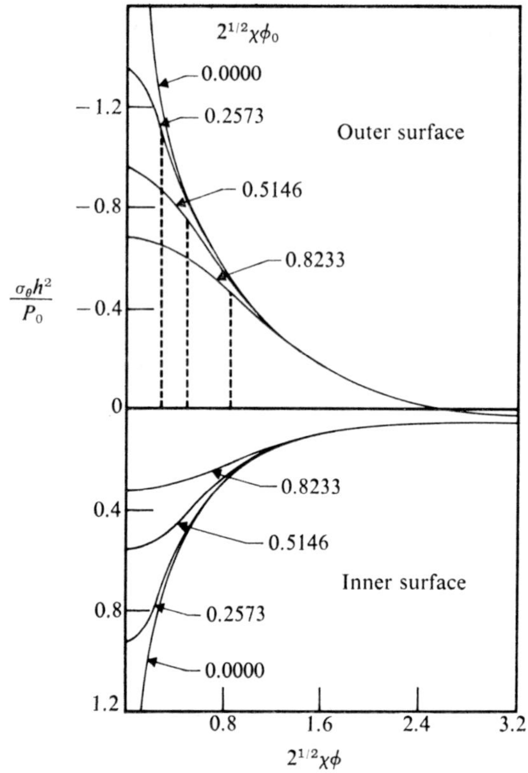

(c) Circumferential stress

Fig. 6.24. Concluded

Measured meridional and circumferential maximum stresses due to bending moments and to middle surface forces are compared with the results of shallow shell theory in Fig. 6.26. As can be seen, the comparison of theory and experiment is quite good.

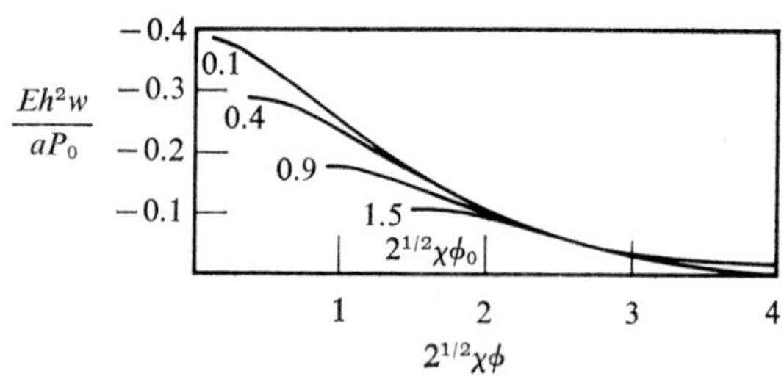

(a) Radical deflection

Fig. 6.25. Deflections and stresses for spheres loaded through rigid inserts at the poles

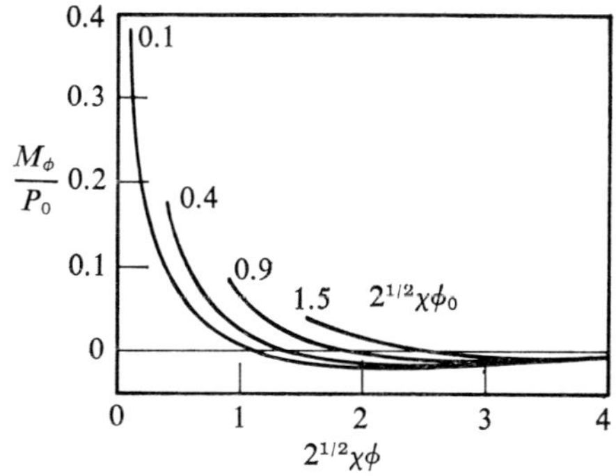

(b) Meridional moment

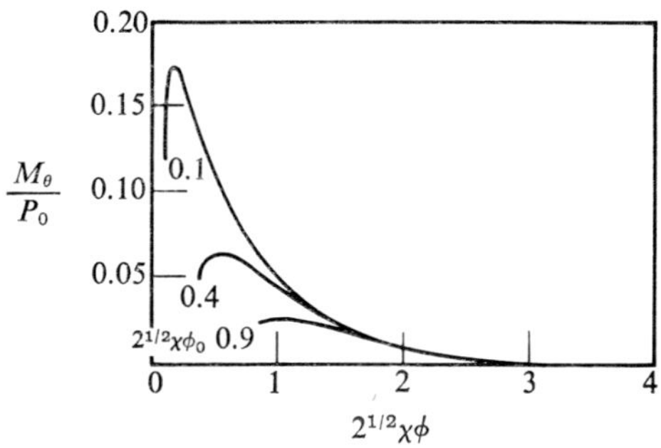

(c) Circumferential moment

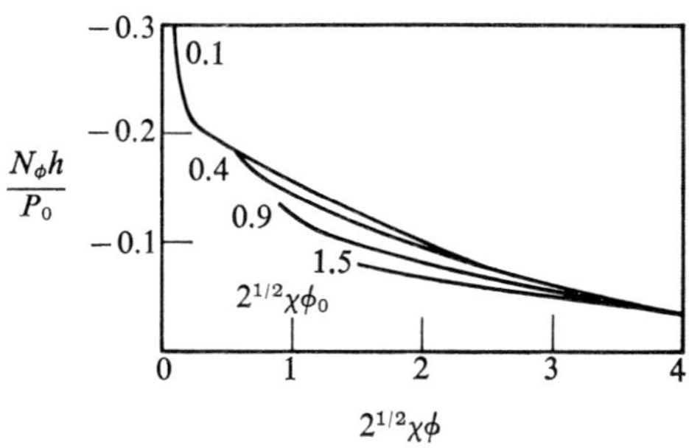

(d) Meridional force

Fig. 6.25. Continued

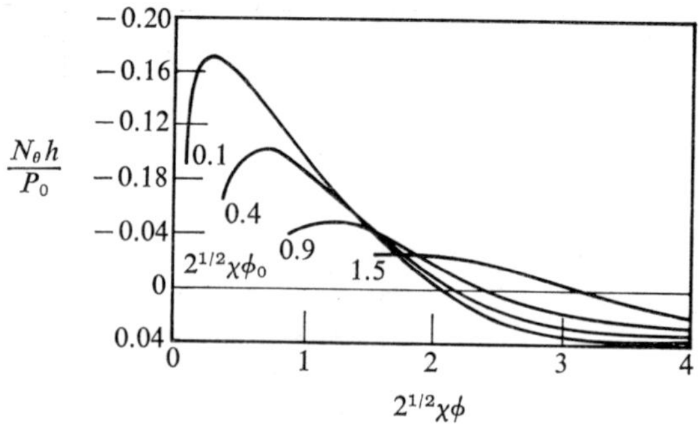

(e) Circumferential force

Fig. 6.25. Concluded

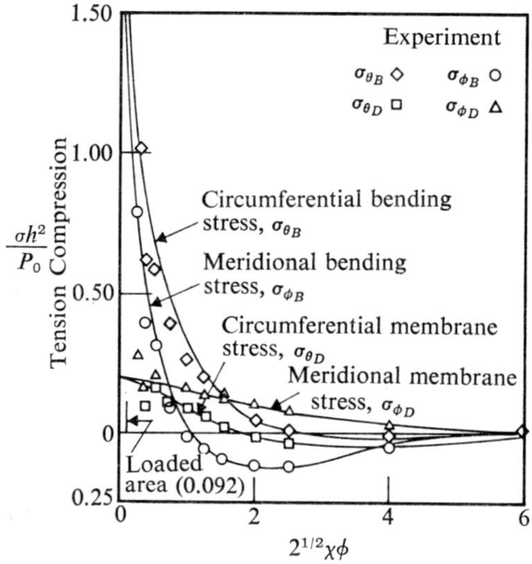

Fig. 6.26. Comparison of theoretical and experimental stresses in a sphere loaded radially at its poles

Supplementary references

[1] Blumenthal, O., "Über asymptotische Integration von Differentialgleichungen mit Anwendung auf die Berechnung von Spannungen in Kugelschalen," *Proc. Fifth Int. Cong. Math.* (Cambridge, England), vol. 2, 1912, pp. 319–327.

[2] Reissner, H., "Spannungen in Kugelschalen (Kuppeln)," *Festschrift Mueller-Breslau*, 1912, pp. 181–193.

[3] Bolle, L., "Festigkeitsberechnung von Kugelschalen," *Schweiz. Bauzeit*, vol. 66, 1915, pp. 105–108, 111–113.

[4] Hetényi, M., "Spherical Shells Subjected to Axial Symmetric Bending," *Publ. Int. Assoc. Bridge and Struct. Eng.*, Zurich, Switzerland, vol. 5, 1937–38, pp. 173–185.

[5] Hoff, N. J., "The Effect of Meridian Curvature on the Influence Coefficients of Thin Spherical Shells," *Problems of Continuum Mechanics*, Soc. Ind. and Appl. Math., Philadelphia, 1961, pp. 178–197.

[6] Watts, G. W., and Lang, H. A., "Stresses in a Pressure Vessel with a Hemispherical Head," *Trans. ASME*, vol. 75, 1953, pp. 83–90.

[7] Hoff, N. J., "General Formulas for Influence Coefficients of Thin Spherical Shells," *J. Aero. Sci.*, vol. 29, no. 2, February 1962, pp. 174–179, 225.

[8] Lur'e, A. I., *Three-Dimensional Problems of the Theory of Elasticity*, Interscience Publishers, 1964, pp. 462–484.

[9] Galletly, G. D., "Analysis of Discontinuity Stresses Adjacent to a Central Circular Opening in a Hemispherical Shell." David Taylor Model Basin Report 870, rev., May 1956.

[10] Morgan, W. C., and Bizon, P. T., "Experimental Evaluation of Theoretical Elastic Stress Distributions for Cylinder-to-Hemisphere and Cone-to-Sphere Junctions in Pressurized Shell Structures," NASA TN D-1565, February 1963.

7

Shells of arbitrary meridian

7.1 General equations for arbitrary shells of revolution

For shells of revolution which have varying meridional curvature it is not possible to express the theory conveniently in terms of the normal displacement.† However, it is possible to express the entire system of equations in terms of another single quantity. For the arbitrary shell of revolution let us take the lines of curvature, i.e., the meridians and the parallel circles normal to the axis of revolution, as our coordinate curves. The coordinates themselves are the angle ϕ, measured between the axis of revolution and the normal to the surface, and the angle θ, measured in the plane of the parallel circle passing through the point in question (Fig. 7.1).

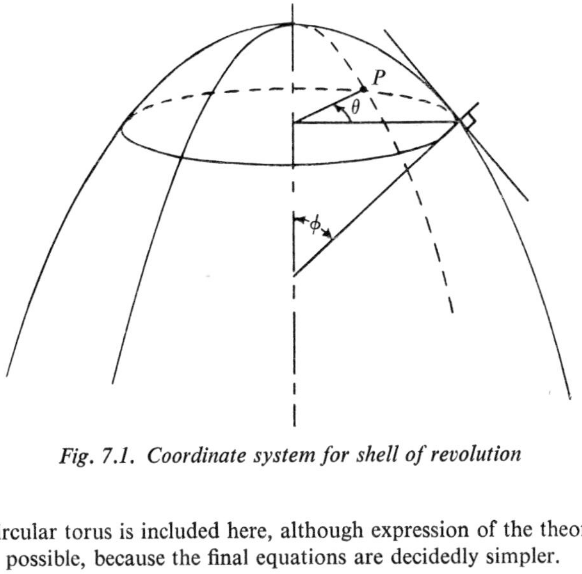

Fig. 7.1. Coordinate system for shell of revolution

† The circular torus is included here, although expression of the theory entirely in terms of w is possible, because the final equations are decidedly simpler.

For axisymmetric loading and deformations we have

$$q_\theta = m_\theta = u_\theta = 0, \tag{7.1.1}$$

while the remaining deformations, surface loads, and surface moments are functions of the angle ϕ only. The strain components and curvature changes are then, from Eqs. (1.5.3) and (1.5.10)

$$\gamma_{\phi\theta} = \bar{\kappa}_{\phi\theta} = 0, \tag{7.1.2a}$$

$$\epsilon_\phi = \frac{1}{R_\phi}\left(\frac{du_\phi}{d_\phi} + w\right), \tag{7.1.2b}$$

$$\epsilon_\theta = \frac{1}{R_\theta}(u_\phi \cot\phi + w), \tag{7.1.2c}$$

$$\kappa_\phi = -\frac{1}{R_\phi}\frac{d\Theta}{d\phi}, \tag{7.1.2d}$$

$$\kappa_\theta = -\frac{\Theta \cot\phi}{R_\theta}, \tag{7.1.2e}$$

where Θ is the rotation of the normal to the shell middle surface, given by

$$\Theta = \frac{1}{R_\phi}\left(\frac{dw}{d\phi} - u_\phi\right). \tag{7.1.2f}$$

Because of the axial symmetry of the problem, we also have

$$\bar{Q}_\theta = \bar{N}_{\phi\theta} = \bar{M}_{\phi\theta} = 0, \tag{7.1.3a}$$

while the remaining stress resultants are functions of ϕ only. Since the twisting moments vanish, we note that

$$\bar{Q}_\phi = Q_\phi. \tag{7.1.3b}$$

These stress resultants are related by the equations of equilibrium of the shell element [Eqs. (2.6.2)] which reduce to

$$\frac{1}{R_\phi}\frac{d}{d\phi}(N_\phi R_\theta \sin\phi) - N_\theta \cos\phi + R_\theta \sin\phi\left(\frac{Q_\phi}{R_\phi} + q_\phi\right) = 0, \tag{7.1.4a}$$

$$\frac{1}{R_\phi}\frac{d}{d\phi}(Q_\phi R_\theta \sin\phi) - R_\theta \sin\phi\left(\frac{N_\phi}{R_\phi} + \frac{N_\theta}{R_\theta} - q_\zeta\right) = 0, \tag{7.1.4b}$$

$$\frac{1}{R_\phi}\frac{d}{d\phi}(M_\phi R_\theta \sin\phi) - M_\theta \cos\phi - R_\theta \sin\phi(Q_\phi - m_\phi) = 0. \tag{7.1.4c}$$

The stress resultants are expressed in terms of strain components and curvature changes by overall force–strain relations which will be assumed, as before, to be given by Eqs. (3.1.4) as

$$N_\phi = K(\epsilon_\phi + \nu\epsilon_\theta), \tag{7.1.5a}$$

$$N_\theta = K(\epsilon_\theta + \nu\epsilon_\phi), \tag{7.1.5b}$$

$$M_\phi = D(\kappa_\phi + \nu\kappa_\theta) = -D\left(\frac{1}{R_\phi}\frac{\mathrm{d}\Theta}{\mathrm{d}\phi} + \frac{\nu\Theta\cot\phi}{R_\theta}\right), \tag{7.1.5c}$$

$$M_\theta = D(\kappa_\theta + \nu\kappa_\phi) = -D\left(\frac{\Theta\cot\phi}{R_\theta} + \frac{\nu}{R_\phi}\frac{d\Theta}{\mathrm{d}\phi}\right). \tag{7.1.5d}$$

We note, from Eqs. (7.1.5c), and (7.1.5d) that the meridional and circumferential bending moments are already expressed in terms of the one quantity Θ which will replace the normal deformation as the dependent variable of the problem. The shear force Q_ϕ can also be expressed in terms of Θ by means of the foregoing equations and Eq. (7.1.4c) as

$$U = R_\theta Q_\phi = -D\left[L(\Theta) - \frac{\nu}{R_\phi}\Theta\right] + U_0, \tag{7.1.6a}$$

where the operator L is defined by

$$L(\ldots) = \frac{1}{R_\phi\sin\phi}\frac{\mathrm{d}}{\mathrm{d}\phi}\left[\frac{R_\theta}{R_\phi}\sin\phi\frac{\mathrm{d}(\ldots)}{\mathrm{d}\phi}\right] - \frac{\cot^2\phi}{R_\theta}(\ldots) \tag{7.1.6b}$$

and

$$U_0 = m_\phi R_\theta. \tag{7.1.6c}$$

We now proceed, as for the cone and sphere, to express the middle surface force components N_ϕ and N_θ in terms of Q_ϕ, and hence in terms of Θ, by integrating Eqs. (7.1.4a) and (7.1.4b) in the following manner. We multiply Eq. (7.1.4a) by $\sin\phi$ and subtract from it Eq. (7.1.4b) multiplied by $\cos\phi$ to yield

$$\frac{\mathrm{d}}{\mathrm{d}\phi}[(N_\phi\sin\phi - Q_\phi\cos\phi)R_\theta\sin\phi] = (q_\zeta\cos\phi - q_\phi\sin\phi)R_\phi R_\theta\sin\phi. \tag{7.1.7a}$$

On integrating this with respect to ϕ we have

$$2\pi(N_\phi\sin\phi - Q_\phi\cos\phi)R_\theta\sin\phi$$
$$= 2\pi\int_{\phi_1}^{\phi}(q_\zeta\cos\phi - q_\phi\sin\phi)R_\phi R_\theta\sin\phi\,\mathrm{d}\phi + P, \tag{7.1.7b}$$

where ϕ_1 is some reference angle and P is a constant of integration. Eq.

(7.1.7b) is the equation of equilibrium of forces parallel to the axis of revolution acting on that portion of the shell of revolution between the angles ϕ_1 and ϕ (see Fig. 7.2). The left side of the equation is the axial resultant of forces acting on the edge defined by the angle ϕ, and the first term on the right side is the axial resultant of forces acting on the shell surfaces. The

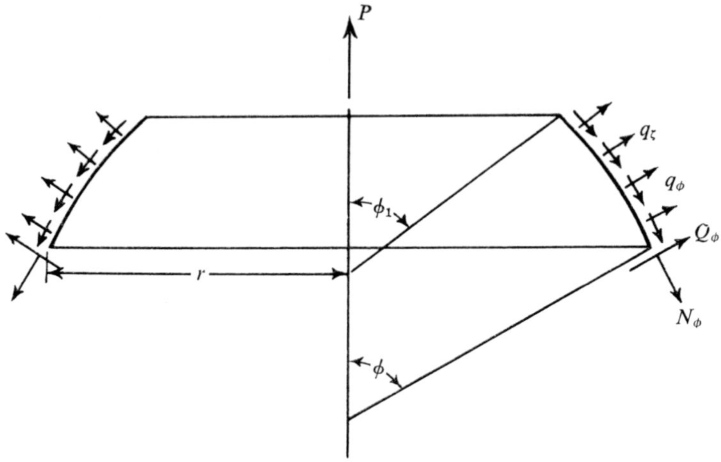

Fig. 7.2. Equilibrium of segment of shell of revolution

constant P can thus be interpreted as the axial resultant of forces acting on the edge defined by the angle ϕ_1. By solving Eq. (7.1.7b) for N_ϕ, we obtain

$$N_\phi = \left(U \cot \phi + \frac{P_\phi}{2\pi \sin^2 \phi}\right) \frac{1}{R_\theta} \tag{7.1.8a}$$

and, from Eqs. (7.1.4a), or (7.1.4b), and (7.1.8a),

$$N_\theta = \frac{1}{R_\phi} \frac{\mathrm{d}U}{\mathrm{d}\phi} + q_\zeta R_\theta - \frac{P_\phi}{2\pi R_\phi \sin^2 \phi}, \tag{7.1.8b}$$

where

$$P_\phi = P + 2\pi \int_{\phi_1}^{\phi} (q_\zeta \cos \phi - q_\phi \sin \phi) R_\phi R_\theta \sin \phi \, \mathrm{d}\phi. \tag{7.1.8c}$$

These can be expressed in terms of Θ by means of Eq. (7.1.6a).

We must now obtain an equation for Θ. With the use of Eqs. (7.1.2a), (7.1.2b), and (7.1.2f) we can verify the following relationship

$$\Theta = \frac{1}{R_\phi \sin \phi} \frac{\mathrm{d}}{\mathrm{d}\phi} (R_\theta \epsilon_\theta \sin \phi) - \epsilon_\phi \cot \phi. \tag{7.1.9}$$

Since, from Eqs. (7.1.5a), (7.1.5b), (7.1.8a), and (7.1.8b), we have

$$\begin{aligned}\epsilon_\phi &= \frac{1}{Eh}(N_\phi - \nu N_\theta)\\ &= \frac{1}{Eh}\left[\frac{\cot\phi}{R_\theta}U + \frac{\nu}{R_\phi}\frac{dU}{d\phi} + \left(\frac{1}{R_\theta} + \frac{\nu}{R_\phi}\right)\frac{P_\phi}{2\pi\sin^2\phi} - \nu q_\zeta R_\theta\right], \quad (7.1.10a)\\ \epsilon_\theta &= \frac{1}{Eh}(N_\theta - \nu N_\phi)\\ &= \frac{1}{Eh}\left[\frac{1}{R_\phi}\frac{dU}{d\phi} - \nu\frac{\cot\phi}{R_\theta}U - \left(\frac{1}{R_\phi} + \frac{\nu}{R_\theta}\right)\frac{P_\phi}{2\pi\sin^2\phi} + q_\zeta R_\theta\right]. \quad (7.1.10b)\end{aligned}$$

Equation (7.1.9) becomes

$$Eh\Theta = L(U) + \frac{\nu}{R_\phi}U - S_\phi, \tag{7.1.11a}$$

with

$$\begin{aligned}S_\phi = &\left\{1 - \left(\frac{R_\theta}{R_\phi}\right)^2 + \tfrac{1}{2}\tan\phi\frac{d}{d\phi}\left[\left(\frac{R_\theta}{R_\phi}\right)^2\right]\right\}\frac{P_\phi\cos\phi}{2\pi R_\theta\sin^3\phi}\\ &- \left[\frac{1}{R_\phi}\frac{d}{d\phi}(q_\zeta R_\theta^2) + \left(\frac{R_\theta}{R_\phi} + \nu\right)q_\phi R_\theta\right].\end{aligned} \tag{7.1.11b}$$

By using Eq. (7.1.6a) for Q_ϕ in terms of Θ we finally obtain

$$\begin{aligned}L^2(\Theta) + \nu\left[\frac{1}{R_\phi}L(\Theta) - L\left(\frac{\Theta}{R_\phi}\right)\right] + \left[\frac{12(1-\nu^2)}{h^2} - \frac{\nu^2}{R_\phi^2}\right]\Theta\\ = \frac{1}{D}\left[L(U_0) + \nu\frac{U_0}{R_\phi} - S_\phi\right].\end{aligned} \tag{7.1.12}$$

We still have the displacements u_ϕ and w to express in terms of Θ. This step can be accomplished through the use of Eqs. (7.1.2a), (7.1.2b), (7.1.10a), and (7.1.10b) to yield

$$\begin{aligned}\sin\phi\frac{d}{d\phi}\left(\frac{u_\phi}{\sin\phi}\right) &= R_\phi\epsilon_\phi - R_\theta\epsilon_\theta\\ &= \frac{1}{Eh}\left[\frac{R_\phi}{R_\theta}U\cot\phi - \frac{R_\theta}{R_\phi}\frac{dU}{d\phi} + \frac{\nu}{\sin\phi}\frac{d}{d\phi}(U\sin\phi)\right.\\ &\quad\left. + \left(\frac{R_\phi}{R_\theta} + 2\nu + \frac{R_\theta}{R_\phi}\right)\frac{P_\phi}{2\pi\sin^2\phi} - \left(\nu + \frac{R_\theta}{R_\phi}\right)R_\phi R_\theta q_\zeta\right],\end{aligned} \tag{7.1.13a}$$

$$\begin{aligned} w &= R_\theta \epsilon_\theta - u_\phi \cot \phi \\ &= \frac{1}{Eh}\left[\frac{R_\theta}{R_\phi}\frac{dU}{d\phi} - \nu U \cot \phi - \left(\frac{R_\theta}{R_\phi} + \nu\right)\frac{P_\phi}{2\pi \sin^2 \phi} + q_\zeta R_\theta^2\right] \\ &\quad - u_\phi \cot \phi, \end{aligned} \tag{7.1.13b}$$

which then can be expressed in terms of Θ through Eq. (7.1.6a).

In many problems involving the joining of shells of different shapes or of different thicknesses, the edge displacements and forces normal and parallel to the axis of revolution are desired and are given by

$$\begin{aligned} u_H &= u_\phi \cos \phi + w \sin \phi = R_\theta \epsilon_\theta \sin \phi \\ &= \frac{\sin \phi}{Eh}\left[\frac{R_\theta}{R_\phi}\frac{dU}{d\phi} - \nu U \cot \phi - \left(\frac{R_\theta}{R_\phi} + \nu\right)\frac{P_\phi}{2\pi \sin^2 \phi} + q_\zeta R_\theta^2\right], \end{aligned} \tag{7.1.14a}$$

$$\begin{aligned} u_V &= w \cos \phi - u_\phi \sin \phi = \int R_\phi(\Theta \cos \phi - \epsilon_\phi \sin \phi)\, d\phi + \text{constant} \\ &= \int\left\{\Theta R_\phi \cos \phi - \frac{1}{Eh}\left[\left(\frac{R_\phi}{R_\theta} + \nu\right)\left(U \cos \phi + \frac{P_\phi}{2\pi \sin \phi}\right)\right.\right. \\ &\quad \left.\left. - \nu q_\zeta R_\phi R_\theta \sin \phi\right]\right\} d\phi + \frac{\nu}{Eh} U \sin \phi + \text{constant}, \end{aligned} \tag{7.1.14b}$$

$$H = N_\phi \cos \phi + Q_\phi \sin \phi = \frac{1}{R_\theta \sin \phi}\left(U + \frac{\cot \phi}{2\pi} P_\phi\right), \tag{7.1.14c}$$

$$V = \frac{P_\phi}{2\pi R_\theta \sin \phi}. \tag{7.1.14d}$$

7.2 An approximate complementary solution of the bending equations

Except for the cylinder, the cone, and the sphere, we cannot solve Eq. (7.1.12) in terms of known functions. We can seek solutions of the equations in the form of an infinite power series of some function of ϕ,† but our previous experience with the series expansions of the Legendre functions in the solution for the spherical shell should be enough to warn us that the solution will very likely prove to be quite unsatisfactory for most practical problems.

Other methods are available, however. It will be recalled that the first

† See, for example, R. Schmidt, 'Analysis of Paraboloidal Shells,' *J. Appl. Mech.*, vol. 26, no. 1, March 1959, pp. 144–145.

term of the asymptotic expansions of the Bessel–Kelvin functions for the cone and of the Legendre functions for the sphere led to adequate representations when the shell was sufficiently thin. For arbitrary shells of revolution, similar approximate solutions are useful when the shells are thin. In this section we shall seek to determine the form of the first term of an asymptotic representation of the solutions of the homogeneous shell equations.

Rather than deal with the fourth-order differential equation for the rotation Θ, Eq. (7.1.12), we shall find it more convenient to work with the two equivalent second-order equations, given by Eqs. (7.1.6a), and (7.1.11a) as

$$L(\Theta) - \frac{\nu}{R_\phi}\Theta + \frac{U}{D} = \frac{U_0}{D}, \tag{7.2.1a}$$

$$L(U) + \frac{\nu}{R_\phi}U - Eh\Theta = S_\phi. \tag{7.2.1b}$$

If we use the convention that $R_\theta \sin\phi$ is always positive, R_θ is negative when $\sin\phi$ is negative. Let us consider Eqs. (7.2.1) when R_θ is either always positive or always negative and introduce the notation

$$d\xi = [3(1-\nu^2)]^{1/4}\frac{R_\phi\, d\phi}{(|R_\theta|h)^{1/2}}, \tag{7.2.2a}$$

$$\overline{\Theta} = \left(\frac{|R_\theta|}{h}\sin^2\phi\right)^{1/4}\Theta, \tag{7.2.2b}$$

$$\overline{U} = \frac{[12(1-\nu^2)]^{1/2}}{Eh^2}\left(\frac{|R_\theta|}{h}\sin^2\phi\right)^{1/4} U \operatorname{sgn} R_\theta, \tag{7.2.2c}$$

$$\overline{U}_0 = \frac{[12(1-\nu^2)]^{1/2}}{Eh^2}\left(\frac{|R_\theta|}{h}\sin^2\phi\right)^{1/4} U_0 \operatorname{sgn} R_\theta, \tag{7.2.2d}$$

$$\overline{\Theta}_0 = \frac{1}{Eh}\left(\frac{|R_\theta|}{h}\sin^2\phi\right)^{1/4}\left[L(U_0) + \frac{\nu}{R_\phi}U_0 - S_\phi\right], \tag{7.2.2e}$$

$$\begin{aligned}\psi = \frac{1}{[3(1-\nu^2)]^{1/2}}\frac{|R_\theta|h}{R_\phi^2}\Bigg\{&\left[\frac{15}{16}\left(\frac{R_\phi}{R_\theta}\right)^2 + \frac{3}{8}\frac{R_\phi}{R_\theta} - \frac{1}{16}\right]\cot^2\phi\\ &- \frac{1}{4}\left(1+\frac{R_\phi}{R_\theta}\right)\csc^2\phi - \frac{1}{4R_\phi}\frac{dR_\phi}{d\phi}\cot\phi\Bigg\},\end{aligned} \tag{7.2.2f}$$

$$\left.\begin{matrix}\theta_1\\ \theta_2\end{matrix}\right\} = \psi \pm \frac{\nu}{[3(1-\nu^2)]^{1/2}}\frac{h}{R_\phi}\operatorname{sgn} R_\theta. \tag{7.2.2g}$$

Then Eqs. (7.2.1a) and (7.2.1b) become

$$\frac{d^2\overline{\Theta}}{d\xi^2} - \theta_1\overline{\Theta} + 2(\overline{U} - \overline{U}_0) = 0, \tag{7.2.3a}$$

$$\frac{d^2(\overline{U} - \overline{U}_0)}{d\xi^2} - \theta_2(\overline{U} - \overline{U}_0) - 2\overline{\Theta} = -2\overline{\Theta}_0. \tag{7.2.3b}$$

A point where R_θ changes sign is treated as a discontinuity in the shell.

In terms of the new dependent variables, the stress resultants and pertinent deformations may be written as

$$\Theta = \left(\frac{|R_\theta|}{h}\sin^2\phi\right)^{-1/4}\overline{\Theta}, \tag{7.2.4a}$$

$$\begin{aligned} u_H &= \frac{R_\theta}{Eh}(N_\theta - \nu N_\phi)\sin\phi = \frac{h}{2[3(1-\nu^2)]^{1/4}}\left(\frac{|R_\theta|}{h}\sin^2\phi\right)^{1/4} \\ &\times \left\{\frac{d\overline{U}}{d\xi}\operatorname{sgn} R_\theta - \frac{\cot\phi}{[3(1-\nu^2)]^{1/4}}\left(\frac{h}{|R_\theta|}\right)^{1/2}\left[\tfrac{1}{4}\left(1 + \frac{R_\theta}{R_\phi}\right) + \nu\right]\overline{U}\right\} \\ &- \frac{P_\phi}{2\pi Eh\sin\phi}\left(\frac{R_\theta}{R_\phi} + \nu\right) \end{aligned} \tag{7.2.4b}$$

$$\begin{aligned} Q_\phi &= \left(N_\phi - \frac{P_\phi}{2\pi R_\theta\sin^2\phi}\right)\tan\phi = \left(H - \frac{P_\phi\cot\phi}{2\pi R_\theta\sin^2\phi}\right)\sin\phi \\ &= \frac{Eh^2}{[12(1-\nu^2)]^{1/2}}\left(\frac{|R_\theta|}{h}\sin^2\phi\right)^{-1/4}\frac{\overline{U}}{|R_\theta|}, \end{aligned} \tag{7.2.4c}$$

$$\begin{aligned} M_\phi &= -\frac{Eh^3}{4[2(1-\nu^2)]^{3/4}}(|R_\theta|h)^{-1/2}\left(\frac{|R_\theta|}{h}\sin^2\phi\right)^{-1/4} \\ &\times \left\{\frac{d\overline{\Theta}}{d\xi} - \frac{\cot\phi}{[3(1-\nu^2)]^{1/4}}\left(\frac{h}{|R_\theta|}\right)^{1/2}\left[\frac{1}{4}\left(1 + \frac{R_\theta}{R_\phi}\right) - \nu\right]\overline{\Theta}\operatorname{sgn} R_\theta\right\}, \end{aligned} \tag{7.2.4d}$$

$$\begin{aligned} M_\theta &= -\frac{Eh^3}{4[3(1-\nu^2)]^{3/4}}(|R_\theta|h)^{-1/2}\left(\frac{|R_\theta|}{h}\sin^2\phi\right)^{-1/4} \\ &\times \left\{\nu\frac{d\overline{\Theta}}{d\xi} - \frac{\cot\phi}{[3(1-\nu^2)]^{1/4}}\left(\frac{h}{R_\theta}\right)^{1/2}\left[\frac{\nu}{4}\left(1 + \frac{R_\theta}{R_\phi}\right) - 1\right]\overline{\Theta}\operatorname{sgn} R_\theta\right\}, \end{aligned} \tag{7.2.4e}$$

$$\begin{aligned} N_\theta &= \frac{Eh^2}{2[2(1-\nu^2)]^{1/4}}(|R_\theta|h)^{-1/2}\left(\frac{|R_\theta|}{h}\sin^2\phi\right)^{-1/4} \\ &\times \left\{\frac{d\overline{U}}{d\xi}\operatorname{sgn} R_\theta - \frac{1}{4}\frac{\cot\phi}{[3(1-\nu^2)]^{1/4}}\left(\frac{h}{|R_\theta|}\right)^{1/2}\left(1 + \frac{R_\theta}{R_\phi}\right)\overline{U}\right\} \\ &+ q_\zeta R_\theta - \frac{P_\phi}{2\pi R_\phi\sin^2\phi}. \end{aligned} \tag{7.2.4f}$$

We note that for the shell shapes we have already examined θ_1 and θ_2 are generally small. For the cylinder, for example, θ_1 and θ_2 vanish identically. For the cone

$$\theta_1 = \theta_2 = \frac{15}{4}\left\{2[3(1-\nu^2)]^{1/4}\left(\frac{s \cot \alpha}{h}\right)^{1/2}\right\}^{-2}. \tag{7.2.5a}$$

The quantity in braces will be recognized as the variable which must be large for the adequacy of the first term of the asymptotic expansions of the Bessel–Kelvin functions and which is usually large in the region of interest of conical shells encountered in practice, so that θ_1 and θ_2 are small. Finally, for the sphere

$$\left.\begin{matrix}\theta_1\\ \theta_2\end{matrix}\right\} = \frac{1}{[3(1-\nu^2)]^{1/2}}\frac{h}{a}\left(\frac{3\csc^2\phi - 5}{4} \pm \nu\right), \tag{7.2.5b}$$

which is small for thin shells provided ϕ is not near 0 degrees or 180 degrees. As an approximation, then let us assume for other shells of revolution that θ_1 and θ_2 are small compared to unity. For this to be so the angle ϕ must not be near 0 degrees or 180 degrees, the radius of curvature of the meridian must not vary rapidly, the shell thickness should be small compared to either radius of curvature and the quantity $R_\phi h/R_\theta^2$ should be small. We also assume that the derivatives of $\overline{\Theta}$ and $\overline{U}$ with respect to the variable ξ are the order of magnitude of the functions themselves. Then if we neglect the terms multiplied by θ_1 and θ_2 and consider only edge loaded shells, Eqs. (7.2.3) reduce to

$$\frac{d^2\overline{\Theta}}{d\xi^2} + 2\overline{U} = 0, \tag{7.2.6a}$$

$$\frac{d^2\overline{U}}{d\xi^2} - 2\overline{\Theta} = 0. \tag{7.2.6b}$$

Equations (7.2.6a) and (7.2.6b) may be combined to yield the single complex equation

$$\left(\frac{d^2}{d\xi^2} - 2i\right)\overline{V} = 0, \tag{7.2.6c}$$

with

$$\overline{V} = \overline{\Theta} + i\overline{U}. \tag{7.2.6d}$$

The solution of Eq. (7.2.6c) may be written as

$$\overline{V} = (A + iB)\,e^{(1+i)\xi} + (C + iD)\,e^{-(1+i)\xi} \tag{7.2.7a}$$

or, equating real and imaginary terms

$$\bar{\Theta} = e^{\xi}(A\cos\xi - B\sin\xi) + e^{-\xi}(C\cos\xi + D\sin\xi), \tag{7.2.7b}$$

$$\bar{U} = e^{\xi}(B\cos\xi + A\sin\xi) + e^{-\xi}(D\cos\xi - C\sin\xi). \tag{7.2.7c}$$

Eqs. (7.2.7) may also be expressed in the form of solutions which decay rapidly as we move into the shell from the edges as

$$\begin{aligned}\bar{\Theta} = {} & e^{-(\xi-\xi_1)}[A_1\cos(\xi-\xi_1) + A_2\sin(\xi-\xi_1)] \\ & + e^{-(\xi_2-\xi)}[A_3\cos(\xi_2-\xi) + A_4\sin(\xi_2-\xi)], \quad \xi_1 \le \xi \le \xi_2,\end{aligned} \tag{7.2.8a}$$

$$\begin{aligned}\bar{U} = {} & e^{-(\xi-\xi_1)}[A_2\cos(\xi-\xi_1) - A_1\sin(\xi-\xi_1)] \\ & + e^{-(\xi_2-\xi)}[A_4\cos(\xi_2-\xi) - A_3\sin(\xi_2-\xi)], \quad \xi_1 \le \xi \le \xi_2,\end{aligned} \tag{7.2.8b}$$

where ξ_1 and ξ_2 are the values of ξ for the shell edges.

Expressions for deformations and stress resultants consistent with the above approximation are given by

$$\left(\frac{|\sin\phi|}{\lambda}\right)^{1/2}\Theta \approx \left(\frac{h}{[3(1-\nu^2)]^{1/4}}\right)^{1/2}\frac{d}{d\xi}\left[\frac{1}{2}\frac{d\bar{U}}{d\xi}\right], \tag{7.2.9a}$$

$$\left(\frac{\lambda}{|\sin\phi|}\right)^{1/2}u_H \approx \left(\frac{h}{[3(1-\nu^2)]^{1/2}}\right)^{1/2}\frac{1}{2}\frac{d\bar{U}}{d\xi}\operatorname{sgn} R_\theta, \tag{7.2.9b}$$

$$\left(\frac{|\sin\phi|}{\lambda}\right)^{3/2}\frac{H}{D\lambda} \approx -\left(\frac{h}{[3(1-\nu^2)]^{1/4}}\right)^{1/2}\frac{d^3}{d\xi^3}\left(\frac{1}{2}\frac{d\bar{U}}{d\xi}\right), \tag{7.2.9c}$$

$$\left(\frac{|\sin\phi|}{\lambda}\right)\frac{M_\phi}{D\lambda} \approx -\left(\frac{h}{[3(1-\nu^2)]^{1/4}}\right)^{1/2}\frac{d^2}{d\xi^2}\left(\frac{1}{2}\frac{d\bar{U}}{d\xi}\right), \tag{7.2.9d}$$

$$M_\theta \approx \nu M_\phi, \tag{7.2.9e}$$

$$N_\theta \approx \frac{Eh}{|R_\theta|}u_H, \tag{7.2.9f}$$

where

$$\lambda = \frac{[3(1-\nu^2)]^{1/4}}{(|R_\theta|h)^{1/2}} \tag{7.2.9g}$$

with $\bar{U}$ given by Eqs. (7.2.7) or (7.2.8). The approximate equations will be recognized as similar to those obtained in the previous chapters for conical and spherical shells, and particularly for cylindrical shells.

The form of the equations suggests, as was found to be true for cones and spheres, that the influence coefficients for sufficiently thin shells for which the minimum edge angles are sufficiently far from 0 degrees or 180 degrees should be approximated by those for a cylindrical shell having an equivalent length parameter λl equal to $\xi_2 - \xi_1$. The influence coefficients appear in edge displacement and slope change equations of the form (exactly)

$$\begin{Vmatrix} \left(\dfrac{\lambda_1}{\sin \phi_1}\right)^{1/2} u_{H_1} \\ \left(\dfrac{\sin \phi_1}{\lambda_1}\right)^{1/2} \Theta_1 \\ \left(\dfrac{\lambda_2}{\sin \phi_2}\right)^{1/2} u_{H_2} \\ \left(\dfrac{\sin \phi_2}{\lambda_2}\right)^{1/2} \Theta_2 \end{Vmatrix} = \begin{Vmatrix} \delta_{H_{11}} & \delta_{M_{11}} & \delta_{H_{12}} & \delta_{M_{12}} \\ \Theta_{H_{11}} & 2\Theta_{M_{11}} & \Theta_{H_{12}} & 2\Theta_{M_{12}} \\ \delta_{H_{21}} & \delta_{M_{21}} & \delta_{H_{22}} & \delta_{M_{22}} \\ \Theta_{H_{21}} & 2\Theta_{M_{21}} & \Theta_{H_{22}} & 2\Theta_{M_{22}} \end{Vmatrix} \begin{Vmatrix} \dfrac{H_1 \sin^{3/2} \phi_1}{2\lambda_1^{5/2} D} \\ \dfrac{M_1 \sin^{1/2} \phi_1}{2\lambda_1^{3/2} D} \\ \dfrac{H_2 \sin^{3/2} \phi_2}{2\lambda_2^{5/2} D} \\ \dfrac{M_2 \sin^{1/2} \phi_2}{2\lambda_2^{3/2} D} \end{Vmatrix}, \tag{7.2.10a}$$

where

$$\lambda_1 |R_{\theta 1}|^{1/2} = \lambda_2 |R_{\theta 2}|^{1/2} = \frac{[3(1-\nu^2)]^{1/4}}{h^{1/2}} \tag{7.2.10b}$$

and the matrix of influence coefficients is symmetric. As an example, some of the edge influence coefficients for the toroidal shells shown in Fig. 7.3, which were obtained by numerical integration of the exact equations,† have been put into the form defined by Eqs. (7.2.10) and are plotted as a function of the parameter $\xi_2 - \xi_1$ in Fig. 7.4. The subscript 2 refers to the edge $\phi = \pm 90°$ while the subscript 1 refers to the edge $\phi < 90°$ or $\phi > -90°$. The results shown are for the thickest toroidal shell which was investigated, one having a value of $a/h = 10$. Despite this low radius/thickness ratio, the results for many of the influence coefficients (primarily those dealing with effects at edge 2, however) practically coincide with the finite cylindrical shell results over the entire range of angle of edge 1 ($75° \leq \phi_1 \leq 15°$) while the remaining coefficients are adequately represented by the cylindrical shell results if the angle ϕ of edge 1 is greater than, say, 60°. The limit on the angle ϕ of edge 1 decreases as the shell becomes thinner.

Eqs. (7.2.7) represent the first terms of an asymptotic expansion of the homogeneous bending equations which are valid when θ_1 and θ_2 are very

† G. D. Galletly, 'Edge Influence Coefficients for Toroidal Shells of Positive Gaussian Curvature,' *J. Engineering for Industry*, vol. 82, no. 3, February 1960, pp. 60–80; 'Edge Influence Coefficients for Toroidal Shells of Negative Gaussian Curvature,' *J. Engineering for Industry*, vol. 82, no. 3, February 1960, pp. 69–75.

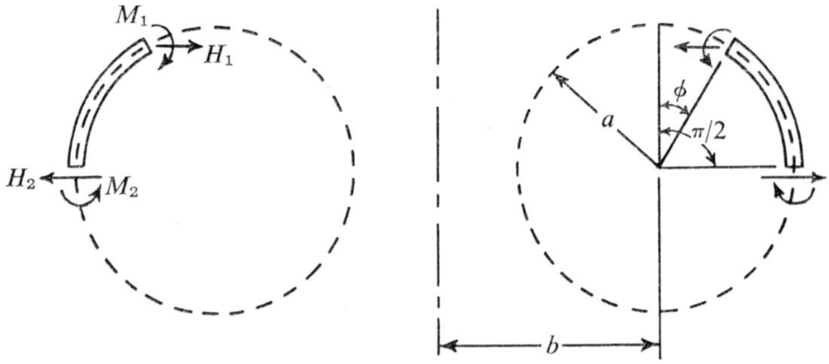

(a) Positive Gaussian curvature

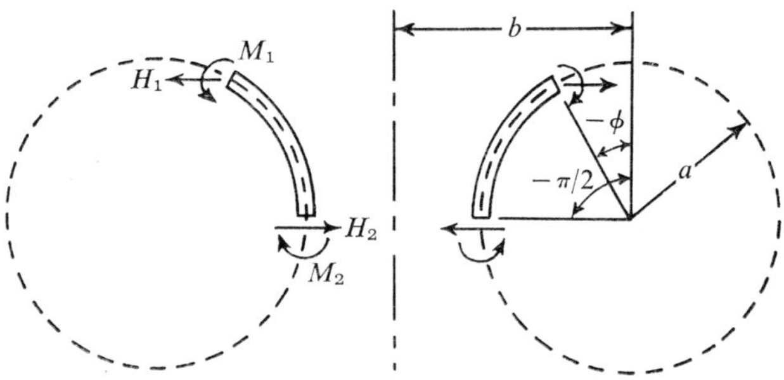

(b) Negative Gaussian curvature

Fig. 7.3. Circular toroidal shell segments under edge loading

small. Additional terms in the series may be obtained† but are rarely used since numerical integration or finite element techniques are much more efficient when these terms are important.

7.3 The variable ξ

In order to make the preceding approximate solution useful we must evaluate the variable ξ as a function of ϕ. For the shells we have considered in previous chapters the independent variable

$$\xi = [3(1 - \nu^2)]^{1/4} \int_0^\phi \frac{R_\phi \, d\phi}{(|R_\theta| \, h)^{1/2}} \tag{7.3.1}$$

† F. B. Hildebrand, 'On Asymptotic Integration in Shell Theory,' *Proc. Third Symposium in Applied Math.*, McGraw-Hill Book Co., 1950, pp. 53–66.

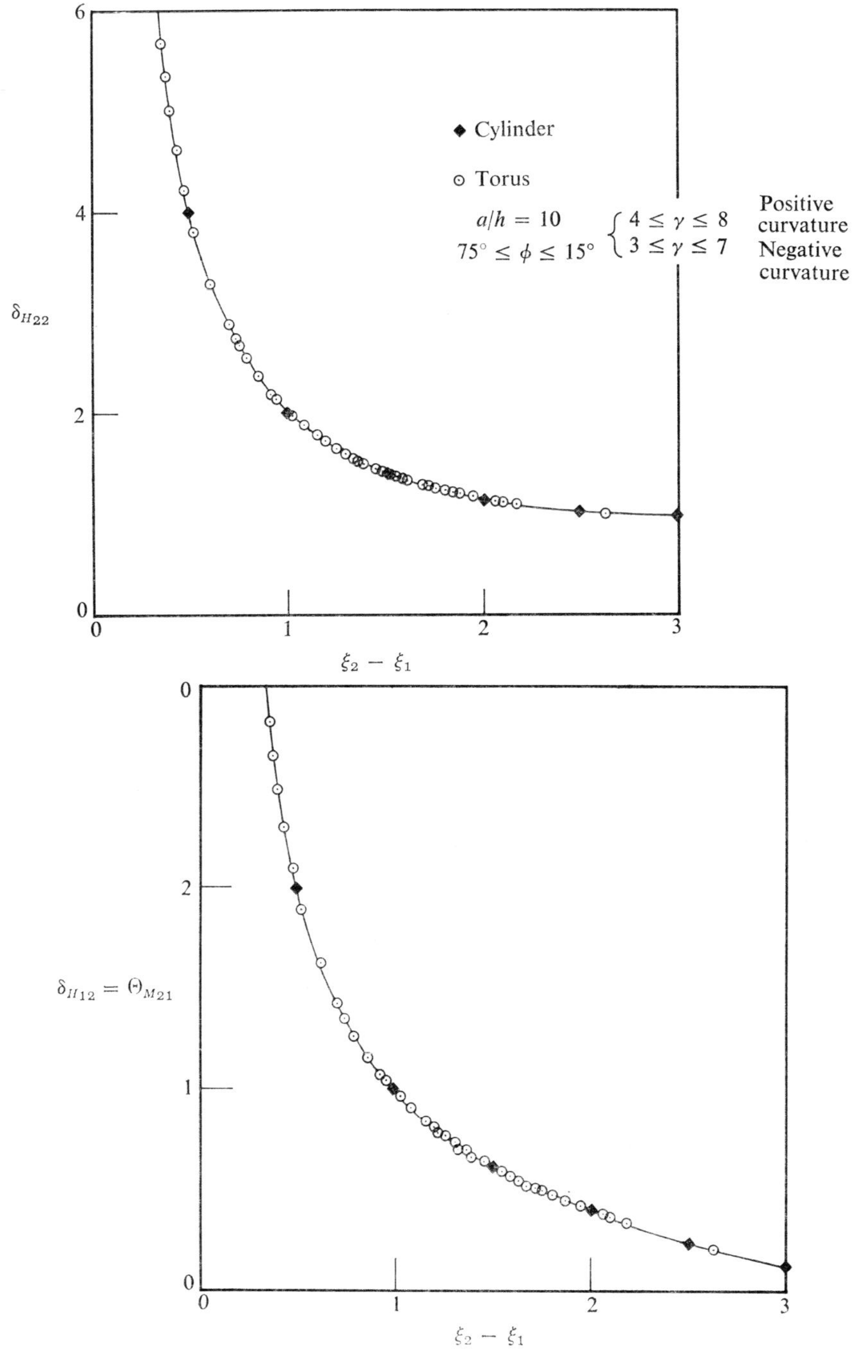

Fig. 7.4. Correlation of influence coefficients for toroidal shell segments

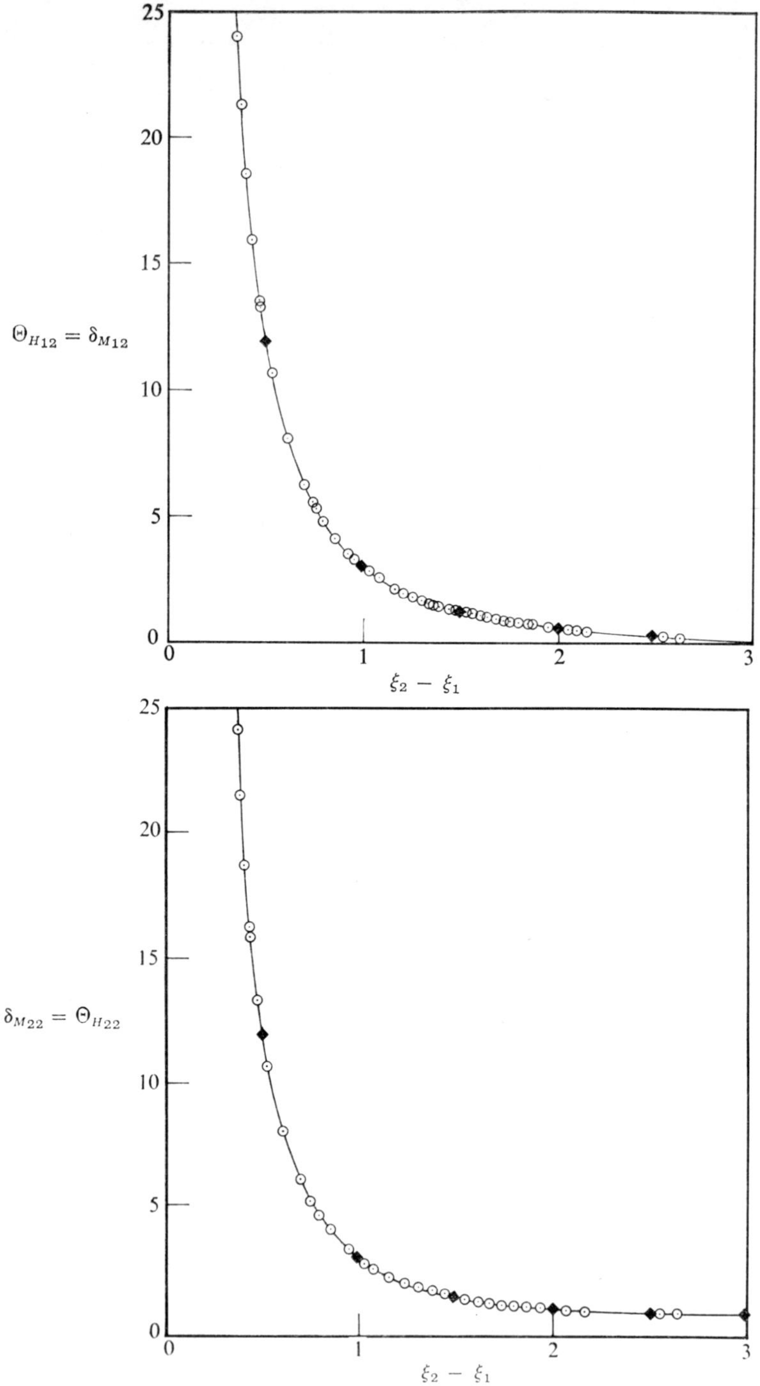

Fig. 7.4. Continued

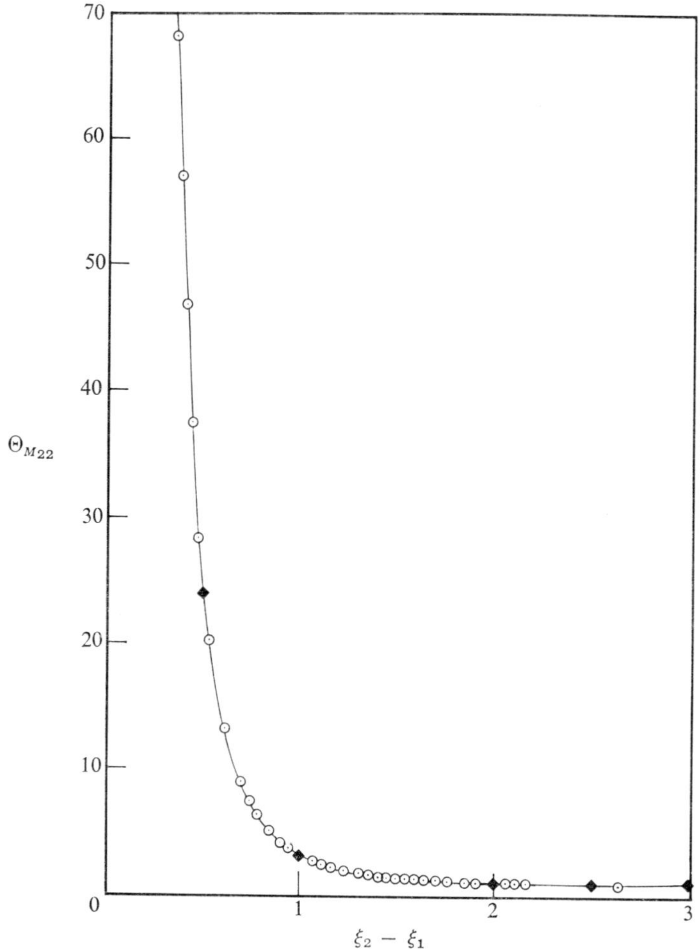

Fig. 7.4. Continued

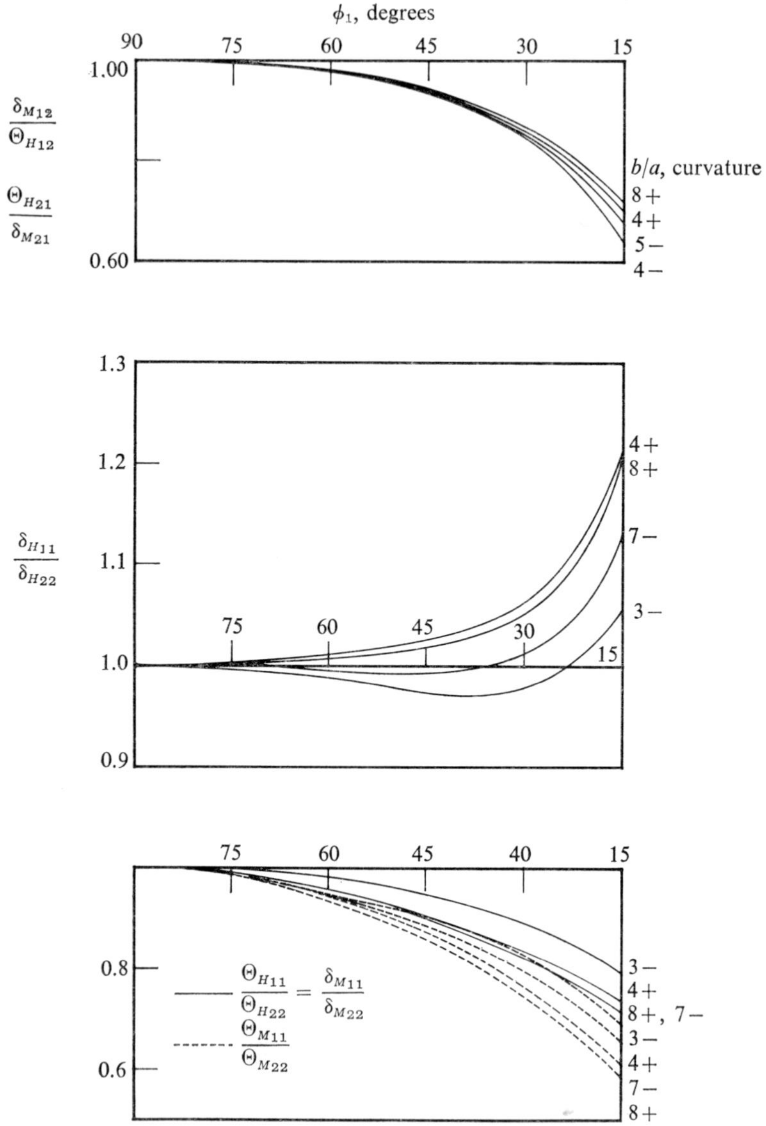

Fig. 7.4. Concluded

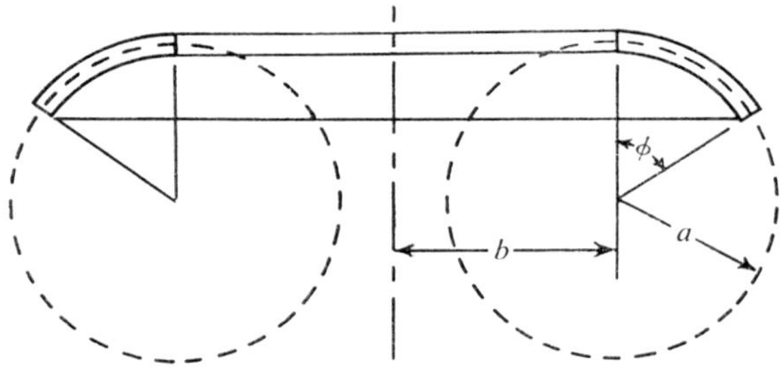

Case I: $\pi \geq \phi \geq 0$, all a/b

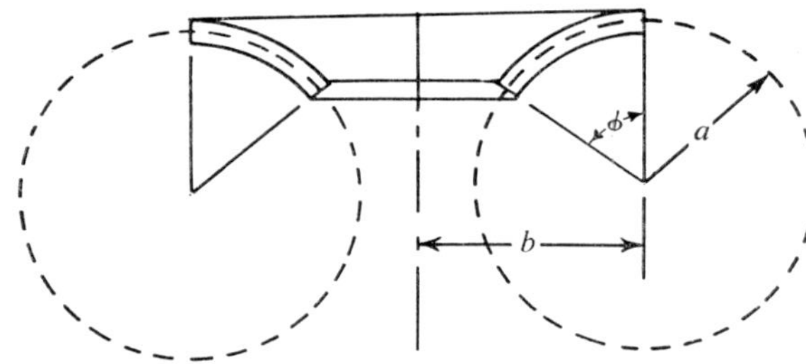

Case II: $a/b < 1$, $-\pi \leq \phi \leq 0$

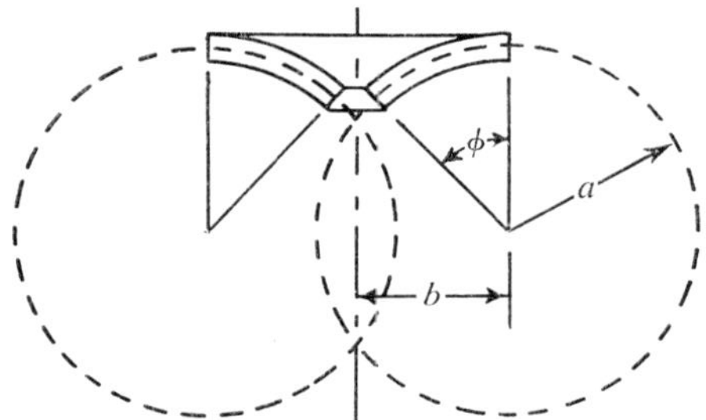

Case III: $a/b > 1$, $-\sin^{-1}(b/a) \leq \phi \leq 0$

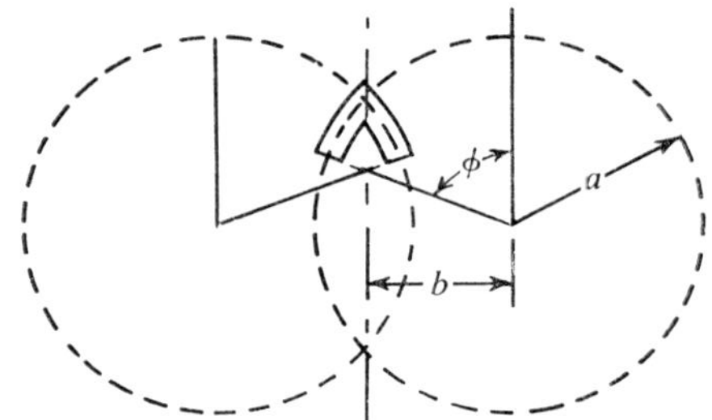

Case IV: $a/b > 1$, $\sin^{-1}(b/a) - \pi \leq \phi \leq -\sin^{-1}(b/a)$

Fig. 7.5. Toroidal shell segments

is a familiar quantity. For the cylinder

$$\xi = \frac{[3(1-\nu^2)]^{1/4}}{(Rh)^{1/2}} \int_0^x \mathrm{d}x = \lambda x. \tag{7.3.2a}$$

For the cone

$$\xi = \frac{[3(1-\nu^2)]^{1/4}}{h \tan \alpha} \int_0^s \frac{\mathrm{d}s}{s^{1/2}} = 2[3(1-\nu^2)]^{1/4} \left(\frac{s \cot \alpha}{h}\right)^{1/2}. \tag{7.3.2b}$$

For the sphere

$$\xi = [3(1-\nu^2)]^{1/4} \left(\frac{a}{h}\right)^{1/2} \int_0^\phi \mathrm{d}\phi = [3(1-\nu^2)]^{1/4} \left(\frac{a}{h}\right)^{1/2} \phi \approx \chi\phi. \tag{7.3.2c}$$

Each of these variables has appeared in asymptotic expansions of the function appearing in the exact solution.

For other shell shapes, in general, numerical integration methods must be used to evaluate Eq. (7.3.1). For certain shell shapes the integral can be evaluated in terms of elliptic functions. For example, for a circular toroidal shell we can express the substitute variable as

$$\xi = [3(1-\nu^2)]^{1/4} \left(\frac{a^2}{bh}\right)^{1/2} \int_0^\phi \left(\frac{\sin|\phi|}{1 + \dfrac{a}{b}\sin\phi}\right)^{1/2} \mathrm{d}\phi, \tag{7.3.3a}$$

for cases (a), (b), and (c) of Fig. 7.5. For toroidal segments of the form shown in Fig. 7.5(d) we write

$$\xi = [3(1-\nu^2)]^{1/4} \left(\frac{a^2}{bh}\right)^{1/2} \int_{\sin^{-1}(b/a)}^\phi \left(\frac{\sin\phi}{(a/b)\sin\phi - 1}\right)^{1/2} \mathrm{d}\phi. \tag{7.3.3b}$$

The integral of Eq. (7.3.3a) is given in closed form by the following expressions†

$$\begin{aligned} \int_0^\phi \left(\frac{\sin|\phi|}{1 + a/b \sin\phi}\right)^{1/2} \mathrm{d}\phi \\ = 2^{1/2}\Bigg\{\Pi\left[\cos^{-1}\tan\frac{1}{2}\left(\frac{\pi}{2} - |\phi|\right), \frac{1}{2}, \left(\frac{1 - a/b \operatorname{sgn}\phi}{2}\right)^{1/2}\right] \\ - F\left[\cos^{-1}\tan\frac{1}{2}\left(\frac{\pi}{2} - |\phi|\right), \left(\frac{1 - a/b \operatorname{sgn}\phi}{2}\right)^{1/2}\right]\Bigg\} \operatorname{sgn}\phi, \\ -\pi/2 \le \phi \le \pi/2, \quad 0 \le a/b \le 1 \end{aligned} \tag{7.3.4a}$$

† The integral is evaluated by transformation to an appropriate form in P. F. Byrd and M. D. Friedman, *Handbook of Elliptic Integrals for Engineers and Physicists*, Second Edition, Springer-Verlag, 1971.

$$= 2\tan^{-1}\left(\frac{\sin\phi}{1-\sin\phi}\right)^{1/2} - 2^{1/2}\tan^{-1}\left(\frac{2\sin\phi}{1-\sin\phi}\right)^{1/2},$$

$$0 \le \phi \le \pi/2,\ a/b = 1 \tag{7.3.4b}$$

$$= -\left[\frac{2^{1/2}}{2}\ln\left(\frac{1+\left(\frac{2\sin|\phi|}{1+\sin|\phi|}\right)^{1/2}}{1-\left(\frac{2\sin|\phi|}{1+\sin|\phi|}\right)^{1/2}}\right) - \ln\left(\frac{1+\left(\frac{\sin|\phi|}{1+\sin|\phi|}\right)^{1/2}}{1-\left(\frac{\sin|\phi|}{1+\sin|\phi|}\right)^{1/2}}\right)\right],$$

$$-\pi/2 \le \phi \le 0,\ a/b = 1 \tag{7.3.4c}$$

$$= \frac{2(b/a)^2}{(1+b/a)^{1/2}}\left\{\Pi\left[\cos^{-1}\left(\frac{1-\sin\phi}{1+a/b\sin\phi}\right)^{1/2}, \frac{1}{1+b/a}, \left(\frac{1-b/a}{1+b/a}\right)^{1/2}\right]\right.$$

$$\left. - F\left[\cos^{-1}\left(\frac{1-\sin\phi}{1+a/b\sin\phi}\right)^{1/2}, \left(\frac{1-b/a}{1+b/a}\right)^{1/2}\right]\right\},$$

$$0 \le \phi \le \pi/2, \quad a/b \ge 1, \tag{7.3.4d}$$

$$= -\frac{2}{(1+b/a)^{1/2}}$$

$$\times\left\{\Pi\left[\cos^{-1}\left(\frac{1-a/b\sin|\phi|}{1+\sin|\phi|}\right)^{1/2}, \frac{b/a}{1+b/a}, \left(\frac{2b/a}{1+b/a}\right)^{1/2}\right]\right.$$

$$\left. - F\left[\cos^{-1}\left(\frac{1-a/b\sin|\phi|}{1+\sin|\phi|}\right)^{1/2}, \left(\frac{2b/a}{1+b/a}\right)^{1/2}\right]\right\},$$

$$\sin^{-1}(-b/a) \le \phi \le 0, \quad a/b \ge 1, \tag{7.3.4e}$$

where $F(\Theta, k)$ and $\Pi(\Theta, \alpha^2, k)$† are elliptic integrals of the first and third kinds, respectively. The integral of Eq. (7.3.3b) is given by

$$\int_{\sin^{-1}b/a}^{\phi}\left(\frac{\sin\phi}{a/b\sin\phi - 1}\right)^{1/2} d\phi = 2\left(\frac{(b/a)^3}{1+b/a}\right)^{1/2}$$

$$\times\,\Pi\left[\cos^{-1}\left(\frac{(b/a)(1-\sin\phi)}{(1-b/a)\sin\phi}\right)^{1/2}, 1-\frac{b}{a}, \left(\frac{1-b/a}{1+b/a}\right)^{1/2}\right]$$

$$\sin^{-1}b/a \le \phi \le \pi/2, \quad a/b > 1. \tag{7.3.5}$$

Some values of the integral are given in Table 7.1 and are shown graphically in Fig. 7.6.

---

† Values of the less common elliptic integral of the third kind are available in R. G. Selfridge and J. E. Maxfield, *A Table of the Incomplete Elliptic Integral of the Third Kind*, Dover Publications, New York, 1958.

Table 7.1 $\int_0^{\phi} \left(\frac{\sin|\phi|}{1 + a/b \sin\phi}\right)^{1/2} d\phi$

ϕ degrees	a/b 0.0	0.1	0.2	0.3	0.4	0.5	1.0
−90	−1.1981	−1.2488	−1.2982	−1.3603	−1.4338	−1.5224	−∞
−85	−1.1109	−1.1527	−1.2004	−1.2558	−1.3210	−1.3997	−3.6478
−80	−1.0240	−1.0613	−1.1034	−1.1526	−1.2096	−1.2782	−2.6711
−75	−0.9378	−0.9705	−1.0075	−1.0499	−1.0993	−1.1580	−2.1032
−70	−0.8527	−0.8809	−0.9128	−0.9492	−0.9912	−1.0407	−1.7043
−65	−0.7688	−0.7930	−0.8200	−0.8505	−0.8856	−0.9264	−1.3989
−60	−0.6866	−0.7068	−0.7293	−0.7545	−0.7833	−0.8163	−1.1537
−55	−0.6065	−0.6231	−0.6414	−0.6619	−0.6849	−0.7111	−0.9507
−50	−0.5287	−0.5420	−0.5567	−0.5728	−0.5908	−0.6111	−0.7795
−45	−0.4538	−0.4640	−0.4756	−0.4880	−0.5016	−0.5167	−0.6332
−40	−0.3822	−0.3901	−0.3985	−0.4077	−0.4177	−0.4286	−0.5071
−35	−0.3141	−0.3199	−0.3259	−0.3325	−0.3395	−0.3471	−0.3981
−30	−0.2501	−0.2540	−0.2581	−0.2625	−0.2672	−0.2722	−0.3040
−25	−0.1908	−0.1932	−0.1959	−0.1987	−0.2016	−0.2046	−0.2221
−20	−0.1369	−0.1383	−0.1398	−0.1415	−0.1430	−0.1446	−0.1545
−15	−0.0892	−0.0898	−0.0907	−0.0915	−0.0921	−0.0929	−0.0972
−10	−0.0486	−0.0488	−0.0490	−0.0493	−0.0496	−0.0499	−0.0514
−5	−0.0172	−0.0172	−0.0173	−0.0173	−0.0174	−0.0174	−0.0177
0	0.0000	0.0000	0.0000	0.0000	0.0000	0.0000	0.0000
5	0.0172	0.0171	0.0171	0.0170	0.0170	0.0170	0.0167
10	0.0486	0.0483	0.0481	0.0478	0.0476	0.0473	0.0462
15	0.0892	0.0885	0.0878	0.0872	0.0866	0.0859	0.0830
20	0.1369	0.1355	0.1342	0.1329	0.1316	0.1305	0.1249
25	0.1908	0.1884	0.1861	0.1839	0.1818	0.1798	0.1707
30	0.2501	0.2463	0.2428	0.2394	0.2362	0.2332	0.2198
35	0.3141	0.3087	0.3036	0.2989	0.2943	0.2900	0.2713
40	0.3822	0.3748	0.3679	0.3614	0.3553	0.3497	0.3250
45	0.4538	0.4441	0.4352	0.4268	0.4190	0.4116	0.3804
50	0.5287	0.5164	0.5050	0.4945	0.4847	0.4756	0.4372
55	0.6065	0.5913	0.5773	0.5644	0.5524	0.5415	0.4953
60	0.6866	0.6682	0.6515	0.6360	0.6218	0.6086	0.5543
65	0.7688	0.7470	0.7274	0.7092	0.6924	0.6771	0.6141
70	0.8527	0.8273	0.8042	0.7834	0.7640	0.7464	0.6746
75	0.9378	0.9086	0.8823	0.8584	0.8364	0.8168	0.7356
80	1.0240	0.9909	0.9611	0.9341	0.9095	0.8873	0.7969
85	1.1109	1.0737	1.0404	1.0103	0.9830	0.9584	0.8584
90	1.1981	1.1570	1.1202	1.0870	1.0568	1.0294	0.9201

For shells of revolution whose meridian is a conic section, we can express the substitute variable in all cases in the form

$$\xi = [3(1-\nu^2)]^{1/4} k_1 \int_0^{\phi} \frac{d\phi}{[1 + k_2 \sin^2\phi]^{5/4}}, \tag{7.3.6}$$

where k_1 and k_2 have the values given in Table 7.2. The integral can be

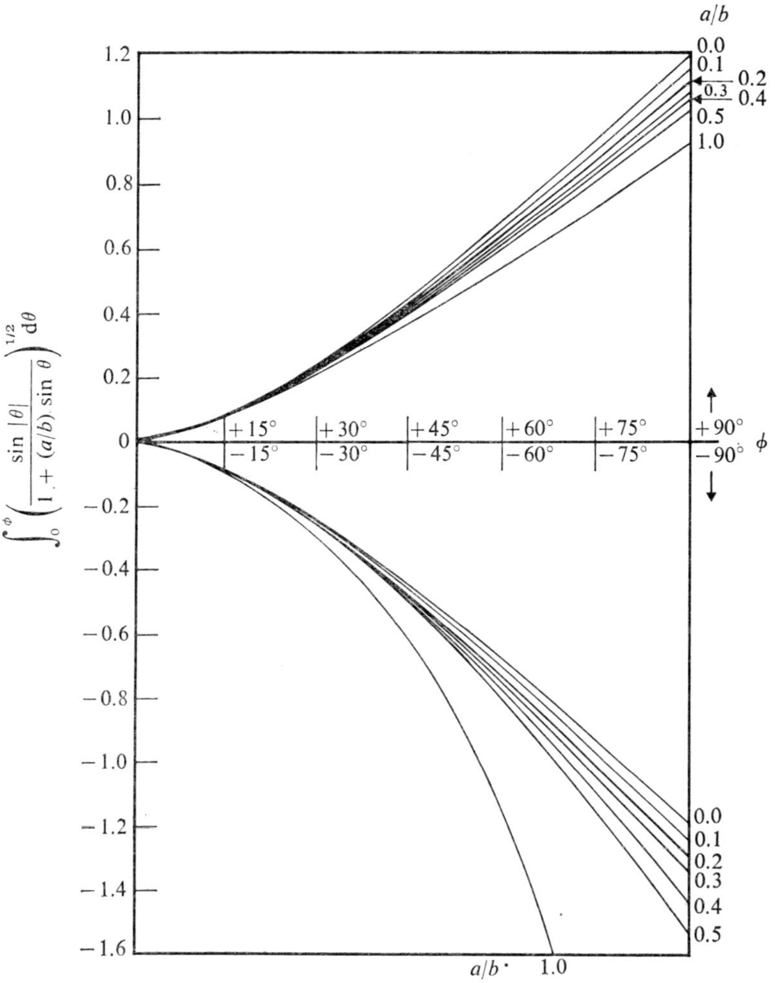

Fig. 7.6. Variation of the integral of eq. (7.4.3a)

Table 7.2 Values of k_1 and k_2 for conic section shells of revolution

Middle Surface		k_1	k_2
	Axis of revolution		
Ellipsoid	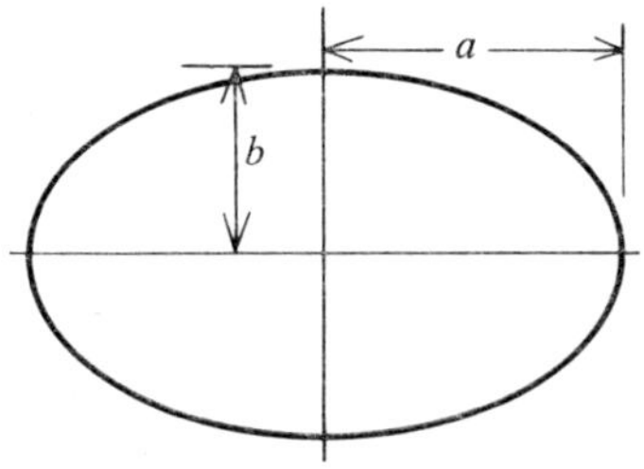	$\left(\frac{a^2}{bh}\right)^{1/2}$	$\frac{a^2}{b^2} - 1$
Paraboloid	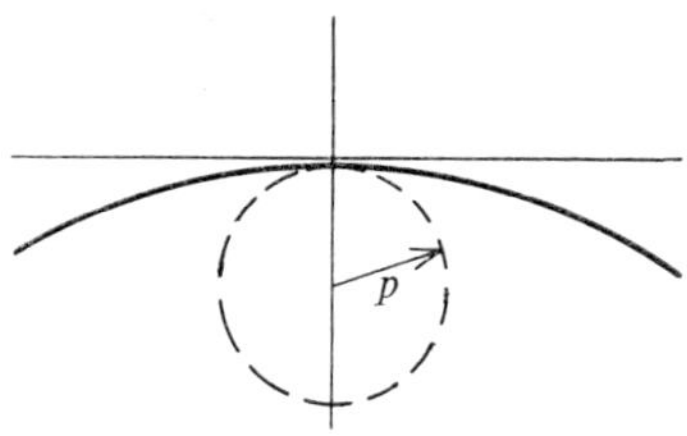	$\left(\frac{p}{h}\right)^{1/2}$	-1
Hyperboloid	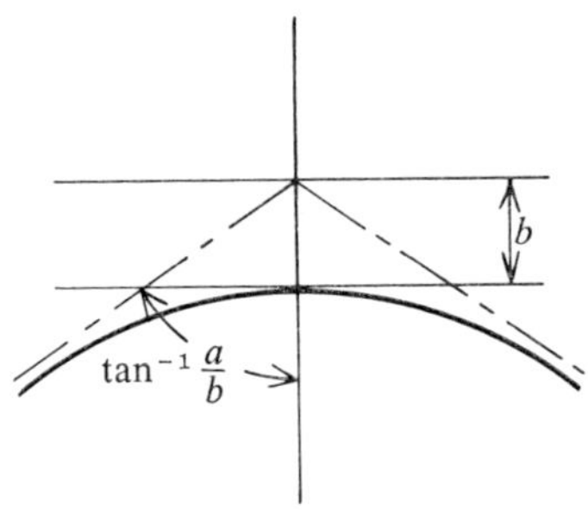	$\left(\frac{a^2}{bh}\right)^{1/2}$	$-\left(\frac{a^2}{b^2} + 1\right)$
Hyperboloid	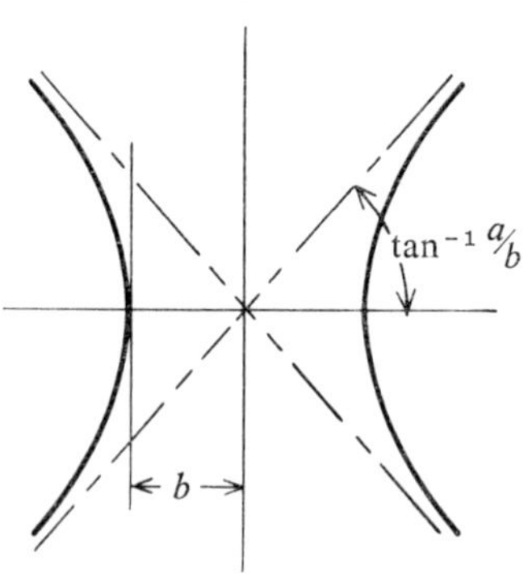	$\left(\frac{a^4}{b^3h}\right)^{1/2}$	$-\left(\frac{a^2}{b^2} + 1\right)$
	Axis of revolution		

Table 7.3 Values of the integral $\int_0^{\phi} \frac{d\theta}{[1 + (a^2/b^2 - 1)\sin^2\theta]^{5/4}}$

ϕ degrees	a/b 0.0	0.1	0.2	0.3	0.4	0.5	0.6	0.7	0.8	0.9
0	0.0000	0.0000	0.0000	0.0000	0.0000	0.0000	0.0000	0.0000	0.0000	0.0000
5	0.0875	0.0875	0.0875	0.0875	0.0875	0.0875	0.0874	0.0874	0.0874	0.0874
10	0.1768	0.1768	0.1767	0.1766	0.1764	0.1762	0.1760	0.1757	0.1753	0.1750
15	0.2695	0.2694	0.2692	0.2688	0.2682	0.2675	0.2667	0.2656	0.2645	0.2632
20	0.3679	0.3677	0.3671	0.3660	0.3646	0.3646	0.3607	0.3583	0.3555	0.3524
25	0.4744	0.4739	0.4726	0.4705	0.4676	0.4639	0.4595	0.4545	0.4489	0.4428
30	0.5923	0.5915	0.5890	0.5849	0.5794	0.5725	0.5644	0.5553	0.5453	0.5347
35	0.7262	0.7247	0.7202	0.7129	0.7030	0.6910	0.6771	0.6618	0.6455	0.6284
40	0.8822	0.8795	0.8717	0.8591	0.8423	0.8222	0.7996	0.7753	0.7499	0.7240
45	1.070	1.065	1.052	1.030	1.002	0.9698	0.9341	0.8968	0.8591	0.8217
50	1.303	1.295	1.272	1.236	1.190	1.138	1.083	1.023	0.9737	0.9217
55	1.607	1.592	1.551	1.489	1.414	1.332	1.250	1.169	1.094	1.024
60	2.022	1.995	1.920	1.811	1.687	1.559	1.436	1.323	1.220	1.129
65	2.631	2.577	2.431	2.234	2.025	1.825	1.646	1.488	1.353	1.236
70	3.611	3.487	3.179	2.805	2.448	2.138	1.880	1.667	1.491	1.344
75	5.424	5.085	4.346	3.596	2.980	2.504	2.140	1.857	1.634	1.455
80	9.683	8.383	6.271	4.689	3.635	2.923	2.423	2.057	1.781	1.567
85	26.99	16.83	9.460	6.137	4.408	3.386	2.724	2.265	1.932	1.680
90	∞	38.38	14.01	7.853	5.256	3.875	3.035	2.478	2.084	1.793

Table 7.3—*contd.*

$a = b$	b/a									
1.0	0.9	0.8	0.7	0.6	0.5	0.4	0.3	0.2	0.1	0.0
0.0000	0.0000	0.0000	0.0000	0.0000	0.0000	0.0000	0.0000	0.0000	0.0000	0.0000
0.0873	0.0872	0.0871	0.0870	0.0868	0.0864	0.0859	0.0846	0.0813	0.0688	0.0000
0.1745	0.1749	0.1733	0.1723	0.1707	0.1683	0.1640	0.1560	0.1384	0.0955	0.0000
0.2618	0.2601	0.2578	0.2545	0.2497	0.2424	0.2303	0.2099	0.1729	0.1059	0.0000
0.3491	0.3451	0.3397	0.3324	0.3221	0.3069	0.2841	0.2488	0.1937	0.1110	0.0000
0.4363	0.4287	0.4188	0.4055	0.3873	0.3624	0.3271	0.2770	0.2071	0.1138	0.0000
0.5236	0.5109	0.4945	0.4735	0.4459	0.4094	0.3613	0.2978	0.2163	0.1157	0.0000
0.6109	0.5912	0.5668	0.5362	0.4978	0.4495	0.3888	0.3137	0.2229	0.1169	0.0000
0.6981	0.6698	0.6360	0.5944	0.5443	0.4838	0.4115	0.3262	0.2279	0.1179	0.0000
0.7854	0.7467	0.7012	0.6482	0.5859	0.5135	0.4305	0.3363	0.2318	0.1186	0.0000
0.8727	0.8215	0.7636	0.6978	0.6232	0.5397	0.4467	0.3448	0.2351	0.1192	0.0000
0.9599	0.8952	0.8237	0.7445	0.6575	0.5629	0.4607	0.3520	0.2377	0.1197	0.0000
1.047	0.9675	0.8810	0.7879	0.6891	0.5837	0.4733	0.3583	0.2401	0.1201	0.0000
1.135	1.038	0.9359	0.8294	0.7181	0.6030	0.4846	0.3639	0.2422	0.1204	0.0000
1.222	1.107	0.9897	0.8687	0.7457	0.6208	0.4950	0.3691	0.2440	0.1207	0.0000
1.309	1.175	1.042	0.9068	0.7719	0.6376	0.5048	0.3739	0.2458	0.1209	0.0000
1.396	1.244	1.093	0.9437	0.7972	0.6539	0.5141	0.3784	0.2474	0.1211	0.0000
1.484	1.311	1.143	0.9798	0.8218	0.6695	0.5231	0.3828	0.2490	0.1212	0.0000
1.571	1.378	1.193	1.016	0.8463	0.6850	0.5319	0.3871	0.2506	0.1214	0.0000

evaluated in terms of elliptic integrals for the ellipsoid by means of the following expressions†

$$\int_0^{\phi} \frac{d\phi}{[1 - k^2 \sin^2 \phi]^{1/4}} = J = \frac{1}{[2(1 + k')]^{1/2}}$$
$$\times \left\{ F\left[\cos^{-1} \frac{[1 - k^2 \sin^2 \phi]^{1/2} + (k')^{1/2}}{[1 + (k')^{1/2}][1 - k^2 \sin^2 \phi]^{1/4}}, \frac{1 + (k')^{1/2}}{[2(1 + k')]^{1/2}}\right] \right.$$
$$\left. + F\left[\cos^{-1} \frac{[1 - k^2 \sin^2 \phi]^{1/2} - (k')^{1/2}}{[1 - (k')^{1/2}][1 - k^2 \sin^2 \phi]^{1/4}}, \frac{1 - (k')^{1/2}}{[2(1 + k')]^{1/2}}\right] \right\}, \tag{7.3.7a}$$

$$\int_0^{\phi} \frac{d\phi}{[1 - k^2 \sin^2 \phi]^{5/4}} = J + 4k^2 \frac{\partial J}{\partial(k^2)}, \tag{7.3.7b}$$

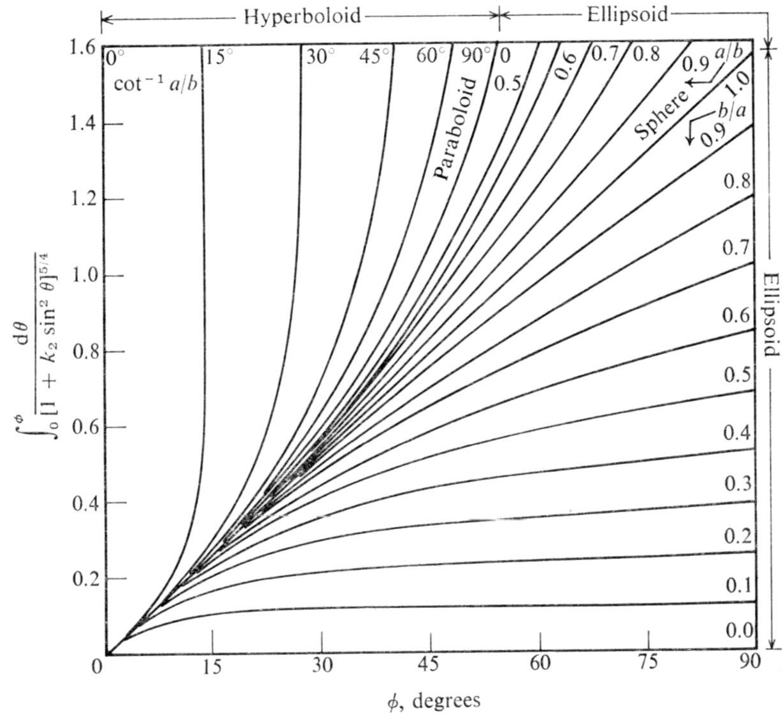

Fig. 7.7. Variation of integral of eq. (7.3.6)

† P. F. Byrd and M. D. Friedman, *Handbook of Elliptic Integrals for Engineers and Physicists*, Second Edition, Springer-Verlag, 1971, p. 243.

with

$$k' = (1 - k^2)^{1/2}. \tag{7.3.7c}$$

When k is equal to unity (the case of the paraboloid), the result can be much simplified to

$$\int_0^{\phi} \frac{d\phi}{(1 - \sin^2\phi)^{5/4}} = \frac{2}{3}\left\{\frac{\tan\phi}{(\cos\phi)^{1/2}} + \frac{2^{1/2}}{2} F\left[\sin^{-1}\left(2^{1/2}\sin\frac{\phi}{2}\right), \frac{2^{1/2}}{2}\right]\right\}. \tag{7.3.7c}$$

Some values of the integral calculated by numerical integration are given in Tables 7.3 and 7.4 and are shown graphically in Fig. 7.7.

Table 7.4 Values of the integral $\int_0^{\phi} \frac{d\theta}{[1 - (a^2/b^2 + 1)\sin^2\theta]^{5/4}}$

ϕ degrees	$\cot^{-1} a/b$, degrees: 90	85	80	75	70	65	60	55
0	0.0000	0.0000	0.0000	0.0000	0.0000	0.0000	0.0000	0.0000
5	0.0875	0.0875	0.0876	0.0876	0.0876	0.0876	0.0876	0.0877
10	0.1768	0.1768	0.1769	0.1769	0.1771	0.1773	0.1776	0.1779
15	0.2695	0.2696	0.2698	0.2701	0.2706	0.2713	0.2723	0.2736
20	0.3679	0.3680	0.3685	0.3694	0.3706	0.3724	0.3749	0.3784
25	0.4744	0.4747	0.4757	0.4775	0.4803	0.4842	0.4897	0.4974
30	0.5923	0.5930	0.5950	0.5984	0.6038	0.6115	0.6226	0.6387
35	0.7262	0.7273	0.7310	0.7374	0.7473	0.7619	0.7835	0.8160
40	0.8822	0.8843	0.8907	0.9022	0.9202	0.9475	0.9893	1.057
45	1.070	1.073	1.085	1.105	1.138	1.190	1.274	1.427
50	1.303	1.310	1.330	1.367	1.428	1.532	1.721	2.164
55	1.607	1.618	1.655	1.725	1.849	2.081	2.631	∞
60	2.022	2.044	2.115	2.258	2.541	3.226	∞	
65	2.631	2.676	2.827	3.166	4.029	∞		
70	3.611	3.716	4.101	5.201	∞			
75	5.424	5.733	7.143	∞				
80	9.683	11.19	∞					
85	26.99	∞						
90	∞							

ϕ degrees	50	45	40	35
0	0.0000	0.0000	0.0000	0.0000
5	0.0877	0.0878	0.0879	0.0881
10	0.1784	0.1791	0.1801	0.1817
15	0.2754	0.2780	0.2819	0.2879
20	0.3833	0.3904	0.4013	0.4194
25	0.5087	0.5257	0.5534	0.6047
30	0.6630	0.7021	0.7734	0.9436
35	0.8684	0.9629	1.186	∞
40	1.178	1.462	∞	
45	1.783	∞		
50	∞			

7.4 Approximate complementary solutions valid at points with horizontal tangents

The preceding exposition indicates asymptotic formulas which are valid for thin shells at points sufficiently far removed from places on the shell at which the tangent to the meridian is horizontal. In many problems, however, the region in the vicinity of horizontal tangents is of interest so that an improved solution is needed. In this section we shall consider an approximate solution which extends the work of Section 7.2 to obtain the dominant terms of the complementary solutions of the bending equations for thin shells. The method of approximation is similar to that used for shallow spherical caps in that we seek to approximate the actual differential equations by one which can be solved in terms of known functions. A particular form of Bessel's equation that is most useful in this regard is the equation

$$\frac{d^2y}{du^2} + \left(1 - \frac{p^2 - \frac{1}{4}}{u^2}\right) y = 0, \tag{7.4.1a}$$

which has the solution

$$y = u^{1/2}Z_p(u), \tag{7.4.1b}$$

where $Z_p(u)$ is the general solution of Bessel's equation,

$$Z_p(u) = AJ_p(u) + BY_p(u). \tag{7.4.1c}$$

We consider the bending equations in the form given by Eqs. (7.2.3). We now neglect the term

$$\frac{\nu}{[3(1 - \nu^2)]^{1/2}} \frac{h}{R_\phi \operatorname{sgn} R_\theta},$$

which distinguishes θ_1 from θ_2, since in the vicinity of an apex it is negligible compared to the other terms in the expression and, from the work of the previous section, it can be neglected away from the apex. With

$$\theta_1 \approx \theta_2 = \psi, \tag{7.4.2}$$

Eqs. (7.2.3a) and (7.2.3b) may now be combined to yield the single complex equation

$$\frac{d^2\bar{V}}{d\bar{\xi}^2} + \left(1 - i\frac{\psi}{2}\right) \bar{V} = \bar{\theta}_0, \tag{7.4.3a}$$

with

$$\bar{\xi} = i(2i)^{1/2}\xi. \tag{7.4.3b}$$

In the homogeneous case, Eq. (7.4.2b) superficially bears a resemblance to Eq. (7.4.1a). Our task is to show that the function $\frac{1}{2}i\psi$ can be approximated by a function of the form $(p^2 - \frac{1}{4})/\bar{\xi}^2$.†

7.4.1 The ellipsoidal shell

Let us consider, as a first example, the case of an ellipsoidal shell for which

$$R_\phi = \frac{a^2}{b}\left[1 + \left(\frac{a^2}{b^2} - 1\right)\sin^2\phi\right]^{-3/2}, \tag{7.4.4a}$$

$$R_\theta = \frac{a^2}{b}\left[1 + \left(\frac{a^2}{b^2} - 1\right)\sin^2\phi\right]^{-1/2}. \tag{7.4.4b}$$

Then

$$\begin{aligned}\xi &= [3(1-\nu^2)]^{1/4}\left(\frac{a^2}{bh}\right)^{1/2}\int_0^\phi \frac{d\phi}{[1 + (a^2/b^2 - 1)\sin^2\phi]^{5/4}} \\ &\approx [3(1-\nu^2)]^{1/4}\left(\frac{a^2}{bh}\right)^{1/2}\sin\phi\left[1 - \frac{5}{12}\left(\frac{a^2}{b^2} - \frac{7}{5}\right)\sin^2\phi\right. \\ &\quad \left. + \frac{9}{32}\left(\frac{a^4}{b^4} - \frac{22}{9}\frac{a^2}{b^2} + \frac{77}{45}\right)\sin^4\phi + \cdots\right]\end{aligned} \tag{7.4.5a}$$

for small ϕ, and

$$\begin{aligned}\psi &= \frac{1}{[3(1-\nu^2)]^{1/2}}\frac{bh}{a^2}\left[1 + \left(\frac{a^2}{b^2} - 1\right)\sin^2\phi\right]^{1/2}\left[\frac{3}{4}\csc^2\phi + \frac{1}{4}\left(\frac{a^2}{b^2} - 6\right)\right. \\ &\quad \left. + \frac{7}{16}\left(\frac{a^2}{b^2} - 1\right)\left(\frac{a^2}{b^2} - \frac{23}{7}\right)\sin^2\phi - \frac{11}{16}\left(\frac{a^2}{b^2} - 1\right)^2\sin^4\phi\right].\end{aligned} \tag{7.4.5}$$

On examination, ψ can be written as

$$\psi \approx \frac{\frac{3}{4}}{\xi^2} = -\frac{\frac{3}{2}i}{\bar{\xi}^2}, \tag{7.4.6}$$

in the vicinity of the apex. Since ψ is assumed to be negligible away from the apex, we may use Eq. (7.4.6) as an approximation over that range of ϕ and write Eq. (7.4.3b) as

$$\frac{d^2\bar{V}}{d\bar{\xi}^2} + \left(1 - \frac{\frac{3}{4}}{\bar{\xi}^2}\right)\bar{V} = 0, \tag{7.4.7}$$

---

† The validity of the argument has been investigated by R. E. Langer, 'On the Asymptotic Solution of Ordinary Differential Equations, with Reference to the Stokes' Phenomenon about a Singular Point,' *Trans. Am. Math. Soc.*, vol. 37, 1935, pp. 397–416.

with

$$p^2 = \tfrac{3}{4} + \tfrac{1}{4} = 1. \tag{7.4.8a}$$

The solution of Eq. (7.5.7) can be expressed in terms of Bessel–Kelvin functions as

$$\bar{\Theta} + i\bar{U} = \xi^{1/2}\{A[\operatorname{ber}'(2^{1/2}\xi) + i\operatorname{bei}'(2^{1/2}\xi)] + [B\operatorname{ker}'(2^{1/2}\xi) + i\operatorname{kei}'(2^{1/2}\xi)]\}, \tag{7.4.8b}$$

where the primes indicate differentiation with respect to $2^{1/2}\xi$, and A and B are arbitrary complex constants of integration. The same analysis applies to paraboloidal shells when the appropriate definition of ξ is used.

Because of the symmetry with respect to the equator of the functions involved in the equations of bending of ellipsoidal shells, the above analysis should be limited, strictly speaking, to the range of angle

$$0 \leq \phi \leq \frac{\pi}{2},$$

while a similar approximation holds for the range of angle

$$\frac{\pi}{2} \leq \phi \leq \pi,$$

with continuity conditions at the equator used to match the solutions for the two ranges. In many problems, however, the decay of the edge solution is so rapid that analytic continuation of the solution may be ignored as unnecessary.

With the solutions given above and the use of Eqs. (7.2.4), approximate expressions for forces and displacements may be obtained. For the ellipsoid we obtain

$$\Theta \approx \left[1 + \left(\frac{a^2}{b^2} - 1\right)\sin^2\phi\right]^{1/8}\left(\frac{I}{\sin\phi}\right)^{1/2}\Big[A_1\operatorname{ber}'(2^{1/2}\xi) - A_2\operatorname{bei}'(2^{1/2}\xi) + A_3\operatorname{ker}'(2^{1/2}\xi) - A_4\operatorname{kei}'(2^{1/2}\xi)\Big], \tag{7.4.9a}$$

$$H\sin\phi = N_\phi\tan\phi \approx \frac{bh}{a^2}\frac{Eh}{[12(1-\nu^2)]^{1/2}} \times \left[1 + \left(\frac{a^2}{b^2} - 1\right)\sin^2\phi\right]^{5/8}\left(\frac{I}{\sin\phi}\right)^{1/2} \times [A_1\operatorname{bei}'(2^{1/2}\xi) + A_2\operatorname{ber}'(2^{1/2}\xi) + A_3\operatorname{kei}'(2^{1/2}\xi) + A_4\operatorname{ker}'(2^{1/2}\xi)], \tag{7.4.9b}$$

$$\frac{u_H}{\sin\phi} \approx \frac{(a^2/bh)^{1/2}h}{[12(1-\nu^2)]^{1/4}}\left[1+\left(\frac{a^2}{b^2}-1\right)\sin^2\phi\right]^{-1/8}\left(\frac{I}{\sin\phi}\right)^{1/2}$$

$$\left\langle A_1\left\{\operatorname{ber}(2^{1/2}\xi) - \frac{1+[1+2\nu+\frac{1}{2}(a^2/b^2-1)\sin^2\phi]\times[1+(a^2/b^2-1)\sin^2\phi]^{1/4}I\cot\phi}{2^{3/2}\xi}\operatorname{bei}'(2^{1/2}\xi)\right\}\right.$$

$$-A_2\left\{\operatorname{bei}(2^{1/2}\xi)\right.$$

$$+\frac{1+[1+2\nu+\frac{1}{2}(a^2/b^2-1)\sin^2\phi]\times[1+(a^2/b^2-1)\sin^2\phi]^{1/4}I\cot\phi}{2^{3/2}\xi}$$

$$\left.\left.\times\operatorname{ber}'(2^{1/2}\xi)\right\}+\cdots\right\rangle, \tag{7.4.9c}$$

$$N_\theta \approx \frac{Eh}{[12(1-\nu^2)]^{1/4}}\left(\frac{bh}{a^2}\right)^{1/2}\left[1+\left(\frac{a^2}{b^2}-1\right)\sin^2\phi\right]^{3/8}\left(\frac{I}{\sin\phi}\right)^{1/2}$$

$$\left\langle A_1\left\{\operatorname{ber}(2^{1/2}\xi)\right.\right.$$

$$-\frac{1+[1+\frac{1}{2}(a^2/b^2-1)\sin^2\phi][1+(a^2/b^2-1)\sin^2\phi]^{1/4}I\cot\phi}{2^{3/2}\xi}$$

$$\left.\times\operatorname{bei}'(2^{1/2}\xi)\right\}-A_2\left\{\operatorname{bei}(2^{1/2}\xi)\right.$$

$$+\frac{1+[1+\frac{1}{2}(a^2/b^2-1)\sin^2\phi][1+(a^2/b^2-1)\sin^2\phi]^{1/4}I\cot\phi}{2^{3/2}\xi}$$

$$\left.\left.\times\operatorname{ber}'(2^{1/2}\xi)\right\}+\cdots\right\rangle, \tag{7.4.9d}$$

$$M_\phi \approx \frac{D}{h}[12(1-\nu^2)]^{1/4}\left(\frac{bh}{a^2}\right)^{1/2}\left[1+\left(\frac{a^2}{b^2}-1\right)\sin^2\phi\right]^{3/8}\left(\frac{I}{\sin\phi}\right)^{1/2}$$

$$\left\langle A_1\left\{\operatorname{bei}(2^{1/2}\xi) + \frac{1+[1-2\nu+\frac{1}{2}(a^2/b^2-1)\sin^2\phi]\times[1+(a^2/b^2-1)\sin^2\phi]^{1/4}I\cot\phi}{2^{3/2}\xi}\right.\right.$$

$$\left.\times\operatorname{ber}'(2^{1/2}\xi)\right\}$$

$$+A_2\left\{\operatorname{ber}(2^{1/2}\xi) - \frac{1+[1-2\nu+\frac{1}{2}(a^2/b^2-1)\sin^2\phi]\times[1+(a^2/b^2-1)\sin^2\phi]^{1/4}I\cot\phi}{2^{3/2}\xi}\right.$$

$$\left.\left.\times\operatorname{bei}'(2^{1/2}\xi)\right\}+\cdots\right\rangle, \tag{7.4.9e}$$

$$M_\theta \approx \frac{\nu D}{h}[12(1-\nu^2)]^{1/4}\left(\frac{bh}{a^2}\right)^{1/2}\left[1+\left(\frac{a^2}{b^2}-1\right)\sin^2\phi\right]^{3/8}\left(\frac{I}{\sin\phi}\right)^{1/2}$$

$$\left\langle A_1\left\{\operatorname{bei}(2^{1/2}\xi)+\frac{\begin{array}{r}1+[1-2/\nu+\frac{1}{2}(a^2/b^2-1)\sin^2\phi]\\ \times[1+(a^2/b^2-1)\sin^2\phi]^{1/4}I\cot\phi\end{array}}{2(2^{1/2}\xi)}\right.\right.$$

$$\left.\times\operatorname{ber}'(2^{1/2}\xi)\right\}$$

$$+A_2\left\{\operatorname{bei}(2^{1/2}\xi)-\frac{\begin{array}{r}1+[1-2/\nu+\frac{1}{2}(a^2/b^2-1)\sin^2\phi]\\ \times[1+(a^2/b^2-1)\sin^2\phi]^{1/4}I\cot\phi\end{array}}{2(2)^{1/2}\xi}\right.$$

$$\left.\left.\times\operatorname{bei}'(2^{1/2}\xi)\right\}+\cdots\right\rangle, \qquad (7.4.9f)$$

where

$$I=\int_0^\phi\frac{d\phi}{[1+(a^2/b^2-1)\sin^2\phi]^{5/4}}, \qquad (7.4.9g)$$

$$\xi=\left(\frac{a^2}{bh}\right)^{1/2}I, \qquad (7.4.9h)$$

and primes indicate differentiation with respect to $2^{1/2}\xi$. These results will be adequate so long as the larger of bh/a^2 or a^3h/b^4 is small compared to unity. For spherical shells, with

$$a=b, \qquad (7.4.10a)$$

$$I=\phi, \qquad (7.4.10b)$$

$$\xi=(a/h)^{1/2}\phi, \qquad (7.4.10c)$$

Eqs. (7.4.9) reduce to their shallow shell counterparts when we put

$$\frac{\phi}{\sin\phi}\approx\phi\cot\phi\approx 1. \qquad (7.4.11)$$

Examination of Eqs. (7.4.9) also indicates that the influence coefficients for most thin ellipsoidal† shells should be adequately approximated by those

† Values calculated more accurately by G. D. Galletly, 'Bending of 2:1 and 3:1 Open-Crown Ellipsoidal Shells,' Welding Research Council Bulletin Series, no. 54, October 1959, verify this statement. These findings also apply for paraboloidal and hyperboloidal shells and, in general, to shells whose meridian is a second degree curve in the vicinity of the apex. See B. R. Baker and G. B. Cline, Jr., 'Influence Coefficients for Thin Smooth Shells of Revolution Subjected to Symmetric Loads,' *J. Appl. Math.*, vol. 29, no. 2, June 1962, pp. 335–339.

for a spherical for which the edge values of $\chi\phi$ are equal to the ellipsoidal shell edge values of ξ. In particular, influence coefficients for shallow ellipsoidal shells can be taken as those for a corresponding shallow spherical cap (see Section 6.4).

7.4.2 The circular toroidal shell

Another example which can be readily investigated by similar methods is the circular toroidal shell for which we have

$$R_\phi = a, \tag{7.4.12a}$$

$$R_\theta = a + \frac{b}{\sin\phi}. \tag{7.4.12b}$$

Then

$$\begin{aligned}\xi &= [3(1-\nu^2)]^{1/4}\left(\frac{a^2}{bh}\right)^{1/2}\int_0^\phi \left(\frac{|\sin\phi|}{1+a/b\sin\phi}\right)^{1/2} \mathrm{d}\phi \\ &\approx \tfrac{2}{3}[2(1-\nu^2)]^{1/4}\left(\frac{a^2}{bh}\right)^{1/2}\sin^{3/2}|\phi| \\ &\quad\times\left[1-\frac{3}{10}\frac{a}{b}\sin\phi+\frac{3}{14}\left(1+\frac{3}{4}\frac{a^2}{b^2}\right)\sin^2\phi+\cdots\right]\operatorname{sgn}\phi\end{aligned} \tag{7.4.12c}$$

for small ϕ and

$$\begin{aligned}\psi &= -\frac{\frac{5}{16}}{[3(1-\nu^2)]^{1/2}}\left(\frac{bh}{a^2}\right) \\ &\quad\times\left[1+\frac{8}{5}\frac{a}{b}\sin\phi-\frac{1}{5}\left(1+12\frac{a^2}{b^2}\right)\sin^2\phi+\frac{4}{5}\frac{a}{b}\sin^3\phi\right. \\ &\quad\left.+4\frac{a^2}{b^2}\sin^4\phi\right]\frac{1}{(1+a/b\sin\phi)\sin^3|\phi|}.\end{aligned} \tag{7.4.12d}$$

For small ϕ, we can ascertain that

$$\psi \approx -\frac{\frac{5}{36}}{\xi^2} \approx \frac{\frac{5}{18}i}{\xi^2}. \tag{7.4.12e}$$

Arguments similar to those used for the ellipsoid enable us to approximate Eq. (7.4.3b) over the applicable range of ϕ by

$$\frac{\mathrm{d}^2\overline{V}}{\mathrm{d}\xi^2} + \left(1+\frac{\frac{5}{36}}{\xi^2}\right)\overline{V} = 0, \tag{7.4.13a}$$

so that, with

$$p^2 = \frac{1}{4} - \frac{5}{36} = \frac{1}{9} \tag{7.4.13b}$$

the solution of Eq. (7.4.13a) can be written as

$$\bar{\Theta} + i\bar{U} = \xi^{1/2}[AH_{1/3}^{(1)}(2^{1/2}i^{3/2}\xi) + BH_{1/3}^{(2)}(2^{1/2}i^{3/2}\xi)], \tag{7.4.14a}$$

where $H_{1/3}^{(1)}(\zeta)$ and $H_{1/3}^{(2)}(\zeta)$ are Hankel functions defined in terms of the more familiar Bessel functions of the first kind as

$$H_{1/3}^{(1)}(\zeta) = \left(1 + \frac{3^{1/2}}{3}i\right)J_{1/3}(\zeta) - \frac{2(3^{1/2})}{3}iJ_{1/3}(\zeta), \tag{7.4.14b}$$

$$H_{1/3}^{(2)}(\zeta) = \left(1 - \frac{3^{1/2}}{3}i\right)J_{1/3}(\zeta) + \frac{2(3^{1/2})}{3}iJ_{-1/3}(\zeta). \tag{7.4.14c}$$

Since the modified Hankel functions (or Airy functions)

$$h_1(z) = (\tfrac{2}{3}z^{3/2})^{1/3}H_{1/3}^{(1)}(\tfrac{2}{3}z^{3/2}), \tag{7.4.15a}$$

$$h_2(z) = (\tfrac{2}{3}z^{3/2})^{1/3}H_{1/3}^{(2)}(\tfrac{2}{3}z^{3/2}), \tag{7.4.15b}$$

are tabulated for complex arguments,† a preferable way of expressing the solution of Eq. (7.4.13a) is

$$\bar{\Theta} + i\bar{U} = x^{1/4}[Ah_1(ix) + Bh_2(ix)], \tag{7.4.15c}$$

with

$$x = (\tfrac{3}{2}\,2^{1/2}\xi)^{2/3}\operatorname{sgn}\xi. \tag{7.4.15d}$$

Values of the real and imaginary parts of the functions and the first derivatives‡ are given in Table 7.5 for positive x. For large positive x we have

$$\left.\begin{matrix} h_1(ix) \\ h_2(ix) \end{matrix}\right\} \sim \frac{2^{1/3}3^{1/6}}{\pi^{1/2}}\frac{1}{(ix)^{1/4}}\exp\left[\mp\left(\tfrac{2}{3}i^{1/2}x^{3/2} + \frac{5\pi}{12}i\right)\right], \tag{7.4.16}$$

so that $h_1(ix)$ decays exponentially with increasing x while $h_2(ix)$ grows exponentially with increasing x. At the apex we have the values

$$\operatorname{Re} h_1(0) = \operatorname{Re} h_2(0) = 0, \tag{7.4.17a}$$

$$\operatorname{Im} h_1(0) = -\operatorname{Im} h_2(0) = -\frac{2^{4/3}}{3^{1/2}\Gamma(\tfrac{2}{3})} = -1.07438, \tag{7.4.17b}$$

† Harvard University Press, Cambridge, Mass., 1945, *Tables of the Modified Hankel Functions of Order One-Third and of Their Derivatives.*

‡ Extensive tables of the various functions required in the solution of circular toroidal shells are given in L. N. Osipova and S. A. Tumarkin, *Tables of the Computation of Toroidal Shells*, P. Noordhoff Ltd., Groningen, The Netherlands, 1965. An inclusive list of references is also given.

Table 7.5 Values of the Airy functions

x	$\mathrm{Re}\, h_1(ix)$	$\mathrm{Im}\, h_1(ix)$	$\frac{\mathrm{d}}{\mathrm{d}x}\mathrm{Re}\, h_1(ix)$	$\frac{\mathrm{d}}{\mathrm{d}x}\mathrm{Im}\, h_1(ix)$	$\mathrm{Re}\, h_2(ix)$	$\mathrm{Im}\, h_2(ix)$	$\frac{\mathrm{d}}{\mathrm{d}x}\mathrm{Re}\, h_2(ix)$	$\frac{\mathrm{d}}{\mathrm{d}x}\mathrm{Im}\, h_2(ix)$
0.0	0.0000	−1.0744	−0.3916	0.6783	0.0000	1.0744	0.3916	0.6783
1	−0.0390	−1.0065	−0.3865	0.6782	0.0309	1.1422	0.3860	0.6784
2	−0.0770	−0.9388	−0.3719	0.6773	0.0768	1.2101	0.3683	0.6793
3	−0.1131	−0.8711	−0.3494	0.6749	0.1122	1.2781	0.3372	0.6817
4	−0.1466	−0.8039	−0.3201	0.6703	0.1437	1.3465	0.2912	0.6862
0.5	−0.1770	−0.7372	−0.2855	0.6630	0.1699	1.4155	0.2290	0.6934
6	−0.2036	−0.6714	−0.2468	0.6524	0.1889	1.4853	0.1492	0.7033
7	−0.2262	−0.6068	−0.2053	0.6384	0.1991	1.5562	0.0503	0.7159
8	−0.2446	−0.5438	−0.1622	0.6207	0.1983	1.6285	−0.0692	0.7309
9	−0.2587	−0.4828	−0.1187	0.5993	0.1845	1.7024	−0.2108	0.7473
1.0	−0.2684	−0.4241	−0.0757	0.5742	0.1553	1.7780	−0.3762	0.7635
1	−0.2738	−0.3681	−0.0342	0.5457	0.1084	1.8551	−0.5670	0.7775
2	−0.2753	−0.3150	0.0050	0.5141	0.0410	1.9333	−0.7849	0.7862
3	−0.2729	−0.2653	0.0412	0.4798	−0.0495	2.0120	−1.0315	0.7859
4	−0.2672	−0.2192	0.0739	0.4433	−0.1663	2.0900	−1.3085	0.7715
1.5	−0.2583	−0.1767	0.1025	0.4052	−0.3123	2.1657	−1.6172	0.7371
6	−0.2468	−0.1381	0.1268	0.3660	−0.4908	2.2365	−1.9585	0.6752
7	−0.2331	−0.1035	0.1466	0.3264	−0.7051	2.2995	−2.3329	0.5768
8	−0.2176	−0.0729	0.1620	0.2870	−0.9585	2.3503	−2.7400	0.4317
9	−0.2008	−0.0461	0.1729	0.2483	−1.2541	2.3838	−3.1782	0.2274
2.0	−0.1832	−0.0232	0.1796	0.2108	−1.5950	2.3934	−3.6445	−0.0499
1	−0.1651	−0.0039	0.1823	0.1751	−1.9838	2.3709	−4.1334	−0.4162
2	−0.1468	0.0119	0.1813	0.1416	−2.4222	2.3066	−4.6370	−0.8893
3	−0.1289	0.0245	0.1772	0.1106	−2.9113	2.1888	−5.1438	−1.4888
4	−0.1115	0.0342	0.1702	0.0824	−3.4506	2.0039	−5.6378	−2.2358
2.5	−0.0949	0.0411	0.1609	0.0572	−4.0377	1.7360	−6.0975	−3.1527
6	−0.0794	0.0457	0.1498	0.0350	−4.6680	1.3670	−6.4952	−4.2623
7	−0.0650	0.0482	0.1373	0.0159	−5.3335	0.8764	−6.7950	−5.5874
8	−0.0520	0.0490	0.1239	−0.0002	−6.0222	0.2416	−6.9517	−7.1490
9	−0.0403	0.0483	0.1100	−0.0133	−6.7172	−0.5619	−6.9097	−8.9651

3.0	−0.0299	0.0464	0.0960	−0.0236	−7.3953	−1.5603	−6.6010	−11.048
1	−0.0210	0.0436	0.0823	−0.0313	−8.0258	−2.7806	−5.9440	−13.402
2	−0.0135	0.0402	0.0691	−0.0367	−8.5693	−4.2495	−4.8424	−16.019
3	−0.0072	0.0364	0.0566	−0.0400	−8.9758	−5.9924	−3.1844	−18.875
4	−0.0021	0.0323	0.0451	−0.0415	−9.1834	−8.0309	−0.8423	−21.923
3.5	0.0019	0.0281	0.0347	−0.0416	−9.1167	−10.381	2.3266	−25.089
6	0.0049	0.0240	0.0255	−0.0403	−8.6853	−13.049	6.4783	−28.261
7	0.0076	0.0201	0.0174	−0.0381	−7.7828	−16.028	11.778	−31.282
8	0.0084	0.0164	0.0106	−0.0352	−6.2858	−19.293	18.395	−33.939
9	0.0092	0.0131	0.0049	−0.0318	−4.0543	−22.795	26.494	−35.953
4.0	0.0094	0.0101	0.0004	−0.0281	−0.9327	−26.451	36.219	−36.968
1	0.0093	0.0074	−0.0031	−0.0244	3.2476	−30.140	47.694	−36.534
2	0.0088	0.0052	−0.0057	−0.0206	8.6638	−33.691	60.940	−34.104
3	0.0081	0.0033	−0.0075	−0.0170	15.494	−36.872	75.955	−29.018
4	0.0073	0.0018	−0.0086	−0.0136	23.908	−39.381	92.573	−20.501
4.5	0.0064	0.0006	−0.0091	−0.0106	34.052	−40.829	110.47	−7.6629
6	0.0055	−0.0003	−0.0092	−0.0078	46.027	−40.736	129.09	10.494
7	0.0046	−0.0010	−0.0089	−0.0055	59.867	−38.517	147.61	35.054
8	0.0038	−0.0015	−0.0083	−0.0035	75.506	−33.474	164.83	67.150
9	0.0030	−0.0017	−0.0075	−0.0019	92.737	−24.798	179.12	107.91
5.0	0.0023	−0.0018	−0.0066	−0.0006	111.16	−11.570	188.32	158.35
1	0.0016	−0.0018	−0.0057	0.0004	130.14	7.2222	189.66	219.29
2	0.0011	−0.0018	−0.0047	0.0011	148.72	32.653	179.68	291.15
3	0.0007	−0.0016	−0.0038	0.0016	165.56	65.812	154.16	373.77
4	0.0004	−0.0015	−0.0030	0.0019	178.89	107.73	108.12	466.12
5.5	0.0001	−0.0013	−0.0023	0.0020	186.31	159.29	35.766	565.97
6	−0.0001	−0.0011	−0.0016	0.0020	184.93	221.05	−69.350	669.47
7	−0.0002	−0.0009	−0.0011	0.0019	171.11	293.11	−214.18	770.72
8	−0.0003	−0.0007	−0.0006	0.0017	140.53	374.84	−405.85	861.21
9	−0.0004	−0.0005	−0.0003	0.0015	88.153	464.61	−651.13	929.25
6.0	−0.0004	−0.0004	−0.0000	0.0013	8.3245	559.43	−955.71	959.39

$$\frac{\mathrm{d}\,\mathrm{Re}\,h_1(0)}{\mathrm{d}x} = -\frac{\mathrm{d}\,\mathrm{Re}\,h_2(0)}{\mathrm{d}x} = -\frac{2^{1/3}}{3^{1/6}\Gamma(\frac{1}{3})} = -0.39167, \tag{7.4.17c}$$

$$\frac{\mathrm{d}\,\mathrm{Im}\,h_1(0)}{\mathrm{d}x} = \frac{\mathrm{d}\,\mathrm{Im}\,h_2(0)}{\mathrm{d}x} = \frac{6^{1/3}}{\Gamma(\frac{1}{3})} = 0.67830, \tag{7.4.17d}$$

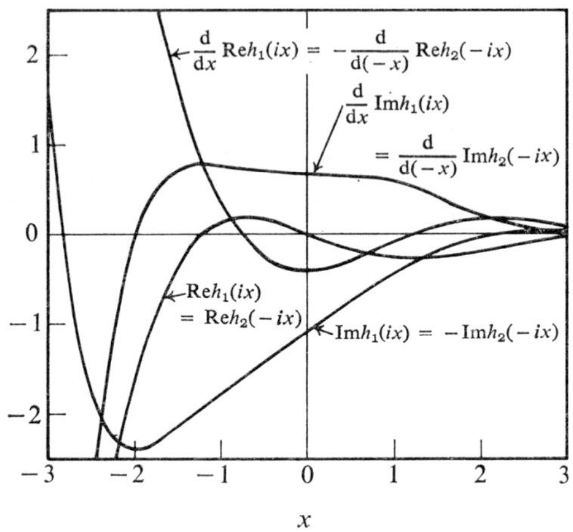

Fig. 7.8. The functions $h_1(ix)$, $h_2(ix)$

which are useful in certain applications. For negative values of x the functions are defined as follows (see Fig. 7.8).

$$\mathrm{Re}\,h_1(ix) = \mathrm{Re}\,h_2(-ix), \tag{7.4.18a}$$

$$\frac{\mathrm{d}\,\mathrm{Re}\,h_1(ix)}{\mathrm{d}x} = -\frac{\mathrm{d}\,\mathrm{Re}\,h_2(-ix)}{\mathrm{d}(-x)}; \tag{7.4.18b}$$

$$\mathrm{Im}\,h_1(ix) = -\mathrm{Im}\,h_2(-ix), \tag{7.4.18c}$$

$$\frac{\mathrm{d}\,\mathrm{Im}\,h_1(ix)}{\mathrm{d}x} = \frac{\mathrm{d}\,\mathrm{Im}\,h_2(-ix)}{\mathrm{d}(-x)} \tag{7.4.18d}$$

$$\mathrm{Re}\,h_2(ix) = \mathrm{Re}\,h_1(-ix), \tag{7.4.18e}$$

$$\frac{\mathrm{d}\,\mathrm{Re}\,h_2(ix)}{\mathrm{d}x} = -\frac{\mathrm{d}\,\mathrm{Re}\,h_1(-ix)}{\mathrm{d}(-x)}; \tag{7.4.18f}$$

$$\mathrm{Im}\,h_2(ix) = -\mathrm{Im}\,h_1(-ix), \tag{7.4.18g}$$

$$\frac{\mathrm{d}\,\mathrm{Im}\,h_2(ix)}{\mathrm{d}x} = \frac{\mathrm{d}\,\mathrm{Im}\,h_1(-ix)}{\mathrm{d}(-x)}. \tag{7.4.18h}$$

With the approximate solution given we may obtain equations for the horizontal displacement, middle surface rotation, horizontal force and bending moment for the calculation of influence coefficients as

$$\Theta = \left[\frac{h}{b}\frac{x}{\sin\phi(1 + a/b\sin\phi)}\right]^{1/4}$$
$$\times\ [A_1 \operatorname{Re} h_1(ix) - A_2 \operatorname{Im} h_1(ix) + B_1 \operatorname{Re} h_2(ix) - B_2 \operatorname{Im} h_2(ix)], \tag{7.4.19a}$$

$$H = N_\phi \sec\phi = Q_\phi \csc\phi = \frac{Eh^2}{[12(1 - \nu^2)]^{1/2} b(1 + a/b\sin\phi)}$$
$$\times \left[\frac{h}{b}\frac{x}{\sin\phi(1 + a/b\sin\phi)}\right]^{1/4}$$
$$\times\ [A_1 \operatorname{Im} B_1(ix) + A_2 \operatorname{Re} h_1(ix) + B_1 \operatorname{Im} h_1(ix) + B_2 \operatorname{Re} h_2(ix)], \tag{7.4.19b}$$

$$u_H = \frac{h}{[12(1 - \nu^2)]^{1/4}}\left[\frac{b}{h}\frac{\sin\phi(1 + a/b\sin\phi)}{x}\right]^{1/4}$$
$$\times \left\langle A_1 \frac{\mathrm{d} \operatorname{Im} h_1(ix)}{\mathrm{d}x} + A_2 \frac{\mathrm{d} \operatorname{Re} h_1(ix)}{\mathrm{d}x} + B_1 \frac{\mathrm{d} \operatorname{Im} h_2(ix)}{dx}\right.$$
$$+ B_2 \frac{\mathrm{d} \operatorname{Re} h_2(ix)}{\mathrm{d}x} + \left\{\frac{1}{4x} - \frac{\cos\phi}{[12(1 - \nu^2)]^{1/4}}\right.$$
$$\times \left.\left[\frac{h}{b}\frac{x}{\sin\phi(1 + a/b\sin\phi)}\right]^{1/2}\left(\frac{1}{2} + \nu + \frac{1}{4}\frac{b/a}{\sin\phi}\right)\right\}$$
$$\times\ [A_1 \operatorname{Im} h_1(ix) + A_2 \operatorname{Re} h_1(ix)$$
$$\left. + B_1 \operatorname{Im} h_2(ix) + B_2 \operatorname{Re} h_2(ix)]\right\rangle, \tag{7.4.19c}$$

$$M_\phi = -\frac{D}{b(1 + a/b\sin\phi)}\left[12(1 - \nu^2)\frac{b}{h}\frac{\sin\phi(1 + a/b\sin\phi)}{x}\right]^{1/4}$$
$$\times \left\langle A_1 \frac{\mathrm{d} \operatorname{Re} h_1(ix)}{\mathrm{d}x} - A_2 \frac{\mathrm{d} \operatorname{Im} h_1(ix)}{\mathrm{d}x} + B_1 \frac{\mathrm{d} \operatorname{Re} h_2(ix)}{\mathrm{d}x}\right.$$
$$- B_2 \frac{\mathrm{d} \operatorname{Im} h_2(ix)}{\mathrm{d}x} + \left\{\frac{1}{4x} - \frac{\cos\phi}{[12(1 - \nu^2)]^{1/4}}\right.$$
$$\times \left.\left[\frac{h}{b}\frac{x}{\sin\phi(1 + a/b\sin\phi)}\right]^{1/2}\left(\frac{1}{2} - \nu + \frac{1}{4}\frac{b/a}{\sin\phi}\right)\right\}$$
$$\times\ [A_1 \operatorname{Re} h_1(ix) - A_2 \operatorname{Im} h_1(ix)$$
$$\left. + B_1 \operatorname{Re} h_2(ix) - B_2 \operatorname{Im} h_2(ix)]\right\rangle, \tag{7.4.19d}$$

which can be used over the range of angle $\pi/2 \leq \phi \leq \pi/2$, with similar expressions for the range of angle $\pi/2 \leq \pi - \phi \leq \pi/2$. We note the following approximations for small ϕ, which are useful in evaluating quantities at or near the apex:

$$\frac{x}{\sin\phi(1 + a/b\sin\phi)} \approx \left[[12(1-\nu^2)]^{1/2}\frac{a^2}{bh}\right]^{1/3} \times \left[1 - \frac{6}{5}\frac{a}{b}\sin\phi + \frac{1}{7}\left(1 + \frac{227}{25}\frac{a^2}{b^2}\right)\sin^2\phi + \cdots\right], \quad (7.4.20a)$$

$$\frac{1}{4}\left\{\frac{1}{x} - \frac{\cos\phi}{[12(1-\nu^2)]^{1/4}}\left[\frac{h}{b}\frac{x}{\sin\phi(1 + a/b\sin\phi)}\right]^{1/2}\frac{b/a}{\sin\phi}\right\} \approx \frac{1}{4}\left[\frac{bh}{[12(1-\nu^2)]^{1/2}a^2}\right]^{1/3}\left[\frac{17}{20}\frac{9}{b} + \frac{1}{4}\left(1 - \frac{461}{200}\frac{a^2}{b^2}\right)\sin\phi + \cdots\right]. \quad (7.4.20b)$$

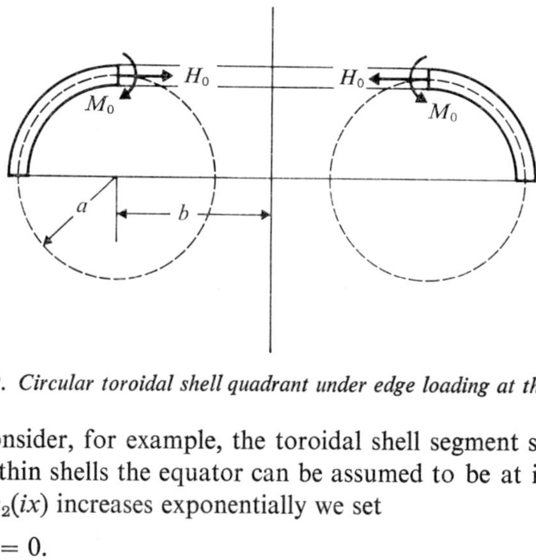

Fig. 7.9. Circular toroidal shell quadrant under edge loading at the apex

Let us consider, for example, the toroidal shell segment shown in Fig. 7.9. For very thin shells the equator can be assumed to be at infinity. Since the function $h_2(ix)$ increases exponentially we set

$$B_1 = B_2 = 0. \quad (7.4.21a)$$

Then at the apex we have, on neglecting terms of the order of $\langle ah/\{[12(1-\nu^2)]^{1/2}b^2\}\rangle^{1/3}$ compared to unity

$$M_0 = \frac{D}{b}[12(1-\nu^2)]^{1/4}\left[\frac{1}{[12(1-\nu^2)]^{1/4}}\frac{b^2}{ah}\right]^{1/6}\frac{6^{1/3}}{\Gamma(\frac{1}{3})}\left(A_2 + \frac{1}{3^{1/2}}A_1\right), \quad (7.4.21b)$$

$$H_0 = -\frac{D}{bh}\left[[12(1-\nu^2)]^{1/4}\frac{ah}{b^2}\right]^{1/6}\frac{2^{4/3}}{3^{1/2}\Gamma(\frac{2}{3})}A_1, \quad (7.4.21c)$$

from which we obtain

$$\Theta_0 = \frac{b}{D}\left\{\frac{H_0 h}{3^{1/2}} + \frac{2\Gamma(\frac{1}{3})}{3^{5/6}\Gamma(\frac{2}{3})}\left[\frac{ah}{[12(1-\nu^2)]^{1/2}}\right]^{1/3} M_0\right\}$$

$$\approx \frac{b}{D}\left\{0.577 H_0 h + 1.584\left[\frac{ah}{[12(1-\nu^2)]^{1/2}}\right]^{1/3} M_0\right\}, \tag{7.4.22a}$$

$$u_{H_0} = \frac{hh}{[12(1-\nu^2)]^{1/2} D}\left\{\frac{3^{5/6}\Gamma(\frac{2}{3})}{2\Gamma(\frac{1}{3})}\left[\frac{b^2[12(1-\nu^2)]^{1/2}}{ah}\right]^{1/3} H_0 h - \frac{M_0}{3^{1/2}}\right\}$$

$$\approx \frac{hb}{[12(1-\nu^2)]^{1/2} D}\left\{0.631\left[\frac{[12(1-\nu^2)]^{1/2} b^2}{ah}\right]^{1/3} H_0 h - 0.577 M_0\right\}. \tag{7.4.22b}$$

Also of interest is the vertical displacement of the apex relative to the equator, given by Eqs. (7.1.14b), (7.2.2), and (7.4.15d) as

$$u_{V_0} \approx -a\left(\frac{h}{b}\right)^{1/4}\int_0^{\pi/2}\left[\frac{x}{\sin\phi(1 + a/b\cos\phi)}\right]^{1/4}$$

$$\times\left\{A_1 \operatorname{Re} h_1(ix) - A_2 \operatorname{Im} h_1(ix) - \frac{h/a}{[12(1-\nu^2)]^{1/2}}\right.$$

$$\left.\times\left(\nu + \frac{a/b\sin\phi}{1 + a/b\sin\phi}\right)[A_1 \operatorname{Im} h_1(ix) + A_2 \operatorname{Re} h_1(ix)]\right\}\cos\phi\, d\phi, \tag{7.4.23a}$$

where the assumed equivalence of $\phi = \pi/2$ and $x = \infty$ has been used. Since for thin shells the functions $\bar{\Theta}$ and $\bar{U}$ decay very rapidly with ϕ, we may assume that the integral is essentially governed by the behavior of the functions near ϕ equal to zero. Then, on deleting terms of the order of h/a or h/b compared to unity, we may approximate the integral by †

$$u_{V_0} = -\left[\frac{ah}{[12(1-\nu^2)]^{1/4}}\right]^{1/2}\int_0^{\infty}[A_1 \operatorname{Re} h_1(ix) - A_2 \operatorname{Im} h_1(ix)]\, dx$$

$$\approx \frac{hb}{[12(1-\nu^2)]^{1/2} D}\left[\frac{[12(1-\nu^2)]^{1/2} a^2}{bh}\right]^{1/3}$$

$$\times\left\{\frac{\Gamma(\frac{1}{3})}{6^{1/3}} M_0 \int_0^{\infty} \operatorname{Im} h_1(ix)\, dx + \frac{3^{1/2}\Gamma(\frac{2}{3})}{2^{4/3}}\left[\frac{[12(1-\nu^2)]^{1/2} b^2}{ah}\right]^{1/3}\right.$$

$$\left.\times H_0 h \int_0^{\infty}\left[\operatorname{Re} h_1(ix) + \frac{1}{3^{1/2}} \operatorname{Im} h_1(ix)\right] dx\right\}$$

$$\approx -\frac{hb}{[12(1-\nu^2)]^{1/2}}\left[\frac{[12(1-\nu^2)]^{1/2} a^2}{bh}\right]^{1/3}$$

$$\times\left\{1.288 M_0 + 0.948\left[\frac{[12(1-\nu^2)]^{1/2} b^2}{ah}\right]^{1/3} H_0 h\right\}, \tag{7.4.23b}$$

† R. A. Clark, 'On the Theory of Thin Elastic Toroidal Shells,' *J. Math. & Phys.*, vol. 29, no. 3, 1950, pp. 146–178.

since numerical integration yields

$$\int_0^\infty h_1(ix)\,dx \approx -(0.504 + 0.874i). \tag{7.4.23c}$$

For less thin shells we can use the complete approximate equations to satisfy the conditions

$$\begin{aligned} M_\phi &= M_0 \quad \text{at } \phi = 0 \\ &= 0 \quad \text{at} \quad \phi = 90°, \\ H &= H_0 \quad \text{at } \phi = 0 \\ &= 0 \quad \text{at} \quad \phi = 90°. \end{aligned}$$

A comparison of the force and moment distribution thus obtained with the results of a series solution† of Eqs. (7.2.3) for the case

$$\begin{aligned} a/b &= 0.1, \\ a/h &= 20, \\ \nu &= 0.3, \end{aligned}$$

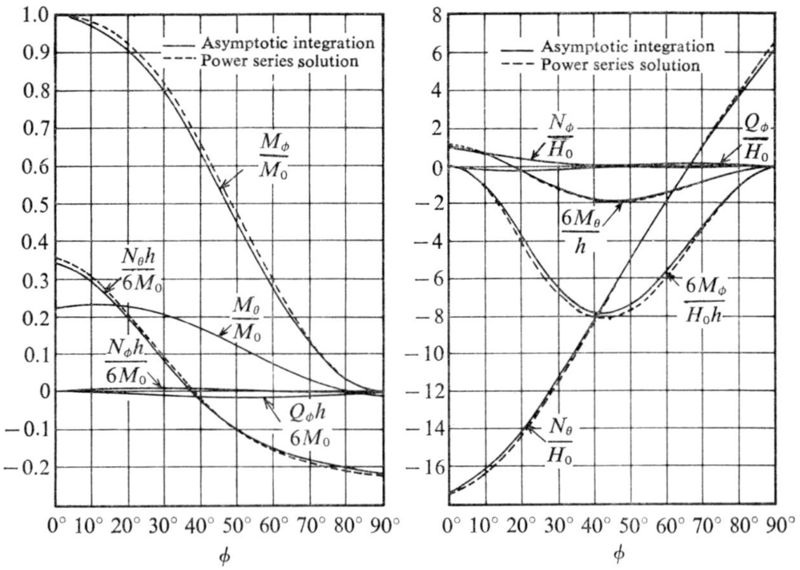

(a) Edge moment (b) Edge force

Fig. 7.10. Stress resultants for applied edge force and moment

† H. Wissler, 'Festigkeitsberechnung von Ringflächenschalen,' Dissertation, E.T.H., Zürich, 1916.

is shown in Fig. 7.10. Although the parameter $[12(1 - \nu^2)]^{1/2}a^2/bh$ has the relatively small value of 6.61, we see that the agreement between the two solutions is quite good.

A similar method can be used to obtain approximate solutions for other shells of revolution.

7.5 Membrane theory as a particular solution of the bending equations

When a shell is loaded on its surfaces or by axial edge loading, we must obtain particular solutions of Eqs. (7.1.6a), (7.1.10c), and (7.1.11a) to be used in conjunction with edge-loading solutions. It is interesting to note that if we put Q_ϕ equal to m_ϕ in Eqs. (7.1.8), (7.1.11a), (7.1.13), and (7.1.14), we obtain precisely the solution for direct stress resultants and displacements as is given by the membrane theory of shells in Section 3.7.

In certain instances these are exact solutions of the bending equations. The conditions under which this is so are obtained by noting that if

$$Q_\phi R_\theta = U_0 = m_\phi R_\theta, \tag{7.5.1a}$$

$$\Theta = \Theta_0 = \frac{1}{Eh}\left[L(m_\phi R_\theta) + \frac{\nu}{R_\phi} m_\phi R_\theta - S_\phi\right], \tag{7.5.1b}$$

the membrane theory solution will be completely exact only if Θ_0 vanishes, for then

$$M_\phi = M_\theta = 0 \tag{7.5.1c}$$

and Q_ϕ is indeed equal to m_ϕ. If Θ_0 is not identically zero but satisfies the equation

$$L(\Theta_0) - \frac{\nu}{R_\phi}\Theta_0 = 0, \tag{7.5.2a}$$

we will have bending moments given by

$$M_\phi = -D\left(\frac{1}{R_\phi}\frac{d\Theta_0}{d\phi} + \frac{\nu}{R_\theta}\Theta_0 \cot\phi\right), \tag{7.5.2b}$$

$$M_\theta = -D\left(\frac{1}{R_\theta}\Theta_0 \cot\phi + \frac{\nu}{R_\phi}\frac{d\Theta_0}{d\phi}\right), \tag{7.5.2c}$$

in addition to the membrane stress resultants. The displacements will be given by the membrane equations. Finally, if Θ_0 satisfies the equation

$$\left(L + \frac{\nu}{R_\phi}\right)\left[L(\Theta_0) - \frac{\nu}{R_\phi}\Theta_0\right] = 0, \tag{7.5.3a}$$

but not

$$L(\Theta_0) - \frac{\nu}{R_\phi}\Theta_0 = 0, \tag{7.5.3b}$$

Θ_0 is an exact particular solution of Eq. (7.1.12), but the expressions for the forces and displacements will generally be modified from the membrane theory results.

For the cone, for example, if we make the substitutions

$$R_\phi \, d\phi = ds, \tag{7.5.4a}$$

$$R_\phi = \infty, \tag{7.5.4b}$$

$$\phi = \frac{\pi}{2} - \alpha, \tag{7.5.4c}$$

$$R_\theta = s \tan \alpha, \tag{7.5.4d}$$

Eq. (7.6.2a) becomes

$$s\frac{d}{ds}s\frac{d\Theta_0}{ds} - \Theta_0 = 0, \tag{7.5.5a}$$

and we find that if Θ_0 is of the form

$$\Theta_0 = \frac{a}{s} + bs, \tag{7.5.5b}$$

the membrane theory solution is exact for displacements and membrane forces but not for moments. On the other hand, if Θ_0 is of the form

$$\Theta_0 = c + ds^2, \tag{7.5.6}$$

satisfying Eqs. (7.5.3), only the membrane normal displacements are exact. Only if Θ_0 vanishes is membrane theory completely exact. These statements coincide with those of Section 5.4.

For shells with slowly varying geometry and surface loading the results of membrane theory can be shown† to yield stresses and middle surface displacements which are certainly adequate, even if not exact, except in the vicinity of edges. In general the surface loading and geometry should not vary rapidly over distances which are of the order of magnitude of $(R_\theta h)^{1/2}$.

7.6 Some particular solutions for circular toroidal shells

For some shells of revolution, in particular the circular torus, the membrane solution is not valid near points having horizontal tangents even for simple

† F. B. Hildebrand, loc. cit.

loadings such as uniform internal pressure. Let us consider a circular toroidal shell subjected to uniform internal pressure and to a vertical force $-P$. Then

$$q_\zeta \approx p, \tag{7.6.1a}$$

$$q_\phi = m_\phi = 0, \tag{7.6.1b}$$

yielding

$$P_\phi = 2\pi pab \sin\phi\left(1 + \frac{1}{2}\frac{a}{b}\sin\phi\right) - P, \tag{7.6.1c}$$

$$S_\phi = \frac{\cot\phi}{1 + a/b\sin\phi}\left[\tfrac{1}{2}pa + \frac{Pb}{\pi a^2}\frac{1 + \frac{3}{2}(a/b)\sin\phi}{\sin^3\phi}\right]. \tag{7.6.1d}$$

We see from Eq. (7.5.1b) that a singularity in the membrane change of slope exists at points having horizontal tangents. Thus the membrane approximation is invalid in the vicinity of those points and a more accurate representation of the particular solution of the approximate bending equation

$$\frac{d^2\overline{V}}{d\bar{\xi}^2} + \left(1 - i\frac{\psi}{2}\right)\overline{V} = -\left(\frac{b}{h}\right)^{1/4}\frac{\cot\phi}{Eh}\left[\sin|\phi|\left(1 + \frac{a}{b}\sin\phi\right)\right]^{1/4}$$
$$\times\frac{\tfrac{1}{2}pa + \dfrac{Pb}{\pi a^2}\left(1 + \dfrac{3}{2}\dfrac{a}{b}\sin\phi\right)\csc^3\phi}{1 + a/b\sin\phi} \tag{7.6.2}$$

must be found, where ψ and $\bar{\xi}$ are given by Eqs. (7.4.12). If we introduce the alternate variable

$$\overline{V} = \overline{V}_1 + i\frac{[3(1-\nu^2)]^{1/2}}{\pi Eh^2}\left[\frac{b}{h}\sin|\phi|\left(1 + \frac{a}{b}\sin\phi\right)\right]^{1/4}\frac{P\cot\phi}{1 + a/b\sin\phi}, \tag{7.6.3a}$$

then Eq. (7.6.2) can be shown to reduce to

$$\frac{d^2\overline{V}_1}{d\bar{\xi}^2} + \left(1 - i\frac{\psi}{2}\right)\overline{V}_1 = -\left(\frac{b}{h}\right)^{1/4}\frac{\cot\phi}{Eh}\left[\sin|\phi|\left(1 + \frac{a}{b}\sin\phi\right)\right]^{1/4}$$
$$\times\frac{\tfrac{1}{2}pa + i\dfrac{[3(1-\nu^2)]^{1/2}P}{\pi h}}{1 + \dfrac{a}{b}\sin\phi}, \tag{7.6.3b}$$

where a small term of the order of magnitude of h/a compared to unity has been deleted from the right side of Eq. (7.6.3b). Thus the problems of determining a particular solution for internal pressure or for axial force are identical.

Let us write the particular solution as

$$\bar{V}_1^{(p)} = -\left(\frac{b}{h}\right)^{1/4} \frac{\cot\phi}{Eh} \left[\sin|\phi|\left(1 + \frac{a}{b}\sin\phi\right)\right]^{1/4}$$

$$\times \frac{\frac{1}{2}pa + i\dfrac{[3(1-\nu^2)]^{1/2}P}{\pi h}}{1 + \dfrac{a}{b}\sin\phi} T(\bar{\xi}). \tag{7.6.4}$$

For large $\bar{\xi}$ (positive or negative) we require that

$$T(\bar{\xi}) \rightarrow 1, \tag{7.6.5a}$$

for then the particular solution of Eq. (7.6.2a) will approach the membrane solution. For small $\bar{\xi}$, on the other hand, the behavior of $T(\bar{\xi})$ can be shown to be described by the equation

$$\frac{d^2T}{d\bar{\xi}^2} - \frac{1}{\bar{\xi}}\frac{dT}{d\bar{\xi}} + \left(1 + \frac{\frac{8}{9}}{\bar{\xi}^2}\right)T = 1, \tag{7.6.5b}$$

obtained by substituting Eq. (7.6.4) into Equation (7.6.2a) and making use of Eq. (7.6.2c). An approximate particular solution of Eq. (7.6.2a), which is valid over the region of interest, is then a particular solution of Eq. (7.6.5b), which approaches unity for large $\bar{\xi}$.

The desired particular solution is given by

$$T = \bar{\xi} S_{0,\,1/3}(\bar{\xi}), \tag{7.6.6}$$

where $S_{\mu,\nu}^{(\bar{\xi})}$ is called a Lommel function.† Lommel functions are particular solutions of the equation

$$\frac{d^2w}{dz^2} + \frac{1}{z}\frac{dw}{dz} + \left(1 - \frac{\nu^2}{z^2}\right)w = z^{\mu-1}. \tag{7.6.7}$$

We may also express the solution as

$$T(\bar{\xi}) = x\bar{T}(x), \tag{7.6.8a}$$

with

$$x = (\tfrac{3}{2}\, 2^{1/2}\xi)^{2/3} \operatorname{sgn} \xi, \tag{7.6.8b}$$

† See, for example, W. Magnus and F. Oberhettinger, *Formulas and Theorems for the Functions of Mathematical Physics*, Chelsea Publishing Co., New York, 1954, pp. 42–43; A. Erdélyi, W. Magnus, F. Oberhettinger, and F. G. Tricomi, *Higher Transcendental Functions*, Bateman Manuscript Project, California Institute of Technology, vol. 2, McGraw-Hill Book Co., Inc., 1953, pp. 40–42, 84–85.

and

$$\bar{T}(x) = i\int_0^\infty e^{-(1/3t^3+ixt)}\,dt$$

$$= \int_0^\infty e^{-1/3t^3}\sin xt\,dt + i\int_0^\infty e^{-1/3t^3}\cos xt\,dt. \qquad (7.7.8c)$$

The function $\bar{T}(x)$ and its first derivative with respect to x are finite at the apex and are given by

$$\bar{T}(0) = \frac{2\pi i}{3^{7/6}\Gamma(\frac{2}{3})} = 1.28789i, \qquad (7.6.9a)$$

$$\frac{d\bar{T}(0)}{dx} = \frac{2\pi}{3^{5/6}\Gamma(\frac{1}{3})} = 0.93889. \qquad (7.6.9b)$$

Some values of the real and imaginary parts of $\bar{T}(x)$ and of its first derivative with respect to x are given in Table 7.6 for positive x and are shown graphically in Fig. 7.11.† For negative x the following relations can be seen to apply

$$\operatorname{Re}\bar{T}(x) = -\operatorname{Re}\bar{T}(-x), \qquad (7.7.10a)$$

$$\frac{d\operatorname{Re}\bar{T}(x)}{dx} = \frac{d\operatorname{Re}\bar{T}(-x)}{d(-x)}; \qquad (7.7.10b)$$

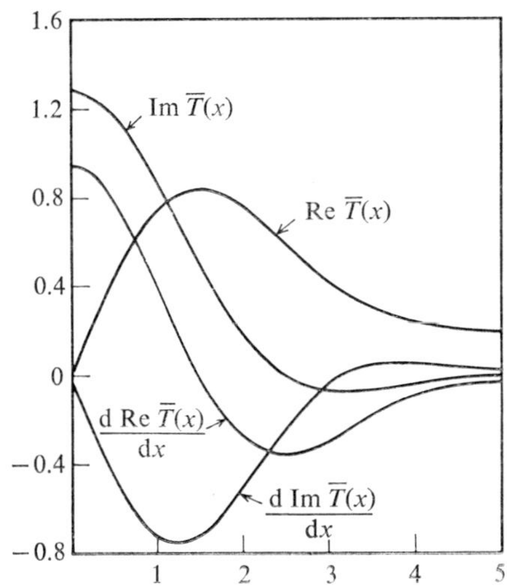

Fig. 7.11. The function $\bar{T}(x)$

† R. A. Clark, loc. cit.

Table 7.6 Values of the function $\overline{T}(x)$

x	Re $\overline{T}(x)$	Im $\overline{T}(x)$	$\frac{d}{dx}$ Re $\overline{T}(x)$	$\frac{d}{dx}$ Im $\overline{T}(x)$	x	Re $\overline{T}(x)$	Im $\overline{T}(x)$	$\frac{d}{dx}$ Re $\overline{T}(x)$	$\frac{d}{dx}$ Im $\overline{T}(x)$
0.00	0.0000	1.2879	0.9389	−0.0000	3.00	0.4281	−0.0704	−0.2967	−0.0387
0.05	0.0469	1.2868	0.9373	−0.0500	3.05	0.4136	−0.0720	−0.2859	−0.0251
0.10	0.0937	1.2829	0.9325	−0.0997	3.10	0.3995	−0.0730	−0.2748	−0.0126
0.15	0.1401	1.2767	0.9245	−0.1489	3.15	0.3861	−0.0733	−0.2634	−0.0012
0.20	0.1861	1.2680	0.9133	−0.1975	3.20	0.3732	−0.0731	−0.2517	+0.0090
0.25	0.2314	1.2570	0.8991	−0.2452	3.25	0.3609	−0.0724	−0.2400	0.0182
0.30	0.2759	1.2435	0.8819	−0.2917	3.30	0.3492	−0.0713	−0.2282	0.0264
0.35	0.3195	1.2278	0.8619	−0.3368	3.35	0.3381	−0.0698	−0.2165	0.0335
0.40	0.3621	1.2099	0.8390	−0.3804	3.40	0.3276	−0.0680	−0.2040	0.0396
0.45	0.4034	1.1898	0.8135	−0.4223	3.45	0.3176	−0.0659	−0.1934	0.0449
0.50	0.4434	1.1677	0.7855	−0.4622	3.50	0.3082	−0.0635	−0.1822	0.0492
0.55	0.4819	1.1436	0.7552	−0.5000	3.55	0.2994	−0.0609	−0.1712	0.0528
0.60	0.5189	1.1177	0.7227	−0.5356	3.60	0.2911	−0.0582	−0.1605	0.0555
0.65	0.5541	1.0901	0.6882	−0.5689	3.65	0.2833	−0.0554	−0.1502	0.0576
0.70	0.5877	1.0609	0.6519	−0.5996	3.70	0.2761	−0.0924	−0.1403	0.0589
0.75	0.6193	1.0302	0.6140	−0.6277	3.75	0.2693	−0.0495	−0.1308	0.0597
0.80	0.6490	0.9981	0.5747	−0.6531	3.80	0.2630	−0.0465	−0.1218	0.0599
0.85	0.6768	0.9649	0.5342	−0.6758	3.85	0.2571	−0.0435	−0.1131	0.0597
0.90	0.7024	0.9206	0.4927	−0.6956	3.90	0.2517	−0.0406	−0.1050	0.0590
0.95	0.7260	0.8954	0.4505	−0.7125	3.95	0.2466	−0.0376	−0.0973	0.0578
1.00	0.7475	0.8594	0.4078	−0.7266	4.00	0.2419	−0.0348	−0.0901	0.0564
1.05	0.7668	0.8228	0.3647	−0.7378	4.05	0.2376	−0.0320	−0.0834	0.0546
1.10	0.7839	0.7857	0.3214	−0.7461	4.10	0.2336	−0.0293	−0.0772	0.0526
1.15	0.7989	0.7482	0.2783	−0.7516	4.15	0.2299	−0.0268	−0.0714	0.0504
1.20	0.8118	0.7106	0.2355	−0.7542	4.20	0.2264	−0.0243	−0.0661	0.0480
1.25	0.8225	0.6729	0.1931	−0.7542	4.25	0.2232	−0.0220	−0.0612	0.0455
1.30	0.8311	0.6352	0.1514	−0.7514	4.30	0.2203	−0.0197	−0.0567	0.0429
1.35	0.8376	0.5978	0.1106	−0.7462	4.35	0.2176	−0.0177	−0.0527	0.0420
1.40	0.8422	0.5606	0.0708	−0.7384	4.40	0.2150	−0.0157	−0.0490	0.0375
1.45	0.8447	0.5239	0.0321	−0.7283	4.45	0.2126	−0.0139	−0.0458	0.0348

1.50	0.8454	0.4878	−0.0052	−0.7160	4.50	0.2104	−0.0122	−0.0428	0.0322
1.55	0.8442	0.4524	−0.0410	−0.7015	4.55	0.2084	−0.0107	−0.0402	0.0296
1.60	0.8413	0.4177	−0.0752	−0.6851	4.60	0.2064	−0.0093	−0.0380	0.0270
1.65	0.8367	0.3839	−0.1078	−0.6670	4.65	0.2046	−0.0080	−0.0360	0.0245
1.70	0.8306	0.3510	−0.1386	−0.6471	4.70	0.2028	−0.0068	−0.0342	0.0221
1.75	0.8229	0.3192	−0.1674	−0.6258	4.75	0.2011	−0.0058	−0.0327	0.0198
1.80	0.8139	0.2885	−0.1944	−0.6032	4.80	0.1995	−0.0048	−0.0315	0.0177
1.85	0.8035	0.2589	−0.2194	−0.5794	4.85	0.1980	−0.0040	−0.0304	0.0156
1.90	0.7920	0.2306	−0.2423	−0.5546	4.90	0.1965	−0.0033	−0.0295	0.0137
1.95	0.7793	0.2035	−0.2632	−0.5290	4.95	0.1950	−0.0026	−0.0288	0.0119
2.00	0.7657	0.1777	−0.2820	−0.5027	5.00	0.1936	−0.0021	−0.0282	0.0102
2.05	0.7512	0.1532	−0.2987	−0.4758	5.05	0.1922	−0.0016	−0.0277	0.0087
2.10	0.7358	0.1301	−0.3134	−0.4487	5.10	0.1908	−0.0012	−0.0274	0.0073
2.15	0.7199	0.1083	−0.3260	−0.4214	5.15	0.1895	−0.0009	−0.0271	0.0060
2.20	0.7033	0.0880	−0.3367	−0.3940	5.20	0.1881	−0.0006	−0.0269	0.0048
2.25	0.6862	0.0689	−0.3454	−0.3667	5.25	0.1868	−0.0004	−0.0268	0.0038
2.30	0.0518	0.0513	−0.3522	−0.3397	5.30	0.1854	−0.0002	−0.0267	0.0029
2.35	0.6510	0.0350	−0.3572	−0.3129	5.35	0.1841	−0.0001	−0.0267	0.0021
2.40	0.6331	0.0200	−0.3605	−0.2867	5.40	0.1828	−0.0000	−0.0266	0.0014
2.45	0.6150	0.0063	−0.3620	−0.2610	5.45	0.1814	+0.0000	−0.0266	0.0008
2.50	0.5969	−0.0061	−0.3620	−0.2361	5.50	0.1801	0.0000	−0.0267	+0.0003
2.55	0.5788	−0.0173	−0.3605	−0.2118	5.55	0.1788	0.0001	−0.0267	−0.0002
2.60	0.5609	−0.0273	−0.3576	−0.1885	5.60	0.1774	0.0000	−0.0267	−0.0005
2.65	0.5431	−0.0362	−0.3535	−0.1660	5.65	0.1761	+0.0000	−0.0267	−0.0008
2.70	0.5255	−0.0440	−0.3481	−0.1446	5.70	0.1748	−0.0000	−0.0267	−0.0010
2.75	0.5083	−0.0507	−0.3416	−0.1242	5.75	0.1734	−0.0001	−0.0267	−0.0012
2.80	0.4914	−0.0564	−0.3342	−0.1048	5.80	0.1721	−0.0002	−0.0266	−0.0013
2.85	0.4749	−0.0612	−0.3259	−0.0866	5.85	0.1708	−0.0002	−0.0266	−0.0014
2.90	0.4588	−0.0651	−0.3168	−0.0695	5.90	0.1694	−0.0003	−0.0265	−0.0014
2.95	0.4432	−0.0681	−0.3070	−0.0535	5.95	0.1681	−0.0004	−0.0264	−0.0014
					6.00	0.1668	−0.0004	−0.0263	−0.0014

$$\operatorname{Im}\bar{T}(x) = \operatorname{Im}\bar{T}(-x), \tag{7.7.10c}$$

$$\frac{\mathrm{d}\operatorname{Im}\bar{T}(x)}{\mathrm{d}x} = -\frac{\mathrm{d}\operatorname{Im}\bar{T}(-x)}{\mathrm{d}(-x)}. \tag{7.7.10d}$$

With the solution given above, the particular solutions for the rotation and shear force become

$$\Theta = -\frac{\cos\phi}{2Eh}\frac{x}{\sin\phi(1 + a/b\sin\phi)} \times \left[pa\operatorname{Re}\bar{T}(x) - \frac{[12(1-\nu^2)]^{1/2}P}{\pi h}\operatorname{Im}\bar{T}(x)\right], \tag{7.6.11a}$$

$$U = -\frac{h\cos\phi}{2[12(1-\nu^2)]^{1/2}\sin\phi(1 + a/b\sin\phi)} \times \left\{pax\operatorname{Im}\bar{T}(x) - \frac{[12(1-\nu^2)]^{1/2}P}{\pi h}[1 - x\operatorname{Re}\bar{T}(x)]\right\}. \tag{7.6.11b}$$

The derivatives of these expressions are useful and are given by

$$\frac{\mathrm{d}\Theta}{\mathrm{d}\phi} = \frac{1}{2Eh}\left[paf_1(\phi) - \frac{[12(1-\nu^2)]^{1/2}P}{\pi h}f_2(\phi)\right], \tag{7.6.12a}$$

$$\frac{\mathrm{d}U}{\mathrm{d}\phi} = \frac{h}{2[12(1-\nu^2)]^{1/2}}\left\{paf_2(\phi) + \frac{[12(1-\nu^2)]^{1/2}P}{\pi h}[g(\phi) - f_1(\phi)]\right\}, \tag{7.6.12b}$$

where

$$f_1(\phi) = \frac{1}{[\sin\phi(1 + a/b\sin\phi)]^2}\left(1 + 2\frac{a}{b}\sin\phi - \frac{a}{b}\sin^3\phi\right)x\operatorname{Re}\bar{T}(x) - \left[[12(1-\nu^2)]^{1/2}\frac{a^2}{bh}\right]^{1/2}\left[\frac{x}{\sin\phi(1 + a/b\sin\phi)}\right]^{3/2} \times \frac{\cos\phi}{1 + a/b\sin\phi}\left[x\operatorname{Re}\bar{T}(x) + x^2\frac{\mathrm{d}}{\mathrm{d}x}\operatorname{Re}\bar{T}(x)\right], \tag{7.6.13a}$$

$$f_2(\phi) = \frac{1}{[\sin\phi(1 + a/b\sin\phi)]^2}\left(1 + 2\frac{a}{b}\sin\phi - \frac{a}{b}\sin^3\phi\right)x\operatorname{Im}\bar{T}(x) - \left[[12(1-\nu^2)]^{1/2}\frac{a^2}{bh}\right]^{1/2}\left[\frac{x}{\sin\phi(1 + a/b\sin\phi)}\right]^{-3/2} \times \frac{\cos\phi}{1 + a/b\sin\phi}\left[x\operatorname{Im}\bar{T}(x) + x^2\frac{\mathrm{d}}{\mathrm{d}x}\operatorname{Im}\bar{T}(x)\right], \tag{7.6.13b}$$

$$g(\phi) = \frac{1 + 2a/b\sin\phi - a/b\sin^3\phi}{[\sin\phi(1 + a/b\sin\phi)]^2}. \tag{7.6.13c}$$

Limiting values at the apex are given by

$$\Theta \to \frac{[12(1-\nu^2)]^{1/2}P}{3^{7/6}\Gamma(\frac{2}{3})Eh^2}\left[[12(1-\nu^2)]^{1/2}\frac{a^2}{bh}\right]^{1/3}, \tag{7.6.14a}$$

$$U \to \frac{P}{2\pi}\left(\frac{1}{\sin\phi}-\frac{a}{b}\right) - \frac{pah}{[12(1-\nu^2)]^{1/2}}\frac{\pi}{3^{7/6}\Gamma(\frac{2}{3})}\left[\frac{[12(1-\nu^2)]^{1/2}a^2}{bh}\right]^{1/3}, \tag{7.6.14b}$$

$$f_1(\phi) \to -\frac{2\pi}{3^{5/6}\Gamma(\frac{1}{3})}\left[[12(1-\nu^2)]^{1/2}\frac{a^2}{bh}\right]^{2/3}, \tag{7.6.14c}$$

$$f_2(\phi) \to \frac{6}{5}\frac{2\pi}{3^{7/6}\Gamma(\frac{2}{3})}\left[[12(1-\nu^2)]^{1/2}\frac{a^2}{bh}\right]^{1/3}\frac{a}{b}, \tag{7.6.14d}$$

$$g(\phi) \to \frac{1}{\sin^2\phi}-\frac{a^2}{b^2}. \tag{7.6.14e}$$

The approximate particular solutions thus obtained can be interpreted as those for the problems shown in Fig. 7.12, a complete torus subjected to an internal pressure p and a slit torus pulled by axial edge loading in such a manner that the edge does not rotate. With these particular solutions and the approximate solutions of the homogeneous bending equations, we may obtain solutions for various problems involving toroidal segments subjected to internal pressure and to axial loading.

As an example of the accuracy to be expected from the particular solutions let us calculate the moment at the apex of an internally pressurized complete toroidal shell for comparisons with more accurate calculations† by means of numerical integration of the bending equations. We have from Eqs. (7.1.5), (7.6.12a), and (7.6.14)

$$M_\phi|_{\phi=0} = -\frac{D}{a}\left(\frac{d\Theta}{d\phi}+\nu\frac{a}{b}\frac{\cos\phi}{1+a/b\sin\phi}\Theta\right)_{\phi=0}$$

$$\approx 0.0896\left[\frac{1}{1-\nu^2}\frac{a^2}{bh}\right]^{2/3}ph^2. \tag{7.6.15}$$

Values for the cases

$a/b = \frac{2}{3}$,

$a/h = 20, 50, 200$,

$\nu = 0.3$

† A. Kalnins, 'Analysis of Shells of Revolution Subjected to Symmetrical and Nonsymmetrical Loads,' *J. Appl. Mech.*, vol. 31, no. 3, September 1964, pp. 467–476.

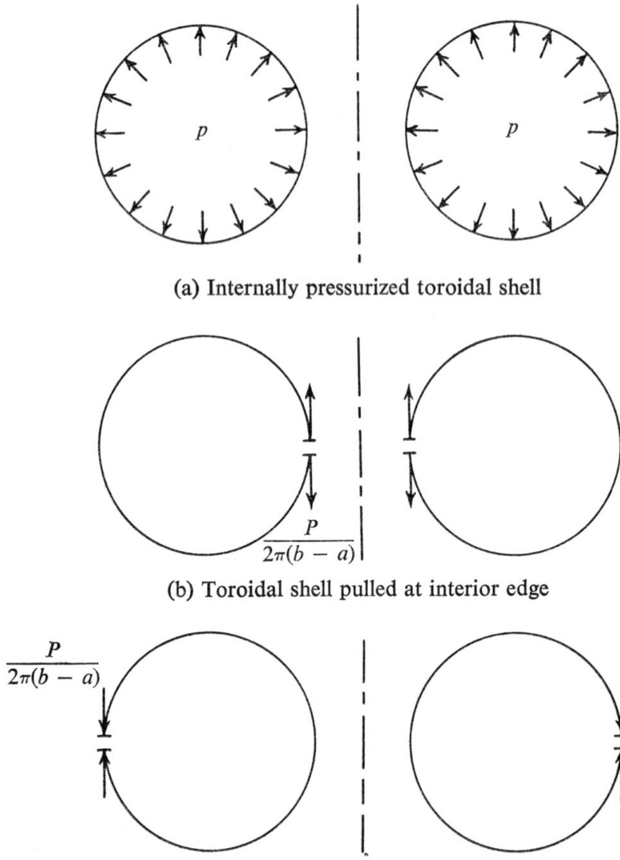

(a) Internally pressurized toroidal shell

(b) Toroidal shell pulled at interior edge

(c) Toroidal shell pulled at exterior edge

Fig. 7.12. Problems treated by particular solutions of Section 7.6

are shown in Table 7.7 and can be seen to be in error by about 4 percent for the worst case. The maximum moment generally occurs not at the apex but

Table 7.7 Bending moment at abex of internally pressurized toroidal shell

	$(M_\phi/ph)_{\phi=0}$		
a/h	Eq. (7.6.15)	Kalnins	% Error
20	0.536	0.515	4.1
50	0.988	0.949	4.1
200	2.490	2.470	0.8

in that part of the shell which has negative Caussian curvature to the left of the apex in the region of negative ϕ as shown in Fig. 7.13. It is interesting to note that although the membrane theory expression for the shear force and rotation have to be modified, the force resultants given by membrane

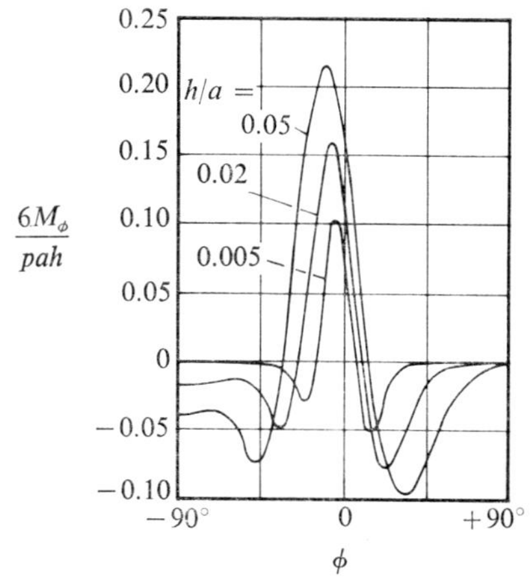

Fig. 7.13. Variation of meridional moment

theory are quite accurate. If we calculate the meridional and circumferential force at the apex we have

$$N_\phi|_{\phi=0} \approx pa\left\{1 - 0.644\left[\frac{ah}{[12(1-\nu^2)]^{1/2}b^2}\right]^{2/3}\right\}, \tag{7.6.16a}$$

$$N_\theta|_{\phi=0} \approx \tfrac{1}{2}pa\left\{1 - 1.546\left[\frac{ah}{[12(1-\nu^2)]^{1/2}b^2}\right]^{2/3}\right\}. \tag{7.6.16b}$$

For thin shells the correction term in the brackets is small compared to unity in each case.

A similar method of approximate analysis may be used for other surface loading on toroidal shells or for other shells of revolution, if necessary, leading to solutions involving Lommel functions of various orders.

Supplementary references

[1] Meissner, E., 'Das Elastizitätsproblem für dünne Schalen von Ringflächen-, Kugel-, oder Kegelform,' *Physik. Zeit.*, vol. 14, 1919, pp. 343–349.

[2] Geckeler, J. W., 'Über die Festigkeit achsensymetrischer Schalen,' *Forsch. Ing. Wes.*, vol. 276, 1926, pp. 1–52.

[3] Clark, R. A., Gilroy, T. I., and Reissner, E., 'Stresses and Deformations of Toroidal Shells of Elliptical Cross-section,' *J. Appl. Mech.*, vol. 19, no. 1, March 1952, pp. 37–48.

[4] Dahl, N. C., 'Toroidal Shell Expansion Joints,' *J. Appl. Mech.*, vol. 20, no. 4, December 1953, pp. 497–503.

[5] Horvay, G., and Clausen, I. M., 'Membrane and Bending Analysis of Axisymmetrically Loaded Axisymmetrical Shells,' *J. Appl. Mech.*, vol. 22, no. 1, March 1955, pp. 25–30.

[6] Horvay, G., Linkous, C., and Born, J. S., 'Analysis of Short, Thin Axisymmetrical Shells under Axisymmetrical Loading,' *J. Appl. Mech.*, vol. 23, no. 1, March 1956, pp. 68–79.

[7] Clark, R. A., and Reissner, E., 'On Axially Symmetric Bending of Nearly Cylindrical Shells of Revolution,' *J. Appl. Mech.*, vol. 23, no. 1, March 1956, pp. 59–67.

[8] Hetenyi, M., and Timms, R. J., 'Analysis of Axially Loaded Annular Shells with Applications to Welded Bellows,' *Trans. ASME,* vol. 82C, 1960, pp. 741–755.

[9] Kraus, H., Bilodeau, G. G., and Langer, B. F., 'Stresses in Thin-Walled Pressure Vessels with Ellipsoidal Heads,' *Trans. ASME,* vol. 82B, 1961, pp. 29–42.

[10] Johns, R. H., and Orange, T. W., 'Theoretical Elastic Stress Distributions Arising from Discontinuities and Edge Loads in Several Shell Type Structures,' NASA TR R-103, 1961.

[11] Steele, C. R., 'On the Asymptotic Solution of Nonhomogeneous Ordinary Differential Equations with a Large Parameter,' *Quart. Appl. Math.*, vol. 23, 1965, pp. 193–201.

[12] Tsui, E. Y. W., and Massard, J. M., 'Bending Behavior of Toroidal Shells,' *J. Eng. Mech. Div., ASCE,* vol. 94, no. EM2, April 1968, pp. 439–464.

8

Torsion and circumferential bending of shells of revolution

8.1 Torsion of shells of revolution

In the preceding chapters we have considered axisymmetric deformations of shells of revolution due to bending. It is also possible to obtain axisymmetric deformations when the surface or edge loading consists of circumferential shear forces which are independent of the angle θ. In this case we have

$$q_\phi = q_\zeta = m_\phi = 0, \tag{8.1.1a}$$

$$q_\theta = q_\theta(\phi), \qquad m_\theta = m_\theta(\phi). \tag{8.1.1b}$$

We assume the deformations to be given by

$$u_\phi = w = 0, \tag{8.1.2a}$$

$$u_\theta = u_\theta(\phi), \tag{8.1.2b}$$

which leads to

$$\epsilon_\phi = \epsilon_\theta = \kappa_\phi = \kappa_\theta = 0, \tag{8.1.3a}$$

$$\gamma_{\phi\theta} = \frac{R_\theta \sin\phi}{R_\phi} \frac{\mathrm{d}}{\mathrm{d}\phi}\left(\frac{u_\theta}{R_\theta \sin\phi}\right), \tag{8.1.3b}$$

$$\bar{\kappa}_{\phi\theta} = \frac{\sin\phi}{R_\phi} \frac{\mathrm{d}}{\mathrm{d}\phi}\left(\frac{u_\theta}{R_\theta \sin\phi}\right) - \frac{1}{4}\left(\frac{1}{R_\phi} + \frac{1}{R_\theta}\right)\gamma_{\phi\theta} = \frac{1}{4}\left(\frac{3}{R_\theta} - \frac{1}{R_\phi}\right)\gamma_{\phi\theta}, \tag{8.1.3c}$$

and, from Eqs. (2.7.2),

$$N_\phi = N_\theta = M_\phi = M_\theta = 0, \tag{8.1.4a}$$

$$N_{\phi\theta} = \frac{1-\nu}{2} K\gamma_{\phi\theta}\left\{1 + \frac{h^2}{16}\left(\frac{1}{R_\phi} - \frac{1}{R_\theta}\right)^2 \times \left[1 + \left(3 - \frac{R_\phi}{R_\theta}\right)\sum_{n=1}^{\infty} \frac{1}{2n+3}\left(\frac{h}{2R_\phi}\right)^{2n}\right]\right\}, \tag{8.1.4b}$$

$$\overline{M}_{\phi\theta} = \frac{1-\nu}{4} D\gamma_{\phi\theta} \left[\frac{3}{R_\theta} - \frac{1}{R_\phi} - R_\phi\left(\frac{1}{R_\phi} - \frac{1}{R_\theta}\right)^2 \times \sum_{n=1}^{\infty} \frac{3}{2n+3}\left(\frac{h}{2R_\phi}\right)^{2n}\right]. \tag{8.1.4c}$$

Equilibrium Eqs. (2.6.6) then reduce to the single equation

$$\frac{\mathrm{d}}{\mathrm{d}\phi}\left[(R_\theta \sin\phi)^2 Gh(1+\delta)\gamma_{\phi\theta}\right] = -R_\phi(R_\theta \sin\phi)^2\left(q_\theta + \frac{m_\theta}{R_\theta}\right), \tag{8.1.5}$$

where G is the shear modulus, given by

$$G = \frac{E}{2(1+\nu)}, \tag{8.1.6a}$$

and

$$\delta = \frac{h^2}{12}\left[\left(\frac{1}{R_\phi^2} - \frac{3}{R_\phi R_\theta} + \frac{3}{R_\theta^2}\right) + R_\phi\left(\frac{1}{R_\phi} - \frac{1}{R_\theta}\right)^3 \sum_{n=1}^{\infty} \frac{3}{2n+3}\left(\frac{h}{2R_\phi}\right)^{2n-1}\right]. \tag{8.1.6b}$$

Eq. (8.1.5) can be integrated to yield

$$\gamma_{\phi\theta} = \frac{1}{2\pi Gh(R_\theta \sin\phi)^2(1+\delta)}\left[T - 2\pi\int_{\phi_1}^{\phi} R_\phi(R_\theta \sin\phi)^2\left(q_\theta + \frac{m_\theta}{R_\theta}\right)\mathrm{d}\phi\right], \tag{8.1.7}$$

where T is a torque applied to the edge $\phi = \phi_1$ and the entire quantity in brackets is the total torque due to circumferential surface stresses, body forces, and edge shear forces at the cross-section defined by the angle ϕ.

From Eqs. (8.1.3b) and (8.1.7) we then can express the angle of twist as

$$\Theta = \frac{u_\theta}{R_\theta \sin\phi} = \frac{1}{Gh}\int_{\phi_2}^{\phi} \frac{R_\phi\,\mathrm{d}\phi}{(1+\delta)(R_\theta \sin\phi)^3} \times \left[\frac{T}{2\pi} - \int_{\phi_1}^{\phi}(R_\theta \sin\phi)^2\left(q_\theta + \frac{m_\theta}{R_\theta}\right)R_\phi\,\mathrm{d}\phi\right], \tag{8.1.8}$$

where the boundary condition of zero circumferential displacement at the section corresponding to ϕ_2 has been imposed. It is obvious that if the ratio of the shell thickness to the least principal radius of curvature is small, the quantity δ in Eqs. (8.1.7) and (8.1.8) may be neglected in comparison with unity, a simplification which is equivalent to using the membrane theory of shells and which yields formulas similar to the familiar Bredt formula for thin tubular sections.

The shear stresses parallel to the middle surface in the shell wall corresponding to Eqs. (8.1.7) and (8.1.8) can be obtained from Eqs. (2.7.1c) and (8.1.3c) as

$$\tau_{\phi\rho}^{(\zeta)} = G\gamma_{\phi\theta}\left(\frac{1+\dfrac{\zeta}{R_\theta}}{1+\dfrac{\zeta}{R_\phi}}\right). \tag{8.1.9}$$

Note that for a spherical shell, with R_ϕ equal to R_θ, the shear stresses are uniform across the wall thickness. The non-zero circumferential displacements are given by

$$u_\theta^{(\zeta)} = u_\theta\left(1+\frac{\zeta}{R_\theta}\right), \tag{8.1.10}$$

which implies that each of the conical surfaces defined by constant ϕ rotates as a rigid body about the axis of revolution of the shell (Fig. 8.1). The remaining non-zero stress is given by Eq. (2.11.1) as

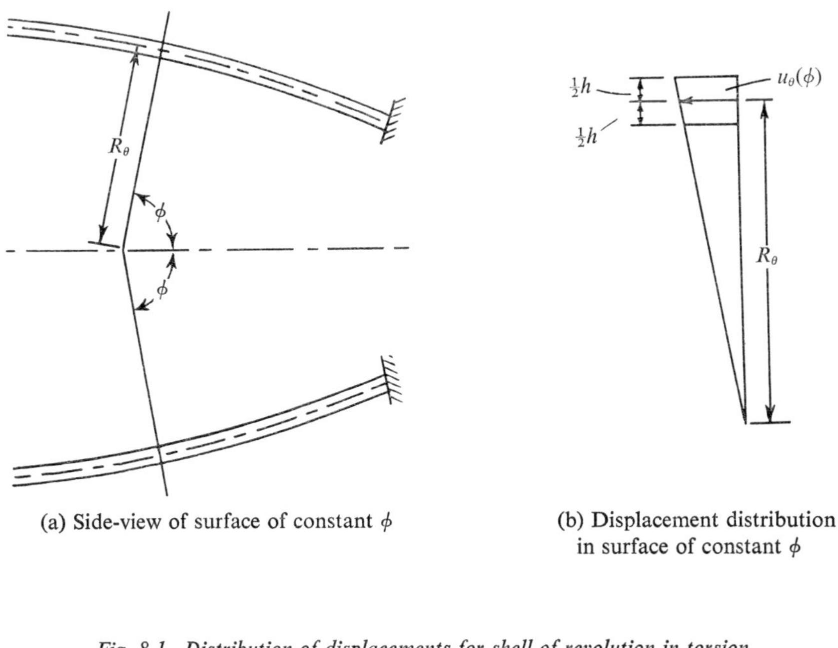

(a) Side-view of surface of constant ϕ

(b) Displacement distribution in surface of constant ϕ

Fig. 8.1. Distribution of displacements for shell of revolution in torsion

$$\left(1+\frac{\zeta}{R_\phi}\right)\left(1+\frac{\zeta}{R_\theta}\right)^2\tau_{\theta\zeta}^{(\zeta)} = \left(1-\frac{h}{2R_\phi}\right)\left(1-\frac{h}{2R_\theta}\right)^2\tau_{\theta\zeta}^{(-h/2)}$$

$$-\frac{1}{R_\phi(R_\theta\sin\phi)^2}\frac{\mathrm{d}}{\mathrm{d}\phi}\left\{\frac{\displaystyle\int_{-h/2}^{\zeta}\frac{(1+\zeta/R_\theta)^2}{(1+\zeta/R_\phi)}\,\mathrm{d}\zeta}{\displaystyle\int_{-h/2}^{h/2}\frac{(1+\zeta/R_\theta)^2}{(1+\zeta/R_\phi)}\,\mathrm{d}\zeta}\right.$$

$$\left.\times\left[\frac{T}{2\pi}-\int_{\phi_1}^{\phi}(R_\theta\sin\phi)^2\left(q_\theta+\frac{m_\theta}{R_\theta}\right)R_\phi\,\mathrm{d}\phi\right]\right\}$$

$$-\int_{-h/2}^{\zeta}\left(1+\frac{\zeta}{R_\phi}\right)\left(1+\frac{\zeta}{R_\theta}\right)^2\rho_\theta^{(\zeta)}\,\mathrm{d}\zeta. \tag{8.1.11a}$$

The last equation can be easily shown to yield the correct surface stresses when we use the relation

$$q_\theta+\frac{m_\theta}{R_\theta} = \left(1+\frac{h}{2R_\phi}\right)\left(1+\frac{h}{2R_\theta}\right)^2\tau_{\theta\zeta}^{(h/2)} - \left(1-\frac{h}{2R_\phi}\right)\left(1-\frac{h}{2R_\theta}\right)^2\tau_{\theta\zeta}^{(-h/2)}$$

$$+\int_{-h/2}^{h/2}\left(1+\frac{\zeta}{R_\phi}\right)\left(1+\frac{\zeta}{R_\theta}\right)^2\rho_\theta^{(\zeta)}\,\mathrm{d}\zeta. \tag{8.1.11b}$$

The above solution is an exact solution of the isotropic shell equations for cylindrical and spherical shells under edge torque alone, provided the edge shear forces are applied as given by the pertinent equations. In this case, with

$$\tau_{\theta\zeta}^{(h/2)} = \tau_{\theta\zeta}^{(-h/2)} = \rho_\theta^{(\zeta)} = 0, \tag{8.1.12a}$$

and the principal radii of curvature independent of the angle ϕ, Eq. (8.1.11a) yields

$$\tau_{\theta\zeta}^{(\zeta)} = 0. \tag{8.1.12b}$$

Table 8.1 Shear stresses and displacements in spheres and cylinders under end torque

Shell	$\tau_{\phi\theta}^{(\zeta)}$	$u_\theta^{(\zeta)}$
Cylinder	$\dfrac{T(1+\zeta/R)}{2\pi hR^2(1+h^2/4R^2)}$	$\dfrac{T(1+\zeta/R)x}{2\pi GhR^2(1+h^2/4R^2)}$
Sphere	$\dfrac{T}{2\pi ha^2\sin^2\phi(1+h^2/12a^2)}$	$\dfrac{T(1+\zeta/a)\sin\phi}{4\pi Gha(1+h^2/12a^2)}\left[\ln\left(\dfrac{\cot\phi_2/2}{\cot\phi/2}\right) - \dfrac{(\cos\phi-\cos\phi_2)(1+\cos\phi\cos\phi_2)}{\sin^2\phi\sin^2\phi_2}\right]$

Thus the shear angle is identically equal to zero. Since both the transverse direct stress and the transverse direct strain vanish also, there is no difference in the behavior of an isotropic shell and a transversely rigid shell in these cases. The shear stresses and displacements in the shell wall are given by Table 8.1.

The comparison of theoretical and experimental results shown in Fig. 8.2† for a spherical cap loaded as shown in Fig. 8.3 indicates the very good agreement that may be expected for shells in torsion.

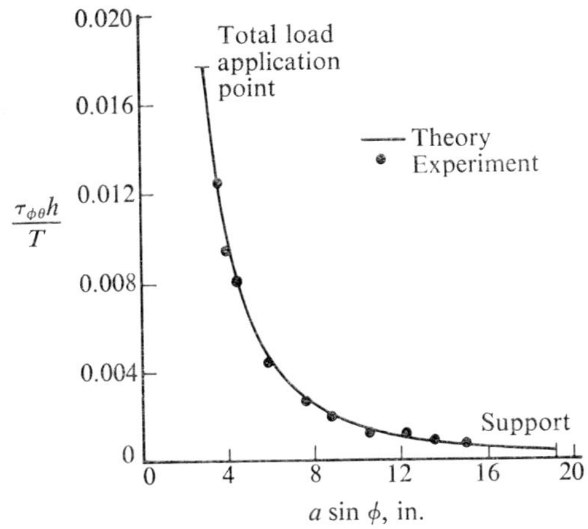

(a) Shear stress

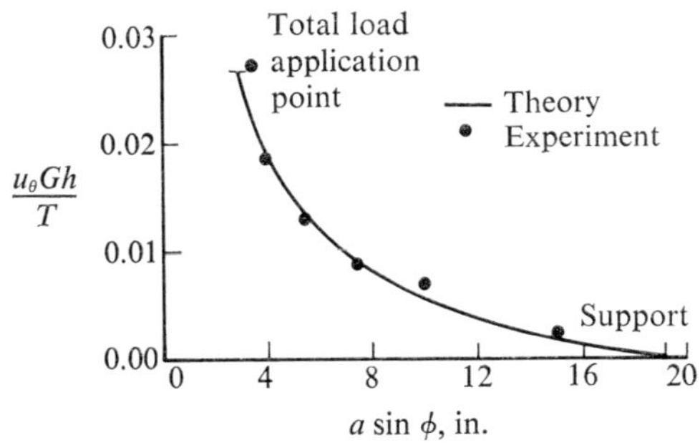

(b) Circumferential displacement

Fig. 8.2. Comparison of theory and experiment for a spherical shell in torsion

† A. S. Tooth and R. M. Kenedi, 'The Influence Line Technique of Shell Analysis,' in *Proc. Colloquium on Simplified Calculation Methods of Shell Structures*, Brussels, September 4–6, 1961, North-Holland Publishing Co., Amsterdam, 1962, pp. 51–56.

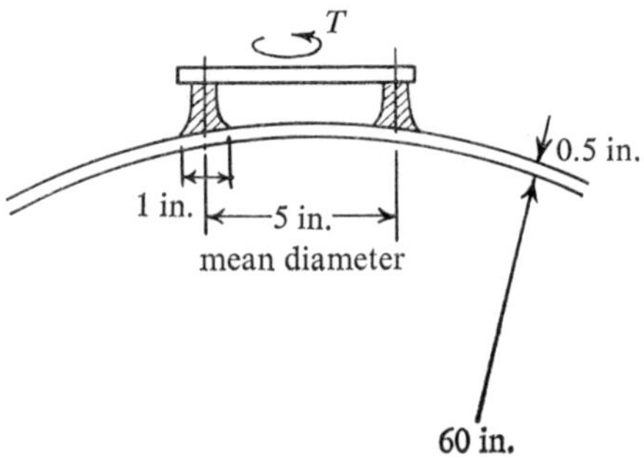

Fig. 8.3. Method of loading of spherical specimen

8.2 Circumferential bending of shells of revolution

A final class of problems which we can consider to be axisymmetric involves strains, and thus stress resultants, which are independent of circumferential location, but not all of the displacements are axisymmetric. Let us assume the following displacement distributions

$$u_\phi = u_\phi(\phi), \tag{8.2.1a}$$

$$w = w(\phi), \tag{8.2.1b}$$

$$u_\theta = \mu\theta R_\theta \sin\phi, \tag{8.2.1c}$$

where μ is an arbitrary constant. Then from Eqs. (1.5.3) and (1.5.10) we have

$$\epsilon_\phi = \frac{1}{R_\phi}\left(\frac{du_\phi}{d\phi} + w\right), \tag{8.2.2a}$$

$$\epsilon_\theta = \frac{1}{R_\theta}(u_\phi \cot\phi + w) + \mu, \tag{8.2.2b}$$

$$\gamma_{\phi\theta} = \kappa_{\phi\theta} = 0, \tag{8.2.2c}$$

$$\kappa_\phi = -\frac{1}{R_\phi}\frac{d\Theta}{d\phi}, \tag{8.2.2d}$$

$$\kappa_\theta = -\frac{1}{R_\theta}(\Theta \cot\phi - \mu), \tag{8.2.2e}$$

where the meridional slope change is given by

$$\Theta = \frac{1}{R_\phi}\left(\frac{dw}{d\phi} - u_\phi\right). \tag{8.2.2f}$$

The contribution of the circumferential displacements can be seen to be a uniform circumferential stretching of the middle surface which remains at the same distance from the axis of revolution. Such a deformation pattern is possible, however, only if the shell is considered to be slit along a generator and the slit permitted to enlarge or the slit edges to overlap (Fig. 8.4) since the displacement pattern is not single valued. When μ vanishes, Eqs. (8.2.2) reduce to those for axisymmetric deformations of shells of revolution.

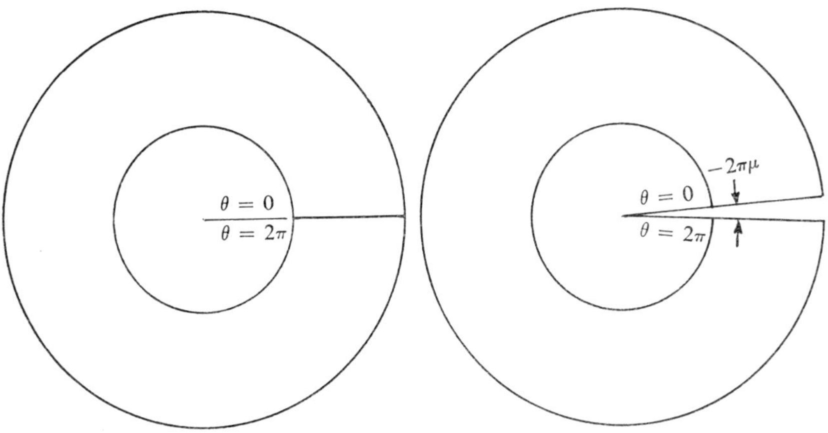

(a) Undeformed slit shell (b) Deformed slit shell

Fig. 8.4. Multivalued circumferential displacement

Since the strains are independent of θ, so are the stress resultants and, for consistency, the surface loads. Since the axisymmetric surface loads are treated in Chapters 5–7, however, we shall not consider them further. Much of the development of Chapter 7 can be taken over, however. The force equations of equilibrium, Eqs. (7.1.4a) and (7.1.4b) are satisfied by

$$Q_\phi = \frac{1}{R_\theta} U, \tag{8.2.3a}$$

$$N_\phi = \frac{\cot\phi}{R_\theta} U, \tag{8.2.3b}$$

$$N_\theta = \frac{1}{R_\phi}\frac{dU}{d\phi}, \tag{8.2.3c}$$

while the moments are given by Eqs. (7.1.5c), (7.1.5d), (8.2.2d) and (8.2.2e) as

$$M_\phi = -D\left[\frac{1}{R_\phi}\frac{d\Theta}{d\phi} + \frac{\nu}{R_\theta}(\Theta\cot\phi - \mu)\right], \tag{8.2.4a}$$

$$M_\theta = -D\left[\frac{1}{R_\theta}(\Theta\cot\phi - \mu) + \frac{\nu}{R_\phi}\frac{d\Theta}{d\phi}\right]. \tag{8.2.4b}$$

The relationship between U and Θ is then given by Eqs. (7.1.4c) and (8.2.4) as

$$U = -D\left[L(\Theta) - \frac{\nu}{R_\phi}\Theta + \mu\left(\frac{1}{R_\theta} - \frac{\nu}{R_\phi}\right)\cot\phi\right]. \tag{8.2.5}$$

Finally, since Eq. (7.1.9) is replaced by

$$\Theta = \frac{1}{R_\phi}\sin\phi\frac{d}{d\phi}[(\epsilon_\theta - \mu)R_\theta\sin\phi] - \epsilon_\phi\cot\phi, \tag{8.2.6a}$$

while Eqs. (7.1.10) are unchanged, we have

$$Eh\Theta = L(U) + \frac{\nu}{R_\phi}U - \mu Eh\cot\phi. \tag{8.2.6b}$$

Thus the equations of bending for the present deformation state are equivalent to those for axisymmetrical deformations of a shell of revolution subjected to surface loading such that

$$\overline{m}_\phi = -\frac{\mu D}{R_\theta}\left[\frac{1}{R_\theta} - \frac{\nu}{R_\phi}\right]\cot\phi, \tag{8.2.7a}$$

$$\overline{S}_\phi = \mu Eh\cot\phi, \tag{8.2.7b}$$

which can be treated by the methods of Chapters 4 to 7.

The above class of solutions is of interest primarily for toroidal shells whose meridional cross-section is closed curve. For those shells of revolution the problem which is solved can be interpreted as the bending of a curved thin-walled tube as shown in Fig. 8.5. The total applied moment is given by

$$\begin{aligned} M &= \oint\left(N_\theta + \frac{M_\theta}{R_\theta}\right)R_\theta R_\phi\sin\phi\, d\phi \\ &= -\oint\left\{\left[UR_\phi - D\left(\nu - \frac{R_\phi}{R_\theta}\right)\Theta\right]\cos\phi - \mu D\frac{R_\phi\sin\phi}{R_\theta}\right\}d\phi. \end{aligned} \tag{8.2.8}$$

From Eq. (8.2.1c) the change of inclination of each meridional cross-section is given by $\mu\theta$. The curvature change of the curved tube can then be defined by

$$\Delta K = \mu \frac{d\theta}{ds} = \frac{\mu}{b}, \tag{8.2.9}$$

where b is the radius of curvature of the centroidal circle of the cross-section. We can now define a stiffness of the curved tube by

$$\frac{M}{\Delta K} = \beta E I_0, \tag{8.2.10}$$

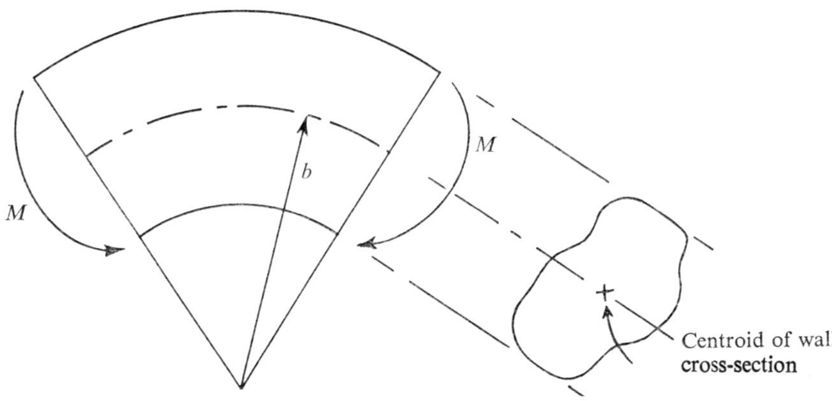

Fig. 8.5. Bending of curved thin-walled tube

where I_0 is the moment of inertia of the shell cross-section about a vertical axis passing through the centroid of the cross-section. For a straight tube β would be equal to unity, while for a curved tube β is less than unity.

For a toroidal shell having a circular cross-section, for example, we have

$$\overline{m}_\phi = \frac{\mu D \cos\phi}{ab[1 + a/b \sin\phi]} \left(\nu - \frac{a/b \sin\phi}{1 + a/b \sin\phi}\right), \tag{8.2.11a}$$

$$\overline{S}_\phi = \mu E h \cot\phi. \tag{8.2.11b}$$

Then Eqs. (8.2.5) and (8.2.6b) can be written as

$$\frac{d^2\overline{\Theta}}{d\xi^2} - \theta_1\overline{\Theta} + 2\overline{U} = 2\left[\frac{b}{h}\sin|\phi|\left(1 + \frac{a}{b}\sin\phi\right)\right]^{1/4} \times \left(\nu + \frac{a/b \sin\phi}{1 + a/b \sin\phi}\right)\frac{h/a}{[12(1-\nu^2)]^{1/2}}\,\mu\cot|\phi|, \tag{8.2.12a}$$

$$\frac{d^2\overline{U}}{d\xi^2} - \theta_2\overline{U} - 2\overline{\Theta} = 2\left[\frac{b}{h}\sin|\phi|\left(1 + \frac{a}{b}\sin\phi\right)\right]^{1/4}\mu\cot\phi, \tag{8.2.12b}$$

where the notation of Eq. (7.2.2) is used. By means of approximations similar to those of Section 7.5, and the deletion of terms small compared to unity, we may rewrite Eq. (8.2.12) as

$$\frac{d^2\overline{V}}{d\xi^2} + \left(1 - i\frac{\psi}{2}\right)\overline{V} = -\left[\frac{b}{h}\sin|\phi|\left(1 + \frac{a}{b}\sin\phi\right)\right]^{1/4}\mu\cot\phi. \tag{8.2.13}$$

But Eq. (8.2.13) is similar to Eq. (7.7.2) for an internally pressurized torus so that an approximate particular solution may be written as

$$\overline{V} \approx -\left[\frac{b}{h}\sin|\phi|\left(1 + \frac{a}{b}\sin\phi\right)\right]^{1/4}\mu\cot\phi[x\overline{T}(x)], \quad -\frac{\pi}{2} \leq \phi \leq \frac{\pi}{2}. \tag{8.2.14a}$$

Then

$$\Theta \approx -\mu x\cot\phi\,\mathrm{Re}\,\overline{T}(x), \tag{8.2.14b}$$

$$U \approx -\frac{Eh^2}{[12(1-\nu^2)]^{1/2}}\mu x\cot\phi\,\mathrm{Im}\,\overline{T}(x), \tag{8.2.14c}$$

and the moment may be written as

$$M \approx \frac{2\mu Eh^2a}{[12(1-\nu^2)]^{1/2}}\int_{-\pi/2}^{\pi/2}\left\{\frac{x}{\sin\phi}\cos^2\phi\left[\mathrm{Im}\,\overline{T}(x) - \frac{h/a}{[12(1-\nu^2)]^{1/2}}\right.\right.$$
$$\left.\left.\times\left(\nu - \frac{a/b\sin\phi}{1 + a/b\sin\phi}\right)\mathrm{Re}\,\overline{T}(x)\right] + \frac{h/b}{[12(1-\nu^2)]^{1/2}}\frac{\sin^2\phi}{1 + a/b\sin\phi}\right\}d\phi$$
$$\approx \frac{2\mu Eh^2a}{[12(1-\nu^2)]^{1/2}}\int_{-\pi/2}^{\pi/2}\frac{\cos^2\phi}{\sin\phi}x\,\mathrm{Im}\,\overline{T}(x)\,d\phi, \tag{8.2.15a}$$

when terms of the order of magnitude of h/a or h/b compared to unity are deleted. We may again argue that the integral is determined primarily by what happens in the vicinity of the apex ($\phi = 0$) so that Eq. (8.2.15a) may be approximated by

$$M \approx \frac{4\mu Eh^2a}{[12(1-\nu^2)]^{1/2}}\int_0^\infty \mathrm{Im}\,\overline{T}(x)\,dx$$
$$\approx \frac{2\pi\mu Eh^2a}{[12(1-\nu^2)]^{1/2}}. \tag{8.2.15b}$$

Then, with

$$I_0 = \pi a^3h, \tag{8.2.16}$$

Eqs. (8.2.9), (8.2.10), and (8.2.15b) yield

$$\beta \approx \frac{2}{[12(1 - \nu^2)]^{1/2}(a^2/bh)}. \tag{8.2.17}$$

More accurate calculations† show that the approximate formula is quite accurate for values of $[12(1 - \nu^2)]^{1/2}(a^2/bh)$ as small as 3.

† See R. A. Clark and E. Reissner, 'Bending of Curved Tubes,' in Advances in Applied Mechanics, vol. II, Academic Press, 1951, pp. 93–122. Other results and references are given in D. H. Cheng and H. J. Thailer, 'In-Plane Bending of Curved Circular Tubes,' Paper No. 68-PVP-12, *ASME 1st Pressure Vessels and Piping Conference*, Dallas, Texas, September 22–26, 1968.

Part III

Asymmetrically loaded shells

9

Asymmetric deformations of spherical shells

9.1 General relations for asymmetrically loaded spherical shells

When we leave the subject of axisymmetric deformations of shells of revolution, we leave that portion of the theory of shells which is more or less concise and developed in depth. In problems of axisymmetric deformations we were able to deal with one or at most two unknown quantities. Now we must be prepared, in general, to deal with three unknowns unless we are willing to be satisfied with Vlasov's equations and the associated assumptions. In the case of the spherical shell, however, we can obtain a representation which is aesthetically satisfying with no simplifying assumptions, a result which might have been anticipated because of the close relation to the theory of flat plates. For this reason we shall start with the theory of spherical shells rather than the slightly more complicated, although easier to calculate, theory of circular cylinders.

The equations defining the forces and deformations (Fig. 9.1) in a spherical shell are the following, obtained from Eqs. (1.5.3), (1.5.10), (2.4.9), (2.6.1), (2.6.2), and (2.7.2). The equations of equilibrium are given by

$$\frac{\partial}{\partial \phi}(N_\phi \sin\phi) - N_\theta \cos\phi + \frac{\partial}{\partial \theta}\left(\bar{N}_{\phi\theta} - \frac{\bar{M}_{\phi\theta}}{a}\right) + (\bar{Q}_\phi + q_\phi a)\sin\phi = 0, \tag{9.1.1a}$$

$$\frac{\partial}{\partial \theta}(N_\theta) + \frac{\partial}{\partial \phi}\left[\left(\bar{N}_{\phi\theta} - \frac{\bar{M}_{\phi\theta}}{a}\right)\sin\phi\right] + \left(\bar{N}_{\phi\theta} + \frac{\bar{M}_{\phi\theta}}{a}\right)\cos\phi$$
$$+ (\bar{Q}_\theta + q_\theta a)\sin\phi = 0, \tag{9.1.1b}$$

$$\frac{\partial}{\partial \phi}(\bar{Q}_\phi \sin\phi) + \frac{\partial \bar{Q}_\theta}{\partial \theta} - \frac{2}{a}\frac{\partial^2 \bar{M}_{\phi\theta}}{\partial \phi\, \partial \theta} - (N_\phi + N_\theta - q_\zeta a)\sin\phi = 0, \tag{9.1.1c}$$

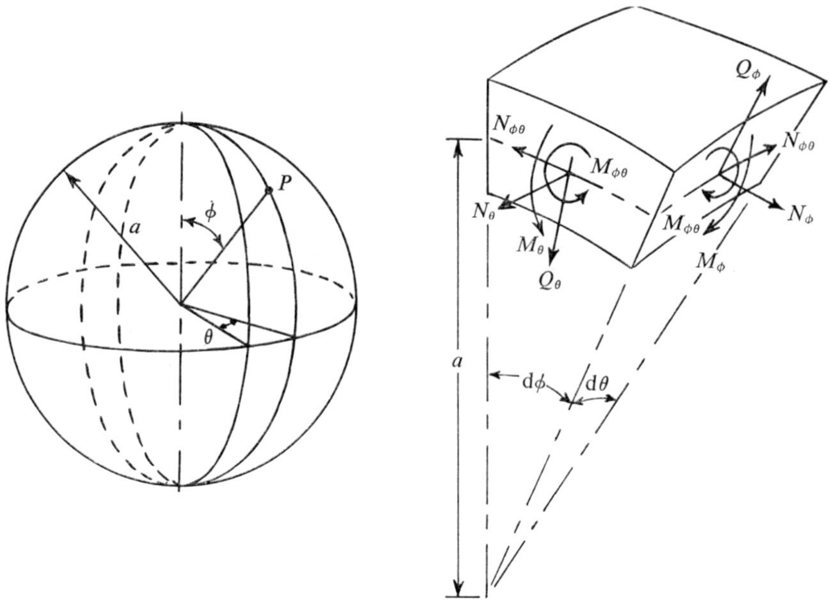

(a) Coordinate system for spherical shell (b) Forces on spherical shell element

Fig. 9.1. Notation for spherical shell

$$\frac{\partial}{\partial \phi}(M_\phi \sin \phi) - M_\theta \cos \phi + 2\frac{\partial \overline{M}_{\phi\theta}}{\partial \theta} - (\overline{Q}_\phi - m_\phi)a \sin \phi = 0, \tag{9.1.1d}$$

$$\frac{\partial}{\partial \theta}(M_\theta) + 2\frac{\partial}{\partial \phi}(\overline{M}_{\phi\theta} \sin \phi) - (\overline{Q}_\theta - m_\theta)a \sin \phi = 0, \tag{9.1.1e}$$

where, for the spherical shell,

$$\overline{M}_{\phi\theta} = M_{\phi\theta} = M_{\theta\phi}, \tag{9.1.1f}$$

$$\overline{N}_{\phi\theta} = N_{\phi\theta} = N_{\theta\phi}. \tag{9.1.1g}$$

The force–strain relations are, with no approximations,

$$N_\phi = K(\epsilon_\theta + \nu\epsilon_\theta), \tag{9.1.2a}$$

$$N_\theta = K(\epsilon_\theta + \nu\epsilon_\phi), \tag{9.1.2b}$$

$$\overline{N}_{\phi\theta} = \frac{1 - \nu}{2} K\gamma_{\phi\theta}, \tag{9.1.2c}$$

$$M_\phi = D(\kappa_\phi + \nu\kappa_\theta), \tag{9.1.2d}$$

$$M_\theta = D(\kappa_\theta + \nu\kappa_\phi), \tag{9.1.2e}$$

$$\overline{M}_{\phi\theta} = D(1-\nu)\bar{\kappa}_{\phi\theta}. \tag{9.1.2f}$$

Finally, the strain–displacement relations are

$$\epsilon_\phi = \frac{1}{a}\left(\frac{\partial u_\phi}{\partial \phi} + w\right), \tag{9.1.3a}$$

$$\epsilon_\theta = \frac{1}{a}\left(\frac{1}{\sin\phi}\frac{\partial u_\theta}{\partial\theta} + u_\phi\cot\phi + w\right), \tag{9.1.3b}$$

$$\gamma_{\phi\theta} = \frac{1}{a}\left[\frac{1}{\sin\phi}\frac{\partial u_\phi}{\partial\theta} + \sin\phi\frac{\partial}{\partial\phi}\left(\frac{u_\theta}{\sin\phi}\right)\right], \tag{9.1.3c}$$

$$\kappa_\phi = -\frac{1}{a^2}\left(\frac{\partial^2 w}{\partial\phi^2} - \frac{\partial u_\phi}{\partial\phi}\right) = -\frac{1}{a^2}\left(\frac{\partial^2 w}{\partial\phi^2} + w\right) + \frac{\epsilon_\phi}{a}, \tag{9.1.3d}$$

$$\begin{aligned}\kappa_\theta &= -\frac{1}{a^2}\left[\frac{1}{\sin^2\phi}\frac{\partial^2 w}{\partial\theta^2} + \frac{\partial w}{\partial\phi}\cot\phi - \frac{1}{\sin\phi}\frac{\partial u_\theta}{\partial\theta} - u_\phi\cot\phi\right], \\ &= -\frac{1}{a^2}\left(\frac{1}{\sin^2\phi}\frac{\partial^2 w}{\partial\theta^2} + \frac{\partial w}{\partial\phi}\cot\phi + w\right) + \frac{\epsilon_\theta}{a},\end{aligned} \tag{9.1.3e}$$

$$\begin{aligned}\bar{\kappa}_{\phi\theta} &= -\frac{1}{a^2\sin\phi}\left[\frac{\partial^2 w}{\partial\phi\,\partial\theta} - \frac{\partial w}{\partial\theta}\cot\phi - \frac{1}{2}\frac{\partial u_\phi}{\partial\theta} - \frac{1}{2}\sin^2\phi\frac{\partial}{\partial\phi}\left(\frac{u_\theta}{\sin\phi}\right)\right] \\ &= -\frac{1}{a^2}\frac{\partial^2}{\partial\phi\,\partial\theta}\left(\frac{w}{\sin\phi}\right) + \frac{1}{2}\frac{\gamma_{\phi\theta}}{a}.\end{aligned} \tag{9.1.3f}$$

These equations may be reduced to a form similar to that of Vlasov's equation for the sphere in the following manner. From Eqs. (9.1.2) and (9.1.3) we can rewrite the force–strain relations for moments as

$$M_\phi = -\frac{D}{a^2}\left[\frac{\partial^2 w}{\partial\phi^2} + w + \nu\left(\frac{1}{\sin^2\phi}\frac{\partial^2 w}{\partial\theta^2} + \frac{\partial w}{\partial\phi}\cot\phi + w\right)\right] + \frac{h^2}{12a}N_\phi, \tag{9.1.4a}$$

$$M_\theta = -\frac{D}{a^2}\left[\frac{1}{\sin^2\phi}\frac{\partial^2 w}{\partial\theta^2} + \frac{\partial w}{\partial\phi}\cot\phi + w + \nu\left(\frac{\partial^2 w}{\partial\phi^2} + w\right)\right] + \frac{h^2}{12a}N_\theta, \tag{9.1.4b}$$

$$\overline{M}_{\phi\theta} = -\frac{D(1-\nu)}{a^2}\frac{\partial^2}{\partial\phi\,\partial\theta}\left(\frac{w}{\sin\phi}\right) + \frac{h^2}{12a}\overline{N}_{\phi\theta}, \tag{9.1.4c}$$

which are similar in form to those used in Section 6.1. By multiplying Eqs.

(9.1.1a) and (9.1.1b) by $h^2/12a$ and subtracting them respectively from Eqs. (9.1.1d) and (9.1.1e) we obtain the modified equilibrium equations

$$\left(1+\frac{h^2}{12a}\right)^{-1}\left\{\frac{\partial}{\partial\phi}\left[\left(M_\phi-\frac{h^2}{12a}N_\phi\right)\sin\phi\right]-\left(M_\theta-\frac{h^2}{12a}N_\theta\right)\cos\phi\right.$$
$$\left.+\frac{\partial}{\partial\theta}\left(\overline{M}_{\phi\theta}-\frac{h^2}{12a}\overline{N}_{\phi\theta}\right)+\left(m_\phi a-\frac{h^2}{12}q_\phi\right)\sin\phi\right\}$$
$$=\overline{Q}_\phi-\frac{1}{a\sin\phi}\frac{\partial\overline{M}_{\phi\theta}}{\partial\theta}=Q_\phi, \tag{9.1.5a}$$

$$\left(1+\frac{h^2}{12a^2}\right)^{-1}\left\{\frac{1}{\sin\phi}\frac{\partial}{\partial\theta}\left(M_\theta-\frac{h^2}{12a}N_\theta\right)+\frac{1}{\sin^2\phi}\frac{\partial}{\partial\phi}\right.$$
$$\left.\times\left[\left(\overline{M}_{\phi\theta}-\frac{h^2}{12a}\overline{N}_{\phi\theta}\right)\sin^2\phi\right]+m_\theta a-\frac{h^2}{12}q_\theta\right\}$$
$$=\overline{Q}_\theta-\frac{\partial\overline{M}_{\phi\theta}}{a\partial\phi}=Q_\theta. \tag{9.1.5b}$$

When Eqs. (9.1.4) are used, the transverse shear forces can be expressed in terms of surface loading and radial displacements as

$$Q_\phi=\left(1+\frac{h^2}{12a^2}\right)^{-1}\left[m_\phi-\frac{h^2}{12a}q_\phi-\frac{D}{a}\frac{\partial}{\partial\phi}\left(\nabla^2+\frac{2}{a^2}\right)w\right], \tag{9.1.6a}$$

$$Q_\theta=\left(1+\frac{h^2}{12a^2}\right)^{-1}\left[m_\theta-\frac{h^2}{12a}q_\theta-\frac{D}{a\sin\phi}\frac{\partial}{\partial\theta}\left(\nabla^2+\frac{2}{a^2}\right)w\right], \tag{9.1.6b}$$

where the operator ∇^2 is defined by

$$\nabla^2(\ldots)=\frac{1}{a^2}\left[\frac{\partial^2(\ldots)}{\partial\phi^2}+\cot\phi\frac{\partial(\ldots)}{\partial\phi}+\frac{1}{\sin^2\phi}\frac{\partial^2(\ldots)}{\partial\theta^2}\right]. \tag{9.1.6c}$$

The substitution of Eqs. (9.1.6) into Eqs. (9.1.1) then yields the following three equations for the determination of the direct forces:

$$\frac{\partial}{\partial\phi}(N_\phi\sin\phi)-N_\theta\cos\phi+\frac{\partial\overline{N}_{\phi\theta}}{\partial\theta}=\left(1+\frac{h^2}{12a^2}\right)^{-1}\sin\phi$$
$$\times\left[\frac{D}{a}\frac{\partial}{\partial\phi}\left(\nabla^2+\frac{2}{a^2}\right)w-(m_\phi+q_\phi a)\right], \tag{9.1.7a}$$

$$\frac{\partial N_\theta}{\partial\theta}+\frac{1}{\sin\phi}\frac{\partial}{\partial\phi}(\overline{N}_{\phi\theta}\sin^2\phi)$$
$$=\left(1+\frac{h^2}{12a^2}\right)^{-1}\sin\phi\left[\frac{D}{a\sin\phi}\frac{\partial}{\partial\theta}\left(\nabla^2+\frac{2}{a^2}\right)w-(m_\theta+q_\theta a)\right], \tag{9.1.7b}$$

$$N_\phi + N_\theta = q_\zeta a + \left(1 + \frac{h^2}{12a^2}\right)^{-1} \left\langle \frac{1}{\sin\phi}\left\{\frac{\partial}{\partial\phi}\left[\left(m_\phi - \frac{h^2}{12a} q_\phi\right)\sin\phi\right] + \frac{\partial}{\partial\theta}\left(m_\theta - \frac{h^2}{12a} q_\theta\right)\right\} - aD\nabla^2\left(\nabla^2 + \frac{2}{a^2}\right) w \right\rangle. \qquad (9.1.7c)$$

The left sides of Eqs. (9.1.7) are identical in form to the equations of the membrane theory of shells. We may satisfy Eqs. (9.1.7a) and (9.1.7b) by expressing the forces in terms of a stress function S, the radial displacement w, and a particular solution as†

$$N_\phi = \frac{1}{a^2}\left(\frac{1}{\sin^2\phi}\frac{\partial^2 S}{\partial\theta^2} + \cot\phi\frac{\partial S}{\partial\phi} + S\right) + \left(1 + \frac{h^2}{12a^2}\right)^{-1}\frac{D}{a}\left(\nabla^2 + \frac{2}{a^2}\right) w + N_\phi^p, \qquad (9.1.8a)$$

$$N_\theta = \frac{1}{a^2}\left(\frac{\partial^2 S}{\partial\phi^2} + S\right) + \left(1 + \frac{h^2}{12a^2}\right)^{-1}\frac{D}{a}\left(\nabla^2 + \frac{2}{a^2}\right) w + N_\theta^p, \qquad (9.1.8b)$$

$$\overline{N}_{\phi\theta} = -\frac{1}{a^2}\frac{\partial^2}{\partial\phi\,\partial\theta}\left(\frac{S}{\sin\phi}\right) + N_{\phi\theta}^p, \qquad (9.1.8c)$$

where N_ϕ^p, N_θ^p, and $\overline{N}_{\phi\theta}^p$ are particular solutions of the membrane force equations‡

$$\frac{\partial}{\partial\phi}(N_\phi^p\sin\phi) - N_\theta^p\cos\phi + \frac{\partial\overline{N}_{\phi\theta}^p}{\partial\theta} = -\left(1 + \frac{h^2}{12a^2}\right)^{-1}(m_\phi + q_\phi a)\sin\phi, \qquad (9.1.9a)$$

$$\frac{\partial N_\theta^p}{\partial\theta} + \frac{1}{\sin\phi}\frac{\partial}{\partial\phi}(\overline{N}_{\phi\theta}^p\sin^2\phi) = -\left(1 + \frac{h^2}{12a^2}\right)^{-1}(m_\theta + q_\theta a)\sin\phi, \qquad (9.1.9b)$$

$$N_\phi^p + N_\theta^p = q_\zeta a + \frac{1}{\sin\phi}\left(1 + \frac{h^2}{12a^2}\right)^{-1} \times \left\{\frac{\partial}{\partial\phi}\left[\left(m_\phi - \frac{h^2}{12a} q_\phi\right)\sin\phi\right] + \frac{\partial}{\partial\theta}\left(m_\theta - \frac{h^2}{12a} q_\theta\right)\right\}. \qquad (9.1.9c)$$

† This representation partially breaks down when the displacements and forces vary as $f_1(\phi)\sin\theta + f_2(\phi)\cos\theta$. This special case will be treated separately.

‡ We can, of course, define the particular solution with q_ζ deleted from the equation as in Vlasov's equations. There does not appear to be any advantage in using this other than a closer resemblance to the theory of plates. On the other hand the above definition corresponds to the membrane theory of shells.

The third of the equations, Eq. (9.1.7c), is satisfied if the radial deformation w and the stress function S are related by the equation

$$\left(\nabla^2 + \frac{2}{a^2}\right)\left[D\left(1 + \frac{h^2}{12a^2}\right)^{-1}\left(\nabla^2 + \frac{2}{a^2}\right) w + \frac{S}{a}\right] = 0. \tag{9.1.10}$$

A second equation relating S and w comes from consideration of compatibility of strains. The functions S and w define the stress resultants N_ϕ, N_θ, and $\bar{N}_{\phi\theta}$. From Eqs. (9.1.2) and (9.1.3) we then have three equations for the determination of the remaining displacements u_ϕ and u_θ. For these three equations to be compatible, the following relationship, obtained by eliminating u_ϕ and u_θ from Eqs. (9.1.2) and (9.1.3), must be satisfied

$$\begin{aligned}
&\frac{1}{a}\left[\frac{1}{\sin^2\phi}\frac{\partial^2}{\partial\phi\,\partial\theta}(\gamma_{\phi\theta}\sin\phi) - \frac{1}{\sin^2\phi}\frac{\partial^2\epsilon_\phi}{\partial\theta^2}\right.\\
&\qquad\left. + \cot\phi\frac{\partial\epsilon_\phi}{\partial\phi} - 2\epsilon_\phi - \frac{\partial^2\epsilon_\theta}{\partial\phi^2} - 2\cot\phi\frac{\partial\epsilon_\theta}{\partial\phi}\right]\\
&\quad = -\left(\nabla^2 + \frac{2}{a^2}\right) w\\
&\quad = -\frac{1}{Eh/a}\left\{\left[\nabla^2 + \frac{1-\nu}{a^2}\right](N_\phi + N_\theta) - \frac{1+\nu}{a^2\sin\phi}\frac{\partial}{\partial\phi}\right.\\
&\qquad \times\left[\frac{\partial(N_\phi\sin\phi)}{\partial\phi} - N_\theta\cos\phi + \frac{\partial\bar{N}_{\phi\theta}}{\partial\theta}\right]\\
&\qquad \left. - \frac{1+\nu}{a^2\sin^2\phi}\frac{\partial}{\partial\theta}\left[\frac{1}{\sin\phi}\frac{\partial}{\partial\phi}(\bar{N}_{\phi\theta}\sin^2\phi) + \frac{\partial N_\theta}{\partial\theta}\right]\right\}.
\end{aligned} \tag{9.1.11a}$$

With the aid of Eqs. (9.1.8), (9.1.9), and (9.1.10) we then may rewrite Eq. (9.1.11a) as

$$\begin{aligned}
&\left(\nabla^2 + \frac{2}{a^2}\right)\left\langle \nabla^2 S - \frac{Eh}{a} w + q_\zeta a + \left(1 + \frac{h^2}{12a^2}\right)^{-1}\frac{1}{\sin\phi}\right.\\
&\qquad\left. \times\left\{\frac{\partial}{\partial\phi}\left[\left(m_\phi - \frac{h^2}{12a}q_\phi\right)\sin\phi\right] + \frac{\partial}{\partial\theta}\left(m_\theta - \frac{h^2}{12a}q_\theta\right)\right\}\right\rangle\\
&\quad = \frac{1+\nu}{a}\left\{q_\zeta - \frac{1}{\sin\phi}\left[\frac{\partial(q_\phi\sin\phi)}{\partial\phi} + \frac{\partial q_\theta}{\partial\theta}\right]\right\}.
\end{aligned} \tag{9.1.11b}$$

Eqs. (9.1.10) and (9.1.11b) are the exact representation of the theory of spherical shells in terms of the radial displacement of the middle surface and a stress function. To complete the theory we can express the two equations as a single complex equation in terms of a single complex variable by multiplying Eq. (9.1.11b) by the constant

$$k = \frac{1 - 2i\chi^2}{4\chi^4 + 1} a \tag{9.1.12}$$

and adding the result to Eq. (9.1.10) to yield

$$\begin{aligned}\left(\nabla^2 + \frac{2}{a^2}\right)\Big\langle D\left(\nabla^2 + \frac{1 + 2i\chi^2}{a^2}\right)\left[w + (1 - 2i\chi^2)\frac{S}{Eha}\right] + a\frac{1 - 2i\chi^2}{4\chi^4 + \nu^2} \\ \times \left\{q_\zeta a + \frac{(1 + h^2/12a^2)^{-1}}{\sin\phi}\left[\frac{\partial}{\partial\phi}\left(m_\phi - \frac{h^2}{12a}q_\phi\right)\sin\phi\right]\right. \\ \left. + \frac{1}{\sin\phi}\frac{\partial}{\partial\theta}\left(m_\theta - \frac{h^2}{12a}q_\theta\right)\right\}\Big\rangle \\ = (1 + \nu)\frac{1 - 2i\chi^2}{4\chi^4 + \nu^2}\left\{q_\zeta - \frac{1}{\sin\phi}\left[\frac{\partial(q_\phi \sin\phi)}{\partial\phi} + \frac{\partial q_\theta}{\partial\theta}\right]\right\},\end{aligned} \tag{9.1.13}$$

which is the desired equation. We shall defer a discussion of the calculation of the displacements u_ϕ and u_θ.

9.2 'Wind' loading of spherical shells

When the surface loading, the middle surface forces, bending moments, and displacements vary as $\cos\theta$ or $\sin\theta$ around the circumference, the case of so-called 'wind-loading,' the foregoing derivation must be modified. We return to Eqs. (9.1.7) and start afresh. Let us assume that the radial displacements, the middle surface forces, and the surface loading are expressed as

$$w = \hat{w}(\phi)\begin{cases}\cos\theta \\ \sin\theta,\end{cases} \tag{9.2.1a}$$

$$N_\phi = \hat{N}_\phi(\phi)\begin{cases}\cos\theta \\ \sin\theta,\end{cases} \tag{9.2.1b}$$

$$N_\theta = \hat{N}_\theta(\phi)\begin{cases}\cos\theta \\ \sin\theta,\end{cases} \tag{9.2.1c}$$

$$\bar{N}_{\phi\theta} = \hat{\bar{N}}_{\theta\phi}(\phi)\begin{cases}\sin\theta \\ \cos\theta,\end{cases} \tag{9.2.1d}$$

$$m_\phi = \hat{m}_\phi(\phi)\begin{cases}\cos\theta \\ \sin\theta,\end{cases} \tag{9.2.1e}$$

$$m_\theta = \hat{m}_\theta(\phi)\begin{cases}\sin\theta \\ \cos\theta,\end{cases} \tag{9.2.1f}$$

$$q_\phi = \hat{q}_\phi(\phi)\begin{cases}\cos\theta\\ \sin\theta,\end{cases} \tag{9.2.1g}$$

$$q_\theta = \hat{q}_\theta(\phi)\begin{cases}\sin\theta\\ \cos\theta.\end{cases} \tag{9.2.1h}$$

Then Eqs. (9.1.7a) and (9.1.7b) become

$$\frac{\mathrm{d}(\hat{N}_\phi \sin\phi)}{\mathrm{d}\phi} - \hat{N}_\theta\cos\phi \pm \hat{\bar{N}}_{\phi\theta}$$
$$= \left(1 + \frac{h^2}{12a^2}\right)^{-1} \sin\phi\left[\frac{D}{a}\frac{\mathrm{d}}{\mathrm{d}\phi}\left(\nabla_\phi^2 + \frac{2}{a^2}\right)\hat{w} - (\hat{m}_\phi + \hat{q}_\phi a)\right], \tag{9.2.2a}$$

$$\mp\hat{N}_\theta + \frac{1}{\sin\phi}\frac{\mathrm{d}}{\mathrm{d}\phi}(\hat{\bar{N}}_{\phi\theta}\sin^2\phi)$$
$$= \left(1 + \frac{h^2}{12a^2}\right)^{-1} \sin\phi\left[\mp\frac{D}{a\sin\phi}\left(\nabla_\phi^2 + \frac{2}{a^2}\right)\hat{w} - (\hat{m}_\theta + \hat{q}_\theta a)\right], \tag{9.2.2b}$$

where the upper and lower signs refer to the upper and lower variations of θ in Eqs. (9.2.1) and

$$\nabla_\phi^2(\ldots) = \frac{1}{a^2}\left[\frac{\mathrm{d}^2(\ldots)}{\mathrm{d}\phi^2} + \cot\phi\frac{\mathrm{d}(\ldots)}{\mathrm{d}\phi} - \frac{(\ldots)}{\sin^2\phi}\right]. \tag{9.2.2c}$$

By eliminating $\hat{N}_\theta$ from Eqs. (9.2.2a) and (9.2.2b) we obtain

$$\frac{\mathrm{d}}{\mathrm{d}\phi}[(\hat{N}_\phi \mp \hat{\bar{N}}_{\phi\theta}\cos\phi)\sin\phi]$$
$$= \left(1 + \frac{h^2}{12a^2}\right)^{-1}\left\{\frac{D}{a}\frac{\mathrm{d}}{\mathrm{d}\phi}\left[\sin\phi\left(\nabla_\phi^2 + \frac{2}{a^2}\right)\hat{w}\right]\right.$$
$$\left. - [\hat{m}_\phi + \hat{q}_\phi a \mp (\hat{m}_\theta + \hat{q}_\theta a)\cos\phi]\sin\phi\right\}, \tag{9.2.3a}$$

which can be integrated to yield

$$\hat{N}_\phi \mp \hat{\bar{N}}_{\phi\theta}\cos\phi = \frac{C_1}{\sin\phi} + \left(1 + \frac{h^2}{12a^2}\right)^{-1}$$
$$\times\left\{\frac{D}{a}\left(\nabla_\phi^2 + \frac{2}{a^2}\right)\hat{w}\right.$$
$$\left. - \frac{1}{\sin\phi}\int_0^\phi [\hat{m}_\phi + \hat{q}_\phi a \mp (\hat{m}_\theta + \hat{q}_\phi a)\cos\phi]\sin\phi\,\mathrm{d}\phi\right\} \tag{9.2.3b}$$

Eq. (9.2.3b) is used in place of Eq. (9.2.2a). The elimination of N_ϕ and N_θ from Eqs. (9.1.7c), (9.2.2b), and (9.2.3b) results in an equation for $\hat{N}_{\phi\theta}$ in terms of the radial deformation w and the surface loading which can be integrated to yield

$$\begin{aligned}\mp \hat{N}_{\phi\theta} = & -\frac{1}{\sin^3\phi}\Bigg\langle C_1\cos\phi + C_2 \\ & + \int_0^\phi (\hat{q}_\zeta \sin\phi + \hat{q}_\phi \cos\phi \mp \hat{q}_\theta) a \sin\phi \, \mathrm{d}\phi \\ & - \left(1 + \frac{h^2}{12a^2}\right)^{-1}\left\{\cos\phi \int_0^\phi [\hat{m}_\phi + \hat{q}_\phi a \right. \\ & \mp (\hat{m}_\theta + \hat{q}_\theta a)\cos\phi] \sin\phi \, \mathrm{d}\phi \\ & \left. - \sin^2\phi \left(\hat{m}_\phi - \frac{h^2}{12a}\hat{q}_\phi\right)\right\}\Bigg\rangle \\ & + \left(1 + \frac{h^2}{12a^2}\right)^{-1}\frac{D}{a}\frac{\mathrm{d}}{\mathrm{d}\phi}\left[\frac{1}{\sin\phi}\left(\nabla_\phi^2 + \frac{2}{a^2}\right)\hat{w}\right]. \end{aligned} \tag{9.2.4a}$$

From Eqs. (9.2.2b) and (9.2.3b) we can obtain similar expressions for N_ϕ and N_θ as

$$\begin{aligned}\hat{N}_\phi = & \frac{1}{\sin^3\phi}\Bigg\langle C_1 + C_2\cos\phi - \left(1 + \frac{h^2}{12a^2}\right)^{-1} \\ & \times \left\{\int_0^\phi [\hat{m}_\phi + \hat{q}_\phi a \mp (\hat{m}_\theta + \hat{q}_\theta a)\cos\phi]\sin\phi\,\mathrm{d}\phi \right. \\ & \left. - \left(\hat{m}_\phi - \frac{h^2}{12a}\hat{q}_\phi\right)\sin^2\phi\cos\phi\right\} \\ & + \cos\phi\int_0^\phi (\hat{q}_\zeta\sin\phi + \hat{q}_\phi\cos\phi \mp \hat{q}_\theta) a \sin\phi\,\mathrm{d}\phi\Bigg\rangle \\ & - \left(1 + \frac{h^2}{12a^2}\right)^{-1}\frac{D}{a}\frac{\mathrm{d}}{\mathrm{d}\phi}\left[\cot\phi\left(\nabla_\phi^2 + \frac{2}{a^2}\right)\hat{w}\right), \end{aligned} \tag{9.2.4b}$$

$$\begin{aligned}\hat{N}_\theta = & \hat{q}_\zeta a - \left(1 + \frac{h^2}{12a^2}\right)^{-1}\left[\frac{D}{a}\frac{\mathrm{d}^2}{\mathrm{d}\phi^2}\left(\nabla_\phi^2 + \frac{2}{a^2}\right)\hat{w} - \frac{\mathrm{d}}{\mathrm{d}\phi}\left(\hat{m}_\phi - \frac{h^2}{12a}\hat{q}_\phi\right)\right] \\ & - \frac{1}{\sin^3\phi}\Bigg\langle C_1 + C_2\cos\phi + \cos\phi\int_0^\phi (\hat{q}_\zeta\sin\phi \\ & + \hat{q}_\phi\cos\phi \mp \hat{q}_\theta) a\sin\phi\,\mathrm{d}\phi - \left(1 + \frac{h^2}{12a^2}\right)^{-1} \\ & \times \int_0^\phi [\hat{m}_\phi + \hat{q}_\phi a \mp (\hat{m}_\theta + \hat{q}_\theta a)\cos\phi]\sin\phi\,\mathrm{d}\phi \\ & + \left(\hat{m}_\theta - \frac{h^2}{12a}\hat{q}_\theta\right)\sin^2\phi\Big\}\Bigg\rangle. \end{aligned} \tag{9.2.4c}$$

The compatibility condition which must be satisfied is somewhat simpler than Eq. (9.1.11a) and is given by

$$\sin\phi \frac{d}{d\phi}(\hat{\epsilon}_\theta) - \hat{\epsilon}_\phi \cos\phi \mp \hat{\gamma}_{\phi\theta} = \frac{1}{a}\sin^2\phi \frac{d}{d\phi}\left(\frac{\hat{w}}{\sin\phi}\right)$$
$$= \frac{1}{Eh}\left\{\sin^2\phi \frac{d}{d\phi}\left(\frac{\hat{N}_\phi + \hat{N}_\theta}{\sin\phi}\right) - (1+\nu)\right.$$
$$\times \left[\frac{d}{d\phi}(\hat{N}_\phi \sin\phi) - \hat{N}_\theta \cos\phi \pm \hat{N}_{\phi\theta}\right.$$
$$\left.\left. - (\hat{N}_\phi \cos\phi \mp \hat{N}_{\phi\theta})\right]\right\}. \tag{9.2.5a}$$

The use of Eqs. (9.1.7c) (9.2.2a), (9.2.4a) and (9.2.4b) then yields, after integration, the following fourth-order differential equation for w

$$\left[\frac{1}{4\chi^4+1} a^2\nabla_\phi^2(a^2\nabla_\phi^2 + 2) + 1\right]\hat{w}$$
$$= \frac{1}{Eh/a}\left\langle \hat{q}_\zeta a + \left(1 + \frac{h^2}{12a^2}\right)^{-1} \frac{1}{\sin\phi}\left\{\frac{d}{d\phi}\left[\left(\hat{m}_\phi - \frac{h^2}{12a}\hat{q}_\phi\right)\sin\phi\right]\right.\right.$$
$$\left.\mp\left(\hat{m}_\theta - \frac{h^2}{12a}\hat{q}_\theta\right)\right\} - (1+\nu)\left[\frac{1}{2}C_2\left(\sin\phi \ln\tan\frac{\phi}{2} - \cot\phi\right)\right.$$
$$- \sin\phi \int_0^\phi \frac{q_\phi a}{\sin\phi}\, d\phi + \sin\phi \int_0^\phi \frac{d\phi}{\sin^3\phi}$$
$$\left.\left.\times \int_0^\phi (\hat{q}_\zeta \sin\phi + \hat{q}_\phi \cos\phi \mp \hat{q}_\theta) a \sin\phi \, d\phi\right]\right\rangle + C_3 \sin\phi, \tag{9.2.5b}$$

which replaces Eq. (9.1.13).

9.3 Edge-loaded spherical shells: inextensional deformations and membrane stress states

For a spherical shell loaded only at its edges or by concentrated surface loads, Eq. (9.1.13) reduces to

$$\left(\nabla^2 + \frac{2}{a^2}\right)\left(\nabla^2 + \frac{1 - 2i\chi^2}{a^2}\right)\left[w + (1 + 2i\chi^2)\frac{S}{Eha}\right] = 0. \tag{9.3.1}$$

Since the operators in parentheses are commutative, the solutions of Eq. (9.3.1) are the sum of the solutions of the two second-order equations

$$\left(\nabla^2 + \frac{2}{a^2}\right)\left[w + (1 + 2i\chi^2)\frac{S}{Eha}\right] = 0, \tag{9.3.2a}$$

$$\left(\nabla^2 + \frac{1 - 2i\chi^2}{a^2}\right)\left[w + (1 + 2i\chi^2)\frac{S}{Eha}\right] = 0. \tag{9.3.2b}$$

In this section we shall deal only with solutions of Eq. (9.3.2a). This equation is interesting in that it reveals the possibility of states of stress associated with no radial deformation or with no middle surface forces. Since the operator $\nabla^2 + 2/a^2$ is real, Eq. (9.3.2a) is equivalent to the two equations

$$\left(\nabla^2 + \frac{2}{a^2}\right) w = \left(\nabla^2 + \frac{2}{a^2}\right) S = 0, \tag{9.3.3}$$

which must both be satisfied simultaneously. This is obviously possible if we have

$$w = 0, \qquad \left(\nabla^2 + \frac{2}{a^2}\right) S = 0, \tag{9.3.4a}$$

or

$$S = 0, \qquad \left(\nabla^2 + \frac{2}{a^2}\right) w = 0. \tag{9.3.4b}$$

The state of stress defined by Eq. (9.3.4b) corresponds to what is called a state of inextensional deformation. If S vanishes, then the middle surface forces N_ϕ, N_θ, and $\overline{N}_{\phi\theta}$ vanish, and from Eqs. (9.1.2) the middle surface strains ϵ_ϕ, ϵ_θ, and $\gamma_{\phi\theta}$ also vanish. Thus deformation occurs with no stretching of the middle surface and only bending and twisting moments are present in the shell wall.

The state of stress defined by Eq. (9.3.4a) has no simple physical interpretation. It is possible to adjust the solutions, however, in such a manner that the state of stress defined by Eq. (9.3.4a) is replaced by a pure membrane stress state. From Eqs. (9.3.3) we have

$$\frac{1}{a^2}\left(\frac{\partial^2 w}{\partial \phi^2} + w\right) = -\frac{1}{a^2}\left(\cot\phi\,\frac{\partial w}{\partial \phi} + \frac{1}{\sin^2\phi}\,\frac{\partial^2 w}{\partial \theta^2} + w\right), \tag{9.3.5a}$$

$$\frac{1}{a^2}\left(\frac{\partial^2 S}{\partial \phi^2} + S\right) = -\frac{1}{a^2}\left(\cot\phi\,\frac{\partial S}{\partial \phi} + \frac{1}{\sin^2\phi}\,\frac{\partial^2 S}{\partial \theta^2} + S\right). \tag{9.3.5b}$$

Then the middle surface forces may be expressed, with the use of Eqs. (9.1.8), as

$$N_\theta = -N_\phi = \frac{1}{a^2}\left(\frac{\partial^2 S}{\partial \phi^2} + S\right) = -\frac{1}{a^2}\left(\cot\phi\,\frac{\partial S}{\partial \phi} + \frac{1}{\sin^2\phi}\,\frac{\partial^2 S}{\partial \theta^2} + S\right), \tag{9.3.6a}$$

$$\overline{N}_{\phi\theta} = -\frac{1}{a^2}\frac{\partial^2}{\partial\phi\,\partial\theta}\left(\frac{S}{\sin\phi}\right). \tag{9.3.6b}$$

Eqs. (9.1.4), (9.3.5), and (9.3.6) then yield

$$\begin{aligned} M_\theta = -M_\phi &= \frac{(1-\nu)D}{a^2}\left(\frac{\partial^2}{\partial\phi^2}+1\right)\left[w+\frac{(1+\nu)S}{Eha}\right] \\ &= -\frac{(1-\nu)D}{a^2}\left(\cot\phi\frac{\partial}{\partial\phi}+\frac{1}{\sin^2\phi}\frac{\partial^2}{\partial\theta^2}+1\right)\left[w+\frac{(1+\nu)S}{Eha}\right] \end{aligned} \tag{9.3.7a}$$

$$\overline{M}_{\phi\theta} = M_{\phi\theta} = -\frac{(1-\nu)D}{a^2}\frac{\partial^2}{\partial\phi\,\partial\theta}\left[\frac{w+(1+\nu)\dfrac{S}{Eha}}{\sin\phi}\right]. \tag{9.3.7b}$$

Obviously, the moments will vanish if the radial deflections are defined by

$$w = -\frac{(1+\nu)S}{Eha}, \tag{9.3.8}$$

which is then the radial deformation corresponding to a pure membrane state of stress.

Let us investigate the nature of the inextensional and membrane solutions. The simplest cases occur when the shell closes upon itself in the circumferential direction, is cut along parallel circles, and the load is applied only at the edges. In this instance, the deformations or the stress function can be expanded in the form

$$w, S = \sum_{n=0}^{\infty} a_n(\phi)\cos n\theta + \sum_{n=1}^{\infty} b_n(\phi)\sin n\theta. \tag{9.3.9}$$

The case $n = 0$ has already been discussed in Chapter 6 while the case $n = 1$ requires special considerations. If we restrict ourselves to n greater than or equal to 2, the substitution of Eq. (9.3.9) into Eq. (9.3.3) yields the following differential equations for the coefficients

$$\frac{d^2p_n}{d\phi^2} + \cot\phi\frac{dp_n}{d\phi} + \left(2 - \frac{n^2}{\sin^2\phi}\right)p_n = 0 \qquad (n \geq 2), \tag{9.3.10a}$$

where p_n stands for either a_n or b_n. The solutions of Eq. (9.3.10a) can be readily obtained when we note that the transformation $x = \cos\phi$ will yield the standard form of the equation for the associated Legendre spherical harmonics $P_1^n(x)$ and $Q_1^n(x)$. Then the solution for p_n may be written as†

† W. Magnus, F. Oberhettinger, and R. P. Soni, *Formulas and Theorems for the Special Functions of Mathematical Physics*, Second Edition, Springer-Verlag, New York, Inc., 1966, pp. 166–169.

$$p_n = p_{1n}(n + \cos\phi)\tan^n\frac{\phi}{2} + p_{2n}(n - \cos\phi)\cot^n\frac{\phi}{2} \qquad (n \geq 2), \tag{9.3.10b}$$

where p_{1n} and p_{2n} are arbitrary constants.

In the case of a pure membrane state of stress let us write the expression for the stress function as

$$S_n = \left[A_n(n + \cos\phi)\tan^n\frac{\phi}{2} + B_n(n - \cos\phi)\cot^n\frac{\phi}{2}\right]\begin{Bmatrix}\cos n\theta\\ \sin n\theta\end{Bmatrix} \qquad (n \geq 2). \tag{9.3.11}$$

From Eq. (9.3.8), the radial displacement is given by

$$w_n = -\frac{(1+\nu)}{Eha}S_n. \tag{9.3.12}$$

The middle surface forces are obtained from Eqs. (9.3.6) and (9.3.11) as

$$N_{\phi n} = -N_{\theta n} = -\frac{n(n^2-1)}{a^2\sin^2\phi}\left(A_n\tan^n\frac{\phi}{2} + B_n\cot^n\frac{\phi}{2}\right)\begin{Bmatrix}\cos n\theta\\ \sin n\theta\end{Bmatrix} \qquad (n \geq 2), \tag{9.3.13a}$$

$$\overline{N}_{\phi\theta n} = \mp\frac{n(n^2-1)}{a^2\sin^2\phi}\left(A_n\tan^n\frac{\phi}{2} - B_n\cot^n\frac{\phi}{2}\right)\begin{Bmatrix}\sin n\theta\\ \cos n\theta\end{Bmatrix} \qquad (n \geq 2), \tag{9.3.13b}$$

and the remaining stress resultants vanish. The tangential displacements are defined by Eqs. (9.1.2a), (9.1.2b), (9.1.3a), and (9.1.3b) as

$$\frac{\partial u_\phi}{\partial\phi} = \frac{(1+\nu)a}{Eh}N_\phi - w, \tag{9.3.14a}$$

$$\frac{\partial u_\theta}{\partial\theta} = -\frac{(1+\nu)aN_\phi\sin\phi}{Eh} - (w\sin\phi + u_\phi\cos\phi), \tag{9.3.14b}$$

which yield

$$u_{\phi n} = -\frac{1+\nu}{Eha}\left\{A_n\left[\frac{n(n+\cos\phi)}{\sin\phi} - \sin\phi\right]\tan^n\frac{\phi}{2} - B_n\left[\frac{n(n-\cos\phi)}{\sin\phi} - \sin\phi\right]\cot^n\frac{\phi}{2}\right\}\begin{Bmatrix}\cos n\theta\\ \sin n\theta,\end{Bmatrix} \tag{9.3.15a}$$

$$u_{\theta n} = \pm\frac{1+\nu}{Eha}\frac{n}{\sin\phi}\left[A_n(n+\cos\phi)\tan^n\frac{\phi}{2} + B_n(n-\cos\phi)\cot^n\frac{\phi}{2}\right]\begin{Bmatrix}\sin n\theta\\ \cos n\theta.\end{Bmatrix} \tag{9.3.15b}$$

The condition that Eqs. (9.3.13b) and (9.3.15) must satisfy Eq. (9.1.2c) indicates that there can be no additional constants of integration. The above solutions represent the distribution of middle surface forces and displacements in spherical caps with the edges loaded by self-equilibrating middle surface forces in the manner defined by Eqs. (9.3.13). The solutions multiplied by A_n decay as ϕ decreases and are valid for a cap which includes the pole $\phi = 0$, while the solutions multiplied by B_n decay as ϕ increases and are valid for the remainder of the sphere. As an example, the variation with ϕ and n of the functions $[\tan^n (\phi/2)]/[\sin^2 \phi]$ entering into the expression for the middle surface force is shown in Fig. 9.2. We see that as n increases the

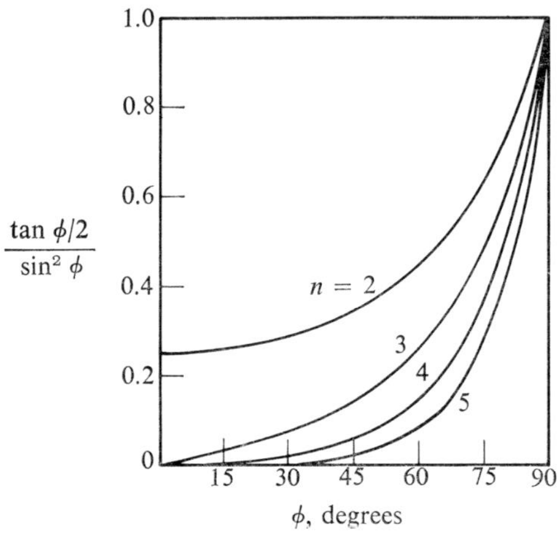

Fig. 9.2. Variation of membrane stress state forces and of inextensional deformation state moments

rate of decay of the functions also increases, so that for large n the effects of an edge-loading membrane state solution are confined, like the edge loading solution for axisymmetric deformations, to a region immediately in the vicinity of the edge. For a shell loaded along two parallel circles, both solutions apply.

For inextensional deformations let us write the expression for the radial deformations in the form

$$w_n = \left[C_n(n + \cos\phi)\tan^n\frac{\phi}{2} + D_n(n - \cos\phi)\cot^n\frac{\phi}{2}\right]\begin{Bmatrix}\cos n\theta\\ \sin n\theta\end{Bmatrix} \quad (n \geq 2). \quad (9.3.16)$$

The tangential displacements are then obtained from the equations

$$\epsilon_\phi = \epsilon_\theta = \gamma_{\phi\theta} = 0, \quad (9.3.17)$$

which yield

$$u_{\phi n} = -\sin\phi\left(C_n \tan^n \frac{\phi}{2} - D_n \cot^n \frac{\phi}{2}\right)\begin{Bmatrix}\cos n\theta \\ \sin n\theta\end{Bmatrix} \qquad (n \geq 2), \qquad (9.3.18a)$$

$$u_{\theta n} = \mp\sin\phi\left(C_n \tan^n \frac{\phi}{2} + D_n \cot^n \frac{\phi}{2}\right)\begin{Bmatrix}\sin n\theta \\ \cos n\theta\end{Bmatrix} \qquad (n \geq 2). \qquad (9.3.18b)$$

The moments in the shell wall are given by

$$\begin{aligned} M_{\phi n} = -M_{\theta n} &= \frac{(1-\nu)D}{a^2}\left(\frac{\partial w_n}{\partial \phi}\cot\phi + w_n + \frac{1}{\sin^2\phi}\frac{\partial^2 w_n}{\partial \theta^2}\right) \\ &= -\frac{n(n^2-1)(1-\nu)D}{a^2\sin^2\phi}\left(C_n \tan^n \frac{\phi}{2} + D_n \cot^n \frac{\phi}{2}\right)\begin{Bmatrix}\cos n\theta \\ \sin n\theta\end{Bmatrix} \\ &\qquad (n \geq 2), \qquad (9.3.19a) \end{aligned}$$

$$\begin{aligned} \overline{M}_{\phi\theta n} = M_{\phi\theta n} &= -\frac{(1-\nu)D}{a^2}\frac{\partial^2}{\partial\phi\,\partial\theta}\left(\frac{w_n}{\sin\phi}\right) \\ &= \pm\frac{n(n^2-1)(1-\nu)D}{a^2\sin^2\phi}\left(C_n \tan^n \frac{\phi}{2} - D_2 \cot^n \frac{\phi}{2}\right)\begin{Bmatrix}\sin n\theta \\ \cos n\theta\end{Bmatrix} \\ &\qquad (n \geq 2) \qquad (9.3.19b) \end{aligned}$$

and the remaining stress resultants vanish. The above solutions represent the distribution of displacements and moments in spherical caps with the edges loaded by self-equilibrating bending and twisting moments in the manner defined by Eqs. (9.3.19).

For comparable deformations of the spherical shell the stresses corresponding to a state of inextensional deformations are much smaller than those corresponding to a membrane state of stress. For instance, suppose edge moments were imposed on a shell subjected to edge forces yielding a membrane stress state in such a manner as to make the radial deformations vanish entirely. From Eqs. (9.3.7) and (9.3.8) we find that the distributions of moments are then given by

$$M_{\phi i} = -M_{\theta i} = \frac{h^2}{12a} N_{\phi m} = -\frac{h^2}{12a} N_{\theta m}, \qquad (9.3.20a)$$

$$\overline{M}_{\phi\theta i} = \frac{h^2}{12a} \overline{N}_{\phi\theta m}. \qquad (9.3.20b)$$

Thus the stresses due to bending moments are $h/2a$ times the stresses due to middle surface forces. Stresses of this order of magnitude are easily possible accidental perturbations of membrane state edge loading so that unless edge deformations are controlled a spherical shell may have deformations that are much larger than expected.

We must treat the case $n = 1$ separately since the equations given above are not completely valid. From Eqs. (9.2.4) we have, for no surface loading,

$$\hat{N}_\phi = \frac{C_1 + C_2 \cos\phi}{\sin^3\phi} - \left(1 + \frac{h^2}{12a^2}\right)^{-1} \frac{D}{a} \frac{\mathrm{d}}{\mathrm{d}\phi} \times \left[\cot\phi \left(\nabla_\phi^2 + \frac{2}{a^2}\right) \hat{w}\right], \tag{9.3.21a}$$

$$\hat{N}_\theta = -\frac{C_1 + C_2 \cos\phi}{\sin^3\phi} - \left(1 + \frac{h^2}{12a^2}\right)^{-1} \frac{D}{a} \frac{\mathrm{d}^2}{\mathrm{d}\phi^2} \left(\nabla_\phi^2 + \frac{2}{a^2}\right) \hat{w}, \tag{9.3.21b}$$

$$\mp \hat{N}_{\phi\theta} = -\frac{C_1 \cos\phi + C_2}{\sin^3\phi} + \left(1 + \frac{h^2}{12a^2}\right)^{-1} \frac{D}{a} \frac{\mathrm{d}}{\mathrm{d}\phi} \times \left[\frac{1}{\sin\phi} \left(\nabla_\phi^2 + \frac{2}{a^2}\right) \hat{w}\right], \tag{9.3.21c}$$

where $\hat{w}$ satisfies the equation

$$\left[\frac{1}{4\chi^4 + 1} \nabla_\phi^2\left(\nabla_\phi^2 + \frac{2}{a^2}\right) + 1\right]\hat{w} = -\frac{1 + \nu}{2Eh/a} C_2\left(\sin\phi \ln \tan\frac{\phi}{2} - \cot\phi\right) + C_3 \sin\phi. \tag{9.3.22}$$

We seek only a particular solution of Eq. (9.3.22) since the complementary solutions are of the type given by Eq. (9.3.2b) which will be discussed shortly. We can easily verify by substitution that the solutions of the equation

$$\left(\nabla_\phi^2 + \frac{2}{a^2}\right) \hat{w} = 0, \tag{9.3.23a}$$

are of the form

$$\hat{w} = A\left(\sin\phi \ln \tan\frac{\phi}{2} - \cot\phi\right) + B \sin\phi. \tag{9.3.23b}$$

Then a particular solution of Eq. (9.3.22) is

$$\hat{w} = -\frac{1 + \nu}{2Eh/a} C_2 \left(\sin\phi \ln \tan\frac{\phi}{2} - \cot\phi\right) + C_3 \sin\phi, \tag{9.3.23c}$$

and Eqs. (9.3.21) are reduced to the terms involving C_1 and C_2. In addition we have bending and twisting moments given by

$$M_\phi = -M_\theta = \frac{h^2}{12a}\frac{C_1}{\sin^3\phi}\begin{cases}\cos\theta\\ \sin\theta,\end{cases} \tag{9.3.24a}$$

$$\overline{M}_{\phi\theta} = \pm\frac{h^2}{12a}\frac{C_1\cos\phi}{\sin^3\phi}\begin{cases}\sin\theta\\ \cos\theta.\end{cases} \tag{9.3.24b}$$

Thus when $n = 1$ we have a pure membrane state of stress, denoted by those terms multiplied by C_2 and a mixed state of stress, denoted by the terms multiplied by C_1 which is almost a pure membrane stress state since the stresses due to the bending moments given by Eqs. (9.3.24) are small compared to those due to the middle surface forces. The tangential deformations of the middle surface can be obtained from Eqs. (9.3.14), (9.3.21), and (9.3.23c) as

$$u_\phi = \left\{\frac{1+\nu}{2Eh/a}\left[C_1\left(\ln\tan\frac{\phi}{2} - \frac{\cos\phi}{\sin^2\phi}\right) - C_2\left(\frac{1}{\sin^2\phi} + \cos\phi\ln\tan\frac{\phi}{2}\right)\right]\right.$$
$$\left. + C_3\cos\phi + C_4\right\}\begin{cases}\cos\theta\\ \sin\theta\end{cases} \tag{9.3.25a}$$

$$u_\theta = \mp\left\langle\frac{1+\nu}{2Eh/a}\left\{C_1\left(\cos\phi\ln\tan\frac{\phi}{2} + 1 + \frac{1}{\sin^2\phi}\right)\right.\right.$$
$$\left. - C_2\left[\ln\tan\frac{\phi}{2} - \cos\phi\left(1 + \frac{1}{\sin^2\phi}\right)\right]\right\}$$
$$\left. + C_3 + C_4\cos\phi\right\rangle\begin{cases}\sin\theta\\ \cos\theta\end{cases} \tag{9.3.25b}$$

where an additional constant of integration is permitted in this case.

The constants of integration C_1, C_2, C_3, and C_4 can be given a physical interpretation. The terms multiplied by the constant C_3 correspond to a rigid body translation of the sphere in a direction perpendicular to the axis of revolution of the shell (Fig. 9.3a) while those multiplied by the constant C_4 correspond to a rotation of the sphere about an axis lying in the plane of the equator of the shell (Fig. 9.3b). The significance of the constants C_1 and C_2 can be found by considering equilibrium of a spherical cap. Let us deal with the upper variation with θ in each of the equations. Then the summation of forces acting on the cut edge indicates a net force acting in the plane of the parallel circle as shown in Fig. 9.4 and given by

$$H = \int_0^{2\pi}(N_\phi\cos\phi\cos\theta - \overline{N}_{\phi\theta}\sin\theta)a\sin\phi\,d\theta = -\pi C_2 a. \tag{9.3.26}$$

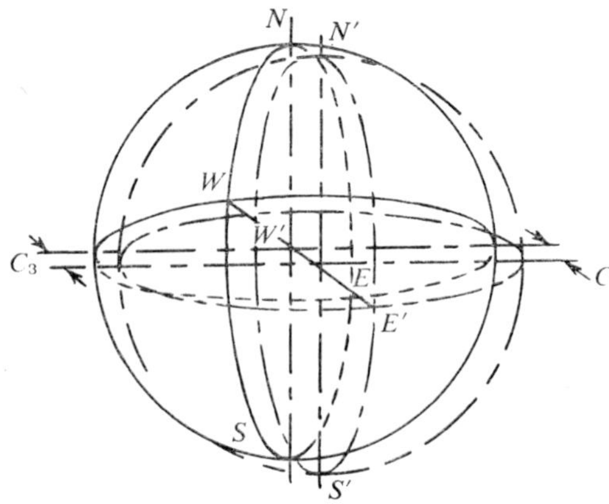

(a) Rigid body translation
(u_ϕ varies as cos θ, u_θ as sin θ)

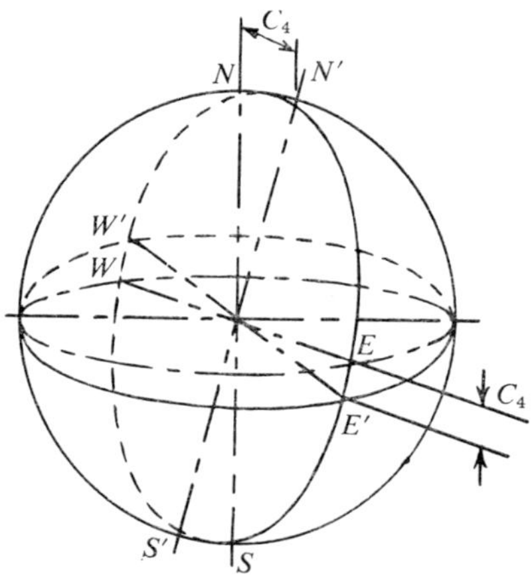

(b) Rigid body rotation
(u_ϕ varies as cos θ, u_θ as sin θ)

Fig. 9.3. Rigid body motions

If we take the moment of all edge forces and moments about an axis passing through the pole (Fig. 9.5) we find a net moment given by

$$M = \int_0^{2\pi} [(N_\phi \sin \phi)(a \sin \phi \cos \theta) + M_\phi \cos \theta - \overline{M}_{\phi\theta} \cos \phi \sin \theta] a \sin \phi \, d\theta - Ha(1 - \cos \phi)$$

$$= \pi a^2 \left[\left(1 + \frac{h^2}{12a^2}\right) C_1 + C_2 \right]. \tag{9.3.27}$$

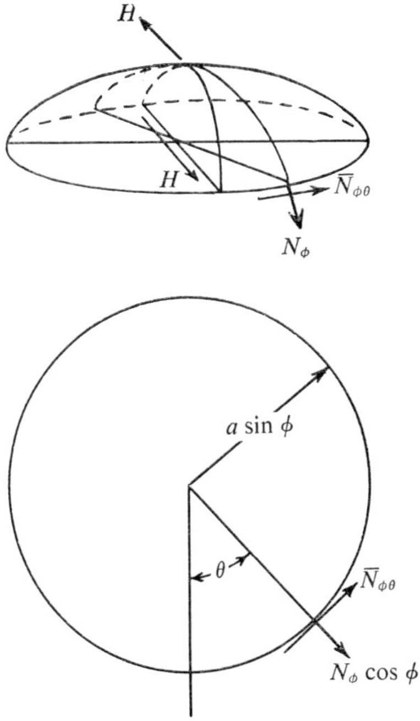

Fig. 9.4. Force acting on spherical cap

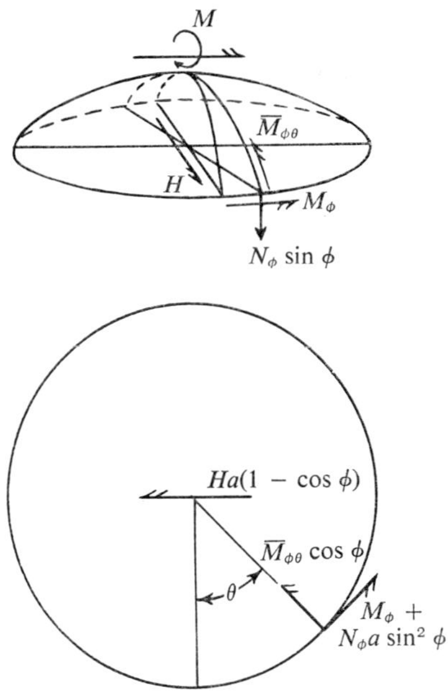

Fig. 9.5. Moments about apex of spherical cap

Thus the terms multiplied by the constant C_1 correspond to the solution for a sphere subjected at its poles to equal but opposite moments (Fig. 9.6a)

$$M_0 = \pi \left(1 + \frac{h^2}{12a^2}\right) C_1 a^2. \tag{9.3.28a}$$

The terms multiplied by the constant C_2 correspond to the solution for a sphere loaded at the poles by equal but opposite tangential forces

$$H_0 = -\pi C_2 a, \tag{9.3.28b}$$

and by equal moments of magnitude $H_0 a$ (Fig. 9.6b). To be more strictly correct, recalling the membrane and bending solutions in Chapter 6 for a sphere subjected to radial concentrated loads at its poles, we should denote these as solutions for portions of spheres, excluding the poles, when the edge forces and moments have the specified distributions.

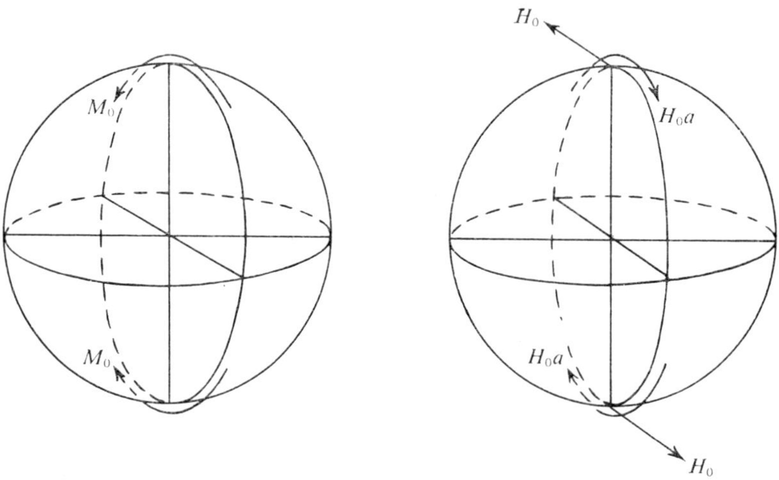

(a) Self-equilibrating moments at the poles

(b) Self-equilibrating forces and moments at the poles

Fig. 9.6. Force and moment systems corresponding to stress states for $n = 1$

9.4 Complex variable representation of membrane stress states

The solutions given above do not exhaust the possible membrane solutions of the bending equations. Other singular solutions which correspond to physically possible concentrated load systems can be obtained which are different from those already discussed. We first note that for membrane stress states the radial deformation satisfies the equation

$$\left(\nabla^2 + \frac{2}{a^2}\right) w = 0. \tag{9.4.1}$$

Then Eqs. (9.1.7) may be written, in the absence of surface loading, as

$$\frac{\partial}{\partial \phi}(N_\phi \sin \phi) - N_\theta \cos \phi + \frac{\partial \bar{N}_{\phi\theta}}{\partial \theta} = 0, \tag{9.4.2a}$$

$$\frac{\partial N_\theta}{\partial \theta} + \frac{1}{\sin \phi} \frac{\partial}{\partial \phi}(\bar{N}_{\phi\theta} \sin^2 \phi) = 0, \tag{9.4.2b}$$

$$N_\phi + N_\theta = 0. \tag{9.4.2c}$$

The elimination of N_θ from Eqs. (9.4.2a) and (9.4.2b) by means of Eq. (9.4.2c) yields

$$\sin \phi \frac{\partial}{\partial \phi}(N_\phi \sin^2 \phi) + \frac{\partial}{\partial \theta}(\bar{N}_{\phi\theta} \sin^2 \phi) = 0, \tag{9.4.3a}$$

$$\frac{\partial}{\partial \theta}(N_\phi \sin^2 \phi) - \sin \phi \frac{\partial}{\partial \phi}(\bar{N}_{\phi\theta} \sin^2 \phi) = 0. \tag{9.4.3b}$$

If now we introduce the notation

$$N_\phi \sin^2 \phi = u, \tag{9.4.4a}$$

$$\bar{N}_{\phi\theta} \sin^2 \phi = v, \tag{9.4.4b}$$

$$\eta = \ln \tan \frac{\phi}{2}, \tag{9.4.4c}$$

Eqs. (9.4.3) become

$$\frac{\partial u}{\partial \eta} + \frac{\partial v}{\partial \theta} = 0, \tag{9.4.5a}$$

$$\frac{\partial u}{\partial \theta} - \frac{\partial v}{\partial \eta} = 0. \tag{9.4.5b}$$

But Eqs. (9.4.5) are the Cauchy–Riemann conditions† for the regularity of the function

$$F(z) = u + iv, \tag{9.4.6a}$$

where

$$z = \theta + i\eta = \theta + i \ln \tan \frac{\phi}{2}. \tag{9.4.6b}$$

† See, for example, G. F. Carrier, M. Krook, and C. E. Pearson, *Functions of a Complex Variable*, McGraw-Hill Book Co., 1966, pp. 25–29.

Thus, solutions of Eqs. (9.4.2) may be written as

$$N_\phi = -N_\theta = \frac{1}{\sin^2\phi}\,\mathrm{Re}\,F(z), \tag{9.4.7a}$$

$$\bar{N}_{\phi\theta} = \frac{1}{\sin^2\phi}\,\mathrm{Im}\,F(z), \tag{9.4.7b}$$

where $F(z)$ is an arbitrary analytic function of the complex variable z.

If $F(z)$ has a real period of 2π the stresses will be periodic around the circumference of the sphere and will therefore be single valued. However, $F(z)$ must be singular at one or more points on the sphere.† Only some of these singular solutions are of interest here, those which yield force or moment resultants acting at the singular points which are different from those already obtained or any combination of them.

Let us consider first the function

$$F(z) = iC\cot\tfrac{1}{2}(z - z_0), \tag{9.4.8a}$$

where

$$z_0 = \theta_0 + i\ln\tan\frac{\phi_0}{2}, \tag{9.4.8b}$$

and C is real. On separating this function into its real and imaginary parts we have from Eqs. (9.4.7)

$$N_\phi = -N_\theta = \frac{C}{\sin^2\phi} \times \frac{\cos\phi_0 - \cos\phi}{1 - \cos\phi_0\cos\phi - \sin\phi_0\sin\phi\cos(\theta-\theta_0)}, \tag{9.4.9a}$$

$$\bar{N}_{\phi\theta} = \frac{C}{\sin\phi}\,\frac{\sin\phi_0\sin(\theta-\theta_0)}{1 - \cos\phi_0\cos\phi - \sin\phi_0\sin\phi\cos(\theta-\theta_0)}. \tag{9.4.9b}$$

We immediately see that the middle surface forces are singular when $\phi = 0$ or π. When $\phi = \phi_0$ and $\theta = \theta_0$ the expressions become indeterminate since both the numerator and the denominator of each equation vanishes. By taking the limit of the expression, however, we can show that the forces become infinite at that point. Therefore there must be concentrated loads at these points on the shell. To determine what these concentrated forces must be, we first investigate the forces at the poles. The forces at the point $\phi = \phi_0$, $\theta = \theta_0$ must then equilibrate the pole forces.

† Liouville's Theorem; see, for example, L. A. Pipes, *Applied Mathematics for Engineers and Physicists*, McGraw-Hill Book Co. Inc., 1946, pp. 466–467.

If we isolate a small spherical cap including the pole (with $\phi < \phi_0$) equilibrium of forces requires (1) that any vertical concentrated force at the pole be given by (Fig. 9.7)

$$V_1 = \int_{-\pi}^{\pi} aN_\phi \sin^2 \phi \, d\theta = -2\pi Ca, \tag{9.4.10a}$$

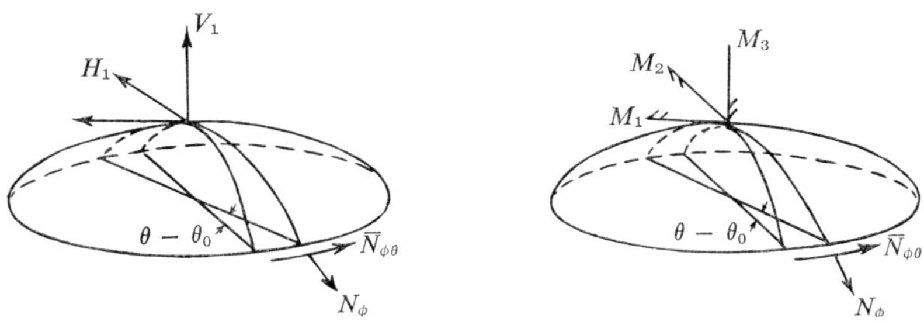

(a) Pole forces (b) Pole moments

Fig. 9.7. Notation for spherical cap ($\phi < \phi_0$)

(2) that any horizontal concentrated forces in the direction parallel and perpendicular to the plane $\theta = \theta_0$ be given by

$$\begin{aligned} H_1 &= \int_{-\pi}^{\pi} [N_\phi \cos \phi \cos (\theta - \theta_0) - N_{\phi\theta} \sin (\theta - \theta_0)] \sin \phi \, d\theta \\ &= -2\pi aC \cot \frac{\phi_0}{2}, \end{aligned} \tag{9.4.10b}$$

$$\begin{aligned} H_2 = \int_{-\pi}^{\pi} & [N_\phi \cos \phi \sin (\theta - \theta_0) \\ & + N_{\phi\theta} \cos (\theta - \theta)] a \sin \phi \, d\theta = 0, \end{aligned} \tag{9.4.10c}$$

and (3) that moments about the apex be given by

$$M_1 = \int_{-\pi}^{\pi} N_\phi \sin^3 \phi \cos (\theta - \theta_0) \, d\theta - H_1 a (1 - \cos \phi) = 0, \tag{9.4.10d}$$

$$M_2 = \int_{-\pi}^{\pi} N_\phi a^2 \sin^3 \phi \sin (\theta - \theta_0) \, d\theta - H_2 a (1 - \cos \phi) = 0, \tag{9.4.10e}$$

$$M_3 = \int_{-\pi}^{\pi} N_{\phi\theta} a^2 \sin^2 \phi \, d\theta = 0. \tag{9.4.10f}$$

A similar procedure for a spherical cap which includes the other pole (with $\phi > \phi_0$) yields a concentrated vertical force given by

$$V_2 = -\int_{-\pi}^{\pi} aN_\phi \sin^2 \phi \, d\theta = V_1, \tag{9.4.11a}$$

$$\begin{aligned} H_3 &= -\int_{-\pi}^{\pi} [N_\phi \cos \phi \cos (\theta - \theta_0) \\ &\qquad - N_{\phi\theta} \sin (\theta - \theta_0)] a \sin \phi \, d\theta \\ &= 2\pi a C \tan \frac{\phi_0}{2}. \end{aligned} \tag{9.4.11b}$$

All other force and moment components vanish. For equilibrium, then, the singularity at the point, $\phi = \phi_0$, $\theta = \theta_0$ must correspond to a concentrated force lying in the plane $\theta = \theta_0$ which has the vertical and horizontal components

$$P_V = -(V_1 + V_2) = 4\pi a C, \tag{9.4.12a}$$

$$P_H = -(H_1 + H_3) = 4\pi a C \cot \phi_0 . \tag{9.4.12b}$$

These are the components of a force tangent to the meridian. If we denote the magnitude of this force by P we have

$$P = (P_V^2 + P_H^2)^{1/2} = \frac{4\pi a C}{\sin \phi_0} \tag{9.4.13a}$$

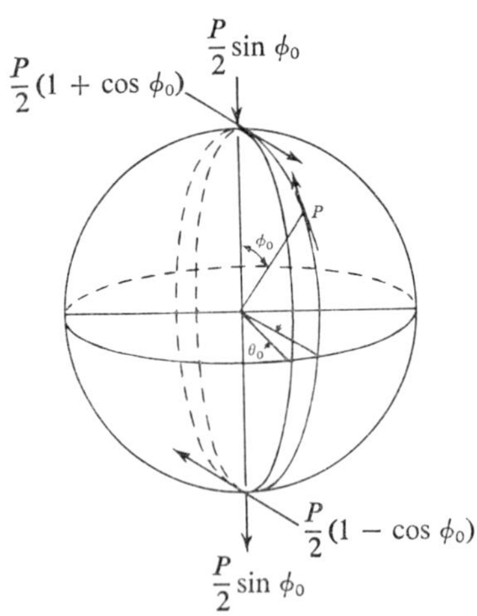

Fig. 9.8. System of forces corresponding to

$$F(z) = i\frac{P \sin \phi_0}{4\pi a} \cot \tfrac{1}{2}(z - z_0)$$

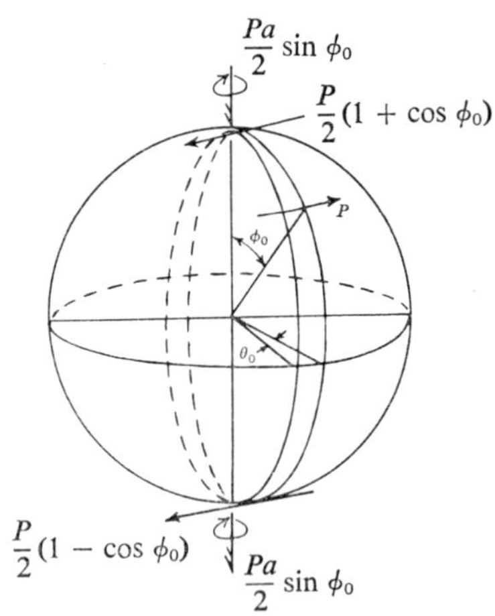

Fig. 9.9. System of forces corresponding to

$$F(z) = \frac{P \sin \phi_0}{4\pi a} \cot \tfrac{1}{2}(z - z_0)$$

or

$$C = \frac{P \sin \phi_0}{4\pi a}. \tag{9.4.13b}$$

The total system of forces to which this membrane solution corresponds is then as shown in Fig. 9.8.

If the constant C is imaginary and we write the function $F(z)$ as

$$F(z) = \frac{P \sin \phi_0}{4\pi a} \cot \tfrac{1}{2}(z - z_0), \tag{9.4.14a}$$

we have

$$N_\phi = -N_\theta = \frac{P \sin^2 \phi_0}{4\pi a \sin \phi} \frac{\sin (\theta - \theta_0)}{1 - \cos \phi_0 \cos \phi - \sin \phi_0 \sin \phi \cos (\theta - \theta_0)}, \tag{9.4.14b}$$

$$\overline{N}_{\phi\theta} = -\frac{P \sin \phi_0}{4\pi a \sin^2 \phi} \frac{\cos \phi_0 - \cos \phi}{1 - \cos \phi_0 \cos \phi - \sin \phi_0 \sin \phi \cos (\theta - \theta_0)}. \tag{9.4.14c}$$

A similar investigation of force and moment equilibrium indicates that the force system is that shown in Fig. 9.9.

A final membrane state of stress which is of interest is that due to edge forces which are equivalent to a normal force at the point $\phi = \phi_0$, $\theta = \theta_0$. We obtain the desired solution by considering two forces tangent to the meridian $\theta = \theta_0$ at the adjacent points corresponding to

$$\phi = \phi_0 + \frac{\Delta\phi_0}{2} \quad \text{and} \quad \phi = \phi_0 - \frac{\Delta\phi_0}{2}$$

as shown in Fig. 9.10. The resultant of these forces is a force in the direction of the radius at $\phi = \phi_0$ whose magnitude is given by

$$P = 2F \sin \frac{\Delta\phi_0}{2} \approx F\,\Delta\phi_0, \tag{9.4.15}$$

in addition to polar forces as shown in Fig. 9.11. Thus the solution for the middle surface forces is the limit of the sum of the solutions for each tangential load as $\Delta\phi_0$ approaches 0 but $F\,\Delta\phi_0$ remains finite and equal to P. We can obtain the pertinent stress function $F(z)$ as

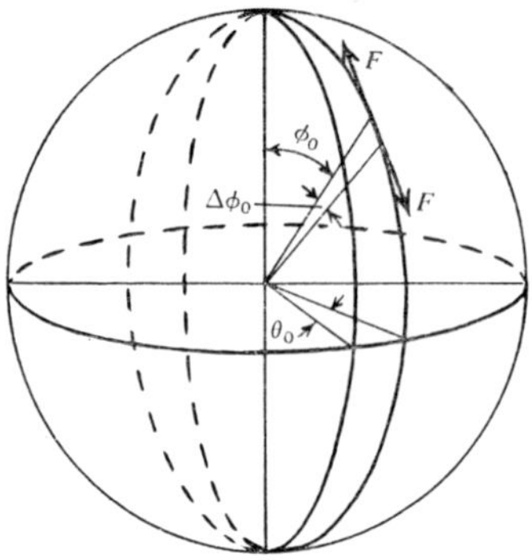

Fig. 9.10. *Radial force as resultant of tangential forces*

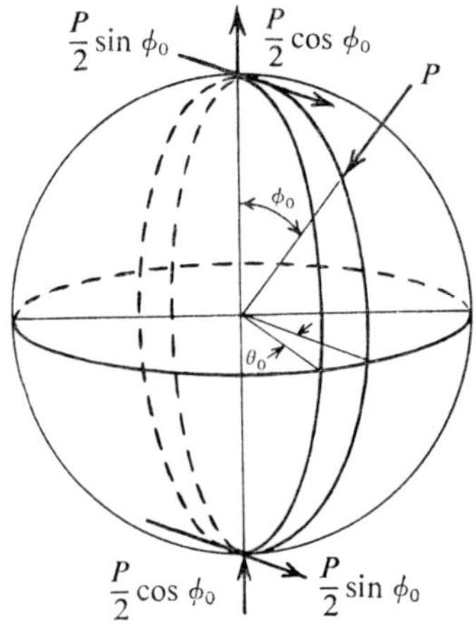

Fig. 9.11. *System of forces corresponding to*

$$F(z) = -\frac{P}{4\pi a}\left[i\cos\phi_0\cot\tfrac{1}{2}(z-z_0) + \tfrac{1}{2}\csc^2\tfrac{1}{2}(z-z_0)\right]$$

$$\begin{aligned} F(z) = \lim_{\Delta\phi_0\to 0}\frac{iF}{4\pi a}\Bigg[&\sin\left(\phi_0-\frac{\Delta\phi_0}{2}\right)\\ &\times\cot\frac{1}{2}\left(z-z_0+i\ln\frac{1-\tan\frac{\Delta\phi_0}{4}\cot\frac{\phi_0}{2}}{1+\tan\frac{\Delta\phi_0}{4}\tan\frac{\phi_0}{2}}\right)\\ &-\sin\left(\phi_0+\frac{\Delta\phi_0}{2}\right)\\ &\times\cot\frac{1}{2}\left(z-z_0+i\ln\frac{1+\tan\frac{\Delta\phi_0}{4}\cot\frac{\phi_0}{2}}{1-\tan\frac{\Delta\phi_0}{4}\tan\frac{\phi_0}{2}}\right)\Bigg]\end{aligned}$$

$$= -\frac{P}{4\pi a}\left[i\cos\phi_0\cot\tfrac{1}{2}(z-z_0)+\tfrac{1}{2}\csc^2\tfrac{1}{2}(z-z_0)\right] \tag{9.4.16a}$$

from which we can obtain the membrane stress distribution as

$$N_\phi = -N_\theta = -\frac{P}{4\pi a \sin^2\phi} \times \left\{\frac{1 - 2\cos\phi\cos\phi_0 + \cos^2\phi_0}{1 - \cos\phi\cos\phi_0 - \sin\phi\sin\phi_0\cos(\theta-\theta_0)} - \left[\frac{\cos\phi_0 - \cos\phi}{1 - \cos\phi\cos\phi_0 - \sin\phi\sin\phi_0\cos(\theta-\theta_0)}\right]^2\right\}, \quad (9.4.16b)$$

$$\bar{N}_{\phi\theta} = -\frac{P\sin^2\phi_0\sin(\theta-\theta_0)[\cos\phi\sin\phi_0 + \sin\phi\cos\phi_0\cos(\theta-\theta_0)]}{4\pi a\sin\phi[1 - \cos\phi\cos\phi_0 - \sin\phi\sin\phi_0\cos(\theta-\theta_0)]^2}. \quad (9.4.16c)$$

The three concentrated load systems given above, together with the two concentrated load systems obtained in Section 9.3 and the two axisymmetric stress states for concentrated radial loads and torques at the poles (Sections 6.2 and 8.1) are sufficient to define any physically possible system of concentrated loads which hold the sphere in equilibrium. Other solutions of the equations will be found, although with difficulty in some cases, to correspond to combinations and/or rotations of the solutions already obtained. For example, the solution for the system of moments shown in Fig. 9.12 can be obtained by superposition of the four other known concentrated load solutions which are shown. The relations necessary for rotation of the coordinate system are

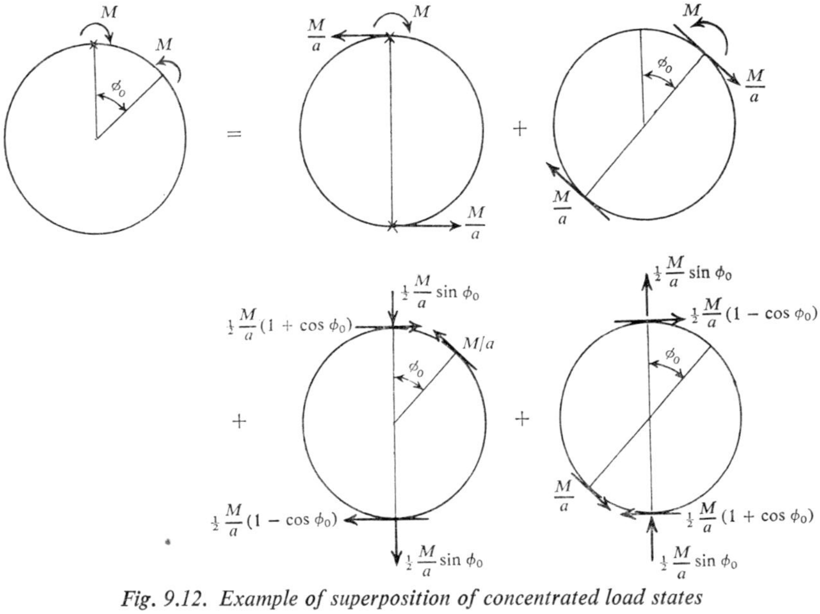

Fig. 9.12. Example of superposition of concentrated load states

$$\sin\phi\cos(\theta-\theta_1)=\cos\bar{\theta}\sin\bar{\phi}\cos\phi_1+\cos\bar{\phi}\sin\phi_1, \tag{9.4.17a}$$

$$\sin\phi\sin(\theta-\theta_1)=\sin\bar{\theta}\sin\bar{\phi}, \tag{9.4.17b}$$

$$\cos\phi=\cos\bar{\phi}\cos\phi_1-\cos\bar{\theta}\sin\bar{\phi}\sin\phi_1, \tag{9.4.17c}$$

where ϕ, θ are a set of spherical coordinates for a point, $\bar{\phi}$, $\bar{\theta}$ are a second set of spherical coordinates for the same point, ϕ_1 is the angle between the polar axes of the two systems, and θ_1 is the rotation of the reference plane of the second system relative to the first (see Fig. 9.13).

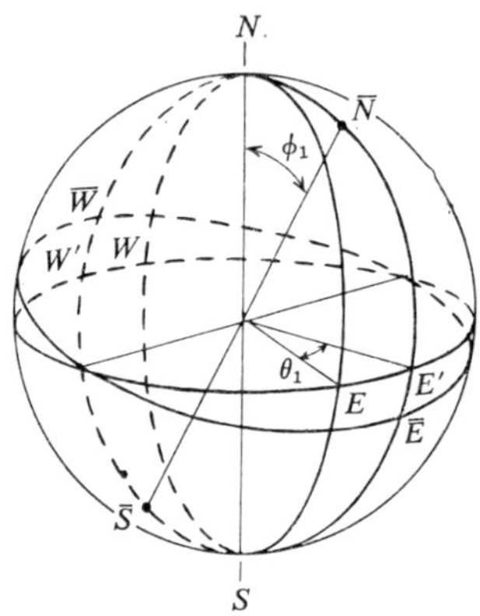

(a) Relation between two coordinate references

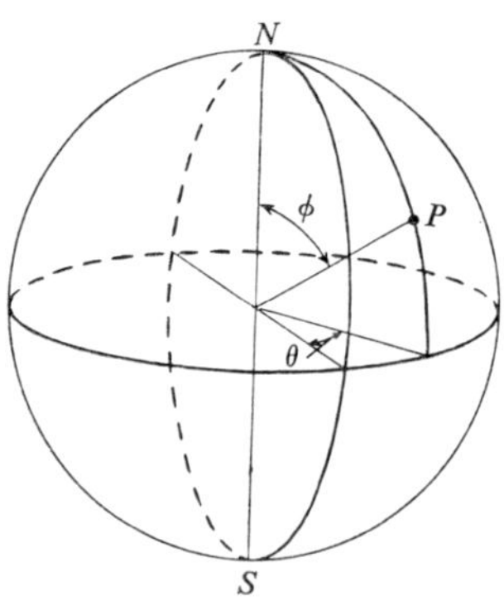

(b) Coordinates of point with respect to one coordinate system

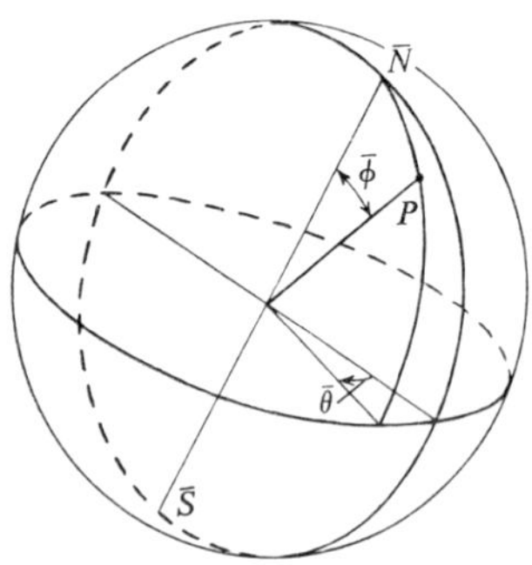

(c) Coordinates of point with respect to other coordinate system

Fig. 9.13. Notation for rotated spherical coordinate systems

The displacements corresponding to the stress states given above are not easy to determine, in general, although formal expressions may be obtained. From Eqs. (9.3.6b) and (9.3.8) can obtain the radial deflection w as

$$w = -\frac{(1+\nu)a}{Eh}\sin\phi \iint \overline{N}_{\phi\theta}\, \mathrm{d}\phi\, \mathrm{d}\theta, \tag{9.4.18a}$$

and from Eqs. (9.1.2a), (9.1.2b), (9.1.3a), (9.1.3b), and (9.3.6a)

$$\frac{\partial u_\phi}{\partial \phi} = \frac{(1+\nu)a}{Eh} N_\phi - w, \tag{9.4.18b}$$

$$\frac{\partial u_\theta}{\partial \theta} = -\left[\frac{\partial}{\partial \phi}(u_\phi \sin\phi) + 2w\sin\phi\right]. \tag{9.4.18c}$$

The integrations involved are not generally possible in terms of known functions.

9.5 Edge-loaded spherical shells: mixed bending–stretching solutions

In addition to the membrane and inextensional solutions of the bending equations for edge-loaded spheres given in Sections 9.3 and 9.4, there are other solutions which are defined by Eq. (9.3.2b) as

$$\left(\nabla^2 + \frac{1+2i\chi^2}{a^2}\right)\left[w + (1-2i\chi^2)\frac{S}{Eha}\right] = 0, \tag{9.5.1}$$

or by the two equivalent equations

$$\frac{S}{Eha} = -\frac{1}{4\chi^4+1}(a^2\nabla^2 + 2)w, \tag{9.5.2a}$$

$$w = a^2\nabla^2 \frac{S}{Eha}. \tag{9.5.2b}$$

It should be noted that this group of equations is valid for spherical shells with axisymmetric or with "wind loading" deformations and yields self-equilibrating groups of forces.

Before investigating the solutions of Eqs. (9.5.1) we obtain some general relationships for the stress resultants and displacements belonging to this group of solutions. From Eqs. (9.1.8) and (9.5.2) we have

$$N_\phi = \frac{1}{a^2}\left(\cot\phi \frac{\partial S}{\partial \phi} + \frac{1}{\sin^2\phi}\frac{\partial^2 S}{\partial \phi^2}\right), \tag{9.5.3a}$$

$$N_\theta = \frac{1}{a^2}\frac{\partial^2 S}{\partial \phi^2} = -N_\phi + \nabla^2 S = -N_\phi + \frac{Eh}{a} w, \tag{9.5.3b}$$

$$\bar{N}_{\phi\theta} = N_{\phi\theta} = -\frac{1}{a^2}\frac{\partial^2}{\partial\phi\,\partial\theta}\left(\frac{S}{\sin\phi}\right). \tag{9.5.3c}$$

With these equations we can obtain explicit relationships for the tangential displacements u_ϕ and u_θ in terms of S. From Eqs. (9.1.2a), (9.1.2b), (9.1.3a), and (9.5.3b) we have

$$\begin{aligned}\frac{\partial u_\phi}{\partial\phi} &= a\epsilon_\phi - w = \frac{a}{Eh}(N_\phi - \nu N_\theta) - w \\ &= -\frac{(1+\nu)a}{Eh}\left(-N_\phi + \frac{Eh}{a}w\right) \\ &= -\frac{1+\nu}{Eha}\frac{\partial^2 S}{\partial\phi^2}.\end{aligned} \tag{9.5.4a}$$

Thus

$$u_\phi = -\frac{1+\nu}{Eha}\frac{\partial S}{\partial\phi}. \tag{9.5.4b}$$

To determine u_θ we utilize Eqs. (9.1.3b) and (9.5.3a) in addition to those already cited. Then

$$\begin{aligned}\frac{\partial u_\theta}{\partial\theta} &= a\epsilon_\theta\sin\phi - u_\phi\cos\phi - w\sin\phi \\ &= \frac{a}{Eh}(N_\theta - \nu N_\phi)\sin\phi - u_\phi\cos\phi - w\sin\phi \\ &= -\left[(1+\nu)\frac{a}{Eh}N_\phi\sin\phi + u_\phi\cos\phi\right] \\ &= -\frac{1+\nu}{Eha\sin\phi}\frac{\partial^2 S}{\partial\theta^2},\end{aligned} \tag{9.5.5a}$$

which yields

$$u_\theta = -\frac{1+\nu}{Eha\sin\phi}\frac{\partial S}{\partial\theta}. \tag{9.5.5b}$$

Any additional functions due to the integration of Eqs. (9.5.4a) and (9.5.5a) can be shown to be limited to rigid body motions which have already been treated.

Similar manipulations with Eqs. (9.1.4) for the bending and twisting moments yields the relationships

$$M_\phi = \frac{S}{a} + \frac{Eh^3}{12(1+\nu)a^2}\left(\cot\phi\frac{\partial}{\partial\phi} + \frac{1}{\sin^2\phi}\frac{\partial^2}{\partial\theta^2} + 1\right)\left[w + (1+\nu)\frac{S}{Eha}\right], \tag{9.5.6a}$$

$$M_\theta = \nu\frac{S}{a} - \frac{Eh^3}{12(1+\nu)a^2}\left(\cot\phi\frac{\partial}{\partial\phi} + \frac{1}{\sin^2\phi}\frac{\partial^2}{\partial\theta^2} - \nu\right)\left[w + (1+\nu)\frac{S}{Eha}\right], \tag{9.5.6b}$$

$$\overline{M}_{\phi\theta} = M_{\phi\theta} = -\frac{Eh^3}{12(1+\nu)a^2}\frac{\partial^2}{\partial\phi\,\partial\theta}\left[\frac{w + (1+\nu)(S/Eha)}{\sin\phi}\right], \tag{9.5.6c}$$

which are free of second derivatives with respect to the coordinate ϕ. Finally we have from Eqs. (9.1.6) and (9.3.2b)

$$\bar{Q}_\phi - \frac{1}{a\sin\phi}\frac{\partial\overline{M}_{\phi\theta}}{\partial\theta} = Q_\phi = \frac{1}{a^2}\frac{\partial S}{\partial\phi} = -\frac{Eh}{(1+\nu)a}u_\phi, \tag{9.5.7a}$$

$$\bar{Q}_\theta - \frac{\partial\overline{M}_{\phi\theta}}{a\,\partial\phi} = Q_\theta = \frac{1}{a^2\sin\phi}\frac{\partial S}{\partial\theta} = -\frac{Eh}{(1+\nu)a}u_\theta. \tag{9.5.7b}$$

Our task now is to find solutions of Eq. (9.5.1). This is most easily accomplished if the deformations and stress resultants can be expanded in a Fourier series around the circumference, which restricts us to spherical shells cut along parallel circles perpendicular to the axis of revolution and which have no discontinuities or cut-outs within the confines of the region considered. Let us then express the complex dependent variable of Eq. (9.5.1a) as

$$\begin{aligned} w + (1 - 2i\chi^2)\frac{S}{Eha} &= \sum_{n=0}^{\infty}\left[w_n(\phi) + (1-2i\chi^2)\frac{S_n(\phi)}{Eha}\right]\cos(n\theta - \theta_n) \\ &= \sum_{n=0}^{\infty} F_n(\phi)\cos(n\theta - \theta_n). \end{aligned} \tag{9.5.8a}$$

The functions $F_n(\phi)$ must be solutions of the following equations

$$\frac{d^2F_n}{d\phi^2} + \cot\phi\frac{dF_n}{d\phi} + \left(1 + 2i\chi^2 - \frac{n^2}{\sin^2\phi}\right)F_n = 0 \quad (n = 0, 1, 2, \ldots). \tag{9.5.8b}$$

We have already encountered a special case of Eq. (9.5.8b) in Chapters 6 and 7 in our investigation of axisymmetric deformations of spherical shells. In this instance we found solutions of the equation in several ways: an exact solution in terms of Legendre functions, an approximate solution for thin shells and small or large angles in terms of Bessel–Kelvin functions, and an asymptotic solution for thin shells and angles near 90° in terms of trigonometric and exponential functions. We can obtain counterparts of these solutions in the general case when n is not equal to zero. Whatever the form of the functions involved, the complete solution of Eq. (9.5.8b) can be expressed in terms of two independent solutions as

$$F_n = (\alpha_n + i\beta_n)(\mathrm{Re}\, F_{n1} + i\, \mathrm{Im}\, F_{n1}) + (\gamma_n + i\delta_n)(\mathrm{Re}\, F_{n2} + i\, \mathrm{Im}\, F_{n2}). \tag{9.5.8c}$$

By separating real and imaginary parts of Eq. (9.5.8c) we can obtain

$$S_n = \bar{A}_n\, \mathrm{Im}\, F_{n1} + \bar{B}_n\, \mathrm{Re}\, F_{n1} + \bar{C}_n\, \mathrm{Im}\, F_{n2} + \bar{D}_n\, \mathrm{Re}\, F_{n2}, \tag{9.5.9a}$$

$$\begin{aligned} w_n = -\frac{2\chi^2}{Eha}\bigg[&\bar{A}_n\left(\mathrm{Re}\, F_{n1} + \frac{1}{2\chi^2}\mathrm{Im}\, F_{n1}\right) - \bar{B}_n\left(\mathrm{Im}\, F_{n1} - \frac{1}{2\chi^2}\mathrm{Re}\, F_{n1}\right) \\ &+ \bar{C}_n\left(\mathrm{Re}\, F_{n2} + \frac{1}{2\chi^2}\mathrm{Im}\, F_{n2}\right) - \bar{D}_n\left(\mathrm{Im}\, F_{n2} - \frac{1}{2\chi^2}\mathrm{Re}\, F_{n2}\right)\bigg] \end{aligned} \tag{9.5.9b}$$

and, for use in Eqs. (9.5.6),

$$\begin{aligned} &\frac{Eh^3}{12(1+\nu)a}\left[w_n + (1+\nu)\frac{S}{Eha}\right] \\ &\quad = -\frac{2(1-\nu)\chi^2}{4\chi^4+\nu^2}\bigg[\bar{A}_n\left(\mathrm{Re}\, F_{n1} - \frac{\nu}{2\chi^2}\mathrm{Im}\, F_{n1}\right) \\ &\quad - \bar{B}_n\left(\mathrm{Im}\, F_{n1} + \frac{\nu}{2\chi^2}\mathrm{Re}\, F_{n1}\right) + \bar{C}_n\left(\mathrm{Re}\, F_{n2} - \frac{\nu}{2\chi^2}\mathrm{Im}\, F_{n2}\right) \\ &\quad - \bar{D}_n\left(\mathrm{Im}\, F_{n2} + \frac{\nu}{2\chi^2}\mathrm{Re}\, F_{n2}\right)\bigg]. \end{aligned} \tag{9.5.9c}$$

The exact solutions of Eq. (9.5.8b) can be expressed in terms of associated Legendre functions as†

$$F_{n1} = P_\mu^n(\cos\phi), \tag{9.5.10a}$$

$$F_{n2} = P_\mu^n[\cos(\pi - \phi)]. \tag{9.5.10b}$$

When χ^2/n is small, a rapidly convergent expression for $P_\mu^n(\cos\phi)$ is given by

$$\begin{aligned} P_\mu^n(\cos\phi) = &\frac{\Gamma(\mu+n+1)}{\Gamma(\mu-n+1)}\frac{(-1)^n \tan^n \phi/2}{n!} \\ &\times \left\{1 - \frac{\sin^2\phi/2}{n+1} - \frac{2(\chi^4+\frac{1}{4})}{(n+1)(n+2)}\sin^4\frac{\phi}{2}\right. \end{aligned}$$

† M. Abramowitz and I. A. Stegun, *Handbook of Mathematical Functions*, National Bureau of Standards, Appl. Math. Ser. no. 55, June 1964, pp. 333, 338.

$$- \frac{5(\chi^4 + \frac{1}{4})}{3(n+1)(n+2)(n+3)} \sin^6 \frac{\phi}{2}$$

$$+ \frac{2(\chi^4 + \frac{1}{4})(\chi^4 - \frac{55}{4})}{3(n+1)(n+2)(n+3)(n+4)} \sin^8 \frac{\phi}{2} + \cdots$$

$$- \frac{2i\chi^2}{n+1} \sin^2 \frac{\phi}{2} \left[1 - \frac{2(\chi^4 + \frac{1}{4})}{3(n+2)(n+3)} \sin^4 \frac{\phi}{2}\right.$$

$$- \frac{8(\chi^4 + \frac{1}{4})}{3(n+2)(n+3)(n+4)} \sin^6 \frac{\phi}{2}$$

$$\left.+ \frac{2(\chi^4 + \frac{1}{4})(\chi^4 - \frac{359}{4})}{15\,(n+2)\,(n+3)\,(n+4)\,(n+5)} \sin^8 \frac{\phi}{2} + \cdots\right]\Bigg\}. \tag{9.5.11}$$

It is interesting to note for the real part of the infinite series in braces that as n becomes large the first term predominates so that this part of the solution approaches the solutions for membrane stress states and states of inextensional deformation.

If χ^2 is large compared to n or is comparable to n the series expression converges slowly and we must depend on asymptotic expansions for more practical computation methods. If we introduce the transformation

$$F_n = \frac{U_n}{(\sin \phi)^{1/2}}, \tag{9.5.12a}$$

into Eq. (9.5.8b) we obtain the equation

$$\frac{d^2 U_n}{d\phi^2} + \left(\frac{5}{4} + 2i\chi^2 - \frac{n^2 - \frac{1}{4}}{\sin^2 \phi}\right) U_n = 0. \tag{9.5.12b}$$

If χ is large and is much greater than n^2, we may then approximate Eq. (9.5.12b) by

$$\frac{d^2 U_n}{d\phi^2} + \left(2i\chi^2 - \frac{n^2 - \frac{1}{4}}{\phi^2}\right) U_n = 0, \tag{9.5.13}$$

which has the solutions†

$$U_{n1} = \phi^{1/2}[\mathrm{ber}_n\,(2^{1/2}\chi\phi) - i\,\mathrm{bei}_n\,(2^{1/2}\chi\phi)], \tag{9.5.14a}$$

$$U_{n2} = \phi^{1/2}[\mathrm{ker}_n\,(2^{1/2}\chi\phi) - i\,\mathrm{kei}_n\,(2^{1/2}\chi\phi)]. \tag{9.5.14b}$$

For solutions near the lower pole we merely replace ϕ by $\pi - \phi$ in the equations. The functions are generalizations of the Bessel–Kelvin functions which

† See H. B. Dwight, *Tables of Integrals and Other Mathematical Data*, 3rd Ed. The Macmillan Co., New York, 1957, pp. 280–282.

entered into our investigation of axisymmetric deformations of shells of revolution.

We have obtained practical forms of the solutions of Eq. (9.5.8b) for the cases when n is large compared to χ^2 and when χ is large compared to n^2. The next logical question is what approximate solution can we obtain which interpolates between these two extremes? There does not appear to be any tabulated function which reduces to Bessel–Kelvin functions of order n when χ is much greater than n^2 and to trigonometric functions when n is much greater than χ^2. However, we can obtain an approximate solution which is useful when χ and n are of the same order of magnitude.

Let us introduce the following transformations for both the dependent and independent variables

$$\xi_n = \xi_n(\phi), \tag{9.5.15a}$$

$$F_n(\phi) = V_n(\phi)U_n(\xi_n), \tag{9.5.15b}$$

where the functions ξ_n and $V_n(\phi)$ are to be determined. Then Eq. (9.5.8b) becomes

$$\left(\frac{\mathrm{d}\xi_n}{\mathrm{d}\phi}\right)^2 \frac{\mathrm{d}^2 U_n}{\mathrm{d}\xi^2} + \left[2\,\frac{\mathrm{d}\xi_n}{\mathrm{d}\phi}\,\frac{V_n/\mathrm{d}\phi}{V_n} + \frac{1}{\sin\phi}\,\frac{\mathrm{d}}{\mathrm{d}\phi}\left(\frac{\mathrm{d}\xi_n}{\mathrm{d}\phi}\sin\phi\right)\right]\frac{\mathrm{d}U_n}{\mathrm{d}\xi_n}$$
$$+ \left[1 + 2i\chi^2 - \frac{n^2}{\sin^2\phi} + \frac{\mathrm{d}/\mathrm{d}\phi(\mathrm{d}V_n/\mathrm{d}\phi\,\sin\phi)}{V_n\sin\phi}\right]U_n = 0. \tag{9.5.15c}$$

We define V_n by requiring that the coefficient of the first derivative of U_n vanish, so that

$$2\,\frac{\mathrm{d}V_n/\mathrm{d}\phi}{V_n} + \frac{\mathrm{d}/\mathrm{d}\phi(\mathrm{d}\xi_n/\mathrm{d}\phi\,\sin\phi)}{\mathrm{d}\xi_n/\mathrm{d}\phi\,\sin\phi} = 0. \tag{9.5.16a}$$

Eq. (9.5.16a) can be integrated to yield

$$V_n = \frac{1}{(|\mathrm{d}\xi_n/\mathrm{d}\phi|\,\sin\phi)^{1/2}}, \tag{9.5.16b}$$

and Eq. (9.5.15c) reduces to

$$\left(\frac{\mathrm{d}\xi_n}{\mathrm{d}\phi}\right)^2 \frac{\mathrm{d}^2 U_n}{\mathrm{d}\xi_n{}^2}$$
$$+ \left[\frac{5}{4} + 2i\chi^2 - \frac{n^2 - \frac{1}{4}}{\sin^2\phi} + \left(\frac{1}{2}\,\frac{\mathrm{d}^2\xi_n/\mathrm{d}\phi^2}{\mathrm{d}\xi_n/\mathrm{d}\phi}\right)^2 - \frac{\mathrm{d}}{\mathrm{d}\phi}\left(\frac{1}{2}\,\frac{\mathrm{d}^2\xi_n/\mathrm{d}\phi^2}{\mathrm{d}\xi_n/\mathrm{d}\phi}\right)\right]U_n = 0. \tag{9.5.16c}$$

We now wish to choose the function ξ so that Eq. (9.5.16c) is of the form

$$\frac{d^2U_n}{d\xi_n^2} + [\lambda^2 + \alpha_n(\phi)]U_n = 0, \tag{9.5.17}$$

where $\alpha_n(\phi)$ is small compared to λ^2. If we define $d\xi_n/d\phi$ by the relationship†

$$\frac{d\xi_n}{d\phi} = -\left(1 + \frac{in^2}{2\chi^2 \sin^2 \phi}\right)^{1/2}, \tag{9.5.18a}$$

we have

$$\lambda^2 = 2i\chi^2, \tag{9.5.18b}$$

$$\alpha_n(\phi) = \frac{\left[\dfrac{in^2}{\chi^2}\left(2 + \dfrac{in^2}{\chi^2}\right) + \left(1 + 3i\dfrac{n^2}{\chi^2}\right)\sin^2\phi + 5\sin^4\phi\right]\sin^2\phi.}{4\left(\dfrac{in^2}{2\chi^2} + \sin^2\phi\right)^3} \tag{9.5.18c}$$

The function $\alpha_n(\phi)$ vanishes at the poles and is generally of the order of unity or less. When n/χ is small and $\sin\phi$ is of the order of n/χ, however, the function is of the order of χ^2/n^2. If we except this case which has been treated previously, the function $\alpha_n(\phi)$ is negligible compared to λ^2 over the entire range of ϕ from 0 degrees to 180 degrees for thin shells. Then Eq. (9.5.17) may be approximated by

$$\frac{d^2U_n}{d\xi_n^2} + 2i\chi^2 U_n = 0, \tag{9.5.19a}$$

which has the simple solution

$$U_n = \alpha_n\, e^{(1-i)\chi\xi_n} + \beta_n\, e^{-(1-i)\chi\xi_n}. \tag{9.5.19b}$$

The function $\xi_n(\phi)$ is obtained by integrating Eq. (9.5.18a) with the aid of the transformation

$$u_n = \frac{\cos\phi}{(\sin^2\phi + in^2/2\chi^2)^{1/2}}, \tag{9.5.20a}$$

by means of which Eq. (9.5.18a) becomes

$$\begin{aligned} \xi_n &= \int \frac{du_n}{1 + u_n^2} + \frac{in^2}{2\chi^2}\int \frac{du_n}{1 - (in^2/2\chi^2)u_n^2} \\ &= \tan^{-1} u_n + \left(\frac{in^2}{2\chi^2}\right)^{1/2} \tanh^{-1}\left(\frac{in^2}{2\chi^2}\right)^{1/2} u_n. \end{aligned} \tag{9.5.20b}$$

† This choice of variables is due to A. Havers, *Asymptotische Biegetheorie der Unbelasteten Kugelschale*, Ingenieur-Archiv, vol. 6, 1935, pp. 282–312.

Table 9.1a ξ_{1n} as a function of ϕ and $n^2/2\chi^2$

$\frac{n^2}{2\chi^2}$	$\phi = 80°$	70°	60°	50°	40°	30°	25°	20°	15°	10°	5°	0°
0.00	0.1745	0.3491	0.5236	0.6981	0.8727	1.0472	1.1345	1.223	1.309	1.396	1.484	1.571
0.01	0.1754	0.3509	0.5265	0.7024	0.8787	1.0559	1.1453	1.236	1.328	1.425	1.544	∞
0.02	0.1763	0.3527	0.5294	0.7066	0.8847	1.0647	1.1563	1.250	1.347	1.455	1.604	
0.03	0.1772	0.3546	0.5323	0.7109	0.8908	1.0736	1.1674	1.264	1.366	1.485	1.661	
0.04	0.1781	0.3564	0.5353	0.7152	0.8969	1.0826	1.1786	1.278	1.385	1.516	1.716	
0.05	0.1790	0.3583	0.5382	0.7195	0.9031	1.0916	1.1899	1.293	1.405	1.547	1.770	
0.06	0.1799	0.3602	0.5412	0.7238	0.9092	1.1007	1.2012	1.308	1.425	1.578	1.821	
0.07	0.1808	0.3620	0.5442	0.7281	0.9154	1.1099	1.2126	1.323	1.445	1.609	1.871	
0.08	0.1817	0.3639	0.5472	0.7325	0.9217	1.1191	1.2241	1.338	1.465	1.639	1.918	
0.09	0.1826	0.3658	0.5502	0.7369	0.9280	1.1283	1.2356	1.353	1.485	1.668	1.964	
0.10	0.1836	0.3677	0.5532	0.7413	0.9343	1.1375	1.2471	1.368	1.505	1.697	2.009	
0.15	0.1883	0.3773	0.5685	0.7636	0.9663	1.1847	1.3059	1.443	1.605	1.837	2.219	
0.20	0.1930	0.3871	0.5841	0.7865	0.9990	1.2329	1.3657	1.519	1.702	1.969	2.409	
0.25	0.1978	0.3970	0.5999	0.8098	1.0321	1.2816	1.4259	1.594	1.797	2.086	2.583	
0.30	0.2027	0.4071	0.6159	0.8333	1.0656	1.3303	1.4858	1.667	1.888	2.202	2.746	
0.35	0.2076	0.4173	0.6321	0.8569	1.0992	1.3787	1.5448	1.739	1.977	2.318	2.903	
0.40	0.2125	0.4276	0.6484	0.8806	1.1329	1.4267	1.6026	1.810	2.063	2.431	3.054	
0.45	0.2175	0.4379	0.6648	0.9043	1.1666	1.4742	1.6592	1.874	2.148	2.541	3.201	
0.50	0.2225	0.4483	0.6813	0.9281	1.2002	1.5212	1.7148	1.946	2.230	2.649	3.342	
0.55	0.2275	0.4586	0.6978	0.9519	1.2336	1.5677	1.7695	2.0123	2.3101	2.750	3.479	
0.60	0.2325	0.4689	0.7143	0.9757	1.2667	1.6136	1.8234	2.0772	2.3881	2.848	3.611	
0.65	0.2375	0.4791	0.7307	0.9994	1.2995	1.6589	1.8766	2.1408	2.4642	2.944	3.739	
0.70	0.2424	0.4893	0.7470	0.1230	1.3320	1.7036	1.9291	2.2032	2.5386	3.037	3.862	
0.75	0.2473	0.4995	0.7632	1.0464	1.3642	1.7477	1.9808	2.2643	2.6114	3.127	3.982	
0.80	0.2522	0.5097	0.7793	1.0696	1.3961	1.7912	2.0318	2.3242	2.6826	3.215	4.097	
0.85	0.2572	0.5199	0.7953	1.0925	1.4276	1.8341	2.0821	2.3830	2.7523	3.300	4.209	
0.90	0.2622	0.5301	0.8112	1.1152	1.4588	1.8764	2.1317	2.4407	2.8206	3.384	4.318	
0.95	0.2672	0.5403	0.8270	1.1376	1.4897	1.9181	2.1805	2.4974	2.8876	3.466	4.423	
1.00	0.2722	0.5505	0.8426	1.1598	1.5203	1.9592	2.2286	2.5532	2.9532	3.5461	4.4568	

1.05	0.2771	0.5605	0.8581	1.1818	1.5505	1.9997	2.2748	2.6081	3.0175	3.6249	4.6287
1.10	0.2819	0.5704	0.8735	1.2035	1.5804	2.0396	2.3204	2.6621	3.0806	3.7023	4.7291
1.15	0.2866	0.5801	0.8887	1.2250	1.6099	2.0789	2.3658	2.7152	3.1427	3.7784	4.8281
1.20	0.2913	0.5897	0.9038	1.2463	1.6390	2.1177	2.4103	2.7674	3.2039	3.8532	4.9257
1.25	0.2959	0.5992	0.9187	1.2574	1.6677	2.1559	2.4543	2.8198	3.2642	3.9268	5.0219
1.30	0.3005	0.6085	0.9335	1.2883	1.6959	2.1935	2.4978	2.8694	3.3237	3.9993	5.1167
1.35	0.3051	0.6179	0.9481	1.3090	1.7237	2.2306	2.5409	2.9192	3.3824	4.0707	5.2102
1.40	0.3096	0.6272	0.9626	1.3295	1.7511	2.2671	2.5835	2.9683	3.4403	4.1410	5.3023
1.45	0.3141	0.6364	0.9769	1.3498	1.7781	2.3030	2.6256	3.0167	3.4975	4.2103	5.3931
1.50	0.3185	0.6455	0.9911	1.3698	1.8047	2.3384	2.6672	3.0644	3.5539	4.2786	5.4825
1.55	0.3229	0.6545	1.0051	1.3896	1.8310	2.3733	2.7082	3.1115	3.6096	4.3460	5.5706
1.60	0.3272	0.6634	1.0190	1.4092	1.8570	2.4078	2.7486	3.1580	3.6654	4.4125	5.6574
1.65	0.3317	0.6723	1.0328	1.4286	1.8828	2.4419	2.7884	3.2039	3.7186	4.4781	5.7428
1.70	0.3360	0.6811	1.0464	1.4478	1.9084	2.4757	2.8276	3.2493	3.7719	4.5428	5.8268
1.75	0.3403	0.6898	1.0599	1.4668	1.9338	2.5092	2.8663	3.2941	3.8244	4.6066	5.9094
1.80	0.3445	0.6984	1.0733	1.4856	1.9590	2.5424	2.9044	3.3384	3.8761	4.6695	5.9907
1.85	0.3487	0.7069	1.0865	1.5042	1.9840	2.5753	2.9420	3.3821	3.9271	4.7315	6.0706
1.90	0.3527	0.7153	1.0996	1.5225	2.0088	2.6079	2.9791	3.5243	3.9773	4.7926	6.1492
1.95	0.3568	0.7237	1.1126	1.5406	2.0334	2.6403	3.0156	3.4680	4.0267	4.8528	6.2264
2.00	0.3609	0.7320	1.1255	1.5585	2.0578	2.6724	3.0516	3.5102	4.0753	4.9121	6.3022

→ ∞

Table 9.1b ξ_{2n} as a function of ϕ and $n^2/2\chi^2$

$\frac{n^2}{2\chi^2}$	$\phi = 80°$	70°	60°	50°	40°	30°	25°	20°	15°	10°	5°	0°
0.00	0.1745	0.3491	0.5236	0.6981	0.8727	1.0472	1.1345	1.222	1.309	1.396	1.484	1.571
0.01	0.1736	0.3472	0.5207	0.6939	0.8667	1.039	1.124	1.208	1.292	1.369	1.430	1.468
0.02	0.1727	0.3454	0.5178	0.6898	0.8608	1.030	1.113	1.195	1.273	1.343	1.390	1.410
0.03	0.1719	0.3436	0.5150	0.6856	0.8550	1.022	1.102	1.182	1.256	1.320	1.357	1.374
0.04	0.1710	0.3419	0.5122	0.6815	0.8492	1.013	1.093	1.169	1.240	1.298	1.329	1.345
0.05	0.1701	0.3401	0.5094	0.6775	0.8434	1.005	1.083	1.157	1.224	1.277	1.306	1.320
0.06	0.1693	0.3383	0.5066	0.6734	0.8377	0.997	1.073	1.145	1.209	1.258	1.285	1.299
0.07	0.1684	0.3366	0.5038	0.6694	0.8321	0.989	1.063	1.133	1.196	1.240	1.266	1.279
0.08	0.1676	0.3348	0.5010	0.6654	0.8265	0.981	1.053	1.121	1.181	1.223	1.249	1.261
0.09	0.1667	0.3331	0.4983	0.6614	0.8209	0.973	1.044	1.110	1.168	1.208	1.233	1.244
0.10	0.1659	0.3314	0.4956	0.6575	0.8154	0.965	1.035	1.098	1.156	1.193	1.218	1.228
0.15	0.1618	0.3229	0.4822	0.6384	0.7887	0.928	0.991	1.046	1.096	1.126	1.148	1.155
0.20	0.1579	0.3147	0.4694	0.6199	0.7633	0.894	0.951	1.001	1.041	1.071	1.089	1.095
0.25	0.1541	0.3068	0.4571	0.6024	0.7393	0.862	0.914	0.959	0.994	1.025	1.039	1.045
0.30	0.1504	0.2992	0.4453	0.5857	0.7166	0.832	0.881	0.922	0.953	0.984	0.996	1.001
0.35	0.1469	0.2919	0.4339	0.5697	0.6952	0.804	0.850	0.888	0.918	0.946	0.957	0.962
0.40	0.1435	0.2849	0.4230	0.5544	0.6750	0.779	0.822	0.858	0.886	0.912	0.922	0.927
0.45	0.1402	0.2782	0.4125	0.5398	0.6560	0.755	0.797	0.831	0.858	0.880	0.891	0.895
0.50	0.1370	0.2718	0.4025	0.5260	0.6381	0.733	0.773	0.806	0.832	0.851	0.862	0.866
0.55	0.1340	0.2657	0.3929	0.5129	0.6213	0.6122	0.7507	0.783	0.808	0.824	0.835	0.839
0.60	0.1311	0.2598	0.3837	0.5005	0.6055	0.6945	0.7301	0.762	0.785	0.798	0.810	0.814
0.65	0.1283	0.2542	0.3750	0.4887	0.5906	0.6786	0.7108	0.742	0.764	0.775	0.788	0.791
0.70	0.1256	0.2488	0.3667	0.4775	0.5766	0.6601	0.6927	0.723	0.744	0.753	0.766	0.770
0.75	0.1230	0.2436	0.3588	0.4669	0.5634	0.6443	0.6757	0.705	0.725	0.734	0.747	0.750
0.80	0.1206	0.2387	0.3514	0.4568	0.5509	0.6293	0.6597	0.688	0.707	0.716	0.729	0.732
0.85	0.1183	0.2340	0.3444	0.4473	0.5390	0.6150	0.6446	0.672	0.691	0.699	0.712	0.715
0.90	0.1161	0.2296	0.3378	0.4383	0.5277	0.6014	0.6304	0.656	0.675	0.684	0.697	0.700
0.95	0.1140	0.2254	0.3316	0.4298	0.5169	0.5885	0.6171	0.642	0.660	0.671	0.682	0.685
1.00	0.1120	0.2214	0.3275	0.4218	0.5065	0.5762	0.6046	0.628	0.647	0.658	0.668	0.671

1.05	0.1100	0.2175	0.3201	0.4141	0.4966	0.5646	0.5928	0.615	0.634	0.646	0.655	0.658
1.10	0.1081	0.2138	0.3147	0.4066	0.4872	0.5536	0.5816	0.603	0.621	0.633	0.642	0.645
1.15	0.1063	0.2102	0.3095	0.3994	0.4783	0.5432	0.5709	0.591	0.601	0.621	0.630	0.633
1.20	0.1045	0.2067	0.3044	0.3925	0.4698	0.5334	0.5607	0.581	0.599	0.610	0.619	0.622
1.25	0.1028	0.2034	0.2995	0.3859	0.4617	0.5242	0.5510	0.570	0.588	0.599	0.608	0.611
1.30	0.1012	0.2002	0.2947	0.3796	0.4541	0.5157	0.5417	0.561	0.578	0.589	0.597	0.600
1.35	0.0996	0.1971	0.2901	0.3736	0.4469	0.5076	0.5328	0.552	0.569	0.580	0.588	0.590
1.40	0.0981	0.1942	0.2856	0.3679	0.4401	0.4998	0.5243	0.543	0.560	0.571	0.578	0.581
1.45	0.0967	0.1914	0.2812	0.3625	0.4337	0.4923	0.5161	0.535	0.551	0.562	0.569	0.571
1.50	0.0954	0.1888	0.2772	0.3575	0.4277	0.4851	0.5083	0.528	0.543	0.554	0.560	0.562
1.55	0.0942	0.1863	0.2732	0.3527	0.4220	0.4782	0.5008	0.520	0.535	0.546	0.552	0.554
1.60	0.0930	0.1839	0.2694	0.3481	0.4165	0.4936	0.4936	0.513	0.527	0.538	0.544	0.546
1.65	0.0918	0.1815	0.2657	0.3436	0.4112	0.4652	0.4868	0.506	0.520	0.531	0.536	0.538
1.70	0.0907	0.1792	0.2622	0.3392	0.4060	0.4590	0.4803	0.499	0.513	0.524	0.529	0.530
1.75	0.0896	0.1770	0.2588	0.3349	0.4009	0.4530	0.4739	0.493	0.507	0.517	0.521	0.523
1.80	0.0885	0.1748	0.2556	0.3306	0.3958	0.4472	0.4678	0.486	0.500	0.510	0.515	0.517
1.85	0.0874	0.1727	0.2525	0.3264	0.3907	0.4416	0.4618	0.480	0.494	0.504	0.508	0.510
1.90	0.8664	0.1706	0.2495	0.3223	0.3857	0.4361	0.4561	0.474	0.488	0.497	0.502	0.504
1.95	0.0854	0.1685	0.2466	0.3183	0.3807	0.4307	0.4507	0.468	0.481	0.491	0.496	0.498
2.00	0.0844	0.1665	0.2439	0.3144	0.3756	0.4254	0.4456	0.462	0.475	0.485	0.490	0.492

The function ξ_n thus defined is complex and is antisymmetric about $\phi = \pi/2$, varying from 0 when $\phi = \pi/2$ to ∞ when $\phi = 0$ and $-\infty$ when $\phi = 180°$. If we write $\xi_n(\phi)$ as

$$\xi_n = \mathrm{Re}\,\xi_n + i\,\mathrm{Im}\,\xi_n. \tag{9.5.21}$$

Eq. (9.5.19b) may be written as

$$U_n = \alpha_n\,\mathrm{e}^{\chi\xi_{1n}}(\cos\chi\xi_{2n} - i\sin\chi\xi_{2n}) + \beta_n\,\mathrm{e}^{-\chi\xi_{1n}}(\cos\chi\xi_{2n} + i\sin\chi\xi_{2n}), \tag{9.5.22a}$$

where

$$\xi_{1n} = \mathrm{Re}\,\xi_n + \mathrm{Im}\,\xi_n, \tag{9.5.22b}$$

$$\xi_{2n} = \mathrm{Re}\,\xi_n - \mathrm{Im}\,\xi_n. \tag{9.5.22c}$$

Numerical values of ξ_{1n} and ξ_{2n} have been calculated as functions of ϕ for various values of $n^2/2\chi^2$ and are given in Table 9.1. For values of ϕ greater than 90 degrees, the values of ξ_{1n} and ξ_{2n} are the negative of the values given for $(180 - \phi)$ degrees.

When we convert back to the expression for F_n, we have

$$F_n = \frac{1}{i[\sin^2\phi + i(n^2/2\chi^2)]^{1/4}}\,U_n. \tag{9.5.23}$$

If now we put

$$\sin^2\phi + i\frac{n^2}{2\chi^2} = \zeta_n^4\,\mathrm{e}^{4i\psi_n}, \tag{9.5.24a}$$

and solve for ζ_n and ψ_n to obtain

$$\zeta_n^8 = \frac{n^4}{4\chi^4} + \sin^4\phi, \tag{9.5.24b}$$

$$\tan 4\psi_n = \frac{n^2}{2\chi^2\sin^2\phi}, \tag{9.5.24c}$$

the expression for F_n can be written as

$$\begin{aligned} F_n &= -\frac{1}{\zeta_n}(\sin\psi_n + i\cos\psi_n)[\alpha_n\,\mathrm{e}^{\chi\xi_{1n}}(\cos\chi\xi_{2n} - i\sin\chi\xi_{2n}) \\ &\quad + \beta_n\,\mathrm{e}^{-\chi\xi_{1n}}(\cos\chi\xi_{2n} + i\sin\chi\xi_{2n})] \\ &= -\frac{1}{\zeta_n}\{\alpha_n\,\mathrm{e}^{\chi\xi_{1n}}[\sin(\chi\xi_{2n} + \psi_n) + i\cos(\chi\xi_{2n} + \psi_n)] \\ &\quad - \beta_n\,\mathrm{e}^{-\chi\xi_{1n}}[\sin(\chi\xi_{2n} - \psi_n) - i\cos(\chi\xi_{2n} - \psi_n)]\}. \end{aligned} \tag{9.5.25}$$

Thus the stress function S_n and the radial deformation w_n can be expressed in terms of the functions

$$\operatorname{Re} F_{n1} = \frac{1}{\zeta_n} e^{\chi\xi_{1n}} \sin(\chi\xi_{2n} + \psi_n), \tag{9.5.26a}$$

$$\operatorname{Im} F_{n1} = \frac{1}{\zeta_n} e^{\chi\xi_{1n}} \cos(\chi\xi_{2n} + \psi_n), \tag{9.5.26b}$$

$$\operatorname{Re} F_{n2} = \frac{1}{\zeta_n} e^{-\chi\xi_{1n}} \sin(\chi\xi_{2n} - \psi_n), \tag{9.5.26c}$$

$$\operatorname{Im} F_{n2} = \frac{1}{\zeta_n} e^{-\chi\xi_{1n}} \cos(\chi\xi_{2n} - \psi_n). \tag{9.5.26d}$$

We also need the first derivatives of these functions with respect to ϕ, a process which is complicated by the fact that ζ_n, ξ_{1n}, ξ_{2n}, and ψ_n are all functions of ϕ. The derivatives of these functions with respect to ϕ can be readily calculated, however, to yield

$$\frac{d \operatorname{Re} F_{n1}}{d\phi} = -\frac{1}{\zeta_n} e^{\chi\xi_{1n}} \times \left[\frac{1}{4\zeta_n^4} \sin 2\phi \sin(\chi\xi_{2n} + 5\psi_n) + \frac{2^{1/2}\chi\zeta_n^2}{\sin\phi} \sin\left(\chi\xi_{2n} - \psi_n + \frac{\pi}{4}\right)\right], \tag{9.5.27a}$$

$$\frac{d \operatorname{Im} F_{n1}}{d\phi} = -\frac{1}{\zeta_n} e^{\chi\xi_{1n}} \times \left[\frac{1}{4\zeta_n^4} \sin 2\phi \cos(\chi\xi_{2n} + 5\psi_n) + \frac{2^{1/2}\chi\zeta_n^2}{\sin\phi} \cos\left(\chi\xi_{2n} - \psi_n + \frac{\pi}{4}\right)\right], \tag{9.5.27b}$$

$$\frac{d \operatorname{Re} F_{n2}}{d\phi} = -\frac{1}{\zeta_n} e^{-\chi\xi_{1n}} \times \left[\frac{1}{4\zeta_n^4} \sin 2\phi \sin(\chi\xi_{2n} - 5\psi_n) - \frac{2^{1/2}\chi\zeta_n^2}{\sin\phi} \sin\left(\chi\xi_{2n} + \psi_n - \frac{\pi}{4}\right)\right], \tag{9.5.27c}$$

$$\frac{d \operatorname{Im} F_{n2}}{d\phi} = -\frac{1}{\zeta_n} e^{-\chi\xi_{1n}} \times \left[\frac{1}{4\zeta_n^4} \sin 2\phi \cos(\chi\xi_{2n} - 5\psi_n) - \frac{2^{1/2}\chi\zeta_n^2}{\sin\phi} \cos\left(\chi\xi_{2n} + \psi_n - \frac{\pi}{4}\right)\right]. \tag{9.5.27d}$$

9.6 Bending of a spherical shell by moments at the poles: exact solution

We are now in a position to solve various problems of interest with the use of the solutions of the bending equation obtained in the preceding sections. The first of these which we shall investigate is the stress distribution in a spherical shell subjected to pure bending moments at its poles (Fig. 9.6a). We have already obtained a solution of the equations for which the resultants of forces and moments in the vicinity of the poles correspond to a pure moment. We cannot accept this solution as complete, however, since it gives discontinuous displacements at the apex and bears no resemblance to the known solution for a flat plate. Thus we must investigate the problem further. We shall first write an exact solution in terms of Legendre functions.

If we restrict ourselves to solutions involving only the circumferential variation $\cos\theta$ or $\sin\theta$, we deal with Eqs. (9.3.21) to (9.3.29), and the pertinent equations of Section 9.5 with $n = 1$. In the former equations we put C_2 equal to 0 since the loads have no horizontal force resultant and we can ignore rigid body displacements so that C_3 and C_4 vanish. From conditions of symmetry of the loading the radial displacements should be symmetrical about the equator. When we note that the function $\{P_\mu^1(\cos\phi) + P_\mu^1[\cos(\pi - \phi)]\}$ satisfies this symmetry condition, we put $\bar{C}_1$ equal to $\bar{A}_1$ and $\bar{D}_1$ equal to $\bar{B}_1$ in Eq. (9.5.9b) and write

$$w = -\frac{2\chi^2}{Eha}\left\{\bar{A}_1\left[\operatorname{Re}\tilde{P}_\mu^1(\phi) + \frac{1}{2\chi^2}\operatorname{Im}\tilde{P}_\mu^1(\phi)\right] - \bar{B}_1\left[\operatorname{Im}\tilde{P}_\mu^1(\phi) - \frac{1}{2\chi^2}\operatorname{Re}\tilde{P}_\mu^1(\phi)\right]\right\}\cos\theta, \tag{9.6.1a}$$

$$S = [\bar{A}_1 \operatorname{Im}\tilde{P}_\mu^1(\phi) + \bar{B}_1 \operatorname{Re}\tilde{P}_\mu^1(\phi)]\cos\theta, \tag{9.6.1b}$$

where

$$\tilde{P}_\mu^1(\phi) = \frac{\pi}{2}\{P_\mu^1(\cos\phi) + P_\mu^1[\cos(\pi - \phi)]\}\csc\pi\mu, \tag{9.6.1c}$$

and the constant $2/\pi \sin\pi\mu$ has been absorbed in A_1 and B_1. The meridional and circumferential displacements are then given by

$$u_\phi = \frac{1+\nu}{2Eha}\left\{\frac{M_0}{\pi(1 + h^2/12a^2)}\left(\ln\tan\frac{\phi}{2} - \frac{\cos\phi}{\sin^2\phi}\right) - 2\left[\bar{A}_1\frac{\mathrm{d}\operatorname{Im}\tilde{P}_\mu^1(\phi)}{\mathrm{d}\phi} + \bar{B}_1\frac{\mathrm{d}\operatorname{Re}\tilde{P}_\mu^1(\phi)}{\mathrm{d}\phi}\right]\right\}\cos\theta \tag{9.6.2a}$$

$$u_\theta = -\frac{1+\nu}{2Eha}\left\{\frac{M_0}{\pi(1+h^2/12a^2)}\left(\cos\phi \ln\tan\frac{\phi}{2} + 1 + \frac{1}{\sin^2\phi}\right) - \frac{2}{\sin\phi}\left[\bar{A}_1 \operatorname{Im}\tilde{P}^1_\mu(\phi) + \bar{B}_1 \operatorname{Re}\tilde{P}^1_\mu(\phi)\right]\right\}\sin\theta. \tag{9.6.2b}$$

If the sphere is to remain continuous and have no holes at the poles, we must have

$$w = 0 \quad \text{at} \quad \phi = 0 \quad \text{and} \quad \pi, \tag{9.6.3a}$$

$$u_\phi\Big|_{\substack{\phi=0,\pi\\ \theta}} = -\,u_\theta\Big|_{\substack{\phi=0,\pi\\ \theta+\pi/2}} \tag{9.6.3b}$$

which are two conditions for the determination of A_1 and B_1. To determine whether or not we can satisfy the conditions, we must investigate the behavior of $\tilde{P}^1_\mu(\phi)$ and its first derivative in the vicinity of the poles. It can be shown that

$$\tilde{P}^1_\mu(\phi) \approx -\frac{1}{\phi} + 0(\phi), \tag{9.6.4a}$$

$$\frac{1}{\sin\phi}\tilde{P}^1_\mu(\phi) \approx -\frac{1}{\phi^2} - \frac{1}{3} - \frac{1+2i\chi^2}{2} \times \left[\frac{\pi}{2}\tan\frac{\pi\mu}{2} - \ln\tan\frac{\phi}{2} - \gamma - \psi(\mu+1) + \frac{1}{2}\right] + 0(\phi^2), \tag{9.6.4b}$$

$$\frac{d\tilde{P}^1_\mu(\phi)}{d\phi} \approx \frac{1}{\phi^2} - \frac{1}{6} - \frac{1+2i\chi^2}{2} \times \left[\frac{\pi}{2}\tan\frac{\pi\mu}{2} - \ln\tan\frac{\phi}{2} - \gamma - \psi(\mu+1) - \frac{1}{2}\right] + 0(\phi^2). \tag{9.6.4c}$$

The behavior of the deflections in the vicinity of the apex is then

$$w \approx \frac{1}{Eha}\left[\frac{2\chi^2\bar{A}_1 + \bar{B}_1}{\phi} + 0(\phi)\right]\cos\theta, \tag{9.6.5a}$$

$$u_\phi \approx \frac{1+\nu}{2Eha}\left\{\left[\frac{M_0}{\pi(1+h^2/12a^2)} - (2\chi^2\bar{A}_1 + \bar{B}_1)\right]\ln\tan\frac{\phi}{2} - \frac{1}{\phi^2}\left[\frac{M_0}{\pi(1+h^2/12a^2)} + 2\bar{B}_1\right] + 0(1)\right\}\cos\theta, \tag{9.6.5b}$$

$$u_\theta \approx -\frac{1+\nu}{2Eha}\left[\left\{\frac{M_0}{\pi(1+h^2/12a^2)} - (2\chi^2\bar{A}_1 + \bar{B}_1)\right]\ln\tan\frac{\phi}{2} + \frac{1}{\phi^2}\left[\frac{M_0}{\pi(1+h^2/12a^2)} + 2\bar{B}_1\right] + 0(1)\right\}\sin\theta. \tag{9.6.5c}$$

For the radial displacement to vanish at the apex, Eq. (9.6.5a) indicates that the coefficient of $1/\phi$ must vanish so that A_1 and B_1 are related by

$$2\chi^2 \bar{A}_1 + \bar{B}_1 = 0. \tag{9.6.6a}$$

The condition relating u_θ and u_ϕ is more difficult. If Eq. (9.6.6a) holds there is no possibility of getting rid of both of the singularities in Eqs. (9.6.5b) and (9.6.5c). We note, however, that if the coefficient of $1/\phi^2$ is made to vanish; i.e.

$$2\bar{B}_1 + \frac{M_0}{\pi(1 + h^2/12a^2)} = 0, \tag{9.6.6b}$$

then the remaining singularity satisfies Eq. (9.6.3b) in the sense that both displacements approach infinity in the required manner. Further investigation indicates that Eq. (9.6.3b) is *not* equivalent to the statement

$$u_\phi \Big|_{\substack{\phi = 0,\, \pi \\ \theta}} + u_\phi \Big|_{\substack{\phi = 0,\, \pi \\ \theta + \pi/2}} = 0, \tag{9.6.7}$$

since the sum has a finite limit.

With A_1 and B_1 given by Eqs. (9.6.6), the displacements have the following expressions

$$w = -\frac{M_0 a}{4\pi\chi^2 D} \operatorname{Im} \tilde{P}_\mu^1(\phi) \cos\theta, \tag{9.6.8a}$$

$$u_\phi = \frac{(1+\nu)M_0}{2\pi Eha(1 + h^2/12a^2)} \times \left[\ln\tan\frac{\phi}{2} - \frac{\cos\phi}{\sin^2\phi} + \frac{\mathrm{d}\operatorname{Re}\tilde{P}_\mu^1(\phi)}{\mathrm{d}\phi} - \frac{1}{2\chi^2}\frac{\mathrm{d}\operatorname{Im}\tilde{P}_\mu^1(\phi)}{\mathrm{d}\phi}\right]\cos\theta, \tag{9.6.8b}$$

$$u_\theta = -\frac{(1+\nu)M_0}{2\pi Eha(1 + h^2/12a^2)} \times \left[\cos\phi\ln\tan\frac{\phi}{2} + 1 + \frac{1}{\sin^2\phi} + \frac{\operatorname{Re}\tilde{P}_\mu^1(\phi) - (1/2\chi^2)\operatorname{Im}\tilde{P}_\mu^1(\phi)}{\sin\phi}\right]\sin\theta. \tag{9.6.8c}$$

Since there are no other coefficients which can be adjusted, we must settle for a solution containing logarithmic displacement singularities. These vanish, however, when the radius of the sphere becomes infinite and thus do not contradict the results of flat plate theory for which only normal deformations occur when moment is applied in a plane normal to the plate. The slope at the apex is likewise infinite, for

$$\Theta\Bigg|_{\phi=\theta=0} = \frac{1}{a}\left(\frac{\partial w}{\partial \phi} - u_\phi\right)\Bigg|_{\phi=\theta=0} \to \quad -\frac{M_0 a}{4\pi D}\left[1 + \frac{2(1+\nu)}{4\chi^4+1}\right] \ln \tan\frac{\phi}{2}. \tag{9.6.9}$$

Singularities also appear in the expressions for the middle surface forces and bending and twisting moments. The middle surface forces are given by

$$N_\phi = \frac{M_0 \cos\theta}{2\pi(1 + h^2/12a^2)a^2} \times \left\langle \frac{2}{\sin^3\phi} - \frac{\mathrm{d}}{\mathrm{d}\phi}\left\{\cot\phi\left[\operatorname{Re}\tilde{P}^1_\mu(\phi) - \frac{1}{2\chi^2}\operatorname{Im}\tilde{P}^1_\mu(\phi)\right)\right]\right\rangle, \tag{9.6.10a}$$

$$N_\theta = -N_\phi + \frac{Eh}{a}w, \tag{9.6.10b}$$

$$N_{\phi\theta} = \frac{M_0 \sin\theta}{2\pi(1 + h^2/12a^2)a^2} \times \left\langle \frac{2\cos\phi}{\sin^3\phi} - \frac{\mathrm{d}}{\mathrm{d}\phi}\left\{\frac{1}{\sin\phi}\left[\operatorname{Re}\tilde{P}^1_\mu(\phi) - \frac{1}{2\chi^2}\operatorname{Im}\tilde{P}^1_\mu(\phi)\right]\right\}\right\rangle, \tag{9.6.10c}$$

When we use Eqs. (9.6.4) we can show that the behavior of the forces in the vicinity of the apex is as follows:

$$N_\phi \approx \frac{M_0 \cos\theta}{2\pi(1 + h^2/12a^2)a^2}\left[\frac{1}{\phi} + 0(\phi)\right], \tag{9.6.11a}$$

$$N_\theta \approx -\frac{M_0 \cos\theta}{2\pi(1 + h^2/12a^2)a^2}\left[\frac{1}{\phi} + 0(\phi)\right], \tag{9.6.11b}$$

$$N_{\phi\theta} \approx \frac{M_0 \sin\theta}{2\pi(1 + h^2/12a^2)a^2}\,0(\phi), \tag{9.6.11c}$$

so that the imposition of the continuity or regularity conditions has reduced the severity of the singularities in the case of the meridional and circumferential forces and has completely eliminated the singularity of the middle surface shear force.

The bending and twisting moments and transverse shear forces are similarly given by

$$M_\phi = -\frac{M_0}{2\pi a}\left\langle \operatorname{Re}\tilde{P}^1_\mu(\phi) - \frac{\nu}{2\chi^2}\operatorname{Im}\tilde{P}^1_\mu(\phi) + \frac{1-\nu}{2\chi^2}\,\frac{4\chi^4-\nu}{4\chi^4+1}\,\frac{\mathrm{d}}{\mathrm{d}\phi}\left[\cot\phi\operatorname{Im}\tilde{P}^1_\mu(\phi)\right] + \frac{1-\nu^2}{4\chi^4+1}\left\{\frac{\mathrm{d}}{\mathrm{d}\phi}\left[\cot\phi\operatorname{Re}\tilde{P}^1_\mu(\phi)\right] - \frac{2}{\sin^3\phi}\right\}\right\rangle \cos\theta, \tag{9.6.12a}$$

$$M_\theta = -\frac{M_0}{2\pi a}\left\langle \nu\left[\operatorname{Re}\tilde{P}^1_\mu(\phi) - \frac{\nu}{2\chi^2}\operatorname{Im}\tilde{P}^1_\mu(\phi)\right] - \frac{1-\nu}{2\chi^2}\frac{4\chi^4-\nu}{4\chi^4+1}\frac{\mathrm{d}}{\mathrm{d}\phi}[\cot\phi\operatorname{Im}\tilde{P}^1_\mu(\phi)] - \frac{1-\nu^2}{4\chi^4+1}\left\{\frac{\mathrm{d}}{\mathrm{d}\phi}[\cot\phi\operatorname{Re}\tilde{P}^1_\mu(\phi)] - \frac{2}{\sin^3\phi}\right\}\right\rangle\cos\theta, \tag{9.6.12b}$$

$$M_{\phi\theta} = -\frac{M_0}{2\pi a}\frac{1-\nu}{4\chi^4+1}\left\langle\frac{4\chi^4-\nu}{2\chi^2}\frac{\mathrm{d}}{\mathrm{d}\phi}\left[\frac{1}{\sin\phi}\operatorname{Im}\tilde{P}^1_\mu(\phi)\right] + (1+\nu)\left\{\frac{\mathrm{d}}{\mathrm{d}\phi}\left[\frac{1}{\sin\phi}\operatorname{Re}\tilde{P}^1_\mu(\phi)\right] - \frac{2\cos\phi}{\sin^3\phi}\right\}\right\rangle, \sin\theta \tag{9.6.12c}$$

$$Q_\phi = -\frac{4\chi^4+\nu^2}{4\chi^4+1}\frac{M_0}{2\pi a^2}\frac{\mathrm{d}}{\mathrm{d}\phi}\left[\operatorname{Re}\tilde{P}^1_\mu(\phi) - \frac{1}{2\chi^2}\operatorname{Im}\tilde{P}^1_\mu(\phi)\right]\cos\theta, \tag{9.6.12d}$$

$$Q_\theta = \frac{4\chi^4+\nu^2}{4\chi^4+1}\frac{M_0}{2\pi a^2}\frac{1}{\sin\phi}\left[\operatorname{Re}\tilde{P}^1_\mu(\phi) - \frac{1}{2\chi^2}\operatorname{Im}\tilde{P}^1_\mu(\phi)\right]\sin\theta, \tag{9.6.12e}$$

and vary in the vicinity of the apex as follows:

$$M_\phi \to \frac{(1+\nu)M_0}{4\pi a\phi}\left[1 + \frac{2(1-\nu)}{4\chi^4+1}\right]\cos\theta, \tag{9.6.13a}$$

$$M_\phi \to \frac{(1+\nu)M_0}{4\pi a\phi}\left[1 - \frac{2(1-\nu)}{4\chi^4+1}\right]\cos\theta, \tag{9.6.13b}$$

$$M_{\phi\theta} \to -\frac{(1-\nu)M_0}{4\pi a\phi}\left[1 - \frac{2(1+\nu)}{4\chi^4+1}\right]\sin\theta, \tag{9.6.13c}$$

$$\frac{Q_\phi}{\cos\theta} \to \frac{Q_\theta}{\sin\theta} \to -\frac{M_0}{2\pi a^2\phi^2}\left(1 - \frac{1-\nu^2}{4\chi^4+1}\right). \tag{9.6.13d}$$

Except for small terms of the order of h^2/a^2 compared to unity, the behavior of the moments and transverse shear forces in the spherical shell is identical to the behavior of the corresponding quantities in a flat circular plate under concentrated moment.

9.7 Bending of a spherical shell by moments at the poles: shallow shell theory

For practical calculation, but only because the Legendre functions are untabulated, it is preferable to make use of the approximate solution of the bending equations given by Eqs. (9.5.12) and (9.5.14) which may also be written as

$$\operatorname{Re} F_{n1} = \left(\frac{\phi}{\sin \phi}\right)^{1/2} \operatorname{ber}' (2^{1/2}\chi\phi), \tag{9.7.1a}$$

$$\operatorname{Im} F_{n1} = -\left(\frac{\phi}{\sin \phi}\right)^{1/2} \operatorname{bei}' (2^{1/2}\chi\phi), \tag{9.7.1b}$$

$$\operatorname{Re} F_{n2} = \left(\frac{\phi}{\sin \phi}\right)^{1/2} \operatorname{ker}' (2^{1/2}\chi\phi), \tag{9.7.1c}$$

$$\operatorname{Im} F_{n2} = -\left(\frac{\phi}{\sin \phi}\right)^{1/2} \operatorname{kei}' (2^{1/2}\chi\phi), \tag{9.7.1d}$$

where primes indicate differentiation with respect to $2^{1/2}\chi\phi$. As in the case of a spherical shell under radial loads at its poles we can treat only the upper half of the sphere and set the coefficients of terms involving the derivatives of the ber and bei functions equal to zero on the basis of the approximation that the equator is far away. Since the significant region of ϕ is very close to the apex for thin shells we may replace $(\phi/\sin \phi)^{1/2}$ by unity. Then Eq. (9.5.9b) for the radial deflection can be written as

$$\begin{aligned} w \approx -\frac{2\chi^2}{Eha} \Bigg[&\left(\bar{C}_1 + \frac{1}{2\chi^2}\bar{D}_1\right) \operatorname{ker}' (2^{1/2}\chi\phi) \\ &+ \left(\bar{D}_1 - \frac{1}{2\chi^2}\bar{C}_1\right) \operatorname{kei}' (2^{1/2}\phi)\Bigg] \cos\theta. \end{aligned} \tag{9.7.2}$$

If w is to vanish at the apex we must have

$$\bar{C}_1 = -\frac{1}{2\chi^2}\bar{D}_1 \tag{9.7.3}$$

since $\operatorname{ker}' (2^{1/2}\chi\phi)$ is infinite there while $\operatorname{kei} (2^{1/2}\chi\phi)$ vanishes. For all practical purposes C_1 vanishes since terms of order $1/2\chi^2$ compared to unity should be omitted as they were in the approximate equation for F_n. We thus can express the radial deflection as

$$w = -\frac{a\bar{D}_1}{2\chi^2 D} \operatorname{kei}' (2^{1/2}\chi\phi) \cos\theta, \tag{9.7.4a}$$

and the meridional and circumferential displacements as

$$\begin{aligned} u_\phi \approx \frac{(1+\nu)}{Eha} \Bigg\{ &\frac{M_0}{2\pi}\left(\ln \tan\frac{\phi}{2} - \frac{\cos\phi}{\sin^2\phi}\right) \\ &+ 2^{1/2}\chi\bar{D}_1 \left[\frac{1}{2^{1/2}\chi\phi} \operatorname{ker}' (2^{1/2}\chi\phi) + \operatorname{kei}' (2^{1/2})\right]\Bigg\} \cos\theta, \end{aligned} \tag{9.7.4b}$$

$$u_\theta \approx -\frac{(1+\nu)}{Eha}\left[\frac{M_0}{2\pi}\left(\cos\phi \ln\tan\frac{\phi}{2} + 1 + \frac{1}{\sin^2\phi}\right) - 2^{1/2}\chi \bar{D}_1 \frac{\text{ker}'(2^{1/2}\chi\phi)}{2^{1/2}\chi\phi}\right]\sin\theta. \tag{9.7.4c}$$

In the vicinity of the apex these become

$$u_\phi \approx \frac{1+\nu}{Eha}\left[\frac{M_0}{2\pi}\ln\tan\frac{\phi}{2} - \left(\frac{\bar{D}_1}{2^{1/2}\chi} + \frac{M_0}{2\pi}\right)\frac{1}{\phi^2} + 0(1)\right]\cos\theta, \tag{9.7.5a}$$

$$u_\theta \approx -\frac{1+\nu}{Eha}\left[\frac{M_0}{2\pi}\ln\tan\frac{\phi}{2} + \left(\frac{\bar{D}_1}{2^{1/2}\chi} + \frac{M_0}{2\pi}\right)\frac{1}{\phi^2} + 0(1)\right]\sin\theta. \tag{9.7.5b}$$

Then for the amplitude of u_θ to be equal to the negative of the amplitude of u_ϕ we must have

$$\bar{D}_1 = -\frac{2^{1/2}M_0\chi}{2\pi} \tag{9.7.5c}$$

The final displacement expressions are, after deleting all terms of order $1/2\chi^2$ or less compared to unity and assuming that ϕ is small,

$$w \approx \frac{2^{1/2}\chi^3 M_0}{\pi Eha}\text{kei}'(2^{1/2}\chi\phi)\cos\theta, \tag{9.7.6a}$$

$$u_\phi \approx -\frac{(1+\nu)\chi^2 M_0}{\pi Eha} \times \left[\frac{1}{2^{1/2}\chi\phi}\text{ker}'(2^{1/2}\chi\phi) + \text{kei}(2^{1/2}\chi\phi) + \frac{1}{(2^{1/2}\chi\phi)^2}\right]\cos\theta, \tag{9.7.6b}$$

$$u_\theta \approx -\frac{(1+\nu)\chi^2 M_0}{\pi Eha}\left[\frac{1}{2^{1/2}\chi\phi}\text{ker}'(2^{1/2}\chi\phi) + \frac{1}{(2^{1/2}\chi\phi)^2}\right]\sin\theta. \tag{9.7.6c}$$

We note that the meridional and circumferential displacements are of the order of $1/\chi$ compared to radial deformation and hence are not of the same order as the terms which have been omitted. In deleting higher order terms we have lost the singularity in the meridional and circumferential displacements. We can show, however, that the singularity of displacement predominates only in a region extremely close to the poles.

Similar deletions of higher order terms for the stress resultants yields

$$N_\phi \approx \frac{2^{1/2}\chi^3 M_0}{\pi a^2}\left[\frac{2}{(2^{1/2}\chi\phi)^3} + \frac{2\,\text{ker}'(2^{1/2}\chi\phi)}{(2^{1/2}\chi\phi)^2} + \frac{\text{kei}(2^{1/2}\chi\phi)}{2^{1/2}\chi\phi}\right]\cos\theta, \tag{9.7.7a}$$

$$N_\theta \approx -\frac{2^{1/2}\chi^3 M_0}{\pi a^2}$$
$$\times \left[\frac{2}{(2^{1/2}\chi\phi)^3} + \frac{2\,\mathrm{ker}'\,(2^{1/2}\chi\phi)}{(2^{1/2}\chi\phi)^2} + \frac{\mathrm{kei}\,(2^{1/2}\chi\phi)}{2^{1/2}\chi\phi} + \mathrm{kei}'\,(2^{1/2}\chi\phi)\right] \cos\theta, \tag{9.7.7b}$$

$$N_{\phi\theta} \approx \frac{2^{1/2}\chi^3 M_0}{\pi a^2}$$
$$\times \left[\frac{2}{(2^{1/2}\chi\phi)^3} + \frac{2\,\mathrm{ker}'\,(2^{1/2}\chi\phi)}{(2^{1/2}\chi\phi)^2} + \frac{\mathrm{kei}\,(2^{1/2}\chi\phi)}{2^{1/2}\chi\phi}\right] \sin\theta, \tag{9.7.7c}$$

$$M_\phi \approx -\frac{2^{1/2} M_0 \chi}{2\pi a}$$
$$\times \left\{\mathrm{ker}'\,(2^{1/2}\chi\phi) - \frac{1+\nu}{2^{1/2}\chi\phi}\left[\mathrm{ker}\,(2^{1/2}\chi\phi) - \frac{2\,\mathrm{kei}'\,(2^{1/2}\chi\phi)}{2^{1/2}\chi\phi}\right]\right\} \cos\theta \tag{9.7.7d}$$

$$M_\theta \approx -\frac{2^{1/2} M_0 \chi}{2\pi a}$$
$$\times \left\{\nu\,\mathrm{ker}'\,(2^{1/2}\chi\phi) + \frac{1-\nu}{2^{1/2}\chi\phi}\left[\mathrm{ker}\,(2^{1/2}\chi\phi) - \frac{2\,\mathrm{kei}'\,(2^{1/2}\chi\phi)}{2^{1/2}\chi\phi}\right]\right\} \cos\theta \tag{9.7.7.e}$$

$$M_{\phi\theta} \approx \frac{2^{1/2}(1-\nu)\chi M_0}{2\pi a}\left[\frac{\mathrm{ker}\,(2^{1/2}\chi\phi)}{2^{1/2}\chi\phi} - \frac{2\,\mathrm{kei}'\,(2^{1/2}\chi\phi)}{2^{1/2}\chi\phi}\right] \sin\theta, \tag{9.7.7f}$$

$$Q_\phi \approx \frac{\chi^2 M_0}{\pi a^2}\left[\frac{1}{2^{1/2}\chi\phi}\,\mathrm{ker}'\,(2^{1/2}\chi\phi) + \mathrm{kei}\,(2^{1/2}\chi\phi)\right] \cos\theta, \tag{9.7.7g}$$

$$Q_\theta \approx \frac{\chi^2 M_0}{\pi a^2}\,\frac{1}{2^{1/2}\chi\phi}\,\mathrm{ker}'\,(2^{1/2}\chi\phi)\sin\theta. \tag{9.7.7h}$$

When we examine the expressions for N_ϕ and N_θ, we find that we have lost the singularity at the pole which would be given by the addition of the term $1/2\chi^2(1/2^{1/2}\chi\phi)$ to the bracketed expressions. While the neglected term is not insignificant, except for very thin shells, the similar singularity in the equations for the bending moments has not been lost and yields stresses of the order of magnitude a/h compared to the stresses due to the middle surface force singularity. Thus the inaccuracy in the middle surface forces is of no real significance for thin shells. The displacements and stress resultants given by Eqs. (9.7.6) and (9.7.7) as a function of the parameter $2^{1/2}\chi\phi$ are shown in Fig. 9.14. The various singularities can be relieved by the application of the

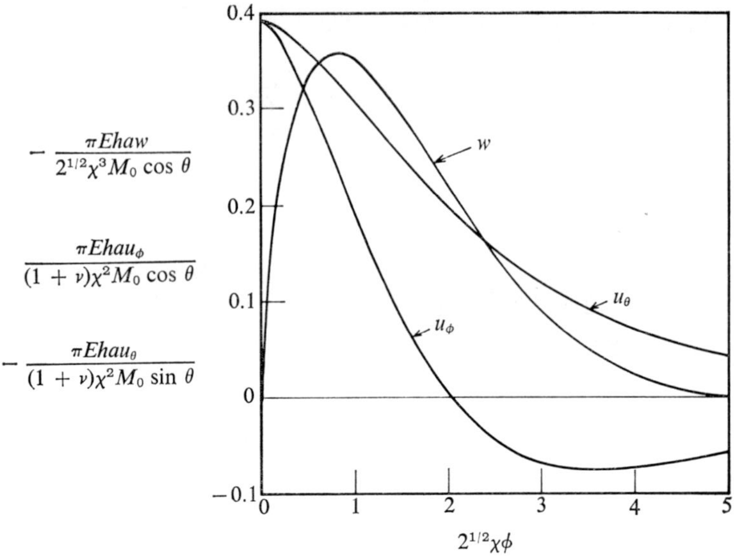

(a) Displacements

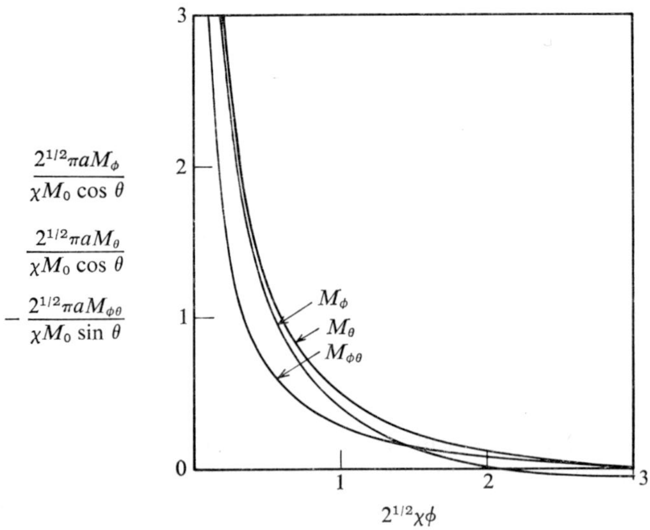

(b) Bending moments

Fig. 9.14. "*Shallow shell*" *results for sphere loaded by bending moments at the poles*

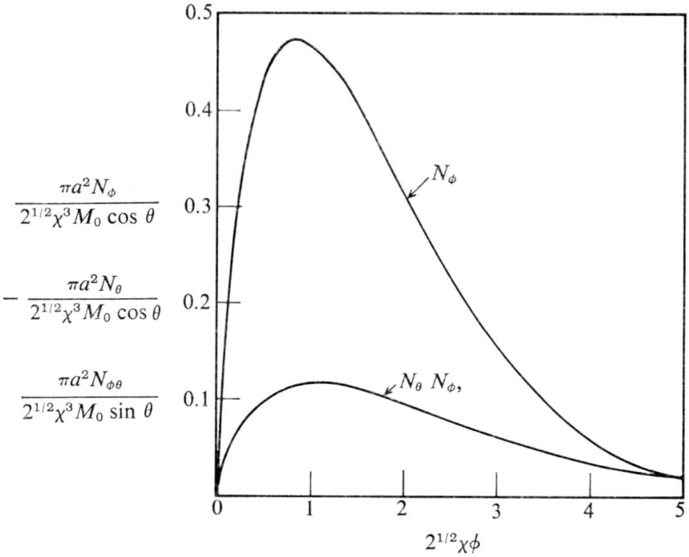

(c) Middle surface forces

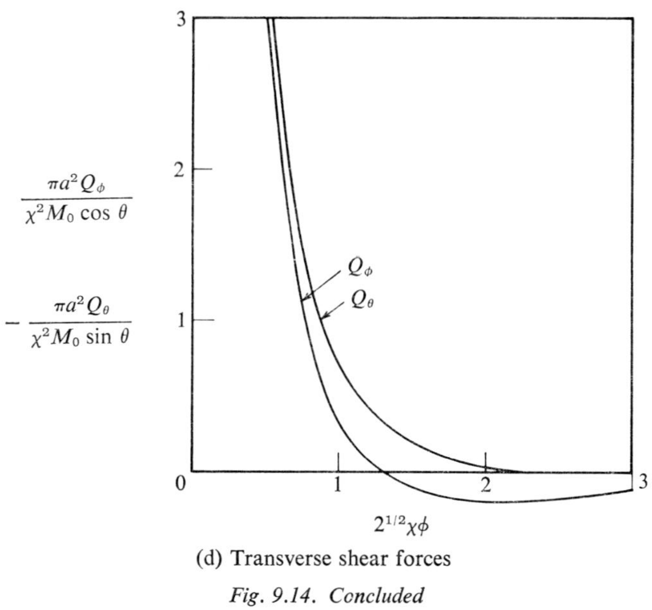

(d) Transverse shear forces

Fig. 9.14. Concluded

moments as the resultant of loads distributed over a small portion of the surface in the vicinity of the poles or by assuming the moment to be applied to a small rigid insert at the shell apex (Fig. 9.15). The latter problem differs from the concentrated moment problem only in the boundary conditions to be imposed at the junction of the shell and the rigid insert. When the shell is rigidly connected to the insert, the boundary conditions may be written as (see Fig. 9.16)

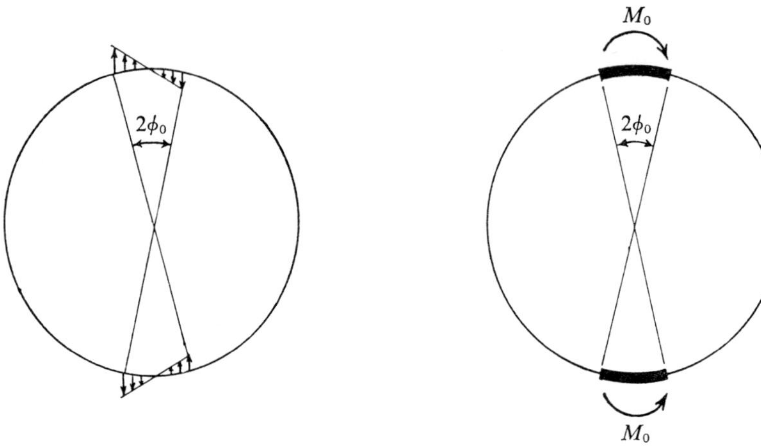

(a) Moment due to distributed load (b) Moment applied to rigid insert

Fig. 9.15. Problems which yield no displacement or force singularities

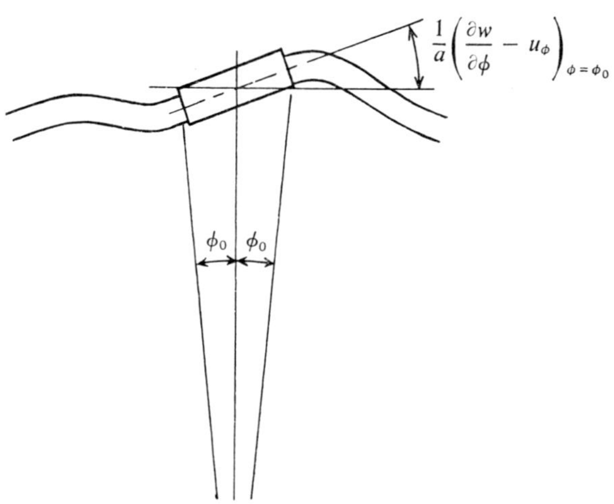

Fig. 9.16. Junction of spherical shell and rigid insert

$$\epsilon_\theta = 0, \tag{9.7.8a}$$

$$\frac{\partial w}{\partial \phi} = w \cot \phi, \tag{9.7.8b}$$

at $\phi = \phi_0$. Deformations and stress resultants calculated by means of shallow shell approximations† for small inserts are shown in Fig. 9.17 and illustrate the alleviation as the rigid insert is increased in size. The rotation and horizontal displacement of the insert relative to the equator are shown in Fig. 9.18 as a function of insert size.

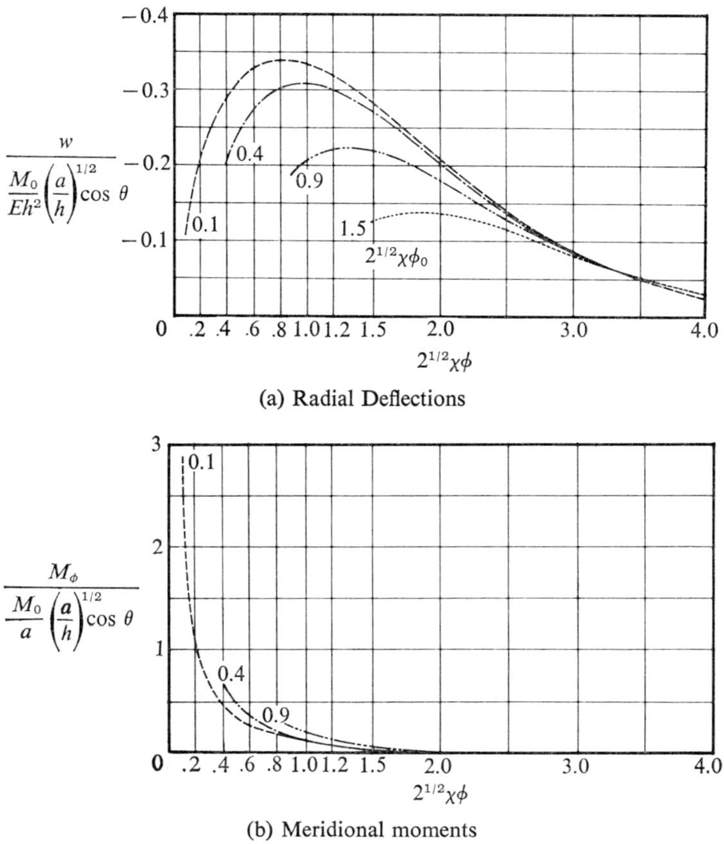

Fig. 9.17. Forces and deformations of a shell loaded by bending moment through a rigid insert

† P. P. Bijlaard; 'On the Stresses from Loads in Spherical Pressure Vessels,' Bulletin U.S. Welding Research Committee 34, 1956.

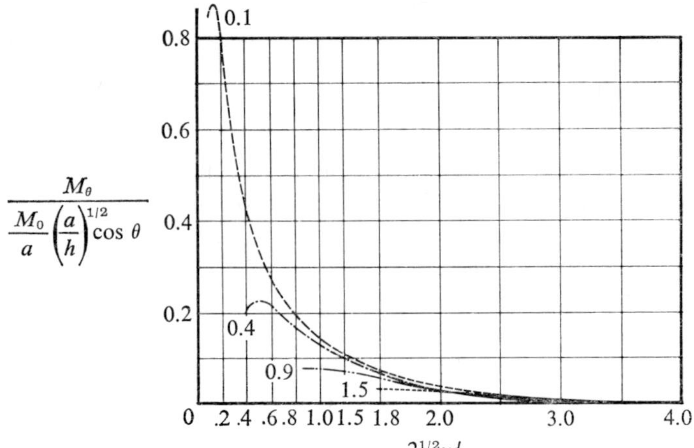

(c) Circumferential moments

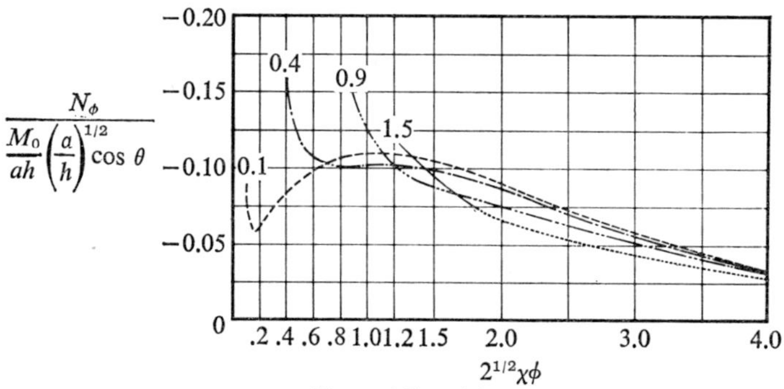

(d) Meridional force

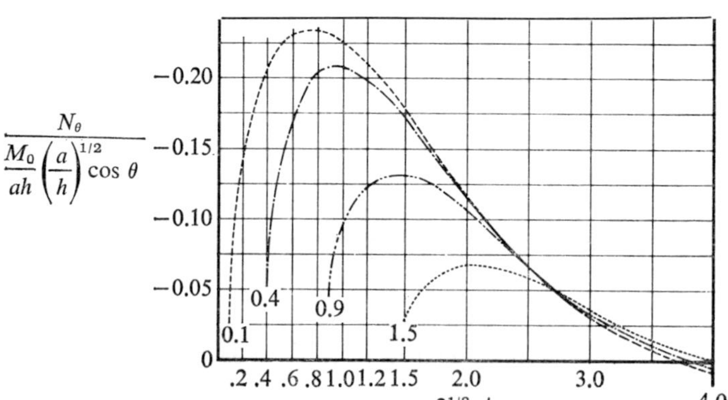

(e) Circumferential force

Fig. 9.17. Concluded

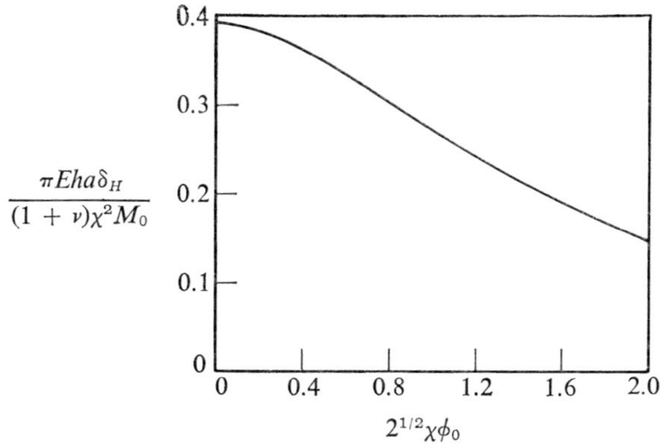

(a) Horizontal displacement

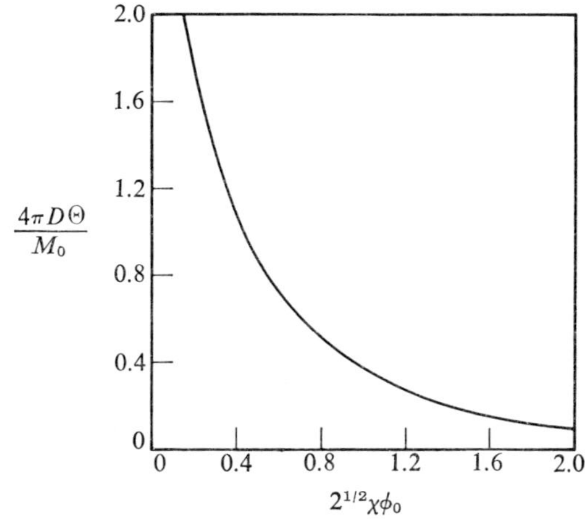

(b) Rotation

Fig. 9.18. Displacement of rigid insert

The accuracy of the shallow shell approximations is indicated in Fig. 9.19 by a comparison of theoretical and experimental† results for a spherical shell to which moment was locally applied as the resultant of equal forces of opposite direction and a fixed distance apart applied to a circular boss as

† A. S. Tooth and R. M. Kenedi; 'The Influence Line Technique of Shell Analysis,' in 'Shell Structures,' *Proc. Colloquium on Simplified Calculation Methods*, Brussels, September 4–6, 1961, North Holland Publishing Co., Amsterdam, the Netherlands, 1962, pp. 44–63.

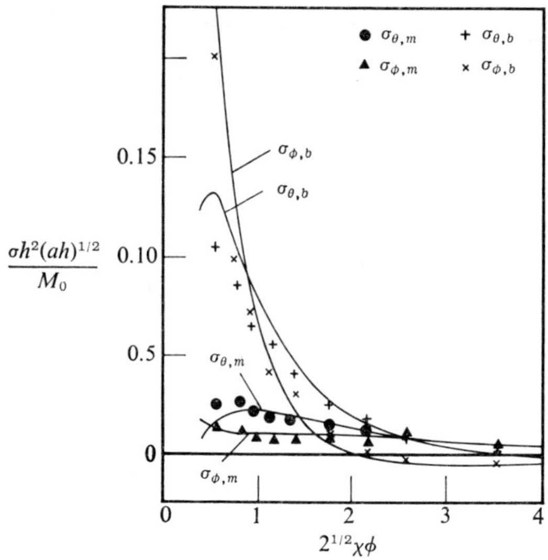

Fig. 9.19. Comparison of theory and experiment for spherical shell loaded by moment through an insert

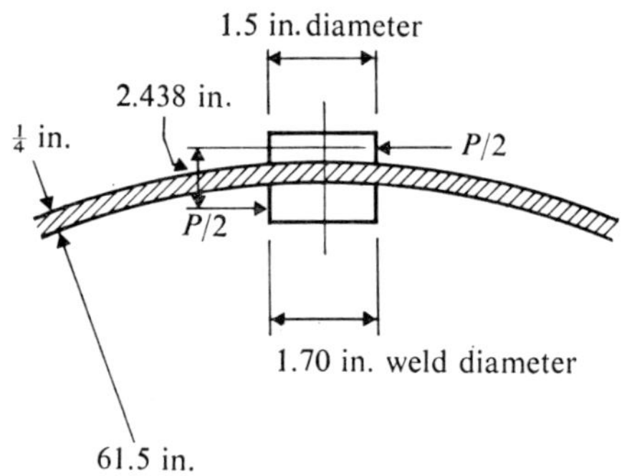

Fig. 9.20. Dimensions of test specimen

shown in Fig. 9.20. The theoretical solution corresponds to the results for a shell with a rigid insert for a value of $2^{1/2}\chi\phi_0$ of 0.394. We see that although discrepancies exist between calculated and experimental values, the agreement is quite satisfactory for all practical purposes.

9.8 Bending of a spherical shell by tangential loads

The second problem we shall investigate is the nature of the deformations and stresses in the vicinity of a tangential load H_0 applied to a spherical shell, which we shall consider to be applied at the apex. We shall assume for convenience that the load is equilibrated at the other pole by a tangential load in the opposite direction, $-H_0$, and by a pure moment $2H_0a$ as in Fig. 9.21. The solution is facilitated if we consider the load system to be the sum of systems which are symmetrical and antisymmetric about the equator of the shell (Fig. 9.22). Since the solution of the symmetrical problem has been investigated, we need consider only the antisymmetric problem in detail, so far as the exact solution is concerned. We note that the function

$$\{P_\mu^1[\cos(\pi - \phi)] - P_\mu^1(\cos\phi)\}$$

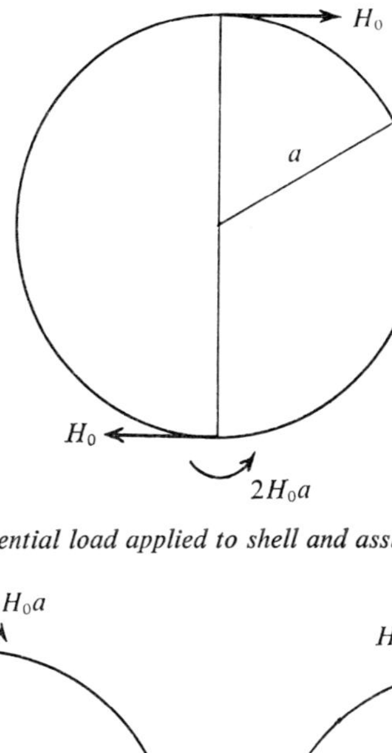

Fig. 9.21. Tangential load applied to shell and assumed resisting forces

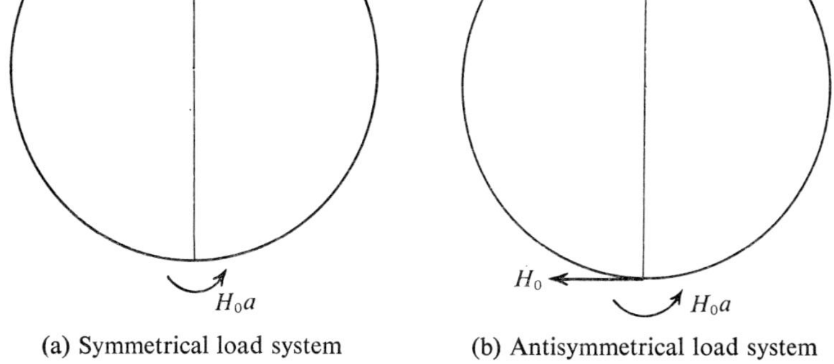

(a) Symmetrical load system (b) Antisymmetrical load system

Fig. 9.22. Symmetrical and antisymmetrical load states equivalent to assumed load system

is antisymmetric about the equator and that a membrane state of stress which yields the antisymmetric force system is given by those terms of Eqs. (9.3.21) to (9.3.29) which involve only the coefficient C_2. Then the displacements can be written as

$$w = \frac{1}{Eha}\left\langle -\frac{(1+\nu)H_0 a}{2\pi}\left(\sin\phi \ln\tan\frac{\phi}{2} - \cot\phi\right) + 2\chi^2\left\{A_1\left[\operatorname{Re}\hat{P}^1_\mu(\phi) + \frac{1}{2\chi^2}\operatorname{Im}\hat{P}^1_\mu(\phi)\right] - \bar{B}_1\left[\operatorname{Im}\hat{P}^1_\mu(\phi) - \frac{1}{2\chi^2}\operatorname{Re}\hat{P}^1_\mu(\phi)\right]\right\}\right\rangle \cos\theta, \tag{9.8.1a}$$

$$u_\phi = \frac{1+\nu}{2Eha}\left\{-\frac{H_0 a}{\pi}\left(\frac{1}{\sin^2\phi} + \cos\phi\ln\tan\frac{\phi}{2}\right) + 2\left[\bar{A}_1\frac{d\operatorname{Im}\hat{P}^1_\mu(\phi)}{d\phi} + \bar{B}_1\frac{d\operatorname{Re}\hat{P}^1_\mu}{d\phi}\right]\right\}\cos\theta, \tag{9.8.1b}$$

$$u_\theta = -\frac{1+\nu}{2Eha}\left\{-\frac{H_0 a}{\pi}\left[\ln\tan\frac{\phi}{2} - \cos\phi\left(1 + \frac{1}{\sin^2\phi}\right)\right] + \frac{2}{\sin\phi} \times [\bar{A}_1\operatorname{Im}\hat{P}^1_\mu(\phi) + \bar{B}_1\operatorname{Re}\hat{P}^1_\mu(\phi)]\right\}\sin\theta, \tag{9.8.1c}$$

where

$$\hat{P}^1_\mu(\phi) = \frac{\pi}{2}\{P^1_\mu[\cos(\pi-\phi)] - P^1_\mu(\cos\phi)\}\csc\pi\mu. \tag{9.8.1d}$$

The behavior of $\hat{P}^1_\mu(\phi)$ and its first derivative in the vicinity of the poles is almost identical to that of $\tilde{P}^1_\mu(\phi)$. In Eqs. (9.6.4) and (9.6.6) we need only replace the term $\pi/2\tan(\pi/2)\mu$ by $-\pi/2\cot\pi\mu/2$. Thus the variation of the displacements in the vicinity of the apex is given by

$$w \approx -\frac{1}{Eha}\left\{\left[2\chi^2\bar{A}_1 + B_1 - \frac{(1+\nu)H_0 a}{2\pi}\right]\frac{1}{\phi} + 0(\phi)\right\}\cos\theta, \tag{9.8.2a}$$

$$u_\phi \approx \frac{1+\nu}{2Eha}\left\{\left[-\frac{H_0 a}{\pi} + 2\chi^2\bar{A} + \bar{B}_1\right]\ln\tan\frac{\phi}{2} + \frac{1}{\phi^2} \times \left[-\frac{H_0 a}{\pi} + 2\bar{B}_1\right] + 0(1)\right\}\cos\theta, \tag{9.8.2b}$$

$$u_\theta \approx -\frac{1+\nu}{2Eha}\left\{\left[-\frac{H_0 a}{\pi} + 2\chi^2\bar{A}_1 + \bar{B}_1\right]\ln\tan\frac{\phi}{2} - \frac{1}{\phi^2} \times \left[-\frac{H_0 a}{\pi} + 2\bar{B}_1\right] + 0(1)\right\}\sin\theta. \tag{9.8.2c}$$

For regularity at the apex we must have [see Eqs. (9.6.3)]

$$2\chi^2\bar{A}_1 + \bar{B}_1 - (1+\nu)\frac{H_0 a}{2\pi} = 0, \tag{9.8.3a}$$

$$-\frac{H_0 a}{\pi} + 2\bar{B}_1 = 0, \tag{9.8.3b}$$

which completely define the solution of the problem. Here again we must allow the displacements u_ϕ and u_θ to be singular at the apex, but in the present problem the singularity is to be expected in analogy with the corresponding flat plate problem.

With $\bar{A}_1$ and $\bar{B}_1$ defined by Eqs. (9.8.3), the displacements are given by

$$w = -\frac{\chi^2 H_0}{\pi E h}\left\{\left(1 - \frac{\nu}{4\chi^4}\right)\operatorname{Im}\hat{P}_\mu^1(\phi) - \frac{1+\nu}{2\chi^2}\right.$$
$$\left.\times\left[\operatorname{Re}\hat{P}_\mu^1(\phi) + \cot\phi - \sin\phi\ln\tan\frac{\phi}{2}\right]\right\}\cos\theta, \tag{9.8.4a}$$

$$u_\phi = \frac{(1+\nu)H_0}{2\pi E h}\left\{\frac{\mathrm{d}}{\mathrm{d}\phi}\left[\operatorname{Re}\hat{P}_\mu^1(\phi) + \frac{\nu}{2\chi^2}\operatorname{Im}\hat{P}_\mu^1(\phi)\right] - \frac{1}{\sin^2\phi}\right.$$
$$\left. - \cos\phi\ln\tan\frac{\phi}{2}\right\}\cos\theta, \tag{9.8.4b}$$

$$u_\theta = -\frac{(1+\nu)H_0}{2\pi E h}\left\{\frac{1}{\sin\phi}\left[\operatorname{Re}\hat{P}_\mu^1(\phi) + \frac{\nu}{2\chi^2}\operatorname{Im}\hat{P}_\mu^1(\phi)\right]\right.$$
$$\left. + \cos\phi\left(1 + \frac{1}{\sin^2\phi}\right) - \ln\tan\frac{\phi}{2}\right\}\sin\theta. \tag{9.8.4c}$$

The stress resultants can be obtained as

$$N_\phi = \left\langle\frac{2\cos\phi}{\sin^3\phi} - \frac{\mathrm{d}}{\mathrm{d}\phi}\left\{\cot\phi\left[\operatorname{Re}\hat{P}_\mu^1(\phi) + \frac{\nu}{2\chi^2}\operatorname{Im}\hat{P}_\mu^1(\phi)\right]\right\}\right\rangle\frac{H_0}{2\pi a}\cos\theta, \tag{9.8.5a}$$

$$N_\theta = -\left\{N_\phi + 2\chi^2\left[\left(1 - \frac{\nu}{4\chi^4}\right)\operatorname{Im}\hat{P}_\mu^1(\phi)\right.\right.$$
$$\left.\left. - \frac{1+\nu}{2\chi^2}\operatorname{Re}\hat{P}_\mu^1(\phi)\right]\frac{H_0}{2\pi a}\cos\theta\right\}, \tag{9.8.5b}$$

$$\bar{N}_{\phi\theta} = \left\langle\frac{2}{\sin^3\phi} - \frac{\mathrm{d}}{\mathrm{d}\phi}\left\{\frac{1}{\sin\phi}\left[\operatorname{Re}\hat{P}_\mu^1(\phi) + \frac{\nu}{2\chi^2}\operatorname{Im}\hat{P}_\mu^1(\phi)\right]\right\}\right\rangle\frac{H_0}{2\pi a}\sin\theta, \tag{9.8.5c}$$

$$Q_\phi = -\frac{d}{d\phi}\left[\operatorname{Re}\hat{P}^1_\mu(\phi) + \frac{\nu}{2\chi^2}\operatorname{Im}\hat{P}^1_\mu(\phi)\right]\frac{H_0}{2\pi a}\cos\theta, \tag{9.8.5d}$$

$$Q_\theta = \frac{1}{\sin\phi}\left[\operatorname{Re}\hat{P}^1_\mu(\phi) + \frac{\nu}{2\chi^2}\operatorname{Im}\hat{P}^1_\mu(\phi)\right]\frac{H_0}{2\pi a}\sin\theta, \tag{9.8.5e}$$

$$M_\phi = -\left\{\operatorname{Re}\hat{P}^1_\mu(\phi) + \frac{1}{2\chi^2}\operatorname{Im}\hat{P}^1_\mu(\phi) + \frac{1-\nu}{2\chi^2} \times \frac{d}{d\phi}[\cot\phi \operatorname{Im}\hat{P}^1_\mu(\phi)]\right\}\frac{H_0}{2\pi}\cos\theta, \tag{9.8.5f}$$

$$M_\theta = -\left\{\nu\left[\operatorname{Re}\hat{P}^1_\mu(\phi) + \frac{1}{2\chi^2}\operatorname{Im}\hat{P}^1_\mu(\phi)\right] - \frac{1-\nu}{2\chi^2} \times \frac{d}{d\phi}[\cot\phi \operatorname{Im}\hat{P}^1_\mu(\phi)]\right\}\frac{H_0}{2\pi}\cos\theta, \tag{9.8.5g}$$

$$\overline{M}_{\phi\theta} = -\frac{1-\nu}{2\chi^2}\frac{d}{d\phi}\left[\frac{1}{\sin\phi}\operatorname{Im}\hat{P}^1_\mu(\phi)\right]\frac{H_0}{2\pi}\sin\theta. \tag{9.8.5h}$$

A study of the behavior of the functions in the vicinity of the apex indicates that the deformations and stress resultants vary as follows:

$$w \to 0, \tag{9.8.6a}$$

$$\frac{u_\phi}{\cos\theta} \to -\frac{u_\theta}{\sin\theta} \to -\frac{(1-\nu^2)H_0}{4\pi Eh}\ln(a\phi), \tag{9.8.6b}$$

$$N_\phi \to N_\theta \to -\frac{(1+\nu)H_0}{4\pi a\phi}\cos\theta, \tag{9.8.6c}$$

$$N_{\phi\theta} \to \frac{(1-\nu)H_0}{4\pi a\phi}\sin\theta, \tag{9.8.6d}$$

$$\frac{Q_\phi}{\cos\theta} \to \frac{Q_\theta}{\sin\theta} \to -\frac{H_0}{2\pi a\phi^2}, \tag{9.8.6e}$$

$$M_\phi \to M_\theta \to \frac{(1+\nu)H_0}{4\pi\phi}\cos\theta, \tag{9.8.6f}$$

$$M_{\phi\theta} \to -\frac{(1-\nu)H_0}{4\pi\phi}\sin\theta. \tag{9.8.6g}$$

We note that the stresses due to middle surface forces are of the order of h/a compared to those due to bending and twisting moments and are thus negligible in the vicinity of the apex for thin shells. We can also determine that transverse shear stresses are large compared to bending stresses within a region about the apex having a radius of the order of the thickness of the sphere.

The solution for the stress and deformation state in the sphere loaded as shown in Fig. 9.21 is now obtained by adding the solution of Section 9.6, with

$$M_0 = -H_0 a, \tag{9.8.7}$$

to the solution just obtained. The behavior of the various quantities in the vicinity of a tangential load can then be obtained as

$$w \to 0, \tag{9.8.8a}$$

$$\frac{u_\phi}{\cos\theta} \to -\frac{u_\theta}{\sin\theta} \to -\frac{(1+\nu)H_0}{4\pi Eh}\left(3-\nu-\frac{h^2/6a^2}{1+h^2/12a^2}\right)\ln(a\phi), \tag{9.8.8b}$$

$$N_\phi \to -\frac{H_0}{4\pi a\phi}\left(3+\nu-\frac{h^2/6a^2}{1+h^2/12a^2}\right)\cos\theta, \tag{9.8.8c}$$

$$N_\theta \to \frac{H_0}{4\pi a\phi}\left(1-\nu-\frac{h^2/6a^2}{1+h^2/12a^2}\right)\cos\theta, \tag{9.8.8d}$$

$$N_{\phi\theta} \to \frac{(1-\nu)H_0}{4\pi a\phi}\sin\theta, \tag{9.8.8e}$$

$$\frac{M_\phi}{\cos\theta} \to -\frac{M_\theta}{\cos\theta} \to \frac{M_{\phi\theta}}{\cos\theta} \to -\frac{h^2/12a^2}{1+h^2/12a^2}\frac{H_0}{2\pi\phi}, \tag{9.8.8f}$$

$$\frac{Q_\phi}{\cos\theta} \to \frac{Q_\theta}{\sin\theta} \to -\frac{h^2/12a^2}{1+h^2/12a^2}\frac{H_0}{2\pi a\phi^2}. \tag{9.8.8g}$$

Aside from small quantities of the order of h^2/a^2 compared to unity, the displacement and middle surface force singularities are identical with those for a force at a point in an infinite flat plate. Although we have moment singularities, we note that the corresponding bending stresses are of the order of h/a compared to the stresses due to middle surface forces and are thus negligible for thin shells. The transverse shear force singularity is such that the corresponding stresses are comparable to those due to middle surface forces in a region around the apex for which the radius is of the order of h/a compared to the plate thickness.

The problem can also be reinvestigated with the aid of the shallow shell approximations. In doing so it is necessary to deal with the complete problem rather than the symmetric and antisymmetric problems since the terms which will be important in the combined problem are those which will have been deleted in the separate problems on the basis of their being of the order of magnitude of $1/2\chi^2$ compared to unity. The stress resultants given by the analysis are

$$N_\phi = -\frac{H_0 2^{1/2}\chi}{2\pi a}\left\{\frac{1}{2^{1/2}\chi\phi} + (1+\nu) \times \left[\frac{2\,\mathrm{kei}'(2^{1/2}\chi\phi)}{(2^{1/2}\chi\phi)^2} - \frac{\mathrm{ker}(2^{1/2}\chi\phi)}{2^{1/2}\chi\phi}\right]\cos\theta\right\}, \tag{9.8.9a}$$

$$N_\theta = \frac{H_0 2^{1/2}\chi}{2\pi a}\left\{\frac{1}{2^{1/2}\chi\phi} + (1+\nu) \times \left[\frac{2\,\mathrm{kei}'(2^{1/2}\chi\phi)}{(2^{1/2}\chi\phi)^2} - \frac{\mathrm{ker}(2^{1/2}\chi\phi)}{2^{1/2}\chi\phi} + \mathrm{ker}'(2^{1/2}\chi\phi)\right]\right\}\cos\theta, \tag{9.8.9b}$$

$$N_{\phi\theta} = \frac{H_0 2^{1/2}\chi}{2\pi a}\left\{\frac{1}{2^{1/2}\chi\phi} - (1+\nu)\left[\frac{2\,\mathrm{kei}'(2^{1/2}\chi\phi)}{(2^{1/2}\chi\phi)^2} - \frac{\mathrm{ker}(2^{1/2}\chi\phi)}{2^{1/2}\chi\phi}\right]\right\}\sin\theta, \tag{9.8.9c}$$

$$M_\phi = \frac{(1+\nu)H_0}{2^{3/2}\pi\chi}\left\{\mathrm{kei}'(2^{1/2}\chi\phi) - (1-\nu) \times \left[\frac{2\,\mathrm{ker}'(2^{1/2}\chi\phi)}{(2^{1/2}\chi\phi)^2} + \frac{\mathrm{kei}(2^{1/2}\chi\phi)}{2^{1/2}\chi\phi} + \frac{2}{(2^{1/2}\chi\phi)^3}\right]\right\}\cos\theta, \tag{9.8.9d}$$

$$M_\theta = \frac{(1+\nu)H_0}{2^{3/2}\pi\chi}\left\{\nu\,\mathrm{kei}'(2^{1/2}\chi\phi) + (1-\nu) \times \left[\frac{2\,\mathrm{kei}'(2^{1/2}\chi\phi)}{(2^{1/2}\chi\phi)^2} + \frac{\mathrm{kei}(2^{1/2}\chi\phi)}{2^{1/2}\chi\phi} + \frac{2}{(2^{1/2}\chi\phi)^3}\right]\right\}\cos\theta, \tag{9.8.9e}$$

$$M_{\phi\theta} = -\frac{(1-\nu^2)H_0}{2^{3/2}\pi\chi}\left[\frac{2\,\mathrm{ker}'(2^{1/2}\chi\phi)}{(2^{1/2}\chi\phi)^2} + \frac{\mathrm{kei}(2^{1/2}\chi\phi)}{2^{1/2}\chi\phi} + \frac{2}{(2^{1/2}\chi\phi)^3}\right]\sin\theta, \tag{9.8.9f}$$

$$Q_\phi = -\frac{(1+\nu)H_0}{2\pi a}\left[\frac{\mathrm{kei}'(2^{1/2}\chi\phi)}{2^{1/2}\chi\phi} - \mathrm{ker}(2^{1/2}\chi\phi)\right]\cos\theta, \tag{9.8.9g}$$

$$Q_\theta = -\frac{(1+\nu)H_0}{2\pi a}\frac{\mathrm{kei}'(2^{1/2}\chi\phi)}{2^{1/2}\chi\phi}\sin\theta. \tag{9.8.9h}$$

A comparison of the behavior near the apex of the approximate stress resultant expressions with that of the exact expressions indicates that we have lost singularities in the bending and twisting moment expressions and have replaced the singularities of $1/\phi^2$ in the exact expressions for Q_ϕ and Q_θ by singularities of $\ln\phi$ in the approximate expressions. As explained previously, however, the difference between the results is not very significant.

The variation of the stress resultants given by Eqs. (9.8.9) is shown in Fig. 9.23. In the vicinity of the tangential load the variation of the middle

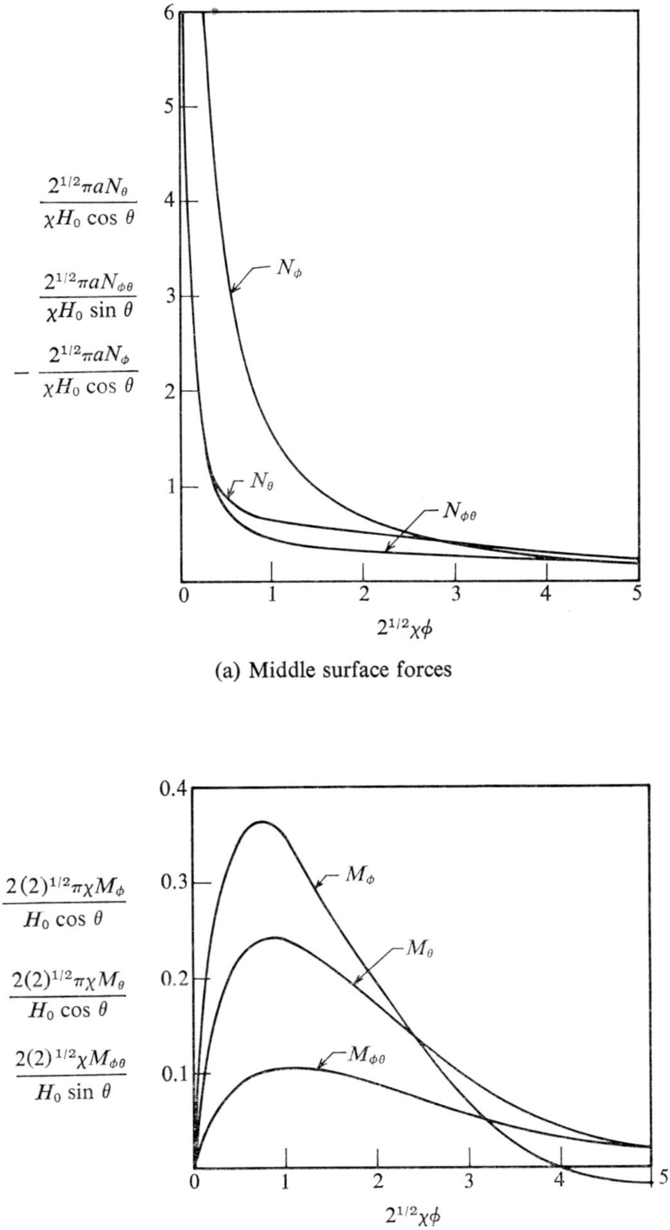

(a) Middle surface forces

(b) Moments

Fig. 9.23. Stress resultants in a shell loaded tangentially at the apex

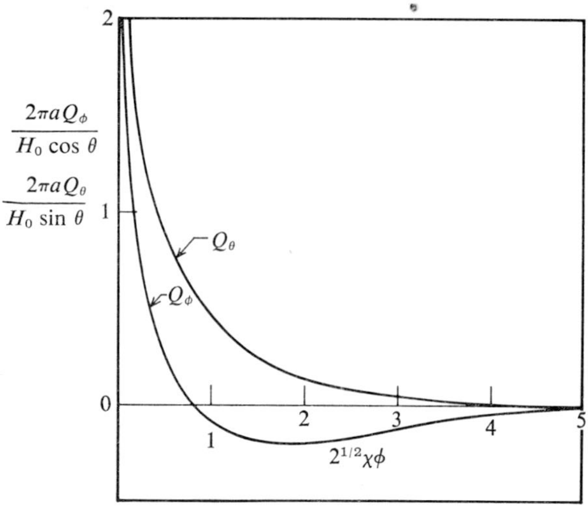

(c) Transverse shear forces

Fig. 9.23. Concluded

surface forces N_ϕ, N_θ, and $N_{\phi\theta}$ is identical to that in an infinite flat plate subjected to a concentrated load in the plane of the plate. Farther away from the load the forces approach those given by the membrane theory of shells. Stresses due to the middle surface forces are everywhere large compared to those due to bending and twisting moments and those due to transverse shear forces. It is interesting to note that despite this circumstance, the complete bending theory must be used rather than just the membrane theory which neglects bending moments and transverse shear forces.

The coincidence between the results for a flat plate subjected to a load in the plane of the plate and those for a sphere subjected to a tangential load has been used to obtain approximate expressions for the stresses in a spherical cap with a small rigid insert to which a tangential load is applied (Fig. 9.24a).† For a flat circular plate bonded to a rigid circular insert at the center and with displacements prevented at the outer radius (Fig. 9.24b), the stresses for an assumed state of plane stress are given by

$$\sigma_r = -\frac{H_0}{4\pi t r}\left[3 + \nu + \frac{(1+\nu)^2 r^2}{(3-\nu)(r_0^2 + r_1^2)} - \frac{(1+\nu) r_0^2 r_1^2}{(r_0^2 + r_1^2) r^2}\right]\cos\theta, \quad (9.8.10a)$$

$$\sigma_\theta = \frac{H_0}{4\pi t r}\left[1 - \nu - \frac{3(1+\nu)^2 r^2}{(3-\nu)(r_0^2 + r_1^2)} - \frac{(1+\nu) r_0^2 r_1^2}{(r_0^2 + r_1^2) r^2}\right]\cos\theta, \quad (9.8.10b)$$

† A. S. Tooth and R. M. Kenedi, op. cit.

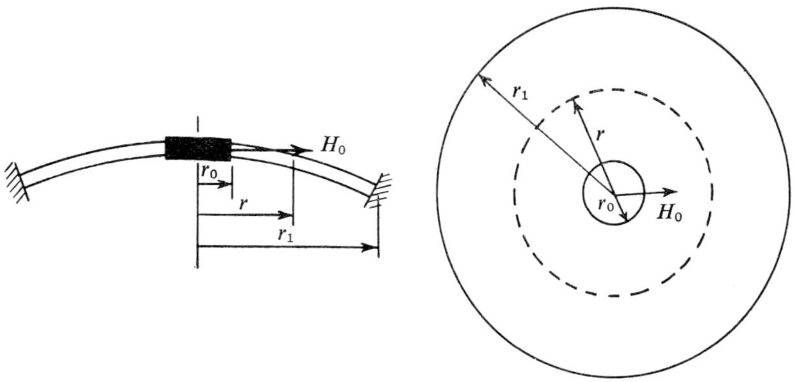

(a) Spherical cap (b) Equivalent flat circular plate

Fig. 9.24. Equivalent problems for approximate analysis of a spherical cap tangentially loaded through a rigid insert at the pole

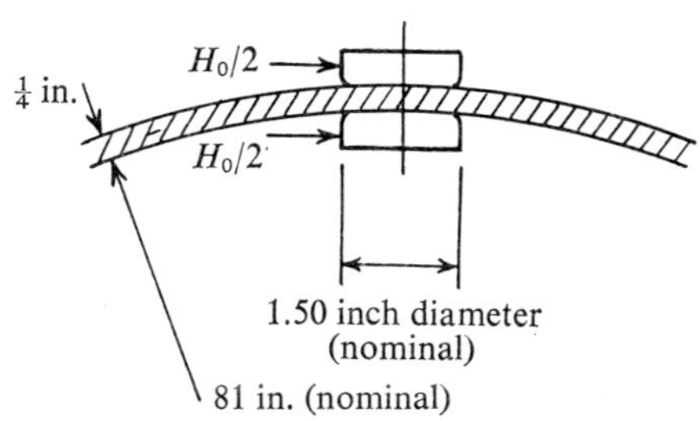

Fig. 9.25. Dimensions of test specimen

$$\tau_{\phi\theta} = \frac{H_0}{4\pi t r}\left[1 - \nu - \frac{(1+\nu)^2 r^2}{(3-\nu)(r_0^2 + r_1^2)} + \frac{(1+\nu) r_0^2 r_1^2}{(r_0^2 + r_1^2) r^2}\right] \sin\theta. \qquad (9.8.10c)$$

The remarkable agreement between the meridional and circumferential stresses given by the above equations (with r_1 equal to infinity) and those obtained experimentally for the spherical shell of Fig. 9.25 is shown in Fig. 9.26. A theoretically more accurate solution may, of course, be derived with the use of the solutions of the bending equations.

9.9 Some other problems of edge-loaded spherical shells

The problems we have discussed above for complete shells involve only the 'wind-loading' equations for the spherical shell. Other problems, however,

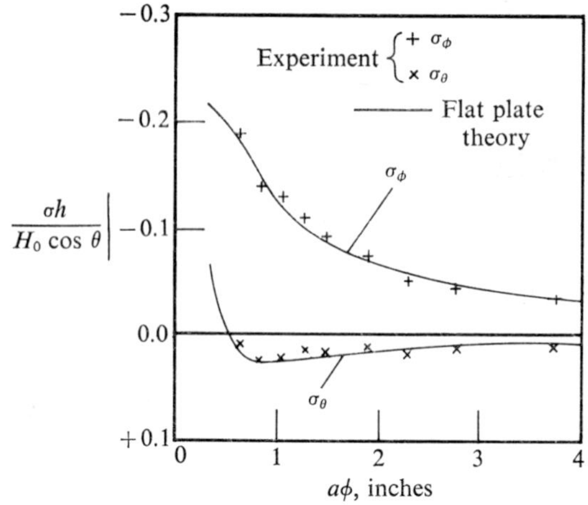

Fig. 9.26. Comparison of theory and experiment for spherical shell under tangential load at an insert

may deal with segments of spherical shells or with spherical caps which are loaded or restrained at their edges in a manner requiring consideration of more than one Fourier component of the solution and stress resultant boundary conditions as well as displacement boundary conditions. For spherical segments we require the eight independent solutions of the bending equations for each Fourier component to satisfy four boundary conditions on each edge. Only the four independent solutions for each Fourier component which are finite at the apex are required to satisfy boundary conditions at the edge of a spherical cap.

An example of the calculations involved in such problems is provided by the problem of determining the stress distribution in a complete hemispherical shell subjected to concentrated edge moments at the ends of a diameter† (Fig. 9.27). The boundary conditions which must be satisfied at the loaded edge ($\phi = \pi/2$) are as follows:

$$N_\phi = N_{\phi\theta} + \frac{1}{a} M_{\phi\theta} = Q_\phi + \frac{1}{a}\frac{\partial M_{\phi\theta}}{\partial \theta} = 0, \tag{9.9.1a}$$

$$M_\phi = \begin{cases} \dfrac{M_0}{a} \quad \text{at} \quad \theta = 0, \pi \\ 0 \text{ elsewhere.} \end{cases} \tag{9.9.1b}$$

† J. G. Berry, 'On Thin Hemispherical Shells Subjected to Concentrated Edge Moments and Forces,' *Proc. 2nd Midwest Conference on Solid Mechanics*, Purdue University, September 1955, pp. 25–44.

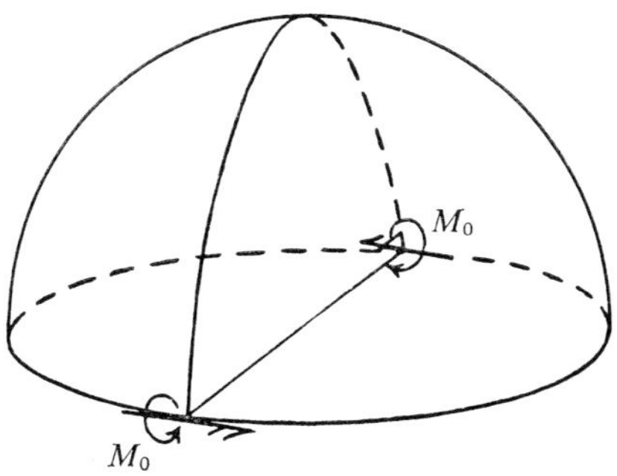

Fig. 9.27. Hemispherical shell subjected to concentrated moments at the ends of a diameter

The concentrated edge moments can be expressed in the form of a Fourier series over the entire perimeter of the hemisphere as

$$M_\phi\big|_{\phi=\pi/2} = \frac{M_0}{\pi a}\left[1 + 2\sum_{n=1}^{\infty}\cos 2n\theta\right]. \tag{9.9.1c}$$

Since no load is applied at the apex, the stress resultants and displacements should be finite there.

When we collect all of the solutions of the equations for spherical shells which are finite at the apex and have circumferential distributions of $\cos 2n\theta$ we have

$$\begin{aligned} a^2 N_\phi = \sum_{n=1}^{\infty} \Bigg\{ & 2n(4n^2-1)A_{2n}\frac{\tan^{2n}\phi/2}{\sin^2\phi} + \bar{A}_{2n} \\ & \times\left[\cot\phi\,\frac{\mathrm{d}\,\mathrm{Im}\,P_\mu^{2n}(\cos\phi)}{\mathrm{d}\phi} - \frac{4n^2}{\sin^2\phi}\,\mathrm{Im}\,P_\mu^{2n}(\cos\phi)\right] \\ & + \bar{B}_{2n}\left[\cot\phi\,\frac{\mathrm{d}\,\mathrm{Re}\,P_\mu^{2n}(\cos\phi)}{\mathrm{d}\phi} - \frac{4n^2}{\sin^2\phi}\,\mathrm{Re}\,P_\mu^{2n}(\cos\phi)\right]\Bigg\}\cos 2n\theta, \end{aligned} \tag{9.9.2a}$$

$$\begin{aligned} a^2\left(N_{\phi\theta} + \frac{1}{a}M_{\phi\theta}\right) = \sum_{n=1}^{\infty} 2n\Bigg\langle & (4n^2-1)\left[(1-\nu)\frac{DC_{2n}}{a} - A_{2n}\right]\frac{\tan^{2n}\phi/2}{\sin^2\phi} \\ & + \frac{\bar{A}_{2n}}{\sin\phi}\Bigg\{\frac{4\chi^4+\nu}{4\chi^4+\nu^2}\left[\frac{\mathrm{d}\,\mathrm{Im}\,P_\mu^{2n}(\cos\phi)}{\mathrm{d}\phi} - \cot\phi\,\mathrm{Im}\,P_\mu^{2n}(\cos\phi)\right] \\ & - \frac{2(1-\nu)\chi^2}{4\chi^4+\nu^2}\left[\frac{\mathrm{d}\,\mathrm{Re}\,P_\mu^{2n}(\cos\phi)}{\mathrm{d}\phi} - \cot\phi\,\mathrm{Re}\,P_\mu^{2n}(\cos\phi)\right]\Bigg\} \\ & + \frac{\bar{B}_{2n}}{\sin\phi}\Bigg\{\frac{4\chi^4+\nu}{4\chi^4+\nu^2}\left[\frac{\mathrm{d}\,\mathrm{Re}\,P_\mu^{2n}(\cos\phi)}{\mathrm{d}\phi} - \cot\phi\,\mathrm{Re}\,P_\mu^{2n}(\cos\phi)\right] \\ & + \frac{2(1-\nu)\chi^2}{4\chi^4+\nu^2}\left[\frac{\mathrm{d}\,\mathrm{Im}\,P_\mu^{2n}(\cos\phi)}{\mathrm{d}\phi} - \cot\phi\,\mathrm{Im}\,P_\mu^{2n}(\cos\phi)\right]\Bigg\}\Bigg\rangle\sin 2n\theta, \end{aligned} \tag{9.9.2b}$$

$$a^2\left(Q_\phi + \frac{1}{a}\frac{\partial M_{\phi\theta}}{\partial\theta}\right) = \sum_{n=0}^{\infty}\left\langle 4n^2(4n^2-1)(1-\nu)\frac{DC_{2n}}{a}\frac{\tan^n\phi/2}{\sin^2\phi} + \bar{A}_{2n}\right.$$

$$\times\left\{\frac{\mathrm{d}\,\mathrm{Im}\,P_\mu^{2n}(\cos\phi)}{\mathrm{d}\phi} - \frac{8(1-\nu)\chi^2n^2}{4\chi^4+\nu^2}\left(\frac{\mathrm{d}}{\mathrm{d}\phi} - \cot\phi\right)\right.$$

$$\left.\times\left[\mathrm{Re}\,P_\mu^{2n}(\cos\phi) - \frac{\nu}{2\chi^2}\,\mathrm{Im}\,P_\mu^{2n}(\cos\phi)\right]\right\}$$

$$+ \bar{B}_{2n}\left\{\frac{\mathrm{d}\,\mathrm{Re}\,P_\mu^{2n}(\cos\phi)}{\mathrm{d}\phi} + \frac{8(1-\nu)\chi^2n^2}{4\chi^4+\nu^2}\left(\frac{\mathrm{d}}{\mathrm{d}\phi} - \cot\phi\right)\right.$$

$$\left.\left.\times\left[\mathrm{Im}\,P_\mu^{2n}(\cos\phi) + \frac{\nu}{2\chi^2}\,\mathrm{Re}\,P_\mu^{2n}(\cos\phi)\right]\right\}\right\rangle \cos 2n\theta, \qquad (9.9.2c)$$

$$M_\phi a = \sum_{n=0}^{\infty}\left\langle -\frac{2n(4n^2-1)(1-\nu)DC_{2n}}{a}\frac{\tan^n\phi/2}{\sin^2\phi}\right.$$

$$+ \bar{A}_{2n}\left\{\mathrm{Im}\,P_\mu^{2n}(\cos\phi) - \frac{2(1-\nu)\chi^2}{4\chi^4+\nu^2}\left(1 - \frac{n^2}{\sin^2\phi} + \cot\phi\frac{\mathrm{d}}{\mathrm{d}\phi}\right)\right.$$

$$\left.\times\left[\mathrm{Re}\,P_\mu^{2n}(\cos\phi) - \frac{\nu}{2\chi^2}\,\mathrm{Im}\,P_\mu^{2n}(\cos\phi)\right]\right\}$$

$$+ \bar{B}_{2n}\left\{\mathrm{Re}\,P_\mu^{2n}(\cos\phi) + \frac{2(1-\nu)\chi^2}{4\chi^4+\nu^2}\left(1 - \frac{n^2}{\sin^2\phi} + \cot\phi\frac{\mathrm{d}}{\mathrm{d}\phi}\right)\right.$$

$$\left.\left.\times\left[\mathrm{Im}\,P_\mu^{2n}(\cos\phi) + \frac{\nu}{2\chi^2}\,\mathrm{Re}\,P_\mu^{2n}(\cos\phi)\right]\right\}\right\rangle \cos 2n\theta. \qquad (9.9.2d)$$

In order for Eq. (9.9.1) to be satisfied, each Fourier component of Eqs. (9.9.2a), (9.9.2b), and (9.9.2c) must vanish at $\phi = \pi/2$ while each component of Eq. (9.9.2d) calculated at $\phi = \pi/2$ must be equal to the corresponding value given by Eq. (9.9.1c). Thus the coefficients A_{2n}, C_{2n}, $\bar{A}_{2n}$, and $\bar{B}_{2n}$ are determined by the solution of the following simultaneous equations when n is equal to or greater than unity

$$(4n^2-1)A_{2n} - 2n[\bar{A}_{2n}\,\mathrm{Im}\,P_\mu^{2n}(0) + \bar{B}_{2n}\,\mathrm{Re}\,P_\mu^{2n}(0)] = 0, \qquad (9.9.3a)$$

$$(4n^2-1)\left[(1-\nu)\frac{D}{a}C_{2n} - A_{2n}\right]$$

$$+ \bar{A}_{2n}\left[\frac{4\chi^4+\nu}{4\chi^4+\nu^2}\frac{\mathrm{d}\,\mathrm{Im}\,P_\mu^{2n}(0)}{\mathrm{d}\phi} - \frac{2(1-\nu)\chi^2}{4\chi^4+\nu^2}\frac{\mathrm{d}\,\mathrm{Re}\,P_\mu^{2n}(0)}{\mathrm{d}\phi}\right]$$

$$+ \bar{B}_{2n}\left[\frac{4\chi^4+\nu}{4\chi^4+\nu^2}\frac{\mathrm{d}\,\mathrm{Re}\,P_\mu^{2n}(0)}{\mathrm{d}\phi} + \frac{2(1-\nu)\chi^2}{4\chi^4+\nu^2}\frac{\mathrm{d}\,\mathrm{Im}\,P_\mu^{2n}(0)}{\mathrm{d}\phi}\right] = 0, \qquad (9.9.3b)$$

$$4n^2(4n^2-1)(1-\nu)\frac{D}{a}C_{2n}$$

$$+\bar{A}_{2n}\left\{\frac{\mathrm{d}\,\mathrm{Im}\,P_\mu^{2n}(0)}{\mathrm{d}\phi}\left[1+\frac{4\nu(1-\nu)n^2}{4\chi^4+\nu^2}\right]-\frac{8(1-\nu)\chi^2n^2}{4\chi^4+\nu^2}\frac{\mathrm{d}\,\mathrm{Re}\,P_\mu^{2n}(0)}{\mathrm{d}\phi}\right\}$$

$$+\bar{B}_{2n}\left\{\frac{\mathrm{d}\,\mathrm{Re}\,P_\mu^{2n}(0)}{\mathrm{d}\phi}\left[1+\frac{4\nu(1-\nu)n^2}{4\chi^4+\nu^2}\right]+\frac{8(1-\nu)\chi^2n^2}{4\chi^4+\nu^2}\frac{\mathrm{d}\,\mathrm{Im}\,P_\mu^{2n}(0)}{\mathrm{d}\phi}\right\}=0, \tag{9.9.3c}$$

$$-2n(4n^2-1)(1-\nu)\frac{D}{a}C_{2n}$$

$$+\bar{A}_{2n}\left\{\frac{4\chi^4+\nu-\nu(1-\nu)n^2}{4\chi^4+\nu^2}\,\mathrm{Im}\,P_\mu^{2n}(0)\right.$$

$$\left.+\frac{2(n^2-1)(1-\nu)\chi^2}{4\chi^4+\nu^2}\,\mathrm{Re}\,P_\mu^{2n}(0)\right]$$

$$+\bar{B}_{2n}\left[\frac{4\chi^4+\nu-\nu(1-\nu)n^2}{4\chi^4+\nu^2}\,\mathrm{Re}\,P_\mu^{2n}(0)\right.$$

$$\left.-\frac{2(n^2-1)(1-\nu)\chi^2}{4\chi^4+\nu^2}\,\mathrm{Im}\,P_\mu^{2n}(0)\right]=2\frac{M_0}{\pi}. \tag{9.9.3d}$$

When n is equal to zero, the coefficients are determined from

$$\bar{A}_0\frac{\mathrm{d}\,\mathrm{Im}\,P_\mu(0)}{\mathrm{d}\phi}+\bar{B}_0\frac{\mathrm{d}\,\mathrm{Re}\,P_\mu(0)}{\mathrm{d}\phi}=0, \tag{9.9.4a}$$

$$\bar{A}_0\left[\frac{4\chi^4+\nu}{4\chi^4+\nu^2}\,\mathrm{Im}\,P_\mu(0)-\frac{2(1-\nu)\chi^2}{4\chi^4+\nu^2}\,\mathrm{Re}\,P_\mu(0)\right]$$

$$+\bar{B}_0\left[\frac{4\chi^4+\nu}{4\chi^4+\nu^2}\,\mathrm{Re}\,P_\mu(0)+\frac{2(1-\nu)\chi^2}{4\chi^4+\nu^2}\,\mathrm{Im}\,P_\mu(0)\right]=\frac{M_0}{\pi}. \tag{9.9.4b}$$

The variation along the meridians at $\theta = 0$ and $\pi/4$ of stresses due to middle surfaces or to bending moments is shown in Fig. 9.28 for a steel shell whose dimensions and material properties are as follows:

$a = 6$ in.
$h = 0.095$ in.
$\nu = 0.3$
$E = 30 \times 10^6$ psi

The calculated stress distributions† for the hemisphere when it is subjected

† The solution given by Berry is in error in its treatment of the axisymmetric portion of the stress distributions but the numerical values are not significantly affected.

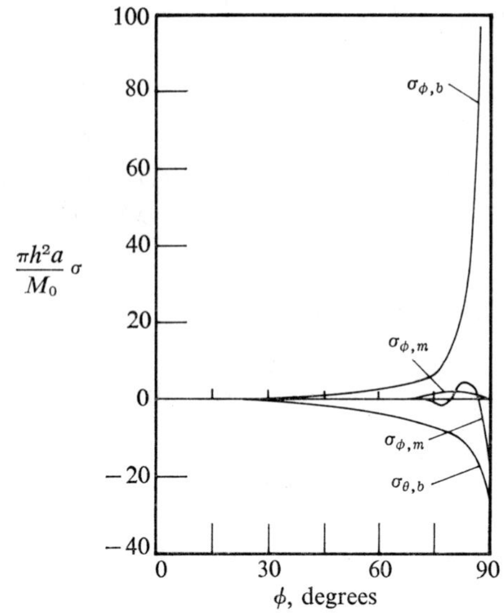

(a) Membrane and bending stresses, $\theta = 0$

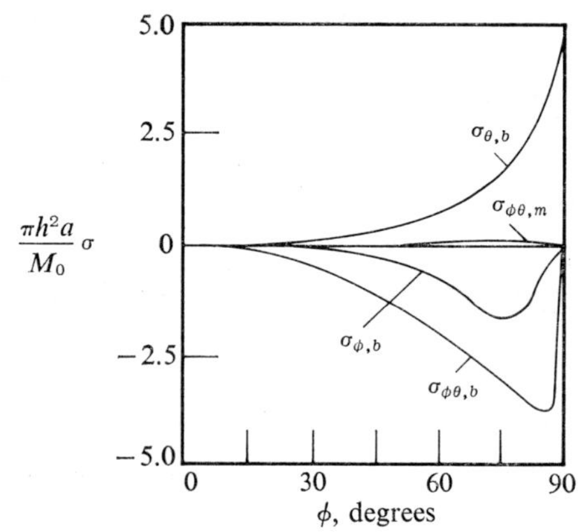

(b) Shear and bending stresses, $\theta = \pi/4$

Fig. 9.28. Calculated stresses for hemisphere loaded by concentrated moments

to a concentrated load at the apex and to vertical concentrated reactive loads (Fig. 9.29) are shown in Fig. 9.30. In both cases it is interesting to note that the stresses due to bending moments are generally more important than those due to middle surface forces over most of the shell.

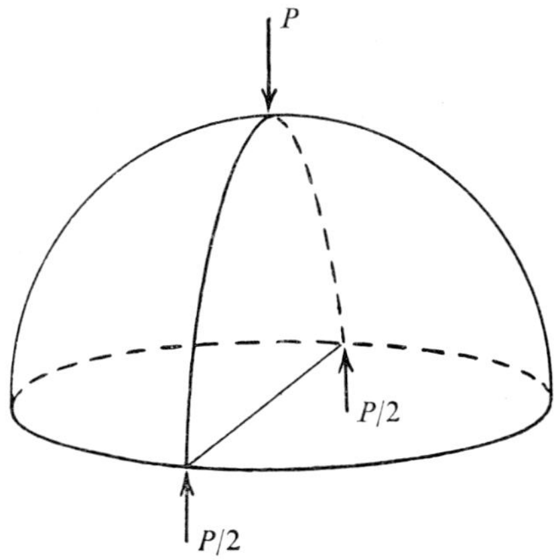

Fig. 9.29. Hemispherical shell subjected to concentrated loads at the pole and at the ends of a diameter

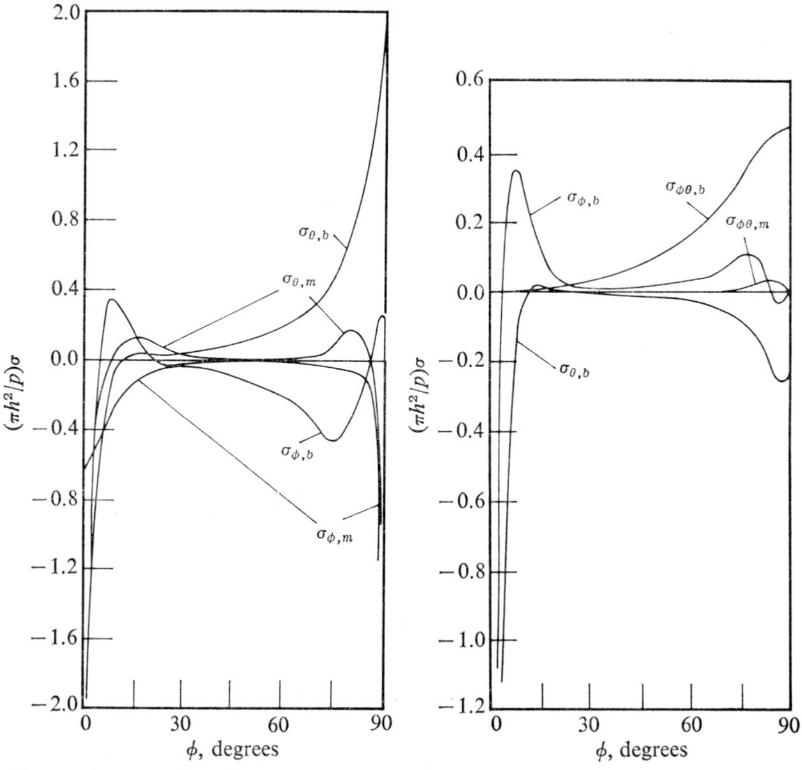

(a) Membrane and bending stresses, $\theta = 0$ (b) Shear and bending stresses, $\theta = \pi/4$

Fig. 9.30. Calculated stresses for hemisphere loaded by concentrated forces

9.10 Surface-loaded spherical shells

When the spherical shell is loaded on its surface we must find particular solutions to be added to the solutions already obtained. The one we can obtain with the least difficulty is a particular solution of Eq. (9.1.9) which is almost the solution we would obtain if the membrane theory of shells were used. Let us assume, for example, that the surface loading is expressible as

$$q_\phi = \sum_{n=0}^{\infty} q_{\phi n}^{(\phi)} \cos n\theta, \tag{9.10.1a}$$

$$q_\theta = \sum_{n=1}^{\infty} q_{\theta n}^{(\phi)} \sin n\theta, \tag{9.10.1b}$$

$$q_\zeta = \sum_{n=0}^{\infty} q_{\zeta n}^{(\phi)} \cos n\theta, \tag{9.10.1c}$$

$$m_\phi = \sum_{n=0}^{\infty} m_{\phi n}^{(\phi)} \cos n\theta, \tag{9.10.1d}$$

$$m_\theta = \sum_{n=1}^{\infty} m_{\theta\phi}^{(\phi)} \sin n\theta, \tag{9.10.1e}$$

and that the particular solutions are expressible as

$$N_\phi^p = \sum_{n=0}^{\infty} N_{\phi n}^{p(\phi)} \cos n\theta, \tag{9.10.2a}$$

$$N_\theta^p = \sum_{n=0}^{\infty} N_{\theta n}^{p(\phi)} \cos n\theta, \tag{9.10.2b}$$

$$N_{\phi\theta}^p = \sum_{n=1}^{\infty} N_{\phi\theta n}^{p(\phi)} \sin n\theta. \tag{9.10.2c}$$

Substitution into Eqs. (9.1.9) yields the following equations for the determination of the coefficients of the particular solution

$$\frac{\mathrm{d}}{\mathrm{d}\phi}(N_{\phi n}^p \sin\phi) - N_{\theta n}^p \cos\phi + nN_{\phi\theta n}^p$$
$$= -\left(1 + \frac{h^2}{12a^2}\right)^{-1}(m_{\phi n} + q_{\phi n}a)\sin\phi, \tag{9.10.3a}$$

$$nN_{\theta n}^p - \frac{1}{\sin\phi}\frac{\mathrm{d}}{\mathrm{d}\phi}(N_{\phi\theta n}^p \sin^2\phi) = \left(1 + \frac{h^2}{12a^2}\right)^{-1}(m_{\theta n} + q_{\theta n}a)\sin\phi, \tag{9.10.3b}$$

$$N_{\phi n}^{p} + N_{\theta n}^{p} = q_{\zeta n}a + \frac{1}{\sin\phi}\left(1 + \frac{h^2}{12a^2}\right)^{-1}$$
$$\times \left\{\frac{\mathrm{d}}{\mathrm{d}\phi}\left[\left(m_{\phi n} - \frac{h^2}{12a}q_{\phi n}\right)\sin\phi\right] + n\left(m_{\theta} - \frac{h^2}{12a}q_{\theta n}\right)\right\}. \tag{9.10.3c}$$

We may solve this system of equations by eliminating $N_{\theta n}^{p}$ and by adding and subtracting the resultant equations to obtain the two independent equations

$$\frac{1}{\sin\phi}\frac{\mathrm{d}}{\mathrm{d}\phi}\left[(N_{\phi n}^{p} + N_{\phi\theta n}^{p})\sin^2\phi\right] + n(N_{\phi n}^{p} + N_{\phi\theta n}^{p})$$
$$= a(n + \cos\phi)\left\{q_{\zeta n} - \frac{1}{\sin\phi}\left[\frac{\mathrm{d}}{\mathrm{d}\phi}(q_{\phi n}\sin\phi) + nq_{\theta n}\right]\right\} + \left(1 + \frac{h^2}{12a^2}\right)^{-1}$$
$$\times \frac{1}{\sin\phi}\left\{\frac{\mathrm{d}}{\mathrm{d}\phi}\left[(n + \cos\phi)(m_{\phi n} + q\phi_{n}a)\sin\phi\right]\right.$$
$$\left. + \left[n(n + \cos\phi) - \sin^2\phi\right](m_{\theta n} + q_{\theta n}a)\right\}, \tag{9.10.4a}$$

$$\frac{1}{\sin\phi}\frac{\mathrm{d}}{\mathrm{d}\phi}\left[(N_{\phi n}^{p} - N_{\phi\theta n}^{p})\sin^2\phi\right] - n(N_{\phi n}^{p} - N_{\phi\theta n}^{p})$$
$$= -\left\langle a(n - \cos\phi)\left\{q_{\zeta n} - \frac{1}{\sin\phi}\left[\frac{\mathrm{d}}{\mathrm{d}\phi}(q_{\phi n}\sin\phi + nq_{\theta n})\right]\right\}\right.$$
$$+ \left(1 + \frac{h^2}{12a^2}\right)^{-1}\frac{1}{\sin\phi}\left\{\frac{\mathrm{d}}{\mathrm{d}\phi}\left[(n - \cos\phi)(m_{\phi n} + q_{\phi n}a)\sin\phi\right]\right.$$
$$\left.\left. - \left[n(n - \cos\phi) - \sin^2\phi\right](m_{\theta n} + q_{\theta n}a)\right\}\right\rangle, \tag{9.10.4b}$$

which can be solved to yield

$$\left.\begin{matrix} N_{\phi n}^{p} \\ N_{\phi\theta n}^{p} \end{matrix}\right\} = \frac{1}{2\sin^2\phi}\left(\cot^n\frac{\phi}{2}\int\left\langle a(n + \cos\phi)\right.\right.$$
$$\times \left[q_{\zeta n}\sin\phi - \frac{\mathrm{d}}{\mathrm{d}\phi}(q_{\phi n}\sin\phi) - nq_{\theta n}\right]$$
$$+ \left(1 + \frac{h^2}{12a^2}\right)^{-1}\left\{\frac{\mathrm{d}}{\mathrm{d}\phi}\left[(n + \cos\phi)(m_{\phi n} + q_{\phi n}a)\sin\phi\right]\right.$$
$$\left.\left. + \left[n(n + \cos\phi) - \sin^2\phi\right](m_{\theta n} + q_{\theta n}a)\right\}\right\rangle$$

(*continued overleaf*)

$$\times \tan^n \frac{\phi}{2}\, \mathrm{d}\phi \mp \tan^n \frac{\phi}{2} \int \Big\langle (n - \cos\phi)$$

$$\times \left[q_{\zeta n} \sin\phi - \frac{\mathrm{d}}{\mathrm{d}\phi}(q_{\phi n} \sin\phi) - n q_{\theta n} \right]$$

$$+ \left(1 + \frac{h^2}{12a^2}\right)^{-1} \left\{ \frac{\mathrm{d}}{\mathrm{d}\phi} [(n - \cos\phi)(m_{\phi n} + q_{\phi n} a) \sin\phi] \right.$$

$$\left. + [n(n - \cos\phi) - \sin^2\phi](m\theta_n + q\theta_n a) \right\} \Big\rangle \cot^n \frac{\phi}{2}\, \mathrm{d}\phi \Big) \; (n \neq 0). \tag{9.10.5}$$

The circumferential direct stress coefficient $N^p_{\theta n}$ is determined by Eq. (9.10.3c).

The other solution which we need is a particular solution of Eqs. (9.1.10) and (9.1.11b), which we recall are

$$(a^2\nabla^2 + 2)\left[D\left(1 + \frac{h^2}{12a^2}\right)^{-1} (a^2\nabla^2 + 2) w^p + aS^p \right] = 0, \tag{9.10.6a}$$

$$(a^2\nabla^2 + 2)\Big\langle a^2\nabla^2 S^p - Ehaw^p + q_\zeta a^3 + \left(1 + \frac{h^2}{12a^2}\right)^{-1} \frac{a^2}{\sin\phi}$$

$$\times \left\{ \frac{\partial}{\partial\phi}\left[\left(m_\phi - \frac{h^2}{12a} q_\phi\right) \sin\phi\right] + \frac{\partial}{\partial\theta}\left(m_\theta - \frac{h^2}{12a} q_\theta\right) \right\} \Big\rangle$$

$$= (1 + \nu) a^3 \left\{ q_\zeta - \frac{1}{\sin\phi}\left[\frac{\partial(q_\phi \sin\phi)}{\partial\phi} + \frac{\partial q_\theta}{\partial\theta} \right] \right\}. \tag{9.10.6b}$$

These may be integrated to yield

$$D\left(1 + \frac{h^2}{12a^2}\right)^{-1} (a^2\nabla^2 + 2) w^p + aS^p = 0, \tag{9.10.7a}$$

$$a^2\nabla^2 S^p - Ehaw^p = - Ehaw^m, \tag{9.10.7b}$$

where w^m is the radial membrane displacement corresponding to the membrane state of stress of Eqs. (9.10.2), given by

$$w^m = \frac{a}{Eh}\left\{ q_\zeta a + \left(1 + \frac{h^2}{12a^2}\right)^{-1} \frac{1}{\sin\phi} \frac{\partial}{\partial\phi}\left[\left(m_\phi - \frac{h^2}{12a} q_\phi\right) \sin\phi\right] \right.$$

$$\left. + \frac{\partial}{\partial\theta}\left(m_\theta - \frac{h^2}{12a} q_\theta\right) - (1 + \nu) a\psi^p \right\}, \tag{9.10.8a}$$

and ψ^p is a particular solution of the equation

$$(a^2\nabla^2 + 2)\psi^p = f(\phi) = q_\zeta - \frac{1}{\sin\phi}\left[\frac{\partial(q_\phi \sin\phi)}{\partial\phi} + \frac{\partial q_\theta}{\partial\theta} \right]. \tag{9.10.8b}$$

When the surface loading varies as Eq. (9.10.1), the method of variation of parameters can be used to obtain

$$\psi^p = \sum_{n=0}^{\infty} \psi_n^p \cos n\theta, \tag{9.10.9a}$$

where

$$2n(n^2 - 1)\,\psi_n^p = (n + \cos\phi)\tan^n\frac{\phi}{2}\int f(\phi)(n - \cos\phi)\cot^n\frac{\phi}{2}\sin\phi\, d\phi$$

$$- (n - \cos\phi)\cot^n\frac{\phi}{2}\int f(\phi)(n + \cos\phi)\tan^n\frac{\phi}{2}\sin\phi\, d\phi, \tag{9.10.9b}$$

$$\psi_1^p = \int\frac{1}{\sin^3\phi}\left[\int f(\phi)\sin^2\phi\, d\phi\right]d\phi, \tag{9.10.9c}$$

$$\psi_0^p = \int\frac{1}{\sin\phi\cos^2\phi}\left[\int f(\phi)\sin\phi\cos\phi\, d\phi\right]d\phi. \tag{9.10.9d}$$

The particular solution w^p must satisfy the differential equation

$$\left[\frac{a^2\nabla^2(a^2\nabla^2 + 2)}{4\chi^4 + 1} + 1\right]w^p = w^m. \tag{9.10.10}$$

If now we write

$$w^p = \sum_{n=0}^{\infty}\frac{w_n^p}{(4\chi^4 + 1)^n}, \tag{9.10.11a}$$

the coefficients of the equation obtained by substituting (9.10.11a) into Eq. (9.10.10) and equating coefficients of like powers of $(4\chi^4 + 1)$ are

$$w_0^p = w^m, \tag{9.10.11b}$$

$$w_n^p = (-1)^n[a^2\nabla^2(a^2\nabla^2 + 2)]^n w^m, \tag{9.10.11c}$$

so that the solution is (formally at least)

$$w^p = w^m + \sum_{n=1}^{\infty}(-1)^n\frac{[a^2\nabla^2(a^2\nabla^2 + 2)]^n w^m}{4\chi^4 + 1}, \tag{9.10.11d}$$

$$S^p = -\frac{Eha}{4\chi^4 + 1}(a^2\nabla^2 + 2)w^p. \tag{9.10.11e}$$

An alternate solution when w^m is expressed in the form

$$w^m = \sum_{n=0}^{\infty} w_n^m(\phi)\cos n\theta, \tag{9.10.12a}$$

can be obtained by expressing w^p as

$$w^p = \sum_{n=0}^{\infty} \sum_{k=0}^{\infty} w_{nk}^p P_k^n(\cos\phi) \cos n\theta, \tag{9.10.12b}$$

where $P_k^n(\cos\phi)$ are associated Legendre polynomials.† With the use of the known orthogonality conditions for these functions we can show that w_{nk}^p is given by

$$w_{nk}^p = \frac{(n+\frac{1}{2})(k-n)!}{(k+n)!}\left[1 + \frac{(k-1)k(k+1)(k+2)}{4\chi^4+1}\right]^{-1}$$
$$\times \int_0^\pi w_n^m(\phi) P_k^n(\cos\phi) \sin\phi \, d\phi. \tag{9.10.13e}$$

If the spherical shell includes one or both poles, the particular solution should be modified by the addition of solutions of the homogeneous bending equations. The polar singularities of the complementary solutions should cancel any singularities of the particular solutions to ensure displacement continuity.

Supplementary references

[1] Reissner, E., 'Stresses and Small Displacements of Shallow Spherical Shells,' *J. Math. Phys.*, vol. 25, no. 1, 1946, pp. 80–85.

[2] Reissner, E., 'On the Determination of Stresses and Displacements for Unsymmetrical Deformations of Shallow Spherical Shells,' *J. Math. and Phys.*, vol. 38, no. 1, April 1959, pp. 16–35.

[3] Zandbergen, P. J., 'Determination of the Stresses in a Not Shallow Spherical Shell with a Hole, Due to an Axial Force, a Bending Moment and a Transverse Force,' Nationaal Luchtvaartlaboratorium, Amsterdam, NLL-TR S. 518, January 1958.

[4] Wilkinson, J. P., and Kalnins, A., 'Deformation of Open Spherical Shells Under Arbitrarily Located Concentrated Loads,' *J. Appl. Mech.*, vol. 33, no. 2, June 1966, pp. 305–312.

† W. Magnus *et al.*, op. cit.

10

Asymmetric deformations of circular cylindrical shells

10.1 General relations for cylindrical shells

The analysis of many circular cylinder shell problems poses no real difficulty other than the tedium of the lengthy calculations involved. The equations of equilibrium of the cylinder are given by Eq. (2.6.6) as

$$\frac{\partial N_x}{\partial \xi} + \frac{\partial}{\partial \theta}\left(\overline{N}_{x\theta} - \frac{1}{2}\frac{\overline{M}_{x\theta}}{R}\right) = -q_x R, \tag{10.1.1a}$$

$$\frac{\partial}{\partial \theta}\left(N_\theta + \frac{M_\theta}{R}\right) + \frac{\partial}{\partial \xi}\left(\overline{N}_{x\theta} + \frac{3}{2}\frac{\overline{M}_{x\theta}}{R}\right) = -\left(q_\theta + \frac{m_\theta}{R}\right)R, \tag{10.1.1b}$$

$$\frac{1}{R}\left(\frac{\partial^2 M_x}{\partial \xi^2} + 2\frac{\partial^2 \overline{M}_{x\theta}}{\partial \xi\, \partial \theta} + \frac{\partial^2 M_\theta}{\partial \theta^2}\right) - N_\theta = -\left(q_\zeta R + \frac{\partial m_x}{\partial \xi} + \frac{\partial m_\theta}{\partial \theta}\right), \tag{10.1.1c}$$

together with the definitions of the equivalent transverse forces as

$$\bar{Q}_x = \frac{1}{R}\left(\frac{\partial M_x}{\partial \xi} + 2\frac{\overline{M}_{x\theta}}{\partial \theta}\right), \tag{10.1.1d}$$

$$\bar{Q}_\theta = \frac{1}{R}\left(\frac{\partial M_\theta}{\partial \theta} + 2\frac{\partial \overline{M}_{x\theta}}{\partial \xi}\right), \tag{10.1.1e}$$

where $\xi = x/R$. (10.1.2)

For the overall force–strain relations, let us use the approximation given by Eqs. (3.1.2), (1.5.3), (1.5.10), and (2.4.9) as

$$N_x = \frac{K}{R}\left[\frac{\partial u_x}{\partial \xi} + \nu\left(\frac{\partial u_\theta}{\partial \theta} + w\right) - k\frac{\partial^2 w}{\partial \xi^2}\right], \tag{10.1.3a}$$

$$N_\theta = \frac{K}{R}\left[\frac{\partial u_\theta}{\partial \theta} + w + \nu\frac{\partial u_x}{\partial \xi} + k\left(\frac{\partial^2 w}{\partial \theta^2} + w\right)\right], \tag{10.1.3b}$$

$$\overline{N}_{x\theta} = \frac{1-\nu}{2}\frac{K}{R}\left(\frac{\partial u_\theta}{\partial \xi} + \frac{\partial u_x}{\partial \theta}\right)(1 + \tfrac{3}{4}k), \tag{10.1.3c}$$

$$M_x = -Kk\left[\frac{\partial^2 w}{\partial \xi^2} - \frac{\partial u_x}{\partial \xi} + \nu\left(\frac{\partial^2 w}{\partial \theta^2} - \frac{\partial u_\theta}{\partial \theta}\right)\right], \tag{10.1.3d}$$

$$M_\theta = -Kk\left(\frac{\partial^2 w}{\partial \theta^2} + w + \nu\frac{\partial^2 w}{\partial \xi^2}\right), \tag{10.1.3e}$$

$$\overline{M}_{x\theta} = -Kk(1-\nu)\left(\frac{\partial^2 w}{\partial \xi\, \partial \theta} + \frac{1}{4}\frac{\partial u_x}{\partial \theta} - \frac{3}{4}\frac{\partial u_\theta}{\partial \xi}\right), \tag{10.1.3f}$$

where the wall-bending stiffness D has been replaced by the equivalent representation KkR^2 with

$$k = \frac{1}{12}\frac{h^2}{R^2}. \tag{10.1.4}$$

Although the overall force–strain relations given by Eqs. (3.1.4) are somewhat simpler, there is no real advantage in using them since they do not yield simpler final equations.

Unlike the spherical shell, the equations for the cylindrical shell do not permit the rigorous introduction of a stress function to take the place of the two tangential displacements. Thus, at this point, we obtain three simultaneous differential equations for the three displacements u_x, u_θ and w by substituting Eqs. (10.1.3) into Eqs. (10.1.1) to yield

$$\left[\frac{\partial^2}{\partial \xi^2} + \frac{1-\nu}{2}(1+k)\frac{\partial^2}{\partial \theta^2}\right]u_x + \frac{1+\nu}{2}\frac{\partial^2 u_\theta}{\partial \xi\, \partial \theta}$$
$$+ \left[\nu - k\left(\frac{\partial \xi^2}{\partial^2} - \frac{1-\nu}{2}\frac{\partial^2}{\partial \theta^2}\right)\right]\frac{\partial w}{\partial \xi} = -\frac{R^2}{K}q_x, \tag{10.1.5a}$$

$$\frac{1+\nu}{2}\frac{\partial u_x}{\partial \xi\, \partial \theta} + \left[\frac{1-\nu}{2}(1+3k)\frac{\partial^2}{\partial \xi^2} + \frac{\partial^2}{\partial \theta^2}\right]u_\theta + \left(1 - \frac{3-\nu}{2}k\frac{\partial^2}{\partial \xi^2}\right)\frac{\partial w}{\partial \theta}$$
$$= -\left(q_\theta + \frac{m_\theta}{R}\right)\frac{R^2}{K}, \tag{10.1.5b}$$

$$\left[\nu - k\left(\frac{\partial^2}{\partial \xi^2} - \frac{1-\nu}{2}\frac{\partial^2}{\partial \theta^2}\right)\right]\frac{u_x}{\partial \xi} + \left(1 - \frac{3-\nu}{2}k\frac{\partial^2}{\partial \xi^2}\right)\frac{\partial u}{\partial \theta}$$
$$+ \left[1 + k\left(\nabla^4 + 2\frac{\partial^2}{\partial \theta^2} + 1\right)\right]w = \frac{R^2}{K}\left[q_\zeta + \frac{1}{R}\left(\frac{\partial m_x}{\partial \xi} + \frac{\partial m_\theta}{\partial \theta}\right)\right], \tag{10.1.5c}$$

where

$$\nabla^2 = \frac{\partial^2}{\partial \xi^2} + \frac{\partial^2}{\partial \theta^2}, \tag{10.1.6a}$$

$$\nabla^4 = \nabla^2\nabla^2 = \frac{\partial^4}{\partial \xi^4} + 2\frac{\partial^4}{\partial \xi^2\,\partial \theta^2} + \frac{\partial^4}{\partial \theta^4}. \tag{10.1.6b}$$

Since these are linear partial differential equations with constant coefficients, all of the operators are commutative. It thus is simple to obtain equations for u_x and u_θ in terms of w and the surface loading from Eqs. (10.1.5a) and (10.1.5b) as

$$L_3(u_x) = -L_1(w) - \frac{R^2}{K}\left\{\left[(1 + 3k)\frac{\partial^2}{\partial \xi^2} + \frac{2}{1 - \nu}\frac{\partial^2}{\partial \theta^2}\right]q_x - \frac{1 + \nu}{1 - \nu}\frac{\partial^2}{\partial \xi\,\partial \theta}\left(q_\theta + \frac{m_\theta}{R}\right)\right\}, \tag{10.1.7a}$$

$$L_3(u_\theta) = -L_2(w) + \frac{R^2}{K}\left\{\frac{1 + \nu}{1 - \nu}\frac{\partial q^2{}_x}{\partial \xi\,\partial \theta} - \left[\frac{2}{1 - \nu}\frac{\partial^2}{\partial \xi^2} + (1 + k)\frac{\partial^2}{\partial \theta^2}\right]\left(q_\theta + \frac{m_\theta}{R}\right)\right\}, \tag{10.1.7b}$$

where the operators L_1, L_2, and L_3 are given by

$$L_1 = \left[(1 + 3k)\left(\nu\frac{\partial^2}{\partial \xi^2} - k\frac{\partial^4}{\partial \xi^4}\right) - \frac{\partial^2}{\partial \theta^2} + k\left(3\frac{1 - \nu}{2}k\frac{\partial^4}{\partial \xi^2\,\partial \theta^2} + \frac{\partial^4}{\partial \theta^4}\right)\right]\frac{\partial}{\partial \xi}, \tag{10.1.8a}$$

$$L_2 = \left\{(2 + \nu)\frac{\partial^2}{\partial \xi^2} + (1 + k)\frac{\partial^2}{\partial \theta^2} - 2k\left[\frac{\partial^4}{\partial \xi^4} + \left(1 + \frac{3 - \nu}{4}k\right)\frac{\partial^4}{\partial \xi^2\,\partial \theta^2}\right]\right\}\frac{\partial}{\partial \theta}, \tag{10.1.8b}$$

$$L_3 = \nabla^4 + k\left[3\frac{\partial^4}{\partial \xi^4} + 2(1 - \nu)(1 + \tfrac{3}{4}k)\frac{\partial^4}{\partial \xi^2\,\partial \theta^2} + \frac{\partial^4}{\partial \theta^4}\right]. \tag{10.1.8c}$$

With the use of Eqs. (10.1.5c), (10.1.7), and (10.1.8) we may obtain an equation for the radial deformation w alone as

$$\frac{D}{R^4}L(w) = L_1(q_x) + L_2\left(q + \frac{m_\theta}{R}\right) + L_3\left[q_\zeta + \frac{1}{R}\left(\frac{\partial m_x}{\partial \xi} + \frac{\partial m_\theta}{\partial \theta}\right)\right], \tag{10.1.9}$$

where

$$L = (1 + 3k)(1 - k)\frac{\partial^8}{\partial\xi^8} + 4\left(1 + \frac{11 - 3\nu}{8}k + 9\frac{1 - \nu}{8}k^2\right)\frac{\partial^8}{\partial\xi^6\,\partial\theta^2}$$

$$+ 6\left(1 + \frac{2 - \nu}{2}k - \frac{\nu^2}{6}k^2\right)\frac{\partial^8}{\partial\xi^4\,\partial\theta^4}$$

$$+ 4\left(1 + \frac{7 - 3\nu}{8}k + 3\frac{1 - \nu}{8}k^2\right)\frac{\partial^8}{\partial\xi^2\,\partial\theta^6}$$

$$+ (1 + k)\frac{\partial^8}{\partial\theta^8} + 2\nu(1 + 3k)\frac{\partial^6}{\partial\xi^6} + 6\left(1 + \frac{2 - \nu + \nu^2}{2}k\right)\frac{\partial^6}{\partial\xi^4\,\partial\theta^2}$$

$$+ 2\left(4 - \nu + \frac{7 - 5\nu}{2}k + 3\frac{1 - \nu}{2}k^2\right)\frac{\partial^6}{\partial\xi^2\,\partial\theta^4}$$

$$+ 2(1 + k)\frac{\partial^6}{\partial\theta^6} + (1 + 3k)\left(\frac{1 - \nu^2}{k} + 1\right)\frac{\partial^4}{\partial\xi^4}$$

$$+ 2\left(2 - \nu + 7\frac{1 - \nu}{4}k + 3\frac{1 - \nu}{4}k^2\right)\frac{\partial^4}{\partial\xi^2\,\partial\theta^2}$$

$$+ (1 + k)\frac{\partial^4}{\partial\theta^4}. \tag{10.1.10}$$

Similar equations for u_x and u_θ may be obtained as

$$\frac{D}{R^4}L(u_x) = -L_4(q_x) + L_5\left(q_\theta + \frac{m_\theta}{R}\right) - L_1\left[q_\zeta + \frac{1}{R}\left(\frac{\partial m_x}{\partial\xi} + \frac{\partial m_\theta}{\partial\theta}\right)\right], \tag{10.1.11a}$$

$$\frac{D}{R^4}L(u_\theta) = L_5(q_x) - L_6\left(q_\theta + \frac{m_\theta}{R}\right) - L_2\left[q_\zeta + \frac{1}{R}\left(\frac{\partial m_x}{\partial\xi} + \frac{\partial m_\theta}{\partial\theta}\right)\right], \tag{10.1.11b}$$

where

$$L_4 = (1 + k)(1 + 3k)\frac{\partial^2}{\partial\xi^2} + k \times \left[(1 + 3k)\frac{\partial^6}{\partial\xi^6} + \frac{2}{1 - \nu}\left(2 - \nu + \frac{3 - 6\nu - \nu^2}{4}k\right)\frac{\partial^6}{\partial\xi^4\,\partial\theta^2} + \left(\frac{5 - \nu}{1 - \nu} + 3k\right)\frac{\partial^6}{\partial\xi^2\,\partial\theta^4} + \frac{2}{1 - \nu}\frac{\partial^2}{\partial\theta^2}\left(\frac{\partial^2}{\partial\theta^2} + 1\right)^2 + 4\left(\frac{2 - \nu}{1 - \nu} + \frac{3}{2}k\right)\frac{\partial^4}{\partial\xi^2\,\partial\theta^2}\right], \tag{10.1.12a}$$

$$L_5 = \left\{1 + \frac{1+\nu}{1-\nu}k\left[\left(1 - \frac{3-\nu}{1+\nu}k\right)\frac{\partial^4}{\partial\xi^4}\right.\right.$$

$$+ 2\left(1 + \frac{3-\nu}{4}\frac{1-\nu}{1+\nu}k\right)\frac{\partial^4}{\partial\xi^2\,\partial\theta^2} + \frac{\partial^4}{\partial\theta^4}$$

$$\left.\left. + \frac{2+3\nu-\nu^2}{1+\nu}\frac{\partial^2}{\partial\xi^2} + \frac{1+3\nu}{1+\nu}\frac{\partial^2}{\partial\theta^2} + 1\right]\right\}\frac{\partial^2}{\partial\xi\,\partial\theta}, \tag{10.1.12b}$$

$$L_6 = \frac{2}{1-\nu}(1-\nu^2+k)\frac{\partial^2}{\partial\xi^2} + (1+k)^2\frac{\partial^2}{\partial\theta^2}$$

$$+ k\left[2\frac{1-k}{1-\nu}\frac{\partial^6}{\partial\xi^6} + \left(\frac{5-\nu}{1-\nu} + 3k\right)\frac{\partial^6}{\partial\xi^4\,\partial\theta^2}\right.$$

$$+ 2\left(\frac{2-\nu}{1-\nu} + \frac{3+\nu}{4}k\right)\frac{\partial^6}{\partial\xi^2\,\partial\theta^4} + (1+k)\frac{\partial^4}{\partial\theta^4}\left(\frac{\partial^2}{\partial\theta^2} + 2\right)$$

$$\left. + 2\frac{2-\nu+\nu^2}{1-\nu}\frac{\partial^4}{\partial\xi^2\,\partial\theta^2} + \frac{4\nu}{1-\nu}\frac{\partial^4}{\partial\xi^4}\right]. \tag{10.1.12c}$$

An alternate representation of the theory in terms of a single displacement function may be had by defining a function Φ such that

$$u_x = -L_1(\Phi) + u_{xp}, \tag{10.1.13a}$$

$$u_\theta = -L_2(\Phi) + u_{\theta p}, \tag{10.1.13b}$$

$$w = L_3(\Phi) + w_p, \tag{10.1.13c}$$

where u_{xp}, $u_{\theta p}$, and w_p are particular solutions of Eq. (10.1.5) or of Eqs. (10.1.9) and (10.1.11). The function Φ must then satisfy the equation

$$L(\Phi) = 0. \tag{10.1.14}$$

The use of Eqs. (10.1.13) and (10.1.14) relieves us of the necessity of having to solve Eqs. (10.1.7) for the tangential displacements u_x and u_θ in terms of the normal displacement w.

10.2 Simplifications of the equations for radial and edge loading

Since the parameter k is a small number, the bending equations may be simplified by deleting terms which are of the order of k compared to other terms in the equations. The first of the simplified sets of equations of cylindrical shell theory may be written as

$$\nabla^4 u_x = -\left[\nu\frac{\partial^2}{\partial\xi^2} - \frac{\partial^2}{\partial\theta^2} + k\left(\frac{\partial^4}{\partial\theta^4} - \frac{\partial^4}{\partial\xi^4} + 3\frac{1-\nu}{2}k\frac{\partial^4}{\partial\xi^2\,\partial\theta^2}\right)\right]\frac{\partial w}{\partial\xi}, \tag{10.2.1a}$$

$$\nabla^4 u_\theta = -\left[(2+\nu)\frac{\partial^2}{\partial\xi^2} + \frac{\partial^2}{\partial\theta^2} - 2k\left(\frac{\partial^4}{\partial\xi^4} + \frac{\partial^4}{\partial\xi^2\,\partial\theta^2}\right)\right]\frac{\partial w}{\partial\theta}, \tag{10.2.1b}$$

$$\frac{D}{R^4}\left[\nabla^4(\nabla^2+1)^2 + 2(1-\nu)\left(\frac{\partial^6}{\partial\xi^2\,\partial\theta^4} + \frac{\partial^4}{\partial\xi^2\,\partial\theta^2} - \frac{\partial^6}{\partial\xi^6}\right)\right.$$
$$\left. + \frac{1-\nu^2}{k}\frac{\partial^4}{\partial\xi^4}\right]w = \nabla^4 q_\zeta, \tag{10.2.1c}$$

where only radial loading has been considered for simplicity. We shall identify these equations as Flügge's approximate equations† as distinguished from the set of equations derived in the preceding section.

A somewhat more drastic reduction of the equations has been suggested by Morley.‡ This second reduction deletes all terms multiplied by k in Eqs. (10.2.1a) and (10.2.1b) so that they read as

$$\nabla^4 u_x = -\left(\nu\frac{\partial^2}{\partial\xi^2} - \frac{\partial^2}{\partial\theta^2}\right)\frac{\partial w}{\partial\xi}, \tag{10.2.2a}$$

$$\nabla^4 u_\theta = -\left[(2+\nu)\frac{\partial^2}{\partial\xi^2} + \frac{\partial^2}{\partial\theta^2}\right]\frac{\partial w}{\partial\theta}. \tag{10.2.2b}$$

In Eq. (10.2.1c) the terms multiplied by the factor $2(1-\nu)$ are deleted so that the equation becomes

$$\left[\nabla^4(\nabla^2+1)^2 + \frac{1-\nu^2}{k}\frac{\partial^4}{\partial\xi^4}\right]w = \nabla^4 q_\zeta. \tag{10.2.2c}$$

The form of the theory most often used, however, is the even more drastic reduction due to Donnell§ which has already been discussed for shells of arbitrary shape in Section 3.3. To obtain these equations we delete the terms multiplied by k in Eqs. (10.1.3a) to (10.1.3c), the terms involving M_x and N_x in Eqs. (10.1.1a) and (10.1.1b), and terms involving tangential displacements u_x and u_θ in Eqs. (10.1.3d) and (10.1.3f) as well as the term w in Eq. (10.1.3e). The resulting equations differ from those given by Eqs. (10.2.2) only in the replacement of Eq. (10.2.2c) by

$$\left(\nabla^8 + \frac{1-\nu^2}{k}\frac{\partial^4}{\partial\xi^4}\right)w = \nabla^4 q_\zeta. \tag{10.2.3}$$

We shall compare the results of the various approximations for several problems of interest in later sections.

† W. Flügge, *Stresses in Shells*, Springer-Verlag, 1960, pp. 218–221.

‡ L. S. D. Morley, 'An Improvement on Donnell's Approximation for Thin-walled Circular Cylinders,' *Quarterly Journal of Mechanics and Applied Math.*, vol. 12, pt. 4, February 1959, pp. 89–99.

§ L. H. Donnell, 'Stability of Thin-walled Tubes Under Torsion,' NACA Report, no. 479, 1933.

10.3 Cylinders loaded along circular edges

When a cylinder is loaded only along circular edges and has no other discontinuities, the solution for the stresses and deformations within the shell wall is facilitated if the edge loading is expressed in the form of a Fourier series in the angular location θ around the circumference. Since the stresses and deformations must have a period of 2π as we traverse the circumference, the deflection-function Φ can also be expressed as a Fourier series. We note that Eq. (10.1.12) then yields ordinary differential equations with constant coefficients. Thus the form of Φ can be readily determined to be

$$\Phi = \sum_{n=0}^{\infty} \sum_{r=1}^{8} e^{\lambda_{nr}\xi}(A_{nr} \cos n\theta + B_{nr} \sin n\theta), \tag{10.3.1}$$

where the values of λ_{nr} are the eight independent solutions of the following equation, obtained by substituting Eq. (10.3.1) into Eq. (10.1.14),

$$\begin{aligned}
&(1 + 3k)(1 - k)\lambda_n^8 \\
&\quad - \left[4\left(1 + \frac{11 - 3\nu}{8} k + 9\frac{1 - \nu}{8} k^2\right)n^2 - 2\nu(1 + 3k)\right]\lambda_n^6 \\
&\quad + \left[\left(\frac{1 - \nu^2}{k} + 1\right)(1 + 3k) - 6\left(1 + \frac{2 - \nu + \nu^2}{2} k\right)n^2\right. \\
&\quad \left. + 6\left(1 + \frac{2 - \nu}{2} k - \frac{\nu^2}{6} k^2\right)n^4\right]\lambda_n^4 - n^2(n^2 - 1) \\
&\quad \times \left\{4\left(1 + \frac{7 - 3\nu}{8} k + 3\frac{1 - \nu}{8} k^2\right)n^2\right. \\
&\quad \left. - \left[2(2 - \nu) + 7\frac{1 - \nu}{2} k + 3\frac{1 - \nu}{2} k^2\right]\right\}\lambda_n^2 \\
&\quad + (1 + k)n^4(n^2 - 1)^2 = 0.
\end{aligned} \tag{10.3.2}$$

10.3.1 *Solutions for $n = 0, 1$*

When n is equal to 0 (axisymmetric deformations) or to 1 ('wind loading'), the solutions of Eq. (10.3.2) are easily obtained since the constant term and the coefficient of λ_n^2 vanish in both cases, yielding

$$(1 + 3k)\lambda_0^4\left[(1 - k)\lambda_0^4 + 2\nu\lambda_0^2 + \frac{1 - \nu^2}{k} + 1\right] = 0, \tag{10.3.3a}$$

$$\begin{aligned}
\lambda_1^4\Bigg\{(1 + 3k)(1 - k)\lambda_1^4 - \left[2(2 - \nu) + \frac{11 - 15\nu}{2} k + 9\frac{1 - \nu}{2} k^2\right]\lambda_1^2 \\
+ \left[\frac{1 - \nu^2}{k} + 4 - 3\nu^2 + 3(1 - \nu^2)k - \nu^2k^2\right]\Bigg\} = 0.
\end{aligned} \tag{10.3.3b}$$

The four zero roots of Eq. (10.3.3a) and of Eq. (10.3.3b) indicate that Eq. (10.1.14) can be integrated four times with respect to ξ when n is equal to 0 or 1. The corresponding solutions for Φ are then of the form

$$\begin{aligned}\Phi = {} & A_{02} + B_{02}\xi + C_{02}\xi^2 + D_{02}\xi^3 \\ & + (A_{12} + B_{12}\xi + C_{12}\xi^2 + D_{12}\xi^3)\cos\theta \\ & + (\bar{A}_{12} + \bar{B}_{12}\xi + \bar{C}_{12}\xi^2 + \bar{D}_{12}\xi^3)\sin\theta. \end{aligned} \tag{10.3.4}$$

The first three terms on the right side of Eq. (10.3.4) are extraneous since they yield no displacements or stress-resultants and hence need not be considered further. The remaining terms yield displacements and stress resultants which can be expressed in the general form

$$\begin{aligned}(A) = {} & a_0 D_{02} + (a_1 A_{12} + b_1 B_{12} + c_1 C_{12} + d_1 D_{12})\cos\theta \\ & + (a_1 \bar{A}_{12} + b_1 \bar{B}_{12} + c_1 \bar{C}_{12} + d_1 \bar{D}_{12})\sin\theta, \end{aligned} \tag{10.3.5a}$$

$$\begin{aligned}(B) = {} & (a_1 A_{12} + b_1 B_{12} + c_1 C_{12} + d_1 D_{12})\sin\theta \\ & - (a_1 \bar{A}_{12} + b_1 \bar{B}_{12} + c_1 \bar{C}_{12} + d_1 \bar{D}_{12})\cos\theta. \end{aligned} \tag{10.3.5b}$$

Values of the coefficients a_0, a_1, b_1, c_1, and d_1 are given in Table 10.1 where the category to which each quantity belongs is also indicated.

The non-zero roots of Eqs. (10.3.3a) and (10.3.3b) may be readily obtained as

$$\begin{aligned}\lambda_{0r} = {} & \pm\left\{\frac{\left[\left(\frac{1-\nu^2}{k}+1\right)(1-k)\right]^{1/2} - \nu}{2(1-k)}\right\}^{1/2} \\ & \pm i\left\{\frac{\left[\left(\frac{1-\nu^2}{k}+1\right)(1-k)\right]^{1/2} + \nu}{2(1-k)}\right\}^{1/2}, \end{aligned} \tag{10.3.6a}$$

$$\begin{aligned}\lambda_{1r} = {} & \\ & \pm\left(\frac{\left\{(1+3k)(1-k)\left[\frac{1-\nu^2}{k} + 4 - 3\nu^2 + 3(1-\nu^2)k - \nu^2k^2\right]\right\}^{1/2} + 2 - \nu + \frac{11-15\nu}{4}k + 9\,\frac{1-\nu}{4}k^2}{2(1+3k)(1-k)}\right)^{1/2} \\ & \pm i\left(\frac{\left\{(1+3k)(1-k)\left[\frac{1-\nu^2}{k} + 4 - 3\nu^2 + 3(1-\nu^2)k - \nu^2k^2\right]\right\}^{1/2} - \left(2 - \nu + \frac{11-15\nu}{4}k + 9\,\frac{1-\nu}{4}k^2\right)}{2(1+3k)(1-k)}\right)^{1/2}. \end{aligned} \tag{10.3.6b}$$

Table 10.1 Coefficients for use in Eq. (10.3.5)

Quantity	Category	a_0	a_1	b_1	c_1	d_1
u_x	A	$-6\nu(1+3k)$	0	$-(1+k)$	$-2(1+k)\xi$	$-[6\nu(1+3k)-9(1-\nu)k^2 + 3(1+k)\xi^2]$
u_θ	B	0	$-(1+k)$	$-(1+k)\xi$	$2[2+\nu+2k+\frac{1}{2}(3-\nu)k^2] - (1+k)\xi^2$	$6[2+\nu+2k+\frac{1}{2}(3-\nu)k^2]\xi - (1+k)\xi^3$
w	A	0	$1+k$	$(1+k)\xi$	$-4[1+(1-\nu)k-\frac{3}{4}(1-\nu)k^2] + (1+k)\xi^2$	$-12[1+(1-\nu)k+\frac{3}{4}(1-\nu)k^2]\xi + (1+k)\xi^3$
$\partial w/\partial\xi$	A	0	0	$1+k$	$2(1+k)\xi$	$-12[1+(1-\nu)k+\frac{3}{4}(1-\nu)k^2] + 3(1+k)\xi^2$
RN_x/K	A	0	0	0	$-2(1-\nu^2)(1+k)^2$	$-6(1-\nu^2)(1+k)^2\xi$
RN_θ/K	A	0	0	0	$4\nu k(1+k)$	$12\nu k(1+k)\xi$
$R\overline{N}_{x\theta}/K$	B	0	0	0	0	$3(1-\nu)(1+k)(1+\frac{3}{4}k) \times [2(1+\nu)+\nu k]$
$M_x/(kK)$	A	0	0	0	$-2(1+k)[2-(1+k)\nu^2]$	$-6(1+k)[2-(1+k)\nu^2]\xi$
$M_\theta/(kK)$	A	0	0	0	$-2\nu(1+k)$	$-6\nu(1+k)\xi$
$\overline{M}_{x\theta}/(kK)$	B	0	0	0	0	$-3(1-\nu)(1+k)(1-\nu-\frac{3}{2}\nu k)$
$R\overline{Q}_x/(kK)$	A	0	0	0	0	$-6(1+k)[3-2\nu-\frac{1}{2}(3-\nu)k]$
$R\overline{Q}_\theta/(kK)$	B	0	0	0	$2\nu(1+k)$	$6\nu(1+k)\xi$

The results for axisymmetric deformation differ somewhat from previous results since the constitutive equations have been altered.

Eq. (10.3.6) can be very much simplified since k is generally very small compared to unity. Then we have

$$\lambda_0 \approx \lambda_1 \approx \pm \left(\frac{1-\nu^2}{4k}\right)^{1/2} (1 \pm i) \tag{10.3.7}$$

with an error of the order of $k^{1/2}$ or of h/R compared to unity, or

$$\lambda_0 \approx \left[\left(\frac{1-\nu^2}{4k}\right)^{1/2} - \frac{\nu}{2}\right]^{1/2} \pm i\left[\left(\frac{1-\nu^2}{4k}\right)^{1/2} + \frac{\nu}{2}\right]^{1/2}, \tag{10.3.8a}$$

$$\lambda_1 \approx \pm\left[\left(\frac{1-\nu^2}{4k}\right)^{1/2} + \frac{2-\nu}{2}\right]^{1/2} \pm i\left[\left(\frac{1-\nu^2}{4k}\right)^{1/2} - \frac{2-\nu}{2}\right]^{1/2}, \tag{10.3.8b}$$

with an error of the order of k or of h^2/R^2 compared to unity.

The displacement function corresponding to the roots given by Eq. (10.3.6) may be written in real form as

$$\begin{aligned}\Phi = {} & (A_0 \cos q_0\xi + B_0 \sin q_0\xi)\, \mathrm{e}^{-p_0\xi} \\ & + [C_0 \cos q_0(\xi_1 - \xi) + D_0 \sin q_0(\xi_1 - \xi)]\, \mathrm{e}^{-p_0(\xi_1 - \xi)} \\ & + \{(A_1 \cos q_1\xi + B_1 \sin q_1\xi)\, \mathrm{e}^{-p_1\xi} \\ & + [C_1 \cos q_1(\xi_1 - \xi) + D_1 \sin q_1(\xi_1 - \xi)]\, \mathrm{e}^{-p_1(\xi - \xi)}\} \cos\theta \\ & + \{(\bar{A}_1 \cos q_1\xi + \bar{B}_1 \sin q_1\xi)\, \mathrm{e}^{-p_1\xi} \\ & + [\bar{C}_1 \cos q_1(\xi_1 - \xi) + D_1 \sin q_1(\xi_1 - \xi)]\, \mathrm{e}^{-p_1(\xi_1 - \xi)}\} \sin\theta,\end{aligned} \tag{10.3.9}$$

where

$$\left.\begin{matrix} p_0 \\ q_0 \end{matrix}\right\} = \left\{\frac{\left[\left(\frac{1-\nu^2}{k} + 1\right)(1-k)\right]^{1/2} \mp \nu}{2(1-k)}\right\}^{1/2}, \tag{10.3.10a}$$

$$\left.\begin{matrix} p_1 \\ q_1 \end{matrix}\right\} = \left[\frac{\left\{(1+3k)(1-k)\left[\frac{1-\nu^2}{k} + 4 - 3\nu^2 + 3(1-\nu^2)k - \nu^2k^2\right]\right\}^{1/2} \pm \left(2 - \nu + \frac{11-15\nu}{4}k + 9\,\frac{1-\nu}{4}k^2\right)}{2(1+3k)(1-k)}\right]^{1/2}, \tag{10.3.10b}$$

$$\xi_1 = \frac{L}{R}. \tag{10.3.10c}$$

We may now obtain expressions, as before, for displacements and stress resultants. In all cases these may be expressed in one of the following forms

$$\begin{aligned}\begin{pmatrix}\mathrm{I}\\ \mathrm{II}\end{pmatrix} = {} & C_0 f_0(\xi_1 - \xi) + D_0 g_0(\xi_1 - \xi) \pm [A_0 f_0(\xi) + B_0 g_0(\xi)] \\ & + (C_1 \cos\theta + \bar{C}_1 \sin\theta) f_1(\xi_1 - \xi) \\ & + (D_1 \cos\theta + \bar{D}_1 \sin\theta) g_1(\xi_1 - \xi) \\ & \pm [(A_1 \cos\theta + \bar{A}_1 \sin\theta) f_1(\xi) \\ & \qquad + (B_1 \cos\theta + \bar{B}_1 \sin\theta) g_1(\xi)],\end{aligned} \tag{10.3.11a}$$

$$\begin{aligned}\begin{pmatrix}\mathrm{III}\\ \mathrm{IV}\end{pmatrix} = {} & (C_1 \sin\theta - \bar{C}_1 \cos\theta) f_1(\xi_1 - \xi) \\ & + (D_1 \sin\theta - \bar{D}_1 \cos\theta) g_1(\xi_1 - \xi) \\ & \pm [(A_1 \sin\theta - \bar{A}_1 \cos\theta) f_1(\xi) \\ & \qquad + (B_1 \sin\theta - \bar{B}_1 \cos\theta) g_1(\xi)],\end{aligned} \tag{10.3.11b}$$

with

$$f_n(u) = (\alpha_n \cos q_n u + \beta_n \sin q_n u)\, \mathrm{e}^{-p_n u}, \tag{10.3.11c}$$

$$g_n(u) = (\alpha_n \sin q_n u - \beta_n \cos q_n u)\, \mathrm{e}^{-p_n u}. \tag{10.3.11d}$$

Equations for α_n and β_n for the various deformations and stress resultants are given in Table 10.2 where the category to which each quantity belongs is also indicated.

Table 10.2 (a) Expressions for α_n for use in Eqs. (10.3.11)

Quantity	α_n
u_x (II)	$-\left\{\left[\nu(1+3k) - 3\dfrac{1-\nu}{2}n^2k^2\right](p_n^2 - 3q_n^2) + n^2(1+kn^2) - k(1+3k) \times (p_n^4 - 10p_n^2q_n^2 + 5q_n^4)\right\}p_n$
u_θ (III)	$\left\{\left[2+\nu+2\left(1+\dfrac{3-\nu}{4}k\right)n^2k\right](p_n^2 - q_n^2) - (1+k)n^2 - 2k(p_n^4 - 6p_n^2q_n^2 + q_n^4)\right\}n$
w (I)	$(1+3k)(p_n^4 - 6p_n^2q_n^2 + q_n^4) - 2n^2[1 + (1-\nu)(1+\frac{3}{4}k)k](p_n^2 - q_n^2) + (1+k)n^4$
$\partial w/\partial\xi$ (II)	$\{(1+3k)(p_n^4 - 10p_n^2q_n^2 + 5q_n^4) - 2n^2[1 + (1-\nu)(1+\frac{3}{4}k)k](p_n^2 - 3q_n^2) + (1+k)n^4\}p_n$
RN_x/K (I)	$(1-\nu)n^2\left\{\left[1+\nu+2\nu(1+\frac{3}{4}k)k + 2k\left(1+\dfrac{2-\nu}{4}k\right)n^2\right](p_n^2 - q_n^2) - 2(1+k)(1+\frac{3}{4}k)k(p_n^4 - 6p_n^2q_n^2 + q_n^4)\right\}$

Table 10.2—*contd.*

Quantity	α_n
RN_θ/K (I)	$\left[(1+3k)(1-\nu^2+k) - 3\left(1+\frac{2-\nu+\nu^2}{2}k\right)kn^2\right](p_n^4 - 6p_n^2q_n^2 + q_n^4) - n^2k\left[2(2-\nu) + \tfrac{1}{2}(1-\nu)(7+3k)k - \left(4-\nu+\frac{7-5\nu}{2}k + 3\,\frac{1-\nu}{2}k^2\right)n^2\right] \times (p_n^2 - q_n^2) + \nu k(1+3k)(p_n^6 - 15p_n^4q_n^2 + 15p_n^2q_n^4 - q_n^6) - k(1+k)n^2(n^2-1)$
$R\overline{N}_{x\theta}/K$ (IV)	$(1-\nu)(1+\tfrac{3}{4}k)np_n\left[\left(1-\nu+\frac{3\nu}{2}k + \frac{2+\nu k}{2}kn^2\right)(p_n^2 - 3q_n^2) + \tfrac{1}{2}kn^2(n^2-1) - \tfrac{3}{2}k(1+k)(p_n^4 - 10p_n^2q_n^2 + 5q_n^4)\right]$
M_x/kK (I)	$(1-k)(1+3k)(p_n^6 - 15p_n^4q_n^2 + 15p_n^2q_n^4 - q_n^6) + \{\nu(1-3k) - [2+\nu+(2-\nu)k + 3(1-\nu)k^2]n^2\} \times (p_n^4 - 6p_n^2q_n^2 + q_4^n) - n^2\{[1+2\nu+2(1-\nu^2)k - \nu^2k^2]n^2 + 1 - 2\nu - \nu^2\} \times (p_n^2 - q_n^2) - \nu(1+k)n^4(n^2-1)$
M_θ/kK (I)	$\nu(1+3k)(p_n^6 - 15p_n^4q_n^2 + 15p_n^2q_n^4 - q_n^6) + \left\{1+3k - \left[1+2\nu+(3-2\nu-2\nu^2)k + 3\,\frac{1-\nu}{2}k^2\right]n^2\right\} \times (p_n^4 - 6p_n^2q_n^2 + q_n^4) + n^2\left\{\left[2+\nu+(2-\nu)k + 3\,\frac{1-\nu}{2}k^2\right]n^2 - \left[2+2(1-\nu)k + 3\,\frac{1-\nu}{2}k^2\right]\right\} \times (p_n^2 - q_n^2) - (1+k)n^4(n^2-1)$
$\overline{M}_{x\theta}/kK$ (IV)	$(1-\nu)n\left\{(1+\tfrac{3}{4}k)(1+k)(p_n^4 - 10p_n^2q_n^2 - 5q_n^4) + \left[\frac{3+\nu}{2} - \frac{3\nu}{4}k - n^2\left(2+\frac{1-4\nu}{2}k - \frac{3\nu}{4}k^2\right)\right] \times (p_n^2 - 3q_n^2) + n^2(n^2-1)(1+\tfrac{3}{4}k)\right\}p_n$
$R\overline{Q}_x/kK$ (II)	$-\left\langle(1-k)(1+3k)(p_n^6 - 21p_n^4q_n^22 + 35p_n^2q_n^4 - 7q_n^6) - \{[4-\nu-\tfrac{1}{2}(11-9\nu)k - 3(1-\nu)k^2]n^2 - \nu(1+3k)\} \times (p_n^4 - 10p_n^2q_n^2 + 5q_n^4) - n^2\left\{[5-2\nu-(1-\nu)(3-2\nu)k - \tfrac{1}{2}\nu(3-\nu)k^2]n^2 - \left(2+3\nu\,\frac{1-\nu}{2}k\right)\right\} \times (p_n^2 - 3q_n^2) - n^4\left[\left(2-\nu+\frac{3-\nu}{2}k\right)n^2 - \left(2-\nu+\frac{3-\nu}{2}k\right)\right]\right\rangle p_n$

Table 10.2—*contd.*

Quantity	α_n
$R\bar{Q}_\theta/kK$ (III)	$n\Big\langle\Big(2-\nu+\frac{7-\nu}{2}k+3\frac{1-\nu}{2}k^2\Big)(p_n^6-15p_n^4q_n^2+15p_n^2q_n^4-q_n^6) - \Big\{[5-2\nu+(4-3\nu+2\nu^2)k-\frac{3}{2}(1-\nu)^2k^2]n^2 - \Big(4-2\nu-\nu^2+3\frac{2-\nu+\nu^2}{2}k\Big)\Big\} \times (p_n^4-6p_n^2q_n^2-q_n^4)+n^2\Big[\Big(4-\nu+\frac{7-5\nu}{2}k+3\frac{1-\nu}{2}k^2\Big)n^2 - \Big(4-2\nu+7\frac{1-\nu}{2}k+3\frac{1-\nu}{2}k^2\Big)\Big] \times (p_n^2-q_n^2)-n^4(n^2-1)(1+k)\Big\rangle$

(b) Expressions for β_n for use in Eqs. (10.3.11)

Quantity	β_n
u_x (II)	$-\Big\{\Big[\nu(1+3k)-3\frac{1-\nu}{2}n^2k^2\Big](3p_n^2-q_n^2)+n^2(1+kn^2) - k(1+3k)(5p_n^4-10p_n^2q_n^2+q_n^4)\Big\}q_n$
u_θ (III)	$2p_nq_nn\Big[2+\nu+2\Big(1+\frac{3-\nu}{4}k\Big)n^2k-2k(p_n^2-q_n^2)\Big]$
w (I)	$4p_nq_n\{(1+3k)(p_n^2-q_n^2)-n^2[1+(1-\nu)(1+\frac{3}{4}k)k]\}$
$\partial w/\partial\xi$ (II)	$\{(1+3k)(5p_n^4-10p_n^2q_n^2+q_n^4) - 2n^2[1+(1-\nu)(1+\frac{3}{4}k)k](3p_n^2-q_n^2)+(1+k)n^4\}q_n$
RN_x/K (I)	$2p_nq_nn^2(1-\nu)\Big[1+\nu+2\nu(1+\frac{3}{4}k)+2k\Big(1+\frac{2-\nu}{4}k\Big)n^2 - 4(1+k)(1+\frac{3}{4}k)k(p_n^2-q_n^2)\Big]$
RN_θ/K (I)	$2p_nq_n\Big\{2\Big[(1+3k)(1-\nu^2+k)-3\Big(1+\frac{2-\nu+\nu^2}{2}k\Big)kn^2(p_n^2-q_n^2) - n^2k[2(2-\nu)+\frac{1}{2}(1-\nu)(7+3k)k - \Big(4-\nu+\frac{7-5\nu}{2}k+3\frac{1-\nu}{2}k^2\Big)n^2\Big] + \nu k(1+3k)(3p_n^4-10p_n^2q_n^2+3q_n^4)\Big\}$

Table 10.2—*contd.*

Quantity	β_n
$R\bar{N}_{x\theta}/K$ (IV)	$\begin{aligned}&(1-\nu)(1+\tfrac{3}{4}k)nq_n\Big[\Big(1+\nu+\frac{3\nu}{2}k+\frac{2+\nu k}{2}kn^2\Big)(3p_n^2-q_n^2)\\&\quad+\tfrac{1}{2}kn^2(n^2-1)-\tfrac{3}{2}k(1+k)(5p_n^4-10p_n^2q_n^2-q_n^4)\Big]\end{aligned}$
M_x/kK (I)	$\begin{aligned}&2p_nq_n\langle(1+3k)(1-k)(3p_n^4-10p_n^2q_n^2+3q_n^4)\\&\quad+2\{\nu(1-3k)-[2+\nu-(2-\nu)k+3(1-\nu)k^2]n^2\}(p_n^2-q_n^2)\\&\quad+n^2\{[1+2\nu+2(1-\nu^2)k-\nu^2k^2]n^2-1-2\nu-\nu^2\}\rangle\end{aligned}$
M_θ/kK (I)	$\begin{aligned}&2p_nq_n\Big\langle\nu(1+3k)(3p_n^4-10p_n^2q_n^2+3q_n^4)\\&\quad+2\Big\{1+3k-\Big[1+2\nu+(3-2\nu-2\nu^2)k+3\frac{1-\nu}{2}k^2\Big]n^2\Big\}\\&\quad\times(p_n^2-q_n^2)+n^2\Big\{\Big[2+\nu-(2-\nu)k+3\frac{1-\nu}{2}k^2\Big]n^2\\&\qquad-\Big[2+2(1-\nu)k+3\frac{1-\nu}{2}k^2\Big]\Big\}\Big\rangle\end{aligned}$
$\bar{M}_{x\theta}/kK$	$\begin{aligned}&(1-\nu)nq_n\Big\{(1+\tfrac{3}{4}k)(1+k)(5p_n^4-10p_n^2q_n^2+q_n^4)\\&\quad+\Big[\frac{3+\nu}{2}-\frac{3\nu}{4}k-n^2\Big(2-\frac{1-4\nu}{2}k-\frac{3\nu}{4}k^2\Big)\Big]\\&\quad\times(3p_n^2-q_n^2)+n^2(n^2-1)(1+\tfrac{3}{4}k)\Big\}\end{aligned}$
$R\bar{Q}_x/kK$ (II)	$\begin{aligned}&-\Big\langle(1-k)(1+3k)(7p_n^6-35p_n^4q_n^2+21p_n^2q_n^4-q_n^6)\\&\quad-\{[4-\nu+\tfrac{1}{2}(11-9\nu)k+3(1-\nu)k^2]n^2-\nu(1+3k)\}\\&\quad\times(5p_n^4-10p_n^2q_n^2-q_n^4)\\&\quad-n^2\Big\{[5-2\nu+(1-\nu)(3-2\nu)k-\tfrac{1}{2}\nu(3-\nu)k^2]n^2\\&\qquad-\Big(2+3\nu\frac{1-\nu}{2}k\Big)\Big\}(3p_n^2-q_n^2)\\&\quad-n^4\Big[\Big(2-\nu+\frac{3-\nu}{2}k\Big)n^2-\Big(2-\nu+\frac{3-\nu}{2}k\Big)\Big]\Big\rangle q_n\end{aligned}$
$R\bar{Q}_\theta/kK$ (III)	$\begin{aligned}&2p_nq_nn\Big\langle\Big(2-\nu+\frac{7-\nu}{2}k+3\frac{1-\nu}{2}k^2\Big)(3p_n^4-10p_n^2q_n^2+3q_n^4)\\&\quad-\Big\{[5-2\nu+(4-3\nu+2\nu^2)k+\tfrac{3}{2}(1-\nu)^2k^2]n^2\\&\qquad-2\Big(4-2\nu-\nu^2+3\frac{2-\nu+\nu^2}{2}k\Big)\Big\}(p_n^2-q_n^2)\\&\quad+n^2\Big[\Big(4-\nu+\frac{7-5\nu}{2}k+3\frac{1-\nu}{2}k^2\Big)n^2\\&\qquad-\Big(4-2\nu+7\frac{1-\nu}{2}k+3\frac{1-\nu}{2}k^2\Big)\Big]\Big\rangle\end{aligned}$

In addition to the above forms of the displacement-function yielding axisymmetric stresses and displacements we need to append

$$\Phi = \frac{(1 + k - \nu^2)E_{02}}{4(1 - \nu)[1 + \nu + 2\nu k(1 + \frac{3}{4}k)]} \times \left[\xi^2\theta^2 - \frac{2 + \nu}{6(1 + k)}\theta^4 - \frac{2(1 - \nu)(1 + k)(1 + \frac{3}{4}k) - \nu}{6(1 + 3k)(1 + k - \nu^2)} k\xi^4\right] + F_{02}\xi\theta^2, \tag{10.3.12}$$

which satisfies the differential equation and yields single valued displacements and stress-resultants. These are needed to supply some terms missing when the Fourier series given by Eq. (10.3.1) is used. The displacements corresponding to the displacement function are given by Eqs. (10.1.8), (10.1.13), and (10.3.12) as

$$u_x = 2F_{02} + (1 + k)E_{02}, \tag{10.3.13a}$$

$$u_\theta = 0, \tag{10.3.13b}$$

$$w = -\nu E_{02}, \tag{10.3.13c}$$

and correspond to a rigid body translation and a uniform stretching of the cylinder. The stress resultants may now be obtained from Eqs. (10.1.3) and (10.3.13) as

$$N_x = \frac{K}{R}(1 + k - \nu^2)E_{02}, \tag{10.3.14a}$$

$$N_\theta = 0, \tag{10.3.14b}$$

$$\bar{N}_{x\theta} = 0, \tag{10.3.14c}$$

$$M_x = Kk(1 + k)E_{02}, \tag{10.3.14d}$$

$$M_\theta = \nu KkE_{02}, \tag{10.3.14e}$$

$$\bar{M}_{x\theta} = 0. \tag{10.3.14f}$$

10.3.2 Solutions for $n \geq 2$

For values of n other than 0 and 1 we have to deal with the full biquartic equation for λ_n given by Eq. (10.3.2). This equation may be solved exactly by means of the formulas of Ferrari and Tartaglia for the solution of the general quartic and cubic equations† or by means of various iteration programs which are available for use on electronic digital computers.‡ Calculations

† M. Abramowitz and I. A. Stegun, *Handbook of Mathematical Functions*, N.B.S. Appl. Math. Series No. 55, June 1964, pp. 17–18.

‡ See, for example, A. Ralston, *A First Course in Numerical Analysis*, McGraw-Hill Book Co., 1965, pp. 350–382

indicate that as n increases the roots of Eq. (10.3.2) are at first complex while for very large values of n four roots are real while the remainder are complex.†

The approximate characteristic equations corresponding to the operators of Eqs. (10.2.1c), (10.2.2c), and (10.2.3) are given respectively by

$$(\lambda_n^2 - n^2)^2[\lambda_n^2 - (n^2 - 1)]^2 - 2(1 - \nu)\lambda_n^2[\lambda_n^4 - n^2(n^2 - 1)] + \frac{1 - \nu^2}{k}\lambda_n^4 = 0, \tag{10.3.15a}$$

$$(\lambda_n^2 - n^2)^2[\lambda_n^2 - (n^2 - 1)]^2 + \frac{1 - \nu^2}{k}\lambda_n^4 = 0, \tag{10.3.15b}$$

$$(\lambda_n^2 - n^2)^4 + \frac{1 - \nu^2}{k}\lambda_n^4 = 0. \tag{10.3.15c}$$

Eq. (10.3.15a) must be solved in the same manner as Eq. (10.3.2) and thus presents no real simplification. The solutions of Eqs. (10.3.15b) and (10.3.15c) can found explicitly and may be expressed in almost identical forms. We may solve Eq. (10.3.15b) to yield

$$\left(\frac{4k}{1 - \nu^2}\right)^{1/4}\lambda_{nr} = (\gamma_{n1} + 1)^{1/2} \times \left\{\pm\left[\frac{(\gamma_{n2}^2 + 1)^{1/2} + \gamma_{n2}}{2}\right]^{1/2} \pm i\left[\frac{(\gamma_{n2}^2 + 1)^{1/2} - \gamma_{n2}}{2}\right]^{1/2}\right\} \quad (r = 1, 2, 3, 4), \tag{10.3.16a}$$

$$\left(\frac{4k}{1 - \nu^2}\right)^{1/4}\lambda_{nr} = (\gamma_{n1} - 1)^{1/2} \times \left\{\pm\left[\frac{(\gamma_{n2}^2 + 1)^{1/2} + \gamma_{n2}}{2}\right]^{1/2} \pm i\left[\frac{(\gamma_{n2}^2 + 1)^{1/2} - \gamma_{n2}}{2}\right]^{1/2}\right\} \quad (r = 5, 6, 7, 8), \tag{10.3.16b}$$

with

$$\left.\begin{matrix}\gamma_{n1}\\ \gamma_{n2}\end{matrix}\right\} = \left\{\frac{\left[\left(1 + \frac{k}{1 - \nu^2}\right)^2 + \psi_n^4\right]^{1/2} \pm \left(1 - \frac{k}{1 - \nu^2}\right)}{2}\right\}^{1/2}, \tag{10.3.16c}$$

$$\psi_n = 2\left[n^2(n^2 - 1)\frac{k}{1 - \nu^2}\right]^{1/4}. \tag{10.3.16d}$$

† P. Seide, 'Roots of the Cylindrical Shell Characteristic Equation for Harmonic Circumferential, Edge Loading', *AIAA J.*, vol. 8, no. 3, March 1970, pp. 452–454 and vol. 10, no. 9, September 1972, pp. 1263–1264.

If we omit the terms involving k explicitly in Eq. (10.3.16c) as being small compared to unity, the solutions may be reduced to the much neater and still adequate form

$$\frac{4k}{1-\nu^2}\lambda_{nr} = \pm\frac{1}{2}\{1 + [(1 + \psi_n^4)^{1/2} + \psi_n^2]^{1/2}\}$$

$$\pm\frac{i}{2}\{1 - [(1 + \psi_n^4)^{1/2} - \psi_n^2]^{1/2}\}, \tag{10.3.17a}$$

$$\frac{4k}{1-\nu^2}\lambda_{nr} = \pm\frac{1}{2}\{(1 + \psi_n^4)^{1/2} + \psi_n^2]^{1/2} - 1\}$$

$$\pm\frac{i}{2}\{1 - [(1 + \psi_n^4)^{1/2} - \psi_n^2]^{1/2}\}. \tag{10.3.17b}$$

Eqs. (10.3.17) are also the solutions of Eq. (10.3.15c) if we neglect unity compared to n^2 in Eq. (10.3.16d) for ψ_n.

The real and imaginary parts of Eqs. (10.3.17) are shown as a function of the single parameter ψ_n in Fig. 10.1. For small values of the parameter ψ_n we determine that the roots given by Eq. (10.3.17a) imply a longitudinal variation similar to that for an axisymmetrically loaded cylinder and hence with an attenuation distance of the order of $(Rh)^{1/2}$. The second set of roots given by $(R/h)/n^2$ times the attenuation length for axisymmetric loading. For large Eq. (10.3.17b) implies an attenuation length which is longer, or the order of values of ψ_n we see that the two sets of roots become nearly identical so that both sets of stress states have almost the same longitudinal variation.

Since the roots of Eqs. (10.3.15b) and (10.3.15c) are always complex, they obviously are not a good approximation to those of Eq. (10.3.2) for large values of n. Calculations indicate that the values of the roots of Eqs. (10.3.15b) and (10.3.15c) are in good agreement with those of Eq. (10.3.2) when n is less than about R/h.† This value of n corresponds to circumferential load variations for which the half-wave length is about 8 shell wall thicknesses. A comparison of the roots of the various equations for some values of R/h and n is shown in Table 10.3. In most problems of interest the terms involving rapidly varying edge loads do not contribute significantly to the total solution. Thus the root approximations given by Eqs. (10.3.16) or (10.3.17) are usually adequate.

If the value of n is such that the roots of Eq. (10.3.2) are all complex we write the roots as

$$\lambda_{nr} = \pm p_{n1} \pm iq_{n1} \qquad (r = 1, 2, 3, 4), \tag{10.3.18a}$$

$$\lambda_{nr} = \pm p_{n2} \pm iq_{n2} \qquad (r = 5, 6, 7, 8). \tag{10.3.18b}$$

† P. Seide, *op. cit.*

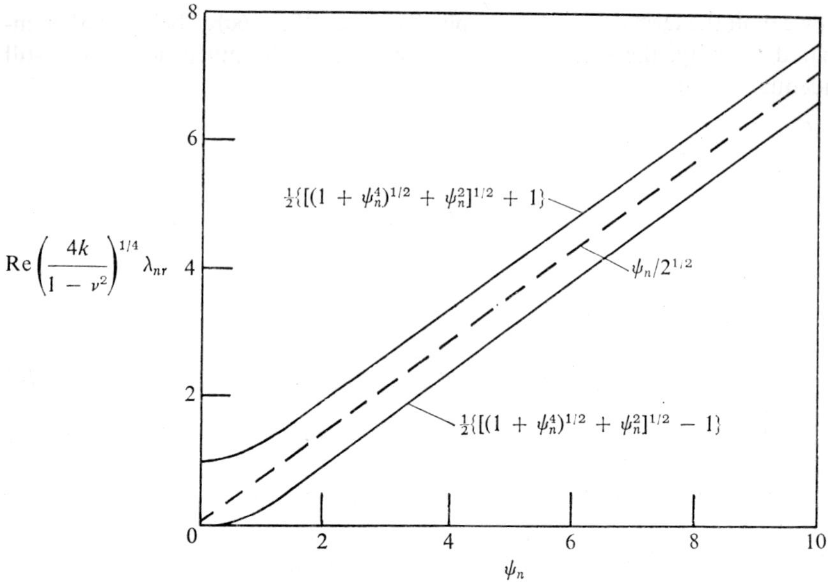

(a) Real part of roots

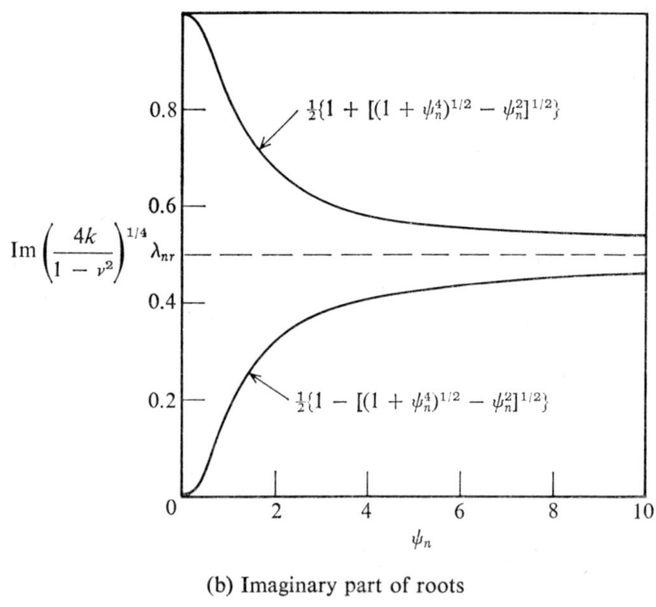

(b) Imaginary part of roots

Fig. 10.1. Variation of roots of Morley and Donnell characteristic equations (Eqs. (10.3.15b) and (10.3.15c))

Then the displacement function may be written as

$$\begin{aligned}\Phi_n = \sum_{j=1}^{2} \{[(A_{nj} \cos n\theta + \bar{A}_{nj} \sin n\theta) \cos q_{nj}\xi \\ + (B_{nj} \cos n\theta + \bar{B}_{nj} \sin n\theta) \sin q_{nj}\xi] \, \mathrm{e}^{-p_{nj}\xi} \\ + [(C_{nj} \cos n\theta + \bar{C}_{nj} \sin n\theta) \cos q_{nj}(\xi_1 - \xi) \\ + (D_{nj} \cos n\theta + \bar{D}_{nj} \sin n\theta) \sin q_{nj}(\xi_1 - \xi)] \, \mathrm{e}^{-p_{nj}(\xi_1 - \xi)}\}. \quad (10.3.19)\end{aligned}$$

Expressions for displacements and stress resultants can now be expressed in one of the following forms, similar to those of Eqs. (10.3.11):

$$\begin{aligned}\begin{pmatrix}\mathrm{I}\\ \mathrm{II}\end{pmatrix} = \sum_{j=1}^{2} \{(C_{nj} \cos n\theta + \bar{C}_{nj} \sin n\theta) f_{nj}(\xi_1 - \xi) \\ + (D_{nj} \cos n\theta + \bar{D}_{nj} \sin n\theta) g_{nj}(\xi_1 - \xi) \\ \pm [(A_{nj} \cos n\theta + \bar{A}_{nj} \sin n\theta) f_{nj}(\xi) \\ + (B_{nj} \cos n\theta + \bar{B}_{nj} \sin n\theta) g_{nj}(\xi)]\}, \quad (10.3.20a)\end{aligned}$$

$$\begin{aligned}\begin{pmatrix}\mathrm{III}\\ \mathrm{IV}\end{pmatrix} = \sum_{j=1}^{2} \{(C_{nj} \sin n\theta - \bar{C}_{nj} \cos n\theta) f_{nj}(\xi_1 - \xi) \\ + (D_{nj} \sin n\theta - \bar{D}_{nj} \cos n\theta) g_{nj}(\xi_1 - \xi) \\ \pm [(A_{nj} \sin n\theta - \bar{A}_{nj} \cos n\theta) f_{nj}(\xi) \\ + (B_{nj} \sin n\theta - \bar{B}_{nj} \cos n\theta) g_{nj}(\xi)]\}, \quad (10.3.20b)\end{aligned}$$

with

$$f_{nj}(u) = (\alpha_{nj} \cos q_{nj}u + \beta_{nj} \sin q_{nj}u) \, \mathrm{e}^{-p_{nj}u}, \quad (10.3.20c)$$

$$g_{nj}(u) = (\alpha_{nj} \sin q_{nj}u - \beta_{nj} \cos q_{nj}u) \, \mathrm{e}^{-p_{nj}u}. \quad (10.3.20d)$$

Expressions for α_{nj} and β_{nj} for each of the deformations and stress resultants can be obtained from Table 10.2 by inserting the value of p_{nj} for p_n and q_{nj} for q_n wherever it appears.

For real roots, part of Eqs. (10.3.19) and (10.3.20) should be changed. If we have four real roots of the form

$$\lambda_{nr} = \pm p_{n1} \qquad (r = 1, 2), \quad (10.3.21a)$$

$$\lambda_{nr} = \pm q_{n1} \qquad (r = 3, 4), \quad (10.3.21b)$$

while the remaining four roots of Eq. (10.3.2) are of the form given by Eq. (10.3.18b), that part of the displacement function involving the subscript $j = 1$ is replaced by

$$\begin{aligned}\Phi_{n1} = (A_{n1} \cos n\theta + \bar{A}_{n1} \sin n\theta) \, \mathrm{e}^{-p_{n1}\xi} \\ + (B_{n1} \cos n\theta + \bar{B}_{n1} \sin n\theta) \, \mathrm{e}^{-q_{n1}\xi} \\ + (C_{n1} \cos n\theta + \bar{C}_{n1} \sin n\theta) \, \mathrm{e}^{-p_{n1}(\xi_1 - \xi)} \\ + (D_{n1} \cos n\theta + \bar{D}_{n1} \sin n\theta) \, \mathrm{e}^{-q_{n1}(\xi_1 - \xi)}. \quad (10.3.22)\end{aligned}$$

Corresponding changes should be made in Eqs. (10.3.20).

Table 10.3 Roots of various characteristic equations

		Complete Flügge equation Eq. (10.3.2)				Simplified e.g.
$\frac{R}{h}$	n	p_{n1}	q_{n1}	p_{n2}	q_{n2}	p_{n1}
	1	4.169 70	3.960 80	—	—	4.168 06
	2	4.568 63	3.638 90	0.455 91	0.396 08	4.567 42
	3	5.257 48	3.272 46	1.147 45	0.748 35	5.256 93
	4	6.130 12	2.984 41	2.021 05	1.037 38	6.130 19
	5	7.079 53	2.781 78	2.971 00	1.240 59	7.080 14
	10	12.039 06	2.320 61	7.932 66	1.703 90	12.042 68
10	15	17.037 44	2.130 81	12.933 97	1.896 44	17.045 21
	25	27.033 86	1.917 77	22.940 49	2.117 00	27.054 08
	50	51.999 84	1.565 93	47.971 56	2.481 43	52.081 62
	75	76.911 22	1.163 60	73.052 60	2.799 62	77.119 01
	100	101.312 52†	102.148 76†	98.224 30	3.135 16	102.164 11
	200	197.070 33†	204.622 90†	199.069 41	5.606 17	200.988 87†
	300	294.708 66†	306.527 28†	299.257 78	8.460 25	300.650 97†
	1	5.822 58	5.674 75	—	—	5.821 98
	2	6.096 11	5.429 62	0.313 85	0.285 55	6.095 58
	3	6.581 09	5.084 77	0.800 68	0.632 28	6.580 68
	4	7.267 21	4.727 50	1.487 40	0.990 16	7.266 97
	5	8.096 35	4.420 86	2.316 84	1.297 10	8.096 28
	10	12.902 34	3.631 80	7.123 45	2.086 79	12.902 96
20	15	17.884 76	3.331 27	12.106 42	2.387 88	17.886 20
	25	27.881 01	3.058 66	22.104 29	2.662 09	27.884 73
	50	52.878 00	2.764 25	47.109 16	2.963 40	52.891 45
	75	77.870 58	2.581 04	72.116 38	3.156 44	77.900 04
	100	102.858 10	2.424 17	97.127 57	3.323 42	102.910 83
	200	202.726 13	1.746 49	197.251 27	3.933 10	202.975 59
	300	301.212 61†	303.482 99†	297.620 01	4.589 41	303.065 33
	1	9.136 00	9.042 48	—	—	9.135 84
	2	9.304 50	8.881 25	0.194 02	0.186 77	9.304 35
	3	9.596 42	8.627 52	0.487 11	0.441 68	9.596 29
	4	10.022 06	8.306 23	0.913 11	0.763 33	10.021 94
	5	10.583 37	7.951 97	1.474 59	1.117 74	10.583 27
	10	14.737 69	6.542 44	5.629 16	2.527 53	14.737 72
50	15	19.603 07	5.875 87	10.494 62	3.194 20	19.603 21
	25	29.559 32	5.308 19	20.451 05	3.762 05	29.559 73
	50	54.550 93	4.842 70	45.443 41	4.228 30	54.552 37
	75	79.550 26	4.652 66	70.444 01	4.419 59	79.553 32
	100	104.549 56	4.531 70	95.445 09	4.542 26	104.554 78
	200	204.543 13	4.221 98	195.451 57	4.863 05	204.562 95
	300	304.530 59	3.982 73	295.463 11	5.118 00	304.575 29

† Real solutions of equation.

Flügge equation (10.3.15a)			Morley equation Eq. (10.3.56)			
q_{n1}	p_{n2}	q_{n2}	p_{n1}	q_{n1}	p_{n2}	q_{n2}
3.958 88	—	—	4.126 78	4.003 78	—	—
3.636 70	0.456 23	0.379 34	4.529 33	3.685 97	0.460 10	0.374 43
3.270 08	1.148 18	0.748 44	5.222 41	3.326 75	1.155 78	0.736 25
2.981 98	2.022 06	1.037 17	6.098 40	3.047 94	2.032 60	1.015 88
2.779 26	2.972 13	1.240 01	7.050 21	2.855 35	2.984 77	1.208 84
2.316 54	7.933 09	1.701 11	12.016 00	2.446 08	7.951 03	1.618 58
2.123 77	12.932 32	1.890 45	17.018 29	2.308 01	12.953 41	1.756 73
1.901 72	22.930 51	2.101 13	27.022 96	2.197 68	22.958 12	1.867 10
1.524 40	47.910 26	2.421 76	52.027 48	2.115 02	47.962 66	1.949 78
1.167 76	72.875 53	2.668 51	77.029 09	2.087 48	72.964 28	1.977 33
0.741 27	97.831 79	2.883 30	102.029 91	2.073 71	97.965 10	1.991 10
203.758 99†	197.624 01	3.560 21	202.031 15	2.053 05	197.966 35	2.011 75
304.510 40†	297.417 94	4.065 24	302.031 56	2.046 16	297.966 78	2.018 62
5.674 10	—	—	5.792 17	5.705 19	—	—
5.428 91	0.313 91	0.285 59	6.067 03	5.461 48	0.315 39	0.283 91
0.083 99	0.800 82	0.632 34	6.553 90	5.119 16	0.804 11	0.628 08
4.726 68	1.487 65	0.990 20	7.242 00	4.765 51	1.492 78	0.982 31
4.420 03	2.317 17	1.297 07	8.072 85	4.463 26	2.32389	1.284 82
3.630 79	7.123 82	2.086 27	12.883 34	3.699 59	7.134 72	2.048 81
3.329 76	12.106 51	2.386 75	17.867 72	3.425 23	12.119 16	2.323 24
3.055 52	22.103 14	2.659 15	27.866 45	2.204 81	22.117 92	2.543 69
2.753 05	47.100 85	2.952 25	52.869 53	3.039 50	47.121 01	2.709 01
2.557 62	72.094 63	3.131 97	77.871 01	2.984 41	72.122 49	2.764 10
2.386 15	97.085 11	3.280 92	102.871 79	2.956 87	97.123 28	2.791 64
1.717 43	197.022 35	3.774 83	202.873 01	2.915 56	197.124 50	2.832 95
0.851 77	296.933 29	4.189 50	302.873 42	1.901 79	297.124 92	2.846 71
9.042 32	—	—	9.116 75	9.061 74	—	—
8.881 08	0.194 03	0.186 78	9.285 59	8.900 86	0.194 42	0.18636
8.627 34	0.487 13	0.441 69	9.578 07	8.647 74	0.488 06	0.440 65
8.306 04	0.913 14	0.763 35	10.004 39	8.327 35	0.914 74	0.761 40
7.951 78	1.474 63	1.117 77	10.566 46	7.974 28	1.476 98	1.114 64
6.542 23	5.629 26	2.527 49	14.723 84	6.573 54	5.634 57	2.515 59
5.875 62	10.494 72	3.194 08	19.590 68	5.917 26	10.501 45	3.171 90
5.307 79	20.451 06	3.761 73	29.548 25	5.370 61	20.459 04	3.718 58
4.841 48	45.442 79	4.227 13	54.541 31	4.957 69	45.452 10	4.131 50
4.650 04	70.442 20	4.417 02	79.541 78	4.819 99	70.452 57	4.269 21
4.527 15	95.441 52	4.537 74	104.542 31	4.751 14	95.453 11	4.338 06
4.204 53	195.435 04	4.845 37	204.543 38	4.647 87	195.454 18	4.441 33
3.945 47	295.423 35	5.079 06	304.543 77	4.613 44	295.454 58	4.475 75

10.4 End-loaded cantilevered cylinder with a rigid end ring

As an example of the calculations involved in the solution of asymmetric problems, let us consider a cylinder which is built into a wall at one end and stiffened by a rigid ring at the other end to which a lateral force and a bending moment are applied (Fig. 10.2). The boundary conditions to be satisfied at the built-in edge are

$$u_x = u_\theta = w = \frac{\partial w}{\partial x} = 0 \quad \text{at} \quad x = 0. \tag{10.4.1}$$

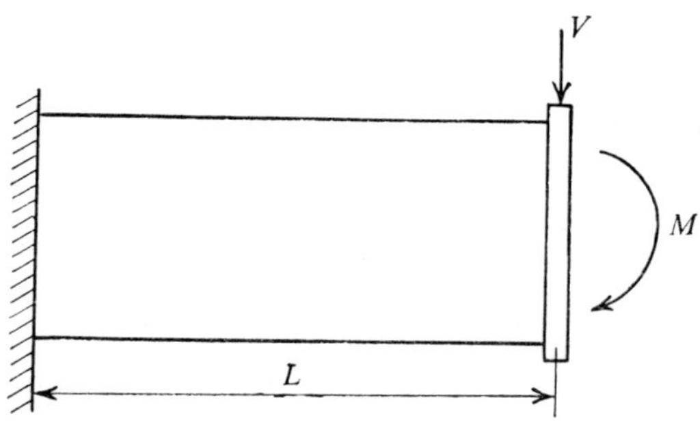

Fig. 10.2. End-loaded cantilevered cylinder

At the loaded edge the boundary conditions can be expressed in terms of the applied loads, but it is somewhat simpler if we use instead the fact that the rigidity of the edge ring implies that it will move as rigid body with a lateral translation δ and a rotation Θ (Fig. 10.3) so that the edge displacements at $x = L$ are given by

$$u_x = \Theta R \cos \theta, \tag{10.4.2a}$$

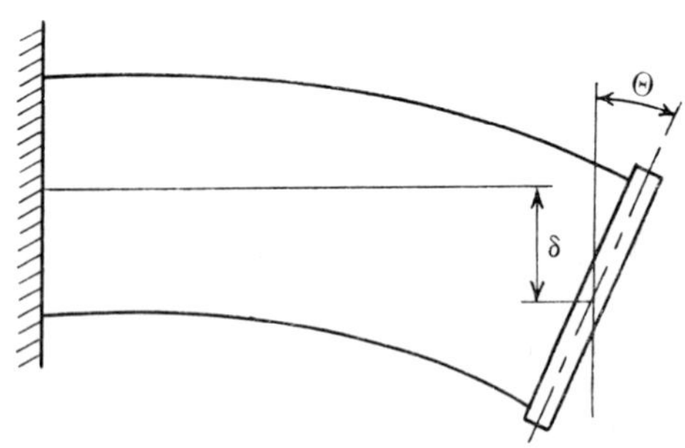

Fig. 10.3. Deflection and rotation of rigid end ring

$$u_\theta = \delta \sin \theta, \tag{10.4.2b}$$
$$w = -\delta \cos \theta. \tag{10.4.2c}$$

In addition we shall assume that the edge of the cylinder is rigidly attached to the edge ring in such a way that the reduction of the cylinder edge is equal to that of the ring or

$$\frac{\partial w}{\partial x} = -\Theta \cos \theta \qquad \text{at} \quad x = L. \tag{10.4.2d}$$

An examination of the solutions of the bending equations and boundary conditions indicates that it is sufficient to limit our investigation to those portions of Φ which vary as $\cos \theta$ in the circumferential direction. Then

$$\begin{aligned}\Phi = \{&A_{12} + B_{12}\xi + C_{12}\xi^2 + D_{12}\xi^3 + \mathrm{e}^{-p_1(\xi_1 - \xi)} \\ &\times [C_1 \cos q_1(\xi_1 - \xi) + D_1 \sin q_1(\xi_1 - \xi)] \\ &+ \mathrm{e}^{-p_1\xi}(A_1 \cos q_1\xi + B_1 \sin q_1\xi)\} \cos \theta.\end{aligned} \tag{10.4.3a}$$

The first four terms correspond to the solution of membrane theory while the remaining terms are associated with edge-bending effects. The eight constants of integration are the solution of the eight simultaneous equations obtained from Eqs. (10.3.5), (10.3.11), (10.4.1), (10.4.2) and Tables 10.1 and 10.2 (see Eqs. (10.4.4) on top of pages 432, 433).

The relationships between the edge deflection and slope are given by the two equations of equilibrium of internal and external loads on a cylinder cross-section which are not automatically satisfied, namely

$$\begin{aligned}&\int_0^{2\pi} (N_{x\theta} \sin \theta - Q_x \cos \theta) R \, \mathrm{d}\theta \\ &\qquad = \int_0^{2\pi} \left[\left(\overline{N}_{x\theta} + \frac{3}{2}\frac{\overline{M}_{x\theta}}{R}\right) \sin \theta - \overline{Q}_x \cos \theta\right] R \, \mathrm{d}\theta = V,\end{aligned} \tag{10.4.5a}$$

$$\int_0^{2\pi} (N_x R + M_x) R \cos \theta \, \mathrm{d}\theta = M + V(L - x), \tag{10.4.5b}$$

which yicld

$$V = 6\pi E h(1 + k)\left[1 + 3k + \frac{\nu^2}{1 - \nu^2} k(1 - k)\right] D_{12}, \tag{10.4.5c}$$

$$M = 2\pi E h R(1 + k)\left[1 + 3k + \frac{\nu^2}{1 - \nu^2} k(1 - k)\right](C_{12} + 3D_{12}\xi_1). \tag{10.4.5d}$$

These equilibrium equations involve only C_{12} and D_{12} since the force systems associated with A_1, B_1, C_1, and D_1 can be shown to be self-equilibrating and there are no stress resultants associated with the rigid body deformations defined by the coefficients A_{12} and B_{12}.

$$\left|\begin{array}{llll}
0 & -1 & 0 & -[6\nu(1+3k)-9(1-\nu)k^2] \\
-1 & 0 & 2\left(2+\nu+2k+\frac{3-\nu}{2}k^2\right) & 0 \\
1 & 0 & -4\left[1+(1-\nu)k+3\frac{1-\nu}{4}k^2\right] & 0 \\
0 & 1 & 0 & -12\left[1+(1-\nu)k+3\frac{1-\nu}{4}k^2\right] \\
0 & -1 & 2(1+k)\xi_1 & -[6\nu(1+3k)-9(1-\nu)k^2+3(1+k)\xi_1^2] \\
-1 & -\xi_1 & 2\left(2+\nu+2k+\frac{3-\nu}{2}k^2\right)-(1+k)\xi_1^2 & 6\left(2+\nu+2k+\frac{3-\nu}{2}k^2\right)\xi_1-(1+k)\xi_1^3 \\
1 & \xi_1 & -4\left[1+(1-\nu)k+3\frac{1-\nu}{4}k^2\right]+(1+k)\xi_1^2 & -12\left[1+(1-\nu)k+3\frac{1-\nu}{4}k^2\right]\xi_1+(1+k)\xi_1^3 \\
0 & 1 & 2(1+k)\xi_1 & -12\left[1+(1-\nu)k+3\frac{1-\nu}{4}k^2\right]+3(1+k)\xi_1^2
\end{array}\right.$$

It is possible to make approximations which simplify Eqs. (10.4.5) considerably and permit their explicit solution. An obvious simplification is obtained by noting that when the cylinder is long compared to the attenuation length of edge-bending effects the coefficients of A_1 and B_1 in the first four equations are very small compared to those in the last four equations and that, conversely, the coefficients of C_1 and D_1 in the last four equations are generally very small compared to those in the first four equations, whereas the coefficients of A_{12}, B_{12}, C_{12}, and D_{12} in all the equations are of the same order of magnitude. By ignoring the coefficients of A_1 and B_1 in the first four equations and the coefficients of C_1 and D_1 in the last four equations it is possible to solve for all the quantities in terms of C_{12} and D_{12} by a series of manipulations. The resulting relationships are given by

$$A_1 = -\frac{1}{\Delta}\{2\nu[\beta_1^{(w')} - \beta_1^{(u_x)}]C_{12} - 6(2+\nu)[\beta_1^{(w)} - \beta_1^{(u_\theta)}]D_{12}\}, \tag{10.4.6a}$$

$$B_1 = -\frac{1}{\Delta}\{2\nu[\alpha_1^{(w')} - \alpha_1^{(u_x)}]C_{12} - 6(2+\nu)[\alpha_1^{(w)} - \alpha_1^{(u_\theta)}]D_{12}\}, \tag{10.4.6b}$$

$$C_1 = -\frac{1}{\Delta}\{2\nu[\beta_1^{(w')} - \beta_1^{(u_x)}](C_{12} + 3D_{12}\xi_1) + 6(2+\nu)[\beta_1^{(w)} + \beta_1^{(u_\theta)}]D_{12}\}, \tag{10.4.6c}$$

$$D_1 = -\frac{1}{\Delta}\{2\nu[\alpha_1^{(w')} - \alpha_1^{(u_x)}](C_{12} + 3D_{12}\xi_1) + 6(2+\nu)[\alpha_1^{(w)} + \alpha_1^{(u_\theta)}]D_{12}\}, \tag{10.4.6d}$$

$$\left\|\begin{array}{cccc} -f_1^{(u_x)}(\xi_1) & -g_1^{(u_x)}(\xi_1) & \alpha_1^{(u_x)} & -\alpha_1^{(u_x)} \\ f_1^{(u_\theta)}(\xi_1) & g_1^{(u_\theta)}(\xi_1) & \alpha_1^{(u_\theta)} & -\alpha_1^{(u_\theta)} \\ f_1^{(w)}(\xi_1) & g_1^{(w)}(\xi_1) & \alpha_1^{(w)} & -\alpha_1^{(w)} \\ f_1^{(w')}(\xi_1) & g_1^{(w')}(\xi_1) & \alpha_1^{(w')} & \alpha_1^{(w')} \\ -\alpha_1^{(u_x)} & \beta_1^{(u_x)} & f_1^{(u_x)}(\xi_1) & g_1^{(u_x)}(\xi_1) \\ \alpha_1^{(u_\theta)} & -\beta_1^{(u_\theta)} & f_1^{(u_\theta)}(\xi_1) & g_1^{(u_\theta)}(\xi_1) \\ \alpha_1^{(w)} & -\beta_1^{(w)} & f_1^{(w)}(\xi_1) & g_1^{(w)}(\xi_1) \\ \alpha_1^{(w')} & -\beta_1^{(w')} & -f_1^{(w')}(\xi_1) & -g_1^{(w')}(\xi_1) \end{array}\right\| \left|\begin{array}{c} (1+k)A_{12} \\ (1+k)B_{12} \\ C_{12} \\ D_{12} \\ C_1 \\ D_1 \\ A_1 \\ B_1 \end{array}\right| = \left|\begin{array}{c} 0 \\ 0 \\ 0 \\ 0 \\ 0 \\ 1 \\ -1 \\ 0 \end{array}\right| \delta + \left|\begin{array}{c} 0 \\ 0 \\ 0 \\ 0 \\ 1 \\ 0 \\ 0 \\ -1 \end{array}\right| \Theta R. \tag{10.4.4}$$

$$\begin{aligned} A_{12} = 2(2+\nu)C_{12} + \frac{1}{\Delta}\langle 2\nu\{[\alpha_1^{(w')} - \alpha_1^{(u_x)}]\beta_1^{(u_\theta)} \\ - [\beta_1^{(w')} - \beta_1^{(u_x)}]\alpha_1^{(u_\theta)}\}C_{12} \\ - 6(2+\nu)[\alpha_1^{(w)}\beta_1^{(u_\theta)} - \alpha_1^{(u_\theta)}\beta_1^{(w)}]D_{12}\rangle, \end{aligned} \tag{10.4.6e}$$

$$\begin{aligned} B_{12} = -6\nu D_{12} + \frac{1}{\Delta}\langle 2\nu[\alpha_1^{(w')}\beta_1^{(u_x)} - \alpha_1^{(u_x)}\beta_1^{(w')}]C_{12} \\ - 6(2+\nu)\{[\alpha_1^{(w)} + \alpha_1^{(u_\theta)}]\beta_1^{(u_x)} - [\beta_1^{(w)} + \beta_1^{(u_\theta)}]\alpha_1^{(u_x)}\}D_{12}\rangle, \end{aligned} \tag{10.4.6f}$$

$$\delta = -\xi_1(\xi_1 + 2\gamma_2)C_{12} - [\xi_1^2 - 12(1+\nu) - 2\gamma_1]D_{12}, \tag{10.4.6g}$$

$$\Theta = -(\xi_1 + 2\gamma_2)(2C_{12} + 3D_{12}\xi_1), \tag{10.4.6h}$$

where

$$\Delta = [\alpha_1^{(w)} + \alpha_1^{(u_\theta)}][\beta_1^{(w')} - \beta_1^{(u_x)}] - [\alpha_1^{(w')} - \alpha_1^{(u_x)}][\beta_1^{(w)} + \beta_1^{(u_\theta)}], \tag{10.4.6i}$$

$$\begin{aligned} \gamma_1 = \frac{3}{\Delta}\Big\langle \frac{2(2+\nu)}{\xi_1}[\alpha_1^{(w)}\beta_1^{(u_\theta)} - \alpha_1^{(u_\theta)}\beta_1^{(w)}] \\ + (2+\nu)\{[\alpha_1^{(w)} + \alpha_1^{(u_\theta)}]\beta_1^{(u_x)} - \alpha_1^{(u_x)}[\beta_1^{(w)} + \beta_1^{(u_\theta)}]\} \\ + \nu\{[\alpha_1^{(w')} - \alpha_1^{(u_x)}]\beta_1^{(u_\theta)} - \alpha_1^{(u_\theta)}[\beta_1^{(w')} - \beta_1^{(u_x)}]\}\Big\rangle, \end{aligned} \tag{10.4.6j}$$

$$\gamma_2 = \frac{\nu}{\Delta}[\alpha_1^{(w')}\beta_1^{(u_x)} - \alpha_1^{(u_x)}\beta_1^{(w')}], \tag{10.4.6k}$$

$$w' = \frac{\partial w}{\partial \xi}, \tag{10.4.6l}$$

and terms of the order of k compared to unity have been omitted. Finally C_{12} and D_{12} are given in terms of M and V by Eqs. (10.4.5c) and (10.4.5d) as

$$C_{12} = -\frac{V}{2\pi Eh}\xi_1 - \frac{M}{2\pi EhR}, \tag{10.4.7a}$$

$$D_{12} = \frac{V}{6\pi Eh}, \tag{10.4.7b}$$

where again small terms of the order of k compared to unity have been omitted.

With these results we can readily calculate deformations and stresses in the cylinder. Of interest are the deflection and rotation of the edge ring in terms of the end loads V and M. From Eqs. (10.4.6g), (10.4.6h), and (10.4.7) we have

$$\delta = \xi_1[\xi_1^2 + 6(1+\nu) + \gamma_1 + 3\xi_1\gamma_2]\frac{V}{3\pi Eh} + \xi_1(\xi_1 + 2\gamma_2)\frac{M}{2\pi EhR}, \tag{10.4.8a}$$

$$\Theta = (\xi_1 + 2\gamma_2)\left(\xi_1\frac{V}{2\pi EhR} + \frac{M}{\pi EhR^2}\right). \tag{10.4.8b}$$

In order to determine the magnitude of γ_1 and γ_2 let us examine the various quantities entering into these expressions. We have from Eqs. (10.3.6)

$$\left.\begin{matrix} p_1 \\ q_1 \end{matrix}\right\} = \left(\frac{1-\nu^2}{4k}\right)^{1/4}\left[1 \pm \frac{2-\nu}{4}\left(\frac{4k}{1-\nu^2}\right)^{1/2} + 0(k)\right], \tag{10.4.9a}$$

$$p_1^2 - q_1^2 = 2 - \nu + 0(k), \tag{10.4.9b}$$

and from Table 10.2 and Eq. (10.4.9)

$$\left.\begin{matrix} \alpha_1^{(u_x)} \\ \beta_1^{(u_x)} \end{matrix}\right\} = \left(\frac{1-\nu^2}{4k}\right)^{3/4}\left[\pm 2\nu + \frac{10 + 6\nu - 11\nu^2}{2}\left(\frac{4k}{1-\nu^2}\right)^{1/2} + 0(k)\right], \tag{10.4.10a}$$

$$\alpha_1^{(u_\theta)} = 5 - 3\nu^2 + 0(k), \tag{10.4.10b}$$

$$\beta_1^{(u_\theta)} = 2\left(\frac{1-\nu^2}{4k}\right)^{1/2}[2 + \nu + 0(k)], \tag{10.4.10c}$$

$$\alpha_1^{(w)} = -\frac{1-\nu^2}{k}[1 + 0(k)], \tag{10.4.10d}$$

$$\beta_1^{(w)} = 4\left(\frac{1-\nu}{4k}\right)^{1/2} [1 - \nu + 0(k)], \tag{10.4.10e}$$

$$\left.\begin{matrix}\alpha_1^{(w')} \\ \beta_1^{(w')}\end{matrix}\right\} = -4\left(\frac{1-\nu^2}{4k}\right)^{5/4} \left[1 \pm \frac{6-5\nu}{4}\left(\frac{4k}{1-\nu^2}\right)^{1/2} + 0(k)\right]. \tag{10.4.10f}$$

With these quantities given we can determine approximate expressions for δ and Θ as

$$\delta = \langle \xi_1^3\{1 - 3[\nu^2\epsilon + \nu(2+\nu)\epsilon^2 + (2+\nu)\epsilon^3]\} + 6(1+\nu)\xi_1\rangle \frac{V}{3\pi Eh}$$
$$+ \xi_1^2(1 - 2\nu^2\epsilon)\frac{M}{2\pi EhR}, \tag{10.4.11a}$$

$$\Theta = \xi_1(1 - 2\nu^2\epsilon)\left(\xi_1 \frac{V}{2\pi EhR} + \frac{M}{\pi EhR^2}\right), \tag{10.4.11b}$$

where now terms of the order of $k^{1/2}$ (or h/R) compared to unity have been omitted. The quantity

$$\epsilon = \frac{1}{\xi_1}\left(\frac{4k}{1-\nu^2}\right)^{1/4} = \frac{1}{[3(1-\nu^2)]^{1/4}} \frac{(Rh)^{1/2}}{L} \tag{10.4.12}$$

will be recognized as the ratio of the attenuation length of edge effects and the cylinder length, a quantity which was encountered in the theory of axisymmetrically loaded cylindrical shells. As we might have expected, the effect of edge restraint (the terms involving ϵ) is to reduce the end deflection and rotation of the cylinder.

When we note that the longitudinal variation of quantities for an edge loading which varies circumferentially as $\cos\theta$ is similar to that for a symmetrically loaded cylindrical shell and recall the results for finite edge-loaded shells obtained in Section 4.1, we may conclude that Eqs. (10.4.11) are valid when ϵ is less than about 1/6. For this value of ϵ the coefficient of ξ_1^3 in Eq. (10.4.11a) is reduced by about 18 percent from unity when ν is equal to 0.3 while the factor $1 - 2\nu^2\epsilon$ is reduced by about 3 percent from unity. However, since the value of ϵ for shells of practical proportions is usually much less than 1/6, the terms involving ϵ may also be neglected and Eqs. (10.4.11) reduce to

$$\delta = \left[1 + \frac{6(1+\nu)R^2}{L^2}\right]\frac{VL^3}{3\pi EhR^3} + \frac{ML^2}{2\pi EhR^3}, \tag{10.4.13a}$$

$$\Theta = \frac{VL^2}{2\pi EhR^3} + \frac{ML}{\pi EhR^3}. \tag{10.4.13b}$$

When we note further that the quantity πhR^3 is approximately the moment of inertia of the circular cross-section of the cylinder about a horizontal axis passing through its centroid, Eqs. (10.4.13) may be recognized as being similar to the results of the Bernoulli–Euler theory of beams, the only difference being the quantity $[6(1 + \nu)R^2]/L^2$ in Eq. (10.4.13a) which is a correction due to shearing of the middle surface of the cylinder and which becomes negligible when the length is large compared to the cross-sectional radius.

Much the same results can be obtained if we introduce into Eq. (10.4.4) the approximation that the tangential deformation u_x and u_θ due to bending can be neglected in comparison with membrane tangential deformations. Thus the first, second, fifth, and sixth of Eqs. (10.4.4) become four equations for A_{12}, B_{12}, δ, and Θ in terms of C_{12} and D_{12}. The expressions for δ and Θ are then identical with Eqs. (10.4.13) while the expressions for A_{12} and B_{12} become

$$A_{12} \approx -(2 + \nu)\left(\frac{M}{\pi EhR} + \frac{V\xi_1}{\pi Eh}\right), \tag{10.4.14a}$$

$$B_{12} \approx -\frac{V}{\pi Eh}, \tag{10.4.14b}$$

as compared to the more exact expressions

$$\begin{aligned} A_{12} \approx &-(2 + \nu)\left[1 - \nu\left(\frac{k}{1 - \nu^2}\right)^{1/2}\right]\left(\frac{M}{\pi EhR} + \frac{V\xi_1}{\pi Eh}\right) \\ &+ (2 + \nu)^2\left(\frac{4k}{1 - \nu^2}\right)^{3/4}\frac{V}{2\pi Eh}, \end{aligned} \tag{10.4.15a}$$

$$\begin{aligned} B_{12} \approx &-\left[\nu - \frac{2 + \nu}{2}\left(\frac{k}{1 - \nu^2}\right)^{1/2}\right]\frac{V}{\pi Eh} \\ &+ \nu^2\left(\frac{4k}{1 - \nu^2}\right)^{1/4}\left[1 - \frac{2 + 5\nu}{2}\left(\frac{k}{1 - \nu^2}\right)^{1/2}\right]\left(\frac{M}{\pi EhR} + \frac{V\xi_1}{\pi Eh}\right). \end{aligned} \tag{10.4.15b}$$

While the approximate expression for A_{12} is accurate, with errors of the order of magnitude of h/R compared to unity, the expression for B_{12} is not, since we are omitting a term of the order of $(h/R)^{1/2}L/R$ compared to unity, a term which can be large. The error in the tangential deformations can be shown to be considerable in the vicinity of the clamped end, but we have already shown that the term in question is negligible at the loaded end of the cylinder where the largest deformations occur. Similar conclusions can be reached about the errors in the approximate expressions for the remaining coefficients, which can be readily obtained with the use of the remaining four equations of (10.4.4) with the length of the cylinder assumed to be large compared to the attenuation length of end effects.

10.5 Infinite cylindrical shell loaded by diametrically opposed concentrated loads

Other problems may be solved with the use of the relationships derived previously. Among these is the problem of determining the effect of the concentrated loads on the stresses and deformations of a cylindrical shell.† We shall assume that the ends are sufficiently far away so that the cylinder can be considered to be infinitely long and loaded at the center. At the center section two diametrically opposed concentrated loads are applied (Fig. 10.4).

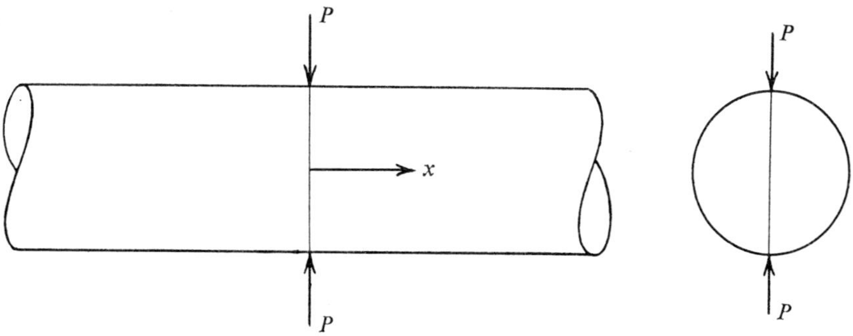

Fig. 10.4. Infinitely long cylinder subjected to diametrically opposed concentrated loads

The solution to this problem is readily obtained if we consider the loads to be applied as discontinuities of the transverse shear force Q_x. From symmetry of loading it is apparent that we need consider only half of the infinitely long shell and that three of the boundary conditions at the station at which the loads are applied are

$$u_x = \frac{\partial w}{\partial x} = \overline{N}_{x\theta} + \frac{3}{2}\frac{\overline{M}_{x\theta}}{R} = 0, \qquad (\xi = 0) \tag{10.5.1}$$

The remaining edge condition is obtained by considering the concentrated loads to be applied as the limit of uniform transverse shear loads acting over a small portion of the circumference of the middle surface (Fig. 10.5). Then, since symmetry of loading implies that $M_{x\theta}$ vanishes at the loading station we have

† The corresponding problem for completely isotropic shells has been treated by J. C. Yao, 'Long Cylindrical Tube subjected to Two Diametrically Opposite Loads,' *Aeronautical Quarterly*, vol. 20, pt. 4, November 1969, pp. 365–381. A related problem using Donnell theory is treated by P. Seide, 'On the Bending of Circular Cylindrical Shells by Equal and Equally Spaced End Radial Shear Forces and Moments,' *J. Appl. Mech.*, vol. 28, no. 1, March 1961, pp. 117–126.

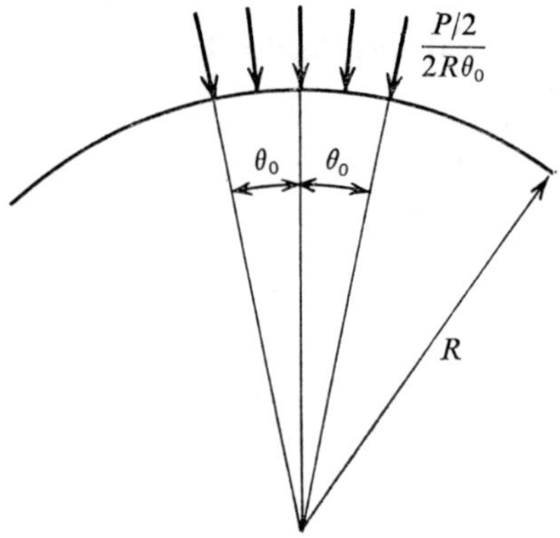

Fig. 10.5. Concentrated transverse shear loading as the limit of uniformly distributed forces

$$\bar{Q}_x = \begin{cases} \operatorname{L}_{\theta_0 \to 0} \dfrac{P}{4R\theta_0} & -\theta_0 < \theta < \theta_0 \\ 0 & \theta_0 < \theta < \pi - \theta_0 \\ \operatorname{L}_{\theta_0 \to 0} \dfrac{P}{4R\theta_0} & \pi - \theta_0 < \theta < \pi + \theta_0 \\ 0 & \pi + \theta_0 < \theta < 2\pi - \theta_0. \end{cases} \tag{10.5.2}$$

A Fourier series expansion of the function given above then yields

$$\bar{Q}_x = \frac{P}{\pi R}\left(\tfrac{1}{2} + \sum_{n=2,4,6,\ldots}^{\infty} \cos n\theta\right), \tag{10.5.3}$$

which is the fourth edge condition and implies that only terms with $\cos n\theta$ where n is even need be used in the displacement function. At infinity, the stresses should vanish and the displacements should vanish or be bounded.

The algebra involved in the solution may be reduced considerably if we note that the last of Eqs. (10.5.1) may be written, with the use of Eqs. (10.1.3a) and (10.1.3f) as

$$\bar{N}_{x\theta} + \frac{3}{2}\frac{\bar{M}_{x\theta}}{R} = \frac{1-\nu}{2}\frac{K}{R}\left[(1+3k)\frac{\partial u_\theta}{\partial \xi} + \frac{\partial u_x}{\partial \theta} - 3k\frac{\partial^2 w}{\partial \xi\,\partial \theta}\right]$$

$$= 0 \qquad \text{at} \quad \xi = 0. \tag{10.5.4a}$$

But since u_x and $\partial w/\partial \xi$ vanish at $\xi = 0$, Eq. (10.5.4a) may be replaced by

$$\frac{\partial u_\theta}{\partial \xi} = 0. \tag{10.5.4b}$$

With the use of Eqs. (10.1.11), the first two of Eqs. (10.5.1) and Eq. (10.5.4b) may be written as

$$\left(\frac{\partial^2}{\partial\theta^2} - k\frac{\partial^4}{\partial\theta^4}\right)\frac{\partial\Phi}{\partial\xi} - \left[\nu(1 + 3k) + 3\frac{1-\nu}{2}k^2\frac{\partial^2}{\partial\theta^2}\right]\frac{\partial^3\Phi}{\partial\xi^3}$$
$$+ k(1 + 3k)\frac{\partial^5\Phi}{\partial\xi^5} = 0, \tag{10.5.5a}$$

$$\left[(1 + k)\frac{\partial^3}{\partial\theta^3}\right]\frac{\partial\Phi}{\partial\xi} + \left[(2 + \nu)\frac{\partial}{\partial\theta} - 2k\left(1 + \frac{3-\nu}{4}k\right)\frac{\partial^3}{\partial\theta^3}\right]\frac{\partial^3\Phi}{\partial\xi^3}$$
$$- \left(2k\frac{\partial}{\partial\theta}\right)\frac{\partial^5\Phi}{\partial\xi^5} = 0, \tag{10.5.5b}$$

$$\left[(1 + k)\frac{\partial^4}{\partial\theta^4}\right]\frac{\partial\Phi}{\partial\xi} + \left\{2[1 + (1 - \nu)k(1 + \tfrac{3}{4}k)]\frac{\partial^2}{\partial\theta^2}\right\}\frac{\partial^3\Phi}{\partial\xi^3}$$
$$+ (1 + 3k)\frac{\partial^5\Phi}{\partial\xi^5} = 0. \tag{10.5.5c}$$

If now we write Φ as

$$\Phi = \sum_{n=2,4,6,\ldots} \Phi_n(\xi)\cos n\theta + D_{02}[\xi^2 + 3\nu(1 + 3k)\theta^2]\xi + F_{02}\xi\theta^2 \tag{10.5.6}$$

Eqs. (10.5.5) imply that the boundary conditions may be further replaced by

$$2F_{02} - \nu(1 + 3k)\frac{d^3\Phi_0(0)}{d\xi^3} = \frac{d^5\Phi_0(0)}{d\xi^5} = 0, \tag{10.5.7a}$$

$$\frac{d\Phi_n(0)}{d\xi} = \frac{d^3\Phi_n(0)}{d\xi^3} = \frac{d^5\Phi_n(0)}{d\xi^5} = 0 \qquad (n = 2, 4, 6, \ldots). \tag{10.5.7b}$$

The fourth boundary condition becomes with the aid of Eqs. (10.5.7) and (10.1.10)

$$\frac{d^7\Phi_0(0)}{d\xi^7} = -\frac{\eta}{2}, \tag{10.5.8a}$$

$$\frac{d^7\Phi_n(0)}{d\xi^7} = -\eta \qquad (n = 2, 4, 6, \ldots), \tag{10.5.8b}$$

where

$$\eta = \frac{1}{(1 + 3k)(1 - k)}\frac{PR^2}{\pi D}, \tag{10.5.8c}$$

From Eqs. (10.3.12), (10.3.19), and (10.3.22) the displacement functions which vanish at infinity can be written as

$$\Phi_0(\xi) = \eta(A_0\cos q_0\xi + B_0\sin q_0\xi)\,e^{-p_0\xi}, \tag{10.5.9a}$$

$$\Phi_n(\xi) = \eta \sum_{j=1}^{2} (A_{nj} \cos q_{nj}\xi + B_{nj} \sin q_{nj}\xi)\, e^{-p_{nj}\xi}, \tag{10.5.9b}$$

(if all roots are complex)

$$= \eta[A_{n1}\, e^{-p_{n1}\xi} + B_{n1}\, e^{-q_{n1}\xi} + (A_{n2} \cos q_{n2}\xi + D_{n2} \sin q_{n2}\xi)\, e^{-p_{n2}\xi}] \tag{10.5.9c}$$

(if four roots are real and four are complex).

Then Eqs. (10.5.7), (10.5.8), and (10.5.9) yield

$$A_0 = -\frac{5p_0^4 - 10p_0^2q_0^2 + q_0^4}{4p_0(p_0^2 + q_0^2)^5}, \tag{10.5.10a}$$

$$B_0 = \frac{p_0^4 - 10p_0^2q_0^2 + 5q_0^4}{4q_0(p_0^2 + q_0^2)^5}, \tag{10.5.10b}$$

$$F_{02} = \frac{\nu(1 + 3k)}{4(p_0^2 + q_0^2)}\, \eta, \tag{10.5.10c}$$

and

$$A_{n1} = \frac{1}{p_{n1}} A(p_{n1}, q_{n1}, p_{n2}, q_{n2}), \tag{10.5.11a}$$

$$B_{n1} = \frac{1}{q_{n1}} A(q_{n1}, p_{n1}, q_{n2}, p_{n2}), \tag{10.5.11b}$$

$$A_{n2} = \frac{1}{p_{n2}} A(p_{n2}, q_{n2}, p_{n1}, q_{n1}), \tag{10.5.11c}$$

$$B_{n2} = \frac{1}{q_{n2}} A(q_{n2}, p_{n2}, q_{n1}, p_{n1}), \tag{10.5.11d}$$

with

$$A(x_1, x_2, x_3, x_4) = -\frac{5x_1^4 - 10x_1^2x_2^2 + x_2^4 - 2(3x_1^2 - x_2^2)(x_3^2 - x_4^2) + (x_3^2 + x_4^2)^2}{2(x_1^2 + x_2^2)[(x_1^2 - x_2^2 - x_3^2 + x_4^2)^2 + 4(x_1x_2 + x_3x_4)^2] \times [(x_1^2 - x_2^2 - x_3^2 + x_4^2)^2 + 4(x_1x_2 - x_3x_4)^2]} \tag{10.5.11e}$$

if all roots are complex, or

$$A_{n1} = \frac{q_{n1}^4 - 2q_{n1}^2(p_{n2}^2 - q_{n2}^2) + (p_{n2}^2 + q_{n2}^2)^2}{p_{n1}(p_{n1}^2 - q_{n1}^2)\Delta_n}, \tag{10.5.12a}$$

$$B_{n1} = \frac{p_{n1}^4 - 2p_{n1}^2(p_{n2}^2 - q_{n2}^2) + (p_{n2}^2 + q_{n2}^2)^2}{q_{n1}(q_{n1}^2 - p_{n1}^2)\Delta_n}, \tag{10.5.12b}$$

$$A_{n2} = -\frac{5p_{n2}^4 - 10p_{n2}^2q_{n2}^2 + q_{n2}^4 - (3p_{n2}^2 - q_{n2}^2)(p_{n1}^2 + q_{n1}^2) + p_{n1}^2q_{n1}^2}{2p_{n2}(p_{n2}^2 + q_{n2}^2)\Delta_n}, \tag{10.5.12c}$$

$$B_{n2} = -\frac{5q_{n2}^4 - 10p_{n2}^2q_{n2}^2 + p_{n2}^4 + (3q_{n2}^2 - p_{n2}^2)(p_{n1}^2 + q_{n1}^2) + p_{n1}^2q_{n1}^2}{2q_{n2}(p_{n2}^2 + q_{n2}^2)\Delta_n}, \tag{10.5.12d}$$

with

$$\Delta_n = [(p_{n2}^2 + q_{n2}^2)^2 - (p_{n2}^2 - q_{n2}^2)(p_{n1}^2 + q_{n1}^2) + p_{n1}^2q_{n1}^2]^2 + 4p_{n2}^2q_{n2}^2(p_{n2}^2 - q_{n2}^2)^2 \tag{10.5.12e}$$

if four roots are real and four are complex. It is interesting to note that the form of the solution is identical had we used either Morley's, Donnell's, or Flügge's approximate equations rather than Flügge's complete equations, with differences between the various results arising from the values of the roots used in the equations.

Calculations of the deflection under the concentrated loads have been made† and indicate that the deflection is adequately represented by the following approximate relationship when ν is equal to 0.3.

$$\left(\frac{Ehw}{P}\right)_{\xi=\theta=0} = 0.746\left(\frac{R}{h}\right)^{3/2}. \tag{10.5.13}$$

Eq. (10.5.13) was obtained by means of the exact analysis given above and with roots of Eq. (10.3.2). The accuracy of other approximations is indicated by a comparison of values of the coefficients of the Fourier series for the deflection at the loading station $\xi = 0$. This is given by

$$\frac{Ehw}{P} = \sum_{n=0,2,4,\ldots}^{\infty} \Upsilon_n \cos n\theta, \tag{10.5.14a}$$

† P. Seide, 'Effect of Internal Pressure on an Infinite Cylindrical Shell subjected to Concentrated Radial Loads,' *AIAA J.*, vol. 7, no. 10, October 1969, pp. 1944–1949; see also S. W. Yuan and L. Ting, 'On Radial Deflections of a Cylinder subjected to Equal and Opposite Concentrated Radial Loads,' *J. Appl. Mech.*, vol. 24, no. 2, June 1957, pp. 278–282; S. W. Yuan, 'Thin Cylindrical Shells subjected to Concentrated Loads,' *Q. Appl. Math.*, vol. 4, no. 1, 1946, pp. 13–26; L. Ting and S. W. Yuan, 'On Radial Deflection of a Cylinder of Finite Length with Various End Conditions,' *J. Aerospace Sciences*, vol. 25, no. 4, April 1958, pp. 230–234; L. S. D. Morley, 'The Thin-walled Cylinder subjected to Concentrated Radial Loads,' *Q. J. Mech. Appl. Math.*, vol. 13, pt. 1, February 1960, pp. 24–37.

where

$$\Psi_0 = K_1[(p_0^4 - 6p_0^2q_0^2 + q_0^4)A_0 - 4p_0q_0(p_0^2 - q_0^2)B_0] \tag{10.5.14b}$$

and

$$\Psi_n = \sum_{j=1}^{2} \Psi_{nj}, \tag{10.5.14c}$$

where

$$\begin{aligned}\Psi_{nj} = {} & [K_1(p_{nj}^4 - 6p_{nj}^2q_{nj}^2 + q_{nj}^4) - 2n^2K_2(p_{nj}^2 - q_{nj}^2) + K_3n^4]A_{nj} \\ & - 4p_{nj}q_{nj}[K_1(p_{nj}^2 - q_{nj}^2) - K_2n^2]B_{nj}\end{aligned} \tag{10.5.14d}$$

if all roots are complex. When four roots are real and four are complex, Ψ_{n1} is replaced by

$$\begin{aligned}\Psi_{n1} = {} & (K_1p_{n1}^4 - 2n^2K_2p_{n1}^2 + K_3n^4)A_{n1} \\ & +(K_1q_{n1}^4 - 2n^2K_2q_{n1}^2 + K_3n^4)B_{n1}\end{aligned} \tag{10.5.14e}$$

while Ψ_{n2} is given by Eq. (10.5.14d). The constants K_1, K_2, and K_3 are defined in the complete solution by

$$K_1 = \frac{1 - \nu^2}{\pi k(1 - k)}, \tag{10.5.15a}$$

$$K_2 = \frac{[1 + (1 - \nu)k(1 + \frac{3}{4}k)](1 - \nu^2)}{\pi(1 - k)(1 + 3k)k}, \tag{10.5.15b}$$

$$K_3 = \frac{(1 + k)(1 - \nu^2)}{\pi(1 - k)(1 + 3k)k}, \tag{10.5.15c}$$

and are replaced by $(1 - \nu^2)/\pi k$ for the approximate solutions. The first few coefficients are given in Table 10.4 for the case $R/h = 100$, $\nu = 0.3$.

Table 10.4 Values of Ψ_n in Eq. (10.5.14a) ($R/h = 100$, $\nu = 0.3$)

n	0	2	4	6	8	10
Flügge (complete)	2	529	107	44	24	14
Flügge (approx.)	2	521	99	44	24	14
Morley	2	523	101	43	24	14
Donnell	2	348	91	43	24	14

For this problem Donnell's approximation is unsatisfactory since the predominant term has the greatest error, leading to an error of about 25 per cent in the total maximum deflection. Morley's equation, on the other hand, leads to results in very good agreement with the exact solution.

The effect of the concentrated loads is not so local as might be expected from previous analyses of axisymmetric local loads. Calculations of the longitudinal variation of the radial deflection along the generators passing through the loads ($\theta = 0°$ or $180°$) for a cylinder having a radius/thickness ratio of 100, shown in Fig. 10.6, indicate that the effect of the loads is still in evidence at distances as much as 20 to 30 cylinder radii away from the points of load application. This result is explained by the fact that the predominant terms in the infinite series can be found to vary with x as $e^{-p_{22}\xi}$ with

$$p_{22} \approx q_{22} \approx (3)^{1/2}\left(\frac{4k}{1-\nu}\right)^{1/4} = \left(\frac{3}{1-\nu^2}\right)^{1/4}\left(\frac{h}{R}\right)^{1/2} \tag{10.5.16a}$$

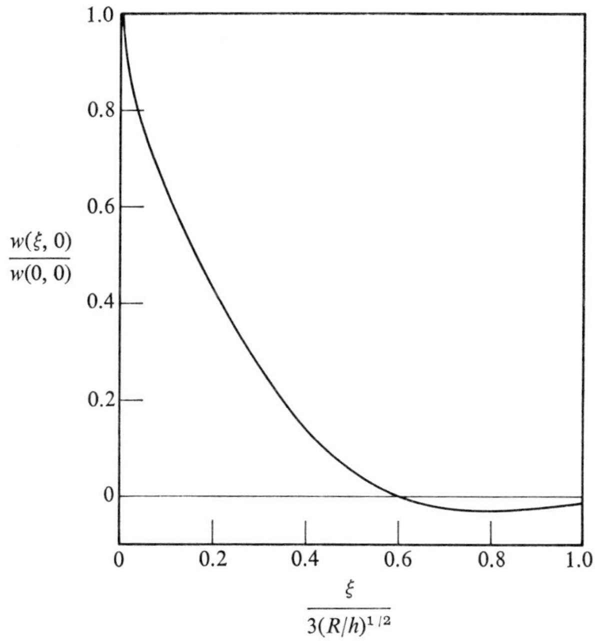

Fig. 10.6. Variation of radial deformation along generator of concentrated load

The exponential becomes negligible in comparison with unity when $p_{22}\xi$ is greater than about 4 so that the attenuation length is given by

$$\frac{x_a}{R} \approx 3\left(\frac{R}{h}\right)^{1/2} \tag{10.5.16b}$$

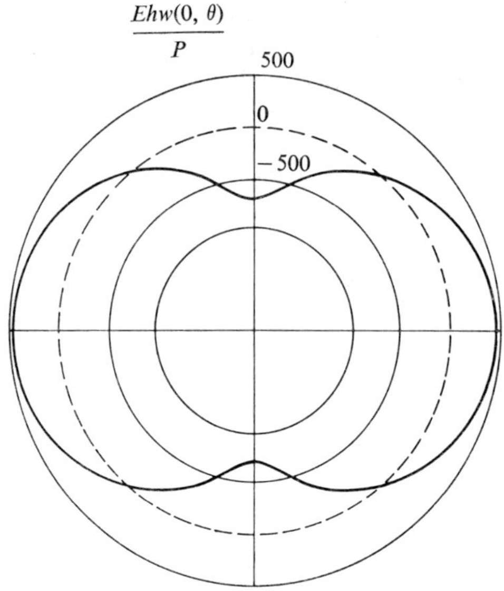

Fig. 10.7. Circumferential variation of radial deformation for loaded section of cylinder

Since the cos 2θ term is predominant for all values of R/h, the concentrated loads affect every part of the circumference near the load positions. The variation of the radial deformation around the circumference for a cylinder with $R/h = 100$ is shown in Fig. 10.7. For other values of R/h, the shape of the radial deformation curve is almost identical.

The solution given above for the infinitely long shell can be used to obtain the solution for a finite shell subjected to diametrically opposed concentrated loads at the center section and having the following boundary conditions at the edges†:

$$w = u_\theta = N_x = M_x = 0 \tag{10.5.17}$$

These conditions are satisfied at the sections midway between the loads of a cylinder subjected to equally spaced diametrically opposed loads which alternate in sign as shown in Fig. 10.8. The solution for the alternating load problem is readily obtained by adding the contributions of all of the loads considered to be acting on an infinite cylinder to the desired quantity at a

† Other boundary conditions are considered in L. Ting and S. W. Yuan, 'On Radial Deflection of a Cylinder of Finite Length with Various End Conditions,' *J. Aerospace Sciences*, vol. 25, no. 4, April 1958, pp. 230–234.

particular point in the shell. The radial deformation, for example, at a distance x from any station at which loads are applied is given by

$$w(x, \theta) = w_\infty(x, \theta) + \sum_{n=1}^{\infty} (-1)^n [w_\infty(nL + x, \theta) + w_\infty(nL - x, \theta)]. \qquad (10.5.18)$$

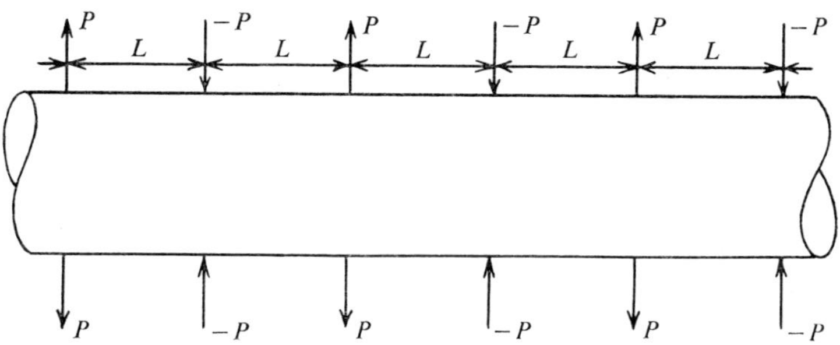

Fig. 10.8. Periodically loaded cylinder

Deflections at the point of application of the loads as a function of shell length have been obtained for a shell having a radius/thickness ratio of 100 with the use of both Flügge's approximate equations and Donnell's equations and are compared in Fig. 10.9. We see that the Donnell approximations become less inaccurate as the shell length decreases. When $R/h = 100$ the two sets of results are in good agreement when the length is less than 10 times the radius.

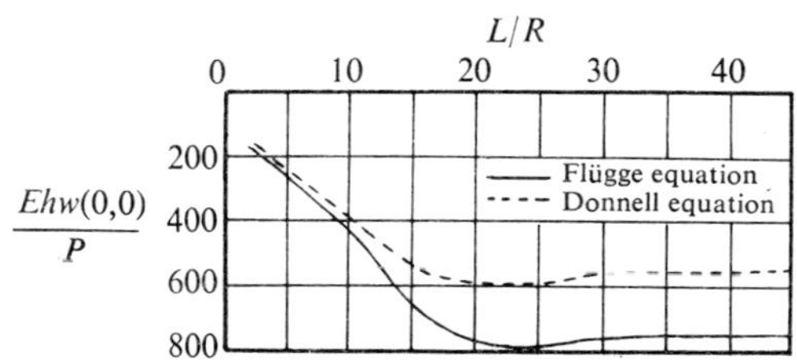

Fig. 10.9. Variation with length of maximum deflection of a simply supported cylinder under central diametrically opposite concentrated loads

10.6 Results for some other concentrated loads

Results for some other concentrated loads may be obtained with the use of the previous concentrated radial load solution. The solution for an infinitely long cylinder under concentrated moments at opposite ends of a diameter is obtained by superimposing the solutions for pairs of equal magnitude but opposite sign concentrated loads an infinitesimal distance δ apart (Fig. 10.10) and taking the limit as δ approaches zero and P becomes infinite in such a way that $P\delta$ remains equal to the concentrated moment M. Then:

$$\Phi_M(\xi,\theta) = \underset{\substack{\delta\to 0\\ P\delta\to M}}{\mathrm{L}} P\left[\Phi_P(\xi,\theta) - \Phi_P\left(\xi - \frac{\delta}{R}, \theta\right)\right]$$

$$= \underset{\substack{\delta\to 0\\ P\delta\to M}}{\mathrm{L}} P\left\{\Phi_P(\xi,\theta) - \left[\Phi_P(\xi,\theta) - \frac{\delta}{R}\frac{\partial\Phi_P(\xi,\theta)}{\partial\xi} + \frac{1}{2!}\left(\frac{\delta}{R}\right)^2\frac{\partial^2\Phi_P(\xi,\theta)}{\partial\xi^2} - \cdots\right]\right\} = \frac{M}{R}\frac{\partial\Phi_P(\xi,\theta)}{\partial\xi}, \tag{10.6.1}$$

where $\Phi_P(\xi, \theta)$ is the displacement function for an infinite cylinder subjected to unit diametrically opposed concentrated loads.

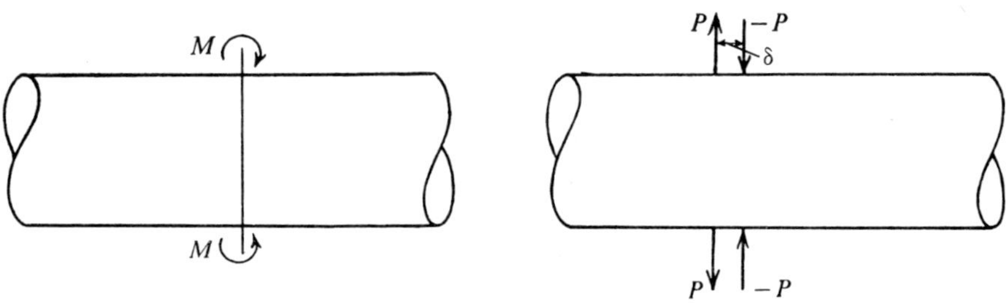

Fig. 10.10 Concentrated longitudinal moment as the limit of concentrated loads

Let us take the solution for a pair of diametrically opposed concentrated loads P and add to it the solution for diametrically opposed loads of opposite sign and rotated through a small angle. The limit as the angle approaches zero while the product of the vertical load components and the horizontal distance between the loads remains equal to a constant T is the solution to the problem pictured in Fig. 10.11. Then

$$\Phi_T(\xi,\theta) = \underset{\substack{\varepsilon\to 0\\ P\varepsilon\to T}}{\mathrm{L}} P[\Phi_P(\xi,\theta) - \Phi_P(\xi,\theta-\epsilon)]$$

$$= \underset{\substack{\varepsilon\to 0\\ P\varepsilon\to T}}{\mathrm{L}} P\left\{\Phi_P(\xi,\theta) - \left[\Phi_P(\xi,\theta) - \epsilon\frac{\partial\Phi_P(\xi,\theta)}{\partial\theta} + \frac{1}{2!}\epsilon^2\frac{\partial^2\Phi_P(\xi,\theta)}{\partial\theta^2} - \cdots\right]\right\} = T\frac{\partial\Phi_P(\xi,\theta)}{\partial\theta}. \tag{10.6.2}$$

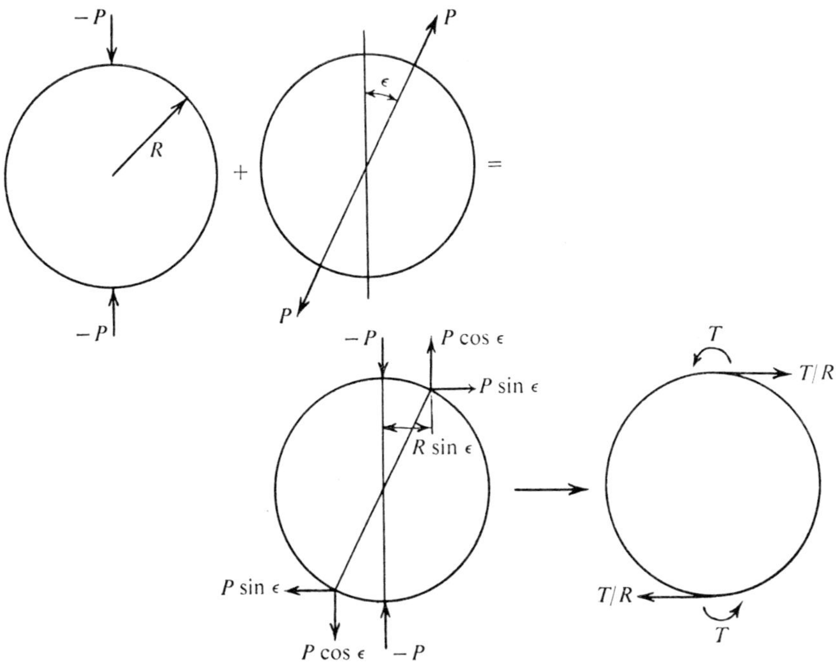

Fig. 10.11. Self-equilibrating system of concentrated loads on a cylindrical shell

10.7 Particular solutions for complete cylindrical shells

When a cylindrical shell is loaded on its surfaces, particular solutions of the bending equations must be obtained in addition to the edge loading solutions of Section 10.3. These particular solutions are obtained most easily when the equivalent surface loads are expressed in the form of an appropriate double Fourier series. For example, let the middle surface loading be expressed as

$$q_x = \sum_{m=0}^{\infty} \sum_{n=0}^{\infty} q_{xmn} \cos m\pi \frac{x}{L} \cos n\theta, \tag{10.7.1a}$$

$$q_\theta + \frac{m_\theta}{R} = \sum_{m=1}^{\infty} \sum_{n=1}^{\infty} q_{\theta mn} \sin \frac{m\pi x}{L} \sin n\theta, \tag{10.7.1b}$$

$$q_\zeta + \frac{1}{R}\left(\frac{\partial m_x}{\partial \xi} + \frac{\partial m_\theta}{\partial \theta}\right) = \sum_{m=1}^{\infty} \sum_{n=0}^{\infty} q_{\zeta mn} \sin m\pi \frac{x}{L} \cos n\theta. \tag{10.7.1c}$$

Then we may expand u_x, u_θ, and w into similar series and use Eqs. (10.1.8) to (10.1.12) to obtain the solution

$$u_{xp} = \frac{D}{R^4} \sum_{m=0}^{\infty} \sum_{n=0}^{\infty} \frac{1}{L_{mn}} \times [A_{mn}^{(1)} q_{xmn} + B_{mn} q_{\theta mn} + C_{mn} q_{\zeta mn}] \cos \frac{m\pi x}{L} \cos n\theta, \tag{10.7.2a}$$

$$u_{\theta p} = \frac{D}{R^4} \sum_{m=1}^{\infty} \sum_{n=1}^{\infty} \frac{1}{L_{mn}} \times [B_{mn} q_{xmn} + A_{mn}^{(2)} q_{\theta mn} + D_{mn} q_{\zeta mn}] \sin \frac{m\pi x}{L} \sin n\theta, \tag{10.7.2b}$$

$$w_p = \frac{D}{R^4} \sum_{m=1}^{\infty} \sum_{n=0}^{\infty} \frac{1}{L_{mn}} \times [C_{mn} q_{xmn} + D_{mn} q_{\theta mn} + A_{mn}^{(3)} q_{\zeta mn}] \sin \frac{m\pi x}{L} \cos n\theta, \tag{10.7.2c}$$

where

$$\begin{aligned} A_{mn}^{(1)} = {} & (1 + k)(1 + 3k)\mu_m^2 \\ & + k\Big\{(1 + 3k)\mu_m^6 + \frac{2}{1 - \nu}\Big(2 - \nu + \frac{3 - 6 - \nu^2}{4} k\Big)\mu_m^4 n^2 \\ & + \Big[\Big(\frac{5 - \nu}{1 - \nu} + 3k\Big)n^2 - 4\Big(\frac{2 - \nu}{1 - \nu} + \frac{3}{2} k\Big)\Big]\mu_m^2 n^2 + \frac{2}{1 - \nu} n^2(n^2 - 1)^2\Big\}, \end{aligned} \tag{10.7.3a}$$

$$\begin{aligned} A_{mn}^{(2)} = {} & 2\Big(1 + \nu + \frac{k}{1 - \nu}\Big)\mu_m^2 + (1 + k)^2 n^2 \\ & + k\Big\{2\frac{1 - k}{1 - \nu}\mu_m^6 + \Big[\Big(\frac{5 - \nu}{1 - \nu} + 3k\Big)n^2 - \frac{4\nu}{1 - \nu}\Big]\mu_m^4 \\ & + 2\Big[\Big(\frac{2 - \nu}{1 - \nu} + \frac{3 + \nu}{4} k\Big)n^2 - \frac{2 - \nu + \nu^2}{1 - \nu}\Big]\mu_m^2 \\ & + (1 + k)n^4(n^2 - 2)\Big\}, \end{aligned} \tag{10.7.3b}$$

$$A_{mn}^{(3)} = (\mu_m^2 + n^2)^2 + k[3\mu_m^4 + 2(1 - \nu)(1 + \tfrac{3}{4}k)\mu_m^2 n^2 + n^4], \tag{10.7.3c}$$

$$\begin{aligned} B_{mn} = {} & \mu_m n\Big\{1 + \frac{1 + \nu}{1 - \nu} k\Big[\Big(1 - \frac{3 - \nu}{1 + \nu} k\Big)\mu_m^4 + 2\Big(1 + \frac{3 - \nu}{4}\frac{1 - \nu}{1 + \nu} k\Big)\mu_m^2 n^2 \\ & + n^4 - \frac{2 + \nu - \nu^2}{1 + \nu}\mu_m^2 - \frac{1 + 3\nu}{1 + \nu} n^2 + 1\Big]\Big\}, \end{aligned} \tag{10.7.3e}$$

$$C_{mn} = \mu_m\left[(1 + 3k)(\nu\mu_m^2 + k\mu_m^4) - n^2 - k\left(n^2 + 3\,\frac{1-\nu}{2}\,k\mu_m^2\right)n^2\right], \tag{10.7.3f}$$

$$D_{mn} = -n\left\{(2 + \nu)\mu_m^2 + (1 + k)n^2 + 2k\left[\mu_m^2 + \left(1 + \frac{3-\nu}{4}\,k\right)n^2\right]\mu_m^2\right\}, \tag{10.7.3g}$$

$$\begin{aligned} L_{mn} = {} & (\mu_m^2 + n^2)^2(\mu_m^2 + n^2 - 1)^2 \\ & + 2(1-\nu)\mu_m^2(\mu_m^4 - n^4 + n^2) + \mu_m^4\,\frac{(1-\nu^2)(1+3k)}{k} \\ & + k\left\{(2 - 3k)\mu_m^8 + \left[\left(\frac{11 - 3\nu}{2} + 9\,\frac{1-\nu}{2}\,k\right)n^2 - 6\nu\right]\mu_m^6\right. \\ & + 3\left[\left(2 - \nu - \frac{\nu^2}{3}\,k\right)n^4 - (2 - \nu + \nu^2)n^2 + 1\right]\mu_m^4 \\ & + \tfrac{1}{2}n^2(n^2 - 1)[(7 + 3k)(1 - \nu)(n^2 - 1) + 4\nu n^2]\mu_m^2 \\ & \left. + n^4(n^2 - 1)^2\right\}, \end{aligned} \tag{10.7.3h}$$

$$\mu_m = \frac{m\pi R}{L}. \tag{10.7.3i}$$

The particular solution given above is the complete solution for simply supported cylindrical shells for which the edge conditions at $x = 0$ and $x = L$ are

$$w = M_x = N_x = u_\theta = 0 \tag{10.7.4}$$

since these edge conditions are satisfied by each term of Eqs. (10.7.2)†. For other edge conditions, different Fourier series expansions may be used if possible and/or appropriate edge solutions may be superimposed to yield the correct edge values of displacements and stress-resultants. It is, of course, possible to use other methods such as the method of variation of parameters to obtain other forms of the particular solution.

† These solutions are used for various problems in P. P. Bijlaard, 'Stresses from Radial Loads in Cylindrical Pressure Vessels,' *Welding J.*, vol. 33, December 1954, pp. 615s–623s, and 'Stresses from Local Loadings in Cylindrical Pressure Vessels.' *Trans. ASME*, vol. 77, no. 6, August 1955, pp. 805–814. Similar solutions of Donnell's equations are the basis of the calculations of J. Kempner, J. Sheng, and F. V. Pohle, 'Tables and Curves for Deformations and Stresses in Circular Cylindrical Shells Under Localized Loadings,' *J. Aero. Sci.*, vol. 24, no. 2, February 1957, pp. 119–129, which has an extensive bibliography of related work.

Let us compare the above results with those obtained by the use of Morley's or Donnell's equations for radial loading. We can obtain from Eqs. (10.2.2) and (10.2.3) the following results for the pertinent coefficients of Eqs. (10.7.2):

$$A_{mn}^{(3)} = (\mu_m^2 + n^2)^2, \tag{10.7.5a}$$

$$C_{mn} = \mu_m(\nu\mu_m^2 - n^2), \tag{10.7.5b}$$

$$D_{mn} = -n[(2 + \nu)\mu_m^2 + n^2], \tag{10.7.5c}$$

$$L_{mn} = \begin{cases} (\mu_m^2 + n^2)^2(\mu_m^2 + n^2 - 1)^2 + \dfrac{1 - \nu^2}{k}\,\mu_m^4 & \text{(Morley)} \quad (10.7.5\text{d}) \\[2ex] (\mu_m^2 + n^2)^4 + \dfrac{1 - \nu^2}{k}\,\mu_m^4 & \text{(Donnell).} \quad (10.7.5\text{e}) \end{cases}$$

The two expressions for $A_{mn}^{(3)}$ given by equation (10.7.3c) and (10.7.5a) differ only by terms of the order of h^2/R^2 compared to unity and are therefore almost identical. The expressions for C_{mn} and D_{mn} given by Eqs. (10.7.3f) and (10.7.3g) and by Eqs. (10.7.5b) and (10.7.5c) do not differ by significant amounts if

$$k\mu_m^2 \ll 1, \tag{10.7.8a}$$

$$kn^2 \ll 1, \tag{10.7.8b}$$

or, equivalently, if the longitudinal and circumferential wave lengths of the Fourier components are large compared to the attenuation length of end effects and the shell thickness, respectively. Thus the primary difference between the various results is the value of L_{mn}. On comparing Eqs. (10.7.3h) and (10.7.5e) we find that the greatest error for the Donnell approximation occurs when μ_m is small. In this case the ratio of the two values of L_{mn} is given

$$\lim_{\mu_m \to 0} \frac{L_{mn}\,(\text{Donnell})}{L_{mn}\,(\text{Flügge})} = \frac{1}{(1 - 1/n^2)^2(1 + k)} \qquad (n \neq 0)$$

$$= \frac{1}{[1 + k/(1 - \nu^2)](1 + 3k)} \qquad (n = 0). \tag{10.7.9}$$

Obviously, the error is greatest when n is equal to unity and decreases as n increases. For given n, the error decreases as μ_m increases but the rate of decrease of the error is slower as n increases. The Morley approximation Eq. (10.7.5d) gives results which do not differ significantly from those given by Eq. (10.7.3h).

On the basis of the above results we conclude that the Donnell approximation should be used with caution, whereas the Morley approximation yields quite accurate results. On the other hand, considering the fact that Eqs. (10.7.3) are known there does not appear to be any real justification for the use of the approximate equations in place of the complete solutions.

The membrane theory of shells is often used to obtain particular solutions. The pertinent equations are given by

$$\frac{\partial N_x}{\partial \xi} + \frac{\partial \overline{N}_{x\theta}}{\partial \theta} = q_x R, \tag{10.7.10a}$$

$$\frac{\partial N}{\partial \theta} + \frac{\partial \overline{N}_{x\theta}}{\partial \xi} = q_\theta R, \tag{10.7.10b}$$

$$N_\theta = q_\zeta R, \tag{10.7.10c}$$

$$\frac{\partial u_x}{\partial \xi} = \frac{R}{Eh}(N_x - \nu N_\theta), \tag{10.7.10d}$$

$$\frac{\partial u_\theta}{\partial \theta} + w = \frac{R}{Eh}(N_\theta - \nu N_x), \tag{10.7.10e}$$

$$\frac{\partial u_\theta}{\partial \xi} + \frac{\partial u_x}{\partial \theta} = 2(1+\nu)\frac{R}{Eh}\overline{N}_{x\theta}. \tag{10.7.10f}$$

These may be solved to yield

$$\frac{Eh}{R^2}u_x = \int_{\xi_0}^{\xi}\left\{(\xi - u)q_x(u,\theta) + \tfrac{1}{2}(\xi - u)^2\left[\frac{\partial^2 q_\zeta(u,\theta)}{\partial \theta^2} - \frac{\partial q_\theta(u,\theta)}{\partial \theta}\right] - \nu q_\zeta(u,\theta)\right\}\mathrm{d}u + A(\theta) + \xi B(\theta) - \tfrac{1}{2}\xi^2\frac{\mathrm{d}C(\theta)}{\mathrm{d}\theta}, \tag{10.7.11a}$$

$$\begin{aligned}\frac{Eh}{R^2}u_\theta = {} & \int_{\xi_0}^{\xi}\left\{(\xi - u)\left[2(1+\nu)q_\theta(u,\theta) - (2+\nu)\frac{\partial q_\zeta(u,\theta)}{\partial \theta}\right]\right. \\ & \left. - \tfrac{1}{2}(\xi - u)^2\frac{\partial q_x(u,\theta)}{\partial \theta} - \tfrac{1}{6}(\xi - u)^3\left[\frac{\partial^3 q_\zeta(u,\theta)}{\partial \theta^3} - \frac{\partial^2 q_\theta(u,\theta)}{\partial \theta^2}\right]\right\}\mathrm{d}u \\ & - \xi\frac{\mathrm{d}A(\theta)}{\mathrm{d}\theta} - \tfrac{1}{2}\xi^2\frac{\mathrm{d}B(\theta)}{\mathrm{d}\theta} + 2(1+\nu)\xi C(\theta) + \tfrac{1}{6}\xi^3\frac{\mathrm{d}^2 C(\theta)}{\mathrm{d}\theta^2} + D(\theta),\end{aligned} \tag{10.7.11b}$$

$$\begin{aligned}\frac{Eh}{R^2}w = q_\zeta + {} & \int_{\xi_0}^{\xi}\left\{(\xi - u)\left[2\frac{\partial^2 q_\zeta(u,\theta)}{\partial \theta^2} - (2+\nu)\frac{\partial q_\theta(u,\theta)}{\partial \theta}\right]\right. \\ & - \nu q_x(u,\theta) + \tfrac{1}{2}(\xi - u)^2\frac{\partial^2 q_x(u,\theta)}{\partial \theta^2} + \tfrac{1}{6}(\xi - u)^3 \\ & \left. \times \left[\frac{\partial^4 q_\zeta(u,\theta)}{\partial \theta^4} - \frac{\partial^3 q_\theta(u,\theta)}{\partial \theta^3}\right]\right\}\mathrm{d}u + \xi\frac{\mathrm{d}^2 A(\theta)}{\mathrm{d}\theta^2} \\ & + \tfrac{1}{2}\xi^2\frac{\mathrm{d}B(\theta)}{\mathrm{d}\theta} - 2(1+\nu)\xi\frac{\mathrm{d}C(\theta)}{\mathrm{d}\theta} - \tfrac{1}{6}\xi^3\frac{\mathrm{d}^3 C(\theta)}{\mathrm{d}\theta^3} - \frac{\mathrm{d}D(\theta)}{\mathrm{d}\theta},\end{aligned} \tag{10.7.11c}$$

$$\frac{N_x}{R} = \int_{\xi_0}^{\xi} \left\{ q_x(u, \theta) + (\xi - u) \left[\frac{\partial^2 q_\zeta(u, \theta)}{\partial \theta^2} - \frac{\partial q_\theta(u, \theta)}{\partial \theta} \right] \right\} \mathrm{d}u$$
$$+ B(\theta) - \xi \frac{\mathrm{d}C(\theta)}{\mathrm{d}\theta}, \tag{10.7.11d}$$

$$\frac{N_\theta}{R} = q_\zeta, \tag{10.7.11e}$$

$$\frac{\overline{N}_{x\theta}}{R} = \int_{\xi_0}^{\xi} \left[q_\theta(u, \theta) - \frac{\partial q_\zeta(u, \theta)}{\partial \theta} \right] \mathrm{d}u + C(\theta). \tag{10.7.11f}$$

The four arbitrary functions of θ can be adjusted to satisfy two boundary conditions for u_x and u_θ only along circular edges of the cylinder. Alternatively, boundary conditions on both forces and displacements along circular edges can be satisfied as discussed in Section 3.8. The validity of the solutions is generally limited to slowly varying loading and to relatively short cylinders for which the inextensional solutions corresponding to vanishing middle surface strains ϵ_x, ϵ_θ, and $\gamma_{x\theta}$ are a reasonably good approximation to the slowly varying exponential solutions given by the bending theory.

10.8 Edge-loaded cylindrical strips

Few relatively simple exact solutions can be obtained for cylindrical shells other than those discussed in the preceding sections for shells complete in the circumferential direction. For finite edge-loaded cylindrical strips (Fig. 10.12), solutions can be obtained for loads which vary harmonically along the straight edges provided certain boundary conditions are satisfied along the curved edges.

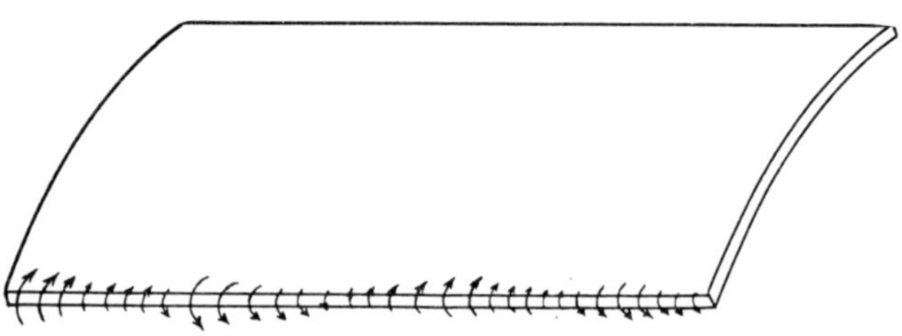

Fig. 10.12. Edge-loaded cylindrical strip

We assume the deflection function Φ to be given by

$$\Phi(\xi, \theta, c) = \sum_{r=1}^{8} A_r \, \mathrm{e}^{\bar{\lambda}_r(c)\theta} \cos(c\xi + \delta_0) \tag{10.8.1}$$

where c is an arbitrary coefficient and the values of $\bar{\lambda}_r(c)$ are the eight independent solutions of the following equation obtained from Eq. (10.3.2) by replacing λ_{nr}^2 by $-c^2$ and n^2 by $-\bar{\lambda}_r^2(c)$:

$$(1 + k)\bar{\lambda}_r^8 + 2\left[1 + k - 2\left(1 + \frac{7 - 3\nu}{8}k + 3\frac{1 - \nu}{8}k^2\right)c^2\right]\bar{\lambda}_r^6$$

$$+ \left\{1 + k - [2(4 - \nu) + (7 - 5\nu)k + 3(1 - \nu)k^2]c^2 + 6\left(1 + \frac{2 - \nu}{2}k - \frac{\nu^2}{6}k^2\right)c^4\right\}\bar{\lambda}_r^4$$

$$- \left[2(2 - \nu) + 7\frac{1 - \nu}{2}k + 3\frac{1 - \nu}{2}k^2 - 6\left(1 + \frac{2 - \nu + \nu^2}{2}k\right)c^2 + 4\left(1 + \frac{11 - 3\nu}{8}k + 9\frac{1 - \nu}{8}k^2\right)c^4\right]c^2\bar{\lambda}_r^2$$

$$+ \left[\frac{1 - \nu^2}{k} + 1 - 2\nu c^2 + (1 - k)c^4\right](1 + 3k)c^4 = 0. \quad (10.8.2)$$

In general all of the roots of Eq. (10.8.2) appear to be complex and may be represented by

$$\bar{\lambda}_r = \pm\bar{p}_1(c) \pm i\bar{q}_1(c) \qquad (r = 1, 2, 3, 4), \quad (10.8.3a)$$

$$\bar{\lambda}_r = \pm\bar{p}_2(c) \pm i\bar{q}_2(c) \qquad (r = 5, 6, 7, 8). \quad (10.8.3b)$$

We may approximate the values of $\bar{p}_1$, $\bar{q}_1$, $\bar{p}_2$, $\bar{q}_2$ by those corresponding to Morley's equation as†

$$\left.\begin{matrix}\bar{p}_1(c)\\ \bar{q}_1(c)\end{matrix}\right\} = [\tfrac{1}{2}(\{[c^2 - \tfrac{1}{2} + \alpha(c)]^2 + \beta^2(c)\}^{1/2} \pm [c^2 - \tfrac{1}{2} + \alpha(c)])]^{1/2}, \quad (10.8.4a)$$

$$\left.\begin{matrix}\bar{p}_2(c)\\ \bar{q}_2(c)\end{matrix}\right\} = [\tfrac{1}{2}(\{[c^2 - \tfrac{1}{2} - \alpha(c)]^2 + \beta^2(c)\}^{1/2} \pm [c^2 - \tfrac{1}{2} - \alpha(c)])]^{1/2}, \quad (10.8.4b)$$

$$\left.\begin{matrix}\alpha(c)\\ \beta(c)\end{matrix}\right\} = \left[\left(\frac{1}{64} + \frac{1 - \nu^2}{4k}c^4\right)^{1/2} \pm \frac{1}{8}\right]^{1/2}. \quad (10.8.4c)$$

† See also J. Moe, *On the Theory of Cylindrical Shells. Explicit Solution of the Characteristic Equation and Discussion of the Accuracy of Various Shell Theories*, Publications of the International Association for Bridge and Structural Engineering (Zürich), vol. 13, 1953, pp. 283–295, where the general solution of the biquartic equation is simplified on the basis of various order of magnitude estimates.

For Donnell's equation we neglect the quantity $\frac{1}{2}$ in comparison with c^2 in Eqs. (10.8.4a) and (10.8.4b) and replace Eqs. (10.8.4c) by

$$\alpha(c) = \beta(c) = \left(\frac{1 - \nu^2}{4k}\right)^{1/4} c. \tag{10.8.4d}$$

Values of $\bar{p}_1, \bar{q}_1, \bar{p}_2$, and $\bar{q}_2$ given by the two sets of equations and by computer solution of Flügge's approximate equation given by

$$(\bar{\lambda}_r^2 - c^2)^2(\bar{\lambda}_r^2 - c^2 - 1)^2 - 2(1 - \nu)c^2(\bar{\lambda}_r^4 + \bar{\lambda}_r^2 - c^4) + \frac{1 - \nu^2}{k} c^4 = 0 \tag{10.8.5}$$

are compared in Table 10.5. We see that Eqs. (10.8.4a) and (10.8.4b) are a quite good approximation to the solution of Eq. (10.8.5). The further simplification afforded by Donnell's approximations is grossly in error when the parameter $[(1 - \nu^2)/4k]^{1/4}c$ is less than about 4 but is reasonably accurate for larger values of c. A graphical representation of the results of Donnell's equation is shown in Fig. 10.13. An approximate representation of the results of Morley's equation valid for $[(1 - \nu^2)/4k]^{1/4}c$ less than 4 is shown in Fig. 10.14.

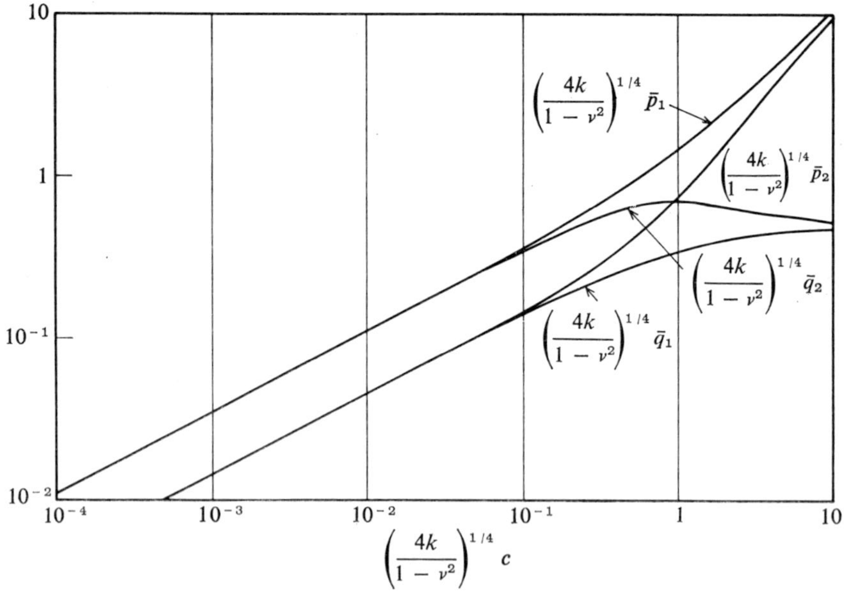

Fig. 10.13. Approximate values of roots

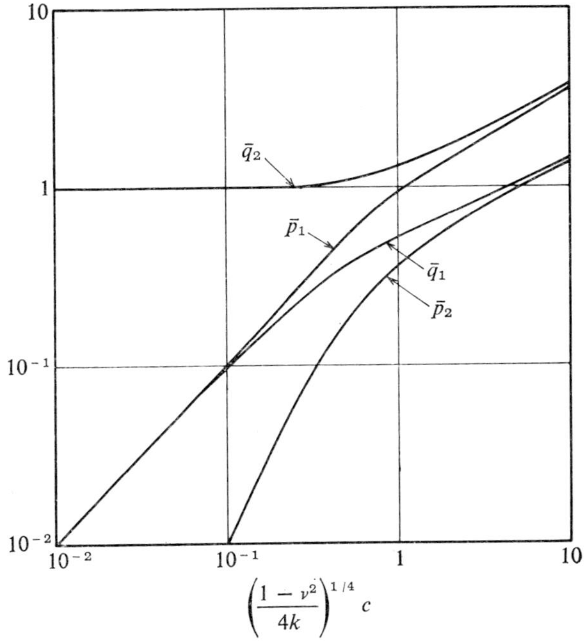

Fig. 10.14. Approximate values of roots for small values of c

The complete solution Φ for a given value of c can be written in real form as

$$\Phi(c) = \cos(c\xi + \delta_0) \sum_{i=1}^{2} \{e^{-\bar{p}_i(\theta-\theta_1)}[A_i \cos \bar{q}_i(\theta - \theta_1) + B_i \sin \bar{q}_i(\theta - \theta_1)] + e^{-\bar{p}_i(\theta_2-\theta)}[C_i \cos \bar{q}_i(\theta_2 - \theta) + D_i \sin \bar{q}_i(\theta_2 - \theta)]\} \quad (10.8.6)$$

which is thus expressed in terms of solutions which decay exponentially as θ increases from θ_1 and solutions which decay as θ decreases from θ_2. From Eqs. (10.1.3), (10.1.8), (10.1.13), and (10.8.6) expressions for the deformations and stress resultants can be obtained in forms similar to those given by Eqs. (10.3.14) for complete cylindrical shells loaded along the curved edges by harmonically varying load. In general the deformations and stress resultants can be expressed either in the form

$$\begin{pmatrix} \mathrm{I} \\ \mathrm{II} \end{pmatrix} = \sum_{i=1}^{2} \langle e^{-\bar{p}_i(\theta-\theta_1)}\{A_i[\bar{\alpha}_i \cos \bar{q}_i(\theta - \theta_1) + \bar{\beta}_i \sin \bar{q}_i(\theta - \theta_1)] + B_i[\bar{\alpha}_i \sin \bar{q}_i(\theta - \theta_1) - \bar{\beta}_i \cos \bar{q}_i(\theta - \theta_1)]\} \pm e^{-\bar{p}_i(\theta_2-\theta)}\{C_i[\bar{\alpha}_i \cos \bar{q}_i(\theta_2 - \theta) + \bar{\beta}_i \sin \bar{q}_i(\theta_2 - \theta)] + D_i[\bar{\alpha}_i \sin \bar{q}_i(\theta_2 - \theta) - \bar{\beta}_i \cos \bar{q}_i(\theta_2 - \theta)]\}\rangle \cos(c\xi + \delta_0) \quad (10.8.7)$$

Table 10.5 Comparison of roots of various characteristic equations for edge-loaded strips

$\left(\frac{1-\nu^2}{k}\right)^{1/2}$	c	Flügge (approx.) $\bar{p}_1$	Flügge (approx.) $\bar{q}_1$	Morley $\bar{p}_1$	Morley $\bar{q}_1$	Donnell $\bar{p}_1$	Donnell $\bar{q}_1$
50	0.01	0.0510	0.0490	0.0506	0.0494	0.2458	0.1017
	0.10	0.5330	0.3662	0.5316	0.3696	0.7824	0.3195
	1.00	2.5469	0.9694	2.5441	0.9801	2.6278	0.9514
	10.00	12.3952	1.9456	12.3921	2.0172	12.4120	2.0142
200	0.01	0.1014	0.0985	0.1012	0.0987	0.3476	0.1439
	0.10	0.9157	0.5121	0.9153	0.5133	1.1026	0.4535
	1.00	3.5371	1.4097	3.5360	1.4131	3.5963	1.3903
	10.00	14.5388	3.4209	14.5372	3.4394	14.5535	3.4356
5000	0.01	0.5254	0.3740	0.5254	0.3741	0.7769	0.3217
	0.10	2.3723	1.0511	2.3723	1.0512	2.4585	1.0169
	1.00	7.7965	3.2063	7.7964	3.2065	7.8237	3.1954
	10.00	26.2695	9.5158	26.2692	9.5168	26.2776	9.5138

$\left(\frac{1-\nu^2}{k}\right)^{1/2}$	c	Flügge (approx.) $\bar{p}_2$	Flügge (approx.) $\bar{q}_2$	Morley $\bar{p}_2$	Morley $\bar{q}_2$	Donnell $\bar{p}_2$	Donnell $\bar{q}_2$
50	0.01	0.0025	1.0000	0.0025	1.0000	0.1018	0.2455
	0.10	0.1812	1.0783	0.1825	1.0767	0.3241	0.7714
	1.00	1.0455	2.3749	1.0520	2.3704	1.0962	2.2807
	10.00	7.7458	3.2942	7.9414	3.2290	7.7689	3.2180
200	0.01	0.0100	1.0002	0.0100	1.0002	0.1440	0.3473
	0.10	0.3611	1.2987	0.3619	1.2982	0.4567	1.0948
	1.00	1.4613	3.4164	1.4640	3.4131	1.4923	3.3506
	10.00	7.0534	7.1011	7.0534	7.0887	7.0711	7.0711
5000	0.01	0.1818	1.0812	0.1818	1.0812	0.3218	0.7768
	0.10	0.9807	2.5426	0.9808	2.5425	1.0183	2.4550
	1.00	3.2290	7.7415	3.2293	7.7414	3.2409	7.7139
	10.00	10.9566	22.8163	10.9573	22.8158	10.9616	22.8069

or in similar forms $\begin{pmatrix}\text{III}\\\text{IV}\end{pmatrix}$ with $\cos(c\xi - \delta_0)$ replaced by $\sin(c\xi - \delta_0)$. The coefficients $\bar{\alpha}_i$ and $\bar{\beta}_i$ are listed in Table 10.6.

If we consider a cylindrical strip of length L and put

$$\delta_0 = -\tfrac{1}{2}\pi, \tag{10.8.8a}$$

Table 10.6 (a) Expressions for $\bar{\alpha}_i$ appearing in Eqs. (10.8.7)

Quantity	$\bar{\alpha}_i$
u_x (III)	$\left[k(\bar p_i^4 - 6\bar p_i^2 q_i^2 + \bar q_i^4) - \left(1 + 3\frac{1-\nu}{2}k^2c^2\right)(\bar p_i^2 - \bar q_i^2) - c^2(\nu + kc^2)(1+3k)\right]c$
u_θ (II)	$-\left\{\left[1 + k + 2k\left(1 + \frac{3-\nu}{4}k\right)c^2\right](\bar p_i^2 - \bar q_i^2) + (2 + \nu + 2kc^2)c^2\right\}\bar p_i$
w (I)	$(1+k)(\bar p_i^4 - 6\bar p_i^2\bar q_i^2 + \bar q_i^4) - 2[1 + (1-\nu)k(1+\frac{3}{4}k)]c^2(\bar p_i^2 - \bar q_i^2) + (1+3k)c^4$
$\partial w/\partial\theta$ (II)	$-\{(1+k)(\bar p_i^4 - 10\bar p_i^2\bar q_i^2 + 5\bar q_i^4) - 2[1+(1-\nu)k(1+\frac{3}{4}k)](3\bar p_i^2 - \bar q_i^2)c^2 + (1+3k)c^4\}\bar p_i$
RN_x/K (I)	$(1-\nu)c^2\left\{2k\left(1+\frac{2-\nu}{4}k\right)(\bar p_i^4 - 6\bar p_i^2\bar q_i^2 + \bar q_i^4) - [1+\nu+2\nu k(1+\frac{3}{4}k) + 2k(1+k)(1+\frac{3}{4}k)c^2](\bar p_i^2-\bar q_i^2)\right\}$
RN_θ/K (I)	$k(1+k)(\bar p_i^2-\bar q_i^2)(\bar p_i^4 - 14\bar p_i^2\bar q_i^2+\bar q_i^4) + k\left[1+k-\left(4-\nu+\frac{7-5\nu}{2}k + 3\frac{1-\nu}{2}k^2\right)c^2\right] \times (\bar p_i^4 - 6\bar p_i^2\bar q_i^2+\bar q_i^4) - 2c^2k\left[2-\nu+\frac{1-\nu}{4}k(7+3k) - \frac{3}{2}\left(1+\frac{2-\nu+\nu^2}{2}k\right)c^2\right] \times (\bar p_i^2-\bar q_i^2) + (1+3k)c^4(1-\nu^2+k-\nu kc^2)$
$R\bar{N}_{x\theta}/K$ (IV)	$\frac{1}{2}(1-\nu)(1+\frac{3}{4}k)\bar p_i c\{k(\bar p_i^4 - 10\bar p_i^2\bar q_i^2 + 5\bar q_i^4) + k[1+(2+\nu k)c^2](\bar p_i^2 - 3\bar q_i^2) - c^2[2(1+\nu)+3\nu k+3(1-k)c^2]$
M_x/kK (I)	$-\nu(1+k)(\bar p_i^2-\bar q_i^2)(\bar p_i^4-14\bar p_i^2\bar q_i^2+\bar q_i^4) - \{\nu(1+k) - c^2[1+2\nu+2(1-\nu^2)k - \nu^2k^2]\}(\bar p_i^4-6\bar p_i^2\bar q_i^2+\bar q_i^4) - c^2\{1-2\nu-\nu^2+c^2[2+\nu+(2-\nu)k+3(1-\nu)k^2]\} \times (\bar p_i^2-\bar q_i^2) + (1+3k)[\nu-(1+k)c^2]c^4$
M_θ/kK (I)	$-(1+k)(\bar p_i^2-\bar q_i^2)(\bar p_i^4-14\bar p_i^2\bar q_i^2+\bar q_i^4) - \left\{1+k-\left[2+\nu+(2-\nu)k+3\frac{1-\nu}{2}k^2\right]c^2\right\} \times (\bar p_i^4-6\bar p_i^2\bar q_i^2+\bar q_i^4) + c^2\left\{2+2(1-\nu)k(1+\frac{3}{4}k) - \left[1+2\nu+(3+2\nu-2\nu^2)k+3\nu\frac{1-\nu}{2}k^2\right]c^2\right\} \times (\bar p_i^2-\bar q_i^2) - (1+3k)(1-\nu c^2)c^4$
$\bar{M}_{x\theta}/kK$ (IV)	$-(1-\nu)\bar p_i c\left\{(1+\frac{3}{4}k)(\bar p_i^4-10\bar p_i^2\bar q_i^2+5\bar q_i^4) + \left[1+\frac{3}{4}k-\left(2+\frac{1-4\nu}{2}k-\frac{3}{4}\nu k^2\right)c^2\right](\bar p_i^2-3\bar q_i^2) - \left[\frac{3+\nu}{2}-\frac{3}{4}\nu k-(1+\frac{3}{4}k)(1+k)c^2\right]\right\}$

Table 10.6 (*Cont.*)

Quantity	$\bar{\alpha}_i$
$\bar{Q}_x R/Kk$ (III)	$c\left\langle\left(2-\nu+\frac{3-\nu}{2}k\right)(\bar{p}_i^2-\bar{q}_i^2)(\bar{p}_i^4-14\bar{p}_i^2\bar{q}_i^2+\bar{q}_i^4) + \left\{2-\nu+\frac{3-\nu}{2}k-\left[5-2\nu+(3-5\nu+2\nu^2)k-\nu\frac{3-\nu}{2}k^2\right]c^2\right\} \times(\bar{p}_i^4-6\bar{p}_i^2\bar{q}_i^2+\bar{q}_i^4)-c^2\left[(1-\nu)(2-\nu-\tfrac{3}{2}\nu k) - \left(4-\nu+\frac{11-9\nu}{2}k+9\frac{1-\nu}{2}k^2\right)c^2\right] \times(\bar{p}_i^2-\bar{q}_i^2)+(1+3k)[\nu-(1-k)c^2]c^4\right\rangle$
$\bar{Q}_\theta R/kK$ (II)	$\bar{p}_i\left\langle(1+k)(\bar{p}_i^6-21\bar{p}_i^4\bar{q}_i^2+35\bar{p}_i^2\bar{q}_i^4-7\bar{q}_i^6) + \left[1+k-\left(4-\nu+\frac{7-5\nu}{2}k+3\frac{1-\nu}{2}k^2\right)c^2\right] \times(\bar{p}_i^4-10\bar{p}_i^2\bar{q}_i^2+5\bar{q}_i^4)-c^2\left\{2(2-\nu)+7\frac{1-\nu}{2}k+3\frac{1-\nu}{2}k^2 - c^2\left[5-2\nu+(4-3\nu+2\nu^2)k+\tfrac{3}{2}(1-\nu)(2-\nu)k^2\right]\right\} \times(\bar{p}_i^2-3\bar{q}_i^2)+c^4\left[4-2\nu-\nu^2+3\frac{2-\nu+\nu^2}{2}k - \left(2-\nu+\frac{7-\nu}{2}k+3\frac{1-\nu}{2}k^2\right)c^2\right]\right\rangle$

(b) Expressions for $\bar{\beta}_i$ appearing in Eqs. (10.8.7)

Quantity	χ_i
u_x (III)	$2\bar{p}_i\bar{q}_i c\left[2k(\bar{p}_i^2-\bar{q}_i^2)-\left(1+3\frac{1-\nu}{2}k^2c^2\right)\right]$
u (II)	$-\bar{q}_i\left\{(2+\nu+2kc^2)c^2+\left[1+k+2k\left(1+\frac{3-\nu}{4}k\right)c^2\right](3\bar{p}_i^2-\bar{q}_i^2)\right\}$
w (I)	$4\bar{p}_i\bar{q}_i\{(1+k)(\bar{p}_i^2-\bar{q}_i^2)-[1+(1-\nu)k(1+\tfrac{3}{4}k)]c^2\}$
$\partial w/\partial\theta$ (II)	$-\bar{q}_i\{(1+k)(5\bar{p}_i^4-10\bar{p}_i^2\bar{q}_i^2+\bar{q}_i^4) - 2[1+(1-\nu)k(1+\tfrac{3}{4}k)]c^2(3\bar{p}_i^2-\bar{q}_i^2)+(1+3k)c^4\}$
RN_x/K (I)	$2(1-\nu)\bar{p}_i\bar{q}_i c^2\left\{4k\left(1+\frac{2-\nu}{4}k\right)(\bar{p}_i^2-\bar{q}_i^2) - [1+\nu+2\nu k(1+\tfrac{3}{4}k)+2k(1+k)(1+\tfrac{3}{4}k)c^2]\right\}$
RN_θ/K (I)	$2\bar{p}_i\bar{q}_i\left\{k(1+k)(3\bar{p}_i^4-10\bar{p}_i^2\bar{q}_i^2+3\bar{q}_i^4) + 2k\left[1+k-\left(4-\nu+\frac{7-5\nu}{2}k+3\frac{1-\nu}{2}k^2\right)c^2\right](\bar{p}_i^2-\bar{q}_i^2) - 2c^2k\left[2-\nu+\frac{1-\nu}{2}k(7+3k)-\frac{3}{2}\left(1+\frac{2-\nu+\nu^2}{2}k\right)c^2\right]\right\}$

Table 10.6 (*Cont.*)

Quantity	χ_i
$R\bar{N}_{x\theta}/K$ (IV)	$\begin{aligned} &\frac{1-\nu}{2}(1+\tfrac{3}{4}k)\bar{q}_i c\{k(5\bar{p}_i^4 - 10\bar{p}_i^2\bar{q}_i^2 + \bar{q}_i^4) \\ &\quad + k[1 + (2+\nu k)c^2](3\bar{p}_i^2 - \bar{q}_i^2) - c^2[2(1+\nu) + 3\nu k + 3(1+k)c^2]\} \end{aligned}$
M_x/kK (I)	$\begin{aligned} &-2\bar{p}_i\bar{q}_i\langle \nu(1+k)(3\bar{p}_i^4 - 10\bar{p}_i^2\bar{q}_i^2 + 3\bar{q}_i^4) \\ &\quad + 2\{\nu(1+k) - c^2[1 + 2\nu + 2(1-\nu^2)k - \nu^2k^2]\}(\bar{p}_i^2 - \bar{q}_i^2) \\ &\quad - c^2\{1 - 2\nu - \nu^2 + c^2[2 + \nu + (2-\nu)k + 3(1-\nu)k^2]\}\rangle \end{aligned}$
M_θ/kK (I)	$\begin{aligned} &-2\bar{p}_i\bar{q}_i\Big\langle (1+k)(3\bar{p}_i^4 - 10\bar{p}_i^2\bar{q}_i^2 + 3\bar{q}_i^4) \\ &\quad + 2\Big\{1 + k - \Big[2 + \nu + (2-\nu)k + 3\frac{1-\nu}{2}k^2\Big]c^2\Big\}(\bar{p}_i^2 - \bar{q}_i^2) \\ &\quad - c^2\Big\{2 + 2(1-\nu)k(1+\tfrac{3}{4}k) \\ &\quad - \Big[1 + 2\nu + (3 + 2\nu - 2\nu^2)k + 3\nu\frac{1-\nu}{2}k^2\Big]c^2\Big\}\Big\rangle \end{aligned}$
$\bar{M}_{x\theta}/kK$ (IV)	$\begin{aligned} &-(1-\nu)\bar{q}_i c\{(1+\tfrac{3}{4}k)(5\bar{p}_i^4 - 10\bar{p}_i^2\bar{q}_i^2 + \bar{q}_i^4) \\ &\quad + \Big[1 + \tfrac{3}{4}k - \Big(2 + \frac{1-4\nu}{2}k - \tfrac{3}{4}\nu k^2\Big)c^2\Big](3\bar{p}_i^2 - \bar{q}_i^2) \\ &\quad - \Big[\frac{3+\nu}{2} - \tfrac{3}{4}\nu k - (1+\tfrac{3}{4}k)(1+k)c^2\Big]\Big\} \end{aligned}$
$\bar{Q}_x R/kK$ (III)	$\begin{aligned} &2\bar{p}_i\bar{q}_i c\Big\langle\Big(2 - \nu + \frac{3-\nu}{2}k\Big)(3\bar{p}_i^4 - 10\bar{p}_i^2\bar{q}_i^2 + 3\bar{q}_i^4) \\ &\quad + 2\Big\{2 - \nu + \frac{3-\nu}{2}k - \Big[5 - 2\nu + (3 - 5\nu + 2\nu^2)k - \nu\frac{3-\nu}{2}k^2\Big]c^2\Big\} \\ &\quad \times (\bar{p}_i^2 - \bar{q}_i^2) - c^2\Big[(1-\nu)(2 - \nu - \tfrac{3}{2}\nu k) \\ &\quad - \Big(4 - \nu + \frac{11-9\nu}{2}k + 9\frac{1-\nu}{2}k^2\Big)c^2\Big]\Big\rangle \end{aligned}$
$\bar{Q}_\theta R/kK$ (II)	$\begin{aligned} &\bar{q}_i\Big\langle(1+k)(7\bar{p}_i^6 - 35\bar{p}_i^4 4\bar{q}_i^2 + 21\bar{p}_i^2\bar{q}_i^4 - \bar{q}_i^6) \\ &\quad + \Big[1 + k - \Big(4 - \nu + \frac{7-5\nu}{2}k + 3\frac{1-\nu}{2}k^2\Big)c^2\Big] \\ &\quad \times (5\bar{p}_i^4 - 10\bar{p}_i^2\bar{q}_i^2 + \bar{q}_i^4) - c^2\Big\{2(2-\nu) + 7\frac{1-\nu}{2}k + 3\frac{1-\nu}{2}k^2 \\ &\quad - c^2[5 - 2\nu + (4 - 3\nu + 2\nu^2)k + \tfrac{3}{2}(1-\nu)(2-\nu)k^2]\Big\}(3\bar{p}_i^2 - \bar{q}_i^2) \\ &\quad + c^4\Big[4 - 2\nu - \nu^2 + 3\frac{2 - \nu + \nu^2}{2}k \\ &\quad - \Big(2 - \nu + \frac{7-\nu}{2}k + 3\frac{1-\nu}{2}k^2\Big)c^2\Big]\Big\rangle \end{aligned}$

$$c = m\pi R/L \qquad (m = 1, 2, 3, \ldots), \tag{10.8.8b}$$

we find that the solutions satisfy the edge conditions for a particular type of simple support

$$w = u_\theta = N_x = M_x = 0. \tag{10.8.9}$$

The eight arbitrary constants of integration are determined by conditions along the straight edges. For arbitrary loading along the straight edges of a finite cylindrical strip with the above type of simple support along the curved edges, an exact solution is obtained by expanding the loading in an appropriate Fourier series in the x-direction and superimposing the above solutions for each Fourier component of loading.

For example, consider a strip of length L which is simply supported in the above manner and which is subjected to equal uniform bending moments along the straight edges (Fig. 10.15). Then the conditions to be satisfied along the straight edges at $\theta_1 = -\frac{1}{2}\theta_0$ and $\theta_2 = \frac{1}{2}\theta_0$ are

$$N_\theta = \frac{\bar{N}_{x\theta} - \frac{1}{2}\bar{M}_{x\theta}}{R} = \bar{Q}_\theta = 0, \tag{10.8.10a}$$

$$M_\theta = M_0. \tag{10.8.10b}$$

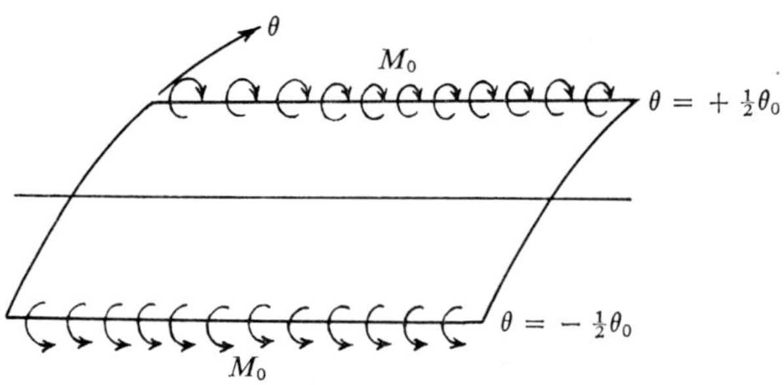

Fig. 10.15. Cylindrical strip with uniform edge moments

The constant value of M_0 may be expanded into a Fourier sine series in the longitudinal direction as

$$M_\theta = \frac{4M_0}{\pi} \sum_{m=1,3,5,\ldots}^{\infty} \frac{1}{m} \sin\frac{m\pi x}{L}. \tag{10.8.10c}$$

Since the loading is symmetrical about the longitudinal axis, we take Φ in a form which is symmetrical about the longitudinal axis by putting A_i equal to

C_i and B_i equal to D_i in Eq. (10.8.7). Thus

$$\Phi = \sum_{m=1,3,5,\dots}^{\infty} \sum_{i=1}^{2} \times \{A_{im}[\mathrm{e}^{-\bar{p}_{im}(\frac{1}{2}\theta_0+\theta)} \cos \bar{q}_{im}(\tfrac{1}{2}\theta_0 + \theta) + \mathrm{e}^{-\bar{p}_{im}(\frac{1}{2}\theta_0-\theta)} \cos \bar{q}_{im}(\tfrac{1}{2}\theta_0 - \theta) + B_{im}[\mathrm{e}^{-\bar{p}_{im}(\frac{1}{2}\theta_0+\theta)} \sin \bar{q}_{im}(\tfrac{1}{2}\theta_0 + \theta) + \mathrm{e}^{-\bar{p}_{im}(\frac{1}{2}\theta_0-\theta)} \sin \bar{q}_{im}(\tfrac{1}{2}\theta_0 - \theta)]\} \sin \frac{m\pi x}{L}. \tag{10.8.11}$$

The equations for the determination of the coefficients are now obtained from Eqs. (10.3.20c), (10.3.20d), (10.8.7), (10.8.10), (10.8.11), and Table 10.6 as

$$\sum_{i=1}^{2} \{A_{im}[\bar{\alpha}_{im}^{(N_\theta)} + \bar{f}_{im}^{(N_\theta)}(\theta_0)] + B_{im}[\bar{g}_{im}^{(N_\theta)}(\theta_0) - \bar{\beta}_{im}^{(N_\theta)}]\} = 0, \tag{10.8.12a}$$

$$\sum_{i=1}^{2} \{A_{im}[\bar{\alpha}_{im}^{(\bar{Q}_\theta)} - \bar{f}_{im}^{(\bar{Q}_\theta)}(\theta_0)] - B_{im}[\bar{g}_{im}^{(\bar{Q}_\theta)}(\theta_0) + \bar{\beta}_{im}^{(\bar{Q}_\theta)}]\} = 0, \tag{10.8.12b}$$

$$\sum_{i=1}^{2} \{A_{im}[\bar{\alpha}_{im}^{(M_\theta)} + \bar{f}_{im}^{(M_\theta)}(\theta_0)] + B_{im}[\bar{g}_{im}^{(M_\theta)}(\theta_0) - \bar{\beta}_{im}^{(M_\theta)}]\} = -\frac{4M_0}{kKm}, \tag{10.8.12c}$$

$$\sum_{i=1}^{2} \langle A_{im}\{\bar{\alpha}_{im}^{(\bar{N}_{x\theta})} - \bar{f}_{im}^{(\bar{N}_{x\theta})}(\theta_0) - \tfrac{1}{2}k[\bar{\alpha}_{im}^{(\bar{M}_{x\theta})} - \bar{f}_{im}^{(\bar{M}_{x\theta})}(\theta_0)]\} - B_{im}\{\bar{\beta}_{im}^{(\bar{N}_{x\theta})} + \bar{g}_{im}^{(\bar{N}_{x\theta})}(\theta_0) - \tfrac{1}{2}k[\bar{\beta}_{im}^{(\bar{M}_{x\theta})} + \bar{g}_{im}^{(\bar{M}_{\theta x})}(\theta_0)]\}\rangle = 0, \tag{10.8.12d}$$

with

$$\bar{f}_{im}(u) = (\bar{\alpha}_{im} \cos \bar{q}_{im}u + \bar{\beta}_{im} \sin \bar{q}_{im}u)\, \mathrm{e}^{-\bar{p}_{im}u}, \tag{10.8.12e}$$

$$\bar{g}_{im}(u) = (\bar{\alpha}_{im} \sin \bar{q}_{im}u - \bar{\beta}_{im} \cos \bar{q}_{im}u)\, \mathrm{e}^{-\bar{p}_{im}u}. \tag{10.8.12f}$$

Other loadings or restraints of the straight edges can be considered in a similar manner. The strip solutions can also be used for complete simply supported cylinders subjected to line loads along generators or to any other loading which is discontinuous in the circumferential direction (Fig. 10.16).

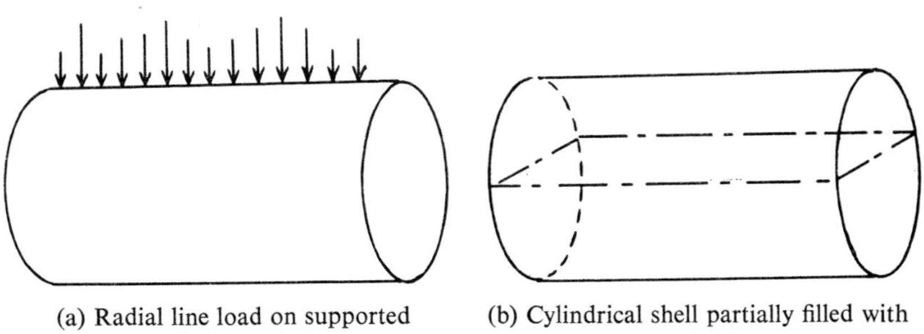

(a) Radial line load on supported cylindrical shell

(b) Cylindrical shell partially filled with fluid

Fig. 10.16. Examples of cylindrical shells with discontinuous circumferential loading

When the curved edges are not simply supported as required by Eqs. (10.8.9) we superimpose solutions for cylinders loaded along the curved edges and simply supported along the straight edges. These may be obtained from Section 10.3 by replacing n there by $2\pi n/\theta_0$ and θ varies from 0 to θ_0. It is necessary to satisfy boundary conditions on all four edges simultaneously, a procedure which generally yields an infinite set of equations in an infinite number of unknowns. The details of the analysis are similar to those for flat plates with arbitrarily supported edges.†

10.9 Infinite cylindrical strips

When the strip is infinitely long, the solutions given in the preceding section can be used as components of a Fourier integral approach to the problem. Since any function of x defined in the interval $-\infty \leq x \leq \infty$ can be expressed as‡

$$f(x) = \frac{1}{2\pi}\int_{-\infty}^{\infty} \mathrm{e}^{icx}\,\mathrm{d}c \int_{-\infty}^{\infty} f(\eta)\,\mathrm{e}^{-ic\eta}\,\mathrm{d}\eta \tag{10.9.1}$$

we take the displacement function to be given by

$$\begin{aligned}\Phi = \int_{-\infty}^{\infty} \sum_{i=1}^{2} \{&\mathrm{e}^{-\bar{p}_i(c)(\theta-\theta_1)}[A_i(c)\cos\bar{q}_i(c)(\theta-\theta_1) \\ &+ B_i(c)\sin\bar{q}_i(c)(\theta-\theta_1)] + \mathrm{e}^{-\bar{p}_i(c)(\theta_2-\theta)}[C_i(c)\cos\bar{q}_i(c)(\theta_2-\theta) \\ &+ D_i(c)\sin\bar{q}_i(c)(\theta_2-\theta)]\}\,\mathrm{e}^{icx}\,\mathrm{d}c.\end{aligned} \tag{10.9.2}$$

The coefficients $A_i(c)$, $B_i(c)$, $C_i(c)$, and $D_i(c)$ are now determined from the boundary conditions on the straight edges by substituting Eq. (10.9.2) into the expressions for the pertinent edge quantities and setting them equal to the desired values also expressed in Fourier integral form. Since each equation must be valid for any value of x, the multipliers of e^{icx} on both sides of the equation must be equal.

For example, consider an infinite strip subjected to equal but opposite bending moments between the limits $\frac{1}{2}a > x > -\frac{1}{2}a$ along the edges $\theta = \pm\frac{1}{2}\theta_0$ (Fig. 10.17). As for the finite strip of the preceding section, we take $A_i(c)$ equal to $B_i(c)$ and $C_i(c)$ equal to $D_i(c)$ for symmetry about the axis = 0. The edge conditions are

$$N_\theta = \bar{N}_{x\theta} - \tfrac{1}{2}\bar{M}_{x\theta}/R = \bar{Q}_\theta = 0, \tag{10.9.3a}$$

† See, for example, S. Timoshenko and S. Woinowsky-Krieger, *Theory of Plates and Shells*, 2nd Edition, McGraw-Hill Book Co., 1959, pp. 181–221.

‡ See, for example, C. Lanczos, *Applied Analysis*, Prentice Hall, Inc., 1956, pp. 248–255.

$$M_\theta = \begin{cases} M_0 & -\tfrac{1}{2}a < x < \tfrac{1}{2}a \\ 0 & \tfrac{1}{2}a < x < \infty \\ 0 & -\infty < x < -\tfrac{1}{2}a. \end{cases} \qquad (10.9.3b)$$

The last edge condition can be expressed as the Fourier integral

$$M = \frac{M_0}{2\pi}\int_{-\infty}^{\infty} \mathrm{e}^{icx}\int_{-1/2a}^{1/2a} \mathrm{e}^{-ic\eta}\,\mathrm{d}\eta = \frac{M_0}{\pi}\int_{-\infty}^{\infty}\frac{1}{c}\,\mathrm{e}^{icx}\sin\tfrac{1}{2}ac\,\mathrm{d}c.$$

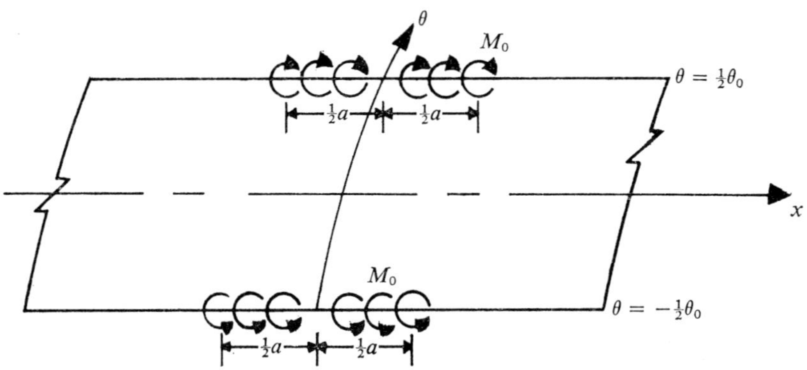

Fig. 10.17. Infinite cylindrical strip subjected to edge moments

The equations for the determination of the coefficients $A_i(c)$ and $B_i(c)$ are identical to Eqs. (10.8.12) with the right side of Eq. (10.8.12c) replaced by $-[M_0/\pi kKc)]\sin\tfrac{1}{2}ac$. The final form of the solution is then obtained by performing the integrations of the various integrals with respect to the variable c.

10.10 Stress concentration around holes

The stress concentration due to holes in shell structures can be as important or even more important than stress concentrations due to edge effects. For example, if a cylindrical shell with a small circular hole is loaded by axial tension which is uniform around the circumference, we would expect the maximum stress to approach the flat plate value of 3 times the stress without the hole† when the cylinder radius is very large compared to the thickness and

† S. Timoshenko and J. N. Goodier, *Theory of Elasticity*, 3rd Edition, McGraw-Hill Book Co., Inc., 1969, pp. 90–92.

the hole radius is very small compared to the cylinder radius. We would like, however, to determine the effect of finite cylinder curvature and finite hole radius on the stress concentration factor. Problems of this sort have not been solved by any exact solution of either Flügge's complete or approximate equations. We can obtain some results, however, if we use the Donnell form of the bending equations for cylindrical shells and if we make certain assumptions as to the shape of the hole and the rapidity of decay of the stress distribution in the vicinity of the hole. In particular we assume that the cylinder is cut in such a manner that the developed shape of the hole is circular (Fig. 10.18). We also assume that the hole is so small compared to the

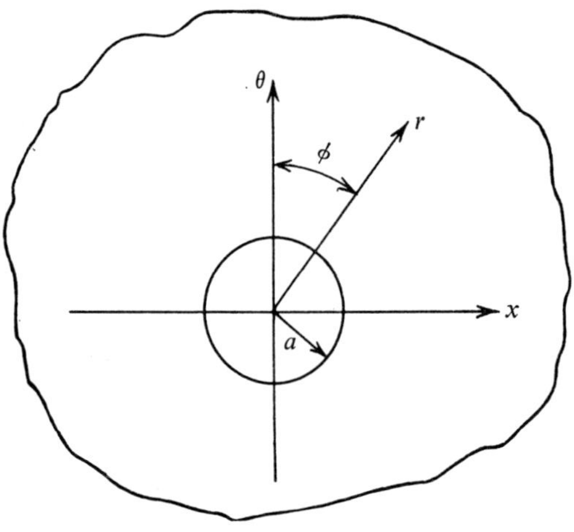

Fig. 10.18. Circular hole in developed cylinder surface

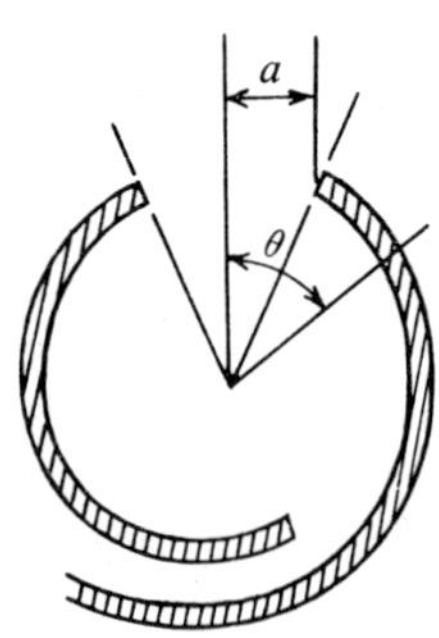

Fig. 10.19. Cross-section of hypothetical spiral shell

circumference and that stresses decay so rapidly that we can consider the cylinder to be unbounded in the circumferential direction rather than closing upon itself, as well as being unbounded in the longitudinal direction. The model chosen is thus an infinite slightly curved spiral shell (Fig. 10.19).

For convenience we use the complex form of Donnell's equations which can be obtained from Eqs. (3.3.19) as

$$\left(\nabla^4 - 4i\mu^2 \frac{\partial^2}{\partial \xi^2}\right)\chi = 0, \tag{10.10.1a}$$

where

$$\chi = w + i\frac{[12(1-\nu^2)]^{1/2}}{Eh^2}\Phi, \tag{10.10.1b}$$

$$\mu = \frac{1}{2}\left(\frac{1-\nu^2}{k}\right)^{1/4}. \tag{10.10.1c}$$

Eq. (10.10.1a) may be factored as

$$\left[\nabla^2 + 2(i)^{1/2}\mu \frac{\partial}{\partial \xi}\right]\left[\nabla^2 - 2(i)^{1/2}\mu \frac{\partial}{\partial \xi}\right]\chi = 0. \tag{10.10.2a}$$

Since the operators are commutative, χ may be taken as the sum of the solutions of the equations

$$\left[\nabla^2 + 2(i)^{1/2}\mu \frac{\partial}{\partial \xi}\right]\chi_1 = 0, \tag{10.10.2b}$$

$$\left[\nabla^2 - 2(i)^{1/2}\mu \frac{\partial}{\partial \xi}\right]\chi_2 = 0. \tag{10.10.2c}$$

When we introduce the transformations

$$\chi_1 = e^{-(i)^{1/2}\mu\xi}\,\bar{\chi}_1, \tag{10.10.3a}$$

$$\chi_2 = e^{(i)^{1/2}\mu\xi}\,\bar{\chi}_2, \tag{10.10.3b}$$

we find that $\bar{\chi}_1$ and $\bar{\chi}_2$ are determined by the single equation

$$(\nabla^2 - i\mu^2)\bar{\chi} = 0. \tag{10.10.4}$$

Finally, when we introduce polar coordinates for the developed surface as (Fig. 10.18)

$$\xi = \eta \sin\phi, \tag{10.10.5a}$$

$$\theta = \eta \cos\phi, \tag{10.10.5b}$$

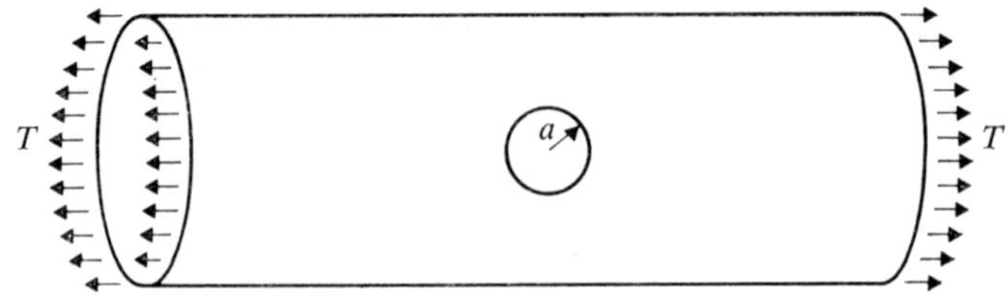

Fig. 10.20. Circular cylinder with hole subjected to uniform axial tension at infinity

Eq. (10.10.4) becomes

$$\left(\frac{\partial^2}{\partial \eta^2} + \frac{1}{\eta}\frac{\partial}{\partial \eta} + \frac{1}{\eta^2}\frac{\partial^2}{\partial \phi^2} - i\mu^2\right)\bar{\chi} = 0. \tag{10.10.5c}$$

We note that the boundary of the hole corresponds to a constant value of η.

We seek solutions of Eq. (10.10.5c) of the form

$$\bar{\chi} = \bar{\psi}_n(\eta)\, e^{ni\phi}. \tag{10.10.6a}$$

Then $\bar{\psi}_n(\eta)$ is determined by the ordinary differential equation

$$\frac{d^2\bar{\psi}_n}{d\eta^2} + \frac{1}{\eta}\frac{d\bar{\psi}_n}{d\eta} - \left(i\mu^2 + \frac{n^2}{\eta^2}\right)\bar{\psi}_n = 0, \tag{10.10.6b}$$

which can be recognized as Bessel's equation of order n. For convenience we write the solution of Eq. (10.10.6b) in terms of Hankel functions† as

$$\bar{\psi}_n = \tilde{A}_n H_n^{(2)}[\mu\eta(-i)^{1/2}] + \tilde{B}_n H_n^{(1)}[\mu\eta(-i)^{1/2}], \tag{10.10.6c}$$

where $(-i)^{1/2}$ is understood to be given by $e^{-1/4i\pi}$. Since Eq. (10.10.6c) is the solution for both $\bar{\chi}_1$ and $\bar{\chi}_2$ for all values of n, we may write the complete solution for χ as

$$\chi = \sum_{n=-\infty}^{\infty} e^{ni\phi}[H_n^{(2)}[\mu\eta(-i)^{1/2}][A_n\, e^{-(i)^{1/2}\mu\xi} + B_n\, e^{(i)^{1/2}\mu\xi}]$$
$$+ H_n^{(1)}[\mu\eta(-i)^{1/2}][C_n\, e^{-(i)^{1/2}\mu\xi} + D_n\, e^{(i)^{1/2}\mu\xi}]\} \tag{10.10.7}$$

For problems in which the loads applied to the edge of the hole are self-equilibrating the displacements and stresses decay rapidly away from the hole. Since the asymptotic behavior of the Hankel functions for large values of the argument is given by

$$H_n^{(2)}(z) = \left(\frac{2}{\pi z}\right)^{1/2} e^{-i[z-\frac{1}{2}\pi(n+\frac{1}{2})]}, \tag{10.10.8a}$$

$$H_n^{(1)}(z) = \left(\frac{2}{\pi z}\right)^{1/2} e^{i[z-\frac{1}{2}\pi(n+\frac{1}{2})]}, \tag{10.10.8b}$$

---

† See M. Abramowitz and I. A. Stegun, *Handbook of Mathematical Functions*, NBS Applied Math. Series No. 55, 1964, pp. 358–369.

reference to Eq. (10.10.7) indicates that the terms involving $H_n^{(1)}[\mu\eta(-i)^{1/2}]$ should be deleted in order for the displacements and stresses to have the proper variation. It can be shown† that the displacements are single-valued and that the stress-resultants are self-equilibrating‡ when the coefficients satisfy the relationship

$$\sum_{n=-\infty}^{\infty} [A_n - (-1)^n B_n] = 0. \tag{10.10.9}$$

For convenience later on we introduce the relationship§

$$e^{iz\sin\phi} = \sum_{k=-\infty}^{\infty} e^{ki\phi} J_k(z) \tag{10.10.10a}$$

so that

$$\begin{aligned} e^{\pm\mu\xi(i)^{1/2}} &= e^{i[\pm\mu\xi(-i)^{1/2}]\sin\phi} \\ &= \sum_{k=-\infty}^{\infty} e^{ki\phi} J_k[\pm\mu\eta(-i)^{1/2}]. \end{aligned} \tag{10.10.10b}$$

Then with the use of the relation

$$J_k(-z) = (-1)^k J_k(z), \tag{10.10.10c}$$

the final form of the solution is given by

$$\begin{aligned} \chi = \sum_{n=-\infty}^{\infty} \sum_{k=-\infty}^{\infty} \Big\{ &(B_n + A_n) J_{2k}[\mu\eta(-i)^{1/2}]\, e^{(n+2k)i\phi} \\ &+ (B_n - A_n) J_{2k+1}[\mu\eta(-i)^{1/2}]\, e^{(n+2k+1)i\phi} \Big\} H_n^{(2)}[\mu\eta(-i)^{1/2}]. \end{aligned} \tag{10.10.11}$$

Simplifications can be made if the displacements and stresses are doubly symmetric or doubly skew symmetric. In the former case χ must contain only terms multiplied by cos $2p\phi$. This can be achieved with

$$B_n + A_n = \begin{cases} 0 & n \text{ odd} \\ \tfrac{1}{2}C_n & n \text{ even} \\ C_0 & n = 0, \end{cases} \tag{10.10.12a}$$

$$B_n - A_n = \begin{cases} \tfrac{1}{2}C_n & n \text{ odd} \\ 0 & n \text{ even}, \end{cases} \tag{10.10.12b}$$

† J. G. Lekkerkerker, 'On the Stress Distributions in Cylindrical Shells Weakened by a Circular Hole,' Doctoral Dissertation, Tech. Univ. Delft, 1965.

‡ Singular solutions of the equations for local concentrated loads and dislocations which involve multi-valued displacements are treated by J. L. Sanders, Jr., and J. G. Simmonds, 'Concentrated Forces on Shallow Cylindrical Shells,' *J. Appl. Mech.*, vol. 37, no. 2, June 1970, pp. 367–373. This paper contains an extensive bibliography of related works.

§ See M. Abramowitz and I. A. Stegun, *Handbook of Mathematical Functions*, NBS Applied Math. Series No. 55, 1964, p. 361.

which satisfies Eq. (10.10.9), and with

$$C_n = C_{-n}. \tag{10.10.12c}$$

Then, with the use of the relations

$$J_{-k}(z) = (-1)^k J_k(z), \tag{10.10.13a}$$

$$H^{(2)}_{-k}(z) = (-1)^k H^{(2)}_k(z), \tag{10.10.13b}$$

we may then write the solution for doubly symmetric loading as

$$\chi = \sum_{n=0}^{\infty} \sum_{p=0}^{\infty} C_n F_{pn}(\mu\eta) \cos 2p\phi, \tag{10.10.14a}$$

where

$$F_{0n}(\mu\eta) = J_{-n}[\mu\eta(-i)^{1/2}] H^{(2)}_n[\mu\eta(-i)^{1/2}], \tag{10.10.14b}$$

$$F_{pn}(\mu\eta) = \{J_{2p-n}[\mu\eta(-i)^{1/2}] + J_{-2p-n}[\mu\eta(-i)^{1/2}]\} H^{(2)}_n[\mu\eta(-i)] \; (p > 0). \tag{10.10.14c}$$

In the latter case only terms multiplied by sin $2p\phi$ must remain. Eqs. (10.10.12a) and (10.10.12b) are still valid but Eq. (10.10.12c) is replaced by

$$C_n = -C_{-n}, \tag{10.10.15}$$

and the solution for doubly skew symmetric loading becomes

$$\chi = \sum_{n=1}^{\infty} \sum_{p=1}^{\infty} C_n G_{pn}(\mu\eta) \sin 2p\phi, \tag{10.10.16a}$$

where

$$G_{pn}(\mu\eta) = \{J_{2p-n}[\mu\eta(-i)^{1/2}] - J_{-2p-n}[\mu\eta(-i)^{1/2}\} H^{(2)}_n[\mu\eta(-i)^{1/2}]. \tag{10.10.16b}$$

In the Donnell theory of cylindrical shells, equivalent stress-resultants are related to the stress function Φ and to the radial displacement w by

$$M_x = -\frac{D}{R^2}\left(\frac{\partial^2 w}{\partial \xi^2} + \nu \frac{\partial^2 w}{\partial \theta^2}\right), \tag{10.10.17a}$$

$$M_\theta = -\frac{D}{R^2}\left(\frac{\partial^2 w}{\partial \theta^2} + \nu \frac{\partial^2 w}{\partial \xi^2}\right), \tag{10.10.17b}$$

$$\overline{M}_{x\theta} = -\frac{D}{R^2}(1-\nu)\frac{\partial^2 w}{\partial \xi\, \partial \theta}, \tag{10.10.17c}$$

$$N_x = \frac{1}{R^2}\frac{\partial^2 \Phi}{\partial \theta^2}, \tag{10.10.17d}$$

$$N_\theta = \frac{1}{R^2}\frac{\partial^2 \Phi}{\partial \xi^2}, \tag{10.10.17e}$$

$$\bar{N}_{x\theta} = -\frac{1}{R^2}\frac{\partial^2\Phi}{\partial\xi\,\partial\theta}, \tag{10.10.17f}$$

$$\bar{Q}_x = -\frac{D}{R^3}\left[\frac{\partial^3 w}{\partial\xi^3} + (2-\nu)\frac{\partial^3 w}{\partial\xi\,\partial\theta^2}\right], \tag{10.10.17g}$$

$$\bar{Q}_\theta = -\frac{D}{R^3}\left[\frac{\partial^3 w}{\partial\theta^3} + (2-\nu)\frac{\partial^3 w}{\partial\theta\,\partial\xi^2}\right]. \tag{10.10.17h}$$

For the directions associated with the transformed coordinate system, the pertinent equivalent stress-resultants can be obtained from Eqs. (A.15), bearing in mind the implications of Eqs. (3.3.1), as

$$\begin{aligned} M_\eta &= M_x \sin^2\phi + M_\theta \cos^2\phi + \bar{M}_{x\theta}\sin 2\phi \\ &= -\frac{D}{R^2}\left[\frac{\partial^2 w}{\partial\eta^2} + \nu\left(\frac{1}{\eta}\frac{\partial w}{\partial\eta} + \frac{1}{\eta^2}\frac{\partial^2 w}{\partial\phi^2}\right)\right], \end{aligned} \tag{10.10.18a}$$

$$\begin{aligned} M_\phi &= M_x \cos^2\phi + M_\theta \sin^2\phi - \bar{M}_{x\theta}\sin 2\phi \\ &= -\frac{D}{R^2}\left(\frac{1}{\eta}\frac{\partial w}{\partial\eta} + \frac{1}{\eta^2}\frac{\partial^2 w}{\partial\phi^2} + \nu\frac{\partial^2 w}{\partial\eta^2}\right), \end{aligned} \tag{10.10.18b}$$

$$\begin{aligned} \bar{M}_{\eta\phi} &= \tfrac{1}{2}(M_\theta - M_x)\sin 2\phi - \bar{M}_{x\theta}\cos 2\phi \\ &= \frac{D}{R^2}\frac{\partial^2}{\partial\eta\,\partial\phi}\left(\frac{w}{\eta}\right), \end{aligned} \tag{10.10.18c}$$

$$\begin{aligned} N_\eta &= N_x \sin^2\phi + N_\theta \cos^2\phi + \bar{N}_{x\theta}\sin 2\phi \\ &= \frac{1}{R^2}\left(\frac{1}{\eta}\frac{\partial\Phi}{\partial\eta} + \frac{1}{\eta^2}\frac{\partial^2\Phi}{\partial\phi^2}\right), \end{aligned} \tag{10.10.18d}$$

$$\begin{aligned} N_\phi &= N_x \cos^2\phi + N_\theta \sin^2\phi - \bar{N}_{x\theta}\sin 2\phi \\ &= \frac{1}{R^2}\frac{\partial^2\Phi}{\partial\eta^2}, \end{aligned} \tag{10.10.18e}$$

$$\begin{aligned} \bar{N}_{\eta\phi} &= \tfrac{1}{2}(N_\theta - N_x)\sin 2\phi - \bar{N}_{x\theta}\cos 2\phi \\ &= \frac{1}{R^2}\frac{\partial^2}{\partial\eta\,\partial\phi}\left(\frac{\Phi}{\eta}\right), \end{aligned} \tag{10.10.18f}$$

$$\begin{aligned} \bar{Q}_\eta &= \left(\bar{Q}_x - \frac{1}{R}\frac{\partial\bar{M}_{x\theta}}{\partial\theta}\right)\sin\phi + \left(\bar{Q}_\theta - \frac{1}{R}\frac{\partial\bar{M}_{x\theta}}{\partial\xi}\right)\cos\phi \\ &\quad + \frac{1}{R\eta}\frac{\partial}{\partial\phi}\left[\bar{M}_{x\theta}\cos 2\phi + \tfrac{1}{2}(M_x - M_\theta)\sin 2\phi\right] \\ &= -\frac{D}{R^3}\left[\frac{\partial}{\partial\eta}(\nabla^2 w) + \frac{1-\nu}{\eta}\frac{\partial^3}{\partial\eta\,\partial\phi^2}\left(\frac{w}{\eta}\right)\right], \end{aligned} \tag{10.10.18g}$$

$$\bar{Q}_\phi = \left(\bar{Q}_x - \frac{1}{R}\frac{\partial \bar{M}_{x\theta}}{\partial \theta}\right)\cos\phi - \left(\bar{Q}_\theta - \frac{1}{R}\frac{\partial \bar{M}_{x\theta}}{\partial \xi}\right)\sin\phi$$

$$+ \frac{1}{R}\frac{\partial}{\partial \eta}\left[\bar{M}_{x\theta}\cos 2\phi + \tfrac{1}{2}(M_x - M_\theta)\sin 2\phi\right]$$

$$= -\frac{D}{R^3}\left[\frac{1}{\eta}\frac{\partial}{\partial \phi}(\nabla^2 w) + (1-\nu)\frac{\partial^3}{\partial \eta^2\,\partial \phi}\left(\frac{w}{\eta}\right)\right]. \tag{10.10.18h}$$

The above relations, together with Eqs. (10.10.1b) and (10.10.11) can be used to obtain solutions for various stress concentration problems.

10.10.1 *Effect of a hole on a circular cylinder under uniform tension*

To illustrate the calculations involved in the preceding analysis, let us consider the case of a cylindrical shell subjected to uniform axial tension far from the hole (Fig. 10.20). For the unweakened cylinder the stress resultants are given by

$$N_x = T, \tag{10.10.19a}$$

$$N_\theta = \bar{N}_{x\theta} = M_x = M_\theta = \bar{M}_{x\theta} = \bar{Q}_x = \bar{Q}_\theta = 0, \tag{10.10.19b}$$

or, for the polar coordinate system,

$$N_\eta = \tfrac{1}{2}T(1 - \cos 2\phi), \tag{10.10.20a}$$

$$N_\phi = \tfrac{1}{2}T(1 + \cos 2\phi), \tag{10.10.20b}$$

$$\bar{N}_{\eta\phi} = -\tfrac{1}{2}T\sin 2\phi, \tag{10.10.20c}$$

$$M_\eta = M_\phi = \bar{M}_{\eta\phi} = \bar{Q}_\eta = \bar{Q}_\phi = 0. \tag{10.10.20d}$$

For the cut cylinder we must nullify the stresses at the edge of the hole. Thus the boundary conditions to be satisfied by the additional stress state are

$$N_\eta = -\tfrac{1}{2}T(1 - \cos 2\phi), \tag{10.10.21a}$$

$$\bar{N}_{\eta\phi} = \tfrac{1}{2}T\sin 2\phi, \tag{10.10.21b}$$

$$M_\eta = \bar{Q}_\eta = 0, \tag{10.10.21c}$$

at the edge of the hole given by

$$\eta = \eta_0 = a/R. \tag{10.10.21d}$$

Since the loading is doubly symmetric we use Eq. (10.10.14). Then the boundary conditions become

$$N_\eta = \frac{\tfrac{1}{4}Eh}{R}\,\mathrm{Im}\left\{\sum_{n=0}^{\infty}\sum_{p=0}^{\infty}(\bar{A}_n + i\bar{B}_n)\left[\frac{1}{\mu\eta}F'_{pn} - \frac{4p^2}{(\mu\eta)^2}F_{pn}\right]_{\eta=\eta_0}\cos 2p\phi\right\}$$

$$= -\tfrac{1}{2}T(1 - \cos 2\phi), \tag{10.10.22a}$$

$$\bar{N}_{\eta\phi} = -\frac{\frac{1}{4}Eh}{R}\,\mathrm{Im}\sum_{n=0}^{\infty}\sum_{p=1}^{\infty}(\bar{A}_n + i\bar{B}_n)2p\left[\frac{1}{\mu\eta}F'_{pn} - \frac{1}{(\mu\eta)^2}F_{pn}\right]_{\eta=\eta_0}\sin 2p\phi$$

$$= \tfrac{1}{2}T\sin 2\phi, \qquad (10.10.22b)$$

$$M_\eta = -\frac{1}{4}\frac{Eh^2}{[12(1-\nu^2)]^{1/2}R}\,\mathrm{Re}\sum_{n=0}^{\infty}\sum_{p=0}^{\infty}(\bar{A}_n + i\bar{B}_n)$$

$$\times\left\{F''_{pn} + \nu\left[\frac{1}{\mu\eta}F'_{pn} - \frac{4p^2}{(\mu\eta)^2}F_{pn}\right]\right\}_{\eta=\eta_0}\cos 2p = 0, \qquad (10.10.22c)$$

$$\bar{Q}_\eta = -\frac{1}{8}\frac{E(h/R)^{3/2}}{[12(1-\nu^2)]^{1/4}}\,\mathrm{Re}\sum_{n=0}^{\infty}\sum_{p=0}^{\infty}(\bar{A}_n + i\bar{B}_n)$$

$$\times\left[F'''_{pn} + \frac{1}{\mu\eta}F''_{pn} - \frac{1+4(2-\nu)p^2}{(\mu\eta)^2}F'_{pn}\right.$$

$$\left. + \frac{4(3-\nu)p^2}{(\mu\eta)^3}F_{pn}\right]_{\eta=\eta_0}\cos 2p = 0, \qquad (10.10.22d)$$

where the primes denote differentiation with respect to $\mu\eta$ and the complex coefficient C_n has been expressed in terms of its real and imaginary parts as $\bar{A}_n - i\bar{B}_n$. Since these equations must be satisfied for all values of ϕ, the equations for the determination of $\bar{A}_n$ and $\bar{B}_n$ can be written in real form as

$$\sum_{n=0}^{\infty}\left[\frac{1}{\mu\eta}(\bar{A}_n\,\mathrm{Im}\,F'_{pn} + B_n\,\mathrm{Re}\,F'_{pn})\right]_{\eta=\eta_0} = \begin{cases} -\,2TR/(Eh) & p = 0, 1 \\ 0 & p \geq 2, \end{cases} \qquad (10.10.23a)$$

$$\sum_{n=0}^{\infty}\left[\frac{1}{(\mu\eta)^2}(\bar{A}_n\,\mathrm{Im}\,F_{pn} + \bar{B}_n\,\mathrm{Re}\,F_{pn})\right]_{\eta=\eta_0} = \begin{cases} -\,TR/(Eh) & p = 1 \\ 0 & p \geq 2, \end{cases} \qquad (10.10.23b)$$

$$\sum_{n=0}^{\infty}\left\{\bar{A}_n\,\mathrm{Re}\left[F''_{pn} + \frac{\nu}{\mu\eta}F'_{pn} - \frac{4\nu p^2}{(\mu\eta)^2}F_{pn}\right]\right.$$

$$\left. - \bar{B}_n\,\mathrm{Im}\left[F''_{pn} + \frac{\nu}{\mu\eta}F'_{pn} - \frac{4\nu p^2}{(\mu\eta)^2}F_{pn}\right]\right\}_{\eta=\eta_0} = 0 \qquad p \geq 0, \qquad (10.10.23c)$$

$$\sum_{n=0}^{\infty}\left\{\bar{A}_n\,\mathrm{Re}\left[F'''_{pn} + \frac{1}{\mu\eta}F''_{pn} - \frac{1+4(2-\nu)p^2}{(\mu\eta)^2}F'_{pn} + \frac{4(3-\nu)p^2}{(\mu\eta)^3}F_{pn}\right]\right.$$

$$- \bar{B}_n\,\mathrm{Im}\left[F'''_{pn} + \frac{1}{\mu\eta}F''_{pn} - \frac{1+4(2-\nu)p^2}{(\mu\eta)^2}F'_{pn}\right.$$

$$\left.\left. + \frac{4(3-\nu)p^2}{(\mu\eta)^3}F_{pn}\right]\right\}_{\eta=\eta_0} = 0 \qquad p \geq 0. \qquad (10.10.23d)$$

These equations cannot be solved exactly so that for practical calculation the infinite number of equations in an infinite number of unknowns must be truncated. If we truncate p at P there are $4P - 3$ equations. From these we must delete Eq. (10.10.23d) for $p = 0$ since the fact that the stress states are self-equilibrating implies that this equation is automatically satisfied. Thus there are $4P - 2$ equations to be satisfied. To achieve the same number of unknowns we truncate n at $2P$ for which we have $2P - 1$ unknown values of A_n and the same number of unknown values of B_n. The procedure is then to choose a truncation value P and to calculate values of the unknowns from the set of truncated equations. The value of P is increased and the new sets of equations solved until the effect of truncation is reduced to the desired small amount. As an example, values of the coefficients for two different truncation numbers ($P = 4, 5$) for the case when $\mu\eta_0$ is equal to 4 are shown in Table 10.7. The corresponding coefficients of $\cos 2p\phi$ in the Fourier series for the circumferential force N_ϕ and the bending moment

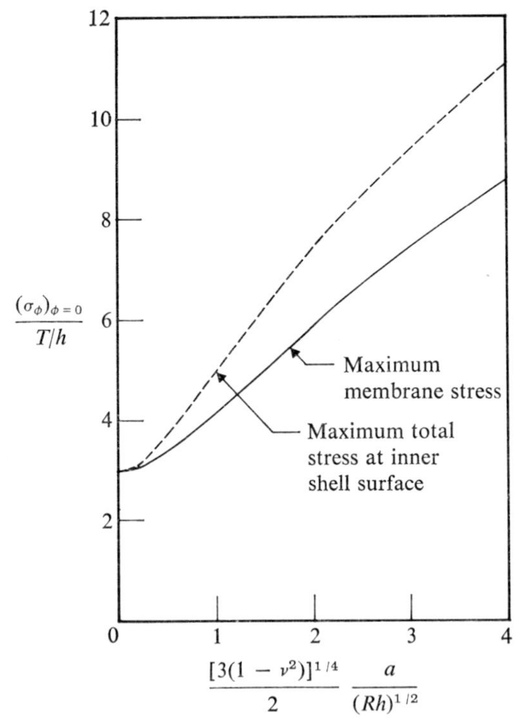

Fig. 10.21. Maximum stress in the cylinder

M_ϕ at the edge of the hole are shown in Table 10.8. We see that for all practical purposes we may terminate the series with $p = 4$ and $n = 8$.

The maximum stresses in the cylinder, which occur at the inner surface at the points $\phi = 0°$ and $180°$ at the edge of the hole, are shown in Fig. 10.21 as a function of the hole size parameter $\mu\eta_0/2^{1/2}$ or $\frac{1}{2}[3(1 - \nu^2)]^{1/4}a/(Rh)^{1/2}$. We see that the stress concentration factor increases considerably above the flat plate value of 3 when the radius of the hole is of the order of magnitude of the attenuation length of axisymmetric effects. The maximum stress due to the stress resultant N_ϕ is also shown and indicates that most of the increase of the stress concentration factor is due to an increase in the membrane force. Calculations indicate that the stress concentration† due to bending is never

Table 10.7 Values of $\bar{A}_n$ and $\bar{B}_n$ for two different truncation numbers ($\mu\eta_0 = 4$, $\nu = 0.3$)

	$2TR\bar{A}_n/(Eh)$		$2TR\bar{B}_n/(Eh)$	
n	$P = 4$	$P = 5$	$P = 4$	$P = 5$
0	−131.064	−131.063	−66.649	−66.649
1	−22.401	−22.399	−31.557	−31.566
2	−33.932	−33.959	−8.585	−8.618
3	84.999	84.875	0.786	0.731
4	6.762	6.486	64.631	64.609
5	−21.134	−21.502	3.819	3.865
6	−0.458	−0.767	−4.482	−4.408
7	0.799	0.638	−0.114	−0.068
8	0.048	−0.000 46	0.060	0.070 8
9	—	−0.006 86	—	−0.000 11
10	—	−0.000 17	—	−0.000 37

Table 10.8 Coefficients of cos $2p$ in expressions for circumferential force and bending moment resultants at the edge of the hole ($\mu\eta_0 = 4$, $\nu = 0.3$)

	N_ϕ/T		M_ϕ/T	
p	$P = 4$	$P = 5$	$P = 4$	$P = 5$
0	0.8372	0.8372	0.3859	0.3859
1	2.7955	2.7955	−2.0068	−2.0067
2	2.2748	2.2748	−0.2768	−0.2769
3	0.2334	0.2332	−0.0564	−0.0557
4	0.0021	0.0037	0.0084	0.0023
5	—	0.0001	—	−0.0002

† P. van Dyke, 'Stresses about a Circular Hole in a Cylindrical Shell,' *AIAA J.*, vol. 3, no. 9, September 1965, pp. 1733–1742.

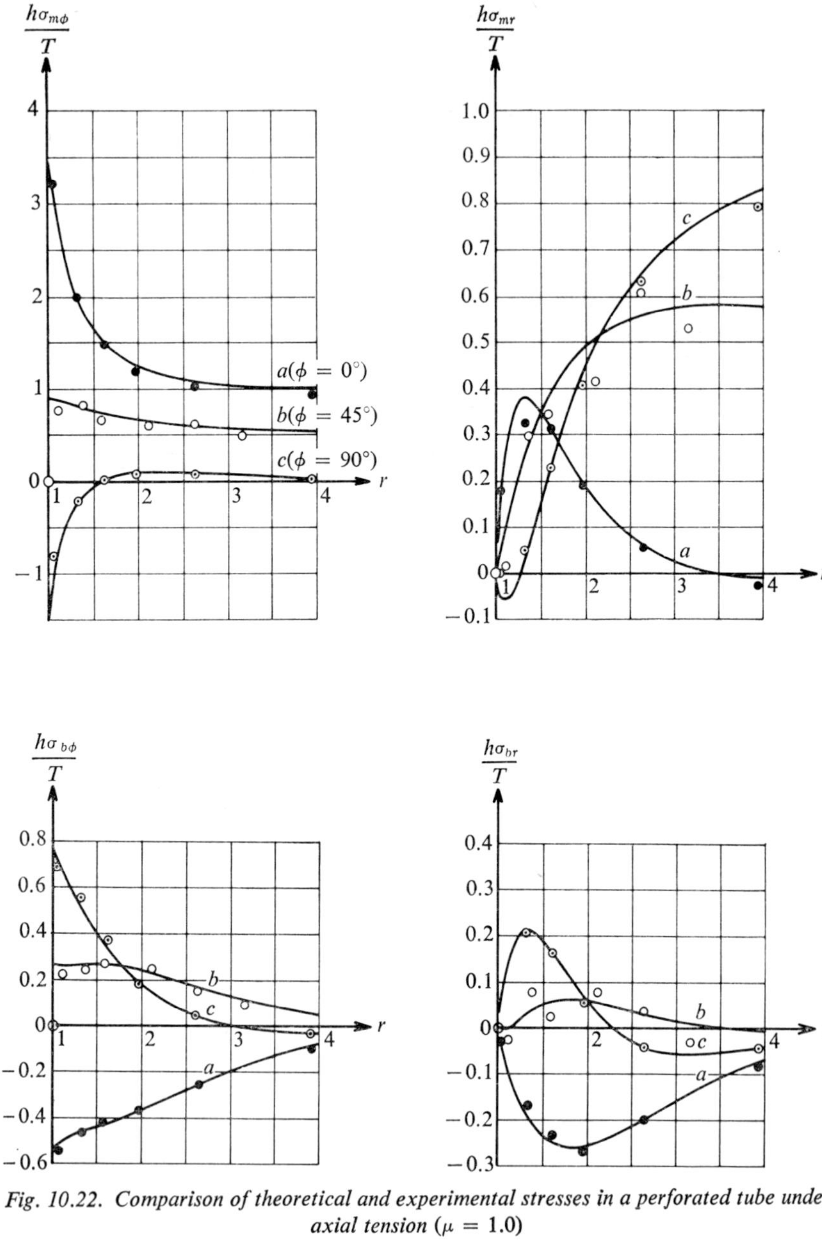

Fig. 10.22. Comparison of theoretical and experimental stresses in a perforated tube under axial tension ($\mu = 1.0$)

more than about 28 percent of that due to membrane forces and decreases as the shell becomes thinner. For large values of the hole size parameter, i.e., for thin shells, the average stress concentration factor has been shown to be approximated by the relationship

$$\frac{N_\phi}{T} \approx 1 + 3{\cdot}05\left\{\frac{[3(1-\nu^2)]^{1/4}}{2}\frac{a}{(Rh)^{1/2}}\right\}^{2/3} \tag{10.10.24}$$

Similar results have been obtained for cylinders under torsion and under pressure loading.†‡ Available experimental results‡ are in quite good agreement with theoretical results, as indicated by the comparison shown in Fig. 10.22 for a tube in axial tension. The analysis given above has also been combined with the analysis of an edge-loaded cylindrical shell (Section 10.3) for the investigation of the behavior of intersecting cylindrical shells.§ Still other results have been obtained for the effect of longitudinal cracks or slits as well as for other hole shapes.||

---

† P. van Dyke, op. cit.

‡ J. G. Lekkerkerker, op. cit.

§ See J. W. Hansberry and N. Jones, 'A Theoretical Study of the Elastic Behavior of Two Normally Intersecting Cylindrical Shells,' Paper No. 68-WA/PVP-1, presented at ASME Winter Annual Meeting, New York, 1968, for some results and for a bibliography of related papers.

|| L. G. Copley, 'A Longitudinal Crack in a Cylindrical Shell,' Ph.D. Dissertation, Harvard Univ., 1965; L. G. Copley and J. L. Sanders, Jr., 'A Longitudinal Crack in a Cylindrical Shell Under Bending and Stretching,' Harvard Univ., Div. Eng. and Appl. Phys., Report SM-19, September 1967; E. S. Folias, 'An Axial Crack in a Pressurized Cylindrical Shell,' *Int. J. Fract. Mech.*, vol. 1, 1965, pp. 104–113; E. S. Folias, 'A Circumferential Crack in a Pressurized Cylindrical Shell,' *Int. J. Fract. Mech.*, vol. 3, 1967, pp. 1–11, M. V. V. Murthy, 'Stresses Around an Elliptic Hole in a Cylindrical Shell,' *J. Appl. Mech.*, vol. 36, no. 1, March 1969, pp. 39–46; M. V. V. Murthy and M. N. Bapu Rao, 'Stresses in a Cylindrical Shell by an Elliptical Hole with Major Axis Perpendicular to Shell Axis,' *J. Appl. Mech.*, vol. 37, no. 4, June 1970, pp. 539–540; O. Tingleff, 'Stress Concentration in a Cylindrical Shell with an Elliptical Cutout,' *AIAA J.*, vol. 9, no. 11, November 1971, pp. 2289–2291; M. E. Duncan-Fama and J. L. Sanders, Jr., "A Circumferential Crack in a Cylindrical Shell Under Tension,' *Int. J. Fract. Mech.*, vol. 8, no. 1, March 1972, pp. 15–20; L. G. Copley and J. L. Sanders, Jr., 'A Longitudinal Crack in a Cylindrical Shell Under Internal Pressure,' *Int. J. Fract. Mech.*, vol. 5, no. 2, June 1969, pp. 117–131. See also the discussion and extensive bibliography in G. N. Savin, 'Stress Distribution Around Holes,' *NASA TT* F-607, November 1970, pp. 791–880.

Supplementary references

[1] Fischer, F. J., 'Stress Diffusion from Axially Loaded Stiffeners into Cylindrical Shells,' *Int. J. Solids Struct.*, vol. 4, pp. 1181–1201.

[2] Kuhn, P., *Stresses in Aircraft and Shell Structures*, McGraw-Hill Book Co., 1956.

[3] Duncan, M. E., and Sanders, J. L., Jr., 'The Effect of a Circumferential Stiffener on the Stress in a Pressurized Cylindrical Shell with a Longitudinal Crack,' *Int. J. Fract. Mech.*, vol. 5, no. 2, June 1969, pp. 133–155.

[4] Biezeno, C. B., and Koch, J. J., 'The Effective Width of Cylinders, Periodically Stiffened by Circular Rings,' *Koninklijke Nederlandsche Akademie van Wetenschappen, Proc.*, vol. XLVIII, 1945, pp. 147–165.

[5] Melform, R. F., Patel, P. D., and Berman, I., 'Shell Theory Solution for Asymmetric Balanced Radial Loads on Long Cylinders,' *J. Eng. Power, Trans. ASME, Ser. A*, April 1968, pp. 177–185.

[6] Patel, P. D., Melform, R. F., and Berman, I., 'Solutions for Distributed Loads on Long Cylinders,' ASME Winter Annual Meeting, New York, N.Y., December 1968, Paper no. 68-WA/PVP-7.

[7] Allentuch, A., Brody, K., and Kempner, J., 'Long Cylindrical Shell Under Concentrated Loads Applied to a Central Reinforcing Ring,' *AIAA J.*, vol. 5, no. 10, October 1967, pp. 1863–1870.

[8] Allentuch, A., Brody, K., Golub, E., and Kempner, J., 'Tests of Cylindrical Shell Under Concentrated Loads Applied to a Reinforcing Ring,' *AIAA J.*, vol. 9, no. 10, October 1971, pp. 2089–2091.

[9] Duncan-Fama, M. E., and Sanders, J. L., Jr., 'The Effect of a Circumferential Stiffener on the Stress in a Pressurized Cylindrical Shell with a Longitudinal Crack,' *Int. J. Fract. Mech.*, vol. 5, no. 2, June 1969, pp. 133–155.

11

Results for other shells

11.1 'Wind' loading of arbitrary shells of revolution

The sphere and the circular cylinder are the only two shell shapes for which exact asymmetric deformation solutions in terms of defined and sometimes tabulated functions can be obtained. When we study asymmetric deformations of other shells of revolution, let alone shells of arbitrary middle surface, we must be content with, at best, asymptotic solutions similar to those for symmetric deformations or with numerical solutions of the various equations. At present, let us investigate what can be learned from analytical study of the theory of shells for surfaces of revolution. We shall limit our investigation here to loads which vary as $\cos\theta$ or as $\sin\theta$ around the circumference, a distribution which is commonly termed 'wind loading'.†

We assume that the shell middle surface is everywhere continuous in the circumferential direction. Then for surface loading varying as $\cos\theta$ or $\sin\theta$ and for appropriate edge loading or deformation conditions, the interior displacements, strains and stress resultants will also vary as $\cos\theta$ or $\sin\theta$. Let us denote the various pertinent quantities as, say,

$$\{q_\phi, q_\zeta, m_\phi, u_\phi, w, \epsilon_\phi, \epsilon_\theta, \kappa_\phi, \kappa_\theta, N_\phi, N_\theta, \bar{Q}_\phi, M_\phi, M_\theta\}$$
$$= \{q_{\phi 1}(\phi), q_{\zeta 1}(\phi), m_{\phi 1}(\phi), u_{\phi 1}(\phi), w_1(\phi), \epsilon_{\phi 1}(\phi), \epsilon_{\theta 1}(\phi),$$
$$\kappa_{\phi 1}(\phi), \kappa_{\theta 1}(\phi), N_{\phi 1}(\phi), N_{\theta 1}(\phi), \bar{Q}_{\phi 1}(\phi), M_{\phi 1}(\phi), M_{\theta 1}(\phi)\} \cos\theta, \quad (11.1.1a)$$

† An approximate analysis for other harmonic components is given in C. R. Steele, 'Nonsymmetric Deformation of Dome-shaped Shells of Revolution,' *J. Appl. Mech.*, vol. 29, no. 2, June 1962, pp. 353–361; and R. D. Schile, 'Asymptotic Solution of Nonshallow Shells of Revolution Subject to Nonsymmetric Loads,' *J. Aero. Sci.*, vol. 29, no. 11, November 1962, pp. 1375–1379. See also C. R. Steele, 'Shells of Revolution with Edge Loads of Rapid Circumferential Variation,' *J. Appl. Mech.*, vol. 29, no. 4, December 1962, pp. 701–705; and 'Shells with Edge Loads of Rapid Variation-II,' *J. Appl. Mech.*, vol. 32, no. 1, March 1965, pp. 335–339. The use of these approximations for the analysis of stresses at multiple shell and ring junctions is discussed in C. R. Steele, 'Juncture of Shells of Revolution,' *J. Spacecraft*, vol. 3, no. 6, June 1966, pp. 881–884.

$$\{q_\theta, m_\theta, u_\theta, \gamma_{\phi\theta}, \bar{\kappa}_{\phi\theta}, \bar{N}_{\phi\theta}, \bar{Q}_\theta, \bar{M}_{\phi\theta}\}$$
$$= \{q_{\theta 1}(\phi), m_{\theta 1}(\phi), u_{\theta 1}(\phi), \gamma_{\theta 1}(\phi), \bar{\kappa}_{\phi\theta 1}(\phi), \bar{N}_{\phi\theta 1}(\phi), \bar{Q}_{\theta 1}(\phi), \bar{M}_{\phi\theta}(\phi)\} \sin\theta. \tag{11.1.1b}$$

Then the equilibrium equations given by Eqs. (2.6.2) become

$$\frac{\mathrm{d}}{\mathrm{d}\phi}(rN_{\phi 1}) - R_\phi N_{\theta 1}\cos\phi + R_\phi\left[\bar{N}_{\phi\theta 1} + \frac{1}{2}\left(\frac{3}{R_\theta} - \frac{1}{R_\phi}\right)\bar{M}_{\phi\theta 1}\right]$$
$$- 2\frac{R_\phi}{R_\theta}\bar{M}_{\phi\theta 1} + r\bar{Q}_{\phi 1} + rR_\phi q_{\phi 1} = 0, \tag{11.1.2a}$$

$$- R_\phi N_{\theta 1} + \frac{1}{r}\frac{\mathrm{d}}{\mathrm{d}\phi}\left\{r^2\left[\bar{N}_{\phi\theta 1} + \frac{1}{2}\left(\frac{3}{R_\theta} - \frac{1}{R_\phi}\right)\bar{M}_{\phi\theta 1}\right]\right\} - \frac{2}{R_\theta}\frac{\mathrm{d}}{\mathrm{d}\phi}(r\bar{M}_{\phi\theta})$$
$$+ R_\phi\bar{Q}_{\theta 1}\sin\phi + rR_\phi q_{\theta 1} = 0, \tag{11.1.2b}$$

$$\frac{\mathrm{d}}{\mathrm{d}\phi}(r\bar{Q}_{\phi 1}) + R_\phi\bar{Q}_{\theta 1} - 2\frac{\mathrm{d}\bar{M}_{\phi\theta 1}}{\mathrm{d}\phi} - rN_{\phi 1} - R_\phi N_\theta \sin\phi$$
$$+ rR_\phi q_{\zeta 1} = 0, \tag{11.1.2c}$$

$$\frac{\mathrm{d}}{\mathrm{d}\phi}(rM_{\phi 1}) - R_\phi M_{\theta 1}\cos\phi + 2R_\phi\bar{M}_{\phi\theta 1} - rR_\phi(\bar{Q}_{\phi 1} - m_{\phi 1}) = 0, \tag{11.1.2d}$$

$$- R_\phi M_{\theta 1} + 2\frac{\mathrm{d}}{\mathrm{d}\phi}(r\bar{M}_{\phi\theta 1}) - rR_\phi(\bar{Q}_{\theta 1} - m_{\theta 1}) = 0. \tag{11.1.2e}$$

We obtain two integrals of the equations by the following means. A first integral is found by eliminating $N_{\theta 1}$ and $Q_{\theta 1}$ from Eqs. (11.1.2a) to yield

$$\frac{\mathrm{d}}{\mathrm{d}\phi}\left\{r\left[N_{\phi 1}\cos\phi + Q_{\phi 1}\sin\phi - \bar{N}_{\phi\theta 1} - \frac{1}{2}\left(\frac{3}{R_\theta} - \frac{1}{R_\phi}\right)\bar{M}_{\phi\theta 1}\right]\right\}$$
$$+ rR_\phi(q_{\phi 1}\cos\phi + q_{\zeta 1}\sin\phi - q_{\theta 1}) = 0, \tag{11.1.3a}$$

or

$$r\left[N_{\phi 1}\cos\phi + \bar{Q}_{\phi 1}\sin\phi - \bar{N}_{\phi\theta 1} - \frac{1}{2}\left(\frac{3}{R_\theta} - \frac{1}{R_\phi}\right)\bar{M}_{\phi\theta 1}\right] = \frac{V(\phi)}{\pi}, \tag{11.1.3b}$$

where

$$V(\phi) = V_0 - \pi\int_{\phi_0}^{\phi} rR_\phi(q_{\phi 1}\cos\phi + q_{\zeta 1}\sin\phi - q_{\theta 1})\,\mathrm{d}\phi, \tag{11.1.3c}$$

and V_0 is the total force acting normal to the axis of revolution on the cross-section $\phi = \phi_0$ (see Fig. 11.1). Eq. (11.1.3b) is the overall equation of equilibrium of transverse forces acting on the shell segment of Fig. 11.1.

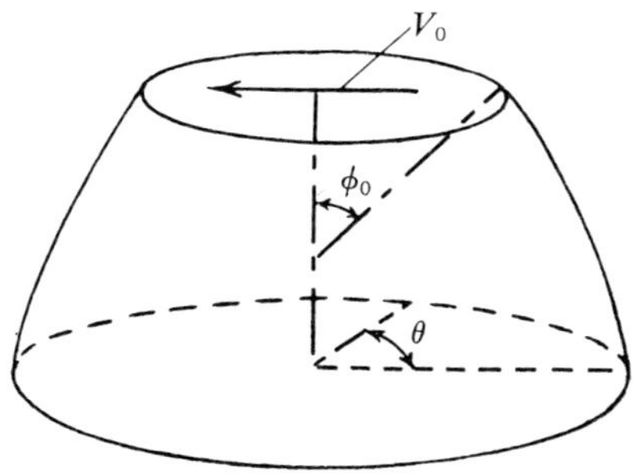

Fig. 11.1. Transverse force on shell of revolution

A second integral is found by eliminating M_θ, $\bar{Q}_\theta$, and N_θ from Eqs. (11.1.2d), (11.1.2e), (11.1.2a), and (11.1.2b) to yield

$$\frac{\mathrm{d}}{\mathrm{d}\phi}\{r[M_{\phi 1} + r(N_{\phi 1}\sin\phi - \bar{Q}_{\phi 1}\cos\phi)]\}$$

$$- rR_\phi\left[N_{\phi 1}\cos\phi + Q_{\phi 1}\sin\phi - \bar{N}_{\phi\theta 1} - \frac{1}{2}\left(\frac{3}{R_\theta} - \frac{1}{R_\phi}\right)\bar{M}_{\phi\theta 1}\right]\sin\phi$$

$$+ rR_\phi[m_\phi - m_\theta\cos\phi + r(q_\phi\sin\phi - q_\zeta\cos\phi)] = 0, \qquad (11.1.4a)$$

or, with the use of Eq. (11.1.3b), the relationship (see Fig. 11.2)

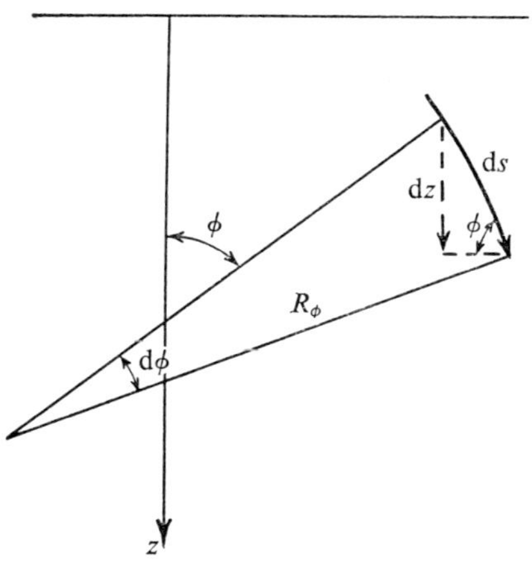

Fig. 11.2. Relationship between meridional length and height increments

$$R_\phi \sin\phi\, \mathrm{d}\phi = \mathrm{d}s \sin\phi = \mathrm{d}z, \qquad (11.1.4b)$$

and integration by parts, we obtain the equation of equilibrium of transverse moments on a shell segment as

$$\pi r[M_{\phi 1} + r(N_{\phi 1}\sin\phi - \bar{Q}_{\phi 1}\cos\phi)] = [M(\phi) + V(\phi)(z - z_0)] \quad (11.1.4c)$$

with

$$M(\phi) = M_0 + \pi \int_{\phi_0}^{\phi} \{(z - z_0)(q_\phi \cos\phi + q_\zeta \sin\phi - q_\theta) \\ - [m_\phi - m_\theta \cos\phi + r(q_\phi \sin\phi - q_\zeta \cos\phi)]\} r R_\phi \, d\phi \qquad (11.1.4d)$$

and M_0 is the moment about an axis lying in the plane of the cross-section at $\phi = \phi_0$ as shown in Fig. 11.3.

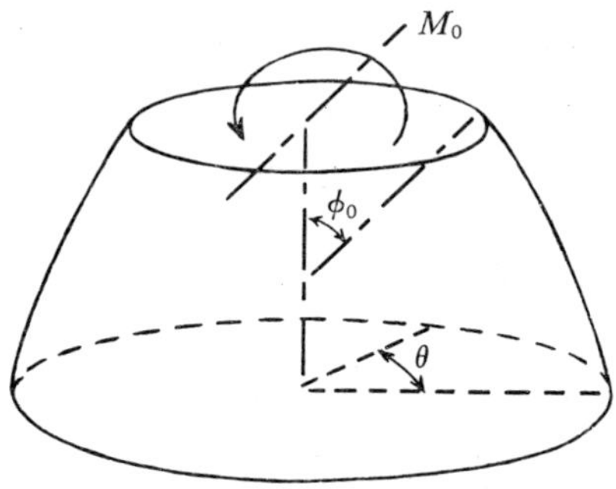

Fig. 11.3. Transverse moment on shell of revolution

The equations of equilibrium may now be taken as the equation obtained by eliminating $\bar{Q}_{\phi 1}$ from Eqs. (11.1.3b) and (11.1.4e):

$$rM_{\phi 1} \sin\phi + r^2\left\{N_{\phi 1} - \left[\bar{N}_{\phi\theta 1} + \frac{1}{2}\left(\frac{3}{R_\theta} - \frac{1}{R_\phi}\right)\bar{M}_{\phi\theta 1}\right] \cos\phi\right\} \\ = [(z - z_0)\sin\phi + r\cos\phi]\frac{V(\phi)}{\pi} + \frac{M(\phi)}{\pi}\sin\phi, \qquad (11.1.5a)$$

the equation obtained by eliminating $\bar{Q}_{\theta 1}$ from Eqs. (11.1.2b) and (11.1.2e):

$$\frac{1}{\sin\phi}\frac{d}{d\phi}\left\{r^2\left[\bar{N}_{\phi\theta 1} + \frac{1}{2}\left(\frac{3}{R_\theta} - \frac{1}{R_\phi}\right)\bar{M}_{\phi\theta 1}\right]\right\} - R_\phi(M_{\theta 1} + N_{\theta 1}R_\theta) \\ + rR_\phi(m_{\theta 1} + q_{\theta 1}R_\theta) = 0, \qquad (11.1.5b)$$

and the equation obtained by eliminating $\bar{Q}_{\phi 1}$ from Eqs. (11.1.2a) and (11.1.3b):

$$\sin\phi\frac{d}{d\phi}(R_\theta N_\phi) + (R_\phi + R_\theta)\left[\bar{N}_{\phi\theta 1} + \frac{1}{2}\left(\frac{3}{R_\theta} - \frac{1}{R_\phi}\right)M_{\phi\theta 1}\right] \\ - N_\theta R_\phi \cos\phi - 2\frac{R_\phi}{R_\theta}\bar{M}_{\phi\theta} + \frac{V(\phi)}{\pi\sin\phi} + rR_\phi q_{\phi 1} = 0. \qquad (11.1.5c)$$

The strain–displacement relations are given by

$$\epsilon_{\phi 1} = \frac{1}{R_\phi}\left(\frac{\mathrm{d}u_{\phi 1}}{\mathrm{d}\phi} + w_1\right), \tag{11.1.6a}$$

$$\epsilon_{\theta 1} = \frac{1}{r}\,(u_{\phi 1}\cos\phi + w_1 \sin\phi + u_{\theta 1}), \tag{11.1.6b}$$

$$\gamma_{\phi\theta 1} = \frac{r}{R_\phi}\frac{\mathrm{d}}{\mathrm{d}\phi}\left(\frac{u_{\theta 1}}{r}\right) - \frac{u_{\phi 1}}{r}, \tag{11.1.6c}$$

$$\kappa_\phi = -\frac{1}{R_\phi}\frac{\mathrm{d}}{\mathrm{d}\phi}\frac{1}{R_\phi}\left(\frac{\mathrm{d}w_1}{\mathrm{d}\phi} - u_{\phi 1}\right), \tag{11.1.6d}$$

$$\kappa_{\theta 1} = \frac{1}{r}\left[\frac{u_{\theta 1}\sin\phi + w_1}{r} - \frac{\cos\phi}{R_\phi}\left(\frac{\mathrm{d}w_1}{\mathrm{d}\phi} - u_{\phi 1}\right)\right], \tag{11.1.6e}$$

$$\bar{\kappa}_{\phi\theta 1} = \frac{1}{rR_\phi}\left(\frac{\mathrm{d}w_1}{\mathrm{d}\phi} - u_{\phi 1}\right) + \frac{1}{4}\left(\frac{1}{R_\phi} + \frac{1}{R_\theta}\right)\frac{u_{\phi 1}}{r} + \frac{1}{4}\frac{r}{R_\phi}\left(\frac{3}{R_\theta} - \frac{1}{R_\phi}\right)\frac{\mathrm{d}}{\mathrm{d}\phi}\left(\frac{u_{\theta 1}}{r}\right). \tag{11.1.6f}$$

The curvature changes may be expressed in terms of strain components by the following relationships which may be verified by substitution:

$$\bar{\kappa}_{\phi\theta 1} = \frac{1}{4}\left(\frac{3}{R_\theta} - \frac{1}{R_\phi}\right)\gamma_{\phi\theta 1} - \frac{\chi_1}{r\sin\phi}, \tag{11.1.7a}$$

$$\kappa_{\phi 1} = \frac{1}{rR_\phi}\frac{\mathrm{d}}{\mathrm{d}\phi}(R_\theta\chi_1) + \frac{\epsilon_{\phi 1}}{R_\theta}, \tag{11.1.7b}$$

$$\kappa_{\theta 1} = \frac{\chi_1\cot\phi}{r} + \frac{\epsilon_{\theta 1}}{R\theta}, \tag{11.1.7c}$$

whcre

$$\chi_1 = \gamma_{\phi\theta 1} + \epsilon_{\phi 1}\cos\phi - \frac{r}{R_\phi}\frac{\mathrm{d}\epsilon_{\theta 1}}{\mathrm{d}\phi}. \tag{11.1.7d}$$

Finally let us use the stress resultant–strain relationships given by Eqs. (3.1.2) as†

$$N_{\phi 1} = K(\epsilon_{\phi 1} + \nu\epsilon_{\theta 1}) - D\left(\frac{1}{R_\phi} - \frac{1}{R_\theta}\right)\left(\kappa_{\phi 1} - \frac{\epsilon_{\phi 1}}{R_\phi}\right), \tag{11.1.8a}$$

$$N_{\theta 1} = K(\epsilon_{\theta 1} + \nu\epsilon_{\phi 1}) - D\left(\frac{1}{R_\theta} - \frac{1}{R_\phi}\right)\left(\kappa_{\theta 1} - \frac{\epsilon_{\theta 1}}{R_\theta}\right), \tag{11.1.8b}$$

† Here, again, we might use the stress resultant–strain relations given by Eqs. (3.1.4) but with no real simplification of the equations.

$$\overline{N}_{\phi\theta 1} = \frac{1-\nu}{2}\left[K + \tfrac{3}{4}D\left(\frac{1}{R_\phi} - \frac{1}{R_\theta}\right)^2\right]\gamma_{\phi\theta 1}, \tag{11.1.8c}$$

$$M_{\phi 1} = D\left[\kappa_{\phi 1} + \nu\kappa_{\theta 1} - \left(\frac{1}{R_\phi} - \frac{1}{R_\theta}\right)\epsilon_{\phi 1}\right], \tag{11.1.8d}$$

$$M_{\theta 1} = D\left(\kappa_{\theta 1} + \nu\kappa_{\phi 1} - \left(\frac{1}{R_\theta} - \frac{1}{R_\phi}\right)\epsilon_{\theta 1}\right], \tag{11.1.8e}$$

$$\overline{M}_{\phi\theta 1} = D(1-\nu)\bar{\kappa}_{\phi\theta 1}. \tag{11.1.8f}$$

We now substitute Eqs. (11.1.7) and (11.1.8) into Eqs. (11.1.5) to obtain

$$\begin{aligned}
&\left[1 + \frac{h^2}{12}\left(\frac{1}{R_\phi^2} - \frac{3}{R_\phi R_\theta} + \frac{3}{R_\theta^2}\right)\right]\left(\epsilon_{\phi 1} - \frac{1-\nu}{2}\gamma_{\phi\theta 1}\cos\phi\right) + \nu\left(1 + \frac{h^2}{12R_\theta^2}\right)\epsilon_{\theta 1}\\
&\quad + \frac{h^2}{12}\left\{\left(\frac{2}{R_\theta} - \frac{1}{R_\phi}\right)\frac{\csc\phi}{R_\phi}\frac{\mathrm{d}\chi_1}{\mathrm{d}\phi} + \left[\frac{7-\nu}{2}\left(\frac{1}{R_\theta} - \frac{1}{R_\phi}\right) + \frac{R_\theta}{R_\theta^2}\right]\frac{\chi_1 \cot\phi}{r}\right\}\\
&\quad = \frac{1}{\pi r^2 K}\{[(z - z_0)\sin\phi + r\cos\phi]V + M\sin\phi\},
\end{aligned} \tag{11.1.9a}$$

$$\begin{aligned}
&\left(1 + \frac{h^2}{12R_\phi R_\theta}\right)\epsilon_{\theta 1} + \nu\left(1 + \frac{h^2}{12R_\theta^2}\right)\epsilon_{\phi 1} - \frac{1-\nu}{2rR_\phi}\\
&\qquad \times \frac{\mathrm{d}}{\mathrm{d}\phi}\left\{\left[1 + \frac{h^2}{12}\left(\frac{1}{R_\phi^2} - \frac{3}{R_\phi R_\theta} + \frac{3}{R_\theta^2}\right)\right]r^2\gamma_{\phi\theta 1}\right\}\\
&\qquad + \frac{h^2}{12rR_\phi}\left\langle\left(\frac{3-\nu}{2} - \frac{1-\nu}{2}\frac{R_\theta}{R_\phi}\right)\frac{\mathrm{d}\chi_1}{\mathrm{d}\phi}\right.\\
&\qquad \left. + \left\{\left[\frac{1-\nu}{2}\left(\frac{R_\theta}{R_\phi} + 1\right) + \nu\frac{R_\phi}{R_\theta}\right]\cot\phi + \frac{1-\nu}{2}\frac{R_\theta}{R_\phi^2}\frac{\mathrm{d}R_\phi}{\mathrm{d}\phi}\right\}\chi_1\right\rangle\\
&\quad = \frac{r}{K}\left(q_{\theta 1} + \frac{m_{\theta 1}}{R_\theta}\right),
\end{aligned} \tag{11.1.9b}$$

$$\begin{aligned}
&\frac{\sin\phi}{R_\phi}\frac{\mathrm{d}}{\mathrm{d}\phi}\left\{\left[1 + \frac{h^2}{12}\left(\frac{1}{R_\theta} - \frac{1}{R_\phi}\right)^2\right]\epsilon_{\phi 1} + \nu\epsilon_{\theta 1}\right\}R_\theta - (\epsilon_{\theta 1} + \nu\epsilon_{\phi 1})\cos\phi\\
&\qquad + \frac{1-\nu}{2}\left(1 + \frac{R_\theta}{R_\phi}\right)\left[1 + \frac{h^2}{24}\frac{3R_\phi + R_\theta}{R_\phi + R_\theta}\left(\frac{1}{R_\phi} - \frac{1}{R_\theta}\right)^2\right]\gamma_{\phi\theta 1}\\
&\qquad + \frac{h^2}{12}\left\{\frac{\sin\phi}{R_\phi}\frac{\mathrm{d}}{\mathrm{d}\phi}\left[\left(1 - \frac{R_\theta}{R_\phi}\right)\frac{1}{rR_\phi}\frac{\mathrm{d}}{\mathrm{d}\phi}(\chi_1 R_\theta)\right]\right.\\
&\qquad \left. + \left[\frac{\cos^2\phi}{R_\theta} + \frac{1-\nu}{2}\left(\frac{1}{R_\theta} - \frac{1}{R_\phi}\right)\right]\left(\frac{1}{R_\theta} - \frac{1}{R_\phi}\right)\frac{\chi_1}{\sin^2\phi}\right\}\\
&\quad = -\frac{1}{K}\left(\frac{V}{\pi R_\phi \sin_\phi} + rq_{\phi 1}\right).
\end{aligned} \tag{11.1.9c}$$

We could proceed further to express the strain components in terms of displacements. This is unnecessary however, since Eqs. (11.1.9) are three equations for the three strain components $\epsilon_{\phi 1}$, $\epsilon_{\theta 1}$, and $\gamma_{\phi\theta 1}$. Displacements may be obtained from the strain components by solving Eqs. (11.1.6a) to (11.1.6c) to yield

$$u_{\phi 1} = [r \sin\phi + (z - z_0)\cos\phi] \times \left\{C_1 + \int_{\phi_0}^{\phi} \frac{(\epsilon_{\phi 1} + \gamma_{\phi\theta 1}\cos\phi)(R_\phi/R_\theta) - \epsilon_{\theta 1}}{\sin^2\phi}\,\mathrm{d}\phi\right\} + \cos\phi \times \left\{C_2 - \int_{\phi_0}^{\phi} \frac{[r\tan\phi + (z - z_0)\gamma_{\phi\theta 1}(R_\phi/R_\theta) + (z - z_0)[\epsilon_{\phi 1}(R_\phi/R_\theta) - \epsilon_{\theta 1}]}{\sin^2\phi}\,\mathrm{d}\phi\right\}, \tag{11.1.10a}$$

$$u_{\theta 1} = -\Bigg\{[R_\phi \sin^2\phi\cos\phi + (z - z_0)] \times \left[C_1 + \int_{\phi_0}^{\phi} \frac{(\epsilon_{\phi 1} + \gamma_{\phi\theta 1}\cos\phi)(R_\phi/R_\theta) - \epsilon_{\theta 1}}{\sin^2\phi}\,\mathrm{d}\phi\right] + C_2 - \int_{\phi_0}^{\phi} \frac{[r\tan\phi + (z - z_0)]\gamma_{\phi\theta 1}(R_\phi/R_\theta) + (z - z_0)[\epsilon_{\phi 1}(R_\phi/R_\theta) - \epsilon_{\theta 1}]}{\sin^2\phi}\,\mathrm{d}\phi \tag{11.1.10b}$$

$$w_1 = \epsilon_{\theta 1}R_\theta + \sin\phi\Bigg\{[(R_\phi - R_\theta)\cos\phi + (z - z_0)] \times \left[C_1 + \int_{\phi_0}^{\phi} \frac{(\epsilon_{\phi 1} + \gamma_{\phi\theta 1}\cos\phi)(R_\phi/R_\theta) - \epsilon_{\theta 1}}{\sin^2\phi}\,\mathrm{d}\phi\right] - C_2 - \int_{\phi_0}^{\phi} \frac{[r\tan\phi + (z - z_0)]\gamma_{\phi\theta 1}(R_\phi/R_\theta) + (z - z_0)[\epsilon_{\phi 1}(R_\phi/R_\theta) - \epsilon_{\theta 1}]}{\sin^2\phi}\,\mathrm{d}\phi\Bigg\} \tag{11.1.10c}$$

where C_1 is a rigid body rotation about an axis passing through the axis of revolution at $z = z_0$ and C_2 is a rigid body translation perpendicular to the axis of revolution (see Fig. 11.4). For conical shells, for which ϕ is constant and equal to $\pi/2 - \alpha$ while R_ϕ is infinite, the solution of Eqs. (11.1.6a) to (11.1.6c) becomes

$$u_{\phi 1} = C_2 \sin\alpha + \int_{s_0}^{s} \epsilon_{\phi 1}\,\mathrm{d}s, \tag{11.1.11a}$$

$$u_{\theta 1} = C_1 s\cos\alpha - C_2 - \left[\int_{s_0}^{s} \epsilon_{\phi 1}\,\mathrm{d}s - s\int_{s_0}^{s} \frac{\epsilon_{\phi 1} + \gamma_{\phi\theta 1}\sin\alpha}{s}\,\mathrm{d}s\right]\csc\alpha, \tag{11.1.11b}$$

$$w_1 = \epsilon_{\theta 1}s\tan\alpha - C_1 s + C_2\cos\alpha + \int_{s_0}^{s} \epsilon_{\phi 1}\,\mathrm{d}s\cot\alpha - \frac{s}{\sin\alpha\cos\alpha}\int_{s_0}^{s} \frac{\epsilon_{\phi 1} + \gamma_{\phi\theta 1}\sin\alpha}{s}\,\mathrm{d}s. \tag{11.1.11c}$$

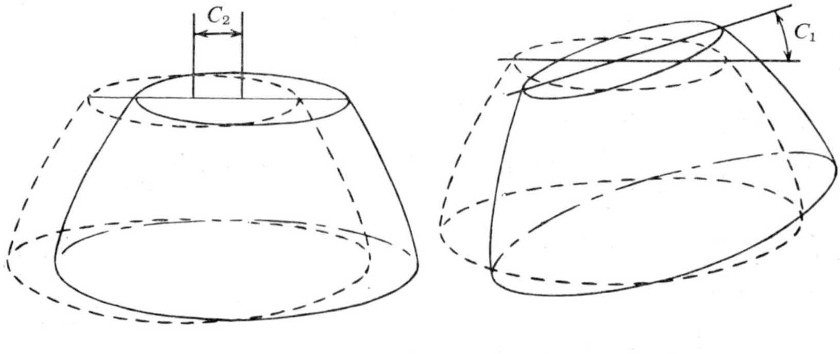

(a) Rigid body translation (b) Rigid body rotation

Fig. 11.4. Rigid body motions of shell segment

A particular solution of Eq. (11.1.9a) is readily obtained if we assume that surface loads are slowly varying and have appropriate behavior in the vicinity of an apex so that the terms involving χ_1 may be omitted. When we also delete terms of the order of h^2/R_ϕ^2, h^2/R_θ^2, $h^2/(R_\phi R_\theta)$ compared to unity in the coefficients of $\epsilon_{\phi 1}$, $\epsilon_{\theta 1}$, and $\gamma_{\phi\theta 1}$, the equations reduce to those of membrane theory and have the solution

$$\epsilon_{\phi 1} + \nu\epsilon_{\theta 1} = \frac{M + V(z - z_0)}{\pi r^2 K \sin\phi}, \tag{11.1.12a}$$

$$\epsilon_{\theta 1} + \nu\epsilon_{\phi 1} = \frac{1}{K}\left[q_\zeta R_\theta - \frac{M + V(z - z_0)}{\pi r R_\phi \sin^2\phi}\right], \tag{11.1.12b}$$

$$\frac{1-\nu}{2}\gamma_{\phi\theta 1} = \frac{1}{\pi r K}\left\{\frac{\cot\phi}{r}[M + V(z - z_0)] - V\right\}. \tag{11.1.12c}$$

In Eqs. (11.1.12) small terms involving $m_{\phi 1}$ and $m_{\theta 1}$ have been omitted as well.

The remaining complementary solutions of Eqs. (11.1.9) are bending solutions which decay away from the edges. We note then that differentiation increases the order of magnitude of the functions while integration decreases the order of magnitude. From these qualitative results we may immediately conclude that the radial displacement w_1 generally predominates since from Eqs. (11.1.10) the expression for w_1 contains the circumferential strain $\epsilon_{\theta 1}$ alone, while the expressions for $u_{\phi 1}$ and $u_{\theta 1}$ are expressed in terms of integrals of the middle surface strain components. To proceed further we retain only the first term in the coefficients of $\epsilon_{\phi 1}$, $\epsilon_{\theta 1}$ and $\gamma_{\phi\theta 1}$ in Eqs. (11.1.9). We then solve Eq. (11.1.9a) for $\epsilon_{\phi 1} + \nu\epsilon_{\theta 1}$ and Eq. (11.1.9b) for $\epsilon_{\theta 1} + \nu\epsilon_{\phi 1}$ in terms of $\gamma_{\phi\theta 1}$ and χ_1. The substitution of the results into Eq. (11.1.9c) then yields, with the deletion of unimportant terms,

$$X_1 = \frac{1}{12}\frac{R_\theta h}{R_\phi^2}\left\{\frac{\mathrm{d}^2\chi_1}{\mathrm{d}\phi^2} - \left[\left(2 - \frac{R_\phi}{R_\theta}\right)\cot\phi + \frac{1}{R_\phi}\frac{\mathrm{d}R_\phi}{\mathrm{d}\phi}\right)\frac{\mathrm{d}\chi_1}{\mathrm{d}\phi} + \chi_1\left(2 - \frac{R_\phi}{R_\theta} - 4\frac{R_\phi}{R_\phi^2}\right)\cot^2\phi\right\}, \tag{11.1.13a}$$

with

$$X_1 = \frac{1-\nu}{2}\frac{R_\theta}{h}\gamma_{\phi\theta 1}\sin^2\phi. \tag{11.1.13b}$$

From Eqs. (11.1.9a), (11.1.9b), and (11.1.13a) expressions for $\epsilon_{\phi 1}$ and $\epsilon_{\theta 1}$ in terms of χ_1 alone may be obtained. The equation which χ_1 must satisfy is then obtained by substituting these into Eq. (11.1.7d). The details are tedious and will not be pursued further here.

Such a procedure indicates, however, that the terms involving χ_1 in Eqs. (11.1.9a) and (11.1.9b) are generally negligible compared to the terms implied by Eqs. (11.1.13a) for thin shells and may be omitted. We now obtain a second approximate equation involving $\gamma_{\phi\theta 1}$ and χ_1 by solving Eqs. (11.1.9a) and (11.1.9b) for $\epsilon_{\phi 1}$ and $\epsilon_{\theta 1}$ in terms of $\gamma_{\phi\theta 1}$ and substituting the results into Eq. (11.1.7d). Then the dominant terms in the equation are

$$(1-\nu^2)\chi_1 + \frac{R_\theta h}{R_\phi^2}\left\{\frac{\mathrm{d}^2 X_1}{\mathrm{d}_\theta^2} - \left[\left(2 - \frac{R_\phi}{R_\theta}\right)\cot\phi + \frac{1}{R_\phi}\frac{\mathrm{d}R_\phi}{\mathrm{d}\phi}\right]\frac{\mathrm{d}X_1}{\mathrm{d}\phi} + X_1\left(2 - \frac{R_\phi}{R_\theta} - 4\frac{R_\phi^2}{R_\theta^2}\right)\cot^2\phi\right\} = 0. \tag{11.1.14}$$

Since Eq. (11.1.14) is of the same form as Eq. (11.1.13a), the two equations may be combined into the single approximate complex equation

$$\frac{1}{[12(1-\nu^2)]^{1/2}}\frac{R_\theta h}{R_\phi^2}\left\{\frac{\mathrm{d}^2 G_1}{\mathrm{d}\phi^2} - \left[\left(2 - \frac{R_\phi}{R_\theta}\right)\cot\phi + \frac{1}{R_\phi}\frac{\mathrm{d}R_\phi}{\mathrm{d}\phi}\right]\frac{\mathrm{d}G_1}{\mathrm{d}\phi} + G_1\left(2 - \frac{R_\phi}{R_\theta} - 4\frac{R_\phi^2}{R_\theta^2}\right)\cot^2\phi\right\} - iG_1 = 0, \tag{11.1.15a}$$

where

$$G_1 = \chi_1 - i\left(3\frac{1-\nu}{1+\nu}\right)^{1/2}\gamma_{\phi\theta 1}\frac{R_\theta}{h}\sin^2\phi. \tag{11.1.15b}$$

When we further introduce the transformations

$$\xi = [12(1-\nu)]^{1/4} \int_0^\phi \frac{R_\phi \, d\phi}{(R_\theta h)^{1/2}}, \tag{11.1.16a}$$

$$\bar{G}_1 = \frac{G_1}{\Psi}, \tag{11.1.16b}$$

$$\Psi = \left(\frac{\sin^2\phi}{R_\theta/h}\right)^{1/4}, \tag{11.1.16c}$$

Eq. (11.1.15a) becomes

$$\frac{d^2\bar{G}_1}{d\xi^2} - \left\{i + \frac{R_\theta h}{[12(1-\nu)]^{1/2} R_\phi^2}\left(\frac{63R_\phi^2}{16R_\theta^2} + \frac{1}{8}\frac{R_\phi}{R_\theta} - \frac{5}{16}\right)\cot^2\phi\right\}\bar{G}_1 = 0. \tag{11.1.17}$$

The substitute variable defined by Eq. (11.1.16a) has been discussed previously in section 7.3.

If $(R_\theta h)/R_\phi^2$, h/R_ϕ, and h/R_θ are small compared to unity, the first term in the coefficient of $\bar{G}_1$ predominates except in the vicinity of an apex. Sufficiently far from the apex Eq. (11.1.17) may be approximated by

$$\frac{d^2\bar{G}_1}{d\xi^2} - i\bar{G}_1 = 0, \tag{11.1.18a}$$

so that $\bar{G}_1$ has the asymptotic form

$$\begin{aligned}\bar{G}_1 = {} & A\, e^{\xi/2^{1/2}}\left(\cos\frac{\xi}{2^{1/2}} + i\sin\frac{\xi}{2^{1/2}}\right) \\ & - B\, e^{-\xi/2^{1/2}}\left(\cos\frac{\xi}{2^{1/2}} - i\sin\frac{\xi}{2^{1/2}}\right).\end{aligned} \tag{11.1.18b}$$

Near an apex we may approximate the second term of the coefficient of $\bar{G}_1$ by a term of the form k/ξ^2 as was done for axisymmetric deformations in section 7.4. For shells of spherical curvature near the apex, for example, we have.

$$R_\phi \approx R_\theta \approx a, \tag{11.1.19a}$$

$$\xi = [12(1-\nu)]^{1/4}\left(\frac{a}{h}\right)^{1/2}\phi, \tag{11.1.19b}$$

$$\cot\phi \approx 1/\phi. \tag{11.1.19c}$$

Eq. (11.1.17) may be approximated by

$$\frac{d^2\bar{G}_1}{d\xi^2} - \left(i + \frac{15}{4\xi^2}\right)\bar{G}_1 = 0, \tag{11.1.20a}$$

which has the solution

$$\bar{G}_1 = \xi^{1/2}[(A_1 + iA_2)(\mathrm{ber}_2\,\xi + i\,\mathrm{bei}_2\,\xi) + (B_1 + iB_2)(\mathrm{ker}_2\,\xi + i\,\mathrm{kei}_2\,\xi)]. \tag{11.1.20b}$$

With $\bar{G}_1$ known, the displacements and stress resultants may be found by means of the preceding equations to an accuracy consistent with the approximations inherent in Eq. (11.1.20b). Similar solutions may be obtained for shells having other behavior at an apex as in section 7.4.†

11.2 Conical frustums subjected to end loads‡

In some problems the desired information is such that solutions of the bending equations are not necessary. As an example, let us consider the problem of a conical frustum which is attached to a wall at the large end and is subjected to end load and moment applied through a rigid ring at the small end (Fig. 11.5). The desired result is a relationship between applied loads and rigid ring motions.

We denote the ring translation δ_0 and its rotation by θ_0 (Fig. 11.6). Then the displacements of the small edge of the cone to which the ring is attached are given by

$$u_s = -(\delta_0 + \theta_0 s_0 \cos\alpha)\sin\alpha\cos\theta, \tag{11.2.1a}$$

$$u_\theta = \delta_0 \sin\theta, \tag{11.2.1b}$$

$$w = -(\delta_0\cos\alpha - \theta_0 s_0 \sin^2\alpha)\cos\theta. \tag{11.2.1c}$$

At the large end we have

$$u_s = u_\theta = w = 0. \tag{11.2.2}$$

† The problem of the conical shell is treated in detail by R. A. Clark and J. F. Garibotti, 'Longitudinal Bending of a Conical Shell,' *Developments in Mechanics*, edited by S. Ostrach and R. H. Scanlon, vol. 2, part 2, Solid Mechanics, Pergamon Press, 1965, pp. 113–130.

‡ The analysis given here is based on results obtained by P. Seide, 'On the Bending of Cantilevered Thin-walled Conical Frustums by End Loads,' Space Technology Laboratories, The Ramo-Wooldridge Corp., Report No. GM-TR-284, 15 December, 1957. The method used is that of asymptotic integration valid far from an apex. The same problem is treated by Clark and Garibotti, but the emphasis there is on stresses rather than displacements. See also, P. Seide, 'Influence Coefficients for End-loaded, Conical Frustums,' *AIAA J.*, vol. 10, no. 12, December 1972, pp. 1717–1718.

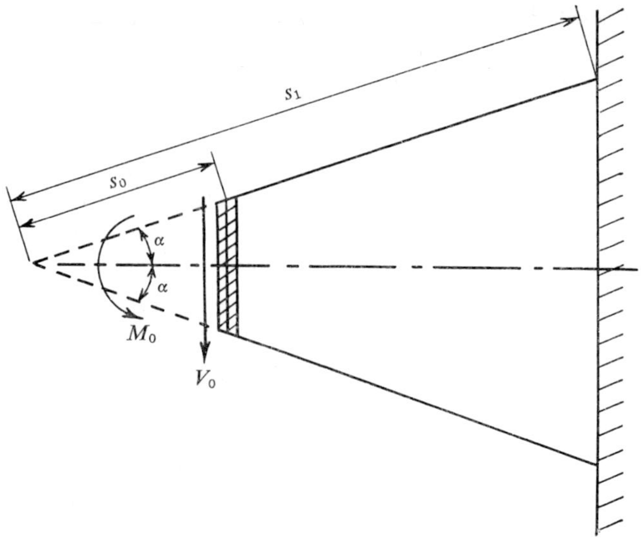

Fig. 11.5. Forces acting on cantilevered conical shell

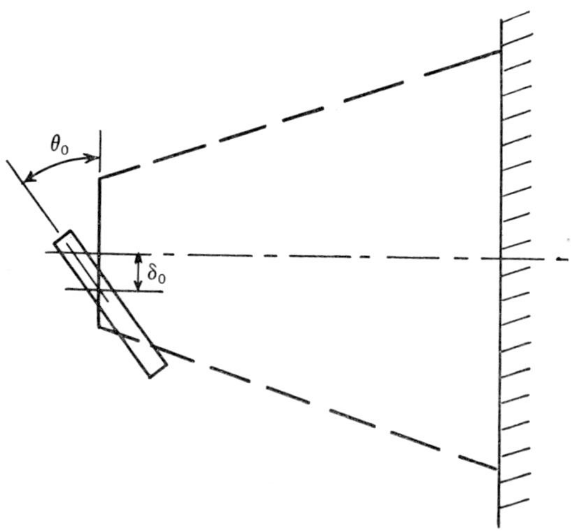

Fig. 11.6. Ring translation and rotation

The form of the boundary conditions suggests that only 'wind' loading solutions are needed. That part of the total deformations given by membrane theory and rigid body motions is obtained from Eqs. (11.1.11) and (11.1.12), with

$$M = M_0 \tag{11.2.3a}$$

$$V = V_0, \tag{11.2.3b}$$

$$R_\phi = \infty, \tag{11.2.3c}$$

$$\phi = \frac{\pi}{2} - \alpha, \tag{11.2.3d}$$

$$r = s \sin \alpha, \tag{11.2.3e}$$

as

$$\frac{u_s}{\cos\theta} = C_2 \sin\alpha + \frac{1}{\pi E h \sin^2\alpha}\left[\frac{M_0 - V_0 s_0 \cos\alpha}{s_0 \cos\alpha}\left(1 - \frac{s_0}{s}\right) + V_0 \ln\frac{s}{s_0}\right], \tag{11.2.4a}$$

$$\begin{aligned}\frac{u_\theta}{\sin\theta} &= C_1 s \cos\alpha - C_2 + \frac{1}{\pi E h \sin^3\alpha}\\ &\times\left\{\frac{M_0 - V_0 s_0 \cos\alpha}{2 s_0 \cos\alpha}\left(1 - \frac{s_0}{s}\right)\left[\frac{s}{s_0} - 1 + 2(1+\nu)\left(1 + \frac{s}{s_0}\right)\sin^2\alpha\right]\right.\\ &\left. - V_0\left(\ln\frac{s}{s_0} - \frac{s}{s_0} + 1\right)\right\},\end{aligned} \tag{11.2.4b}$$

$$\begin{aligned}\frac{w}{\cos\theta} &= C_2 \cos\alpha - C_1 s - \frac{1}{\pi E h \sin\alpha\cos\alpha}\\ &\times\left\{\frac{M_0 - V_0 s_0 \cos\alpha}{s_0\cos\alpha}\left[1 - 2\frac{s_0}{s} + (1+\nu)\frac{s}{s_0} + \frac{1+\nu}{\sin^2\alpha}\frac{s_0}{s}\left(\frac{s}{s_0} - 1\right)^2\right]\right.\\ &\left. + V_0\left[\nu + \left(\frac{s}{s_0} - 1\right)\csc^2\alpha - \ln\frac{s}{s_0}\cot^2\alpha\right]\right\}.\end{aligned} \tag{11.2.4c}$$

We note that in general the displacements u_s, u_θ, and w given by Eq. (11.2.1) are all of the same order of magnitude. Similarly the displacements given by Eqs. (11.2.4) are generally of the same order of magnitude. In the case of the bending solutions, however, the normal displacements w predominate. This same situation was found to be the case for cylindrical shells in section 10.4. We proceed then by requiring that Eq. (11.2.4a) and (11.2.4b) for u_s and u_θ satisfy the boundary conditions given by Eqs. (11.2.1a), (11.2.1b), and the first two of Eqs. (11.2.2). Any discrepancies in w and its various derivatives, depending on whether we prescribe rotationless or moment-free edges, are assumed to be remedied by the bending solutions. Then from Eqs. (11.2.4a), (11.2.4b), (11.2.1a), and (11.2.1b) we have

$$C_2 = -(\delta_0 + \theta_0 s_0 \cos\alpha), \tag{11.2.5a}$$

$$C_1 = -\theta_0. \tag{11.2.5b}$$

The satisfaction of the first two of Eqs. (11.2.2) gives the required relationship between loads and ring deflections as

$$\theta_0 = \frac{1 - s_0/s_1}{\pi E h s_0 \sin^3 \alpha \cos \alpha} \left[(1 - \beta)V_0 + \beta \frac{M_0}{s_0 \cos \alpha}\right], \tag{11.2.6a}$$

$$\delta_0 = \frac{1 - s_0/s_1}{\pi E h \sin^3 \alpha} \left\{\left[\frac{\ln (s_1/s_0)}{1 - s_0/s_1} - 2 + \beta\right] V_0 + (1 - \beta) \frac{M_0}{s_0 \cos \alpha}\right\}, \tag{11.2.6b}$$

with

$$\beta = \left(1 + \frac{s_0}{s_1}\right)[\tfrac{1}{2} + (1 + \nu) \sin^2 \alpha]. \tag{11.2.6c}$$

These may be inverted to yield

$$M_0 = \frac{\pi E h s_0 \sin^3 \alpha \cos \alpha}{\beta \ln (s_1/s_0) - (1 - s_0/s_1)} \times \left\{\left[\frac{\ln (s_1/s_0)}{1 - s_0/s_1} - 2 + \beta\right] \theta_0 s_0 \cos \alpha - (1 - \beta)\, \delta_0\right\}, \tag{11.2.7a}$$

$$V_0 = \frac{\pi E h \sin^3 \alpha}{\beta \ln (s_1/s_0) - (1 - s_0/s_1)} [\beta\, \delta_0 - (1 - \beta)\theta_0 s_0 \cos \alpha]. \tag{11.2.7b}$$

The equations reduce to those given by Eqs. (10.4.14) for the cylindrical shell when the angle α approaches zero and s_0 becomes infinite while $s_0 \sin \alpha$ remains equal to the cylinder radius R and $(s_1 - s_0) \cos \alpha$ is equal to the cylinder length L. For shallow conical shells ($\alpha \to 90°$) the bending–stretching terms become important and should be included.

It is interesting to note that the rotation θ_0 due to transverse load V_0 and the translation δ_0 due to movement M_0 can be negative for sufficiently short cones ($s_0/s_1 \to 1$) or for sufficiently large semi-vertex angles α. The deformation patterns would then appear as in Fig. 11.7. The combinations of s_0/s_1 and α for which the coefficients of Eqs. (11.2.6) become negative ($\beta > 1$) are shown in Fig. 11.8 for $\nu = 0.3$.

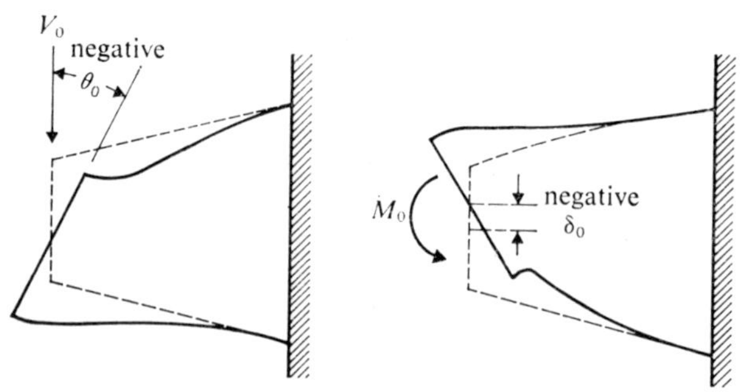

Fig. 11.7. Deformation patterns for negative influence coefficients

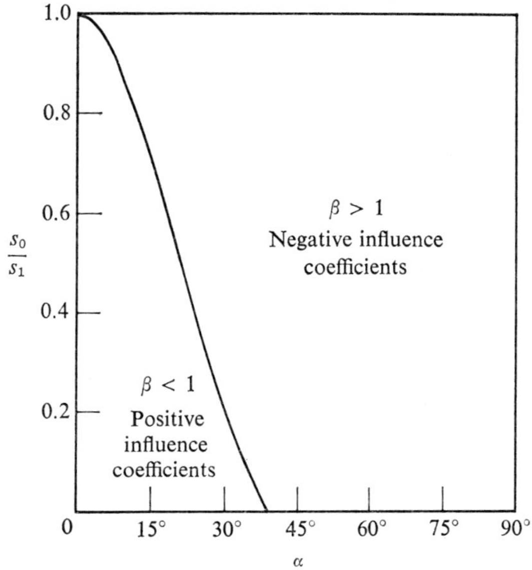

Fig. 11.8. Combinations of conical shell parameters yielding negative influence coefficients ($\nu = 0.3$)

If we interchange s_0 and s_1 in Eqs. (11.2.6) and (11.2.7) we obtain the solution for the problem shown in Fig. 11.9. Deformation patterns of the type shown in Fig. 11.7 can not occur in this case.

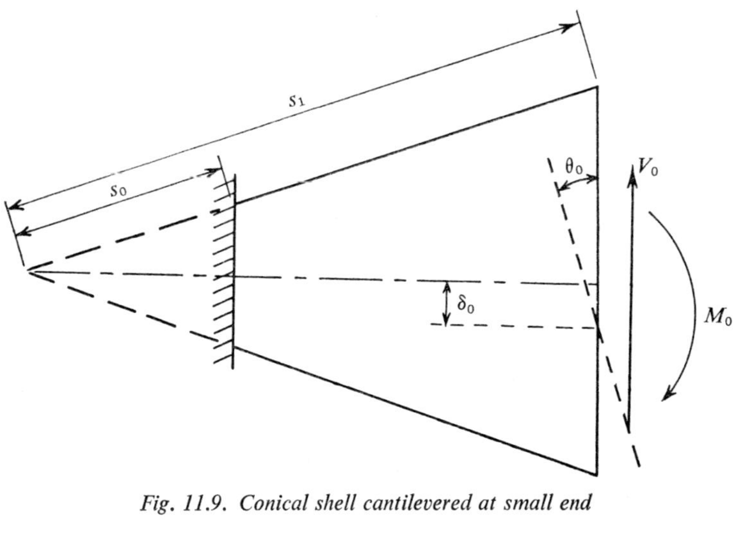

Fig. 11.9. Conical shell cantilevered at small end

11.3 Numerical solution of the bending equations for shells of revolution

The analysis of Section 11.1 and, indeed, all of the preceeding investigations indicate that the analytic solution of shell bending problems generally involves a great deal of tedious manipulation. In general the analyses are inexact since the equations can be solved only approximately, even when the membrane theory of shells is used. With the use of large electronic digital computers the rapid routine calculation of solutions of the shell equations is possible if we abandon our attempts to obtain analytic solutions and instead divert our efforts toward obtaining numerical solutions.

In most programs of numerical investigation the need for solving differential equations is replaced by a requirement that linear simultaneous algebraic equations be solved. The end product of these investigations is an efficient computer program for obtaining the solutions of these equations and desired quantities such as deformations and/or stresses with a minimum of required input data. These goals can be reached by a variety of paths. Two of these paths involve standard methods of numerical analysis, the method of finite differences and the method of step-by-step integration.† These methods are discussed in detail in the numerous volumes devoted to numerical

† Some of the many papers devoted to the description of the application of finite difference and numerical integration methods to the analysis of symmetrical or asymmetrical deformations of arbitrary shells of revolution are:

(a) R. E. Hubka, 'A Generalized Finite Difference Solution of Axisymmetric Stress States in Thin Shells of Revolution,' Report EM-11-19, Space Technology Laboratories (now TRW Systems), Los Angeles, California, 1961.
(b) W. K. Sepetoski, C. E. Pearson, I. W. Dingwell, and A. W. Adkins, 'A Digital Computer Program for the General Axially Symmetric Thin Shell Problem,' *J. Appl. Mech.*, vol. 29, no. 4, December 1962, pp. 655–661.
(c) R. K. Penny, 'Symmetric Bending of the General Shell of Revolution by Finite Difference Methods,' *J. Mech. Eng. Sci.*, vol. 3, 1961, pp. 369–377.
(d) P. P. Radkowski, R. M. Davis, and M. R. Bolduc, 'Numerical Analysis of Equations of Thin Shells of Revolution,' *ARS Jour.*, vol. 32, 1962, pp. 36–44.
(e) B. Budiansky and P. P. Radkowski, 'Numerical Analysis of Unsymmetrical Bending of Shells of Revolution,' *AIAA J.*, vol. 1, no. 8, August 1963, pp. 1833–1842.
(f) A. Kalnins, 'Analysis of Shells of Revolution Subjected to Symmetrical and Unsymmetrical Loads,' *J. Appl. Mech.*, vol. 31, no. 3, September 1964, pp. 467–476.
(g) H. G. Schaeffer, 'Computer Program for Finite-Difference Solutions of Shells of Revolution Under Asymmetric Loads,' NASA Tech. Note D-3926, May 1967.
(h) A. P. Cappelli, T. S. Nishimoto, and P. P. Radkowski, 'Analysis of Shells of Revolution having Arbitrary Stiffness Distributions,' *AIAA J.*, vol. 7, no. 10, October 1969, pp. 1909–1915.

analysis† and will not be examined further. A less standard method of analysis known as the finite element method or the direct stiffness method is fast becoming recognized as the basis for the most efficient analysis programs, especially for arbitrary shells, and will be treated in some detail.

The finite element method is a variant of the Rayleigh–Ritz approach to the analysis of differential equations and usually involves the principle of minimum potential energy. For shells of revolution complete in the circumferential direction we write the principle of minimum potential energy as

$$\delta\left\{\int_{\phi_1}^{\phi_2}\int_0^{2\pi}[W-(q_\phi^* u_\phi+q_\theta^* u_\theta+q_\zeta^* w-m_\phi^*\Theta_\phi-m_\theta^*\Theta_\theta]rR_\phi\,d\phi\,d\theta\right.$$
$$\left.-\int_0^{2\pi}(N_\phi^* u_\phi+N_{\phi\theta}^* u_\theta+Q_\phi^* w-M_\phi^*\Theta_\phi-M_{\phi\theta}^*\Theta_0)r\Big|_{\phi_1}^{\phi_2}d\theta\right\}=0, \tag{11.3.1a}$$

with, say,

$$W=\tfrac{1}{2}K\left(\epsilon_\phi^2+2\nu\epsilon_\phi\epsilon_\theta+\epsilon_\theta^2+\frac{1-\nu}{2}\gamma_{\phi\theta}^2\right)$$
$$+\tfrac{1}{2}D[\kappa_\phi^2+2\nu\kappa_\phi\kappa_\theta+\kappa_\theta^2+2(1-\nu)\bar{\kappa}_{\phi\theta}^2]$$
$$+\tfrac{1}{2}D\left(\frac{1}{R_\theta}-\frac{1}{R_\phi}\right)\left[\epsilon_\phi\left(2\kappa_\phi-\frac{\epsilon_\phi}{R_\phi}\right)-\epsilon_\theta\left(2\kappa_\theta-\frac{\epsilon_\theta}{R_\theta}\right)\right.$$
$$\left.+\tfrac{3}{8}(1-\nu)\left(\frac{1}{R_\theta}-\frac{1}{R_\phi}\right)\gamma_{\phi\theta}^2\right], \tag{11.3.1b}$$

and

$$\epsilon_\phi=\frac{1}{R_\phi}\left(\frac{\partial u_\phi}{\partial\phi}+w\right), \tag{11.3.1c}$$

$$\epsilon_\theta=\frac{1}{r}\left(\frac{\partial u_\phi}{\partial\theta}+u_\phi\cos\phi+w\sin\phi\right), \tag{11.3.1d}$$

$$\gamma_{\phi\theta}=\frac{1}{r}\frac{\partial u_\phi}{\partial\theta}+\frac{1}{R_\phi}\frac{\partial u_\theta}{\partial\phi}-\frac{\cos\phi}{r}u_\theta, \tag{11.3.1e}$$

$$\kappa_\phi=-\frac{1}{R_\phi}\frac{\partial}{\partial\phi}\frac{1}{R_\phi}\left(\frac{\partial w}{\partial\phi}-u_\phi\right), \tag{11.3.1f}$$

† See, for example, F. B. Hildebrand, *Introduction to Numerical Analysis*, McGraw-Hill Book Co., 1962; J. Todd, ed., *Survey of Numerical Analysis*, McGraw-Hill Book Co., 1962; M. L. James, G. M. Smith, and J. C. Wolford, *Applied Numerical Methods for Digital Computation with FORTRAN*, International Textbook Co., 1967; and R. Beckett and J. Hurt, *Numerical Calculations and Algorithms*, McGraw-Hill Book Co., 1967.

$$\kappa_\theta = -\frac{1}{r^2}\frac{\partial}{\partial\theta}\left(\frac{\partial w}{\partial\theta} - u_\theta \sin\phi\right) - \frac{\cos\phi}{rR_\phi}\left(\frac{\partial w}{\partial\phi} - u_\phi\right), \tag{11.3.1g}$$

$$\bar{\kappa}_{\phi\theta} = -\frac{1}{rR_\phi}\frac{\partial^2 w}{\partial\phi\,\partial\theta} + \frac{\cos\phi}{r^2}\frac{\partial w}{\partial\theta} + \frac{1}{4r}\left(\frac{3}{R_\phi} - \frac{\sin\phi}{r}\right)\frac{\partial u_\phi}{\partial\theta}$$
$$+ \frac{1}{4}\left(\frac{3\sin\phi}{r} - \frac{1}{R_\phi}\right)\left(\frac{1}{R_\phi}\frac{\partial u_\theta}{\partial\phi} - \frac{\cos\phi}{r}u_\theta\right), \tag{11.3.1h}$$

$$\Theta_\phi = \frac{1}{R_\phi}\left(\frac{\partial w}{\partial\phi} - u_\phi\right), \tag{11.3.1i}$$

$$\Theta_\theta = \frac{1}{r}\left(\frac{\partial w}{\partial\theta} - u_\theta \sin\phi\right). \tag{11.3.1j}$$

In Eq. (11.3.1b) the stiffnesses K and D may be functions of ϕ. We expand displacements, surface forces and applied edge stress-resultants in a Fourier series in the circumferential direction in the form

$$(\;) = \sum_{n=0}^{\infty} (\;)_n \cos n\theta + \sum_{n=1}^{\infty} \widetilde{(\;)}_n \sin n\theta, \tag{11.3.2}$$

where the coefficients are functions of the meridional coordinate ϕ. On substituting the Fourier series for the various quantities into Eqs. (11.3.1) and performing the integration with respect to θ we obtain

$$\delta \sum_{n=0}^{\infty} (U_n + \tilde{U}_n) = 0, \tag{11.3.3a}$$

where

$$U_n = \int_{\phi_1}^{\phi_2} [W_n - (q_{\phi n}^* u_{\phi n} + \tilde{q}_{\theta n}^* \tilde{u}_{\theta n} + q_{\zeta n}^* w_n - m_{\phi n}^* \Theta_{\phi n} - \tilde{m}_{\theta n}^* \Theta_{\theta n})]$$
$$\times rR_\phi\, \mathrm{d}\phi - (N_{\phi n}^* u_{\phi n} + \tilde{N}_{\phi\theta n}^* \tilde{u}_{\theta n} + Q_{\phi n}^* w_n$$
$$- M_{\phi n}^* \Theta_{\phi n} - \tilde{M}_{\phi\theta n}^* \tilde{\Theta}_{\theta n}) r|_{\phi_1}^{\phi_2}, \tag{11.3.3b}$$

$$\tilde{U}_n = \int_{\phi_1}^{\phi_2} [\tilde{W}_n - (\tilde{q}_{\phi n}^* \tilde{u}_{\phi n} + q_{\theta n}^* u_{\theta n} + \tilde{q}_{\zeta n}^* w_n^* - \tilde{m}_{\phi n}^* \Theta_{\phi n} - m_{\theta n}^* \Theta_{\theta n})]$$
$$\times rR_\phi\, \mathrm{d}\phi - (\tilde{N}_{\phi n}^* \tilde{u}_{\phi n} + N_{\phi\theta n}^* u_{\theta n} + \tilde{Q}_{\phi n}^* w_n$$
$$- \tilde{m}_{\phi n}^* \Theta_{\phi n} - M_{\phi\theta n}^* \Theta_{\theta n}) r|_{\phi_1}^{\phi_2}\} = 0 \qquad (n = 0, 1, 2, \ldots), \tag{11.3.3c}$$

$$W_n = \tfrac{1}{2}K\left(\epsilon^2_{\phi n} + 2\nu\epsilon_{\phi n}\epsilon_{\theta n} + \epsilon^2_{\theta n} + \frac{1-\nu}{2}\gamma^2_{\phi\theta n}\right)$$

$$+ \tfrac{1}{2}D[\kappa^2_{\phi n} + 2\nu\kappa_{\phi n}\kappa_{\theta n} + \kappa^2_{\theta n} + 2(1-\nu)\bar{\kappa}^2_{\phi\theta n}] + \tfrac{1}{2}D\left(\frac{1}{R_\theta} - \frac{1}{R_\phi}\right)$$

$$\times\left[\epsilon_{\phi n}\left(2\kappa_{\phi n} - \frac{\epsilon_{\phi n}}{R_\phi}\right) - \epsilon_{\theta n}\left(2\kappa_{\theta n} - \frac{\epsilon_{\theta n}}{R_\phi}\right)\right.$$

$$\left. + \tfrac{3}{8}(1-\nu)\left(\frac{1}{R_\theta} - \frac{1}{R_\phi}\right)\gamma^2_{\phi\theta n}\right], \tag{11.3.3d}$$

$$\tilde{W}_n = \tfrac{1}{2}K\left(\tilde{\epsilon}^2_{\theta n} + 2\nu\tilde{\epsilon}_{\phi n}\tilde{\epsilon}_{\theta n} + \tilde{\epsilon}^2_{\theta n} + \frac{1-\nu}{2}\tilde{\gamma}^2_{\phi\theta n}\right)$$

$$+ \tfrac{1}{2}D[\tilde{\kappa}^2_{\phi n} + 2\nu\tilde{\kappa}_{\phi n}\tilde{\kappa}_{\theta n} + \tilde{\kappa}^2_{\theta n} + 2(1-\nu)\tilde{\bar{\kappa}}^2_{\phi\theta n}] + \tfrac{1}{2}D\left(\frac{1}{R_\theta} - \frac{1}{R_\phi}\right)$$

$$\times\left[\tilde{\epsilon}_{\phi n}\left(2\tilde{\kappa}_{\phi n} - \frac{\tilde{\epsilon}_{\phi}}{R_\phi}\right) - \tilde{\epsilon}_{\theta n}\left(2\tilde{\kappa}_{\theta n} - \frac{\tilde{\epsilon}_{\theta n}}{R_\theta}\right)\right.$$

$$\left. + \tfrac{3}{8}(1-\nu)\left(\frac{1}{R_\theta} - \frac{1}{R_\phi}\right)\tilde{\gamma}^2_{\phi\theta n}\right], \tag{11.3.3e}$$

$$\epsilon_{\phi n} = \frac{1}{R_\phi}\left(\frac{\mathrm{d}\tilde{u}_{\phi n}}{\mathrm{d}\phi} + w_n\right), \tag{11.3.3f}$$

$$\tilde{\epsilon}_{\phi n} = \frac{1}{R_\phi}\left(\frac{\mathrm{d}u_{\phi n}}{\mathrm{d}\phi} + w_n\right), \tag{11.3.3g}$$

$$\epsilon_{\theta n} = \frac{1}{r}(u_{\phi n}\cos\phi + w_n\sin\phi + n\tilde{u}_{\theta n}), \tag{11.3.3h}$$

$$\tilde{\epsilon}_{\theta n} = \frac{1}{r}(\tilde{u}_{\phi n}\cos\phi + \tilde{w}_n\sin\phi - nu_{\theta n}), \tag{11.3.3i}$$

$$\gamma_{\phi\theta n} = \frac{1}{R_\phi}\frac{\mathrm{d}\tilde{u}_{\theta n}}{\mathrm{d}\phi} - \frac{\cos\phi}{r}\tilde{u}_{\theta n} - \frac{n}{r}u_{\phi n}, \tag{11.3.3j}$$

$$\tilde{\gamma}_{\phi\theta n} = \frac{1}{R_\phi}\frac{\mathrm{d}u_{\theta n}}{\mathrm{d}\phi} - \frac{\cos\phi}{r}u_{\theta n} + \frac{n}{r}\tilde{u}_{\phi n}, \tag{11.3.3k}$$

$$\kappa_{\phi n} = -\frac{1}{R_\phi}\frac{\mathrm{d}}{\mathrm{d}\phi}\left[\frac{1}{R_\phi}\left(\frac{\mathrm{d}w_n}{\mathrm{d}\phi} - u_{\phi n}\right)\right], \tag{11.3.3l}$$

$$\tilde{\kappa}_{\phi n} = -\frac{1}{R_\phi}\frac{\mathrm{d}}{\mathrm{d}\phi}\left[\frac{1}{R_\phi}\left(\frac{\mathrm{d}\tilde{w}_n}{\mathrm{d}\phi} - \tilde{u}_{\phi n}\right)\right], \tag{11.3.3m}$$

$$\kappa_{\theta n} = \frac{n}{r^2}(nw_n + \tilde{u}_{\theta n}\sin\phi) - \frac{\cos\phi}{rR_\phi}\left(\frac{\mathrm{d}w_n}{\mathrm{d}\phi} - u_{\phi n}\right), \tag{11.3.3n}$$

$$\kappa_{\theta n} = \frac{n}{r^2}(n\tilde{w}_n - u_{\theta n}\sin\phi) - \frac{\cos\phi}{rR_\phi}\left(\frac{\mathrm{d}\tilde{w}_n}{\mathrm{d}\phi} - \tilde{u}_{\phi n}\right), \tag{11.3.3o}$$

$$\tilde{\kappa}_{\phi\theta n} = \frac{n}{r}\left[\frac{1}{R_\phi}\frac{\mathrm{d}w_n}{\mathrm{d}\phi} - \frac{\cos\phi}{r}w_n + \frac{1}{4}\left(\frac{\sin\phi}{r} - \frac{3}{R_\phi}\right)u_{\phi n}\right] + \frac{1}{4}\left(\frac{3\sin\phi}{r} - \frac{1}{R_\phi}\right)\left(\frac{1}{R_\phi}\frac{\mathrm{d}\tilde{u}_{\theta n}}{\mathrm{d}\phi} - \frac{\cos\phi}{r}\tilde{u}_{\theta n}\right), \tag{11.3.3p}$$

$$\tilde{\tilde{\kappa}}_{\phi\theta n} = -\frac{n}{R}\left[\frac{1}{R_\phi}\frac{\mathrm{d}\tilde{w}_n}{\mathrm{d}\phi} - \frac{\cos\phi}{r}\tilde{w}_n + \frac{1}{4}\left(\frac{\sin\phi}{r} - \frac{3}{R_\phi}\right)\tilde{u}_{\phi n}\right] + \frac{1}{4}\left(\frac{3\sin\phi}{r} - \frac{1}{R_\phi}\right)\left(\frac{1}{R_\phi}\frac{\mathrm{d}u_{\theta n}}{\mathrm{d}\phi} - \frac{\cos\phi}{r}u_{\theta n}\right), \tag{11.3.3q}$$

$$\Theta_{\phi n} = \frac{1}{R_\phi}\left(\frac{\mathrm{d}w_n}{\mathrm{d}\phi} - u_{\phi n}\right), \tag{11.3.3r}$$

$$\tilde{\Theta}_{\phi n} = \frac{1}{R_\phi}\left(\frac{\mathrm{d}\tilde{w}_n}{\mathrm{d}\phi} - \tilde{u}_{\phi n}\right), \tag{11.3.3s}$$

$$\Theta_{\theta n} = \frac{1}{r}(n\tilde{w}_n - u_{\theta n}\sin\phi), \tag{11.3.3t}$$

$$\tilde{\Theta}_{\theta n} = -\frac{1}{r}(nw_n + \tilde{u}_{\theta n}\sin\phi). \tag{11.3.3u}$$

Since U_n contains only deformation functions $u_{\phi n}$, $\tilde{u}_{\theta n}$, and w_n, while $\tilde{U}_n$ contains only the functions $\tilde{u}_{\phi n}$, $u_{\theta n}$, and $\tilde{w}_n$, Eq. (11.3.3a) may be written as

$$\delta U_n = \delta\tilde{U}_n = 0 \qquad (n = 0, 1, 2 \ldots). \tag{11.3.4}$$

In the standard Rayleigh–Ritz procedure we would assume each of the displacement components to be expandible in an infinite series of the form

$$a_n(\phi) = \sum_{j=1}^{N} a_{nj}\Phi_j(\phi), \tag{11.3.5}$$

where the arbitrary coefficients a_{nj} are constant and the functions $\Phi_j(\phi)$ are known functions each of which satisfies required geometric boundary conditions at the ends of the shell. The substitution of Eqs. (11.3.5) into Eqs. (11.3.4), integration, and minimization with respect to each of the coefficients a_{nj} then yields a set of N linear simultaneous equations for the coefficients. If the set of functions $\Phi_j(\phi)$ is complete, the coefficients a_{nj} converge to definite values and the series converges to the exact value of the function $a_n(\phi)$ as N becomes larger and larger. The disadvantage of this method of analysis, however, is that a great many terms in the series are usually required to describe the behavior of the deformations and stress-resultants in the vicinity of the ends of the shell. In addition it is not always easy to choose functions satisfying prescribed geometric end constraints. The finite element method sidesteps these difficulties by replacing the functions $\Phi_j(\phi)$ which are continuous and with continuous derivatives over the entire length of the shell by a set of simple functions which are piecewise continuous. Each function is defined only over a certain segment of the shell. At the common ends of adjoining segments the corresponding functions and certain of their derivatives are required to be continuous. The unknown quantities to be determined are deformations and rotations at the segment ends rather than series coefficients. Convergence then occurs as the number of segments becomes infinite.

For convenience we use the distance s along the shell generator as the coordinate so that

$$R_\phi \, \mathrm{d}\phi = \mathrm{d}s. \tag{11.3.6}$$

We consider a typical finite segment of the generator, the jth segment, and measure s from one end of the segment. The deformations within the segment will be described in terms of values at the segment ends as, say,

$$u_{\phi nj} = u_{\phi nj}^{(0)}(1 - \xi) + u_{\phi nj}^{(1)}\xi, \tag{11.3.7a}$$

$$\tilde{u}_{\theta nj} = \tilde{u}_{\theta nj}^{(0)}(1 - \xi) + \tilde{u}_{\theta nj}^{(1)}\xi, \tag{11.3.7b}$$

$$\begin{aligned} w_{nj} &= w_{nj}^{(0)}(1 - \xi)^2(1 + 2\xi) + w_{nj}^{(1)}\xi^2(3 - 2\xi) \\ &= + s_j\left[\frac{\mathrm{d}w_{nj}^{(0)}}{\mathrm{d}s}\,\xi(1 - \xi)^2 - \frac{\mathrm{d}w_{nj}^{(1)}}{\mathrm{d}s}\,\xi^2(1 - \xi)\right], \end{aligned} \tag{11.3.7c}$$

$$\xi = s/s_j, \tag{11.3.7d}$$

where the superscript (0) denotes quantities at one end ($\xi = 0$) of the segment while the superscript (1) denotes quantities at the other end ($\xi = 1$). The length of the segment has been denoted by s_j. The tangential displacements have been assumed to vary linearly with the segment while the variation of the radial displacement is described by a cubic curve in terms of the radial displacements and their first derivatives at the end points. These choices are governed by the general requirement for convergence of the Rayleigh–Ritz process that the functions and their derivatives of order up to one less than the highest order derivative in the potential energy expression be continuous. Then within each segment we have

$$\begin{aligned} \epsilon_{\phi nj} = {} & \frac{u_{\phi nj}^{(1)} - u_{\phi nj}^{(0)}}{s_j} + \frac{s_j}{R_\phi}\left[\frac{w_{nj}^{(0)}}{s_1}(1 - \xi)^2(1 + 2\xi) + \frac{w_{nj}^{(1)}}{s_j}\,\xi^2(3 - 2\xi)\right. \\ & \left. + \frac{\mathrm{d}w_{nj}^{(0)}}{\mathrm{d}s}\,\xi(1 - \xi)^2 - \frac{\mathrm{d}w_{nj}^{(1)}}{\mathrm{d}s}\,\xi^2(1 - \xi)\right], \end{aligned} \tag{11.3.8a}$$

$$\epsilon_{\theta nj} = \frac{u^{(0)}_{\phi nj}\cos\phi + n\tilde{u}^{(0)}_{\theta nj}}{r}(1-\xi) + \frac{u^{(1)}_{\phi nj}\cos\phi + n\tilde{u}^{(1)}_{\theta nj}}{r}\,\xi$$

$$+ \frac{s_j\sin\phi}{r}\left[\frac{w^{(0)}_{nj}}{s_j}(1-\xi)^2(1+2\xi) + \frac{w^{(1)}_{nj}}{s_1}\,\xi^2(3-2\xi)\right.$$

$$\left. + \frac{\mathrm{d}w^{(0)}_{nj}}{\mathrm{d}s}\,\xi(1-\xi)^2 - \frac{\mathrm{d}w^{(1)}_{nj}}{\mathrm{d}s}\,\xi^2(1-\xi)\right], \tag{11.3.8b}$$

$$\gamma_{\phi\theta nj} = -\frac{nu^{(0)}_{\phi nj} + \tilde{u}_{\theta nj}\cos\phi}{r}(1-\xi) - \frac{nu^{(1)}_{\phi nj} + \tilde{u}^{(1)}_{\theta nj}\cos\phi}{r}\,\xi$$

$$+ \frac{\tilde{u}^{(1)}_{\theta nj} - \tilde{u}^{(0)}_{\theta nj}}{s_j}, \tag{11.3.8c}$$

$$\kappa_{\phi nj} = -\frac{2}{s_j}\left\{3(1-2\xi)\left[\frac{w^{(1)}_{nj} - w^{(0)}_{nj}}{s_1}\right]\right.$$

$$\left. - \left[(2-3\xi)\frac{\mathrm{d}w^{(0)}_{nj}}{\mathrm{d}s} + (1-3\xi)\frac{\mathrm{d}w^{(1)}_{nj}}{\mathrm{d}s}\right]\right\}$$

$$+ \frac{u^{(1)}_{\phi nj} - u^{(0)}_{\phi nj}}{R_\phi s_j} - [u^{(0)}_{\phi nj}(1-\xi) + u^{(1)}_{\phi nj}\xi]\frac{\mathrm{d}}{\mathrm{d}s}\left(\frac{1}{R_\phi}\right), \tag{11.3.8d}$$

$$\kappa_{\theta nj} = \frac{n}{r^2}\left\{[\tilde{u}^{(0)}_{\theta nj}(1-\xi) + \tilde{u}^{(1)}_{\theta nj}\xi]\sin\phi + s_j n\left[\frac{w^{(0)}_{n}}{s_j}(1-\xi)^2(1+2\xi)\right.\right.$$

$$\left.\left. + \frac{w^{(1)}_{nj}}{s_j}\,\xi^2(3-2\xi) + \frac{\mathrm{d}w^{(0)}_{nj}}{\mathrm{d}s}\,\xi(1-\xi)^2 - \frac{\mathrm{d}w^{(1)}_{nj}}{\mathrm{d}s}\,\xi^2(1-\xi)\right]\right\}$$

$$+ \frac{\cos\phi}{r}\left\{\frac{1}{R_\phi}[u^{(0)}_{\phi nj}(1-\xi) + u^{(1)}_{\phi nj}\xi] - \frac{w^{(1)}_{nj} - w^{(0)}_{nj}}{s_j}\,6\xi(1-\xi)\right.$$

$$\left. - \frac{\mathrm{d}w^{(0)}_{nj}}{\mathrm{d}s}(1-\xi)(1-3\xi) + \frac{\mathrm{d}w^{(1)}_{nj}}{\mathrm{d}s}\,\xi(2-3\xi)\right\}, \tag{11.3.8e}$$

$$\kappa_{\phi\theta nj} = \frac{n}{r}\left\langle\frac{w^{(1)}_{nj} - w^{(0)}_{nj}}{s_j}\,6\xi(1-\xi) + \frac{\mathrm{d}w^{(0)}_{nj}}{\mathrm{d}s}(1-\xi)(1-3\xi)\right.$$

$$- \frac{\mathrm{d}w^{(1)}_{nj}}{\mathrm{d}s}\,\xi(2-3\xi) - \frac{s_j\cos\phi}{r}\left[\frac{w^{(0)}_{nj}}{s_j}(1-\xi)^2(1+2\xi)\right.$$

$$\left. + \frac{w^{(1)}_{nj}}{s_j}\,\xi^2(3-2\xi) + \frac{\mathrm{d}w^{(0)}_{nj}}{\mathrm{d}s}\,\xi(1-\xi)^2 - \frac{\mathrm{d}w^{(1)}_{nj}}{\mathrm{d}s}\,\xi^2(1-\xi)\right]$$

$$\left. + \frac{1}{4}\left(\frac{\sin\phi}{r} - \frac{3}{R_\phi}\right)[u^{(1)}_{\phi nj}(1-\xi) + u^{(0)}_{\phi nj}\xi]\right\rangle$$

$$+ \frac{1}{4}\left(\frac{3\sin\phi}{r} - \frac{3}{R_\phi}\right)\left\{\frac{\tilde{u}^{(1)}_{\theta nj} - \tilde{u}^{(0)}_{\theta nj}}{s_j} - \frac{\cos\phi}{r}[\tilde{u}^{(0)}_{\theta nj}(1-\xi) + \tilde{u}^{(1)}_{\theta nj}\xi]\right\} \tag{11.3.8f}$$

$$\Theta_{\phi nj} = \frac{w_{nj}^{(1)} - w_{nj}^{(0)}}{s_j}\, 6\xi(1-\xi) + \frac{\mathrm{d}w_{nj}^{(0)}}{\mathrm{d}s}(1-\xi)(1-3\xi) - \frac{\mathrm{d}w_{nj}^{(1)}}{\mathrm{d}s}\,\xi(2-3\xi)$$

$$- \frac{1}{R_\phi}\left[u_{\phi nj}^{(0)}(1-\xi) + u_{\phi nj}^{(1)}\xi\right], \tag{11.3.8g}$$

$$\Theta_{\theta nj} = -\frac{ns_j}{r}\left[\frac{w_{nj}^{(0)}}{s_j}(1-\xi)^2(1+2\xi) + \frac{w_{nj}^{(1)}}{s_j}\,\xi^2(3-2\xi)\right.$$

$$\left. + \frac{\mathrm{d}w_{nj}^{(0)}}{\mathrm{d}s}\,\xi(1-\xi)^2 - \frac{\mathrm{d}w_{nj}^{(1)}}{\mathrm{d}s}\,\xi^2(1-\xi)\right] - \frac{\sin\phi}{r}\left[\tilde{u}_{\theta nj}^{(0)}(1-\xi) + \tilde{u}_{\theta nj}^{(1)}\xi\right]\cdot \tag{11.3.8h}$$

The potential energy for the segment is now obtained as a function of end point deformations by substituting Eqs. (11.3.8) into Eqs. (11.3.3a) and (11.3.3b) and performing the integrations. We obtain derivatives of the segment potential energy with respect to the end point deformations

$$\frac{\partial U_{nj}}{\partial a_{nj}} = \int_0^1 \left[\frac{\partial W_{nj}}{\partial a_{nj}} - \left(q_{\phi n}^* \frac{\partial u_{\phi nj}}{\partial a_{nj}} + q_{\theta n}^* \frac{\partial \tilde{u}_{\theta nj}}{\partial a_{nj}} + q_{\zeta n}^* \frac{\partial w_{nj}}{\partial a_{nj}}\right.\right.$$

$$\left.\left. - m_{\phi n}^* \frac{\partial \Theta_{\phi nj}}{\partial a_{nj}} - \tilde{m}_{\theta n}^* \frac{\partial \Theta_{\theta nj}}{\partial a_{nj}}\right)\right] r s_j\, \mathrm{d}\xi$$

$$- \left(N_{\phi n}^* \frac{\partial u_{\phi nj}}{\partial a_{nj}} + \tilde{N}_{\phi\theta n}^* \frac{\partial \tilde{u}_{nj}}{\partial a_{nj}} + Q_{\phi n}^* \frac{\partial w_{nj}}{\partial a_{nj}} - M_{\phi n}^* \frac{\partial \Theta_{\phi nj}}{\partial a_{nj}}\right.$$

$$\left. - \tilde{M}_{\phi\theta}^* \frac{\partial \Theta_{\theta nj}}{\partial a_{nj}}\right) r\Big|_0^1, \tag{11.3.9a}$$

with

$$\frac{\partial W_{nj}}{\partial a_{nj}} = \left[K(\epsilon_{\phi nj} + \nu\epsilon_{\theta nj}) + D\left(\frac{1}{R_\theta} - \frac{1}{R_\phi}\right)\left(\kappa_{\phi nj} - \frac{\epsilon_{\phi nj}}{R_\phi}\right)\right]\frac{\partial \epsilon_{\phi nj}}{\partial a_{nj}}$$

$$+ \left[K(\epsilon_{\theta nj} + \nu\epsilon_{\phi nj}) + D\left(\frac{1}{R_\phi} - \frac{1}{R_\theta}\right)\left(\kappa_{\theta nj} - \frac{\epsilon_{\theta nj}}{R_\theta}\right)\right]\frac{\partial \epsilon_{\theta nj}}{\partial a_{nj}}$$

$$+ \frac{1-\nu}{2}\left[K + \tfrac{3}{8}(1-\nu)D\left(\frac{1}{R_\theta} - \frac{1}{R_\phi}\right)^2\right]\gamma_{\phi\theta nj}\frac{\partial \gamma_{\phi\theta nj}}{\partial a_{nj}}$$

$$+ D\left[\kappa_{\phi nj} + \nu\kappa_{\theta nj} + \left(\frac{1}{R_\theta} - \frac{1}{R_\phi}\right)\epsilon_{\phi nj}\right]\frac{\partial \kappa_{\phi nj}}{\partial a_{nj}}$$

$$+ D\left[\kappa_{\theta nj} + \nu\kappa_{\phi nj} + \left(\frac{1}{R_\phi} - \frac{1}{R_\theta}\right)\epsilon_{\theta nj}\right]\frac{\partial \kappa_{\theta nj}}{\partial a_{nj}}$$

$$+ 2(1-\nu)D\bar{\kappa}_{\phi nj}\frac{\partial \bar{\kappa}_{\phi\theta nj}}{\partial a_{nj}}. \tag{11.3.9b}$$

For interior segments the terms involving edge applied forces are not present since these forces can be prescribed only at shell boundaries. The integrals which must be calculated may in general be obtained by means of numerical integration methods. It is obvious that the length of the element need not be constant so that smaller segments may be used in the vicinity of regions of rapid deformation change than in regions of slowly varying deformation. The derivative equations may be cast in matrix form as

$$\left\{\frac{\partial U_{nj}}{\partial a_{nj}}\right\} = [k_{ij}]\{a_{nj}\} + \{q_{nj}\}, \tag{11.3.10}$$

where the matrix of the coefficients k_{ij} of the end point deformations a_{nj} is symmetrical. This matrix depends only on the shell geometry and is called a stiffness matrix since the coefficients may be interpreted as approximate values of the influence coefficients for the short segment subjected to edge deformations. The elements q_{nj} of the column matrix depend on the surface and edge load distribution as well as shell geometry and thus vary from problem to problem.

Since the total potential energy is given by the sum of the segment potential energies

$$U_n = \sum_{j=1}^{N} U_{nj}, \tag{11.3.11a}$$

the equations for the determination of the nodal point deformations are

$$\sum_{j=1}^{N} \frac{\partial U_{nj}}{\partial a_{nj}} = 0, \tag{11.3.11b}$$

subject to the conditions that corresponding quantities at junctions of adjacent segments are identical, i.e.,

$$a_{nj}^{(1)} = a_{n,j+1}^{(0)}. \tag{11.3.11c}$$

If an edge value of any of the deformation quantities is prescribed, the equation corresponding to differentiation with respect to that quantity is deleted from Eqs. (11.3.11b) and the prescribed value is substituted for the equation and in any other equation in which the quantity appears. Special considerations must be given to boundary conditions at an apex of the shell. These conditions are as follows:†

$$n = 0: \qquad u_{\phi n} = \tilde{u}_{\theta n} = \frac{\mathrm{d}w_n}{\mathrm{d}s} = 0, \tag{11.3.12a}$$

† See G. A. Greenbaum, 'Comments on Numerical Analysis of Unsymmetrical Bending of Shells of Revolution,' *AIAA J.*, vol. 2, no. 3, March 1964, pp. 590–591, and B. Budiansky and P. P. Radkowski, 'Reply by Authors to G. A. Greenbaum,' *AIAA J.*, vol. 2, no. 3, March 1964, p. 592.

$$n = 1: \qquad u_{\phi n} + \tilde{u}_{\theta n} = w_n = 0, \tag{11.3.12b}$$

$$n \geq 2: \qquad u_{\phi n} = \tilde{u}_{\theta n} = w_n = \frac{\mathrm{d}w_n}{\mathrm{d}s} = 0. \tag{11.3.12c}$$

Sometimes another approximation is introduced for ease of computation. This involves approximating the curved shell by a series of conical segments and using the above analysis for each cone element.† Since the element changes direction at each node, it is necessary to transform edge displacements of adjacent segments to a common coordinate system before imposing continuity conditions and minimizing with respect to edge displacements. We may, for example, use a direction which bisects the angle between the two cones and the perpendicular direction. Then

$$\tilde{u}^{(1)}_{\theta n,j-1} = \tilde{u}^{(0)}_{\theta nj}, \tag{11.3.13a}$$

$$\frac{\mathrm{d}w^{(1)}_{n,j-1}}{\mathrm{d}s} = \frac{\mathrm{d}w^{(0)}_{nj}}{\mathrm{d}s} \tag{11.3.13b}$$

$$u^{(0)}_{\phi nj} = \hat{u}^{(0)}_{nj} \cos \frac{\phi_{j-1} - \phi_j}{2} - \hat{w}^{(0)}_{nj} \sin \frac{\phi_{j-1} - \phi_j}{2}, \tag{11.3.13c}$$

$$w^{(0)}_{nj} = \hat{u}^{(0)}_{nj} \sin \frac{\theta_{j-1} - \phi_j}{2} + \hat{w}^{(0)}_{nj} \frac{\cos \phi_{j-1} - \phi_j}{2}, \tag{11.3.13d}$$

$$u^{(1)}_{\phi n,j-1} = \hat{u}^{(1)}_{n,j-1} \frac{\cos \phi_{j-1} - \phi_j}{2} + \hat{w}^{(1)}_{n,j-1} \sin \frac{\phi_{j-1} - \phi_j}{2}, \tag{11.3.13e}$$

$$w^{(1)}_{n,j-1} = -\hat{u}^{(1)}_{n,j-1} \sin \frac{\phi_{j-1} - \phi_j}{2} + \hat{w}^{(1)}_{n,j-1} \frac{\cos \phi_{j-1} - \phi_j}{2}, \tag{11.3.13f}$$

$$\hat{u}^{(1)}_{n,j-1} = \hat{u}^{(0)}_{nj}, \tag{13.3.13g}$$

$$\hat{w}^{(1)}_{n,j-1} = \hat{w}^{(0)}_{nj}. \tag{13.3.13h}$$

These relations would be used in Eqs. (11.3.8) prior to carrying out the operations of Eqs. (11.3.9).

In either case the end result is a set of linear simultaneous equations for the nodal deformation components which may then be solved. Stress resultants may be found by means of Eqs. (11.3.8) and the stress-resultant/strain relationships. Since the first derivatives of $u_{\phi n}$ and $\tilde{u}_{\theta n}$ and the second derivatives of w_n are not continuous, at the segment ends, the stress-resultants calculated in this manner will also be discontinuous. A continuous variation may be obtained by drawing a continuous curve through the values at say, the center of each segment or through the average of the discontinuous end point

† See, for example, J. H. Percy, T. H. H. Pian, S. Klein, and D. R. Navaratna, 'Application of the Matrix Displacement Method to Linear Elastic Analysis of Shells of Revolution,' *AIAA J.*, vol. 3, no. 11, November 1965, pp. 2138–2145.

values. As the number of segments increases, the stress-resultant discontinuities diminish. An alternate method is to use Eqs. (11.3.10) to calculate forces and moments at each end of the segment since the application of Castigliano's principle enables us to write

$$\left\{\frac{\partial U_{nj}}{\partial a_{nj}}\right\}^T = \left\{\frac{\partial U_{nj}}{\partial u^{(0)}_{\phi nj}}, \frac{\partial U_{nj}}{\partial u^{(0)}_{\theta nj}}, \frac{\partial U_{nj}}{\partial W^{(0)}_{nj}}, \frac{\partial U_{nj}}{\partial \frac{\mathrm{d}w^{(0)}_{nj}}{\mathrm{d}s}}, \frac{\partial U_{nj}}{\partial u^{(1)}_{\phi nj}}, \frac{\partial U_{nj}}{\partial u^{(1)}_{\theta nj}}, \frac{\partial U_{nj}}{\partial w^{(1)}_{nj}}, \frac{\partial U_{nj}}{\partial \frac{\mathrm{d}w^{(1)}_{nj}}{\mathrm{d}s}}\right\}$$

$$= \left\{-N^{(0)}_{\phi nj}, -\left[N^{(0)}_{\phi\theta nj} + \tfrac{1}{2}\left(\frac{3}{R_\theta} - \frac{1}{R_\phi}\right)M^{(0)}_{\phi\theta nj}\right], -\bar{Q}^{(\theta)}_{\phi nj}, -M^{(\theta)}_{\phi nj},\right.$$

$$\left. N^{(1)}_{\phi nj}, \left[N^{(1)}_{\phi\theta nj} + \tfrac{1}{2}\left(\frac{3}{R_\theta} - \frac{1}{R_\phi}\right)M^{(1)}_{\phi\theta nj}\right], Q^{(1)}_{\phi nj}, M^{(1)}_{\phi nj}\right\} \quad (11.3.14)$$

Thus with the deformations at each end of the segment known the end force components are determined. Eqs. (11.3.11b) imply that the corresponding stress-resultants calculated from the equations for adjoining segments will

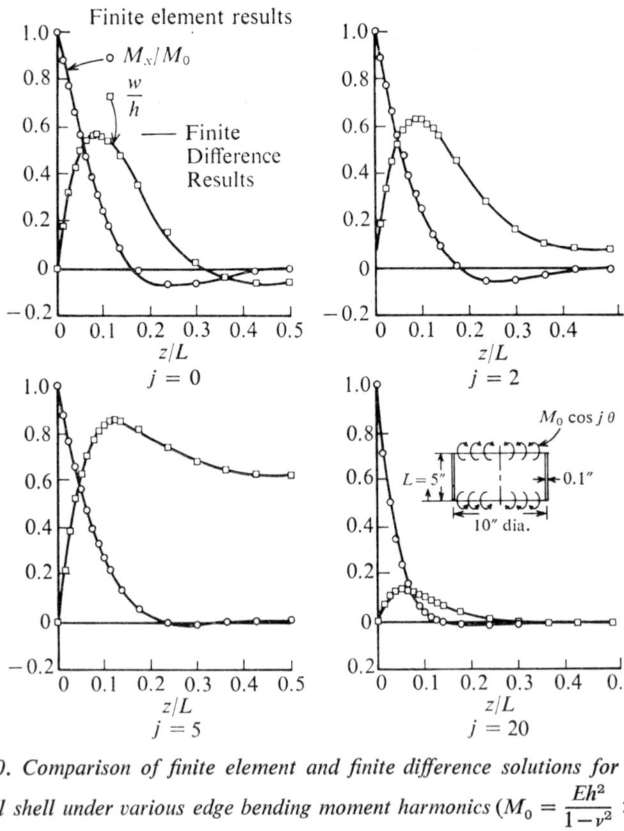

Fig. 11.10. Comparison of finite element and finite difference solutions for a supported cylindrical shell under various edge bending moment harmonics ($M_0 = \frac{Eh^2}{1-\nu^2} \times 10^{-2}$)

be identical. In Fig. 11.10 a comparison of the results of a finite element analysis and of a finite difference analysis of the bending of a cylindrical shell with supported edges subjected to various harmonics of meridional bending moment indicates the accuracy which may be obtained. Similar good agreement between finite element† and exact results‡ for a clamped spherical cap under external pressure is shown in Fig. 11.11.

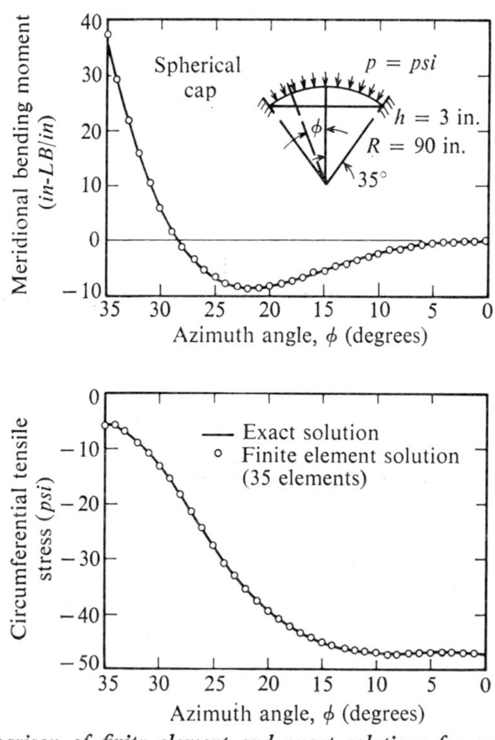

Fig. 11.11. Comparison of finite element and exact solutions for spherical cap under external pressure

It is possible to extend the accuracy of the analysis by including additional terms multiplied by arbitrary constants in Eqs. (11.3.7). The constants may be obtained by minimizing the strain energy (or the potential energy) of each segment with respect to the additional arbitrary coefficients, subject to the restriction that the deflection and rotation at each edge be arbitrary. Such a procedure is essentially an application of the standard Rayleigh–Ritz

† J. H. Percy *et al.*, op. cit.

‡ S. Timoshenko and S. Woinowsky Krieger, *Theory of Plates and Shells*, 2nd Edition, McGraw-Hill Book Co., Inc., 1959, p. 475.

method to a shell segment with prescribed edge displacements and rotations and yields improved equations.† The results of an investigation on the effect of increasing the number of terms in the Taylor's series for the displacements for the case of axisymmetric edge loading ($n = 0$) of a circular cylindrical shell are shown in Fig. 11.12.

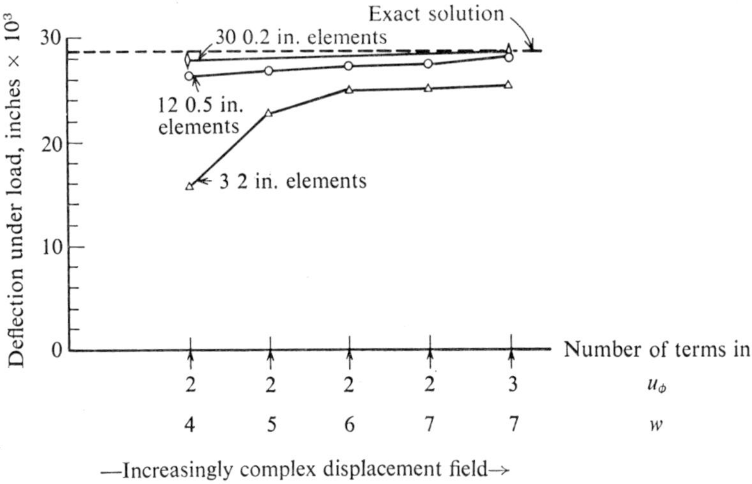

Fig. 11.12. Convergence of end displacements of a cylindrical shell with increasingly complex displacements fields and increasing number of elements

11.4 Shallow translational shells

Few problems of shells for which the middle surface is not a surface of revolution have been solved with any degree of accuracy by means of the exact bending theory of shells.‡ Approximate solutions may be obtained for some shells with the use of the shallow shell equations given by Eq. (3.4.8). We restrict ourselves to shallow shells whose middle surface is described by a second-order equation of the form

† J. H. Percy *et al.*, op. cit.

‡ Some special cases of loading of helicoidal shells have been investigated. See, for example, J. W. Cohen, 'The Inadequacy of the Classical Stress–Strain Relations for the Helicoidal Shell,' *Proc. IUTAM Symposium on the Theory of Thin Elastic Shells*, North Holland Publishing Co., Amsterdam, 1960, pp. 415–433; and E. Reissner, 'On Twisting and Stretching of Helicoidal Shells,' ibid., pp. 434–466. Ring-stiffened oval cylinders have been studied in a series of papers by J. Kempner and his associates. Some results and bibliographies are available in W. P. Vafakos, N. Nissel, J. Kempner, 'Pressurized Oval Cylinders with Closely Spaced Rings,' *AIAA J.*, vol. 4, no. 21, February 1960, pp. 338–348. See also G. B. J. Mah, 'Numerical Analysis of Noncircular Cylindrical Shells,' *J. Eng. Mech. Div. ASCE*, vol. 93, no. EM-3, June 1967, pp. 219–237.

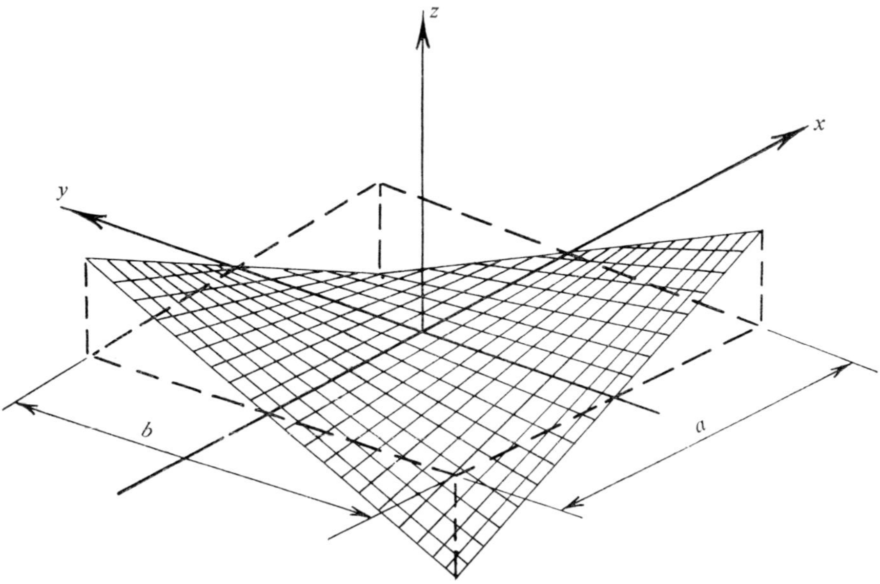

Fig. 11.13. Hyperbolic paraboloid $z = -xy/R_{xy}$

$$z = -\left(\frac{x^2}{2R_x} + \frac{xy}{R_{xy}} + \frac{y^2}{2R_y}\right). \tag{11.4.1}$$

Then Eq. (3.4.8) becomes

$$\left\{\left(\frac{\partial^2}{\partial x^2} + \frac{\partial^2}{\partial y^2}\right)^2 - \frac{[12(1-\nu^2)]^{1/2}}{h}\, i \left(\frac{1}{R_y}\frac{\partial^2}{\partial x^2} - \frac{2}{R_{xy}}\frac{\partial^2}{\partial x\,\partial y} + \frac{1}{R_x}\frac{\partial^2}{\partial y^2}\right)\right\}$$
$$\times \left\{w + i\,\frac{[12(1-\nu)]^{1/2}}{Eh^2}\,\Phi\right\} = \frac{\bar{q}_\zeta}{D}. \tag{11.4.2}$$

Next we assume that the shell is bounded by planes parallel to the $x-z$ and $y-z$ planes and seek complementary solutions of the form

$$w + i\,\frac{[12(1-\nu^2)]^{1/2}}{Eh^2}\,\Phi = F_n(y)\,\mathrm{e}^{\pm i(n\pi x)/a}. \tag{11.4.3a}$$

Then $F_n(y)$ must satisfy the equation

$$\frac{\mathrm{d}^4F_n}{\mathrm{d}y^4} - \left(\frac{n\pi}{a}\right)^2\left\{2 + i\,\frac{[12(1-\nu^2)]^{1/2}}{R_xh}\left(\frac{a}{n\pi}\right)^2\right\}\frac{\mathrm{d}^2F_n}{\mathrm{d}y^2}$$
$$\mp 2\,\frac{[12(1-\nu^2)]^{1/2}}{R_{xy}h}\,\frac{n\pi}{a}\,\frac{\mathrm{d}F_n}{\mathrm{d}y}$$
$$+ \left(\frac{n\pi}{a}\right)^4\left\{1 + i\,\frac{[12(1-\nu^2)]^{1/2}}{R_yh}\left(\frac{a}{n\pi}\right)^2\right\}F_n = 0. \tag{11.4.3b}$$

The solutions of Eqs. (11.4.3b) are given by

$$F_n(y) = \sum_{j=1}^{4} A_{nj}\, e^{\rho_{nj}(n\pi y)/L}, \tag{11.4.3c}$$

where the constants of integration A_{nj} may be complex and the values of ρ_{nj} are the roots of the quartic equation

$$\rho_{nj}^4 - \left\{2 + i\,\frac{[12(1-\nu^2)]^{1/2}}{R_x h}\left(\frac{a}{n\pi}\right)^2\right\}\rho_{nj}^2 \mp 2\,\frac{[12(1-\nu^2)]^{1/2}}{R_{xy} h}\left(\frac{a}{n\pi}\right)^2\rho_{nj}$$

$$+ \left\{1 + i\,\frac{[12(1-\nu^2)]^{1/2}}{R_y h}\left(\frac{a}{n\pi}\right)^2\right\} = 0. \tag{11.4.3d}$$

We note that the solutions of the equation containing one sign of the coefficient preceding ρ_{nj} are the negative of the solutions for the equation containing the other sign. Then the complete complementary solution of the form of Eq. (11.4.3a) may be written as

$$w + i\,\frac{[12(1-\nu^2)]^{1/2}}{Eh^2}\,\Phi = \sum_{j=1}^{4} [A_{nj}\, e^{\rho_{nj}(n\pi y)/a}\, e^{i(n\pi x)/a}$$

$$+ B_{nj}\, e^{-\rho_{nj}(n\pi y)/a}\, e^{-i(n\pi x)/a}] \tag{11.4.4}$$

and may be separated into real and imaginary parts to determine w and ϕ. Eq. (11.4.4) is still valid if the twist $1/R_{xy}$ vanishes.

Let us consider several causes in more detail to determine permissible edge conditions. For the hyperbolic paraboloid bounded by a rectangular set of generators (Fig. 11.13) we have

$$\frac{1}{R_x} = \frac{1}{R_y} = 0, \tag{11.4.5}$$

so that Eq. (11.4.3d) becomes

$$(\rho_{nj}^2 - 1)^2 \mp 2\epsilon_n \rho_{nj} = 0, \tag{11.4.6a}$$

with

$$\epsilon_n = \frac{[12(1-\nu^2)]^{1/2}}{R_{xy} h}\left(\frac{a}{n\pi}\right)^2. \tag{11.4.6b}$$

The solutions of Eq. (11.5.6a) with the upper sign and with

$$\epsilon_n \geq 0, \tag{11.4.7}$$

are given by

$$\rho_{nj} = \tfrac{1}{2}(r_n - 2)^{1/2}\left\{\left(\frac{r_n+2}{r_n-2}\right)^{1/2} \pm \left[2\left(\frac{r_n+2}{r_n-2}\right)^{1/2} - 1\right]^{1/2}\right\} \quad (j=1,2), \tag{11.4.8a}$$

$$\rho_{nj} = \tfrac{1}{2}(r_n - 2)^{1/2}\left\{ - \left(\frac{r_n + 2}{r_n - 2}\right)^{1/2} \pm i\left[2\left(\frac{r_n + 2}{r_n - 2}\right)^{1/2} + 1\right]^{1/2}\right\}$$

$$(j = 3, 4), \quad (11.4.8b)$$

where

$$r_n = \{[(\tfrac{4}{3})^3 + \epsilon_n^2]^{1/2} + \epsilon_n\}^{2/3} + \{[(\tfrac{4}{3})^3 + \epsilon_n^2]^{1/2} - \epsilon_n\}^{2/3} - \tfrac{2}{3}. \qquad (11.4.8c)$$

Examination of Eq. (11.4.4) then indicates that if the expression for w is multiplied by $\cos (n\pi x)/a$, then the expression for Φ is multiplied by $\sin (n\pi x)/a$, and vice versa. The boundary conditions which are satisfied on the edges $x = \pm (a/2)$ are then either those for the special type of simple support

$$w = M_x = N_{xy} = u_x = 0, \qquad (11.4.9a)$$

or those for an edge with fixed slope but free to move transversely

$$\frac{\partial w}{\partial x} = \bar{Q}_x = N_x = u_y = 0. \qquad (11.4.9b)$$

Along the edges $y = \pm (b/2)$ the conditions to be satisfied are arbitrary.

When the twist vanishes the shell is of the translational type shown in Fig. 11.14. The roots of Eq. (11.4.3d) are now given by

$$\rho_{nj} = \pm \left\{\left[\frac{(1 + a_n\alpha_n + 2\alpha_n^2 - 2\beta_n)^{1/2} + (1 - \beta_n)}{2}\right]^{1/2} + i\left[\frac{(1 + a_n\alpha_n + 2\alpha_n^2 - 2\beta_n)^{1/2} - (1 - \beta_n)}{2}\right]^{1/2}\right\} \quad (j = 1, 2),$$

$$(11.4.10a)$$

$$\rho_{nj} = \pm \left\{\left[\frac{(1 - a_n\alpha_n + 2\alpha_n^2 + 2\beta_n)^{1/2} + (1 + \beta_n)}{2}\right]^{1/2} - i\left[\frac{(1 - a_n\alpha_n + 2\alpha_n^2 + 2\beta_n)^{1/2} - (1 + \beta_n)}{2}\right]^{1/2}\right\} \quad (j = 3, 4)$$

$$(11.4.10b)$$

with

$$\alpha_n = \left\{\frac{\left[\left(\frac{a_n}{2}\right)^4 + (b_n - a_n)^2\right]^{1/2} + \left(\frac{a_n}{2}\right)^2}{2}\right\}^{1/2}, \qquad (11.4.10c)$$

$$\beta_n = \left\{\frac{\left[\left(\frac{a_n}{2}\right)^4 + (b_n - a_n)^2\right]^{1/2} - \left(\frac{a_n}{2}\right)^2}{2}\right\}^{1/2} \operatorname{sgn}(b_n - a_n) \qquad (11.4.10d)$$

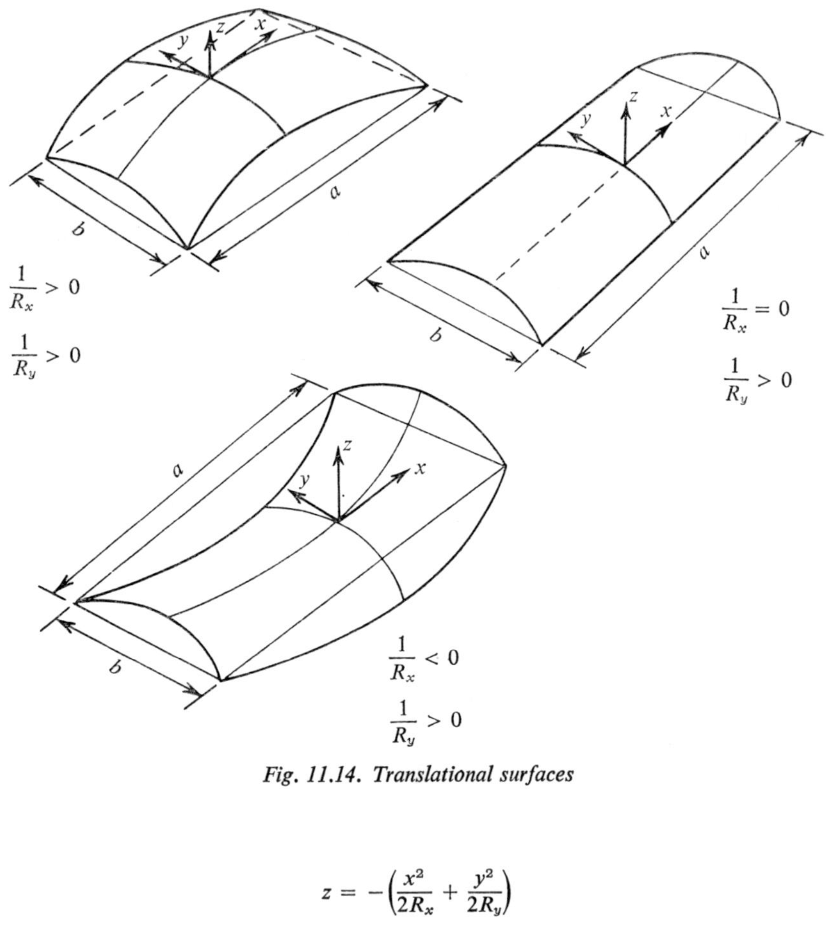

Fig. 11.14. Translational surfaces

$$z = -\left(\frac{x^2}{2R_x} + \frac{y^2}{2R_y}\right)$$

$$a_n = \frac{[12(1 - \nu^2)]^{1/2}}{R_x h} \left(\frac{a}{n\pi}\right)^2, \tag{11.4.10e}$$

$$b_n = \frac{[12(1 - \nu^2)]^{1/2}}{R_y h} \left(\frac{a}{n\pi}\right)^2. \tag{11.4.10f}$$

Examination of Eq. (11.4.4) then indicates that both the expressions for w and Φ are multiplied by either $\cos (n\pi x)/a$ or by $\sin (n\pi x)/a$. The boundary conditions at the edges $x = \pm(a/2)$ are then either those for a more usual type of simple support

$$w = M_x = N_x = u_y = 0, \tag{11.4.11a}$$

or those for another type of edge with fixed slope but free to move transversely

$$\frac{\partial w}{\partial x} = \bar{Q}_x = u_x = \bar{N}_{xy} = 0. \tag{11.4.11b}$$

Along the edges $y = \pm(b/2)$ the conditions to be satisfied are arbitrary.

The solutions given above are appropriate for Lévy-type† solutions of various problems of surface-loaded shells, provided the conditions presented along the edges $x = \pm(a/2)$ are of the form of Eqs. (11.4.9) for the hyperbolic paraboloid bounded by a rectangular set of generators or of Eqs. (11.4.11) for translational shells. A particular solution of Eq. (11.4.2) may be obtained by expanding the function $\bar{q}_\zeta/D$ in an appropriate Fourier series in the x direction. For example, if the edges $x = \pm(a/2)$ are simply supported we may write

$$\frac{\bar{q}_\zeta(x, y)}{D} = \frac{1}{D}\left[\sum_{n=1,3,5,\cdots}^{\infty} \bar{q}_{n1}(y) \cos\frac{n\pi x}{a} + \sum_{n=2,4,6,\cdots}^{\infty} \bar{q}_{n2}(y) \sin\frac{n\pi x}{a}\right], \tag{11.4.12a}$$

with

$$\bar{q}_{n1} = \frac{2}{a}\int_{-a/2}^{a/2} \bar{q}_\zeta \cos\frac{n\pi x}{a}\,dx, \tag{11.4.12b}$$

$$\bar{q}_{n2} = \frac{2}{a}\int_{-a/2}^{a/2} \bar{q}_\zeta \sin\frac{n\pi x}{a}\,dx. \tag{11.4.12c}$$

For translational shells we then find

$$w_p + i\frac{[12(1-\nu^2)]^{1/2}}{Eh^2}\Phi_p = \sum_{n=1,3,5,\cdots}^{\infty} F_{n1}\cos\frac{n\pi x}{a} + \sum_{n=2,4,6,\cdots}^{\infty} F_{n2}\sin\frac{n\pi x}{a}, \tag{11.4.13a}$$

where F_{n1} and F_{n2} are particular solutions of the ordinary differential equations

$$\frac{d^4F_{nj}}{dy^4} - \left(\frac{n\pi}{a}\right)^2\left\{2 + i\frac{[12(1-\nu^2)]^{1/2}}{R_x h}\left(\frac{a}{n\pi}\right)^2\right\}\frac{d^2F_{nj}}{dy^2} + \left(\frac{n\pi}{a}\right)^4\left\{1 + i\frac{[12(1-\nu^2)]^{1/2}}{R_y h}\left(\frac{a}{n\pi}\right)^2\right\}F_{nj} = \frac{\bar{q}_{nj}(y)}{D}. \tag{11.4.13b}$$

† See, for instance, E. H. Mansfield, *The Bending and Stretching of Plates*, Pergamon Press, 1964, pp. 25–30.

For hyperbolic paraboloids bounded by a rectangular set of generators the procedure is somewhat more complicated in that both summations in Eq. (11.4.13a) are taken over all values of n and the equations for F_{n1} and F_{n2} are coupled.

Calculations made for various problems† indicate the necessity for the use of the bending solutions rather than membrane solutions for shallow shells, particularly those of negative Gaussian curvature.

11.5 Finite element analysis of shells of arbitrary middle surface

Shell structures of arbitrary middle surface can, in principle, be analysed numerically in a number of different ways. One procedure would be to approximate the differential equations by a set of finite difference equations relating displacements at discrete points on the shell middle surface.‡ A variation on such a procedure has been used successfully for the analysis of circumferential segments of shells of revolution.§ For arbitrary shells the method has certain disadvantages. It is difficult to write the finite difference equations for an arbitrary network of points on the shell middle surface. It is also difficult to keep track of the equations for each point since they depend on the behavior and the distribution of the network in the vicinity of the point and on the closeness of the point to boundaries. We are therefore led to the finite element method in hopes of reducing the complexity of obtaining equations for the analysis of arbitrary shells.

The application of the finite element method to arbitrary shell structures is not without problems to be surmounted. Not least of these is how to subdivide the shell into finite elements. If we choose, say, three nodal points for a 'triangular' element on the shell middle surface (Fig. 11.15) we still have not defined the element completely since the curves connecting any two points must also be defined. In flat plate analysis the shortest curve between two points, a straight line, is used. For curved shell analysis we could use the geodesics passing through the points, but the amount of analysis required to

† A. L. Bouma, 'Some Applications of the Bending Theory Regarding Doubly Curved Shells,' *Proc. IUTAM Symposium on the Theory of Thin Elastic Shells*, North-Holland Publishing Co., Amsterdam, 1960, pp. 202–234; and K. Apeland and E. P. Popov, 'Analysis of Bending Stresses in Translational Shells,' *Proc. Colloquium on Simplified Calculation Methods of Shell Structures*, North-Holland Publishing Co., Amsterdam, 1962, pp. 9–30.

‡ Such an analysis is described in B. O. Almroth, F. A. Brogan and M. B. Marlowe, 'Collapse Analysis for Elliptic Cones,' *AIAA J.*, vol. 9, no. 1, January 1971, pp. 32–37.

§ A. Kalnins, 'Analysis of Curved Thin-walled Shells of Revolution,' *AIAA J.*, vol. 6, no. 4, April 1968, pp. 584–588.

define these curves is generally more than we would like to do. We also need to consider that for a triangular flat plate element the location of the three nodal points is all we need to define the geometry of the element whereas for a curved finite element a detailed description of the interior surface would generally be needed.

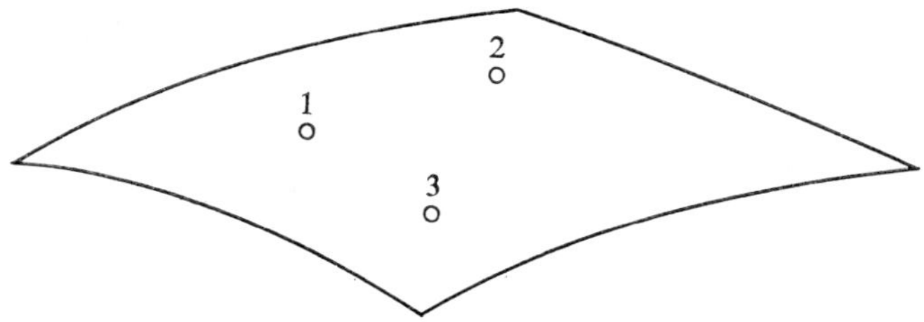

Fig. 11.15. Nodal points on surface

In order to avoid the extensive description needed for a curved element, we introduce an approximation. We replace the continuously curved middle surface by an assembly of flat elements. For an arbitrary shell these are generally flat triangular elements, each of which passes through three nodal points on the middle surface (Fig. 11.16). Such assemblies are not unique,

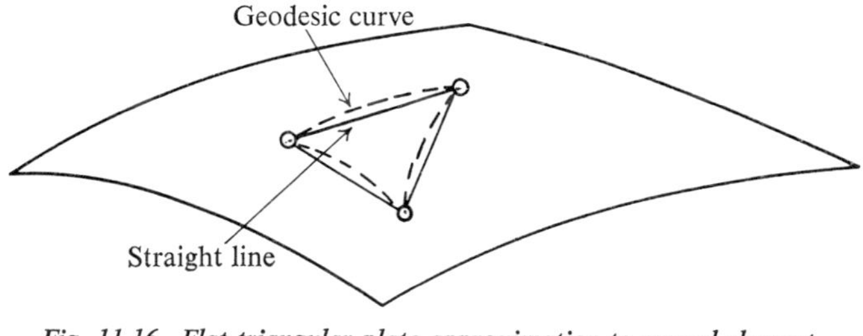

Fig. 11.16. Flat triangular plate approximation to curved element

even when all of the nodal points are prescribed. Also obvious is the fact that the closer the spacing of the middle surface points and the smaller the triangular finite elements, the better the curved surface will be approximated.

The next modification of our previous description of the finite element method is that of the variational principle itself. In the application of the Rayleigh–Ritz approach to variational problems of plate and/or shell analysis it is generally required that deformations and changes of slope be continuous at the junction of integrally connected elements. This requirement is easy to satisfy for one-dimensional problems such as the analysis of each Fourier component of deformation of a shell of revolution, where the slope and displacements must be continuous at the *point* common to adjacent segments of the meridional curve. For arbitrary shells, however, the continuity conditions should be satisfied along the *line* common to adjacent flat plate elements. The literature attests to the difficulty and complexity involved in trying to meet the requirement.† While there is reason to believe that displacement and slope change continuity along the entire boundary of the element are *sufficient* conditions for convergence rather than *necessary* conditions, complete analytic proof of this statement has yet to be devised. In addition, calculations indicate that restrictions must be placed on the chosen displacement functions.

We can surmount the complexities involved in applying the principle of minimum potential energy by turning to an alternate variational formulation of the theory of bending of flat plates‡ subjected to normal surface forces and bending and transverse shell loads:

$$\delta U_B = 0,$$

where

$$\begin{aligned} U_B = \int_A \int \Big\{ & \left(\frac{\partial M_x}{\partial x} + \frac{\partial M_{xy}}{\partial y}\right)\frac{\partial w}{\partial x} + \left(\frac{\partial M_{xy}}{\partial x} + \frac{\partial M_y}{\partial y}\right)\frac{\partial w}{\partial y} \\ & - q_\zeta w - \frac{6}{Eh^3}\left[(M_x + M_y)^2 + 2(1+\nu)(M_{xy} - M_x M_y)\right]\Big\}\, \mathrm{d}x\, \mathrm{d}y \\ & - \oint_S \left(\frac{\partial w}{\partial s} M_{ns} + \bar{Q}_n^* w + M_n \frac{\partial w^*}{\partial n}\right) \mathrm{d}s, \end{aligned} \tag{11.5.1}$$

and $\bar{Q}_n^*$ is the specified effective transverse shear force and $\partial w^*/\partial n$ the specified normal rotation on the boundary S. These quantities are set equal to zero on portions of the boundary where they are not specified. The edge bending and twisting moments per unit length are given by (Fig. 11.17).

† See the exposition in O. C. Zienkiewicz and Y. K. Cheung, *The Finite Element Method in Structural and Continuum Mechanics*, McGraw-Hill Publishing Co., Ltd., 1967, chapter 7, pp. 89–123; see also S. Utku, 'Stiffness Matrices for Thin Triangular Elements of Non-Zero Gaussian Curvature,' *AIAA J.*, vol. 5, no. 9, September 1967, pp. 1650–1657; and G. R. Cowper, E. Kosko, G. M. Lindberg, and M. D. Olson, 'Static and Dynamic Applications of a High-precision Triangular Plate Bending Element," *AIAA J.*, vol. 7, no. 10, October 1969, pp. 1957–1965; G. M. Lindberg and M. D. Olson, 'A High-precision Triangular Cylindrical Shell Finite Element,' *AIAA J.*, vol. 9, no. 3, March 1971, pp. 530–532.

‡ L. R. Herrmann, 'Finite Element Bending Analysis for Plates,' *J. Eng. Mech. Div., ASCE*, vol. 93, no. EM-5, October 1967, pp. 13–26. An improved model is studied by W. Visser, 'A Refined Mixed-type Plate-bending Element,' *AIAA J.*, vol. 7, no. 9, September 1969, pp. 1801–1803.

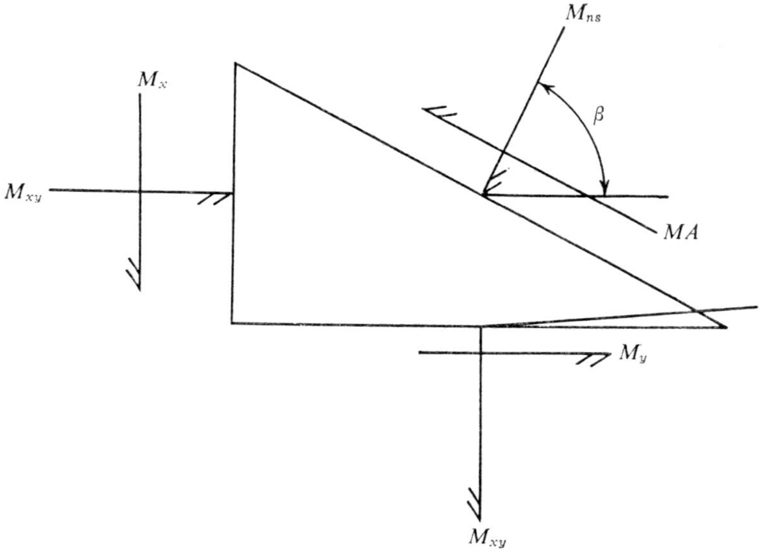

Fig. 11.17. Edge normal and twisting moments

$$M_n = \frac{M_x + M_y}{2} + \frac{M_x - M_y}{2} \cos 2\beta + M_{xy} \sin 2\beta, \tag{11.5.2a}$$

$$M_{ns} = \frac{M_y - M_x}{2} \sin 2\beta + M_{xy} \cos 2\beta, \tag{11.5.2b}$$

where β is the angle between the outwardly directed normal to the edge and the x-axis. In applying the variational principle, the moments M_x, M_y, and M_{xy} as well as the normal deflection w are arbitrary within the restrictions of satisfaction of prescribed edge conditions on normal moment and normal deflection, continuity of normal moment and normal deflection across element boundaries within the plate, and of continuity of second derivatives of moments and normal deflection within each element. Application of the methods of variational calculus as in section 2.7 yields the differential equations to be solved as

$$\frac{\partial^2 w}{\partial x\, \partial y} + \frac{12(1 + \nu)}{Eh^3} M_{xy} = 0, \tag{11.5.3a}$$

$$\frac{\partial^2 w}{\partial x^2} + \frac{12}{Eh^3} (M_x - \nu M_y) = 0, \tag{11.5.3b}$$

$$\frac{\partial^2 w}{\partial y^2} + \frac{12}{Eh^3} (M_y - \nu M_x) = 0, \tag{11.5.3c}$$

$$\frac{\partial^2 M_x}{\partial x^2} + 2\frac{\partial^2 M_{xy}}{\partial x\, \partial y} + \frac{\partial^2 M_y}{\partial y^2} + q = 0, \tag{11.5.3d}$$

which are the usual constitutive equations and equilibrium equation for a flat plate.

In order to apply the 'mixed' variational theorem outlined above to the analysis of shells considered as a conglomeration of triangular flat plate elements† we add to U_B the potential energy of stretching of a plate, i.e.

$$U_S = \int_A \int \left[\frac{K}{2}\left(\epsilon_x^2 + 2\nu\epsilon_x\epsilon_y + \epsilon_y^2 + \frac{1-\nu}{2}\gamma_{xy}^2\right) - (q_x^* u_x + q_y^* u_y)\right] \mathrm{d}x\, \mathrm{d}y - \oint_S (N_n^* u_n + N_{sn}^* u_s)\, \mathrm{d}s, \tag{11.5.4}$$

where it is understood that the first derivatives of u_x and u_y are continuous within the element. For interior triangular flat plate elements the line integral vanishes since the normal and tangential stress resultants N_n^* and N_{sn}^* are not prescribed. For an arbitrary triangle as shown in Fig. 11.18 we choose the displacement field to be a linear function of position

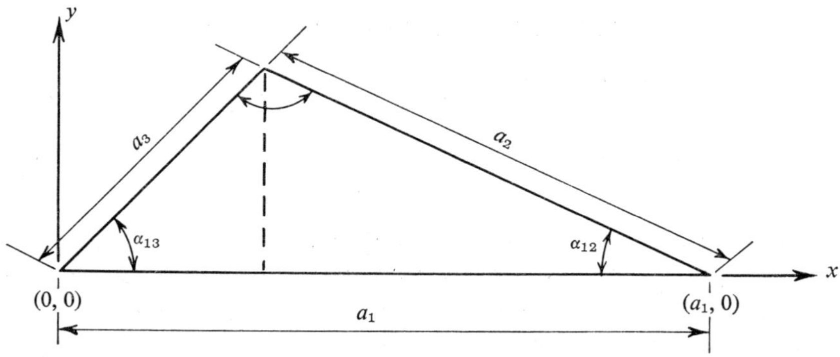

Fig. 11.18. Local coordinate system for arbitrary triangle

$$u_x = u_{x1}\left(1 - \frac{x}{a_1} - \frac{y}{a_1}\cot\alpha_{12}\right) + \frac{u_{x2}}{a_1}(x - y\cot\alpha_{13}) + \frac{u_{x3}}{a_3}y\csc\alpha_{13}, \tag{11.5.5a}$$

$$u_y = u_{y1}\left(1 - \frac{x}{a_1} - \frac{y}{a_1}\cot\alpha_{12}\right) + \frac{u_{y2}}{a_1}(x - y\cot\alpha_{13}) + \frac{u_{y3}}{a_3}y\csc\alpha_{13}, \tag{11.5.5b}$$

† L. R. Herrmann and D. M. Campbell, 'A Finite-element Analysis for Thin Shells,' *AIAA J.*, vol. 6, no. 10, October 1968, pp. 1842–1847.

$$w = w_1\left(1 - \frac{x}{a_1} - \frac{y}{a_1}\cot\alpha_{12}\right) + \frac{w_2}{a_1}(x - y\cot\alpha_{13}) + \frac{w_3}{a_3}y\csc\alpha_{13}, \tag{11.5.5c}$$

yielding the constant strain distribution

$$\epsilon_x = \frac{\partial u_x}{\partial x} = \frac{u_{x2} - u_{x1}}{a_1}, \tag{11.5.6a}$$

$$\epsilon_y = \frac{\partial u_y}{\partial y} = \frac{u_{y3}}{a_3\sin\alpha_{13}} - \frac{u_{y2}}{a_1\tan\alpha_{13}} - \frac{u_{y1}}{a_1\tan\alpha_{12}}, \tag{11.5.6b}$$

$$\gamma_{xy} = \frac{\partial u_x}{\partial y} + \frac{\partial u_y}{\partial x} = \frac{u_{x3}}{a_3\sin\alpha_{13}} - \frac{u_{x2}}{a_1\tan\alpha_{13}} - \frac{u_{x1}}{a_1\tan\alpha_{12}} + \frac{u_{y2} - u_{y1}}{a_1}. \tag{11.5.6c}$$

The rotation along each edge is also needed and is given by

$$\left(\frac{\partial w}{\partial s}\right)_1 = \frac{w_2 - w_1}{a_1}, \tag{11.5.7a}$$

$$\left(\frac{\partial w}{\partial s}\right)_2 = \frac{w_3 - w_2}{a_2}, \tag{11.5.7b}$$

$$\left(\frac{\partial w}{\partial s}\right)_3 = \frac{w_1 - w_3}{a_3}. \tag{11.5.7c}$$

The moments within the triangle are also assumed to be constant. We note then that terms involving derivatives of the moments vanish in Eq. (11.5.1). If we assume

$$M_x = C_x, \tag{11.5.8a}$$

$$M_y = C_y, \tag{11.5.8b}$$

$$M_{xy} = C_{xy}, \tag{11.5.8c}$$

the normal moments at the three edges of the triangle are given by

$$M_{n1} = C_y, \tag{11.5.9a}$$

$$M_{n2} = \frac{C_x + C_y}{2} - \frac{C_x - C_y}{2}\cos 2\alpha_{12} + C_{xy}\sin 2\alpha_{12}, \tag{11.5.9b}$$

$$M_{n3} = \frac{C_x + C_y}{2} - \frac{C_x - C_y}{2}\cos 2\alpha_{13} - C_{xy}\sin 2\alpha_{13}. \tag{11.5.9c}$$

We may solve for C_x, C_y, and C_{xy} in terms of the three normal moments to yield

$$M_x = \frac{-2M_{n1}(\cos\alpha_{12}\cos\alpha_{13}\sin\alpha_{23}) + M_{n2}\sin 2\alpha_{13} + M_{n3}\sin 2\alpha_{12}}{2\sin\alpha_{12}\sin\alpha_{23}\sin\alpha_{31}}, \tag{11.5.10a}$$

$$M_y = M_{n1}, \tag{11.5.10b}$$

$$M_{xy} = \frac{M_{n1}(\sin^2\alpha_{12} - \sin^2\alpha_{13}) + M_{n2}\sin^2\alpha_{13} - M_{n3}\sin^2\alpha_{12}}{2\sin\alpha_{12}\sin\alpha_{23}\sin\alpha_{31}}. \tag{11.5.10c}$$

The normal moments are now the independent variables in the analysis. We may use the moments given by Eqs. (11.5.10) in Eq. (11.5.2b) to determine the edge twisting moment M_{ns} as

$$M_{ns1} = \frac{M_{n1}(\sin^2\alpha_{13} - \sin^2\alpha_{12}) - M_{n2}\sin^2\alpha_{13} + M_{n3}\sin^2\alpha_{12}}{2\sin\alpha_{12}\sin\alpha_{23}\sin\alpha_{31}}, \tag{11.5.11a}$$

$$M_{ns2} = \frac{M_{n1}\sin^2\alpha_{23} + M_{n2}(\sin^2\alpha_{12} - \sin^2\alpha_{23}) - M_{n3}\sin^2\alpha_{12}}{2\sin\alpha_{12}\sin\alpha_{23}\sin\alpha_{31}}, \tag{11.5.11b}$$

$$M_{ns3} = \frac{-M_{n1}\sin^2\alpha_{23} + M_{n2}\sin^2\alpha_{13} + M_{n3}(\sin^2\alpha_{23} - \sin^2\alpha_{13})}{2\sin\alpha_{12}\sin\alpha_{23}\sin\alpha_{31}}. \tag{11.5.11c}$$

All of the quantities which are required in Eqs. (11.5.1) and (11.5.4) have now been determined in terms of three displacements at each of the three corners of the triangular plate element and three constant edge normal moments. Then with the help of matrix notation we have

$$U_B = \{\delta_B'\}^T \left(\frac{1}{2} \left[\begin{array}{c|c} 0 & k_A \\ \hline k_A & k_A' \end{array} \right] \left\{ \delta_B' \right\} - \left\{ \begin{array}{c} k_B \\ \hline 0 \end{array} \right\} \right), \tag{11.5.12a}$$

where

$$\{\delta_B'\} = \begin{Bmatrix} w_1 \\ w_2 \\ w_3 \\ M_{n2} \\ M_{n3} \\ M_{n1} \end{Bmatrix}, \tag{11.5.12b}$$

$$[k_A] = \frac{1}{2 \sin \alpha_{12} \sin \alpha_{23} \sin \alpha_{31}}$$

$$\times \begin{bmatrix} -\sin^2 \alpha_{31} & \sin^2 \alpha_{12} - \sin^2 \alpha_{23} + \sin^2 \alpha_3 & \sin^2 \alpha_{31} - \sin^2 \alpha_{12} + \sin^2 \alpha_{23} \\ \sin^2 \alpha_{12} - \sin^2 \alpha_{23} + \sin^2 \alpha_{31} & -2 \sin^2 \alpha_{12} & \sin^2 \alpha_{23} - \sin^2 \alpha_{31} + \sin^2 \alpha_{12} \\ \sin^2 \alpha_{31} - \sin^2 \alpha_{12} + \sin^2 \alpha_{23} & \sin^2 \alpha_{23} - \sin^2 \alpha_{31} + \sin^2 \alpha_{12} & -2 \sin^2 \alpha_{23} \end{bmatrix}, \tag{11.5.12c}$$

$$[k_A'] = -\left(\iint_A \frac{12\,\mathrm{d}x\,\mathrm{d}y}{Eh^3}\right)[M]^T \begin{bmatrix} 1 & -\nu & 0 \\ -\nu & 1 & 0 \\ 0 & 0 & 2(1+\nu) \end{bmatrix} [M], \tag{11.5.12d}$$

$$[M] = \frac{1}{2 \sin \alpha_{12} \sin \alpha_{23} \sin \alpha_{31}}$$

$$\times \begin{bmatrix} \sin 2\alpha_{13} & \sin 2\alpha_{12} & -2 \cos \alpha_{12} \cos \alpha_{13} \sin \alpha_{23} \\ 0 & 0 & 2 \sin \alpha_{12} \sin \alpha_{13} \sin \alpha_{23} \\ \sin^2 \alpha_{13} & -\sin^2 \alpha_{12} & \sin^2 \alpha_{12} - \sin^2 \alpha_{13} \end{bmatrix}, \tag{11.5.12e}$$

$$\{k_B\} = \begin{Bmatrix} \displaystyle\iint_A q_\zeta\left(1 - \frac{x + y \cot \alpha_{12}}{a_1}\right) \mathrm{d}x\,\mathrm{d}y \\ \displaystyle\iint_A q_\zeta\left(\frac{x - y \cot \alpha_{13}}{a_1}\right) \mathrm{d}x\,\mathrm{d}y \\ \displaystyle\iint_A q_\zeta \frac{y \csc \alpha_{13}}{a_3}\, \mathrm{d}x\,\mathrm{d}y \end{Bmatrix}. \tag{11.5.12f}$$

We also have

$$U_S = \{\delta_s'\}^T([k_C]\{\delta_s'\} - \{k_D\}) \tag{11.5.13a}$$

with

$$\{\delta_s'\} = \begin{Bmatrix} u_{x1} \\ u_{y1} \\ u_{x2} \\ u_{y2} \\ u_{x3} \\ u_{y3} \end{Bmatrix}, \tag{11.5.13b}$$

$$[k_C] = \tfrac{1}{2}\left(\int_A\int K \, dx \, dy\right)[k_C']^T \begin{bmatrix} 1 & \nu & 0 \\ \nu & 1 & 0 \\ 0 & 0 & \dfrac{1-\nu}{2} \end{bmatrix} [k_C'], \tag{11.5.13c}$$

$$[k_C'] = \begin{bmatrix} -\dfrac{1}{a_1} & 0 & \dfrac{1}{a_1} & 0 & 0 & 0 \\ 0 & -\dfrac{1}{a_1 \tan\alpha_{12}} & 0 & -\dfrac{1}{a_1 \tan\alpha_{13}} & 0 & \dfrac{1}{a_3 \sin\alpha_{13}} \\ -\dfrac{1}{a_1 \tan\alpha_{12}} & -\dfrac{1}{a_1} & -\dfrac{1}{a_1 \tan\alpha_{13}} & \dfrac{1}{a_1} & \dfrac{1}{a_3 \sin\alpha_{13}} & 0 \end{bmatrix}, \tag{11.5.13d}$$

$$\{k_D\} = \begin{Bmatrix} \int_A\int q_x^*\left(1 - \dfrac{x + y\cot\alpha_{12}}{a_1}\right) dx\,dy \\ \int_A\int q_y^*\left(1 - \dfrac{x + y\cot\alpha_{12}}{a_1}\right) dx\,dy \\ \int_A\int q_x^*\left(\dfrac{x - y\cot\alpha_{13}}{a_1}\right) dx\,dy \\ \int_A\int q_y^*\left(\dfrac{x - y\cot\alpha_{13}}{a_1}\right) dx\,dy \\ \int_A\int q_x^* \dfrac{y\csc\alpha_{13}}{a_3}\, dx\,dy \\ \int_A\int q_y^* \dfrac{y\csc\alpha_{13}}{a_3}\, dx\,dy \end{Bmatrix}. \tag{11.5.13e}$$

We may combine Eqs. (11.5.12) and (11.5.13) into the equivalent expression

$$U_B + U_S = \{\delta'\}^T(\tfrac{1}{2}[K']\{\delta'\} - \{p'\}), \tag{11.5.14a}$$

where

$$\{\delta'\} = \begin{Bmatrix} u_{x1} \\ u_{y1} \\ w_1 \\ u_{x2} \\ u_{y2} \\ w_2 \\ u_{x3} \\ u_{y3} \\ w_3 \\ M_{n2} \\ M_{n3} \\ M_{n1} \end{Bmatrix} \tag{11.5.14b}$$

and $[K']$ is a symmetrical matrix.

We note that continuity of normal moments will be assured if the edge normal moments for the common edges of adjacent triangles are equal. Since the displacements are linear along edges, continuity of displacements will be assured if the displacement vector at a corner common to adjacent elements is the same for both elements. To expedite matters it is convenient to express the corner displacement components for all elements in terms of a common Cartesian coordinate system. If the Cartesian coordinates of each corner of a triangle are given by (x_i, y_i, z_i) (Fig. 11.19), then the required transformation is

$$\begin{bmatrix} \bar{u}_x \\ \bar{u}_y \\ \bar{u}_z \end{bmatrix} = \begin{bmatrix} \dfrac{x_2 - x_1}{a_1} & \left[\dfrac{x_3 - x_1}{a_3} - \dfrac{(x_2 - x_1)\cos\alpha_{13}}{a_1}\right]\csc\alpha_{13} & \dfrac{(y_3 - y_1)(z_2 - z_1) - (z_3 - z_1)(y_2 - y_1)}{a_1 a_3 \sin\alpha_{13}} \\ \dfrac{y_2 - y_1}{a_1} & \left[\dfrac{y_3 - y_1}{a_3} - \dfrac{(y_2 - y_1)\cos\alpha_{13}}{a_1}\right]\csc\alpha_{13} & \dfrac{(z_3 - z_1)(x_2 - x_1) - (x_3 - x_1)(z_2 - z_1)}{a_1 a_3 \sin\alpha_{13}} \\ \dfrac{z_2 - z_1}{a_1} & \left[\dfrac{z_3 - z_1}{a_3} - \dfrac{(z_2 - z_1)\cos\alpha_{13}}{a_1}\right]\csc\alpha_{13} & \dfrac{(x_3 - x_1)(y_2 - y_1) - (y_3 - y_1)(x_2 - x_1)}{a_1 a_3 \sin\alpha_{13}} \end{bmatrix} \begin{bmatrix} u_x \\ u_y \\ w \end{bmatrix}, \tag{11.5.15}$$

which may be inverted to give the local coordinate system corner displacements in terms of the common coordinate system displacements. The inverse transformation is used to express Eq. (11.5.14a) in terms of displacements referred to a common coordinate system. Then Eq. (11.5.14a) becomes

$$U_B + U_S = \{\delta\}^T(\tfrac{1}{2}[K]\{\delta\} - \{p\}), \tag{11.5.16a}$$

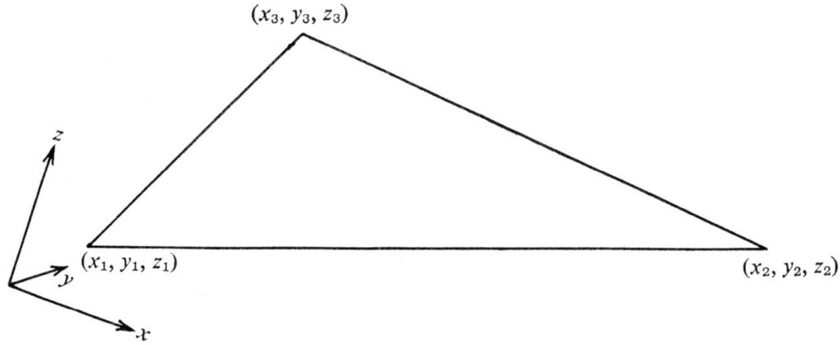

Fig. 11.19. Global coordinate system for arbitrary triangle

where

$$\{\delta\} = \begin{Bmatrix} \bar{u}_{x1} \\ \bar{u}_{y1} \\ \bar{u}_{z1} \\ \bar{u}_{x2} \\ \bar{u}_{y2} \\ \bar{u}_{z2} \\ \bar{u}_{3} \\ \bar{u}_{y3} \\ \bar{u}_{z3} \\ M_{n2} \\ M_{n3} \\ M_{n1} \end{Bmatrix} \tag{11.5.16b}$$

and $[K]$ is a symmetrical matrix.

The variational principle for the entire shell is given in terms of the summation of $U_B + U_S$ for all of the flat triangles making up the shell. Thus

$$\delta \sum_j (U_B + U_S)_j = 0, \tag{11.5.17a}$$

so that the required equations are

$$\sum_j \frac{\partial (U_B + U_S)_j}{\partial a_k} = 0, \tag{11.5.17b}$$

where a_k is a particular nodal displacement component or edge normal moment subject to the restriction that edge corner displacements for corners common to one or more elements are identical, while the sum of normal moment vectors for lines common to one or more elements must vanish. For two adjacent elements this requirement implies that the common normal moments for the two elements are equal. For more than two elements having a common edge one edge normal moment can be expressed in terms of the others.

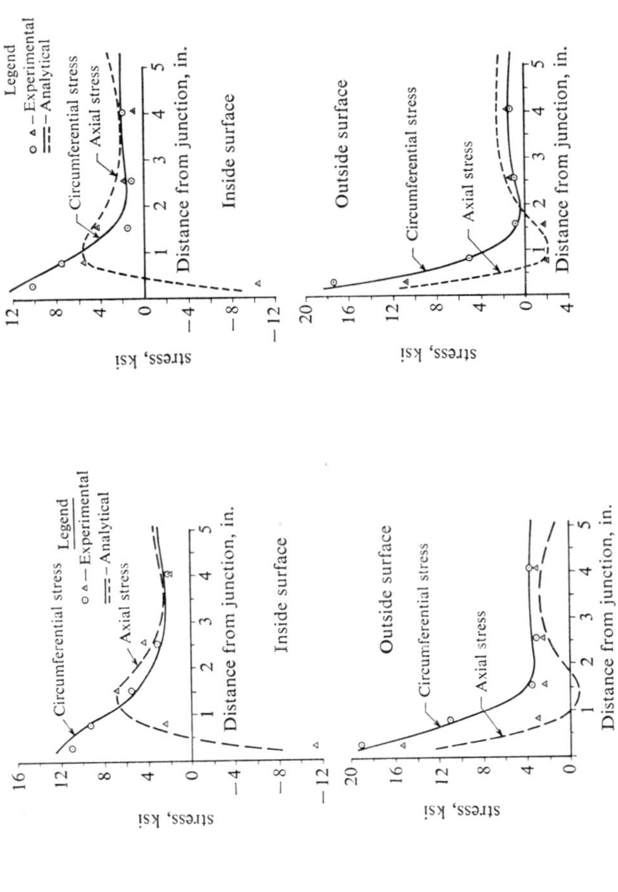

(a) Large cylinder (b) Small cylinder

Fig. 11.20. Comparison of experimental and theoretical surface stresses on the longitudinal plane (internal pressure = 30 psi)

An indication of the accuracy of the method is indicated by the good agreement between theoretical stresses obtained by the use of the method above and experimental values for pressurized cylinders which intersect at right angles (Fig. 11.20).† The grid used for the analysis is shown in Fig. 11.21. In this application four triangular elements were combined to form a quadrilateral shell element (Fig. 11.22) with the interior normal moments and corner displacements eliminated by solving the pertinent equations of Eqs. (11.5.17b) for these quantities in terms of the others.

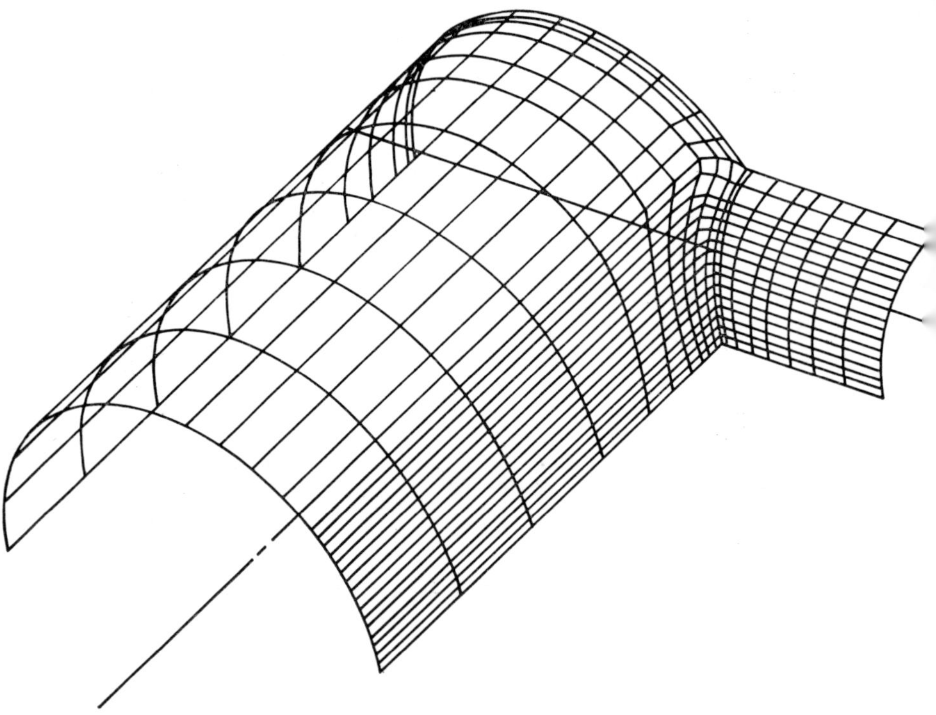

Fig. 11.21. Grid system for intersecting cylinders

† L. R. Herrmann, op. cit.

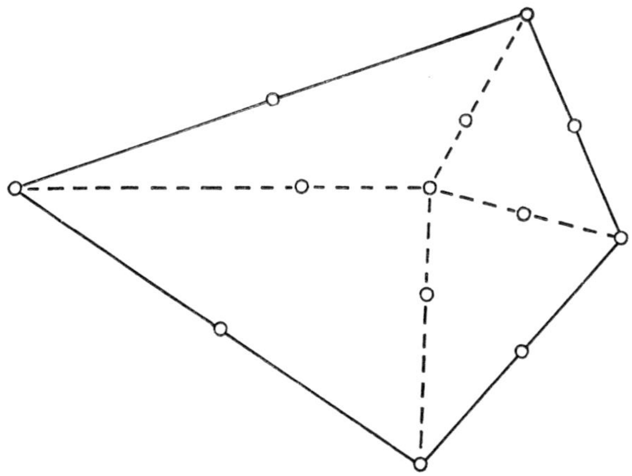

Fig. 11.22. Quadrilateral shell element consisting of four plate elements

Supplementary references

[1] Forsberg, K., and Flügge, W., 'Point Load on a Shallow Elliptic Paraboloid,' *J. Appl. Mech.*, vol. 33, no. 3, September 1966, pp. 575–585.
[2] Sanders, J. L., Jr., 'Singular Solutions to the Shallow Shell Equations,' *J. Appl. Mech.*, vol. 37, no. 2, June 1970, pp. 361–366.
[3] Sanders, J. L., Jr., 'Cutouts in Shallow Shells,' *J. Appl. Mech.*, vol. 37, no. 2, June 1970, pp. 374–382.
[4] Lucas, S. L., and Daugherty, R. L., 'Accuracy of Donnell's Equations for Noncircular Cylinders,' *AIAA J.*, vol. 9, no. 11, November 1971, pp. 2276–2278.

12

Nonuniform anisotropic shells

12.1 Necessity for modifications of the theory of shells

In many cases of practical interest we have to deal with shell structures composed of a material which is anisotropic and/or having a thickness which is not uniform but varies as a function of middle surface location. The shell wall may be of composite construction due to stiffening of one sort or another. The analysis of all of these structures differing from the model of a shell which is uniformly thick and composed of a material which is basically isotropic requires that we adjust the theory of shells to fit the new conditions. Yet another factor which requires the introduction of modifications of the theory is the effect of thermal loading of the shell structure, that is, the effect of subjecting the shell to a temperature field which differs from the implied initial (uniform) temperature distribution in the unstressed state.

Fortunately, not all of the theory must be changed. In all cases the equations of equilibrium are valid and thus the overall equilibrium equations derived by integrating over the thickness of the shell wall. However, the detailed definitions of the equivalent applied middle surface loads and moments and the definitions of the overall stress resultants require slight modification when the shell is no longer uniformly thick and homogeneous. The components of shell theory requiring the most change are the overall force–strain relations since the stress–strain relations of the material differ from those assumed previously if the shell material is anisotropic or if the shell is subjected to thermal loading, while the equations defining overall forces in terms of stresses are different when the shell is nonuniformly thick or a composite. Despite the fact that it is possible to consider the changes required in full generality, it is simpler and more instructive to investigate the modifications introduced when one factor is changed at a time. Although numerical solution aspects of modified shell theory are not emphasized or discussed here, it should be noted that the methods of Sections 11.3 and 11.5

can readily be extended to the modified shell theories once the corresponding strain energy function is defined.

The discussion herein generally does not apply to sandwich shells. A shell of sandwich construction consists of two thin face layers of material separated by a thicker layer of low-density material. As is the case for wide-flange beams in building construction, the purpose of sandwich construction is to achieve increased bending stiffness with little increase in weight. The sandwich shell might be considered to be a special case of a layered shell, were it not for the fact that the low-density core usually has a transverse shear stiffness which is so low that the Love–Kirchhoff assumption of infinite transverse shear stiffness leads to significant error. The analysis of sandwich shells thus requires modifications of the entire theory rather than just changes of the overall force–strain relations of the theory of transversely rigid shells. The reader is referred to the excellent monograph of Plantema† for discussion of the theory and of available results as well as for an extensive bibliography on the subject. A related field is that of extensions of the theory of homogeneous shells to include transverse shear and extensional strains. Such investigations are beyond the scope of this volume.

12.2 Effect of thermal loading

If a shell is subjected to a heat field yielding a temperature distribution $T(\alpha, \beta, \zeta)$ which differs from the uniform temperature T_0 under which the structure is stress free, the stress–strain relations for the material given by Eqs. (2.3.2) are no longed valid and must be replaced by more suitable ones. If we assume that the temperature difference $T - T_0$ is small enough so that the material properties of the shell wall are unchanged, then the only consequence of the heating will be the appearance of additional strains due to thermal expansion of the material which are superimposed upon the strains due to elastic stresses. We further assume, as before, that the stresses are not so large that plasticity effects have to be considered so that the relationship between strain and stress is a linear one.

If we retain the assumption of infinite transverse stiffness of the shell material and further assume that the shell thickness is unchanged by temperature variations then the stress–strain relations for a material isotropic in layers parallel to the middle surface may be written as

$$\epsilon_\alpha^{(\zeta)} = \frac{\sigma_\alpha^{(\zeta)} - \nu\sigma_\beta^{(\zeta)}}{E} + \alpha_T(T - T_0), \tag{12.2.1a}$$

† F. J. Plantema, *Sandwich Construction, the Bending and Buckling of Sandwich Beams, Plates and Shells*, John Wiley & Sons, Inc., 1966. See also, R. Schmidt, 'Sandwich Shells of Arbitrary Shape,' *J. Appl. Mech.*, vol. 31, no. 2, June 1964, pp. 239–244.

$$\epsilon_\beta^{(\zeta)} = \frac{\sigma_\beta^{(\zeta)} - \nu\sigma_\alpha^{(\zeta)}}{E} + \alpha_T(T - T_0), \tag{12.2.1b}$$

$$\gamma_{\alpha\beta}^{(\zeta)} = \frac{2(1+\nu)}{E}\,\tau_{\alpha\beta}^{(\zeta)}, \tag{12.2.1c}$$

where α_T is the coefficient of thermal expansion and is assumed to be independent of temperature. On inverting these, we have

$$\sigma_\alpha^{(\zeta)} = \frac{E}{1-\nu^2}\,[\epsilon_\alpha^{(\zeta)} + \nu\epsilon_\beta^{(\zeta)}] - \frac{E}{1-\nu}\,\alpha_T(T - T_0), \tag{12.2.2a}$$

$$\sigma_\beta^{(\zeta)} = \frac{E}{1-\nu^2}\,[\epsilon_\beta^{(\zeta)} + \nu\epsilon_\alpha^{(\zeta)}] + \frac{E}{1-\nu}\,\alpha_T(T - T_0), \tag{12.2.2b}$$

$$\tau_{\alpha\beta}^{(\zeta)} = \frac{E}{2(1+\nu)}\,\gamma_{\alpha\beta}^{(\zeta)}, \tag{12.2.2c}$$

where $\epsilon_\alpha^{(\zeta)}$, $\epsilon_\beta^{(\zeta)}$, and $\gamma_{\alpha\beta}^{(\zeta)}$ are given by Eqs. (2.4.7) and (2.4.10). When Eqs. (12.2.2) are substituted into Eqs. (2.5.1), (2.5.2), and (2.6.1) giving stress resultants in terms of stresses we find that the following terms are added to the overall force–strain relations given by Eqs. (2.7.2)

$$N_\alpha = \cdots - \frac{E\alpha_T}{1-\nu}\int_{-h/2}^{h/2}\left(1 + \frac{\zeta}{R_\beta}\right)(T - T_0)\,d\zeta, \tag{12.2.3a}$$

$$N_\beta = \cdots - \frac{E\alpha_T}{1-\nu}\int_{-h/2}^{h/2}\left(1 + \frac{\zeta}{R_\alpha}\right)(T - T_0)\,d\zeta, \tag{12.2.3b}$$

$$M_\alpha = \cdots - \frac{E\alpha_T}{1-\nu}\int_{-h/2}^{h/2}\left(1 + \frac{\zeta}{R_\beta}\right)(T - T_0)\,d\zeta, \tag{12.2.3c}$$

$$M_\beta = \cdots - \frac{E\alpha_T}{1-\nu}\int_{-h/2}^{h/2}\zeta\left(1 + \frac{\zeta}{R_\alpha}\right)(T - T_0)\,d\zeta, \tag{12.2.3d}$$

while the equations for $\overline{N}_{\alpha\beta}$ and $\overline{M}_{\alpha\beta}$ are unchanged.† When these are substituted into Eq. (2.6.2) we find that the only changes in the equations may be interpreted as modifications of the surface loads and moments which are now given by

$$\bar{q}_\alpha = q_\alpha - \frac{E\alpha_T}{1-\nu}\frac{1}{g_\alpha}\frac{\partial}{\partial\alpha}\int_{-h/2}^{h/2}\left(1 + \frac{\zeta}{R_\beta}\right)(T - T_0)\,d\zeta, \tag{12.2.4a}$$

† By using the variational principle for an elastic material with thermal stresses and by making suitable approximations in the corresponding strain–energy expression, it is possible to reduce Eqs. (12.2.3) to forms corresponding to Eqs. (3.1.2) and (3.1.4).

$$\bar{q}_\beta = q_\beta - \frac{E\alpha_T}{1-\nu}\frac{1}{g_\beta}\frac{\partial}{\partial\beta}\int_{-h/2}^{h/2}\left(1+\frac{\zeta}{R_\alpha}\right)(T-T_0)\,d\zeta, \tag{12.2.4b}$$

$$\bar{q}_\zeta = q_\zeta + \frac{E\alpha_T}{1-\nu}\int_{-h/2}^{h/2}\left(\frac{1}{R_\alpha}+\frac{1}{R_\beta}+\frac{2\zeta}{R_\alpha R_\beta}\right)(T-T_0)\,d\zeta, \tag{12.2.4c}$$

$$\bar{m}_\alpha = m_\alpha - \frac{E\alpha_T}{1-\nu}\frac{1}{g_\alpha}\frac{\partial}{\partial\alpha}\int_{-h/2}^{h/2}\zeta\left(1+\frac{\zeta}{R_\beta}\right)(T-T_0)\,d\zeta, \tag{12.2.4d}$$

$$\bar{m}_\beta = m_\beta - \frac{E\alpha_T}{1-\nu}\frac{1}{g_\beta}\frac{\partial}{\partial\beta}\int_{-h/2}^{h/2}\zeta\left(1+\frac{\zeta}{R_\alpha}\right)(T-T_0)\,d\zeta, \tag{12.2.4e}$$

where q_α, q_β, q_ζ, m_α, and m_β are expressed by Eqs. (2.5.4) and (2.5.6).

The edge conditions expressed in terms of stress resultants or displacements are unchanged. Thus the deformation of a shell subjected to temperature variations are identical to those for a shell subjected to the equivalent surface loads given by Eqs. (12.2.4) if edge displacements are prescribed. If edge force resultants are prescribed, the shell can be considered to be also subjected to the additional edge force resultants given by the negative of the terms in Eqs. (12.2.3). In both cases, methods previously discussed for the analysis of shells under static surface or edge loading may be used unchanged.†

As an example, let us consider a finite circular cylindrical shell which is built into a rigid wall at both ends and which is heated uniformly from a temperature T_0 to a temperature T_1. From Eqs. (12.2.4) we see that the temperature increase is equivalent to an internal pressure normal to the middle surface such that

$$q_\zeta = \frac{\alpha_T E h}{(1-\nu)R}(T-T_0), \tag{12.2.5}$$

for which the analysis of Section 4.4(c) is applicable directly. The resultants calculated for this problem must be modified, however, by subtracting the quantity $[\alpha_T Eh/(1-\nu)](T-T_0)$ from the middle surface forces and the small quantity $\alpha_T Eh^3/[12(1-\nu)R]$ from the longitudinal bending moment.

When the temperature difference is so large that the material properties themselves are functions of temperature the analysis of shell structures is complicated by the fact that Young's modulus E, Poisson's ratio ν, and the

† See, for example, C. N. De Silva, 'Thermal Stresses in the Bending of Ogival Shells,' *J. Aero. Sci.*, vol. 29, no. 2, February 1962, pp. 207–212; G. S. Stern, 'Thermoelastic Analysis of a Parabolic Shell,' Jet Propulsion Laboratory, Technical Report 32-479, August 1963; and J. H. Huth, 'Thermal Stresses in Conical Shells,' *J. Aero. Sci.*, vol. 20, no. 9, September 1953, pp. 613–616.

coefficient of thermal expansion α_T are functions of position and thus have to be included under the integral sign when Eqs. (12.2.2) are substituted into Eqs. (2.5.1), (2.5.2), and (2.6.1). While this does not change the thermal loading terms significantly, the portion of the expressions involving the middle surface strain and curvature change components will be more complex, with the coefficients of the strain and curvature change components now given by functions of α and β rather than by constant expressions. Thus we have

$$N_\alpha = C_{11}\epsilon_\alpha + C_{12}\epsilon_\beta + C_{13}\kappa_\alpha + C_{14}\kappa_\beta - C_1, \tag{12.2.6a}$$

$$N_\beta = C_{21}\epsilon_\alpha + C_{22}\epsilon_\beta + C_{23}\kappa_\alpha + C_{24}\kappa_\beta - C_2, \tag{12.2.6b}$$

$$M_\alpha = C_{31}\epsilon_\alpha + C_{32}\epsilon_\beta + C_{33}\kappa_\alpha + C_{34}\kappa_\beta - C_3, \tag{12.2.6c}$$

$$M_\beta = C_{41}\epsilon_\alpha + C_{42}\epsilon_\beta + C_{43}\kappa_\alpha + C_{44}\kappa_\beta - C_4, \tag{12.2.6d}$$

$$\overline{N}_{\alpha\beta} = C_{55}\gamma_{\alpha\beta} + C_{56}\bar{\kappa}_{\alpha\beta}, \tag{12.2.6e}$$

$$\overline{M}_{\alpha\beta} = C_{65}\gamma_{\alpha\beta} + C_{66}\bar{\kappa}_{\alpha\beta}, \tag{12.2.6f}$$

where the coefficients are functions of α and β given by

$$C_{11} = \int_{-h/2}^{h/2} \frac{1 + \zeta/R_\beta}{1 + \zeta/R_\alpha} \frac{E}{1 - \nu^2}\,\mathrm{d}\zeta, \tag{12.2.7a}$$

$$C_{22} = \int_{-h/2}^{h/2} \frac{1 + \zeta/R_\alpha}{1 + \zeta/R_\beta} \frac{E}{1 - \nu^2}\,\mathrm{d}\zeta, \tag{12.2.7b}$$

$$C_{33} = \int_{-h/2}^{h/2} \zeta^2 \frac{1 + \zeta/R_\beta}{1 + \zeta/R_\alpha} \frac{E}{1 - \nu^2}\,\mathrm{d}\zeta, \tag{12.2.7c}$$

$$C_{44} = \int_{-h/2}^{h/2} \zeta^2 \frac{1 + \zeta/R_\alpha}{1 + \zeta/R_\beta} \frac{E}{1 - \nu^2}\,\mathrm{d}\zeta, \tag{12.2.7d}$$

$$C_{12} = C_{21} = C_{34} = C_{43} = \int_{-h/2}^{h/2} \frac{\nu E}{1 - \nu^2}\,\mathrm{d}\zeta, \tag{12.2.7e}$$

$$C_{13} = C_{31} = \int_{-h/2}^{h/2} \zeta \frac{1 + \zeta/R_\beta}{1 + \zeta/R_\alpha} \frac{E}{1 - \nu^2}\,\mathrm{d}\zeta, \tag{12.2.7f}$$

$$C_{14} = C_{41} = C_{23} = C_{32} = \int_{-h/2}^{h/2} \zeta \frac{E}{1 - \nu^2}\,\mathrm{d}\zeta, \tag{12.2.7g}$$

$$C_{24} = C_{42} = \int_{-h/2}^{h/2} \zeta \frac{1 + \zeta/R_\alpha}{1 + \zeta/R_\beta} \frac{E}{1 - \nu^2}\,\mathrm{d}\zeta, \tag{12.2.7h}$$

$$C_{55} = \frac{1}{2}\int_{-h/2}^{h/2} \frac{\left[1 + \frac{\zeta}{2}\left(\frac{1}{R_\alpha} + \frac{1}{R_\beta}\right) + \frac{\zeta^2}{4}\left(\frac{1}{R_\alpha} - \frac{1}{R_\beta}\right)^2\right]^2}{\left(1 + \frac{\zeta}{R_\alpha}\right)\left(1 + \frac{\zeta}{R_\beta}\right)} \frac{E}{1+\nu}\, \mathrm{d}\zeta, \tag{12.2.7i}$$

$$C_{66} = \int_{-h/2}^{h/2} \zeta^2 \frac{\left[1 + \frac{\zeta}{2}\left(\frac{1}{R_\alpha} + \frac{1}{R_\beta}\right)\right]^2}{\left(1 + \frac{\zeta}{R_\alpha}\right)\left(1 + \frac{\zeta}{R_\beta}\right)} \frac{E}{1+\nu}\, \mathrm{d}\zeta, \tag{12.2.7j}$$

$$C_{56} = C_{65} = 2\int_{-h/2}^{h/2} \zeta \times \frac{\left[1 + \frac{\zeta}{2}\left(\frac{1}{R_\alpha} + \frac{1}{R_\beta}\right) + \frac{\zeta^2}{4}\left(\frac{1}{R_\alpha} - \frac{1}{R_\beta}\right)^2\right]\left[1 + \frac{\zeta}{2}\left(\frac{1}{R_\alpha} + \frac{1}{R_\beta}\right)\right]}{\left(1 + \frac{\zeta}{R_\alpha}\right)\left(1 + \frac{\zeta}{R_\beta}\right)} \frac{E}{1+\nu}\, \mathrm{d}\zeta, \tag{12.2.7k}$$

and

$$C_1 = C_2 = \int_{-h/2}^{h/2} \left(1 + \frac{\zeta}{R_\beta}\right) \frac{E\alpha_T}{1-\nu}(T - T_0)\, \mathrm{d}\zeta, \tag{12.2.7l}$$

$$C_3 = C_4 = \int_{-h/2}^{h/2} \left(1 + \frac{\zeta}{R_\alpha}\right) \frac{E\alpha_T}{1-\nu}(T - T_0)\, \mathrm{d}\zeta. \tag{12.2.7m}$$

The substitution of these expressions into the equations of equilibrium will now yield substantially different differential equations for the displacements of the middle surface which generally require the use of a digital computer numerical analysis scheme for their solution. As we shall see, similar types of equations are obtained when the shell wall is of nonuniform thickness, or when the material is orthotropic, so that any general method of shell analysis for uniformly thick shells with temperature dependent material properties can have wide applications.

12.3 Effect of nonuniform wall thickness

12.3.1 Derivation of equations for shells of nonuniform wall thickness

For a shell with a nonuniform wall thickness, let us assume that the equation of the middle surface, although in general defined as indicated in Section 2.1, is known explicitly and that the equations of the inner and outer bounding surfaces are defined by known distances along the normal to the middle surface. Thus if the vectorial equation of the middle surface in terms of lines of curvature coordinates α, β is

$$\mathbf{r} = \mathbf{r}(\alpha, \beta), \tag{12.3.1.1a}$$

then the equation of the outer bounding surface is

$$\mathbf{r}_o = \mathbf{r}(\alpha, \beta) + \mathbf{e}_\zeta h_o(\alpha, \beta), \tag{12.3.1.1b}$$

and that of the inner surface is

$$\mathbf{r}_i = \mathbf{r}(\alpha, \beta) - \mathbf{e}_\zeta h_i(\alpha, \beta), \tag{12.3.1.1c}$$

where $\vec{\mathbf{e}}_\zeta$ is the outwardly directed vector normal to the middle surface. For a uniformly thick shell

$$h_o(\alpha, \beta) = h_i(\alpha, \beta) = h/2. \tag{12.3.1.2}$$

We further assume that the shell is rigid in the direction of the normal to the middle surface so that the stress–strain relations and the strain–displacement relations given by Eqs. (2.3.2), (2.4.7), and (2.4.10) are still valid.

Now let us define force and moment resultants as

$$N_\alpha = \int_{-h_i}^{h_o} \sigma_\alpha^{(\zeta)} \left(1 + \frac{\zeta}{R_\beta}\right) d\zeta, \tag{12.3.1.3a}$$

$$N_\beta = \int_{-h_i}^{h_o} \sigma_\beta^{(\zeta)} \left(1 + \frac{\zeta}{R_\alpha}\right) d\zeta, \tag{12.3.1.3b}$$

$$N_{\alpha\beta} = \int_{-h_i}^{h_o} \tau_{\alpha\beta}^{(\zeta)} \left(1 + \frac{\zeta}{R_\beta}\right) d\zeta, \tag{12.3.1.3c}$$

$$N_{\beta\alpha} = \int_{-h_i}^{h_o} \tau_{\alpha\beta}^{(\zeta)} \left(1 + \frac{\zeta}{R_\alpha}\right) d\zeta, \tag{12.3.1.3d}$$

$$M_\alpha = \int_{-h_i}^{h_o} \zeta\sigma_\alpha^{(\zeta)} \left(1 + \frac{\zeta}{R_\beta}\right) d\zeta, \tag{12.3.1.3e}$$

$$M_\beta = \int_{-h_i}^{h_o} \zeta\sigma_\beta^{(\zeta)} \left(1 + \frac{\zeta}{R_\alpha}\right) d\zeta, \tag{12.3.1.3f}$$

$$M_{\alpha\beta} = \int_{-h_i}^{h_o} \zeta\tau_{\alpha\beta}^{(\zeta)} \left(1 + \frac{\zeta}{R_\beta}\right) d\zeta, \tag{12.3.1.3g}$$

$$M_{\beta\alpha} = \int_{-h_i}^{h_o} \zeta\tau_{\alpha\beta}^{(\zeta)} \left(1 + \frac{\zeta}{R_\alpha}\right) d\zeta, \tag{12.3.1.3h}$$

$$Q_\alpha = \int_{-h_i}^{h_o} \tau_{\alpha\zeta}^{(\zeta)} \left(1 + \frac{\zeta}{R_\beta}\right) d\zeta, \tag{12.3.1.3i}$$

$$Q_\beta = \int_{-h_i}^{h_o} \tau_{\beta\zeta}^{(\zeta)} \left(1 + \frac{\zeta}{R_\alpha}\right) d\zeta, \tag{12.3.1.3j}$$

where the limits of integration are functions of α and β. We can show that with these definitions of stress resultants, the equilibrium Eqs. (2.2.5) can be integrated to yield overall equilibrium equations similar to Eqs. (2.5.3) and (2.5.5). We can further introduce the equivalent forces

$$\overline{N}_{\alpha\beta} = \int_{-h_i}^{h_o} \tau_{\alpha\beta}^{(\zeta)} \left[1 + \frac{\zeta}{2}\left(\frac{1}{R_\alpha} + \frac{1}{R_\beta}\right) + \frac{\zeta^2}{4}\left(\frac{1}{R_\alpha} - \frac{1}{R_\beta}\right)^2\right] d\zeta, \tag{12.3.1.4a}$$

$$\overline{M}_{\alpha\beta} = \int_{-h_i}^{h_o} \zeta\tau_{\alpha\beta}^{(\zeta)} \left(1 + \frac{\zeta}{2}\left(\frac{1}{R_\alpha} + \frac{1}{R_\beta}\right)\right] d\zeta, \tag{12.3.1.4b}$$

$$\overline{Q}_\alpha = Q_\alpha + \frac{1}{g_\beta}\frac{\partial M_{\alpha\beta}}{\partial \beta}, \tag{12.3.1.4c}$$

$$\overline{Q}_\beta = Q_\beta + \frac{1}{g_\beta}\frac{\partial M_{\beta\alpha}}{\partial \alpha}, \tag{12.3.1.4d}$$

and rewrite the equations of equilibrium in terms of equivalent forces as before. The edge conditions also remain the same. Thus the basic set of equations for the theory of shells of nonuniform thickness is identical to that for a uniformly thick shell and the primary modifications which must be made are those which enter the overall force–strain relations.

We may obtain the overall force–strain relationships and approximations to them most easily by investigating the expressions for the strain energy of an element of the shell. The strain energy per unit middle surface area is given by

$$\begin{aligned}
W &= \frac{E}{1-\nu^2}\int_{-h_i}^{h_o} \left\{[\epsilon_\alpha^{(\zeta)}]^2 + 2\nu\epsilon_\alpha^{(\zeta)}\epsilon_\beta^{(\zeta)} + [\epsilon_\beta^{(\zeta)}]^2 + \frac{1-\nu}{2}[\gamma_{\alpha\beta}^{(\zeta)}]^2\right\} \\
&\quad \times \left(1 + \frac{\zeta}{R_\alpha}\right)\left(1 + \frac{\zeta}{R_\beta}\right) d\zeta \\
&= \frac{E}{1-\nu^2}\int_{-h_i}^{h_o} \left\langle \frac{1 + \zeta/R_\alpha}{1 + \zeta/R_\beta}(\epsilon_\alpha + \zeta\kappa_\alpha)^2 + 2\nu(\epsilon_\alpha + \zeta\kappa_\alpha)(\epsilon_\beta + \zeta\kappa_\beta)\right. \\
&\quad + \frac{1 + \zeta R_\beta}{1 + \zeta/R_\alpha}(\epsilon_\beta + \zeta\kappa_\beta)^2 + \frac{1-\nu}{2} \\
&\quad \left. \times \frac{\left\{\left[1 + \frac{\zeta}{2}\left(\frac{1}{R_\alpha} + \frac{1}{R_\beta}\right)\right](\gamma_{\alpha\beta} + 2\zeta\overline{\kappa}_{\alpha\beta}) + \frac{\zeta^2}{4}\left(\frac{1}{R_\alpha} - \frac{1}{R_\beta}\right)^2\gamma_{\alpha\beta}\right\}^2}{\left(1 + \frac{\zeta}{R_\alpha}\right)\left(1 + \frac{\zeta}{R_\beta}\right)} \right\rangle d\zeta,
\end{aligned} \tag{12.3.1.5}$$

and can be evaluated readily as it stands or with various truncation approximations. With the strain–energy expression known, the overall force–strain relations can be obtained by the use of Eqs. (2.9.2). In general, the form of the overall force–strain relations is similar to that of Eqs. (12.2.6), even if approximations such as deleting the terms ζ/R_α and ζ/R_β in comparison to unity are made, since integrals such as $\int_{-h_i}^{h_o} \zeta \, d\zeta$ do not necessarily vanish. In the special case when the middle surface bisects the normals along which the thicknesses are measured, so that we have

$$h_o(\alpha, \beta) = h_i(\alpha, \beta) = \tfrac{1}{2}h(\alpha, \beta), \tag{12.3.1.7}$$

the overall force–strain relations are identical in form with Eqs. (2.7.2) for a uniformly thick shell or, if the appropriate approximations are made, with Eqs. (3.1.2) or (3.1.4), with the thickness now a function of middle surface location. In the applications below we shall confine our attention to this special case.

12.3.2 *Axisymmetric deformations of nonuniformly thick shells of revolution*

As might be expected, the theory of nonuniformly thick shells has been investigated primarily for shells whose middle surface is a surface of revolution. We restrict the thickness variations to those which are a function of meridional location only and the loading and deformations to be axisymmetric. We shall, as in Chapter 7, take the overall force–strain relations to be the simplest ones of the bending theory of shells, namely

$$N_\phi = K(\epsilon_\phi + \nu\epsilon_\theta), \tag{12.3.2.1a}$$

$$N_\theta = K(\epsilon_\theta + \nu\epsilon_\phi), \tag{12.3.2.1b}$$

$$M_\phi = D(\kappa_\phi + \nu\kappa_\theta), \tag{12.3.2.1c}$$

$$M_\theta = D(\kappa_\theta + \nu\kappa_\phi), \tag{12.3.2.1d}$$

where

$$K = \frac{Eh}{1 - \nu^2}, \tag{12.3.2.2a}$$

$$D = \frac{Eh^3}{12(1 - \nu)^2}, \tag{12.3.2.2b}$$

$$h = h(\phi). \tag{12.3.2.2c}$$

These, together with the equations of equilibrium

$$\frac{1}{R_\phi}\frac{\mathrm{d}}{\mathrm{d}\phi}(N_\phi R_\theta \sin\phi) - N_\theta \cos\phi + R_\theta \sin\phi\left(\frac{Q_\phi}{R_\phi} + q_\phi\right) = 0, \qquad (12.3.2.3a)$$

$$\frac{1}{R_\phi}\frac{\mathrm{d}}{\mathrm{d}\phi}(Q_\phi R_\theta \sin\phi) - R_\theta \sin\phi\left(\frac{N_\phi}{R_\phi} + \frac{N_\theta}{R_\theta} - q_\zeta\right) = 0, \qquad (12.3.2.3b)$$

$$\frac{1}{R_\phi}\frac{\mathrm{d}}{\mathrm{d}\phi}(M_\phi R_\theta \sin\phi) - M_\theta \cos\phi - R_\theta \sin\phi(Q_\phi - m_\phi) = 0, \qquad (12.3.2.3c)$$

and the strain–displacement relations

$$\epsilon_\phi = \frac{1}{R_\phi}\left(\frac{\mathrm{d}u_\phi}{\mathrm{d}\phi} + w\right), \qquad (12.3.2.4a)$$

$$\epsilon_\theta = \frac{1}{R_\theta \sin\phi}(u_\phi \cos\phi + w \sin\phi), \qquad (12.3.2.4b)$$

$$\kappa_\phi = -\frac{1}{R_\phi}\frac{\mathrm{d}}{\mathrm{d}\phi}\left[\frac{1}{R_\phi}\left(\frac{\mathrm{d}w}{\mathrm{d}\phi} - u_\phi\right)\right], \qquad (12.3.2.4c)$$

$$\kappa_\theta = -\frac{\cot\phi}{R_\phi R_\theta}\left(\frac{\mathrm{d}w}{\mathrm{d}\phi} - u_\phi\right), \qquad (12.3.2.4d)$$

define the solution for the displacements and stresses.

We proceed in a manner similar to that of Section 7.1 and introduce two new dependent variables, defined for the nonuniformly thick shell as

$$\Theta = \frac{1}{R_\phi}\left(\frac{\mathrm{d}w}{\mathrm{d}\phi} - u_\phi\right), \qquad (12.3.2.5a)$$

$$U = Q_\phi R_\theta \frac{[12(1-\nu^2)]^{1/2}}{Eh^2}. \qquad (12.3.2.5b)$$

Then from Eqs. (12.3.2.3c), (12.3.2.1d), (12.3.2.4c), (12.3.2.4d), and (12.3.2.5) we have a relationship between U and Θ given by

$$L(\Theta) + \Phi_1 \Theta + U = 0, \qquad (12.3.2.7)$$

where

$$L(\ldots) = \frac{R_\theta h}{[12(1-\nu^2)]^{1/2}}\left\{\frac{1}{R_\phi}\frac{\mathrm{d}}{\mathrm{d}\phi}\left[\frac{1}{R_\phi}\frac{\mathrm{d}(\ldots)}{\mathrm{d}\phi}\right.\right.$$

$$\left.\left. + \left(\frac{3}{h}\frac{1}{R_\phi}\frac{\mathrm{d}h}{\mathrm{d}\phi} + \frac{\cot\phi}{R_\theta}\right)\frac{1}{R_\phi}\frac{\mathrm{d}(\ldots)}{\mathrm{d}\phi}\right] - \left(\frac{\cot\phi}{R_\theta}\right)^2(\ldots)\right\}, \qquad (12.3.2.8a)$$

$$\Phi_1 = -\frac{\nu h/R_\phi}{[12(1-\nu^2)]^{1/2}}\left(1 - 3\frac{\mathrm{d}h/\mathrm{d}\phi}{h}\cot\phi\right). \qquad (12.3.2.8b)$$

From Eqs. (12.3.2.3a) and (12.3.2.3b) we can obtain the following relations between N_ϕ and U by eliminating N_θ and integrating the resulting expression

$$N_\phi = \frac{Eh^2}{[12(1-\nu^2)]^{1/2}} \frac{U}{R_\theta} \cot \phi + \frac{1}{R_\theta \sin^2 \phi} \left[P + \int_{\phi_0}^{\phi} (q_\zeta \cos \phi - q_\phi \sin \phi) R_\phi R_\theta \sin \phi \, \mathrm{d}\phi\right], \tag{12.3.2.9a}$$

where P is the total axial force acting on the cross-section $\phi = \phi_0$. We may next obtain an expression for N_θ in terms of U as

$$N_\theta = \frac{1}{R_\phi} \frac{\mathrm{d}}{\mathrm{d}\phi} \left[\frac{Eh^2 U}{[12(1-\nu^2)]^{1/2}}\right] + q_\zeta R_\theta - \frac{1}{R_\phi \sin^2 \phi} \left[P + \int_{\phi_0}^{\phi} (q_\zeta \cos \phi - q_\phi \sin \phi) R_\phi R_\theta \sin \phi \, \mathrm{d}\phi\right], \tag{12.3.2.9b}$$

The compatibility condition for strains and curvature changes is

$$\frac{\mathrm{d}}{\mathrm{d}\phi} (\epsilon_\theta R_\theta \sin \phi) - \epsilon_\phi R_\phi \cos \phi = \Theta R_\phi \sin \phi = \frac{\mathrm{d}}{\mathrm{d}\phi} \left(\frac{N_\theta - \nu N_\phi}{Eh} R_\theta \sin \phi\right) - \frac{N_\phi - \nu N_\theta}{Eh} R_\phi \cos \phi, \tag{12.3.2.10}$$

which yields with the use of Eqs. (12.3.2.9)

$$L(U) + \Phi_2 U - \Theta = f_1, \tag{12.3.2.11}$$

where

$$\Phi_2 = \frac{h/R_\phi}{[12(1-\nu^2)]^{1/2}} \left\{\nu\left(1 + \frac{\mathrm{d}h/\mathrm{d}\phi}{h} \cot \phi\right) + 2\frac{R_\theta}{R_\phi}\left[\frac{\mathrm{d}^2 h/\mathrm{d}\phi^2}{h} + \frac{(\mathrm{d}/\mathrm{d}\phi)[(R_\phi \sin \phi)/R_\phi]}{(R_\theta \sin \phi)/R_\phi} \frac{\mathrm{d}h/\mathrm{d}\phi}{h}\right]\right\}, \tag{12.3.2.12a}$$

$$\begin{aligned} f_1 = {} & \frac{1}{Eh}\left\{(1+\nu) q_\zeta R_\theta + \frac{1}{R_\phi \sin^2 \phi}\left(\frac{R_\theta}{R_\phi} - \frac{R_\phi}{R_\theta}\right) \times \left[P + \int_{\phi_0}^{\phi} (q_\zeta \cos \phi - q_\phi \sin \phi) R_\phi R_\theta \sin \phi \, \mathrm{d}\phi\right]\right\} \\ & + \frac{R_\theta}{R_\phi} \tan \phi \frac{\mathrm{d}}{\mathrm{d}\phi}\left(\frac{qR_\theta}{Eh}\right) - \frac{1}{R_\phi \sin \phi \cos \phi} \\ & \times \frac{\mathrm{d}}{\mathrm{d}\phi}\left\{\frac{\nu + R_\theta/R_\phi}{Eh}\left[P + \int_{\phi_0}^{\phi} (q_\zeta \cos \phi - q_\phi \sin \phi) R_\phi R_\theta \sin \phi \, \mathrm{d}\phi\right]\right\}. \end{aligned} \tag{12.3.2.12b}$$

Let us now consider the case when the shell is edge loaded only and is subjected to no axial force so that f_1 vanishes. Then by eliminating U from Eqs. (12.3.2.7) and (12.3.2.11) we find that Θ must satisfy the equation

$$LL(\Theta) + L(\Phi_1\Theta) + \Phi_2 L(\Theta) + (1 + \Phi_1\Phi_2)\Theta = 0. \tag{12.3.2.13a}$$

Similarly, U must satisfy the equation

$$LL(U) + L(\Phi_2 U) + \Phi_1 L(U) + (1 + \Phi_1\Phi_2)U = 0. \tag{12.3.2.13b}$$

In general, these equations are not identical since

$$L(\Phi_1\Theta) + \Phi_2(\ldots) \neq L(\Phi_2\ldots) + \Phi_1 L(\ldots). \tag{12.3.2.14}$$

For certain variations of the thickness h, however, it is possible for the two operators to be identical. In this case the operator

$$\begin{aligned} &LL(\ldots) + L(\Phi_1\ldots) + \Phi_2 L(\ldots) + (1 + \Phi_1\Phi_2)(\ldots) \\ &\quad = LL(\ldots) + L(\phi_2\ldots) + \Phi_1 L(\ldots) + (1 + \Phi_1\Phi_2)(\ldots) \end{aligned} \tag{12.3.2.15}$$

can be factored into two commutative factors, thus simplifying the solution since we need to deal with two second-order ordinary differential equations, as for the cylinder, cone, or sphere of constant thickness, rather than a fourth-order equation.

Suppose that the inequality sign in (12.3.2.14) is replaced by a sign of equality. Then we must have

$$(\Phi_2 - \Phi_1)L(\ldots) - L[(\Phi_2 - \Phi_1)(\ldots)] = 0. \tag{12.3.2.16}$$

This relationship will be satisfied, in general, if $\Phi_2 - \Phi_1$ is a constant, C. The use of Eqs. (12.3.2.8b) and (12.3.2.12a) then yields the relationship which the thickness h must satisfy as

$$\frac{d}{d\phi}\left(\frac{R_\theta \sin\phi}{R_\phi}\frac{dh}{d\phi} - \nu h\cos\phi\right) = [3(1-\nu^2)]^{1/2} C R_\phi \sin\phi, \tag{12.3.2.17}$$

which can be integrated to yield

$$h = (R_\theta \sin\phi)^\nu \left\{ C_1 + \int \frac{C_2 + [3(1-\nu^2)]^{1/2} C \int R_\phi \sin\phi \, d\phi}{(R_\theta \sin\phi)^{\nu+1}} R_\phi \, d\phi \right\}. \tag{12.3.2.18}$$

When h is so defined, Eq. (12.3.2.13a) or Eq. (12.3.2.13b) can be factored in the form

$$(L + a_1 + \Phi_1)(L + a_2 + \Phi_1)(\ldots) = 0, \tag{12.3.2.19a}$$

where

$$a_1, a_2 = \frac{-C \pm (C^2 - 4)^{1/2}}{2}. \tag{12.3.2.19b}$$

Thus the solutions of Θ or U are the sum of the solutions of the second-order differential equations

$$L(\ldots) + (a_1 + \Phi_1)(\ldots) = 0, \tag{12.3.2.19c}$$

$$L(\ldots) + (a_2 + \Phi_1)(\ldots) = 0, \tag{12.3.2.19d}$$

for which exact solutions are obtainable in some cases.

If $|C|$ is less than 2, and this is generally so if the thickness–radius of curvature ratio is to be small, a_1 and a_2 are complex conjugates. We can then show that Eqs. (12.3.2.7) and (12.3.2.11) can be written as one complex equation

$$L(W) + (\Phi_1 - a_1)W = a_1 f_1, \tag{12.3.2.20a}$$

where

$$W = \Theta + a_1 U. \tag{12.3.2.20b}$$

Asymptotic solutions of Eq. (12.3.2.20) can be obtained in a manner similar to that of Section 7.4.†

12.3.3 *Cylindrical shell of linearly varying thickness*

For the cylindrical shell, the thickness variation which permits Eqs. (12.3.2.13) to be factored can be written as

$$h = C_1 + C_2 x + \frac{[3(1 - \nu^2)]^{1/2}}{2} \frac{C}{R} x^2. \tag{12.3.3.1}$$

Let us consider the case of linearly varying thickness‡ so that

$$C = 0, \tag{12.3.3.2}$$

† See C. N. DeSilva and P. M. Naghdi, 'Asymptotic Solutions of a Class of Elastic Shells of Revolution with Variable Thickness,' *Quarterly Appl. Math.*, vol. 15, no. 2, July 1957, pp. 169–182.

‡ The case of parabolically varying thickness ($C_1 = C_2 = 0$, $C \neq 0$) is treated by D. Bushnell and N. J. Hoff, 'Influence Coefficients of a Circular Cylindrical Shell with Rapidly Varying Parabolic Wall Thickness,' *AIAA J.*, vol. 2, no. 12, December 1968, pp. 2167–2173. Exponentially varying thickness is treated approximately by E. Reissner and M. B. Sledd, 'Bounds on Influence Coefficients for Circular Cylindrical Shells,' *J. Math. Phys.*, vol. 36, no. 1, January 1957, pp. 1–19. Other power law thickness variations are treated by Y. C. Pao, 'Influence Coefficients of Short Circular Cylindrical Shells with

and a_1 and a_2 are complex conjugates

$$\begin{Bmatrix} a_1 \\ a_2 \end{Bmatrix} = \pm i. \tag{12.3.3.3}$$

With no loss of generality we shall put C_1 equal to zero as well. For the cylinder we also have

$$\Phi_1 \equiv 0, \tag{12.3.3.4}$$

so that we seek solutions of the equation

$$\frac{RC_2}{[12(1-\nu^2)]^{1/2}}\left(x\frac{\mathrm{d}^2W}{\mathrm{d}x^2} + 3\frac{\mathrm{d}W}{\mathrm{d}x}\right) - iW = 0, \tag{12.3.3.5a}$$

where

$$W = \frac{\mathrm{d}W}{\mathrm{d}x} + i\frac{Q_\phi R}{Eh^2}[12(1-\nu^2)]^{1/2}. \tag{12.3.3.5b}$$

When we introduce the notation

$$x = \frac{RC_2 i}{4[12(1-\nu^2)]^{1/2}}u^2, \tag{12.3.3.6a}$$

$$W = \frac{4[12(1-\nu^2)]^{1/2}}{RC_2 i}\frac{1}{u^2}\overline{W}. \tag{12.3.3.6b}$$

Eq. (12.3.3.5a) becomes Bessel's equation for Bessel functions of second order

$$\frac{\mathrm{d}^2\overline{W}}{\mathrm{d}u^2} + \frac{1}{u}\frac{\mathrm{d}\overline{W}}{\mathrm{d}u} + \left(1 - \frac{4}{u^2}\right)\overline{W} = 0. \tag{12.3.3.7}$$

Thus the solution is

$$\overline{W} = \overline{A}_1 J_2(u) + \overline{A}_2 Y_2(u), \tag{12.3.3.8a}$$

Varying Wall Thickness,' *AIAA J.*, vol. 6, no. 8, August 1968, pp. 1613–1616, and H. C. Wang and Y. C. Pao, 'Radial Deformations of Cylindrical Shells with Variable Wall Thickness,' *AIAA J.*, vol. 6, no. 9, September 1968, pp. 1779–1782. See also H. Reissner, 'Uber die Spannungsverteilung in Zylinderschen Behalterwanden,' *Beton u. Eisen*, vol. 7, 1908, pp. 150–155; K. Federhofer, 'Berechnung der kreiszylindrischen Flussigkeitsbehalter mit quadratisch Wandstarke,' *Oesterreichs Ingenieur Archiv*, vol. 6, 1952, pp. 43–64; A. Cattin, 'Serbatio Cilindrico a Sezione Meridiana di Spessore Variabile,' *Ricerca Ingegneria*, vol. 7, 1939, pp. 80–87; T. J. Lardner, 'Symmetric Deformation of a Circular Cylindrical Shell of Variable Wall Thickness,' *ZAMP*, vol. 19, 1968, pp. 270–277.

where $\overline{A}_1$ and $\overline{A}_2$ are complex constants. We may also use the alternate forms in terms of Bessel–Kelvin functions such as were encountered in the analysis of axisymmetric deformations in conical shells of constant thickness

$$\overline{W} = (A_1 + iA_2)(\text{ber}_2\,\xi + i\,\text{bei}_2\,\xi) + (A_3 + iA_4)(\text{ker}_2\,\xi + \text{kei}_2\,\xi), \tag{12.3.3.8b}$$

with

$$\xi = 2\frac{[12(1-\nu^2)]^{1/4}}{(Rh)^{1/2}}\,x, \tag{12.3.3.9a}$$

$$\text{ber}_2\,\xi = \frac{2\,\text{bei}'\,\xi}{\xi} - \text{ber}\,\xi, \tag{12.3.3.9b}$$

$$\text{bei}_2\,\xi = \frac{2\,\text{ber}'\,\xi}{\xi} - \text{bei}\,\xi, \tag{12.3.3.9c}$$

where primes indicate differentiation with respect to ξ. From Eqs. (12.3.3.5b), (12.3.37), and (12.3.3.8b) we obtain

$$x\frac{dw}{dx} = A_1\,\text{ber}_2\,\xi - A_2\,\text{bei}_2\,\xi + A_3\,\text{ker}_2\,\xi - A_4\,\text{kei}_2\,\xi, \tag{12.3.3.10a}$$

$$[12(1-\nu^2)]^{1/2}\frac{Q_x Rx}{Eh^2} = A_1\,\text{bei}_2\,\xi + A_2\,\text{ber}_2\,\xi + A_3\,\text{kei}_2 + A_4\,\text{ker}_2\,\xi. \tag{12.3.3.10b}$$

We may integrate Eq. (12.3.3.10a) to yield the deflection function

$$w = -\frac{2}{\xi}(A_1\,\text{bei}'\,\xi + A_2\,\text{ber}'\,\xi + A_3\,\text{kei}'\,\xi + A_4\,\text{ker}'\,\xi). \tag{12.3.3.10c}$$

By differentiating Eq. (12.3.3.10a) we can obtain the bending moment expression

$$\begin{aligned}\frac{[3(1-\nu^2)]^{1/2}M_xR}{Eh^2} &= A_1\left(\text{ber}'\,\xi - \frac{4\,\text{ber}\,\xi}{\xi} + \frac{8\,\text{bei}'\,\xi}{\xi^2}\right)\\ &- A_2\left(\text{bei}'\,\xi - \frac{4\,\text{bei}\,\xi}{\xi} - \frac{8\,\text{ber}'\,\xi}{\xi^2}\right) + A_3\left(\text{ker}'\,\xi - \frac{4\,\text{ker}\,\xi}{\xi} + \frac{8\,\text{kei}'\,\xi}{\xi^2}\right)\\ &- A_4\left(\text{kei}'\xi - \frac{4\,\text{kei}\,\xi}{\xi} - \frac{8\,\text{ker}'\,\xi}{\xi^2}\right).\end{aligned} \tag{12.3.3.10d}$$

Let us investigate the effect of subjecting the thick end of a cylinder with thickness tapering to zero to a bending moment and a shear force

(Fig. 12.1). Then, since the deformations and stresses at the edge of zero thickness should be regular, we set

$$A_3 = A_4 = 0. \tag{12.3.3.11}$$

The conditions to be satisfied at the thick edge, $x = x_0$, are

$$M_x = M_0, \tag{12.3.3.12a}$$

$$Q_x = Q_0. \tag{12.3.3.12b}$$

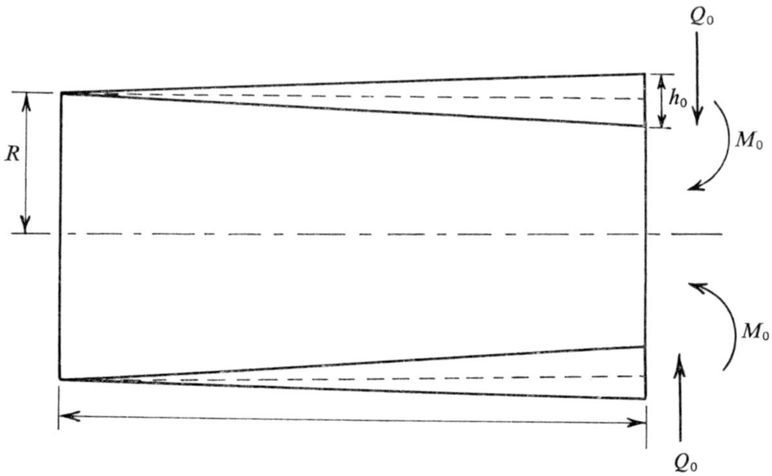

Fig. 12.1. Cylinder with linearly varying thickness subjected to edge loads

Then the coefficients A_1 and A_2 are determined by the equations

$$A_1 \operatorname{bei}_2 \xi_0 + A_2 \operatorname{ber}_2 \xi_0 = \frac{[12(1-\nu^2)]^{1/2} Q_0 R x_0}{E h_0^2}, \tag{12.3.3.13a}$$

$$A_1 \frac{1}{\xi_0}\left(\operatorname{ber}' \xi_0 - \frac{4 \operatorname{ber} \xi_0}{\xi_0} + \frac{8 \operatorname{bei}' \xi_0}{\xi_0^2}\right) - A_2 \frac{1}{\xi_0}\left(\operatorname{bei}' \xi_0 - \frac{4 \operatorname{bei} \xi_0}{\xi_0} - \frac{8 \operatorname{ber}' \xi_0}{\xi_0^2}\right) = \frac{[3(1-\nu^2)]^{1/2} M_0 R}{E h_0^2}, \tag{12.3.3.13b}$$

yielding

$$\Delta_0 A_1 = -\left\{\frac{[3(1-\nu^2)]^{1/2} M_0 R}{E h_0^2} \xi_0 \operatorname{ber}_2 \xi_0 + \frac{[12(1-\nu^2)]^{1/2} Q_0 R x_0}{E h_0^2} \times \left(\operatorname{bei}' \xi_0 - \frac{4 \operatorname{bei} \xi_0}{\xi_0} - \frac{8 \operatorname{ber}' \xi_0}{\xi_0^2}\right)\right\}, \tag{12.3.3.14a}$$

$$\Delta_0 A_2 = \frac{[3(1-\nu^2)]^{1/2} M_0 R}{Eh_0^2} \xi_0 \operatorname{bei}_2 \xi_0 - \frac{[12(1-\nu^2)]^{1/2} Q_0 R x_0}{Eh_0^2} \times \left(\operatorname{ber}' \xi_0 - \frac{4 \operatorname{ber} \xi_0}{\xi_0} + \frac{8 \operatorname{bei}' \xi_0}{\xi_0^2}\right), \quad (12.3.3.14b)$$

where

$$\Delta_0 = \operatorname{ber} \xi_0 \operatorname{ber}' \xi_0 + \operatorname{bei} \xi_0 \operatorname{bei}' \xi_0 - \frac{4}{\xi_0}(\operatorname{ber}^2 \xi_0 + \operatorname{bei}^2 \xi_0) + \frac{16}{\xi_0^2}(\operatorname{ber} \xi_0 \operatorname{bei}' \xi_0 - \operatorname{bei} \xi_0 \operatorname{ber}' \xi_0) - \frac{16}{\xi_0^3}[(\operatorname{ber}' \xi_0)^2 + (\operatorname{bei}' \xi_0)^2]. \quad (12.3.3.14c)$$

The radial deflection and rotation at the loaded edge are thus given by

$$w_0 = \delta_{Q_0}^0 \frac{Q_0}{2\lambda_0^3 D_0} - \delta_{M_0}^0 \frac{M_0}{2\lambda_0^2 D_0}, \quad (12.3.3.15a)$$

$$\frac{\mathrm{d}w_0}{\mathrm{d}x} = \Theta_{Q_0}^0 \frac{Q_0}{2\lambda_0^3 D_0} - \Theta_{M_0}^0 \frac{M_0}{\lambda_0^2 D_0}, \quad (12.3.3.15b)$$

where

$$\delta_{Q_0}^0 = \frac{1}{2^{1/2}\Delta_0}\left[(\operatorname{ber}' \xi_0)^2 + (\operatorname{bei}' \xi_0)^2 - \frac{4}{\xi_0}(\operatorname{ber} \xi_0 \operatorname{ber}' \xi_0 + \operatorname{bei} \xi_0 \operatorname{bei}' \xi_0)\right], \quad (12.3.3.16a)$$

$$\delta_{M_0}^0 = \Theta_{Q_0}^0 = \frac{1}{\Delta_0}\left\{\operatorname{ber} \xi_0 \operatorname{bei}' \xi_0 - \operatorname{bei} \xi_0 \operatorname{ber}' \xi_0 - \frac{2}{\xi_0}[(\operatorname{ber}' \xi_0)^2 + (\operatorname{bei}' \xi_0)^2]\right\}, \quad (12.3.3.16b)$$

$$\Theta_{M_0}^0 = \frac{1}{2^{1/2}\Delta_0}[(\operatorname{ber}_2 \xi_0)^2 + (\operatorname{bei}_2 \xi_0)^2], \quad (12.3.3.16c)$$

and

$$\lambda_0 = \frac{[3(1-\nu^2)]^{1/4}}{(Rh_0)^{1/2}}, \quad (12.3.3.17a)$$

$$D_0 = \frac{Eh_0^3}{12(1-\nu^2)}, \quad (12.3.3.17b)$$

$$h_0 = C_2 x_0. \quad (12.3.3.17c)$$

The edge deflections and slopes in the form of Eqs. (12.3.3.15) have been chosen so that the influence coefficients for the cylinder of linearly varying wall thickness can be directly compared to those for a cylinder of constant thickness equal to the largest thickness. We note that the results for the cylinder of linearly varying thickness can be expressed in terms of the parameter used for the constant thickness cylinder of finite length through the relationship

$$\xi_0 = 2^{3/2}\lambda_0 x_0. \tag{12.3.3.18}$$

The influence coefficients for the loaded edge are shown in Fig. 12.2. As

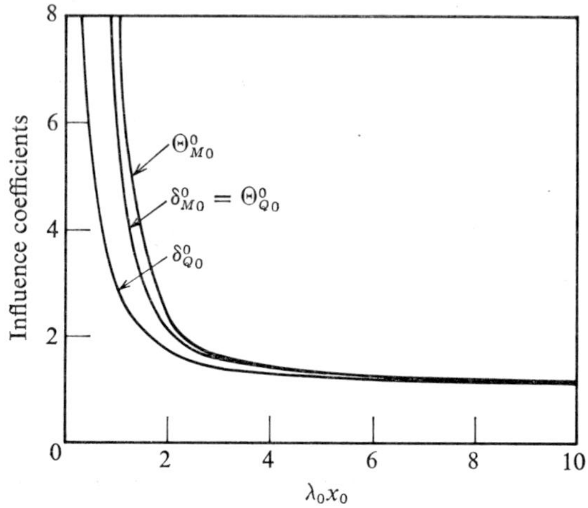

Fig. 12.2. Influence coefficients for cylindrical shells with linearly varying wall thickness

should be expected, the edge influence coefficients for the cylinder of linearly varying thickness are greater than those for the constant thickness cylinder, the difference being most pronounced when the cylinder length is short, so that the thickness tapers rapidly. When the parameter $\lambda_0 x_0$, or ξ_0, is small we may use the series expressions for the Bessel–Kelvin functions to obtain the approximations

$$\delta^0_{Q_0} \approx \frac{3}{\lambda_0 x_0}, \tag{12.3.3.19a}$$

$$\delta^0_{M_0} = \Theta^0_{Q_0} \approx \frac{6}{(\lambda_0 x_0)^2}, \tag{12.3.3.19b}$$

$$\Theta_{M_0} \approx \frac{9}{(\lambda_0 x_0)^3}, \tag{12.3.3.19c}$$

as compared to the corresponding values for the constant thickness cylinder

$$\delta^0_{Q_0} \approx \frac{2}{\lambda_0 x_0}, \tag{12.3.3.20a}$$

$$\delta^0_{M_0} = \Theta^0_{Q_0} \approx \frac{3}{(\lambda_0 x_0)^2}, \tag{12.3.3.20b}$$

$$\Theta_{M_0} \approx \frac{3}{(\lambda_0 x_0)^3}. \tag{12.3.3.20c}$$

When we investigate deformations at the unloaded edge we find that there are four rather than three independent influence coefficients. Thus the radial deflection and slope can be expressed as

$$w|_{\xi=0} = -A_1 = \delta'_{Q_0} \frac{Q_0}{2\lambda_0^3 D_0} - \delta'_{M_0} \frac{M_0}{2\lambda_0^2 D_0}, \tag{12.3.3.21a}$$

$$\left.\frac{\mathrm{d}w}{\mathrm{d}x}\right|_{\xi=0} = \frac{[3(1-\nu^2)]^{1/2}}{RC_2} A_2 = \Theta'_{Q_0} \frac{Q_0}{2\lambda_0^2 D} - \Theta'_{M_0} \frac{M_0}{\lambda_0 D}, \tag{12.3.3.21b}$$

where

$$\delta'_{Q_0} = \frac{\xi_0}{2^{3/2}\Delta_0}\left(\operatorname{bei}'\xi_0 - \frac{4\operatorname{bei}\xi_0}{\xi_0} - \frac{8\operatorname{ber}'\xi_0}{\xi_0^2}\right), \tag{12.3.3.22a}$$

$$\delta'_{M_0} = -\frac{\xi_0}{2\Delta_0}\operatorname{ber}_2\xi_0, \tag{12.3.3.22b}$$

$$\Theta'_{Q_0} = -\frac{1}{8\Delta_0}\xi_0^2\left(\operatorname{ber}'\xi_0 - \frac{4\operatorname{ber}\xi_0}{\xi_0} + \frac{8\operatorname{bei}'\xi_0}{\xi_0^2}\right), \tag{12.3.3.22c}$$

$$\Theta'_{M_0} = -\frac{1}{8(2)^{1/2}\Delta_0}\xi_0^2\operatorname{bei}_2\xi_0. \tag{12.3.3.22d}$$

These are not generally of interest. When ξ_0 is small, however, we can obtain results applicable to tapered rings as

$$\delta'_{Q_0} \approx -\frac{3}{\lambda_0 x_0}, \tag{12.3.3.24a}$$

$$\delta'_{M_0} \approx -\frac{12}{(\lambda_0 x_0)^2}, \tag{12.3.3.24b}$$

$$\Theta'_{Q_0} \approx \frac{6}{(\lambda_0 x_0)^2}, \tag{12.3.3.24c}$$

$$\Theta'_{M_0} \approx \frac{9}{(\lambda_0 x_0)^3}, \tag{12.3.3.24d}$$

which, together with Eqs. (12.3.3.20), can be shown to imply that under an edge moment M_0 the longitudinal cross-section of the short tapered cylinder or ring rotates as a rigid body around a point one-third of the distance from the thick edge and under an edge transverse shear Θ_0 behaves as if the transverse shear were applied at this point together with a moment $\frac{1}{3}Q_0 x_0$ about the point (see Fig. 12.3).

When the cylinder thickness does not taper to a point all of the terms in Eq. (12.3.3.11) must be retained. The calculations are, of course, much more complicated in this case.† For short tapered structures (not necessarily linear thickness variations) such as may be used as transition sections between shells of different constant thickness, it is possible to simplify the calculations by treating the short tapered shells as rings under radial forces passing through the centroid of the longitudinal cross-section and under twisting moments. However, the meridional length of the structure must be very short, of the order of magnitude of the attenuation length of end effects or less, for this simplification to be valid.

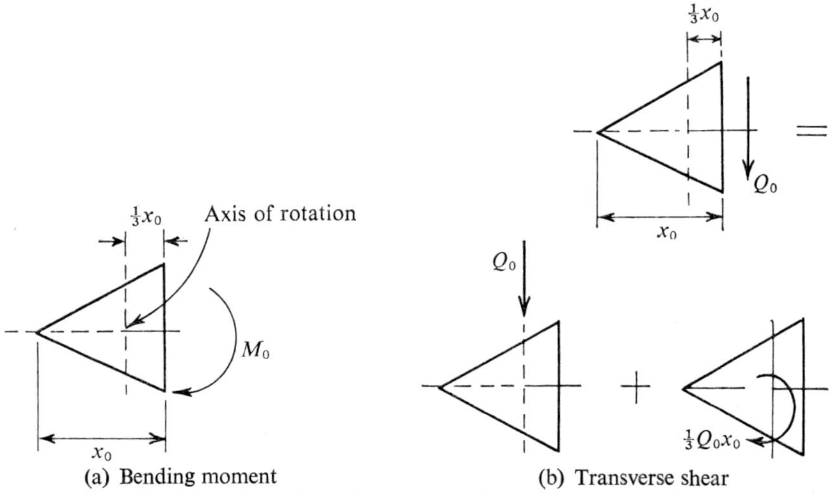

Fig. 12.3. Axis of rotation for short linearly tapered cylinders

12.3.4 Conical shell of linearly varying thickness

For the conical shell, the thickness variation for which Eq. (12.3.2.13) may be factored can be written as

$$h = \begin{cases} C_1 s^\nu + \left[\left(3\dfrac{1+\nu}{1-\nu}\right)^{1/2} C \cot\alpha\right] s + C_2 & (\nu \neq 0), \quad (12.3.4.1a) \\ C_1 + [(3)^{1/2} C \cot\alpha] + C_2 \ln s & (\nu = 0). \quad (12.3.4.1b) \end{cases}$$

We shall assume that the desired thickness variation is linear and write

$$h = a + \left[\left(3\frac{1+\nu}{1-\nu}\right)^{1/2} C \cot\alpha\right] s, \tag{12.3.4.2}$$

† See E. C. Rodabaugh and T. J. Atterbury, 'Stresses in Tapered Transition Joints in Pipelines and Pressure Vessels,' *J. Engng Industry*, vol. 83, no. 3, August 1962, pp. 321–328; and L. S. Wolfson, 'Edge Influence Coefficient for Cylinders with Linearly Varying Wall Thickness,' JPL Technical Report No. 32-435, June 1, 1963.

for which we have, assuming that C is less than 2,

$$a_1 = -\frac{C}{2} + i\left[1 - \left(\frac{C}{2}\right)^2\right]^{1/2}, \tag{12.3.4.3a}$$

$$\Phi_1 = \frac{3\nu(\mathrm{d}h/\mathrm{d}s)\tan\alpha}{[12(1-\nu^2)]^{1/2}} = \frac{3\nu C}{2(1-\nu)}. \tag{12.3.4.3b}$$

Then the equation we wish to solve, Eq. (12.3.2.20), is

$$s\left[a + \left(3\frac{1+\nu}{1-\nu}\right)^{1/2} Cs\cot\alpha\right]\frac{\mathrm{d}^2W}{\mathrm{d}s^2} + \left[a + 4\left(3\frac{1+\nu}{1-\nu}\right)^{1/2} Cs\cot\alpha\right]\frac{\mathrm{d}W}{\mathrm{d}s}$$

$$-\left(\frac{a}{s} - 2\nu\left(3\frac{1+\nu}{1-\nu}\right)^{1/2} C\cot\alpha + i\left\{12(1-\nu^2)\left[1-\left(\frac{C}{2}\right)^2\right]\right\}^{1/2}\cot\alpha\right)W = 0. \tag{12.3.4.5a}$$

where

$$W = \frac{\mathrm{d}w}{\mathrm{d}s} + \left\{-\frac{C}{2} + i\left[1-\left(\frac{C}{2}\right)^2\right]\right\}^{1/2}\frac{[12(1-\nu^2)]^{1/2}Q_s s\tan\alpha}{Eh^2}. \tag{12.3.4.5b}$$

The case when

$$a = 0, \tag{12.3.4.6}$$

for which the thickness tapers to a point at the apex is the simplest to consider. Then Eq. (12.3.4.5a) becomes

$$s^2\frac{\mathrm{d}^2W}{\mathrm{d}s^2} + 4s\frac{\mathrm{d}W}{\mathrm{d}s} + 2\left\{\nu - i\frac{1-\nu}{C}\left[1-\left(\frac{C}{2}\right)^2\right]^{1/2}\right\}W = 0, \tag{12.3.4.7}$$

for which the solution may be readily found as†

$$W = \bar{A}_1 u^{n_1} + \bar{A}_2 u^{n_2}, \tag{12.3.4.8a}$$

$$u = \frac{s}{s_0}, \tag{12.3.4.8b}$$

where

$$n_{1,2} = -\tfrac{3}{2} \pm (\gamma + i\delta), \tag{12.3.4.8c}$$

† Similar solutions may also be obtained for asymmetrical loading of conical shells with thickness proportional to the distance from the vertex. See E. Honegger, 'Festigkeitsberechnung von Kegelschalen mit linear veränderlicher wandstärke,' Dissertation, Luzern, 1919; and W. Flügge, *Stresses in Shells*, Springer-Verlag, Berlin, 1960, pp. 47, 287.

$$\gamma = \left\{\left[\left(\frac{9}{8} - \nu\right)^2 + \left(\frac{1 - \nu}{2\chi}\right)^2\right]^{1/2} + \left(\frac{9}{8} - \nu\right)\right\}^{1/2}, \tag{12.3.4.8d}$$

$$\delta = \left\{\left[\left(\frac{9}{8} - \nu\right)^2 + \left(\frac{1 - \nu}{2\chi}\right)^2\right]^{1/2} - \left(\frac{9}{8} - \nu\right)\right\}^{1/2}, \tag{12.3.4.8e}$$

$$\chi = \frac{C/2}{[1 - (C/2)^2]^{1/2}}, \tag{12.3.4.8f}$$

s_0 is some representative length, and $\bar{A}_1$, $\bar{A}_2$ are complex constants of integration. We may also write Eq. (12.3.4.8a) in the alternate expanded form

$$\begin{aligned} W = {} & u^{\gamma - 3/2}[A_1 \cos \ln u^\delta + A_2 \sin \ln u^\delta] \\ & + u^{-\gamma - 3/2}[A_3 \cos \ln u^\delta + A_4 \sin \ln u^\delta] \\ & + i\{u^{\gamma - 3/2}[A_1 \sin \ln u^\delta - A_2 \cos \ln u^\delta] \\ & - u^{-\gamma - 3/2}[A_3 \sin \ln u^\delta - A_4 \cos \ln u^\delta]\}, \end{aligned} \tag{12.3.4.9a}$$

from which we may obtain

$$\begin{aligned} \left[1 - \left(\frac{C}{2}\right)^2\right]^{1/2} \frac{Q_s C s^2}{2(1 - \nu)D} = {} & u^{\gamma - 3/2}(A_1 \sin \ln u^\delta - A_2 \cos \ln u^\delta) \\ & - u^{-\gamma - 3/2}(A_3 \sin \ln u^\delta - A_4 \cos \ln u^\delta), \end{aligned} \tag{12.3.4.9b}$$

$$\begin{aligned} \frac{\mathrm{d}w}{\mathrm{d}s} = {} & u^{\gamma - 3/2}\{A_1[\cos \ln u^\delta + \chi \sin \ln u^\delta] + A_2[\sin \ln u^\delta - \chi \cos \ln u^\delta]\} \\ & + u^{-\gamma - 3/2}\{A_3 [\cos \ln u^\delta - \chi \sin \ln u^\delta] + A_4[\sin \ln u^\delta + \chi \cos \ln u^\delta]\}. \end{aligned} \tag{12.3.4.9c}$$

From Eqs. (12.3.2.1c), (12.3.2.4c), (12.3.2.4d), and (12.3.4.9c), an expression for the edge bending moment may be written as

$$\begin{aligned} -\frac{M_s s}{D} = {} & u^{\gamma - 3/2}\langle A_1\{[\gamma - \tfrac{3}{2} + \nu + \chi\delta] \cos \ln u^\delta \\ & \qquad + [\chi(\gamma - \tfrac{3}{2} + \nu) - \delta] \sin \ln u^\delta\} \\ & + A_2\{[\gamma - \tfrac{3}{2} + \nu + \chi\delta] \sin \ln u^\delta - [\chi(\gamma - \tfrac{3}{2} + \nu) - \delta] \cos \ln u^\delta\}\rangle \\ & - u^{-\gamma - 3/2}\langle A_3\{[\gamma + \tfrac{3}{2} - \nu + \chi\delta] \cos \ln u^\delta - [\chi(\gamma + \tfrac{3}{2} - \nu) - \delta] \sin \ln u^\delta\} \\ & + A_4\{[\gamma + \tfrac{3}{2} - \nu + \chi\delta] \sin \ln u^\delta + [\chi(\gamma + \tfrac{3}{2} - \nu) - \delta] \cos \ln u^\delta\}\rangle. \end{aligned} \tag{12.3.4.10}$$

Finally, the displacement normal to the longitudinal axis of the conical shell is

$$\begin{aligned} \delta_H & = s\epsilon_\theta \sin \alpha = \frac{N_\theta - \nu N_\phi}{Eh} s \sin \alpha \\ & = \left[\frac{\mathrm{d}}{\mathrm{d}s}(Q_s s) - \nu Q_s\right] \frac{s \sin \alpha \tan \alpha}{Eh}. \end{aligned} \tag{12.3.4.11a}$$

With the use of Eq. (12.3.4.9b) we then have

$$\frac{(1-\nu)\delta_H}{\chi s \cos\alpha} = u^{\gamma-3/2}\{A_1[(\gamma+\tfrac{1}{2}-\nu)\sin\ln u^\delta + \delta\cos\ln u^\delta] - A_2[(\gamma+\tfrac{1}{2}-\nu)\cos\ln u^\delta - \delta\sin\ln u^\delta]\} + u^{-\gamma-3/2}\{A_3[(\gamma-\tfrac{1}{2}+\nu)\sin\ln u^\delta - \delta\cos\ln u^\delta] - A_4[(\gamma-\tfrac{1}{2}+\nu)\cos\ln u^\delta + \delta\sin\ln u^\delta]\}. \quad (12.3.4.11b)$$

Let us now consider a complete cone, subjected to edge-bending moments and shear forces normal to the longitudinal axis (Fig. 12.4). Then, noting

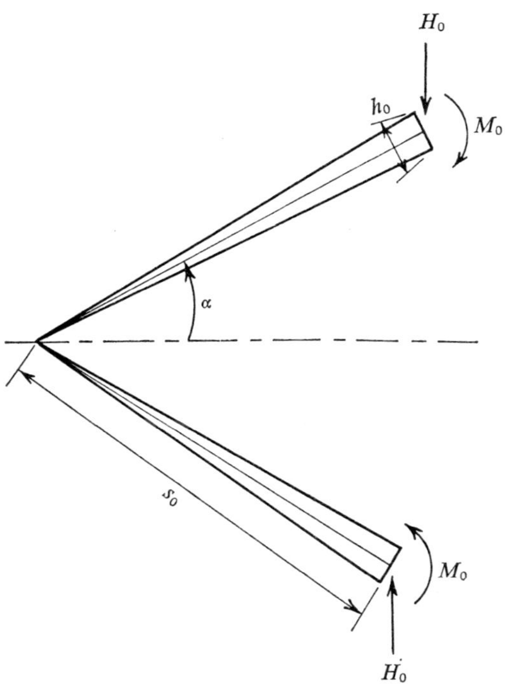

Fig. 12.4. Edge-loaded conical shell with thickness proportional to distance from the vertex

that γ is greater than 1 for all values of C and all usual values of Poisson's ratio ν, it follows that for Q_s, M_s, and δ_H to be regular at the apex we must put

$$A_3 = A_4 = 0, \quad (12.3.4.12)$$

in the equations. The expression for dw/ds will then be regular at the apex if γ is also greater than 3/2, or if

$$C < \frac{2(1-\nu)}{(1+16\nu+\nu^2)^{1/2}}, \quad (12.3.4.13)$$

a condition which is generally satisfied for thin shells. The coefficients A_1 and A_2 are determined by the equations

$$M_s = M_0, \tag{12.3.4.14a}$$

$$\Theta_s = H_0 \cos \alpha, \tag{12.3.4.14b}$$

at $s = s_0$ or $u = 1$ where s_0 is the meridional distance from the vertex to the large end of the cone. With the use of Eqs. (12.3.4.9b) and (12.3.4.10) we have

$$A_1 = -[\gamma - \tfrac{3}{2} + \nu + \chi\delta]^{-1} \times \left\{[\chi(\gamma - \tfrac{3}{2} + \nu) - \delta]\left[1 - \left(\frac{C}{2}\right)^2\right]^{1/2} \frac{H_0 C s_0^2 \cos \alpha}{2(1-\nu)D_0} + \frac{M_0 s_0}{D_0}\right\}, \tag{12.3.4.15a}$$

$$A_2 = -\left[1 - \left(\frac{C}{2}\right)^2\right]^{1/2} \frac{H_0 C s_0^2 \cos \alpha}{2(1-\nu)D_0}. \tag{12.3.4.15b}$$

The deformation normal to the cone axis and the slope at the large end can now be expressed in a form similar to that for the conical shell of uniform thickness as

$$\left(\frac{\delta_H}{\cos \alpha}\right)_0 = \delta_{H0}^0 \frac{H_0 \cos \alpha}{2\lambda_0^3 D_0} - \delta_{M0}^0 \frac{M_0}{2\lambda_0^2 D_0}, \tag{12.3.4.16a}$$

$$\left(\frac{dw}{ds}\right)_0 = \Theta_{H0}^0 \frac{H_0 \cos \alpha}{2\lambda_0^2 D_0} - \Theta_{M0}^0 \frac{M_0}{\lambda_0 D_0}, \tag{12.3.4.16b}$$

where

$$\begin{aligned}\delta_{H0}^0 = \frac{1}{\Delta(1+\chi^2)^{1/4}}\Bigg\{&\left[1 + \left(\frac{\frac{9}{4} - 2\nu}{1-\nu}\chi\right)^2\right]^{1/2} \\ &- \left(\frac{\chi}{2(1-\nu)}\left\{\left[1 + \left(\frac{\frac{9}{4} - 2\nu}{1-\nu}\chi\right)^2\right]^{1/2} + \frac{\frac{9}{4} - 2\nu}{1-\nu}\chi\right\}\right)^{1/2} \\ &- (\tfrac{1}{2} - \nu)(\tfrac{3}{2} - \nu)\frac{\chi}{1-\nu} + 2\chi \\ &\times \left(\frac{1-\nu}{2}\chi\left\{\left[1 + \left(\frac{\frac{9}{4} - 2\nu}{1-\nu}\chi\right)^2\right]^{1/2} - \frac{\frac{9}{4} - 2\nu}{1-\nu}\chi\right\}\right)^{1/2}\Bigg\},\end{aligned} \tag{12.3.4.17a}$$

$$\delta_{M0}^0 = \Theta_{H0}^0 = \frac{(1+\chi^2)^{1/2}}{\Delta}\left\{\left[1 + \left(\frac{\frac{9}{4} - 2\nu}{1-\nu}\chi\right)^2\right]^{1/2} - \frac{\frac{9}{4} - 2\nu}{1-\nu}\chi\right\}^{1/2}, \tag{12.3.4.17b}$$

$$\Theta_{M0}^0 = \frac{(1+\chi^2)^{1/4}}{\Delta}, \tag{12.3.4.17c}$$

$$\Delta = \left\{\left[1 + \left(\frac{\frac{9}{4} - 2\nu}{1 - \nu}\chi\right)^2\right]^{1/2} + \frac{\frac{9}{4} - 2\nu}{1 - \nu}\chi\right\}^{1/2} - (\tfrac{3}{2} - \nu)\left(\frac{2\chi}{1 - \nu}\right)^{1/2}$$
$$+ \left\{\left[1 + \left(\frac{\frac{9}{4} - 2\nu}{1 - \nu}\chi\right)^2\right]^{1/2} - \frac{\frac{9}{4} - 2\nu}{1 - \nu}\chi\right\}^{1/2}. \qquad (12.3.4.17d)$$

The influence coefficients are shown as a function of C in Fig. 12.5. It is interesting to observe that the influence coefficients given above are entirely

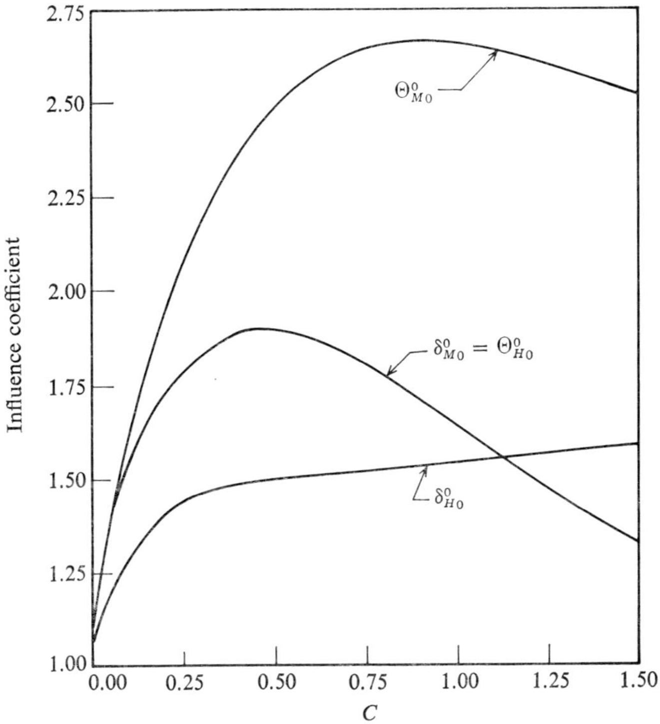

Fig. 12.5. Influence coefficients for edge-loaded conical shells with thickness proportional to the distance from the vertex

independent of the length of the conical shell, thus resembling those for the semi-infinite cylinder. The dependence on the semi-vertex angle is implicit, for if the thickness is expressed in the form

$$h = bs \tan \alpha = bR_\theta, \qquad (12.3.4.18a)$$

the value of C is then

$$C = \left(\frac{1}{3}\frac{1 - \nu}{1 + \nu}\right)^{1/2} b \tan^2 \alpha. \qquad (12.3.4.18b)$$

For conical shell frustums the calculations are again more difficult since all four constants of integration must be considered. The case when a in Eq. (12.3.4.2) is not equal to zero is still more difficult since the solution of Eq. (12.3.4.5a) can no longer be expressed in terms of tabulated or easily calculable functions. If we introduce the notation

$$s = -\frac{b}{a}t, \tag{12.3.4.19a}$$

$$W = \frac{V}{t}, \tag{12.3.4.19b}$$

Eq. (12.3.4.5a) becomes

$$t(1 - t)\frac{d^2V}{dt^2} - (1 + 2t)\frac{dV}{dt} - 2(1 - \nu)\left(1 + \frac{i}{2\chi}\right)V = 0, \tag{12.3.4.20a}$$

or, alternatively,

$$t(1 - t)\frac{d^2V}{dt^2} - [1 + (\bar{\alpha} + \bar{\beta} + 1)t]\frac{dV}{dt} - \bar{\alpha}\bar{\beta}V = 0, \tag{12.3.4.20b}$$

where

$$\left.\begin{matrix}\bar{\alpha}\\ \bar{\beta}\end{matrix}\right\} = \tfrac{1}{2} \pm (\gamma + i\delta). \tag{12.3.4.20c}$$

Eq. (12.3.4.20b) is in the standard form of the hypergeometric equation.† One of the solutions is

$$V_1 = t^2F(\bar{\alpha} + 2, \bar{\beta} + 2; 3, t), \tag{12.3.4.21a}$$

where

$$F(\bar{\alpha} + 2, \bar{\beta} + 2; 3, t) = 1 + \frac{(\bar{\alpha} + 2)(\bar{\beta} + 2)}{3(1!)}t + \frac{(\bar{\alpha} + 2)(\bar{\alpha} + 3)(\bar{\beta} + 2)(\bar{\beta} + 3)}{(3)(4)(2!)}t^2 + \frac{(\bar{\alpha} + 2)(\bar{\alpha} + 3)(\bar{\alpha} + 4)(\bar{\beta} + 2)(\bar{\beta} + 3)(\bar{\beta} + 4)}{(3)(4)(5)(3!)}t^3 + \dots \tag{12.3.4.21b}$$

The other can be expressed in the integral form

$$V_2 = V_1 \int \frac{t\,dt}{V_1^2|1 - t|^{\bar{\alpha} + \bar{\beta} + 2}}. \tag{12.3.4.21c}$$

† See, for example, M. Abramowitz and I. A. Stegun, *Handbook of Mathematical Functions*, NBS Appl. Math. Ser. no. 55, 1964, pp. 556–566.

The general solution thus formed can be used to obtain expressions for deformations and stress resultants as well as edge influence coefficients† in the conical shell with linearly varying wall thickness.

12.4 Effects of anisotropy

When the material of the shell is anisotropic, it is customary to retain the assumption of infinite transverse stiffness of the material and to assume that the stress–strain relations of the material in layers parallel to the middle surface are replaced, in general, by

$$\epsilon_\alpha^{(\zeta)} = s_{11}\sigma_\alpha^{(\zeta)} + s_{12}\sigma_\beta^{(\zeta)} + s_{13}\tau_{\alpha\beta}^{(\zeta)}, \tag{12.4.1a}$$

$$\epsilon_\beta^{(\zeta)} = s_{21}\sigma_\alpha^{(\zeta)} + s_{22}\sigma_\beta^{(\zeta)} + s_{23}\tau_{\alpha\beta}^{(\zeta)}, \tag{12.4.1b}$$

$$\gamma_{\alpha\beta}^{(\zeta)} = s_{31}\sigma_\alpha^{(\zeta)} + s_{32}\sigma_\beta^{(\zeta)} + s_{33}\tau_{\alpha\beta}^{(\zeta)}, \tag{12.4.1c}$$

where the matrix of constant coefficients of the stresses is symmetric. By inverting these to obtain stresses in terms of strains and on using the usual definitions of the stress-resultants we find that the modified overall force–strain relations are of the form

$$\begin{Bmatrix} N_\alpha \\ N_\beta \\ \bar{N}_{\alpha\beta} \\ M_\alpha \\ M_\beta \\ \bar{M}_{\alpha\beta} \end{Bmatrix} = \begin{vmatrix} C_{11} & C_{12} & C_{13} & C_{14} & C_{15} & C_{16} \\ C_{21} & C_{22} & C_{23} & C_{24} & C_{25} & C_{26} \\ C_{31} & C_{32} & C_{33} & C_{34} & C_{35} & C_{36} \\ C_{41} & C_{42} & C_{43} & C_{44} & C_{45} & C_{46} \\ C_{51} & C_{52} & C_{53} & C_{54} & C_{55} & C_{56} \\ C_{61} & C_{62} & C_{63} & C_{64} & C_{65} & C_{66} \end{vmatrix} \begin{Bmatrix} \epsilon_\alpha \\ \epsilon_\beta \\ \gamma_{\alpha\beta} \\ \kappa_\alpha \\ \kappa_\beta \\ \bar{\kappa}_{\alpha\beta} \end{Bmatrix}, \tag{12.4.2}$$

where the matrix of coefficients C_{ij} is symmetric and, in general, contains no vanishing elements.

In many cases of practical interest, the material used is curvilinearly orthotropic with the axes of orthotropy oriented to coincide with the lines of curvature of the middle surface. In this case, we have in Eqs. (12.4.1)

$$s_{13} = s_{31} = s_{23} = s_{32} = 0, \tag{12.4.3a}$$

$$s_{11} = \frac{1}{E_\alpha}, \tag{12.4.3b}$$

† Some of these calculations have been carried out by E. Y. W. Tsui, 'Analysis of Tapered Conical Shells,' *Proc. Fourth U.S. National Congress of Applied Mechanics*, University of California, Berkeley, 1962, pp. 807–816.

$$s_{12} = s_{21} = \frac{\nu_\beta}{E_\alpha} = \frac{\nu_\alpha}{E_\beta}, \tag{12.4.3c}$$

$$s_{22} = \frac{1}{E_\beta}, \tag{12.4.3d}$$

$$s_{33} = \frac{1}{G_{\alpha\beta}}, \tag{12.4.3e}$$

so that the stress–strain relations can be written as

$$\sigma_\alpha^{(\zeta)} = \frac{E_\alpha}{1 - \nu_\alpha \nu_\beta}\left[\epsilon_\alpha^{(\zeta)} + \nu_\alpha \epsilon_\beta^{(\zeta)}\right], \tag{12.4.4a}$$

$$\sigma_\alpha^{(\zeta)} = \frac{E_\beta}{1 - \nu_\alpha \nu_\beta}\left[\epsilon_\beta^{(\zeta)} + \nu_\beta \epsilon_\alpha^{(\zeta)}\right], \tag{12.4.4b}$$

$$\tau_{\alpha\beta}^{(\zeta)} = G_{\alpha\beta}\gamma^{(\zeta)}. \tag{12.4.4c}$$

The overall force–strain relations are then similar to those given by Eqs. (2.7.2) and can readily be written as

$$N_\alpha = K_\alpha(\epsilon_\alpha + \nu_\alpha \epsilon_\beta) + D_\alpha\left(\frac{1}{R_\beta} - \frac{1}{R_\alpha}\right)\left(\kappa_\alpha - \frac{\epsilon_\alpha}{R_\alpha}\right)(1 + \psi_\alpha), \tag{12.4.5a}$$

$$N_\beta = K_\beta(\epsilon_\beta + \nu_\beta \epsilon_\alpha) + D_\beta\left(\frac{1}{R_\alpha} - \frac{1}{R_\beta}\right)\left(\kappa_\beta - \frac{\epsilon_\beta}{R_\beta}\right)(1 + \psi_\beta), \tag{12.4.5b}$$

$$M_\alpha = D_\alpha\left\{\kappa_\alpha + \nu_\alpha \kappa_\beta + \left(\frac{1}{R_\beta} - \frac{1}{R_\alpha}\right)\left[\epsilon_\alpha - \left(\kappa_\alpha - \frac{\epsilon_\alpha}{R_\alpha}\right)R_\alpha \psi_\alpha\right]\right\}, \tag{12.4.5c}$$

$$M_\beta = D_\beta\left\{\kappa_\beta + \nu_\beta \kappa_\alpha + \left(\frac{1}{R_\alpha} - \frac{1}{R_\beta}\right)\left[\epsilon_\beta - \left(\kappa_\beta - \frac{\epsilon_\beta}{R_\beta}\right)R_\beta \psi_\beta\right]\right\}, \tag{12.4.5d}$$

$$\begin{aligned}\overline{N}_{\alpha\beta} = K_{\alpha\beta}\gamma_{\alpha\beta} + \tfrac{1}{4}D_{\alpha\beta}\Bigg\{ & 3\left(\frac{1}{R_\alpha} - \frac{1}{R_\beta}\right)^2 \gamma_{\alpha\beta} \\ & + \left(3 - \frac{R_\alpha}{R_\beta}\right)\left(\frac{1}{R_\beta} - \frac{1}{R_\alpha}\right)\left[\bar{\kappa}_{\alpha\beta} - \tfrac{1}{4}\gamma_{\alpha\beta}\left(\frac{3}{R_\alpha} - \frac{1}{R_\beta}\right)\psi_\alpha\right] \\ & + \left(3 - \frac{R_\beta}{R_\alpha}\right)\left(\frac{1}{R_\alpha} - \frac{1}{R_\beta}\right)\left[\bar{\kappa}_{\alpha\beta} - \tfrac{1}{4}\gamma_{\alpha\beta}\left(\frac{3}{R_\beta} - \frac{1}{R_\alpha}\right)\psi_\beta\right]\Bigg\},\end{aligned} \tag{12.4.5e}$$

$$\begin{aligned}\overline{M}_{\alpha\beta} = 2D_{\alpha\beta}\Bigg\{ & \bar{\kappa}_{\alpha\beta} + \tfrac{1}{4}\left(1 - \frac{R_\alpha}{R_\beta}\right)\left[\bar{\kappa}_{\alpha\beta} - \tfrac{1}{4}\gamma_{\alpha\beta}\left(\frac{3}{R_\alpha} - \frac{1}{R_\beta}\right)\right]\psi_\alpha \\ & + \tfrac{1}{4}\left(1 - \frac{R_\beta}{R_\alpha}\right)\left[\bar{\kappa}_{\alpha\beta} - \tfrac{1}{4}\gamma_{\alpha\beta}\left(\frac{3}{R_\beta} - \frac{1}{R_\alpha}\right)\right]\psi_\beta\Bigg\},\end{aligned} \tag{12.4.5f}$$

where

$$K_\alpha = \frac{E_\alpha h}{12(1 - \nu_\alpha \nu_\beta)}, \tag{12.4.6a}$$

$$K_\beta = \frac{E_\beta h}{12(1 - \nu_\alpha \nu_\beta)}, \tag{12.4.6b}$$

$$K_{\alpha\beta} = G_{\alpha\beta} h, \tag{12.4.6c}$$

$$D_\alpha = \frac{E_\alpha h^3}{12(1 - \nu_\alpha \nu_\beta)}, \tag{12.4.6d}$$

$$D_\beta = \frac{E_\beta h^3}{12(1 - \nu_\alpha \nu_\beta)}, \tag{12.4.6e}$$

$$D_{\alpha\beta} = \tfrac{1}{12} G_{\alpha\beta} h^3. \tag{12.4.6f}$$

With appropriate approximations these may be reduced to forms similar to those of Eqs. (3.1.2) or (3.1.4).

As an example of the effects of anisotropy let us consider the problem of the analysis of an orthotropic circular cylindrical shell, the edges of which are clamped by rigid end plates that can move freely in the longitudinal direction, which is subjected to uniform internal pressure on its curved surface (Fig. 12.6). We take the overall force–strain relations in the form

$$N_x = \frac{E_x h}{1 - \nu_x \nu_\theta}\left(\frac{du}{dx} + \nu_x \frac{w}{R}\right), \tag{12.4.7a}$$

$$N_\theta = \frac{E_\theta h}{1 - \nu_x \nu_0}\left(\frac{w}{R} + \nu_\theta \frac{du}{dx}\right), \tag{12.4.7b}$$

$$M_x = -\frac{E_x h^3}{12(1 - \nu_x \nu_\theta)} \frac{d^2 w}{dx^2}, \tag{12.4.7c}$$

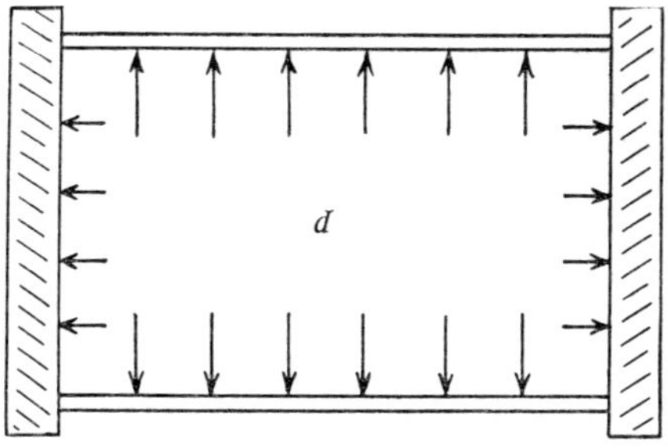

Fig. 12.6. Internally pressurized orthotropic cylindrical shell

$$M_\theta = -\frac{\nu_x E_x h^3}{12(1-\nu_x\nu_\theta)}\frac{d^2w}{dx^2}. \tag{12.4.7d}$$

The equilibrium equations, as for the isotropic shell, are given approximately by

$$N_x = p\frac{R}{2}, \tag{12.4.8a}$$

$$\frac{d^2M_x}{dx^2} - \frac{N_\theta}{R} + p = 0. \tag{12.4.8b}$$

From Eqs. (12.4.7a) and (12.4.8a) we have

$$\frac{du_x}{dx} = -\nu_x\frac{w}{R} + \frac{(1-\nu_x\nu_\theta)pR}{2E_xh}, \tag{12.4.9a}$$

which leads to

$$N_\theta = E_\theta\frac{h}{r}w + \tfrac{1}{2}\nu_x pR. \tag{12.4.9b}$$

Then Eqs. (12.4.7c), (12.4.8b), and (12.4.9b) yield the differential equations for the radial deformation as

$$\frac{d^4w}{dx^4} + \frac{12(1-\nu_x\nu_\theta)}{R^2h^2}\frac{E_\theta}{E_x}w = \frac{p(1-\nu_x/2)}{D_x}. \tag{12.4.10}$$

We thus see that the axisymmetric deflections of an orthotropic cylinder are described by an equation which is practically identical with that for the isotropic cylinder, the main difference being that the attenuation length $1/\lambda$ is replaced by the quantity

$$\frac{1}{\lambda_{\text{orth}}} = (Rh)^{1/2}\left(\frac{E_x/E_\theta}{3(1-\nu_x\nu_\theta)}\right)^{1/4}$$

which is greater than or less than that for the isotropic shell depending on whether the quantity E_x/E_θ is greater than or less than unity.

If we carry through the various algebraic manipulations in a manner similar to that of Sections 4.2 and 4.4, we find that the edge radial deflection and slope of a cylinder of length l for deflections symmetrical about the center can be expressed as

$$w_0 = -\left[\delta_H(\lambda_{\text{orth}}l)\frac{H_0}{2\lambda^3_{\text{orth}}D_x} + \delta_M(\lambda_{\text{orth}}l)\frac{M_0}{2\lambda^2_{\text{orth}}D_x}\right] + \frac{p(1\times\nu_x/2)R^2}{E_\theta h}, \tag{12.4.11a}$$

$$\frac{dw_0}{dx} = \Theta_H(\lambda_{\text{orth}}l)\frac{H_0}{2\lambda^3_{\text{orth}}D_x} + \Theta_M(\lambda_{\text{orth}}l)\frac{M_0}{\lambda_{\text{orth}}D_x}. \tag{12.4.11b}$$

Both of these must vanish for clamped edges, yielding

$$\frac{H_0}{2\lambda_{\mathrm{orth}}^3 D_x} = 2\Theta(\lambda_{\mathrm{orth}} l)\frac{p(1 - \nu_x/2)R^2}{E_\theta h}, \tag{12.4.12a}$$

$$\frac{M_0}{2\lambda_{\mathrm{orth}}^2 D_x} = -\Theta_H(\lambda_{\mathrm{orth}} l)\frac{p(1 - \nu_x/2)R^2}{E_\theta h}. \tag{12.4.12b}$$

By comparing the above expressions to the corresponding ones for an isotropic cylinder, given by Eqs. (4.4.13), we find that for the same internal pressure and cylinder dimensions

$$\frac{(H_0)_{\mathrm{orth}}}{(H_0)_{\mathrm{isotropic}}} = \left(\frac{E_x}{E_\theta}\frac{1 - \nu^2}{1 - \nu_x\nu_\theta}\right)^{1/4}\frac{1 - \nu_x/2}{1 - \nu/2}\frac{\Theta_M(\lambda_{\mathrm{orth}} l)}{\Theta_M(\lambda l)}, \tag{12.4.13a}$$

$$\frac{(M_0)_{\mathrm{orth}}}{(M_0)_{\mathrm{isotropic}}} = \left(\frac{E_x}{E_\theta}\frac{1 - \nu^2}{1 - \nu_x\nu_\theta}\right)^{1/2}\frac{1 - \nu_x/2}{1 - \nu/2}\frac{\Theta_H(\lambda_{\mathrm{orth}} l)}{\Theta_H(\lambda l)}, \tag{12.4.13b}$$

For a long cylinder

$$\frac{\Theta_M(\lambda_{\mathrm{orth}} l)}{\Theta_M(\lambda l)} = \frac{\Theta_H(\lambda_{\mathrm{orth}} l)}{\Theta_H(\lambda l)} = 1. \tag{12.4.14}$$

Thus the edge or maximum stresses in the orthotropic cylinder are generally less than those in the isotropic cylinder if the material used is stiffer in the circumferential direction than in the longitudinal direction and larger if the reverse is true.

Other problems of analysis of orthotropic shells can be treated by the methods used previously for isotropic shells.†

12.5 Layered shells

The simplest type of composite shell is one composed of several bonded layers of different materials. These materials may be isotropic or anisotropic. A familiar example of a layered construction is plywood, which may consist of layers of the same or different woods with the grain direction of alternate layers at right angles to that of the other layers. More recently, layered materials consisting of fiber-reinforced composite materials have been used to great advantage.‡

† See, for example, C. R. Steele and R. F. Hartung, 'Symmetric Loading of Orthotropic Shells of Revolution,' *J. Appl. Mech.*, vol. 32, no. 2, June 1965, pp. 337–345.

‡ With suitably defined material properties, the analysis for fiber-reinforced layered shell walls is identical to that for layered anisotropic shell walls. See, for example, J. M. Whitney, 'Elastic Properties of Fiber-Reinforced Shells,' *AIAA J.*, vol. 5, no. 5, May 1967, pp. 966–968.

The theory for such layered shells is not significantly different from that for homogeneous shells. As in the previous sections of this chapter, all we need change are the overall force–strain relations of the shell which now involve summation of integrals over the thickness of each layer rather than a single integral over the entire thickness of the shell. Let us assume, for simplicity, that each layer is of constant thickness and that the material in each layer is orthotropic, with the axes of orthotropy in the direction of the lines of curvature of the middle surface of the shell. The middle surface is defined in the same manner as for the homogeneous shell. Let us denote the thickness of the nth layer by h_n and the distance from the inner shell surface to the top of the nth layer by x_n (Fig. 12.7). We have

$$x_n = \sum_{m=1}^{n} h_m, \tag{12.5.1a}$$

$$x_0 = 0. \tag{12.5.1b}$$

The total thickness of the shell we denote by h. The stress–strain relations for the nth layer are taken as

$$\sigma_{\alpha n}^{(\zeta)} = \frac{E_{\alpha n}}{1 - \nu_{\alpha n}\nu_{\beta n}} [\epsilon_\alpha^{(\zeta)} + \nu_{\alpha n}\epsilon_\beta^{(\zeta)}], \tag{12.5.2a}$$

$$\sigma_{\beta n}^{(\zeta)} = \frac{E_{\beta n}}{1 - \nu_{\alpha n}\nu_{\beta n}} [\epsilon_\beta^{(\zeta)} + \nu_{\beta n}\epsilon_\alpha^{(\zeta)}], \tag{12.5.2b}$$

$$\tau_{\alpha\beta n}^{(\zeta)} = G_{\alpha\beta n}\gamma_{\alpha\beta}^{(\zeta)}. \tag{12.5.2c}$$

Then the stress-resultants are given by

$$N_\alpha = \sum_{n=1}^{N} \int_{-[(h/2)-x_{n-1}]}^{-[(h/2)-x_n]} \sigma_{\alpha n}^{(\zeta)}\left(1 + \frac{\zeta}{R_\beta}\right) d\zeta, \tag{12.5.3a}$$

$$N_\beta = \sum_{n=1}^{N} \int_{-[(h/2)-x_{n-1}]}^{-[(h/2)-x_n]} \sigma_{\beta n}^{(\zeta)}\left(1 + \frac{\zeta}{R_\alpha}\right) d\zeta, \tag{12.5.3b}$$

$$M_\alpha = \sum_{n=1}^{N} \int_{-[(h/2)-x_{n-1}]}^{-[(h/2)-x_n]} \sigma_{\alpha n}^{(\zeta)}\zeta\left(1 + \frac{\zeta}{R_\beta}\right) d\zeta, \tag{12.5.3c}$$

$$M_\beta = \sum_{n=1}^{N} \int_{-[(h/2)-x_{n-1}]}^{-[(h/2)-x_n]} \sigma_{\beta n}^{(\zeta)}\zeta\left(1 + \frac{\zeta}{R_\alpha}\right) d\zeta, \tag{12.5.3d}$$

$$\bar{N}_{\alpha\beta} = \sum_{n=1}^{N} \int_{-[(h/2)-x_{n-1}]}^{-[(h/2)-x_n]} \tau_{\alpha\beta n}^{(\zeta)} \left[1 + \tfrac{1}{2}\zeta\left(\frac{1}{R_\alpha} + \frac{1}{R_\beta}\right) + \frac{\zeta^2}{4}\left(\frac{1}{R_\alpha} - \frac{1}{R_\beta}\right)^2\right] d\zeta, \tag{12.5.3e}$$

$$\bar{M}_{\alpha\beta} = \sum_{n=1}^{N} \int_{-[(h/2)-x_{n-1}]}^{-[(h/2)-x_n]} \zeta\tau_{\alpha\beta n}^{(\zeta)} \left[1 + \tfrac{1}{2}\zeta\left(\frac{1}{R_\alpha} + \frac{1}{R_\beta}\right)\right] d\zeta, \tag{12.5.3f}$$

where N is the total number of layers. Upon introducing Eqs. (12.5.2) into Eqs. (12.5.3), expanding the integrand into a power series in ζ and retaining only terms up to $0(\zeta^3)$, and performing the indicated integrations, we have the layered shell equivalents of Eqs. (3.1.4), given by

$$N_\alpha + K_\alpha \epsilon_\alpha + K_\nu \epsilon_\beta + A_\alpha\left[\kappa_\alpha - \frac{\epsilon_\alpha}{R_\alpha}\left(1 - \frac{R_\alpha}{R_\beta}\right)\right] + A_\nu \kappa_\beta - \frac{D_\alpha}{R_\alpha}\left(\kappa_\alpha - \frac{\epsilon_\alpha}{R_\alpha}\right)\left(1 - \frac{R_\alpha}{R_\beta}\right), \tag{12.5.4a}$$

$$N_\beta = K_\nu \epsilon_\alpha + K_\beta \epsilon_\beta + A_\nu \kappa_\alpha + A_\beta\left[\kappa_\beta - \frac{\epsilon_\beta}{R_\beta}\left(1 - \frac{R_\beta}{R_\alpha}\right)\right] - \frac{D_\beta}{R_\beta}\left(\kappa_\beta - \frac{\epsilon_\beta}{R_\beta}\right)\left(1 - \frac{R_\beta}{R_\alpha}\right), \tag{12.5.4b}$$

$$M_\alpha = D_\alpha\left[\kappa_\alpha - \left(1 - \frac{R_\alpha}{R_\beta}\right)\frac{\epsilon_\alpha}{R_\alpha}\right] + D_\nu \kappa_\beta + A_\alpha \epsilon_\alpha + A_\nu \epsilon_\beta, \tag{12.5.4c}$$

$$M_\beta = D_\nu \kappa_\alpha + D_\beta\left[\kappa_\beta - \left(1 - \frac{R_\beta}{R_\alpha}\right)\frac{\epsilon_\beta}{R_\beta}\right] + A_\nu \epsilon_\alpha + A_\beta \epsilon_\beta, \tag{12.5.4d}$$

$$\overline{N}_{\alpha\beta} = \left[K_{\alpha\beta} + \frac{3}{4}\left(\frac{1}{R_\alpha} - \frac{1}{R_\beta}\right)^2 D_{\alpha\beta}\right]\gamma_{\alpha\beta} - 2A_{\alpha\beta}\bar{\kappa}_{\alpha\beta}, \tag{12.5.4e}$$

$$\overline{M}_{\alpha\beta} = -A_{\alpha\beta}\gamma_{\alpha\beta} + 2D_{\alpha\beta}\bar{\kappa}_{\alpha\beta}, \tag{12.5.4f}$$

with

$$\begin{Bmatrix} K_\alpha \\ K_\beta \\ K_\nu \end{Bmatrix} = \sum_{n=1}^{N} \begin{Bmatrix} E_{\alpha n} \\ E_{\beta n} \\ \nu_{\alpha n} E_{\alpha n} = \nu_{\beta n} E_{\beta n} \end{Bmatrix} \frac{h_n}{1 - \nu_{\alpha n}\nu_{\beta n}}, \tag{12.5.5a–c}$$

$$\begin{Bmatrix} D_\alpha \\ D_\beta \\ D_\nu \end{Bmatrix} = \sum_{n=1}^{N} \begin{Bmatrix} E_{\alpha n} \\ E_{\beta n} \\ \nu_{\alpha n} E_{\alpha n} = \nu_{\beta n} E_{\beta n} \end{Bmatrix} \frac{h_n}{1 - \nu_{\alpha n}\nu_{\beta n}} \times \left[\frac{h^2}{4} - (x_n + x_{n-1})\frac{h}{2} + \tfrac{1}{3}(x_n^2 + x_n x_{n-1} + x_{n-1}^2)\right], \tag{12.5.5d–f}$$

$$\begin{Bmatrix} A_\alpha \\ A_\beta \\ A_\nu \end{Bmatrix} = \frac{1}{2}\sum_{n=1}^{N} \begin{Bmatrix} E_{\alpha n} \\ E_{\beta n} \\ \nu_{\alpha n} E_{\alpha n} = \nu_{\beta n} E_{\beta n} \end{Bmatrix} \frac{h_n}{1 - \nu_{\alpha n}\nu_{\beta n}}(x_n + x_{n-1} - h), \tag{12.5.5g–i}$$

$$K_{\alpha\beta} = \sum_{n-1}^{N} G_{\alpha\beta n} h_n, \tag{12.5.5j}$$

$$D_{\alpha\beta} = \sum_{n=1}^{N} G_{\alpha\beta n} h_n \left[\frac{h^2}{4} - (x_n + x_{n-1})\frac{h}{2} + \tfrac{1}{3}(x_n^2 + x_n x_{n-1}) + x_{n-1}^2)\right], \tag{12.5.5k}$$

$$A_{\alpha\beta} = \sum_{n=1}^{N} G_{\alpha\beta n} h_n (x_n + x_{n-1} - h). \tag{12.5.5l}$$

For layer arrangements with the thicknesses and material properties symmetrical about the shell middle surface, the coefficients A_α, A_β, A_ν, and $A_{\alpha\beta}$ will vanish. Eqs. (12.5.4) are then similar to Eqs. (12.4.5) for the single-layered orthotropic shell. In many other cases, however, the coupling coefficients A_α, A_β, $A_{\alpha\beta}$, and A_ν do not vanish and are not negligible.

When the material of each layer is isotropic, and the Poisson's ratios of all the layers are identical Eqs. (12.5.4) can be reduced to a form identical to that for the single-layer isotropic shell. This can be achieved by choosing a new reference surface for the coordinate system which is parallel to the middle surface but displaced from it by a distance $\bar{x}$ (see Fig. 12.7). If we use the

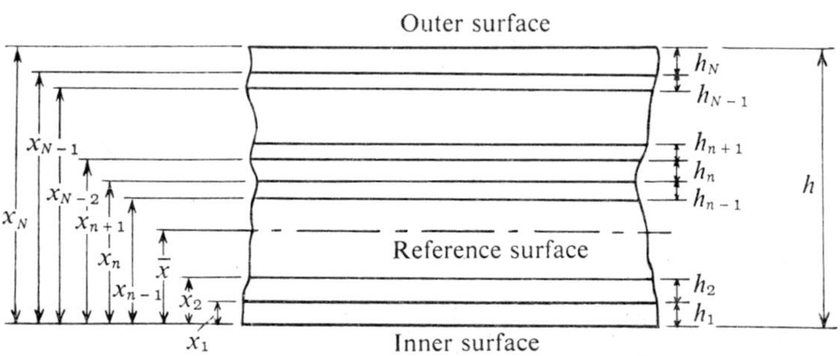

Fig. 12.7. Notation for layered shells

metric quantities and the displacements, strains and changes of curvature of this new surface, and define tangential forces, transverse forces, and bending and twisting moments per unit length of the new surface as well as referring the bending and twisting moments to the new axes, we find that the strain–displacement relationships, equilibrium equations, and force–strain relationships retain their previous form. What does change are the coefficients in the overall force–strain relationships in which we must replace $h/2$ by $(h/2) + \bar{x}$. We then have

$$A_\alpha = A_\beta = A = \frac{1}{2(1 - \nu^2)} \sum_{n=1}^{N} E_n h_n (x_n + x_{n-1} - 2\bar{x} - h), \tag{12.5.6a}$$

$$A_\nu = \nu A, \tag{12.5.6b}$$

$$A_{\alpha\beta} = (1 - \nu)A. \tag{12.5.6c}$$

If we define $\bar{x}$ by

$$\bar{x} = \frac{1}{2}\frac{\sum_{n=1}^{N} E_n h_n (x_n + x_{n-1} - h),}{\sum_{n=1}^{N} E_n h_n} \tag{12.5.7}$$

the coupling constants A_α, A_β, A_ν, and $A_{\alpha\beta}$ vanish. The overall force–strain relationships then become

$$N_\alpha = K_l(\epsilon_\alpha + \nu\epsilon_\beta) - \frac{D_l}{R_\alpha}\left(1 - \frac{R_\alpha}{R_\beta}\right)\left(\kappa_\alpha - \frac{\epsilon_\alpha}{R_\alpha}\right), \tag{12.5.8a}$$

$$N_\beta = K_l(\epsilon_\beta + \nu\epsilon_\alpha) - \frac{D_l}{R_\beta}\left(1 - \frac{R_\beta}{R_\alpha}\right)\left(\kappa_\beta - \frac{\epsilon_\beta}{R_\beta}\right), \tag{12.5.8b}$$

$$M_\alpha = D_l\left[\kappa_\alpha - \left(1 - \frac{R_\alpha}{R_\beta}\right)\frac{\epsilon_\alpha}{R_\alpha} + \nu\kappa_\beta\right], \tag{12.5.8c}$$

$$M_\beta = D_l\left[\kappa_\beta - \left(1 - \frac{R_\beta}{R_\alpha}\right)\frac{\epsilon_\beta}{R_\beta} + \nu\kappa_\alpha\right], \tag{12.5.8d}$$

$$\overline{N}_{\alpha\beta} = \frac{1 - \nu}{2}\left[K_l + \frac{3}{4}\left(\frac{1}{R_\alpha} - \frac{1}{R_\beta}\right)^2 D_l\right]\gamma_{\alpha\beta}, \tag{12.5.8e}$$

$$\overline{M}_{\alpha\beta} = D_l(1 - \nu)\bar{\kappa}_{\alpha\beta}, \tag{12.5.8f}$$

where

$$K_l = \frac{1}{1 - \nu^2}\sum_{n=1}^{N} E_n h_n, \tag{12.5.9a}$$

$$D_l = \frac{1}{1 - \nu^2}\sum_{n=1}^{N}\left\{\frac{1}{3}\sum_{n=1}^{k} E_n h_n (x_n^2 + x_n x_{n-1} + x_{n-1}^2) - \frac{1}{4}\frac{\left[\sum_{n=1}^{N} E_n h_n (x_n + x_{n-1})\right]^2}{\sum_{n=1}^{N} E_n h_n}\right\}. \tag{12.5.9b}$$

If now we define an equivalent thickness $\bar{h}$ and an equivalent Young's modulus $\overline{E}$ such that

$$K_l = \frac{\overline{E}\bar{h}}{1 - \nu^2}, \tag{12.5.10a}$$

$$D_l = \frac{\bar{E}\bar{h}^3}{12(1 - \nu^2)}, \tag{12.5.10b}$$

or

$$\bar{h} = \left(12\frac{D_l}{K_l}\right)^{1/2}, \tag{12.5.10c}$$

$$\bar{E} = (1 - \nu^2)\frac{K_l}{\bar{h}} = (1 - \nu^2)\left(\frac{K_l^3}{12D_l}\right)^{1/2}, \tag{12.5.10d}$$

the correspondence between the shell with different isotropic layers and the shell with a single isotropic layer is complete, for Eqs. (12.5.8) are then identical with those for a shell with a single isotropic layer whose thickness is $\bar{h}$ and whose Young's modulus is $\bar{E}$. Thus all the results previously derived for the single-layer shell are applicable for this special case, remembering, however, that all metric quantities, displacement, forces, and moments are now referred to the displaced reference surface and that the distance from this reference surface is used in the stress–strain relations.

12.6 Stiffened shells

In many applications of shell construction, particularly for the maintenance of a stable structure, it may be more efficient from a weight standpoint to stiffen the structure by means of discrete stringers and ribs than to use a sufficiently thick single-layered construction. In general, we must analyze the shell elements and the discrete stiffeners separately, combining the equations for the different structure elements by means of suitable continuity conditions at the juncture of the shell and the stiffeners. When the stiffening elements are numerous and close, it becomes desirable to simplify the analysis by considering them to be approximated by a stiffening sheet with certain bending, twisting, and extensional properties. By considering this stiffener sheet in conjunction with the basic shell of the structure we can apply the usual continuum equations of the theory of shells to the analysis of the composite structure.

In order to obtain a theory that remains within the framework of the theory of shells given in Chapter 2 we must make certain assumptions about the stiffeners.† We assume that the stiffeners are oriented along the lines of

† The assumptions in various papers in the literature differ to a varying degree. Those used herein are essentially those of M. Baruch, 'Equilibrium and Stability Equations for Stiffened Shells,' *Sixth Annual Conf. Aviation and Astronautics*, February 24–25, 1964, Tel Aviv and Haifa, pp. 117–124.

curvature of the middle surface of the shell and are integral with the shell along a contact line of curvature (Fig. 12.8). The stiffener dimensions are assumed small compared to the principal radii of curvature of the shell. The displacements within the stiffeners are assumed to be given by those along the normal to the deformed middle surface of the shell which passes through the point of contact. The resistance of the stiffener to bending parallel to the tangent plane of the plate is neglected. The effects of restrained warping and interactions with the shell on the torsional resistance are neglected, as are any effects of curvature of the stiffener centroid.

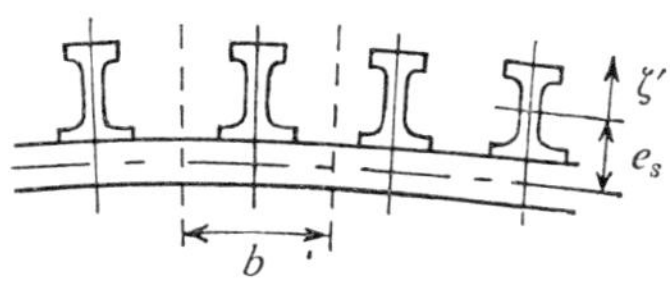

Fig. 12.8. Notation for stiffened shells

With these assumptions, the necessary modification to shell theory is, again, the overall force–strain relationships. Because we have rewritten the equilibrium equations in terms of an equivalent tangential shearing force and an equivalent twisting moment and since the assumed material properties of the stiffeners are those of the theory of beams rather than those of the theory of shells, it is more convenient to obtain these new overall force–strain relationships by means of the energy principle of Section 2.7 rather than by the definitions of forces and moments in terms of stresses. The strain energy per unit middle surface length of a single stiffener oriented in the α direction, say, can be approximated by

$$U_s \approx \tfrac{1}{2}E_{s\alpha}\int_{A_{s\alpha}} (\epsilon_\alpha + e_{s\alpha}\kappa_\alpha + \zeta'\kappa_\alpha)^2\,\mathrm{d}A + \tfrac{1}{2}(GJ)_{s\alpha}\bar{\kappa}_{\alpha\beta}^2, \tag{12.6.1a}$$

where $e_{s\alpha}$ is the distance measured along the normal to the shell middle surface from the middle surface to the centroid of the stiffener cross-section, ζ' is the distance measured along the normal to the shell middle surface from an axis passing through the centroid of the stiffener cross-section and parallel to the middle surface to a point in the stiffener cross-section (Fig. 12.8), and $(GJ)_{s\alpha}$ is the St. Venant torsional stiffness of the stiffener. The integration is over the stiffener cross-section. On carrying out the integrations, remembering that choosing the centroidal axis as a reference axis implies that

$$\int_{A_{s\alpha}} \zeta'\,\mathrm{d}A = 0, \tag{12.6.1b}$$

we have

$$U_s \approx \tfrac{1}{2}E_{s\alpha}[(e_\alpha + e_{s\alpha}\kappa_\alpha)^2 A_{s\alpha} + I_{s\alpha}\kappa_\alpha^2] + \tfrac{1}{2}(GJ)_{s\alpha}\bar{\kappa}_{\alpha\beta}^2, \tag{12.6.1c}$$

where $I_{s\alpha}$ is the moment of inertia of the stiffener about the reference centroidal axis.

Now let us divide U_s by the stiffener spacing, or by half the combined distance to the surrounding stiffeners if the spacing is not uniform, to obtain

$$W_{s\alpha} \approx \frac{1}{2}\left(\frac{EA}{b}\right)_{s\alpha}\epsilon_\alpha^2 + \left(\frac{EAe}{b}\right)_{s\alpha}\epsilon_\alpha\kappa_\alpha + \frac{1}{2}\left(\frac{EAe^2}{b} + \frac{EI}{b}\right)_{s\alpha}\kappa_\alpha^2 + \frac{1}{2}\left(\frac{GJ}{b}\right)_{s\alpha}\bar{\kappa}_{\alpha\beta}^2, \tag{12.6.2}$$

and assume that the coefficients of the strain and curvature-change components are continuous functions of the coordinates α and β. Then the strain and curvature change components can also be assumed to be continuous, in which case $W_{s\alpha}$ can be taken as the stiffener strain energy per unit middle surface area. A similar expression can be written for stiffeners in the β direction, so that the total stiffener strain energy per unit middle surface area is

$$\begin{aligned} W_s \approx{}& \frac{1}{2}\left(\frac{EA}{b}\right)_{s\alpha}\epsilon_\alpha^2 + \frac{1}{2}\left(\frac{EA}{b}\right)_{s\beta}\epsilon_\beta^2 + \left(\frac{EAe}{b}\right)_{s\alpha}\epsilon_\alpha\kappa_\alpha + \left(\frac{EAe}{b}\right)_{s\beta}\epsilon_\beta\kappa_\beta \\ &+ \frac{1}{2}\left(\frac{EA^2}{b} + \frac{EI}{b}\right)_{s\alpha}\kappa_\alpha^2 + \frac{1}{2}\left(\frac{EAe^2}{b} + \frac{EI}{b}\right)_{s\beta}\kappa_\beta^2 \\ &+ \frac{1}{2}\left[\left(\frac{GJ}{b}\right)_{s\alpha} + \left(\frac{GJ}{b}\right)_{s\beta}\right]\bar{\kappa}_{\alpha\beta}^2. \end{aligned} \tag{12.6.3}$$

By adding the stiffener strain energy to the strain energy of the shell and using Eqs. (2.9.2) we obtain the new overall force–strain relations as

$$N_\alpha = \frac{\partial W}{\partial \epsilon_\alpha} = \cdots + \left(\frac{EA}{b}\right)_{s\alpha}\epsilon_\alpha + \left(\frac{EAe}{b}\right)_{s\alpha}\kappa_\alpha, \tag{12.6.4a}$$

$$N_\beta = \frac{\partial W}{\partial \epsilon_\beta} = \cdots + \left(\frac{EA}{b}\right)_{s\beta}\epsilon_\beta + \left(\frac{EAe}{b}\right)_{s\beta}\kappa_\beta, \tag{12.6.4b}$$

$$M_\alpha = \frac{\partial W}{\partial \kappa_\alpha} = \cdots + \left(\frac{EAe}{b}\right)_{s\alpha}\epsilon_\alpha + \left[\frac{E}{b}(I + Ae^2)\right]_{s\alpha}\kappa_\alpha, \tag{12.6.4c}$$

$$M_\beta = \frac{\partial W}{\partial \kappa_\beta} = \cdots + \left(\frac{EAe}{b}\right)_{s\beta}\epsilon_\beta + \left[\frac{E}{b}(I + Ae^2)\right]_{s\beta}\kappa_\beta, \tag{12.6.4d}$$

$$\bar{N}_{\alpha\beta} = \frac{\partial W}{\partial \gamma_{\alpha\beta}} = \cdots, \tag{12.6.4e}$$

$$2\bar{M}_{\alpha\beta} = \frac{\partial W}{\partial \bar{\kappa}_{\alpha\beta}} = \cdots + \left[\left(\frac{GJ}{b}\right)_{s\alpha} + \left(\frac{GJ}{b}\right)_{s\beta}\right]\bar{\kappa}_{\alpha\beta}, \tag{12.6.4f}$$

where the dots denote the usual terms of the theory of shells arising from the strain energy of the shell.†

It is possible, of course, to treat the effect of the stiffeners somewhat more satisfactorily than in the elementary treatment given above. One method is to express the overall force strain relations as

$$N_\alpha = C_{11}\epsilon_\alpha + C_{12}\epsilon_\beta + C_{13}\kappa_\alpha + C_{14}\kappa_\beta, \tag{12.6.5a}$$

$$N_\beta = C_{21}\epsilon_\alpha + C_{22}\epsilon_\beta + C_{23}\kappa_\alpha + C_{24}\kappa_\beta, \tag{12.6.5b}$$

$$M_\alpha = C_{31}\epsilon_\alpha + C_{32}\epsilon_\beta + C_{33}\kappa_\alpha + C_{34}\kappa_\beta, \tag{12.6.5c}$$

$$M_\beta = C_{41}\epsilon_\alpha + C_{42}\epsilon_\beta + C_{43}\kappa_\alpha + C_{44}\kappa_\beta, \tag{12.6.5d}$$

$$\bar{N}_{\alpha\beta} = C_{55}\gamma_{\alpha\beta} + C_{56}\bar{\kappa}_{\alpha\beta}, \tag{12.6.5e}$$

$$\bar{M}_{\alpha\beta} = C_{65}\gamma_{\alpha\beta} + C_{66}\bar{\kappa}_{\alpha\beta}, \tag{12.6.5f}$$

where the matrix of the coefficients of the strain components is symmetrical, and to determine the coefficients experimentally. Since experiments on actual shell elements are quite difficult, it is usual to use equivalent flat-plate elements in these experiments,‡ although some of the effects of curvature are lost in so doing. The above form of the overall force–strain relationships is suitable for other types of stiffening such as waffle stiffening.§

Supplementary references

[1] Flaherty, J. E., and Vafakos, W. P., 'Asymptotic Solution for Pressurized Noncircular Cylinders with Nonuniform Rings,' *AIAA J.*, vol. 9, no. 9, September 1971, pp. 1725–1732.

[2] Wang, J. T.-S., 'Orthogonally Stiffened Cylindrical Shell Subjected to Internal Pressure,' *AIAA J.*, vol. 8, no. 3, March 1970, pp. 455–461.

† To be consistent with the neglect of stiffener curvature effects we should use the terms of Eqs. (3.1.4).

‡ See for example, W. H. Hoppmann II, 'Flexural Vibrations of Orthogonally Stiffened Cylindrical Shells,' *Proc. Ninth International Congress of Theoretical and Applied Mechanics*, Brussels 1956.

§ See N. D. Dow, C. L. Libove, and R. E. Hubka, 'Formulas for the Elastic Constants of Plates with Integral Waffle-like Stiffening,' NACA RM L53E13a, August 10, 1950; R. F. Crawford and C. Libove, 'Shearing Effectiveness of Integral Stiffening,' NACA TN 3443, June 1955; and T. C. Soong, 'Buckling of Cylindrical Shells with Eccentric Spiral-type Stiffeners,' *AIAA J.*, vol. 7, no. 1, January 1969, pp. 65–72.

Part IV

Dynamics of shells

13

Free vibrations of shells

13.1 Equations of motion

Until now we have concentrated on the stresses and deformations of shells to which static loads, loads independent of time, were applied. The area of applicability of static analysis can also be stretched to include slowly varying loads as well. In many applications, however, the loads vary so rapidly that the effect of inertia of the shell material cannot be neglected. The theory of shells derived in Chapter 2 can be extended to permit inertia forces to be taken into account merely by expressing the body forces in the various equations in terms of accelerations of the shell material.† We note that the displacements in the shell wall are given by Eqs. (2.4.5) as

$$u_\alpha^{(\zeta)} = u_\alpha - \zeta\Theta_\alpha, \tag{13.1.1a}$$

$$u_\beta^{(\zeta)} = u_\beta - \zeta\Theta_\beta, \tag{13.1.1b}$$

$$w^{(\zeta)} = w. \tag{13.1.1c}$$

Then the acceleration of a small element of mass within the shell wall is given by

$$\ddot{u}_\alpha^{(\zeta)} = \ddot{u}_\alpha - \zeta\ddot{\Theta}_\alpha, \tag{13.1.2a}$$

$$\ddot{u}_\beta^{(\zeta)} = \ddot{u}_\beta - \zeta\ddot{\Theta}_\beta, \tag{13.1.2b}$$

$$\ddot{w}^{(\zeta)} = \ddot{w}, \tag{13.1.2c}$$

where the dots indicate differentiation with respect to time. The terms involving the shell middle surface rotations in Eqs. (13.1.2) are called rotatory inertia terms. They are the result of rotation of the rigid normals to the shell middle surface.

† Dissipative forces arising from internal friction and the like are not considered.

From d'Alembert's principle the body forces per unit volume acting in the shell are

$$\hat{p}_\alpha^{(\zeta)} = -\rho \ddot{u}_\alpha^{(\zeta)}, \tag{13.1.3a}$$

$$\hat{p}_\beta^{(\zeta)} = -\rho \ddot{u}_\beta^{(\zeta)}, \tag{13.1.3b}$$

$$\hat{p}_\zeta^{(\zeta)} = -\rho \ddot{w}^{(\zeta)}, \tag{13.1.3c}$$

where ρ is the density of the shell material. These enter into the equations of equilibrium of the shell as equivalent surface forces per unit middle surface area and as equivalent surface moments per unit middle surface area. From Eqs. (2.5.4) and (2.5.6) these are given by

$$\begin{aligned}\hat{q}_\alpha &= \int_{-1/2h}^{1/2h} \left(1 + \frac{\zeta}{R_\alpha}\right)\left(1 + \frac{\zeta}{R_\beta}\right)\hat{p}_\alpha^{(\zeta)}\,\mathrm{d}\zeta \\ &= -\rho h\left[\left(1 + \frac{h^2}{R_\alpha R_\beta}\right)\ddot{u}_\alpha - \frac{h^2}{12}\left(\frac{1}{R_\alpha} + \frac{1}{R_\beta}\right)\ddot{\Theta}_\alpha\right],\end{aligned} \tag{13.1.4a}$$

$$\begin{aligned}\hat{q}_\beta &= \int_{-1/2h}^{1/2h} \left(1 + \frac{\zeta}{R_\alpha}\right)\left(1 + \frac{\zeta}{R_\beta}\right)\hat{p}_\beta^{(\zeta)}\,\mathrm{d}\zeta \\ &= -\rho h\left[\left(1 - \frac{h^2}{R_\alpha R_\beta}\right)\ddot{u}_\beta - \frac{h^2}{12}\left(\frac{1}{R_\alpha} + \frac{1}{R_\beta}\right)\ddot{\Theta}_\beta\right],\end{aligned} \tag{13.1.4b}$$

$$\begin{aligned}\hat{q}_\zeta &= \int_{-1/2h}^{1/2h} \left(1 + \frac{\zeta}{R_\alpha}\right)\left(1 + \frac{\zeta}{R_\beta}\right)\hat{p}_\zeta^{(\zeta)}\,\mathrm{d}\zeta \\ &= -\rho h\left(1 + \frac{h^2}{R_\alpha R_\beta}\right)\ddot{w},\end{aligned} \tag{13.1.4c}$$

$$\begin{aligned}\hat{m}_\alpha &= \int_{-1/2h}^{1/2h} \zeta\left(1 + \frac{\zeta}{R_\alpha}\right)\left(1 + \frac{\zeta}{R_\beta}\right)\hat{p}_\alpha^{(\zeta)}\,\mathrm{d}\zeta \\ &= -\frac{\rho h^3}{12}\left[\left(\frac{1}{R_\alpha} + \frac{1}{R_\beta}\right)\ddot{u}_\alpha - \left(1 + \frac{3h^2}{20R_\alpha R_\beta}\right)\ddot{\Theta}_\alpha\right],\end{aligned} \tag{13.1.4d}$$

$$\begin{aligned}\hat{m}_\beta &= \int_{-1/2h}^{1/2h} \rho\left(1 + \frac{\zeta}{R_\alpha}\right)\left(1 + \frac{\zeta}{R_\beta}\right)\hat{p}_\beta^{(\zeta)}\,\mathrm{d}\zeta \\ &= -\frac{\rho h^3}{12}\left[\left(\frac{1}{R_\alpha} + \frac{1}{R_\beta}\right)\ddot{u}_\beta - \left(1 + \frac{3h^2}{20R_\alpha R_\beta}\right)\ddot{\Theta}_\beta\right].\end{aligned} \tag{13.1.4e}$$

The equations of motion of an overall element of the shell wall are then, from Eqs. (2.6.2)

$$\begin{aligned}&\frac{\partial}{\partial\alpha}(g_\beta N_\alpha) - \frac{\partial g_\beta}{\partial\alpha}N_\beta + g_\alpha\frac{\partial}{\partial\beta}\left[\overline{N}_{\alpha\beta} - \frac{1}{2}\left(\frac{1}{R_\alpha} + \frac{1}{R_\beta}\right)\overline{M}_{\alpha\beta}\right] \\ &\qquad + \left[2\overline{N}_{\alpha\beta} + \left(\frac{1}{R_\beta} - \frac{1}{R_\alpha}\right)\overline{M}_{\alpha\beta}\right]\frac{\partial g_\alpha}{\partial\beta} + g_\alpha g_\beta\left(\frac{\overline{Q}_\alpha}{R_\alpha} + q_\alpha^* + \hat{q}_\alpha\right) = 0,\end{aligned} \tag{13.1.5a}$$

$$\frac{\partial}{\partial \beta}(g_\alpha N_\beta) - \frac{\partial g_\alpha}{\partial \beta} N_\alpha + g_\beta \frac{\partial}{\partial \alpha}\left[\overline{N}_{\alpha\beta} - \frac{1}{2}\left(\frac{1}{R_\beta} + \frac{1}{R_\alpha}\right)\overline{M}_{\alpha\beta}\right]$$
$$+ \left[2\overline{N}_{\alpha\beta} + \left(\frac{1}{R_\alpha} - \frac{1}{R_\beta}\right)\overline{M}_{\alpha\beta}\right]\frac{\partial g_\beta}{\partial \alpha} + g_\alpha g_\beta\left(\frac{\overline{Q}_\beta}{R_\beta} + q_\beta^* + \hat{q}_\beta\right) = 0, \tag{13.1.5b}$$

$$\frac{\partial}{\partial \alpha}(g_\beta \overline{Q}_\alpha) + \frac{\partial}{\partial \beta}(g_\alpha \overline{Q}_\beta) - 2\frac{\partial^2 \overline{M}_{\alpha\beta}}{\partial \alpha\, \partial \beta} - g_\alpha g_\beta\left(\frac{N_\alpha}{R_\alpha} + \frac{N_\beta}{R_\beta} - q_\zeta^* - \hat{q}_\zeta\right) = 0, \tag{13.1.5c}$$

$$\frac{\partial}{\partial \alpha}(g_\beta M_\alpha) - \frac{\partial g_\beta}{\partial \alpha} M_\beta + 2\frac{\partial}{\partial \beta}(g_\alpha \overline{M}_{\alpha\beta}) - g_\alpha g_\beta(\overline{Q}_\alpha - m_\alpha^* - \hat{m}_\alpha) = 0, \quad (13.1.5$$

$$\frac{\partial}{\partial \beta}(g_\alpha M_\beta) - \frac{\partial g_\alpha}{\partial \beta} M_\alpha + 2\frac{\partial}{\partial \alpha}(g_\beta \overline{M}_{\alpha\beta}) - g_\alpha g_\beta(\overline{Q}_\beta - m_\beta^* - \hat{m}_\beta) = 0, \tag{13.1.5e}$$

where q_α^*, q_β^*, q_ζ^*, m_α^*, and m_β^* are equivalent surface forces and moments per unit middle surface area due to the applied loading which is now assumed to be an explicit function of middle surface position and time.†

The equations of equilibrium are the only portions of the theory which change. We still use the overall force–strain relations given by Eqs. (2.7.2), (3.1.2), or (3.1.4) depending on what accuracy we wish to attain. Boundary conditions for dynamic problems similarly are still given by Eqs. (2.8.7) to (2.8.9). If we were to substitute the force–strain relations and strain–displacement relations into Eqs. (13.1.5) we would obtain three equations in terms of the displacements of the form

$$L_{11}(u_\alpha) + L_{12}(u_\beta) + L_{13}(w)$$
$$- \rho h\left\{\left[1 + \frac{h^2}{12R_\alpha}\left(\frac{1}{R_\alpha} + \frac{2}{R_\beta}\right)\right]\ddot{u}_\alpha\right.$$
$$\left. - \frac{h^2}{12}\left(\frac{2}{R_\alpha} + \frac{1}{R_\beta} + \frac{3h^2}{20R_\alpha^2 R_\beta}\right)\ddot{\Theta}_\alpha\right\} = -\left(q_\alpha^* - \frac{m_\alpha^*}{R_\alpha}\right), \tag{13.1.6a}$$

$$L_{21}(u_\alpha) + L_{22}(u_\beta) + L_{23}(w) - \rho h\left\{\left[1 + \frac{h^2}{12R_\beta}\left(\frac{2}{R_\alpha} - \frac{1}{R_\beta}\right)\right]\ddot{u}_\beta\right.$$
$$\left. - \frac{h^2}{12}\left(\frac{1}{R_\alpha} + \frac{2}{R_\beta} + \frac{3h^2}{20R_\alpha R_\beta^2}\right)\ddot{\Theta}_\beta\right\} = -\left(q_\beta^* - \frac{m_\beta^*}{R_\beta}\right), \tag{13.1.6b}$$

† We thus exclude problems such as the response of a shell restrained by a medium of any sort where external forces are a function of displacements and velocities of the shell wall.

$$L_{31}(u_\alpha) + L_{32}(u_\beta) + L_{33}(w) - \rho h\left(1 + \frac{h^2}{12R_\alpha R_\beta}\right)\ddot{w}$$

$$- \frac{1}{g_\alpha g_\beta}\left\{\frac{\partial}{\partial\alpha}\frac{\rho h^3}{12} g_\beta \left[\left(\frac{1}{R_\alpha} + \frac{1}{R_\beta}\right)\ddot{u}_\alpha - \left(1 + \frac{3h^2}{20R_\alpha R_\beta}\right)\ddot{\Theta}_\alpha\right]\right.$$

$$\left. + \frac{\partial}{\partial\beta}\frac{\rho h^3}{12} g_\alpha \left[\left(\frac{1}{R_\alpha} + \frac{1}{R_\beta}\right)\ddot{u}_\beta - \left(1 + \frac{3h^2}{20R_\alpha R_\beta}\right)\ddot{\Theta}_\beta\right]\right\}$$

$$= -\left\{q_\zeta^* + \frac{1}{g_\alpha g_\beta}\left[\frac{\partial}{\partial\alpha}(g_\beta m_\alpha^*) + \frac{\partial}{\partial\beta}(g_\alpha m_\beta^*)\right]\right\} \tag{13.1.6c}$$

The operators L_{ij} are linear differential operators involving only space derivatives and are identically those of the static equations. We note then that the set of equations is sixth order with respect to time. This circumstance implies that there are six arbitrary functions of α and β in the general solution of the equations. These six functions are determined by the specification of values of the displacements u_α, u_β, and w and the velocities $\dot{u}_\alpha$, $\dot{u}_\beta$, and $\dot{w}$ at every point of the middle surface at some specified time, the initial state.

The equations are often simplified by neglecting rotatory inertia terms and associated terms which are of the order of h^2/R^2 compared to unity. These simplify the appearance of the equations but usually do not significantly simplify the method of solution. A further simplification is afforded for problems involving predominantly normal loading if we recall that in such static problems the displacements tangential to the shell middle surface are generally small compared to the normal displacements. By assuming this to be true also for dynamic problems we simplify the equations by neglecting inertia forces due to tangential acceleration. The set of equations is then reduced to second order with respect to time, considerably reducing the algebraic details of solution of many problems but permitting us to specify only the normal displacement w and the normal velocity $\dot{w}$ at the initial time. We shall investigate the effect of these simplifications later on.

13.2 Free harmonic vibrations

If a shell is disturbed initially by some distribution of impulses or is initially distorted to some predetermined shape and is suddenly released, the shell will continue in motion with no additional loading. For special distributions of initial disturbances the motions of the unloaded shell will be such that the shape of the spatial distribution of each displacement is independent of time and the time dependence of the amplitudes of the distributions is the same for all three displacements. Such motions are called free vibrations and are defined by displacements such that

$$u_\alpha = \overline{U}_\alpha(\alpha, \beta)f(t), \tag{13.2.1a}$$

$$u_\beta = \overline{U}_\beta(\alpha, \beta)f(t), \tag{13.2.1b}$$

$$w = \overline{W}(\alpha, \beta)f(t). \tag{13.2.1c}$$

If now we substitute Eqs. (13.2.1) into Eqs. (13.1.6), with

$$q_\alpha^* = q_\beta^* = q_\zeta^* = m_\alpha^* = m_\beta^* = 0, \tag{13.2.2}$$

we obtain a set of equations of the form

$$\frac{L_{11}(\overline{U}_\alpha) + L_{12}(\overline{U}_\beta) + L_{13}(\overline{W})}{\rho h\left\{\left[1 + \frac{h^2}{12R_\alpha}\left(\frac{1}{R_\alpha} + \frac{2}{R_\beta}\right]\right)\overline{U}_\alpha - \frac{h^2}{12}\left(\frac{2}{R_\alpha} + \frac{1}{R_\beta} + \frac{3h^2}{20R_\alpha^2 R_\beta}\right)\overline{\Theta}_\alpha\right\}} = \frac{\ddot{f}}{f}, \tag{13.2.3a}$$

$$\frac{L_{21}(\overline{U}_\alpha) + L_{22}(\overline{U}_\beta) + L_{23}(\overline{W})}{\rho h\left\{\left[1 + \frac{h^2}{12R_\beta}\left(\frac{2}{R_\alpha} + \frac{1}{R_\beta}\right)\right]\overline{U}_\beta - \frac{h^2}{12}\left(\frac{1}{R_\alpha} + \frac{2}{R_\beta} + \frac{3h^2}{20R_\alpha R_\beta^2}\right)\overline{\Theta}_\beta\right\}} = \frac{\ddot{f}}{f}, \tag{13.2.3b}$$

$$\frac{L_{31}(\overline{U}_\alpha) + L_{32}(\overline{U}_\beta) + L_{33}(\overline{W})}{\begin{aligned}\rho h\left(1 + \frac{h^2}{12R_\alpha R_\beta}\right)\overline{W} + \frac{1}{g_\alpha g_\beta}\left\{\frac{\partial}{\partial\alpha}\frac{\rho h^3}{12}g_\beta\left[\left(\frac{1}{R_\alpha} + \frac{1}{R_\beta}\right)\overline{U}_\alpha - \left(1 + \frac{3h^2}{20R_\alpha R_\beta}\right)\overline{\Theta}_\alpha\right]\right.\\ \left. + \frac{\partial}{\partial\beta}\frac{\rho h^3}{12}g_\alpha\left[\left(\frac{1}{R_\alpha} + \frac{1}{R_\beta}\right)\overline{U}_\beta - \left(1 + \frac{3h^2}{20R_\alpha R_\beta}\right)\overline{\Theta}_\beta\right]\right\}\end{aligned}} = \frac{\ddot{f}}{f}, \tag{13.2.3c}$$

with

$$\overline{\Theta}_\alpha = \frac{1}{g_\alpha}\frac{\partial\overline{W}}{\partial\alpha} - \frac{\overline{U}_\alpha}{R_\alpha}, \tag{13.2.3d}$$

$$\overline{\Theta}_\beta = \frac{1}{g_\beta}\frac{\partial\overline{W}}{\partial\beta} - \frac{\overline{U}_\beta}{R_\beta}. \tag{13.2.3e}$$

Since the left side of each equation is a function only of the space coordinates α and β while the right side is a function only of time, the equations can be satisfied only if each side is a constant. The form of the equations indicates that the constant is the same for all three equations. We denote it by $-\omega^2$. Then the time function $f(t)$ is determined by the equation

$$\ddot{f} + \omega^2 f = 0, \tag{13.2.4a}$$

which has the harmonic solution

$$f(t) = A \cos \omega t + B \sin \omega t, \tag{13.2.4b}$$

with A and B arbitrary constants.

The circular frequency of motion ω is not arbitrary. The right sides of Eqs. (13.2.3) set equal to $-\omega^2$ are three homogeneous linear partial differential equations for the determination of $\overline{U}_\alpha$, $\overline{U}_\beta$, and $\overline{W}$ as functions of α, β, and ω. The edge conditions are also homogeneous and can be considered to define a transcendental equation for the determination of an infinite sequence of values of ω, called eigenvalues or characteristic values, for which both the equations of motion and the edge conditions are satisfied. Associated with each value of ω are spatial distributions of relative values of $\overline{U}_\alpha$, $\overline{U}_\beta$, and $\overline{W}$ describing the form of the harmonic vibration. These are called eigenfunctions or characteristic functions. We shall see later on that these functions play a role similar to trigonometric functions in Fourier analysis in determining the solution of the equations of motion for arbitrary motions.

From Eqs. (13.2.1) and (13.2.4b) we note that harmonic motions with a circular frequency ω_n will occur if the initial distribution of displacements and velocities is given by

$$u_\alpha|_{t=0} = A_n \overline{U}_{\alpha n}, \tag{13.2.5a}$$

$$u_\beta|_{t=0} = A_n \overline{U}_{\beta n}, \tag{13.2.5b}$$

$$w|_{t=0} = A_n \overline{W}_n, \tag{13.2.5c}$$

$$\dot{u}_\alpha|_{t=0} = \omega_n B_n \overline{U}_{\alpha n}, \tag{13.2.5d}$$

$$\dot{u}_\beta|_{t=0} = \omega_n B_n \overline{U}_{\beta n}, \tag{13.2.5e}$$

$$\dot{w}|_{t=0} = \omega_n B_n \overline{W}_n. \tag{13.2.5f}$$

The determination of the circular frequencies and associated mode shapes depends on the solution of the partial differential equations of motion. In most cases exact solutions of the equations are not known, or if known would be inconvenient to use, so that approximate methods of analysis must be employed. The numerical methods described previously for static shell analysis are applicable to the solution of eigenvalue problems as well. The various step-by-step numerical integration methods, for example, have been applied primarily to shells of revolution for which expansion of the deformations in Fourier series in the circumferential direction makes it necessary to consider only ordinary differential equations. These are an eighth-order set of ordinary differential equations requiring the satisfaction of four boundary conditions at each end of the shell frustum. The usual approach has been to solve the differential equations numerically for a given frequency and for four independent sets of arbitrary values of the quantities which are not specified at one end of the shell. The four specified quantities are set equal to their prescribed value of zero. By requiring that boundary conditions be satisfied at the other end of the shell we obtain a four-by-four determinant which we evaluate. We repeat the procedure with different values of the frequency until the determinant vanishes. Except for the fact that numerical integration is used to determine the solutions of the equations, the procedure is essentially equivalent to the exact transcendental equation approach.

The other numerical approaches which have been used, the method of finite differences, the finite element method, and the Rayleigh–Ritz method, lead to another type of procedure. The result of the various methods is the transformation of the homogeneous partial differential equations and boundary conditions into a set of homogeneous linear alegebraic equations for certain unknown quantities. In the finite difference or finite element methods these quantities are generally physical quantities associated with the stress and deformation states. In the Rayleigh–Ritz approach the unknowns are arbitrary coefficients of sets of prescribed functions. In any case, however, the end result is a set of equations which may be written as

$$[A - \omega^2 B]\{a_i\} = 0, \tag{13.2.6a}$$

where A and B are square arrays of known coefficients and the a_i are the unknown nodal physical quantities or function coefficients. Since the a_i must not all vanish, the determinant of their coefficients must vanish. Thus the equation for the determination of the frequencies ω is given by

$$|A - \omega^2 B| = 0. \tag{13.2.6b}$$

Many numerical techniques exist for the determination of the eigenvalues and the associated eigenvectors or sets of values of a_i for problems defined by Eqs. (13.2.6a) or (13.2.6b).†

It is also possible to combine the methods of finite differences and of numerical integration by using finite difference expansions in one direction of the coordinate system to obtain coupled sets of ordinary differential equations which are integrated numerically. The usefulness of one or the other of the methods outlined above depends to a large extent upon the structure under consideration and upon the computation facilities available.

13.3 Orthogonality conditions for vibration mode shapes

The mode shapes described above satisfy a relationship called an orthogonality condition which is very useful. We may obtain this condition most

† J. H. Wilkinson, *The Algebraic Eigenvalue Problem*, Clarendon Press, Oxford, 1965.

readily by utilizing the principle of virtual work, Eq. (2.8.3) which can be written in terms of the strain energy function W as

$$\int_{\beta_1}^{\beta_2}\int_{\alpha_1}^{\alpha_2} \delta W\, g_\alpha g_\beta \, d\alpha\, d\beta$$

$$= \int_{\beta_1}^{\beta_2}\int_{\alpha_1}^{\alpha_2} (\hat{q}_\alpha \delta u_\alpha + \hat{q}_\beta\, \delta u_\beta + \hat{q}_\zeta\, \delta w + \hat{m}_\alpha\, \delta\Theta_\alpha + \hat{m}_\beta\, \delta\Theta_\beta) g_\alpha g_\beta \, d\alpha\, d\beta$$

$$= -\rho h \int_{\beta_1}^{\beta_2}\int_{\alpha_1}^{\alpha_2} \left[\left(1 + \frac{h^2}{12R_\alpha R_\beta}\right)(\ddot{u}_\alpha\, \delta u_\alpha + \ddot{u}_\beta\, \delta u_\beta + \ddot{w}\, \delta w)\right.$$

$$- \frac{h^2}{12}\left(\frac{1}{R_\alpha} + \frac{1}{R_\beta}\right)(\ddot{\Theta}_\alpha\, \delta u_\alpha + \ddot{u}_\alpha\, \delta\Theta_\alpha + \ddot{\Theta}_\beta\, \delta u_\beta + \ddot{u}_\beta\, \delta\Theta_\beta)$$

$$\left. + \frac{h^2}{12}\left(1 + \frac{3h^2}{20RR}\right)(\ddot{\Theta}_\alpha\, \delta\Theta_\alpha + \ddot{\Theta}_\beta\, \delta\Theta_\beta)\right] g_\alpha g_\beta \, d\alpha\, d\beta. \qquad (13.3.1)$$

Now let the displacements be given by the components of one free vibration mode and the virtual displacements by the components of any other free vibration mode. Then

$$u_\alpha = \overline{U}_{\alpha n} f_n(t), \qquad (13.3.2a)$$
$$u_\beta = \overline{U}_{\beta n} f_n(t), \qquad (13.3.2b)$$
$$w = \overline{W}_n f_n(t), \qquad (13.3.2c)$$
$$\Theta_\alpha = \overline{\Theta}_{\alpha n} f_n(t), \qquad (13.3.2d)$$
$$\Theta_\beta = \overline{\Theta}_{\beta n} f_n(t), \qquad (13.3.2e)$$
$$\delta u_\alpha = \overline{U}_{\alpha m} f_m(t), \qquad (13.3.2f)$$
$$\delta u_\beta = \overline{U}_{\alpha m} f_m(t), \qquad (13.3.2g)$$
$$\delta w = \overline{W}_m f_m(t), \qquad (13.3.2h)$$
$$\delta\Theta_\alpha = \overline{\Theta}_{\alpha m} f_m(t), \qquad (13.3.2i)$$
$$\delta\Theta_\beta = \overline{\Theta}_{\beta m} f_m(t), \qquad (13.3.2j)$$

with

$$f_n(t) = A_n \cos \omega_n t + B_n \sin \omega_n t, \qquad (13.3.2k)$$

$$f_m(t) = A_m \cos \omega_m t + B_m \sin \omega_m t. \qquad (13.3.2l)$$

Also let W_n be the value of the strain energy W and $\bar{e}_{ni}$ ($i = 1, 2, \ldots, 6$) be the values of the strain and curvature change components when the displacements are given by Eqs. (13.3.2a–e) while W_m and $\bar{e}_{mi}$ are the same quantities with the displacements given by Eqs. (13.3.2f–j). The substitution of Eqs. 13.3.2) into Eq. (13.3.1) then yields

$$\int_{\beta_1}^{\beta_2}\int_{\alpha_1}^{\alpha_2} \delta W g_\alpha g_\beta \, d\alpha \, d\beta = \int_{\alpha_1}^{\alpha_2}\int_{\beta_1}^{\beta_2} \sum_{i=1}^{i=6} \frac{\partial W_n}{\partial \bar{e}_{ni}} \bar{e}_{mi} g_\alpha g_\beta \, d\alpha \, d\beta$$

$$= \rho h f_n(t) f_m(t)\omega_n^2 \int_{\beta_1}^{\beta_2}\int_{\alpha_1}^{\alpha_2} \Big[\Big(1 + \frac{h^2}{12R_\alpha R_\beta}\Big)(\bar{U}_{\alpha n}\bar{U}_{\alpha m}$$

$$+ \bar{U}_{\beta n}\bar{U}_{\beta m} + \bar{W}_n\bar{W}_m) - \frac{h^2}{12}\Big(\frac{1}{R_\alpha} + \frac{1}{R_\beta}\Big)(\bar{\Theta}_{\alpha n}\bar{U}_{\alpha m} + \bar{\Theta}_{\alpha m}\bar{U}_{\alpha n}$$

$$+ \bar{\Theta}_{\beta n}\bar{U}_{\beta m} + \bar{\Theta}_{\beta m}\bar{U}_{\beta n}) + \frac{h^2}{12}\Big(1 + \frac{3h^2}{20R_\alpha R_\beta}\Big)(\bar{\Theta}_{\alpha n}\bar{\Theta}_{\alpha m}$$

$$+ \bar{\Theta}_{\beta n}\bar{\Theta}_{\beta m})\Big] g_\alpha g_\beta \, d\alpha \, d\beta. \tag{13.3.3}$$

Because W, the strain energy function, is a quadratic form

$$W = \sum_{i=1}^{6}\sum_{j=1}^{6} a_{ij}e_i e_j, \tag{13.3.4a}$$

we have

$$\sum_{i=1}^{6} \frac{\partial W_n}{\partial \bar{e}_{ni}} \bar{e}_{mi} = \sum_{i=1}^{6}\Big[\sum_{j=1}^{6}(a_{ij} + a_{ji})\bar{e}_{nj}\Big]\bar{e}_{mj}$$

$$= \sum_{j=1}^{6}\Big[\sum_{i=1}^{6}(a_{ji} + a_{ij})\bar{e}_{mi}\Big]\bar{e}_{nj} = \sum_{i=1}^{6}\frac{\partial W_m}{\partial \bar{e}_{mi}}\bar{e}_{ni}. \tag{13.3.4b}$$

Thus if we reverse m and n in Eqs. (13.3.2) and substitute again in Eq. (13.3.1), the left side of the resulting equation similar to Eq. (13.3.3) remains unchanged while the only change in the right side is the replacement of ω_n^2 by ω_m^2. By subtracting the two equations we then have

$$(\omega_n^2 - \omega_m^2)\int_{\beta_1}^{\beta_2}\int_{\alpha_1}^{\alpha_2}\Big[\Big(1 + \frac{h^2}{12R_\alpha R_\beta}\Big)(\bar{U}_{\alpha m}\bar{U}_{\alpha n} + \bar{U}_{\beta m}\bar{U}_{\beta n} + \bar{W}_m\bar{W}_n)$$

$$- \frac{h^2}{12}\Big(\frac{1}{R_\alpha} + \frac{1}{R_\beta}\Big)(\bar{\Theta}_{\alpha m}\bar{U}_{\alpha n} + \bar{\Theta}_{\alpha n}\bar{U}_{\alpha m} + \bar{\Theta}_{\beta m}\bar{U}_{\beta n} + \bar{\Theta}_{\beta n}\bar{U}_{\beta m})$$

$$+ \frac{h^2}{12}\Big(1 + \frac{3h^2}{20R_\alpha R_\beta}\Big)(\bar{\Theta}_{\alpha n}\bar{\Theta}_{\alpha m} + \bar{\Theta}_{\beta n}\bar{\Theta}_{\beta m})\Big] g_\alpha g_\beta \, d\alpha \, d\beta, \tag{13.3.5}$$

which implies that if the two values of ω are distinct ($m \neq n$), the factor multiplying $(\omega_n^2 - \omega_m^2)$ must vanish. This is the desired orthogonality condition. In the case when $n = m$ the factor does not vanish and is denoted by L_n. Thus

$$L_n = \int_{\beta_1}^{\beta_2}\int_{\alpha_1}^{\alpha_2}\Big[\Big(1 + \frac{h^2}{12R_\alpha R_\beta}\Big)(\overline{U}_{\alpha n}^2 + \overline{U}_{\beta n}^2 + \overline{W}_n^2) - \frac{h^2}{6}\Big(\frac{1}{R_\alpha} + \frac{1}{R_\beta}\Big)(\overline{\Theta}_{\alpha n}\overline{U}_{\alpha n} + \overline{\Theta}_{\beta n}\overline{U}_{\beta n}) + \frac{h^2}{12}\Big(1 + \frac{3h^2}{20R_\alpha R_\beta}\Big)(\overline{\Theta}_{\alpha n}^2 + \overline{\Theta}_{\beta n}^2)\Big] g_\alpha g_\beta \, \mathrm{d}\alpha \, \mathrm{d}\beta. \tag{13.3.6}$$

We note that the following relation holds from Eqs. (13.3.3) and (13.3.6)

$$\int_{\beta_1}^{\beta_2}\int_{\alpha_1}^{\beta_2} \sum_{i=1}^{6} \frac{\partial W_n}{\partial \bar{e}_{ni}} \bar{e}_{ni} g_\alpha g_\beta \, \mathrm{d}\alpha \, \mathrm{d}\beta = 2\int_{\beta_1}^{\beta_2}\int_{\alpha_1}^{\beta_2} W_n g_\alpha g_\beta \, \mathrm{d}\alpha \, \mathrm{d}\beta = \rho h \omega_n^2 L_n f_n^2(t). \tag{13.3.7}$$

The orthogonality condition is very much simplified if we neglect rotatory inertia and terms of the order of h^2/R^2 compared to unity. We then have

$$\int_{\beta_1}^{\beta_2}\int_{\alpha_1}^{\beta_2} (\overline{U}_{\alpha n}\overline{U}_{\alpha m} + \overline{U}_{\beta n}\overline{U}_{\beta m} + \overline{W}_n\overline{W}_m) g_\alpha g_\beta \, \mathrm{d}\alpha \, \mathrm{d}\beta = 0. \tag{13.3.8a}$$

If tangential inertia is neglected as well, Eq. (13.3.8a) then reduces to

$$\int_{\beta_1}^{\beta_2}\int_{\alpha_1}^{\alpha_2} \overline{W}_n\overline{W}_m g_\alpha g_\beta \, \mathrm{d}\alpha \, \mathrm{d}\beta = 0. \tag{13.3.8b}$$

The quantities L_n must similarly be modified. It should be noted that corresponding deletions have to be made in Eqs. (13.2.3) describing the space variation of $\overline{U}_{\alpha n}$, $\overline{U}_{\beta n}$, and $\overline{W}_n$.

13.4 Free vibrations of cylindrical shells

The study of the harmonic motions of circular cylindrical shells is the simplest to undertake. As a result more is known about the characteristics of cylindrical shell vibrations than of any other type of shell. The equations of motion can easily be expressed in terms of displacements merely by adding the appropriate inertia terms to Eqs. (10.1.5) to obtain

$$\Big[\frac{\partial^2}{\partial \xi^2} + \frac{1-\nu}{2}(1+k)\frac{\partial^2}{\partial \theta^2} - \frac{(1-\nu^2)\rho R^2}{E}\frac{\partial^2}{\partial t^2}\Big]u_x - \frac{1+\nu}{2}\frac{\partial^2 u_\theta}{\partial \xi \, \partial \theta} + \Big\{\nu - k\Big[\frac{\partial^2}{\partial \xi^2} - \frac{1-\nu}{2}\frac{\partial^2}{\partial \theta^2} - \frac{(1-\nu^2)\rho R^2}{E}\frac{\partial^2}{\partial t^2}\Big]\Big\}\frac{\partial w}{\partial \xi} = 0, \tag{13.4.1a}$$

$$\frac{1+\nu}{2}\frac{\partial^2 u_x}{\partial\xi\,\partial\theta} + \left[\frac{1-\nu}{2}(1+3k)\frac{\partial^2}{\partial\xi^2} + \frac{\partial^2}{\partial\theta^2} - \frac{(1-\nu^2)\rho R^2}{E}(1+3k)\frac{\partial^2}{\partial t^2}\right]u_\theta$$

$$+ \left[1 - \frac{3-\nu}{2}k\frac{\partial^2}{\partial\xi^2} + 2k\frac{(1-{}_b{}^2)\rho R^2}{E}\frac{\partial^2}{\partial t^2}\right]\frac{\partial w}{\partial\theta} = 0, \quad (13.4.1b)$$

$$\left\{\nu - k\left[\frac{\partial^2}{\partial\xi^2} - \frac{1-\nu}{2}\frac{\partial^2}{\partial\theta^2} - \frac{(1-\nu^2)\rho R^2}{E}\frac{\partial^2}{\partial t^2}\right]\right\}\frac{\partial u_x}{\partial\xi}$$

$$+ \left[1 - \frac{3-\nu}{2}k\frac{\partial^2}{\partial\xi^2} + 2k\frac{(1-\nu^2)\rho R^2}{E}\frac{\partial^2}{\partial t^2}\right]\frac{\partial u_\theta}{\partial\theta}$$

$$+ \left\{1 + \frac{(1-\nu^2)\rho R^2}{E}\frac{\partial^2}{\partial t^2} + k\right.$$

$$\left.\times\left[\nabla^4 + 2\frac{\partial^2}{\partial\theta^2} + 1 - \frac{(1-\nu^2)\rho R^2}{E}\frac{\partial^2}{\partial t^2}\nabla^2\right]\right\}w = 0. \quad (13.4.1c)$$

When the shell is closed in the circumferential direction the deformations of a free harmonic vibration mode may be expressed as

$$u_x = \overline{U}_{xn}(\xi)(\alpha_n\cos n\theta + \beta_n\sin n\theta)(\gamma_n\cos\omega_n t + \delta_n\sin\omega_n t), \quad (13.4.2a)$$

$$u_\theta = \overline{U}_n(\xi)(\alpha_n\cos n\theta + \beta_n\sin n\theta)(\gamma_n\cos\omega_n t + \delta_n\sin\omega_n t), \quad (13.4.2b)$$

$$w = \overline{W}_n(\xi)(\alpha_n\cos n\theta + \beta_n\sin n\theta)(\gamma_n\cos\omega_n t + \delta_n\sin\omega_n t), \quad (13.4.2c)$$

$$(n = 0, 1, 2, \ldots)$$

Substitution of Eqs. (13.4.2) into Eqs. (13.4.1) then yields the following set of ordinary differential equations for $\overline{U}_{xn}$, $\overline{U}_{\theta n}$, and $\overline{W}_n$

$$\left[\frac{d^2}{d\xi^2} - \frac{1-\nu}{2}(1+k)n^2 + \Omega_n^2\right]\overline{U}_{xn} + \frac{1+\nu}{2}n\frac{d\overline{U}_{\theta n}}{d\xi}$$

$$+ \left[\nu - k\left(\frac{d^2}{d\xi^2} + \frac{1-\nu}{2}n^2 + \Omega_n^2\right)\right]\frac{d\overline{W}_n}{d\xi^2} = 0, \quad (13.4.3a)$$

$$\frac{1+\nu}{2}n\frac{d\overline{U}_{xn}}{d\xi} - \left[\frac{1-\nu}{2}(1+3k)\frac{d^2}{d\xi^2} - n^2 - (1+3k)\Omega_n^2\right]\overline{U}_{\theta n}$$

$$+ \left(1 - \frac{3-\nu}{2}k\frac{d^2}{d\xi^2} - 2k\Omega_n^2\right)n\overline{W}_n = 0, \quad (13.4.3b)$$

$$\left[\nu - k\left(\frac{d^2}{d\xi^2} + \frac{1-\nu}{2}n^2 + \Omega_n^2\right)\right]\frac{d\overline{U}_{xn}}{d\xi} + \left(1 - \frac{3-\nu}{2}k\frac{d^2}{d\xi^2} - 2k\Omega_n^2\right)n\overline{U}_{\theta n}$$

$$+ \left\{1 - \Omega_n^2 + k\left[\frac{d^4}{d\xi^4} - 2n^2\frac{d^2}{d\xi^2} + (n^2-1)^2 + \Omega_n^2\left(\frac{d^2}{d\xi^2} - n^2\right)\right]\right\}\overline{W}_n = 0,$$

$$(13.4.3c)$$

where

$$\Omega_n^2 = \frac{(1 - \nu^2)\rho R^2}{E}\,\omega_n^2. \tag{13.4.3d}$$

13.4.1 *Simply supported cylindrical shells*

Perhaps the simplest case which can be investigated is that of a cylindrical shell which is simply supported at both ends in such a manner that the boundary conditions are given by

$$u_\theta = w = M_x = N_x = 0, \tag{13.4.4}$$

at $x = 0, L$. This type of simple support can be assumed to be supplied by hypothetical end rings which are rigid in their own plane but which offer no resistance to movement out of their plane or to rotation of the edges. All of the boundary conditions are satisfied if the displacement functions are of the form

$$\overline{U}_{xn} = A_{mn} \cos \frac{m\pi x}{L}, \tag{13.4.5a}$$

$$\overline{U}_n = B_{mn} \sin \frac{m\pi x}{L}, \tag{13.4.5b}$$

$$\overline{W}_n = C_{mn} \sin \frac{m\pi x}{L} \qquad (m = 0, 1, 2, \ldots), \tag{13.4.5c}$$

which are purely sinusoidal variations in the longitudinal direction. Some of the mode shapes are shown in Fig. 13.1.

Substitution of Eqs. (13.4.5) into Eqs. (13.4.3) then indicates that the following equations must be satisfied:

$$\left\{\Omega_{mn}^2 - \left[\lambda_m^2 + \frac{1-\nu}{2}(1+k)n^2\right]\right\}A_{mn} + \frac{1+\nu}{2}n\lambda_m B_{mn}$$
$$+ \left[\nu - k\left(\Omega_{mn}^2 - \lambda_m^2 + \frac{1-\nu}{2}n^2\right)\right]\lambda_m C_{mn} = 0, \tag{13.4.6a}$$

$$-\frac{1+\nu}{2}n\lambda_m A_{mn} - \left\{(1+3k)\Omega_{mn}^2 - \left[\frac{1-\nu}{2}(1+3k)\lambda_m^2 + n^2\right]\right\}B_{mn}$$
$$+ \left[1 - k\left(2\Omega_{mn}^2 - \frac{3-\nu}{2}\lambda_m^2\right)\right]nC_{mn} = 0, \tag{13.4.6b}$$

$$- \left[\nu - k\left(\Omega_{mn}^2 - \lambda_m^2 + \frac{1-\nu}{2} n^2\right)\right] \lambda_m A_{mn}$$
$$+ \left[1 - k\left(2\Omega_{mn}^2 - \frac{3-\nu}{2} \lambda_m^2\right)\right] n B_{mn}$$
$$+ \{1 + k[\lambda_m^4 - 2n^2\lambda_m^2 + (n^2-1)^2] - \Omega_{mn}^2[1 + k(\lambda_m^2 + n^2)]\} C_{mn} = 0, \tag{13.4.6c}$$

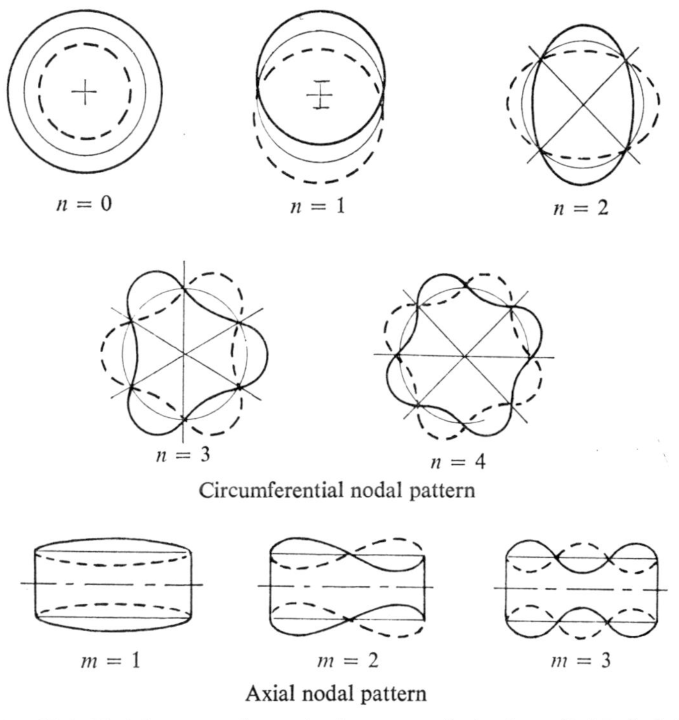

Fig. 13.1. Nodal patterns for a simply supported circular cylindrical shell

with

$$\lambda_m = \frac{m\pi R}{L} \qquad (m = 0, 1, 2, \ldots). \tag{13.4.6d}$$

Eqs. (13.4.6) are a set of homogeneous linear equations for the constants A_{mn}, B_{mn}, and C_{mn}. If these are to have values other than zero the determinant of their coefficients must vanish. When the determinantal equation is expanded we obtain an equation of the form

$$h_{0mn}\Omega_{mn}^6 + h_{1mn}\Omega_{mn}^4 + h_{2mn}\Omega_{mn}^2 + h_{3mn} = 0, \tag{13.4.7a}$$

where the coefficients are functions of k, ν, λ_m, and n. Corresponding values of the constants A_{mn} and B_{mn} relative to C_{mn} may be obtained by solving any two of Eqs. (13.4.6), say the first two, to yield

$$\frac{A_{mn}}{C_{mn}} = \lambda_m \frac{\left[\nu - k\left(\Omega_{mn}^2 - \lambda_m^2 + \frac{1-\nu}{2} n^2\right)\right]\left[(1+3k)\left(\Omega_{mn}^2 - \frac{1+\nu}{2}\lambda_m^2\right) - n^2\right] + \frac{1+\nu}{2} n^2\left[1 - k\left(2\Omega_{mn}^2 - \frac{3-\nu}{2}\lambda_m^2\right)\right]}{\left[\Omega_{mn}^2 - \lambda_m^2 - \frac{1-\nu}{2}(1+k)n^2\right] \times \left[(1+3k)\left(\Omega_{mn}^2 - \frac{1+\nu}{2}\lambda_m^2\right) - n^2\right] - \left(\frac{1+\nu}{2} n\lambda_m\right)^2} \tag{13.4.7b}$$

$$\frac{B_{mn}}{C_{mn}} = n \frac{\left[\Omega_{mn}^2 - \lambda_m^2 - \frac{1-\nu}{2}(1+k)n^2\right]\left[1 - k\left(2\Omega_{mn}^2 - \frac{3-\nu}{2}\lambda_m^2\right)\right] + \frac{1+\nu}{2}\lambda_m^2\left[\nu - k\left(\Omega_{mn}^2 - \lambda_m^2 + \frac{1-\nu}{2} n^2\right)\right]}{\left[\Omega_{mn}^2 - \lambda_m^2 - \frac{1-\nu}{2}(1+k)n^2\right] \times \left[(1+3k)\left(\Omega_{mn}^2 - \frac{1+\nu}{2}\lambda_m^2\right) - n^2\right] - \left(\frac{1+\nu}{2} n\lambda_m\right)^2}. \tag{13.4.7c}$$

(a) $m = 0$ or $n = 0$.

In some cases it is easy to obtain explicit expressions for the frequency parameters. If m is equal to zero, for example, we have the case of purely longitudinal shearing motion for which

$$\Omega_{0n} = n\left[\frac{1-\nu}{2}(1+k)\right]^{1/2}. \tag{13.4.8}$$

If m is not equal to zero but n is equal to zero, the case of axisymmetric deformations, values of the frequency parameter and of the relative magnitudes of the displacements are easily found since Eqs. (13.4.6) separate into the two sets

$$\left(\frac{1-\nu}{2}\lambda_m^2 - \Omega_{m0}^2\right)B_{m0} = 0, \tag{13.4.9a}$$

corresponding to purely torsional motions, and

$$(\lambda_m^2 - \Omega_{m0}^2)A_{m0} - [\nu + k(\lambda_m^2 - \Omega_{m0}^2)]\lambda_m C_{m0} = 0, \tag{13.4.9b}$$

$$[\nu + k(\lambda_m^2 - \Omega_{m0}^2)]\lambda_m A_{m0} - [1 + k(1 + \lambda_m^4) - (1 + k\lambda_m^2)\Omega_{m0}^2]C_{m0} = 0, \tag{13.4.9c}$$

corresponding to coupled radial and longitudinal vibrations.

The frequency of torsional vibrations, characterized by

$$A_{m0} = C_{m0} = 0, \tag{13.4.10a}$$

$$B_{m0} \neq 0, \tag{13.4.10b}$$

is given by Eq. (13.4.9a) as

$$\Omega_{m0} = \left(\frac{1-\nu}{2}\right)^{1/2} \lambda_m. \tag{13.4.10c}$$

From Eqs. (13.4.9b) and (13.4.9c) and the condition that A_{m0} and C_{m0} have values other than zero we find that the frequency of coupled radial and longitudinal vibrations is

$$\Omega_{m0} = \left\{\lambda_m^2 + \frac{[1+k-(1+2\nu k)\lambda_m^2] \pm \{[1+k-(1+2\nu k)\lambda_m^2]^2+4\nu^2\lambda_m^2[1+k(1-k)\lambda_m^2]\}^{1/2}}{2[1+k(1-k)\lambda_m^2]}\right\}^{1/2}, \tag{13.4.11a}$$

and that the ratio of the amplitudes of longitudinal and radial motions is

$$\frac{A_{m0}}{C_{m0}} = \frac{1+k-\lambda_m^2 \mp \{[1+k-(1+2\nu k)\lambda_m^2]^2+4\nu^2\lambda_m^2[1+k(1-k)\lambda_m^2]\}^{1/2}}{2\nu\lambda_m}. \tag{13.4.11b}$$

The circumferential motion vanishes in this case.

The variation with λ_m of the roots of the characteristic frequency equation, given by Eqs. (13.4.10c) and (13.4.11a), is shown by the solid curves of Fig. 13.2. In the range of λ_m shown the effects of bending and of rotatory inertia are negligible so that the frequency parameters are practically independent of wall thickness. The effects of tangential inertia are important over a significant portion of this range of λ_m. If we were to neglect tangential inertia as well as bending stiffness and rotatory inertia, the value of the frequency parameter can be shown to be given by

$$\Omega_{m0} = (1-\nu^2)^{1/2}, \tag{13.4.12}$$

which can be seen to be an adequate representation of the lowest root if λ_m is greater than about 2. For values of λ_m less than 2, however, the approximate value is seriously in error for precisely the reason, as indicated by the amplitude ratios shown in Fig. 13.3., that the longitudinal amplitude is then not negligible compared to the radial amplitude of vibration.

In the preceding discussion the calculations have been restricted to values of λ_m such that $k\lambda_m^4 \ll 1$ or to longitudinal wavelengths of vibration

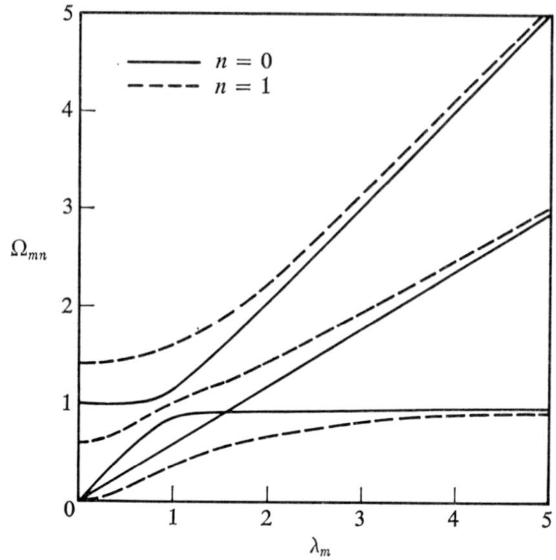

Fig. 13.2. Roots of frequency equation for n = 0 and 1 ($\nu = 0.3$, $\lambda_m^4 \ll 1/k$)

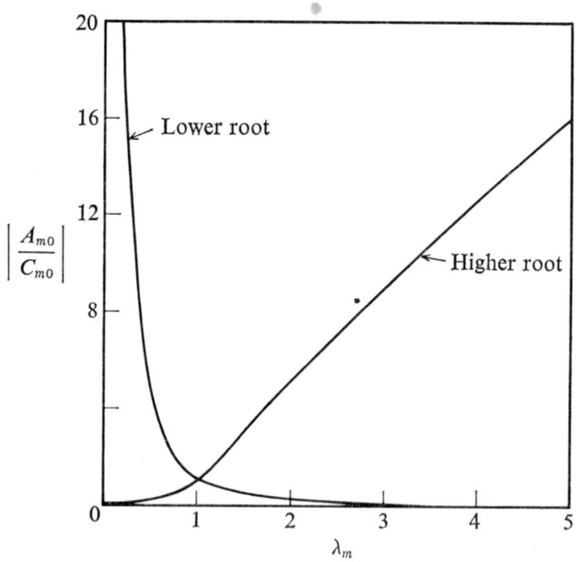

Fig. 13.3. Variation of amplitude ratio with wave length parameter ($n = 0$, $\nu = 0.3$)

which are considerably larger than the attenuation length for static axisymmetric end loads. As λ_m increases the effect of radius–thickness ratio becomes significant. The effect of bending stiffness and rotatory inertia can be calculated most readily when Poisson's ratio ν vanishes. In this case Eqs. (13.4.11) take on the limiting forms

$$\Omega_{m0} = \lambda_m, \qquad A_{m0} \neq 0, \qquad B_{m0} = C_{m0} = 0, \tag{13.4.13a}$$

$$\Omega_{m0} = \left[\frac{1 + k + k(1-k)\lambda_m^4}{1 + k(1-k)\lambda_m^2}\right]^{1/2}, \qquad \frac{A_{m0}}{C_{m0}} = k\lambda_m \qquad B_{m0} = 0. \tag{13.4.13b}$$

We see that this special case corresponds to practically independent radial and longitudinal motions since Eq. (13.4.13a) obviously denotes purely longitudinal motion and Eq. (13.4.13b) indicates radial motion with negligible longitudinal motion unless the wavelength of vibration is extremely small compared to the thickness of the shell. The variation of the frequency parameters is shown in Fig. 13.4 for large values of λ_m. In each case the stiffening effect of bending, indicated by that portion of the curve for a particular value of R/h which is concave upward, becomes important first as λ_m

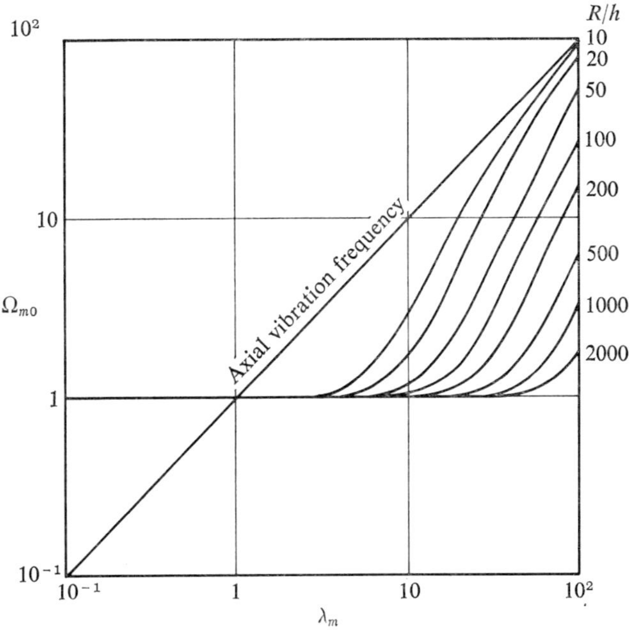

Fig. 13.4. Effect of rotatory inertia on axisymmetric vibration frequencies ($\nu = 0$)

increases. For still larger values of λ_m the increased flexibility effect of rotatory inertia becomes important and the curves tend to approach the curve for purely axial vibrations. The applicability of these last results for the transversely rigid cylinder to the completely isotropic cylinder is doubtful, however, since for such short cylinders or for such short wavelengths of motion the effects of transverse shear and transverse direct stress, which do not enter the investigation of a transversely rigid material, would be important.

(b) $n = 1$.

When n is equal to unity the frequency parameters are more difficult to obtain since the characteristic equation cannot be factored. We thus have to solve the cubic equation for the frequencies. Calculations† indicate that again the effects of bending and of rotatory inertia are negligible for the practical range of values of λ_m. The calculated frequency parameters are shown by the dashed curves of Fig. 13.2. It is interesting to note that despite the difference in vibration mode shape the three values of the frequency parameter approach those for axisymmetric vibrations for the larger values of λ_m. For small values of $\lambda_m (\lambda_m < 0.16$ or $L/R > 20$ for $m = 1)$ the lowest frequency parameter can be shown to be adequately approximated by

$$\Omega_{m1} = \left(\frac{1 - \nu^2}{2}\right)^{1/2} \lambda_m^2, \tag{13.4.14}$$

which is identical with the result we obtain with the Bernoulli–Euler theory of beams. Since beam theory assumes the cylinder to have rigid cross-sections which remain normal to the cylinder axis, we deduce that the cross-sections of flexible cylinders vibrating as beams remain practically circular and normal to the longitudinal axis provided the longitudinal wavelength of motion is large. For greater values of λ_m the effects of middle surface shear deformation, cross-sectional deformation, and longitudinal inertia become important and the vibration frequencies given by shell theory are less than those given by beam theory. The importance of tangential inertia in the determination of the lowest frequency can be investigated by deleting tangential inertia terms as well as bending and rotatory inertia terms from the formulation of the problem. We can easily obtain the approximation

$$\Omega_{m1} = (1 - \nu^2)^{1/2} \frac{\lambda_m^2}{\lambda_m^2 - 1}, \tag{13.4.15}$$

which is compared with the more accurate values of the frequency in Fig. 13.5.

---

† K. Forsberg, 'Axisymmetric and Beam-type Vibrations of Thin Cylindrical Shells,' *AIAA J.*, vol. 7, no. 2, February 1969, pp. 221–227.

While the effect of tangential inertia is not so pronounced as for axisymmetric vibrations, it is still significant, especially when the wavelength parameter λ_m is small. By comparing Eqs. (13.4.14) and (13.4.15) we see that the error in the lowest frequency is about 30 percent for small λ_m.

(c) $n \geq 2$.

When n is greater than or equal to two, one of the three frequencies given by Eq. (13.4.7a) is considerably smaller than the others over much of the interesting range of λ_m and generally corresponds to a vibration mode for

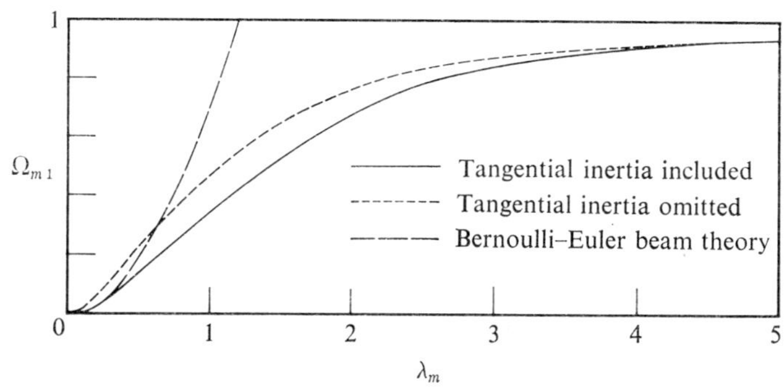

Fig. 13.5. Effect of tangential inertia on the lowest frequency of a simply supported cylindrical shell ($n = 1, \nu = 0.3$)

which radial deformations predominate. For example, for very long cylinders vibrating with very long longitudinal wavelengths so that $\lambda_m \to 0$, the three values of the frequency parameter and the relative deformation amplitudes can be shown to approach the following values:

1. Flexural vibrations of a ring

$$\Omega_n = \frac{n(n^2 - 1)}{(n^2 + 1)^{1/2}} k^{1/2}, \tag{13.4.16a}$$

$$A_n = 0, \qquad B_n/C_n = 1/n. \tag{13.4.16b}$$

2. Axial shear vibrations

$$\Omega_n = \left(\frac{1 - \nu}{2}\right)^{1/2} n, \tag{13.4.16c}$$

$$A_n \neq 0, \qquad B_n = C_n = 0. \tag{13.4.16d}$$

3. Extensional vibrations of a ring

$$\Omega_n = (n^2 + 1)^{1/2}, \tag{13.4.16e}$$

$$A_n = 0, \qquad B_n/C_n = -n. \tag{13.4.16f}$$

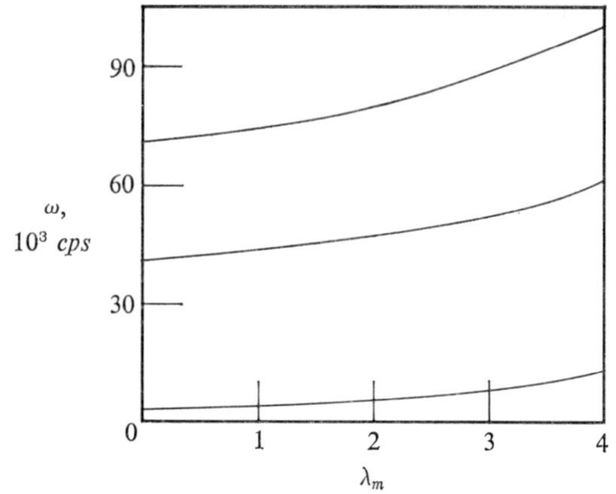

Fig. 13.6. Calculated frequencies for a more complex nodal pattern ($R = 1.925''$, $h = 0.101''$, $n = 4$)

The first of the frequencies is much smaller than the others provided the circumferential wavelength of sinusoidal deformation is considerably greater than the shell wall thickness. The displacement is greatest in the radial direction, becoming predominant as the number of circumferential waves increases. Typical variations of the three frequencies with λ_m is shown in Fig. 13.6.† Some values of amplitude ratio are given in Table 13.1 below.

The variation with λ_m of the lowest frequency parameter for various values of the circumferential wave number n and the thickness–radius ratio h/R is shown in Fig. 13.7.‡ We see that the effect of shell thickness, primarily

Table 13.1 Component amplitude ratios ($k = 2.294 \times 10^{-4}$, $n = 4$, $\nu = 0.29$)

	f_1		f_2		f_3	
λ_m	A_{14}/C_{14}	B_{14}/C_{14}	A_{14}/C_{14}	B_{14}/C_{14}	A_{14}/C_{14}	B_{14}/C_{14}
0.000	0.000	0.250	—	—	0.00	4.00
0.387	0.024	0.252	27.5	1.82	0.36	4.00
1.547	0.072	0.258	6.75	2.14	1.60	4.32
2.707	0.073	0.246	4.53	3.02	3.28	5.05
3.867	0.052	0.220	4.10	3.66	5.79	5.96

† R. N. Arnold and G. B. Warburton, 'Flexural Vibrations of the Walls of Thin Cylindrical Shells having Freely Supported Ends,' *Proc. Roy. Soc., London, Series A*, no. 1049, vol. 197, June 7, 1949, pp. 238–256.

‡ R. N. Arnold and G. B. Warburton, 'The Flexural Vibrations of Thin Cylinders,' *Proc. Inst. Mech. Eng.* (*London*), *Series A*, vol. 167, no. 1, 1953, pp. 62–74.

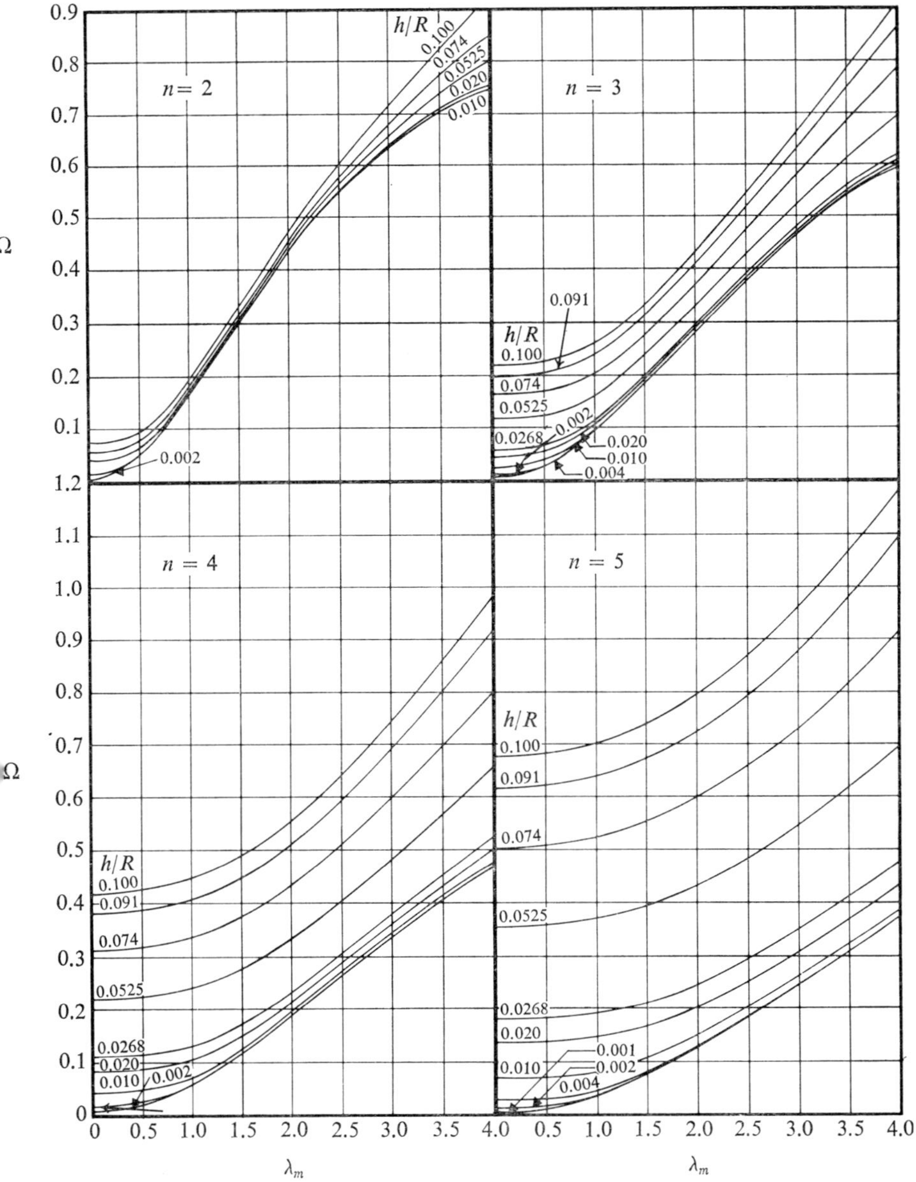

Fig. 13.7. Variation of lowest vibration frequency with circumferential wave number and shell thickness ($\nu = 0.29$)

the terms due to bending of the cylinder wall, becomes more important as n increases since each portion of the cylinder between longitudinal behaves more and more like a simply supported flat plate vibrating in the transverse direction. For sufficiently thin cylinders, however, we see that the effect of wall thickness or bending can be negligible. In that case the kinetic energy of the cylinder is predominantly balanced by the energy of stretching of the cylinder, which thus behaves like a membrane.

Since the deformation pattern leading to the lowest frequency is essentially radial, we would expect an approximation based on the assumption of negligible tangential inertia to be adequate, especially for large values of n. If in Eqs. (13.4.6) we delete the terms due to tangential inertia and those due to rotatory inertia which are negligible for using values of λ_m and k, expand the determinantal equation, and make approximations similar to those leading to Morley's characteristic equation for static deformations of cylinders we obtain the following estimate for the value of the lowest frequency

$$\Omega_n = \frac{1}{\lambda_m^2 + n^2}\,[(1 - \nu^2)\lambda_m^4 + (\lambda_m^2 + n^2)^2(\lambda_m^2 + n^2 - 1)^2 k]^{1/2}. \tag{13.4.19}$$

For very small values of λ_m we have

$$\Omega_n \approx (n^2 - 1)k^{1/2}, \tag{13.4.20}$$

which is in error by 11.8 percent if n is equal to 2, by 5 percent if n is equal to 3, and by less than 5 percent if n is greater than 3. The error for larger values of λ_m is generally of the same order of magnitude, as indicated in Fig. 13.8 by the comparison of the results of Eq. (13.4.19) with those of Fig. 13.7 for the case $n = 2$, $R/h = 100$, $\nu = 0.29$.

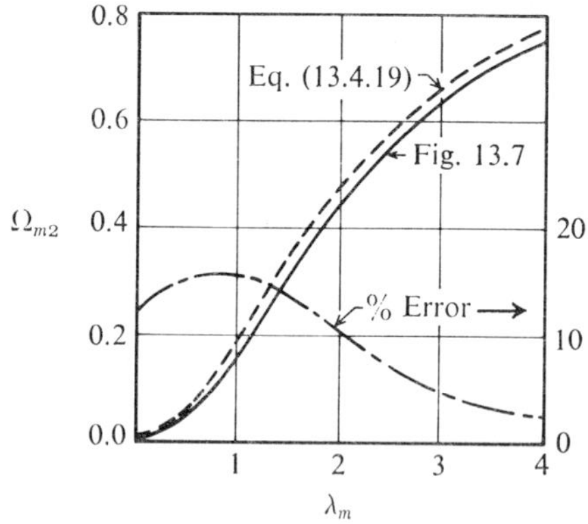

Fig. 13.8. Comparison of exact and approximate values of lowest frequency parameter ($n = 2$, $R/h = 100$, $\nu = 0.29$)

13.4.2 Minimum frequency of vibration

In many practical design problems the minimum frequency of vibration of a given cylinder might be desired. From previous experience with the vibrations of beams or flat plates we would be prepared to find that this frequency is associated with the least complex mode shape, presumably the frequency obtained when we put n equal to zero and m equal to unity in the equations. But a glance at Fig. 13.2 immediately shows us that we would be wrong in one important respect. The monotonic increase of the frequency parameter with λ_m insures that for a given values of the length–radius ratio L/R the frequency corresponding to $m = 1$ is less than that for any larger value of m. We see, however, that contrary to our expectations the frequency parameter for $n = 1$ is less than that for $n = 0$.

If we investigate further for a given value of L/R we find that the lowest frequency generally decreases as n increases until a minimum value is obtained for some particular value of n and then increases with increasing values of n. Typical variations of frequency with n are shown in Fig. 13.9.

A representation of the variation of the minimum frequency with cylinder length can be obtained by superimposing the curves of frequency parameter versus length parameter λ_m for a given value of R/h as in Fig. 13.10† and retaining only the minimum value of the frequency for each value of L/R. Several such envelope curves are shown in Fig. 13.11.‡ We note that the number of circumferential waves associated with the minimum frequency increases with decreasing length and thickness. For sufficiently thin shells and for sufficiently short wavelengths of vibration, the number of circumferential waves associated with the minimum frequency will be large and the frequency parameter curve will vary only slowly with n. We may then assume n to be a continuous rather than a discrete function and take the minimum of the frequency parameter versus *continuous* n curve as the desired minimum value. For large n we may also use Eq. (13.4.19) which we modify to read as

$$\Omega_n \approx \left[\frac{(1-\nu^2)\lambda_m^4}{(\lambda_m^2+n^2)^2} + (\lambda_m^2+n^2)^2 k\right]^{1/2}. \tag{13.4.21a}$$

† R. N. Arnold and G. B. Warburton, 'Flexural Vibrations of the Walls of Thin Cylindrical Shells having Freely Supported Ends,' op. cit.

‡ K. Forsberg, 'Influence of Boundary Conditions on the Modal Characteristics of Thin Cylindrical Shells,' *AIAA J.*, vol. 2, no. 12, December 1964, pp. 2150–2157. See also, G. B. Warburton, 'Vibration of Thin Cylindrical Shells,' *J. Mech. Eng. Sci.*, vol. 7, no. 4, 1965, pp. 399–407.

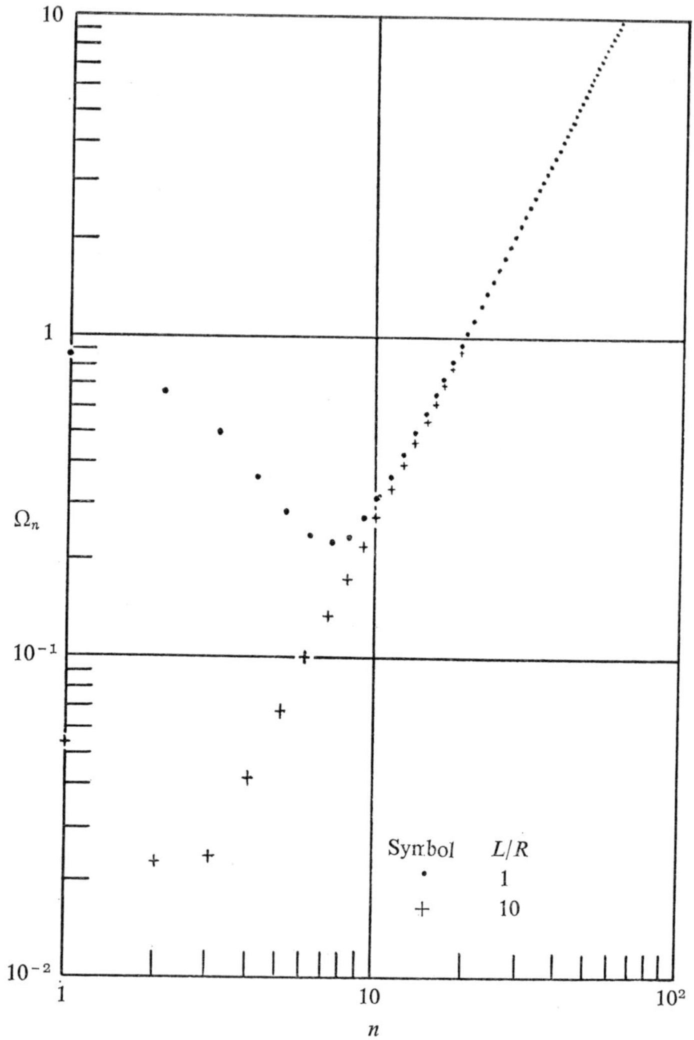

Fig. 13.9. Variation of lowest frequency parameter with circumferential wave number ($\nu = 0.0$, $R/h = 100$, $m = 1$)

Then the minimization of Eq. (13.4.21a) with respect to n yields

$$(\lambda_m^2 + n^2)^2 = \left(\frac{1 - \nu^2}{k}\right)^{1/2} \lambda_m^2, \tag{13.4.21b}$$

and

$$\Omega_{n\min} = [4(1 - \nu^2)k]^{1/4}\lambda_m. \tag{13.4.21c}$$

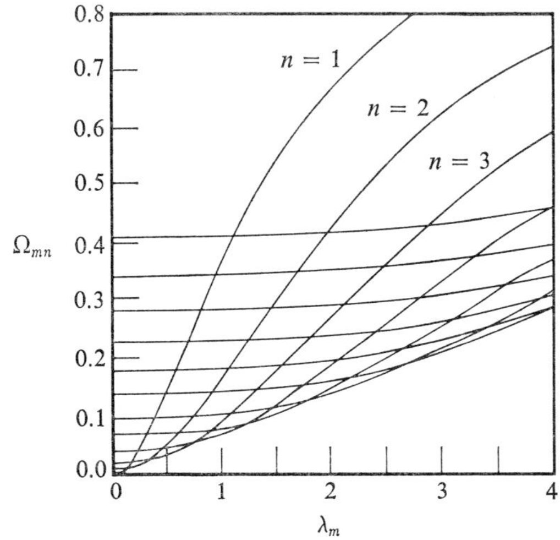

Fig. 13.10. Composite frequency chart (R/h = 100, ν = 0.29)

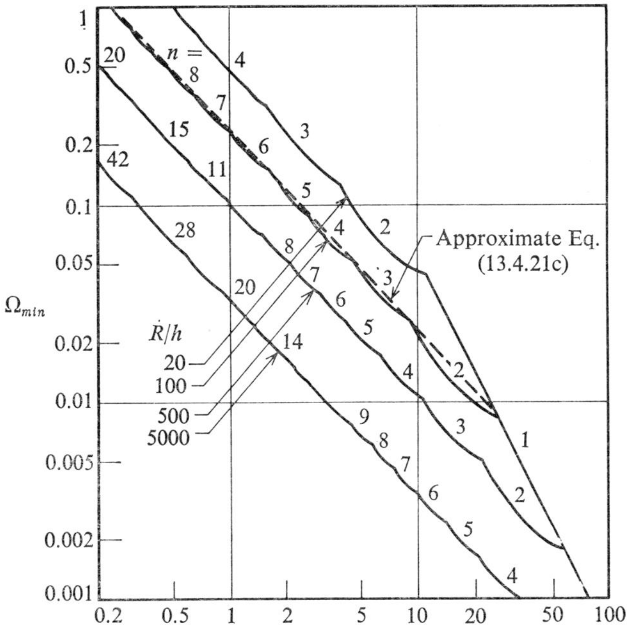

Fig. 13 11. Variation of minimum frequency with cylinder length and thickness (ν = 0.3)

The results of Eq. (13.4.21c) for a cylinder with $R/h = 100$ are shown by the dashed curve of Fig. 13.11. We see that the accuracy of the approximate expression becomes quite good as n increases, as is to be expected.

The anomalous behavior of the natural frequencies of cylindrical shells is attributable to a change in the deformation state from one involving primarily stretching of the shell middle surface to one involving primarily bending of the shell wall as the number of circumferential waves increases. That this change occurs can be determined by comparing the strain energy of stretching and of bending as n increases. If we use the strain energy expression given by Eq. (3.1.3), which is most convenient for our purposes, we have

$$U_t = U_s + U_b, \tag{13.4.22a}$$

where the stretching energy and bending energy are respectively given by

$$U_s = \frac{KR}{2}\int_0^{2\pi}\int_0^L \left[\left(\frac{\partial u_x}{\partial x}\right)^2 + 2\nu\left(\frac{\partial u_x}{\partial x}\right)\frac{1}{R}\left(\frac{\partial u_\theta}{\partial \theta} + w\right) + \frac{1}{R^2}\left(\frac{\partial u_\theta}{\partial \theta} + w\right)^2 + \frac{1-\nu}{2}\left(\frac{\partial u_\theta}{\partial x} + \frac{1}{R}\frac{\partial u_x}{\partial \theta}\right)^2\right] \mathrm{d}x\,\mathrm{d}\theta, \tag{13.4.22b}$$

$$U_b = \frac{DR}{2}\int_0^{2\pi}\int_0^L \left[\left(\frac{\partial^2 w}{\partial x^2}\right)^2 + \frac{2\nu}{R}\frac{\partial^2 w}{\partial x^2}\left(\frac{\partial^2 w}{\partial \theta^2} - \frac{\partial u_\theta}{\partial \theta}\right) + \frac{1}{R^4}\left(\frac{\partial^2 w}{\partial \theta^2} - \frac{\partial u_\theta}{\partial \theta}\right)^2 + \frac{2(1-\nu)}{R^2}\left(\frac{\partial^2 w}{\partial x\,\partial \theta} + \frac{1}{4R}\frac{\partial u_x}{\partial \theta} - \frac{3}{4}\frac{\partial u_\theta}{\partial x}\right)^2\right] \mathrm{d}x\,\mathrm{d}\theta. \tag{13.4.22c}$$

If now we put

$$u_x = A_{mn}\cos\frac{m\pi x}{L}\cos n\theta\cos\omega_{mn}t, \tag{13.4.23a}$$

$$u = B_{mn}\sin\frac{m\pi x}{L}\sin n\theta\cos\omega_{mn}t, \tag{13.4.23b}$$

$$w = C_{mn}\sin\frac{m\pi x}{L}\cos n\theta\cos\omega_{mn}t, \tag{13.4.23c}$$

we have

$$U_s = \frac{\pi EhLC_{mn}^2}{4(1-\nu^2)R}\eta_s\cos^2\omega_{mn}t, \tag{13.4.24a}$$

$$U_b = \frac{\pi EhLC_{mn}^2}{4(1-\nu^2)R}\eta_b\cos^2\omega_{mn}t, \tag{13.4.24b}$$

$$U_t = \frac{\pi EhLC_{mn}^2}{4(1-\nu^2)R}\eta_t\cos^2\omega_{mn}t, \tag{13.4.24c}$$

where

$$\eta_s = \left(1 + n\frac{B_{mn}}{C_{mn}}\right)^2 + 2\nu\frac{A_{mn}}{C_{mn}}\left(1 + n\frac{B_{mn}}{C_{mn}}\right)$$

$$+ \lambda_m^2\left(\frac{A_{mn}}{B_{mn}}\right)^2 + \frac{1-\nu}{2}\left(\lambda_m\frac{B_{mn}}{C_{mn}} - n\frac{A_{mn}}{C_{mn}}\right)^2, \tag{13.4.25a}$$

$$\eta_b = k\left[\left(1 + n\frac{B_{mn}}{C_{mn}}\right)^2 + 2\nu\lambda_m^2\left(1 + n\frac{B_{mn}}{C_{mn}}\right) + \lambda_m^4\right.$$

$$\left. + 2(1-\nu)\left(n\lambda_m + \tfrac{1}{4}n\frac{A_{mn}}{C_{mn}} + \tfrac{3}{4}m\frac{B_{mn}}{C_{mn}}\right)^2\right]. \tag{13.4.25b}$$

$$\eta_t = \eta_b + \eta_s. \tag{13.4.25c}$$

The values of A_{mn}/C_{mn} and B_{mn}/C_{mn} used in Eqs. (13.4.25) can be taken as those given by Eqs. (13.4.7).

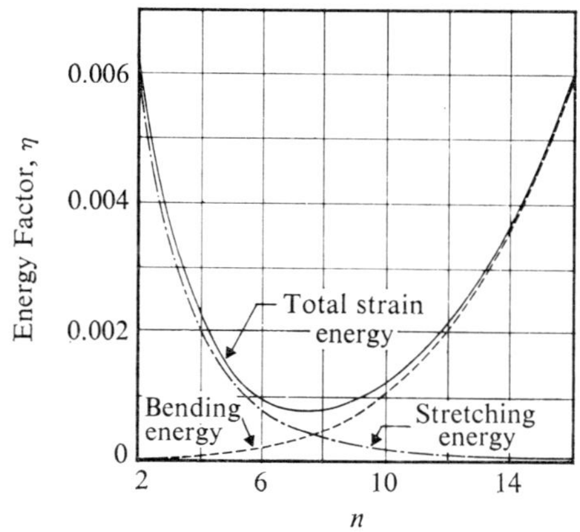

Fig. 13.12. Strain energy due to bending and stretching ($R/h = 100$, $L/R = 0.82$)

For a given cylinder having a given maximum radial amplitude of vibration C_{mn}, the maximum values of the stretching, bending, and total strain energy factors η_s, η_b, and η_t may be plotted as a function of n. A particular variation of the strain energy factors is shown in Fig. 13.12.† We see that the stretching energy factor decreases as n increases while the bending energy factor increases. The minimum frequency of vibration occurs when both

† R. N. Arnold and G. B. Warburton, 'Flexural Vibrations of the Walls of Thin Cylindrical Shells having Freely Supported Ends,' op. cit.

energies are comparable, in which case the total strain energy is a minimum or nearly a minimum.

13.4.3 Effects of other edge conditions

Until recently it was assumed that since tangential displacements are often small compared to radial displacements, the lowest frequencies for problems involving tangential edge conditions other than those of the previous sections would differ little from those already obtained. Calculations indicate, however, that contrary to our expectations, boundary conditions on tangential displacements can have a significant effect on the frequencies of vibration.†

To obtain the frequency equation for cylinders having edge conditions other than the special type of simple support already considered, we must return to the differential equations for the longitudinal variations of displacements, Eqs. (13.4.3). These are an eighth-order set of ordinary differential equations with constant coefficients. We may then express the solution for the displacements in the form

$$\overline{U}_{xn} = \sum_{s=1}^{s=8} A_{ns}\, e^{\lambda_{ns}\xi}, \tag{13.4.26a}$$

$$\overline{U}_{\theta n} = \sum_{s=1}^{s=8} B_{ns}\, e^{\lambda_{ns}\xi}, \tag{13.4.26b}$$

$$\overline{W}_{n} = \sum_{s=1}^{s=8} C_{ns}\, e^{\lambda_{ns}\xi}, \tag{13.4.26c}$$

where A_{ns}, B_{ns}, and C_{ns} are arbitrary constants. The substitution of Eqs. (13.4.26) into Eqs. (13.4.3) indicates that the equations will be satisfied if the following equations are satisfied:

$$\left[\lambda_{ns}^2 - \frac{1-\nu}{2}(1+k)n^2 + \Omega_n^2\right]A_{ns} + \frac{1+\nu}{2}n\lambda_{ns}B_{ns} + \left[\nu - k\left(\lambda_{ns}^2 + \frac{1-\nu}{2}n^2 + \Omega_n^2\right)\right]\lambda_{ns}C_{ns} = 0, \tag{13.4.27a}$$

$$\frac{1+\nu}{2}n\lambda_{ns}A_{ns} - \left[(1+3k)\left(\frac{1-\nu}{2}\lambda_{ns}^2 + \Omega_n^2\right) - n^2\right]B_{ns} + \left[1 - k\left(\frac{3-\nu}{2}\lambda_{ns}^2 + 2\Omega_n^2\right)\right]nC_{ns} = 0, \tag{13.4.27b}$$

† K. Forsberg, 'Influence of Boundary Conditions on the Modal Characteristics of Thin Cylindrical Shells,' op. cit.

$$\left[\nu - k\left(\lambda_{ns}^2 + \frac{1-\nu}{2} n^2 + \Omega_n^2\right)\right]\lambda_{ns}A_{ns} + \left[1 - k\left(\frac{3-\nu}{2}\lambda_{ns}^2 + \Omega_n^2\right)\right]nB_{ns}$$
$$+ \{1 - \Omega_n^2 + k[\lambda_{ns}^4 + \lambda_{ns}^2(\Omega_n^2 - 2n^2) + (n^2-1)^2 - n^2\Omega_n^2]\}C_{ns} = 0$$
$$(s = 1, 2, \ldots, 8). \qquad (13.4.27c)$$

Again, for the constants of integration to have values other than zero the determinant of their coefficients must vanish, yielding an eighth-degree algebraic equation for λ_{ns} of the form

$$g_{n0}\lambda_{ns}^8 + g_{n1}\lambda_{ns}^6 + g_{n2}\lambda_{ns}^4 + g_{n3}\lambda_{ns}^2 + g_{n4} = 0, \qquad (13.4.28a)$$

where the coefficients g_{ni} are functions of Ω_n, h/R, n, and ν. In addition, Eqs. (13.4.27) yield relations between two of the constants, say A_{ns} and B_{ns}, and the third C_{ns} since we may solve any two as was done to obtain Eqs. (13.4.7). In fact the equations may be obtained from Eqs. (13.4.7) by replacing λ_m^2 by $-\lambda_{ns}^2$ wherever it appears and multiplying the left side of Eq. (13.4.7b) by -1.

The resulting solution for the deflection functions is thus given in terms of eight arbitrary constants of integration and the as yet unknown vibration frequency parameter Ω_n. The final step in the calculation is the satisfaction of boundary conditions. In the absence of elastic edge restraints these are of the form

$$M_x = -\frac{D}{R^2}\left[\frac{\partial^2 w}{\partial \xi^2} - \frac{\partial u_x}{\partial \xi} + \nu\left(\frac{\partial^2 w}{\partial \theta^2} - \frac{\partial u_\theta}{\partial \theta}\right)\right] = 0 \quad \text{or} \quad \frac{\partial w}{\partial \xi} = 0,$$
$$(13.4.29a)$$

$$N_x = \frac{K}{R}\left[\frac{\partial u_x}{\partial \xi} + \nu\left(\frac{\partial u_\theta}{\partial \theta} + w\right) - k\frac{\partial^2 w}{\partial \xi^2}\right] = 0 \quad \text{or} \quad u_x = 0, \quad (13.4.29b)$$

$$\bar{N}_{x\theta} = \frac{1-\nu}{2}\frac{K}{R}(1 + \tfrac{3}{4}k)\left(\frac{\partial u_\theta}{\partial \xi} + \frac{\partial u_x}{\partial \theta}\right) = 0 \quad \text{or} \quad u_\theta = 0, \qquad (13.4.29c)$$

$$\bar{Q}_x = -\frac{D}{R^3}\left[\frac{\partial^3 w}{\partial \xi^3} + (2-\nu)\frac{\partial^3 w}{\partial \xi\,\partial \theta^2} - \frac{3-\nu}{2}\frac{\partial^2 u_\theta}{\partial \xi\,\partial \theta} - \frac{\partial^2 u_x}{\partial \xi^2}\right.$$
$$\left. + \frac{1-\nu}{2}\frac{\partial^2 u_x}{\partial \theta^2}\right] = 0 \quad \text{or} \quad w = 0. \qquad (13.4.29d)$$

Since there are four conditions at each edge, we have a total of eight conditions to be satisfied, yielding eight linear homogeneous simultaneous equations for the eight constants of integration for each value of n. The condition that these constants have values other than zero, and thus that the determinant of their coefficients must vanish, yields an equation for the harmonic frequencies. There are 16 possible sets of boundary conditions which may be imposed independently at each edge and 256 possible sets of conditions at both edges. Of these, 136 sets will yield different results. Obviously a complete investigation of the vibration frequencies of cylindrical shells would involve a tremendous amount of computation and graph plotting to represent the results adequately.

The frequency equation is transcendental in general so that there are an infinite number of these frequencies associated with each value of n, each one corresponding to a different mode shape. Because of the transcendental nature of the frequency equation the solutions must usually be found by a trial and error method of calculation. The procedure involves choosing a value of Ω_n, solving Eq. (13.4.28) for the corresponding values of λ_{ns}, deter-

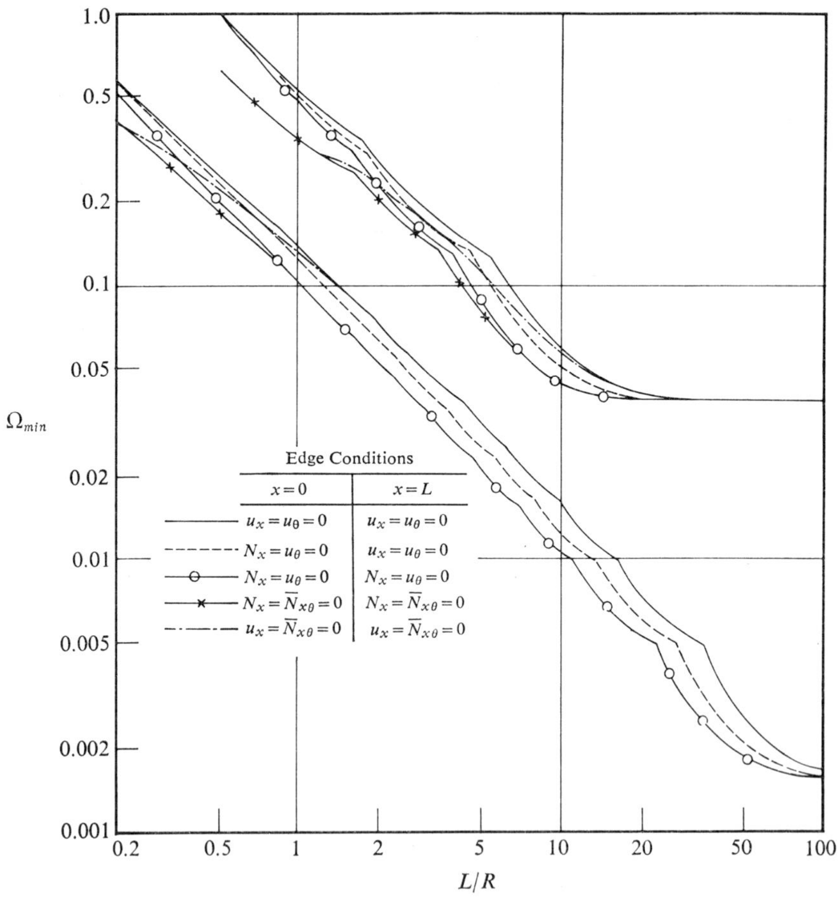

Fig. 13.13. Minimum frequency envelopes for simply supported cylinders with various tangential edge conditions

mining the coefficients A_{ns} and B_{ns} in terms of C_{ns}, and finally obtaining the value of the determinant of the coefficients of the edge condition equations. This procedure is repeated by varying Ω_n until the determinant vanishes or until an accurate value of the frequency parameter can be found by interpolation. The calculation procedure can be programmed for a digital computer† so that frequencies and associated mode shapes can be found readily.

The effect of edge conditions for tangential displacements on the minimum vibration frequency of simply supported cylinders ($w = M_x = 0$ at $x = 0, L$) is shown in Fig. 13.13 where some minimum frequency envelopes

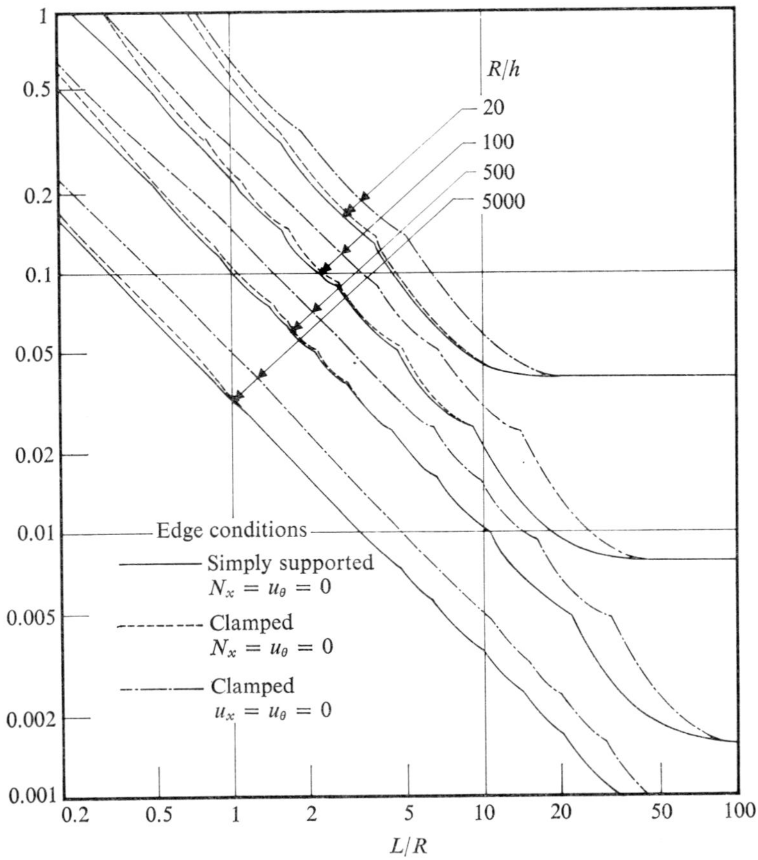

Fig. 13.14. Effect of longitudinal edge conditions on the minimum frequency envelopes for clamped shells

† K. Forsberg, 'Influence of Boundary Conditions on the Modal Characteristics of Thin Cylindrical Shells,' op. cit.

are plotted. The minimum frequencies are largest when both tangential displacements are restrained at both edges; the lowest frequencies are obtained when both tangential forces vanish at both edges. The spread of frequency values given by these two cases, which thus bound all other cases for simply supported cylindrical shells, is 50 percent or more. Some results for clamped shells are shown in Fig. 13.14 where we see that longitudinal displacement restraint can be much more significant in raising the minimum frequency than restraint on edge rotation.

What we may conclude from these comparisons is that we must be quite careful in ascertaining the boundary conditions for a given structure in so far as the determination of vibration frequencies is concerned. Experimental results† usually mirror the uncertainty of edge conditions since precise agreement with theory is not obtained unless extreme care is taken to achieve the edge conditions assumed in the analysis. While the effect of edge conditions on the response of structures is largely unexplored, the free vibration results would appear to indicate that as much care should be taken to ascertain the correct boundary conditions for response problems.

13.5 Free vibrations of spherical shells

The pertinent equations for harmonic vibrations of a spherical shell can be obtained from Eqs. (9.1.1) to (9.1.3) and (13.1.4) as

$$\begin{aligned}\left(a^2\nabla^2 + 1 - \nu - \frac{1+\nu}{2}\frac{1}{\sin^2\phi}\frac{\partial^2}{\partial\theta^2} - \frac{1}{\sin^2\phi} + k_1\Omega_j^2\right)\overline{U}_{\phi j} \\ + \frac{1}{\sin\phi}\left(\frac{1+\nu}{2}\frac{\partial}{\partial\phi} - \frac{3-\nu}{2}\cot\phi\right)\overline{U}_{\theta j} \\ + \frac{\partial}{\partial\phi}\left\{1 + \nu - \frac{k}{1+k}[a^2\nabla^2 + 2 + (1 - \tfrac{9}{5}k)\Omega_j^2]\right\}\overline{W}_j = 0,\end{aligned} \tag{13.5.1a}$$

$$\begin{aligned}\frac{\partial}{\partial\theta}\left(\frac{1+\nu}{2}\frac{\partial}{\partial\phi} + \frac{3-\nu}{2}\cot\phi\right)\overline{U}_{\phi j} + \sin\phi\left[\frac{1-\nu}{2}\left(a^2\nabla^2 + 2 - \frac{1}{\sin^2\phi}\right)\right. \\ \left. + \frac{1+\nu}{2}\frac{1}{\sin^2\phi}\frac{\partial^2}{\partial\theta^2} + k_1\Omega_j^2\right]\overline{U}_{\theta j} + \frac{\partial}{\partial\theta}\left\{1 + \nu\right. \\ \left. - \frac{k}{1+k}\left[a^2\nabla^2 + 2 + (1 + \tfrac{9}{5}k)\Omega_j^2\right]\right\}\overline{W}_j = 0,\end{aligned} \tag{13.5.1b}$$

† R. N. Arnold and G. B. Warburton, 'Flexural Vibrations of the Walls of Thin Cylindrical Shells having Freely Supported Ends,' op. cit.; R. N. Arnold and G. B. Warburton, 'The Flexural Vibrations of Thin Cylinders,' op. cit.; V. I. Weingarten, 'Free Vibration of Thin Cylindrical Shells,' *AIAA J.*, vol. 2, no. 4, April 1964, pp. 717–722.

$$(1 - k_2\Omega_j^2)\frac{1}{\sin\phi}\left[\frac{\partial}{\partial\phi}(\bar{U}_{\phi j}\sin\phi) + \frac{\partial \bar{U}_{\theta j}}{\partial\theta}\right] + \left\{\frac{k}{1+\nu}\left[a^4\nabla^4 + \left(2 + \frac{1 - \frac{1}{5}k}{1+k}\Omega_j^2\right)a^2\nabla^2\right] + 2 - \frac{1+k}{1+\nu}\Omega_j^2\right\}\bar{W}_j = 0,$$ (13.5.1c)

where

$$a^2\nabla^2 = \frac{\partial^2}{\partial\phi^2} + \cot\phi\frac{\partial}{\partial\phi} + \frac{1}{\sin^2\phi}\frac{\partial^2}{\partial\theta^2}, \tag{13.5.2a}$$

$$k = \frac{h^2}{12a^2}, \tag{13.5.2b}$$

$$k_1 = \frac{1 + 6k + \frac{9}{5}k^2}{1+k}, \tag{13.5.2c}$$

$$k_2 = \frac{2k(1 - \frac{3}{5}k)}{(1+\nu)(1+k)}, \tag{13.5.2d}$$

$$\Omega_j = \omega_j a\left[\frac{(1-\nu^2)\rho}{E}\right]^{1/2}, \tag{13.5.2e}$$

and the displacements have been expressed as

$$u_\phi = \bar{U}_{\phi j}(\phi, \theta)(a_j\cos\omega_j t + b_j\sin\omega_j t), \tag{13.5.3a}$$

$$u_\theta = \bar{U}_{\theta j}(\phi, \theta)(a_j\cos\omega_j t + b_j\sin\omega_j t), \tag{13.5.3b}$$

$$w = \bar{W}_j(\phi, \theta)(a_j\cos\omega_j t + b_j\sin\omega_j t). \tag{13.5.3c}$$

The solution of the equations is made much simpler if they can be uncoupled. To this end we eliminate w from Eqs. (13.5.1a) and (13.5.1b) to obtain, after much manipulation,

$$\left[\frac{1-\nu}{2}\left(a^2\nabla^2 + 2 + \frac{1}{\sin^2\phi} - 2\cot\phi\frac{\partial}{\partial\phi}\right) + k_1\Omega_j^2\right] \times \left[\frac{\partial\bar{U}_{\phi j}}{\partial\theta} - \frac{\partial(\bar{U}_{\theta j}\sin\phi)}{\partial\phi}\right] = 0. \tag{13.5.4}$$

We now introduce the auxiliary variables U_j and Ψ_j such that

$$\bar{U}_{\phi j} = \frac{\partial U_j}{\partial\phi} - \Psi_j\sin\phi, \tag{13.5.5a}$$

$$\bar{U}_{\theta j} = \frac{1}{\sin\phi}\frac{\partial U_j}{\partial\theta}. \tag{13.5.5b}$$

Then the substitution of Eqs. (13.5.5) into Eqs. (13.5.4) and integration with respect to θ yields the following equation, apart from an arbitrary function of ϕ and t which can be set equal to zero without loss of generality

$$\left(a^2\nabla^2 + 2 + \frac{2}{1-\nu} k_1\Omega_j^2\right)\Psi_j = 0. \tag{13.5.6}$$

To obtain a second uncoupled equation we substitute Eqs. (13.5.5) into Eq. (13.5.1b) and integrate the result with respect to θ to yield

$$(a^2\nabla^2 + 1 - \nu + k_1\Omega_j^2)U_j - \left(\frac{1+\nu}{2}\sin\phi\frac{\partial}{\partial\phi} + 2\cos\phi\right)\Psi_j + \left\{1 + \nu - \frac{k}{1+k}[a^2\nabla^2 + 2 + (1 + \tfrac{9}{5}k)\Omega_j^2]\right\}\overline{W}_j = 0. \tag{13.5.7a}$$

The substitution of Eqs. (13.5.5) into Eq. (13.5.1c) yields

$$(1 - k_2\Omega_j^2)\left[a^2\nabla^2U_j - \frac{1}{\sin\phi}\frac{\partial(\Psi_j\sin^2\phi)}{\partial\phi}\right] + \left\{\frac{k}{1+\nu}\left[a^4\nabla^4 + \left(2 + \frac{1-\frac{1}{5}k}{1+k}\Omega_j^2\right)a^2\nabla^2\right] + 2 - \frac{1+k}{1+\nu}\Omega_j^2\right\}\overline{W}_j = 0. \tag{13.5.7b}$$

By eliminating $a^2\nabla^2U_j$ from Eqs. (13.5.7a) and (13.5.7b) we then obtain an expression for U_j in terms of Ψ_j and $\overline{W}_j$ as

$$U_j = \frac{[k(1+k)a^4\nabla^4 + \beta_{1j}ka^2\nabla^2 + \beta_{2j}]\overline{W}_j}{(1 - \nu + k_1\Omega_j^2)(1 - k_2\Omega_j^2)(1+\nu)(1+k)} - \frac{\sin\phi(\partial\Psi_j/\partial\phi)}{2\{1 + [k/(1-\nu)]\Omega_j^2\}}, \tag{13.5.8a}$$

with

$$\beta_{1j} = 3 + \nu + 2k + \frac{1 - \frac{6}{5}k + k^2}{1+k}\Omega_j^2, \tag{13.5.8b}$$

$$\beta_{2j} = (1+\nu)[1 - \nu + (3-\nu)k] - \frac{1 - 3\nu k + [(17 - 18\nu)/5]k^2 - [(10 + 3\nu)/5]k^3}{1-k}\Omega_j^2 - \frac{2k^2(1 - \frac{3}{5}k)(1 + \frac{9}{5}k)}{1+k}\Omega_j^4. \tag{13.5.8c}$$

We now take note of the identity

$$a^2\nabla^2\left(\sin\phi\frac{\partial\Psi_j}{\partial\phi}\right) \equiv \frac{1}{\sin\phi}\frac{\partial}{\partial\phi}(\sin^2\phi\, a^2\nabla^2\Psi_j). \tag{13.5.9a}$$

But, from Eq. (13.5.6), we have

$$a^2\nabla^2\Psi_j = -2\left(1 + \frac{k_1}{1-\nu}\Omega_j^2\right)\Psi_j. \tag{13.5.9b}$$

Then

$$a^2\nabla^2\left(\sin\phi\,\frac{\partial\Psi_j}{\partial\phi}\right) = -\frac{2\{1 + [k_1/(1-\nu)]\Omega_j^2\}}{\sin\phi}\,\frac{\partial(\Psi_j\sin^2\phi)}{\partial\phi}. \tag{13.5.9c}$$

Thus if we operate on Eq. (13.5.8a) with the operator $a^2\nabla^2$ we have

$$a^2\nabla^2 U_j - \frac{1}{\sin\phi}\,\frac{\partial(\Psi_j\sin^2\phi)}{\partial\phi} = \frac{[k(1+k)a^6\nabla^6 + \beta_{1j}ka^4\nabla^4 + \beta_{2j}a^2\nabla^2]\overline{W}_j}{(1-\nu+k_1\Omega_j^2)(1-k_2\Omega_j^2)(1+\nu)(1+k)}. \tag{13.5.10}$$

On noting that the left side of Eq. (13.5.10) is of the same form as the terms involving U_j and Ψ_j in Eq. (13.5.7b), we may eliminate these terms to obtain an equation for $\overline{W}_j$ alone as

$$[k(1-k)a^6\nabla^6 + \alpha_{1j}ka^4\nabla^4 + \alpha_{2j}a^2\nabla^2 + \alpha_{3j}]\overline{W}_j = 0, \tag{13.5.11a}$$

where

$$\alpha_{1j} = 4 + (3-\nu)k + \frac{2 + \frac{29}{5}k + \frac{44}{5}k^2 + \frac{9}{5}k^3}{1+k}\,\Omega_j^2, \tag{13.5.11b}$$

$$\begin{aligned}\alpha_{2j} = {} & 1 - \nu^2 + (5-\nu^2)k + 2(1-\nu)k^2 \\ & - \frac{1 - (3+2\nu)k - [(57+14\nu)/5]k^2 - [(87+4\nu)/5]k^3 - \frac{18}{5}k^4}{1-k}\,\Omega_j^2 \\ & + \frac{k(1 + \frac{19}{5}k - \frac{9}{5}k^2 + \frac{9}{5}k^3)}{1+k}\,\Omega_j^4,\end{aligned} \tag{13.5.11c}$$

$$\begin{aligned}\alpha_{3j} = {} & 2(1-\nu^2)(1+k) + \left[1 + 3\nu + 2(5+7\nu)k + \frac{13+23\nu}{5}k^2\right]\Omega_j^2 \\ & - (1+k)(1+6k+\tfrac{9}{5}k^2)\Omega_j^4.\end{aligned} \tag{13.5.11d}$$

Since the coefficients of Eq. (13.5.11a) are constants, the equation can be factored into the form

$$(a^2\nabla^2 - \gamma_{1j})(a^2\nabla^2 - \gamma_{2j})(a^2\nabla^2 - \gamma_{3j})\overline{W}_j = 0, \tag{13.5.12a}$$

where γ_{1j}, γ_{2j}, and γ_{3j} are the three independent solutions of the cubic equation

$$k(1+k)\gamma_j^3 + \alpha_{1j}k\gamma_j^2 + \alpha_{2j}\gamma_j + \alpha_{3j} = 0. \tag{13.5.12b}$$

Then the general solution for $\overline{W}_j$ is the sum of the solutions of the three equations

$$(a^2\nabla^2 - \gamma_{1j})\overline{W}_{1j} = 0, \tag{13.5.13a}$$

$$(a^2\nabla^2 - \gamma_{2j})\overline{W}_{2j} = 0, \tag{13.5.13b}$$

$$(a^2\nabla^2 - \gamma_{3j})\overline{W}_{3j} = 0, \tag{13.5.13c}$$

which are of the same form as Eq. (13.5.6) for Ψ_j. The expression for U_j can then be written as

$$U_j = \sum_{i=1}^{i=3} \epsilon_{ij}\overline{W}_{ij} - \frac{\sin\phi}{2\{1 + [k_1/(1-\nu)]\Omega_j^2\}} \frac{\partial \Psi_j}{\partial \phi}, \tag{13.5.14a}$$

where

$$\epsilon_{ij} = \frac{k(1+k)\gamma_{ij}^2 + \beta_{1j}k\gamma_{ij} + \beta_{2j}}{(1 - \nu + k_1\Omega_j^2)(1 - k_2\Omega_j^2)(1+\nu)(1+k)}. \tag{13.5.14b}$$

With the use of Eqs. (13.5.5), (13.5.6), and (13.5.14) we can obtain expressions for the tangential displacements as

$$\overline{U}_{\phi j} = \sum_{i=1}^{i=3} \epsilon_{ij} \frac{\partial \overline{W}_{ij}}{\partial \phi} + \frac{1}{2\{1 + [k_1/(1-\nu)]\Omega_j^2\}\sin\phi} \frac{\partial^2 \Psi_j}{\partial \theta^2}, \tag{13.5.15a}$$

$$\overline{U}_{\theta j} = \frac{1}{\sin\phi}\sum_{i=1}^{i=3} \epsilon_{ij} \frac{\partial \overline{W}_{ij}}{\partial \theta} - \frac{1}{2\{1 + [k_1/(1-\nu)]\Omega_j^2\}} \frac{\partial^2 \Psi_j}{\partial \phi\, \partial \theta}. \tag{13.5.15b}$$

The simplification of the equations for the harmonic vibrations of spherical shells has been completed since we have reduced the problem to that of solving four partial differential equations of the Poisson type. Once the displacements are known stress-resultants can be found from Eqs. (9.1.2) and (9.1.3). These can be used to satisfy any prescribed homogeneous edge conditions on stress-resultants. The requirement that edge conditions be satisfied yields an equation for the frequency parameter Ω_j. The literature abounds with similar equations† which differ from those given above only in the definitions of α_{ij} and ϵ_{ij}. The cause of these discrepancies is the various approximations which have been made, generally the neglect of rotatory inertia forces and modifications of the overall force-strain relations as well as the neglect of various terms of the order of k compared to unity.

13.5.1 Vibrations of spherical shell segments

When the edges of the middle surface are parallel circumferential lines of curvature so that the shell is closed in the circumferential direction, the ex-

† See for example, A. Kalnins, 'Effect of Bending on Vibrations of Spherical Shells,' *J. Acoust. Soc. Amer.*, vol. 36, no. 1, January 1964, pp. 74–81, which contains an extensive bibliography.

pressions for the space dependent portions of the displacements and the auxiliary variables can be taken as

$$\bar{U}_{\phi n}(\phi, \theta) = U_{\phi n}(\phi)(c_n \cos n\theta + d_n \sin n\theta), \tag{13.5.16a}$$

$$\bar{U}_{\theta n}(\phi, \theta) = U_{\theta n}(\phi)(c_n \sin n\theta - d_n \cos n\theta), \tag{13.5.16b}$$

$$\bar{W}_n(\phi, \theta) = W_n(\phi)(c_n \cos n\theta + d_n \sin n\theta), \tag{13.5.16c}$$

$$U_n(\phi, \theta) = \bar{U}_n(\phi)(c_n \cos n\theta + d_n \sin n\theta), \tag{13.5.16d}$$

$$\Psi_n(\phi, \theta) = \bar{\Psi}_n(\phi)(c_n \cos n\theta + d_n \sin n\theta). \tag{13.5.16e}$$

Then Eqs. (13.5.6) and (13.5.13) may be written as

$$\frac{\mathrm{d}^2 W_{in}}{\mathrm{d}\phi^2} + \cot\phi \frac{\mathrm{d}W_{in}}{\mathrm{d}\phi} + \left[m_{in}(m_{in} + 1) - \frac{n^2}{\sin^2\phi}\right] W_{in} = 0, \quad (i = 1, 2, 3) \tag{13.5.17a}$$

$$\frac{\mathrm{d}^2\bar{\Psi}_n}{\mathrm{d}\phi^2} + \cot\phi \frac{\mathrm{d}\bar{\Psi}_n}{\mathrm{d}\phi} + \left[m_{4n}(m_{4n} + 1) - \frac{n^2}{\sin^2\phi}\right]\bar{\Psi}_n = 0 \tag{13.5.17b}$$

where

$$m_{in} = (\tfrac{1}{4} - \gamma_{in})^{1/2} - \tfrac{1}{2} \qquad (i = 1, 2, 3), \tag{13.5.17c}$$

$$m_{4n} = \{\tfrac{9}{4} - [2k_1/(1 - \nu)]\Omega_n^2\}^{1/2} - \tfrac{1}{2}. \tag{13.5.17d}$$

We have encountered equations of this form previously in the analysis of the static deformations of spherical shells and thus can immediately write the solutions in terms of associated Legendre functions as

$$W_n = \sum_{i=1}^{i=3} A_{in}P^n_{m_{in}}(\cos\phi) + B_{in}P^n_{m_{in}}[\cos(\pi - \phi)], \tag{13.5.18a}$$

$$\bar{\Psi}_n = A_{4n}P^n_{m_{4n}}(\cos\phi) + B_{4n}P^n_{m_{4n}}[\cos(\pi - \phi)], \tag{13.5.18b}$$

from which we obtain, with the use of Eq. (13.5.14),

$$\bar{U}_n = \sum_{i=1}^{i=3} \epsilon_{in}\{A_{in}P^n_{m_{in}}(\cos\phi) + B_{in}P^n_{m_{in}}[\cos(\pi - \phi)]\} - \frac{\sin\phi}{2\{1 + [k_1/(1 - \nu)]\Omega_n^2\}}\left\{A_{4n}\frac{\mathrm{d}P^n_{m_{4n}}(\cos\phi)}{\mathrm{d}\phi} + B_{4n}\frac{\mathrm{d}P^n_{m_{in}}[\cos(\pi - \phi)]}{\mathrm{d}\phi}\right\}. \tag{13.5.18c}$$

The tangential displacements can now be obtained from Eqs. (13.5.15) as

$$U_{\phi n} = \sum_{i=1}^{i=3} \epsilon_{in}\left\{A_{in}\frac{\mathrm{d}P^n_{m_{in}}(\cos\phi)}{\mathrm{d}\phi} + B_{in}\frac{\mathrm{d}P^n_{m_{in}}[\cos(\pi - \phi)]}{\mathrm{d}\phi} - \frac{n^2}{2\{1 + [k_1/(1 - \nu)]\Omega_n^2\}\sin\phi}\bar{\Psi}_n\right\}, \tag{13.5.18d}$$

$$\bar{U}_{\theta n} = -\frac{n}{\sin\phi}\,\bar{U}_n. \tag{13.5.18e}$$

A solution not included in the above analysis is that of pure circumferential motion. If we take

$$\bar{\Psi}_0'(\phi,\,\theta) = \theta\bar{\Psi}_0'(\phi), \tag{13.5.19a}$$

we find that $\bar{\Psi}_0'(\phi)$ satisfies the same equation as $\bar{\Psi}_0(\phi)$ so that we have

$$\bar{\Psi}_0'(\phi) = A_{40}'P_{m_{40}}(\cos\phi) + B_{m_{40}}'P_{m_{40}}[\cos(\pi-\phi)]. \tag{13.5.19b}$$

The expression for $\bar{U}_0'(\phi,\theta)$ is then given by $\theta\bar{U}_0'(\phi)$ where

$$\bar{U}_0'(\phi) = -\frac{\sin\phi}{2\{1+[k_1/(1-\nu)]-\Omega_0^2\}}\,A_{40}'\,\frac{\mathrm{d}P_{m_{40}}(\cos\phi)}{\mathrm{d}\phi} + B_{40}'\,\frac{\mathrm{d}P_{m_{40}}[\cos(\pi-\phi)]}{\mathrm{d}\phi}, \tag{13.5.19c}$$

and the corresponding space variation of the displacements can be found to be

$$\bar{U}_{\phi 0} = \bar{W}_0 = 0, \tag{13.5.20a}$$

$$\bar{U}_{\theta 0} = \frac{1}{\sin\phi}\,\bar{U}_0'(\phi). \tag{13.5.20b}$$

The associated Legendre functions $P_m^n(\cos\phi)$ and $P_m^n[\cos(\pi-\phi)]$ are singular at one or the other pole except when m is an integer. In this case the two functions differ only in sign so that another independent solution of Legendre's differential equation is needed. The deficiency is remedied by the associated Legendre function of the second kind $Q_m^n(\cos\phi)$, which is singular at both poles. If both poles are excluded from the region of interest, both sets of functions, either $P_m^n(\cos\phi)$ and $P_m^n[\cos(\pi-\phi)]$ or $P_m^n(\cos\phi)$ and $Q_m^n(\cos\phi)$, must be used. For spherical caps only functions regular at the poles can be used so that the coefficients B_{in} ($i = 1, 2, 3, 4$) are set equal to zero. The characteristic equation for the determination of the free vibration frequencies is determined by the satisfaction of edge conditions at the two edges of a spherical segment or at the single edge of a spherical cap.

As an example, let us consider the case of a clamped spherical cap. The boundary conditions at $\phi = \phi_0$ are

$$\bar{U}_{\phi n} = \bar{U}_{\theta n} = \bar{W}_n = \frac{\mathrm{d}\bar{W}_n}{\mathrm{d}\phi} = 0. \tag{13.5.21}$$

The substitution of Eqs. (13.5.18) into Eqs. (13.5.21) then yields

$$\begin{bmatrix} P^n_{m_{1n}}(x_0) & P^n_{m_{2n}}(x_0) & P^n_{m_{3m}}(x_0) & 0 \\ \dfrac{\mathrm{d}P^n_{m_{1n}}(x_0)}{\mathrm{d}\phi} & \dfrac{\mathrm{d}P^n_{m_{2n}}(x_0)}{\mathrm{d}\phi} & \dfrac{\mathrm{d}P^n_{m_{3n}}(x_0)}{\mathrm{d}\phi} & 0 \\ \epsilon_{1n}\dfrac{\mathrm{d}P^n_{m_{1n}}(x_0)}{\mathrm{d}\phi} & \epsilon_{2n}\dfrac{\mathrm{d}P^n_{m_{2n}}(x_0)}{\mathrm{d}\phi} & \epsilon_{3n}\dfrac{\mathrm{d}P^n_{m_{3n}}(x_0)}{\mathrm{d}\phi} & -\dfrac{n}{\sin\phi_0}P^n_{m_{4n}}(x_0) \\ \dfrac{n\epsilon_{1n}P^n_{m_{1n}}(x_0)}{\sin\phi_0} & \dfrac{n\epsilon_{2n}P^n_{m_{2n}}(x_0)}{\sin\phi_0} & \dfrac{n\epsilon_{3n}P^n_{m_{3n}}(x_0)}{\sin\phi_0} & -\dfrac{[\mathrm{d}P^n_{m_{4n}}(x_0)]/\mathrm{d}\phi}{2\{1+[k_1/(1-\nu)]\Omega^2_n\}} \end{bmatrix} \begin{bmatrix} A_{1n} \\ A_{2n} \\ A_{3n} \\ A_{4n} \end{bmatrix}$$

$$= 0, \qquad (13.5.22)$$

with $x_0 = \cos\phi_0$. The condition which determines the frequency parameters is the vanishing of the determinant of the coefficients of A_{1n}, A_{2n}, A_{3n}, and A_{4n}. Similar determinantal equations may be obtained readily for other edge conditions. For a spherical segment the determinant would be of eighth order since four conditions would have to be satisfied at each edge.

In most cases the determinantal equation for the natural frequencies must be satisfied by trial and error. The solution of any problem is then dependent upon the ability to calculate values of the associated Legendre functions with real or complex indices since these are generally untabulated. Some calculations of this type have been made for the axisymmetric vibrations of a spherical cap with various edge conditions, with the use of a slightly simplified set of expressions for the quantities γ_{in} and ϵ_{in}.† Vibration frequencies for a roller-hinged spherical cap ($w = N_\phi = M_\phi = 0$ at $\phi = \phi_0$, Fig. 13.15) are given in Table 13.2 for the case $\phi_0 = 60°$, $a/h = 20$, $\nu = 0.3$.

Table 13.2 Natural frequencies for a roller-hinged cap ($\phi_0 = 60°$, $a/h = 20$, $\nu = 0.3$, $n = 0$)

Axisymmetric mode number i	$\Omega_i/(1-\nu^2)^{1/2}$	η_i
1	0.995	0.233
2	1.381	0.656
3	2.110	0.875
4	2.546	0.005
5	3.183	0.953
6	4.563	0.985
7	5.530	0.001
8	6.235	0.990

† A. Kalnins, 'Effect of Bending on Vibrations of Spherical Shells,' op. cit.

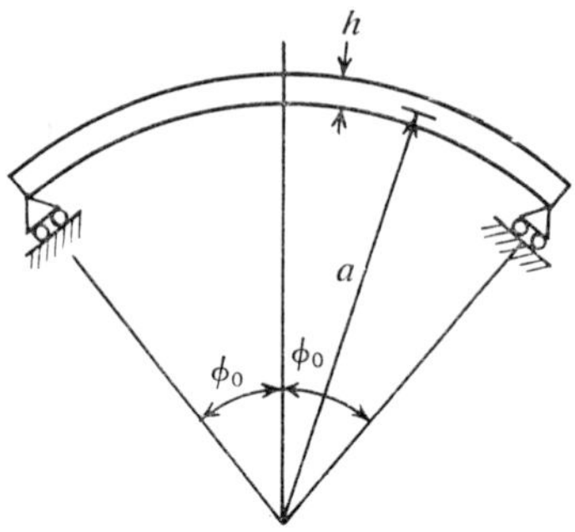

Fig. 13.15. Spherical cap with roller-hinged edge

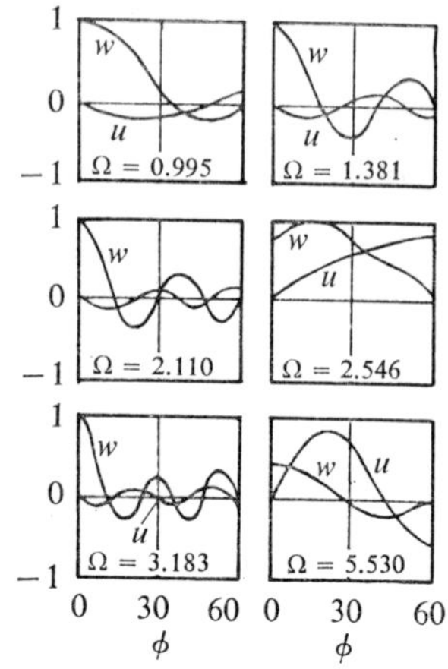

Fig. 13.16. Relative displacements of modes of vibration of a spherical cap with a roller-hinged edge ($h/a = 0.05$, $\nu = 0.3$, $\phi_0 = 60°$)

Associated mode shapes are shown in Fig. 13.16. We see from the figure that there are modes in which radial displacements predominate and the number of crossings of the $w = 0$ axis increases monotonically. There are also modes, however, for which the radial and meridional displacements are of the same order of magnitude, for $\Omega_i/(1 - \nu^2)^{1/2} = 2.546, 5.530$.

Some insight into the essential differences between the two types of modes is obtained if we calculate the ratio of the energy of bending and the total energy of the shell, given by

$$\eta = \frac{V_b}{V_s + V_b}, \tag{13.5.23a}$$

where the energy of bending is given by

$$V_b = \frac{\pi E h^3 a^2}{12(1 - \nu^2)} \int_0^{\phi_0} (\kappa_\phi^2 + 2\nu\kappa_\phi\kappa_\theta + \kappa_\theta^2) \sin\phi \, \mathrm{d}\phi, \tag{13.5.23b}$$

and the energy of stretching is given by

$$V_s = \frac{\pi E h a^2}{1 - \nu^2} \int_0^{\phi_0} (\epsilon_\phi^2 + 2\nu\epsilon_\phi\epsilon_\theta + \epsilon_\theta^2) \sin\phi \, \mathrm{d}\phi, \tag{13.5.23c}$$

with

$$\epsilon_\phi = \frac{1}{a}\left(\frac{\mathrm{d}u_\phi}{\mathrm{d}\phi} + w\right), \tag{13.5.24a}$$

$$\epsilon_\theta = \frac{1}{a}(u_\phi \cot \phi + w), \tag{13.5.24b}$$

$$\kappa_\phi = -\frac{1}{a^2}\left(\frac{d^2 w}{d\phi^2} - \frac{du_\phi}{d\phi}\right), \tag{13.5.24c}$$

$$\kappa_\theta = -\frac{1}{a^2}\left(\frac{dw}{d\phi} - u_\phi\right)\cot \phi. \tag{13.5.24d}$$

For the roller-hinged cap under discussion the values of η are also given in Table 13.2. We see that the modes of vibration with comparable radial and meridional displacements are characterized by bending energy which is small in comparison to stretching energy and are thus primarily membrane modes of vibration.† The other modes of vibration are generally characterized by bending energy which is large compared to stretching energy and can be called bending modes of vibration. We should note, however, that for this example of a relatively thick shell the modes corresponding to the lowest values of the frequency parameter are mixed modes in that the bending and stretching energies are comparable. That the classifications of bending and membrane vibrations are consistent is indicated by an investigation of the variation of the frequency parameters with the radius–thickness ratio for a spherical cap with a free edge. If a mode can be described as a membrane mode its frequency should be relatively independent of shell thickness, whereas the bending mode frequencies should be quite dependent on shell thickness. The results for the first three bending modes and the first membrane mode of spherical caps with angles ϕ_0 of 60° and 90°, respectively, which are shown in Fig. 13.17 indicate that this is so. Asymmetric modes of vibration should exhibit characteristics which are similar to those discussed above, although no exact calculations are available.

13.5.2 *Torsional vibrations of spherical caps*

A special case of vibrations for which some exact results may be obtained readily is that of pure torsional vibrations. For a spherical cap with a fixed edge ($u_\theta = 0$) the frequency equation is given by

$$P^1_{m_{40}}(\cos \phi_0) = 0, \tag{13.5.25a}$$

while for a spherical cap with a free edge ($\gamma_{\phi\theta} = 0$) the corresponding equation can be found to be given by

$$P^2_{m_{40}}(\cos \phi_0) = 0, \tag{13.5.25b}$$

† The small relative values of meridional deformations for bending modes obtained for the case $\phi_0 = 60°$ are fortuitous since for deeper shells the relative meridional displacements are considerably larger.

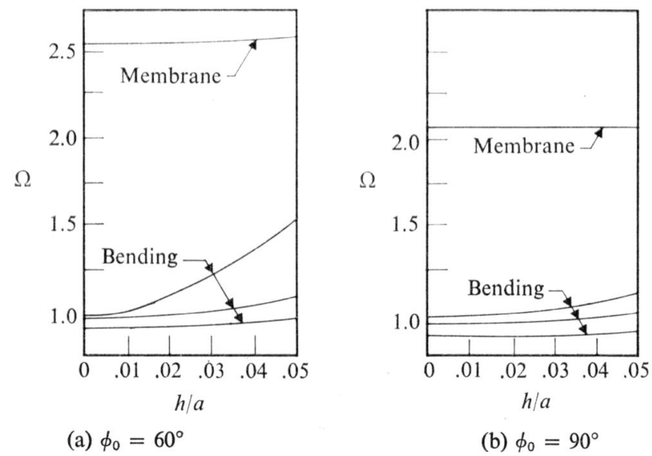

Fig. 13.17. Variation with thickness of the lowest axisymmetric vibration frequency for a free spherical cap ($\nu = 0.3$)

where the frequency of vibration is given by

$$\Omega_0 = \left[\frac{1-\nu}{2k_1}(m_{40}+2)(m_{40}-1)\right]^{1/2}. \tag{13.5.25c}$$

Some solutions of Eqs. (13.5.25a) and (13.5.25b) are given in Table 13.3. These were obtained by interpolation in tables of the associated Legendre function of integral order.†

Some additional values are also obtainable. For small values of ϕ_0 the values of m_{40} for which Eq. (13.5.25a) is satisfied are given by‡

$$m_{40} \approx \frac{x_0}{2\sin\frac{1}{2}\phi_0} - \frac{1}{2} - \frac{1}{12}x_0\left(1 - \frac{3}{x_0^2}\right)\sin\tfrac{1}{2}\phi_0 + \cdots, \tag{13.5.26a}$$

where x_0 are the roots of

$$J_1(x_0) = 0, \tag{13.5.26b}$$

the first few values of which are 3.8317, 7.0156, 10.1735, 13.3237, ...§ For

† *Tables of Associated Legendre Functions*, Mathematical Tables Project, NBS, Columbia University Press, 1945. See also Z. Mursi, *Tables of Legendre Associated Functions*, Fouad I Univ., Faculty of Science, E. & R. Schindler, 1941.

‡ E. W. Hobson, *The Theory of Spherical and Spheroidal Harmonics*, Cambridge Univ. Press, 1931, pp. 404–409.

§ E. Jahnke and F. Emde, *Tables of Functions with Formulae and Curves*, 4th Edition, Dover Publications, New York, 1945, pp. 166–168.

Table 13.3 Corresponding values of m_{40} and ϕ_0 for torsional vibrations of spherical caps
(a) Fixed edge

	Mode number								
m_{40}	1	2	3	4	5	6	7	8	9
	ϕ_0 degrees								
1	180.00								
2	90.00	180.00							
3	63.44	116.56	180.00						
4	49.10	90.00	130.90	180.00					
5	40.08	73.43	106.57	139.92	180.00				
6	33.88	62.04	90.00	117.96	146.12	180.00			
7	29.34	53.72	77.92	102.08	126.28	150.66	180.00		
8	25.87	47.38	68.71	90.00	111.29	132.62	154.13	180.00	
9	23.24	42.37	61.45	80.49	99.51	118.55	137.63	156.76	180.00
10	20.93	38.33	55.59	72.80	90.00	108.20	124.41	141.67	159.07

(b) Free edge

	Mode number								
m_{40}	1	2	3	4	5	6	7	8	9
	ϕ_0, degrees								
1	All angles								
	ϕ_0, degrees								
2		180.00							
3		90.00	180.00						
4		67.79	112.21	180.00					
5		54.74	90.00	125.26	180.00				
6		45.99	75.49	104.51	134.01	180.00			
7		39.70	65.11	90.00	114.89	140.30	180.00		
8		34.93	57.27	79.09	100.91	122.73	145.07	180.00	
9		31.20	51.14	70.63	90.00	109.37	128.86	148.80	180.00
10		28.19	46.21	63.81	81.28	98.72	116.19	133.79	151.81

values of ϕ_0 near 180°, the corresponding values of m_{40} are

$$m_{40} \approx (m+1)[1+(m+2)\tan^2 \tfrac{1}{2}(\pi-\phi_0)] \qquad (m = 0, 1, 2, \ldots). \tag{13.5.27}$$

Finally when ϕ_0 is equal to 90° the values of m_{40} are given by

$$m_{40} = 2(m-1), \qquad m = 0, 1, 2, \ldots. \tag{13.5.28}$$

Similarly for Eq. (13.5.25b) the values of m_{40} for small values of ϕ_0 are given by‡

$$m_{40} \approx \frac{\bar{x}_0}{2\sin\frac{1}{2}\phi_0} - \frac{1}{2} - \frac{1}{12}\bar{x}_0\left(1 - \frac{15}{\bar{x}_0^2}\right)\sin\tfrac{1}{2}\phi_0 + \cdots, \tag{13.5.29a}$$

where $\bar{x}_0$ are the roots of the equation

$$J_2(\bar{x}_0) = 0, \tag{13.5.29b}$$

the first few values of which are 5.136, 8.417, 11.620, 14.796, ...† When ϕ_0 is near 180° the values of m_{40} are

$$m_{40} \approx (m + 2)[1 + \tfrac{1}{2}(m + 4)(m + 3)(m + 1)\tan^4 \tfrac{1}{2}(\pi - \phi_0)] \qquad (m = 0, 1, 2, \ldots), \tag{13.5.30}$$

and when ϕ_0 is equal to 90° the values of m_{40} are

$$m_{40} = 2m + 1, \qquad m = 0, 1, 2, \ldots. \tag{13.5.31}$$

As indicated in Table 13.3b the rigid body rotation solution

$$m_{40} = 1 \quad \text{or} \quad \Omega_0 = 0, \tag{13.5.32}$$

is valid for all angles ϕ_0 of the spherical cap when the edge is free.

13.5.3 Approximate solutions for shallow spherical caps

For shallow spherical shells it is possible to make significant simplifications in the equations as was done for static deformations of spherical shells. When the angle ϕ is small and n is much smaller than γ_{in}, we may use the shallow shell approximations of replacing $\cot \phi$ by $1/\phi$ and $\sin^2 \phi$ by ϕ^2 in Eqs. (13.5.17) to obtain Bessel's equation. Then the associated Legendre functions in the determinantal equation are replaced by Bessel functions of nth order of real or complex argument, for which some tables exist.‡ An alternate simplification is to replace $\cot \phi$ by $1/\sin \phi$, ϕ by $\sin \phi$, and to take the independent variable as

$$r = a \sin \phi. \tag{13.5.33}$$

This procedure leads to the same equations, with the exception that the argument of the Bessel functions involves the sin of the angle rather than the angle. The attractiveness of this version of the theory is its resemblance to the theory of flat circular plates.

Although the two shallow shell simplifications are similar, the first appears to be valid over a much wider range of angles. We can show this extension of the range of validity by reinvestigating the torsional vibrations of a spherical cap with either a fixed edge or a free edge. For a spherical cap with a fixed edge the shallow shell approximations lead to the following torsional frequency equation, the solutions of which have been previously discussed:

$$J_1(x_0) = 0, \tag{13.5.34a}$$

† E. Jahnke and F. Emde, ibid.

‡ See M. Abramowitz and I. A. Stegun, editors, *Handbook of Mathematical Functions*, National Bureau of Standards Applied Maths., Series No. 55, 1964, for some tables and an extensive bibliography.

where

$$x_0 = \left[2\left(1 + \frac{k_1}{1-\nu}\Omega_0^2\right)\right]^{1/2}\begin{cases}\phi_0 \\ \text{or} \\ \sin\phi_0.\end{cases} \tag{13.5.34b}$$

Although either of the definitions of x_0 is valid for small ϕ_0, the comparison of exact and approximate results shown in Fig. 13.18 indicates that the use of the first definition of x_0 gives fundamental frequencies which are accurate for ϕ_0 as large as 90°, where the error is less than 1 percent, while the use of the second definition gives torsional frequencies which are accurate only for values of ϕ_0 less than about 30°. The agreement between the exact and approximate results is similar for the more complex modes of torsional vibration of a spherical cap with a fixed edge.

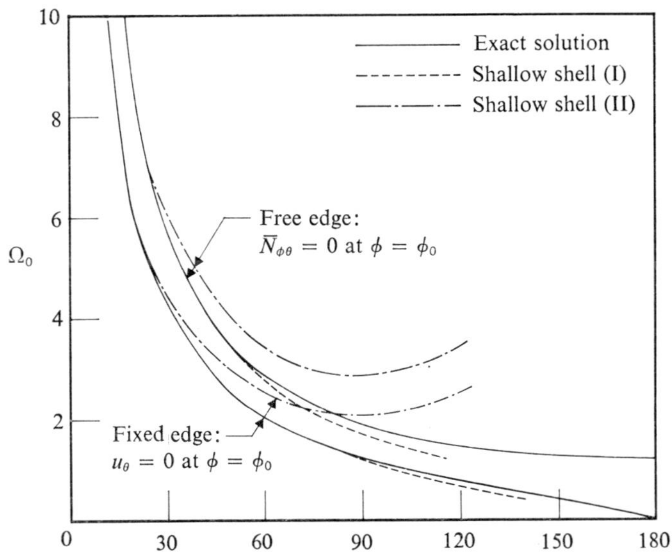

Fig. 13.18. Comparison of exact and approximate lowest torsional frequencies for a spherical cap ($\nu = 0.3$)

When the spherical cap has a free edge the shallow shell approximations can be shown to yield the torsional frequency equation

$$J_2(x_0) = 0, \tag{13.5.35}$$

whose solutions have also been given previously. The comparison of exact and approximate torsional frequencies shown in Fig. 13.18 for this case again indicates that the accuracy of the first definition of x_0 is better. It is interesting to note, however, that both approximations fail to yield the zero rigid body rotation frequency since a rigid body rotation is not a solution of the approximate shallow shell equations as outlined above.

Other natural frequencies and mode shapes for shallow spherical caps with a clamped edge have also been calculated.† The shallow shell approximations used neglect various small terms in Eqs. (13.5.11b) to (13.5.11d) and in Eqs. (13.5.14b) as well as using the second of the independent variable approximations. The results of the investigation are given in Table 13.4 in a form which uses the semi-chord length r_0 as a representative length in the

Table 13.4 Values of $\dfrac{\Omega_n\, r_0/a}{(1-\nu^2)^{1/2}}$ for clamped spherical shells ($\nu = 0.3$)

	$r_0/h = 5$		10				20	
n	$r_0/a = 0$	0.5	0.0	0.1	0.3	0.5	0.0	0.5
	0.6182	0.9291	0.3091	0.3386	0.5205	0.7515	0.1545	0.6561
	2.376	2.376	1.204	1.205	1.247	1.331	0.6020	0.8675
	2.408	2.457	2.376	2.376	2.376	2.376	1.348	1.449
0	4.017	4.058	2.696	2.698	2.711	2.737	2.376	2.376
	4.351	4.351	4.017	4.018	4.032	4.060	2.394	2.444
	5.392	5.410	4.351	4.351	4.351	4.351	3.738	3.757
	6.309	6.309	4.787	4.789	4.796	4.814	4.017	4.077
	1.287	1.346	0.6434	0.6520	0.7140	0.8195	0.3217	0.6161
	2.052	2.110	1.841	1.841	1.844	1.852	0.9205	1.047
	3.332	3.333	2.052	2.054	2.087	2.151	1.816	1.857
1	3.682	3.716	3.332	3.332	3.332	3.334	2.052	2.135
	5.250	5.254	3.632	3.637	3.648	3.667	3.012	3.056
	5.628	5.655	5.250	5.251	5.253	5.255	3.332	3.336
	7.261	7.264	5.628	5.629	5.634	5.650	4.506	4.529
	2.112	2.144	1.056	1.060	1.099	1.168	0.5280	0.7361
	3.195	3.222	2.560	2.562	2.571	2.589	1.280	1.368
2	4.283	4.288	3.195	3.196	3.208	3.230	2.327	2.373
	5.120	5.137	4.283	4.283	4.284	4.288	3.195	3.221
	6.142	6.144	4.655	4.657	4.664	4.683	3.672	3.713
	3.096	3.103	1.548	1.550	1.572	1.622	0.7740	0.9293
	4.154	4.172	3.360	3.361	3.369	3.383	1.680	1.751
3	5.266	5.275	4.154	4.155	4.160	4.173	2.879	2.919
	6.720	6.724	5.266	5.267	5.269	5.273	4.154	4.160
	7.011	7.019	5.759	5.760	5.766	5.780	4.374	4.410

† A. Kalnins, 'Free Nonsymmetric Vibrations of Shallow Spherical Shells,' *Proc. Fourth U.S. Nat. Cong. Appl. Mech.*, 1962, pp. 225–233.

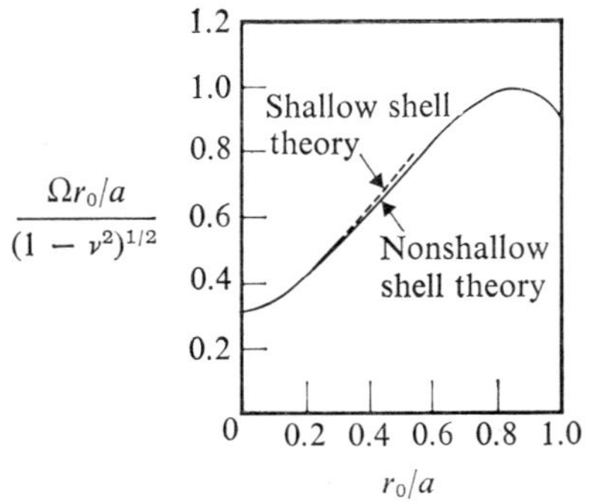

Fig. 13.19. Lowest axisymmetric vibration frequency for a spherical cap with a clamped edge ($r_0/h = 10$, $\nu = 0.3$)

frequency parameter and enables us to compare the calculated value with that for a flat circular plate having a radius r_0. A comparison with more exact values of the lowest vibration frequency thus calculated for a spherical cap with $r_0/h = 10$ and with the cap angle ϕ_0 varying from 0° to 90°, shown in Fig. 13.19, indicates the accuracy of the approximate solution.

Some of the calculated mode shapes for shallow shells are shown in Fig. 13.20. We note that there are a number of modes for which the tangential displacements are relatively small compared to the radial displacement, a

Table 13.5 Effect of tangential inertia on the calculated frequency parameters $\frac{\Omega_0 r_0/a}{(1-\nu^2)^{1/2}}$ for shallow spherical caps ($r_0/a = 0.5$, $r_0/h = 10$, $\nu = 0.3$)

	Tangential inertia			Tangential inertia	
n	Included	Omitted	n	Included	Omitted
	0.7515	0.7580		1.168	1.182
	1.331	1.339		2.589	2.611
	2.376	—	2	3.230	—
0	2.737	2.749		4.288	—
	4.060	—		4.683	4.683
	4.351	—			
	4.814	4.815			
	0.8195	0.8412		1.622	1.633
	1.852	1.913		3.383	3.399
	2.151	—	3	4.173	—
1	3.334	—		5.273	—
	3.667	3.670		5.780	5.783
	5.255	—			
	5.650	—			

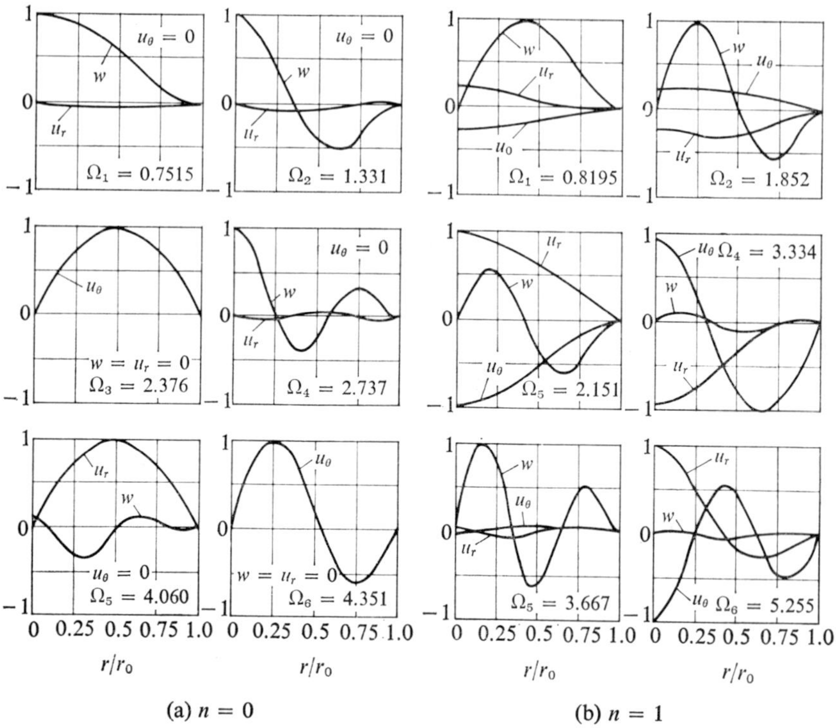

Fig. 13.20. Relative displacements for various vibration modes ($\nu = 0.3$, $r_0/a = 0.5$, $r_0/h = 10$)

circumstance which suggests the further simplification of omitting tangential inertia terms from the formulation of the problem of calculating these modes. A comparison of some natural frequencies calculated with and without tangential inertia terms is shown in Table 13.5.† We see that the modes obtained when tangential inertia is omitted do correspond to those for which the tangential deformations are small and that the error in the calculated frequencies is usually less than 1 percent. For the first two modes for $n = 1$ the error is larger, on the order of 3 percent or so, since the tangential displacements for these modes are not so insignificant as for the others. It should be noted, however, that the results shown are for a spherical cap having the relatively large included angle of about 60°. For shallower spherical caps the agreement would be considerably better.

† A. Kalnins, 'Free Nonsymmetric Vibrations of Shallow Spherical Shells,' op. cit.

13.5.4 *Vibrations of complete spherical shells*

The simplest problem of spherical shell theory is that of the free vibrations of a spherical shell which is unrestrained at every point and has no openings at all (see Fig. 13.21). For this body the displacements must be finite at both

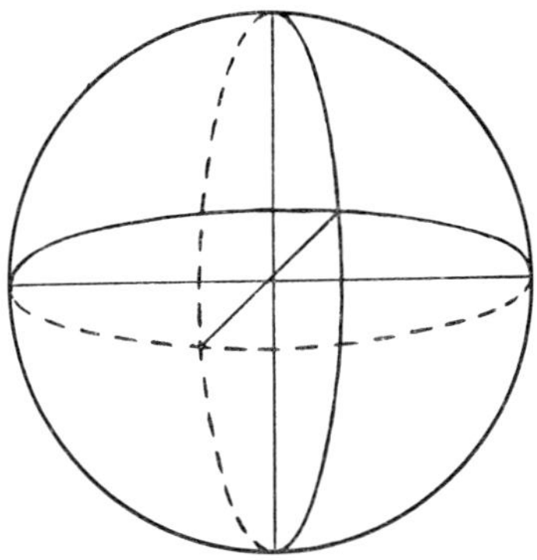

Fig. 13.21. Complete spherical shell

poles. The condition is satisfied for the radial displacements if the order m_{in} of one of the associated Legendre functions in Eq. (13.5.18a) is an integer m and the other functions vanish. We then have

$$\overline{W}_{mn} = A_{mn}P_m^n(\cos\phi). \qquad (13.5.36)$$

The corresponding value of γ_{mn} is

$$\gamma_{mn} = -\, m(m-1) = -\,\eta, \qquad (13.5.37)$$

and Eq. (13.5.12b) becomes an equation for Ω_{mn}, for if we rearrange terms and use Eq. (13.5.37), Eq. (13.5.12b) may be written as

$$\left[1 + (\eta+8)k + \frac{19\eta+74}{5}k^2 - \frac{3(3\eta-16)}{5}k^3 + \frac{9(\eta+1)}{5}k^4\right]\Omega_{mn}^4$$
$$-\left\{\eta + 1 + 3\nu + [2\eta^2 - (3+2\nu)\eta + 11 + 17\nu]k\right.$$
$$+\frac{29\eta^2 - (57+14\nu)\eta + 3(21+31\nu)}{5}k^2$$
$$\left.+\frac{44\eta^2 - (87+4\nu)\eta + 13 + 23\nu}{5}k^3 + \frac{9\eta(\eta-2)}{5}k^4\right\}\Omega_{mn}^2$$
$$+ (1+k)(\eta-2)$$
$$\times [1 - \nu^2 + (\eta-1+\nu)(\eta-1-\nu)k + \eta(\eta-1+\nu)k^2] = 0. \qquad (13.5.38)$$

It remains to be seen if the other displacements are regular at the poles. If we assume in addition that only the constant of integration A_{mn} is different from zero in the expression for Ψ_n, the remaining displacements are given by

$$\bar{U}_{\phi mn} = \epsilon_{mn} A_{mn} \frac{\mathrm{d}P_m^n(\cos\phi)}{\mathrm{d}\phi}, \tag{13.5.39a}$$

$$\bar{U}_{\theta mn} = -\, n \csc\phi\, \epsilon_{mn} A_{mn} P_m^n(\cos\phi). \tag{13.5.39b}$$

For integral m the function $P_m^n(\cos\phi)$ can be expressed as

$$P_m^n(z) = \frac{(-1)^m}{2^m\, m!} (1 - z^2)^{(1/2)n} \frac{\mathrm{d}^{m+n}(1 - z^2)^m}{\mathrm{d}z^{m+n}}. \tag{13.5.40a}$$

The derivative with respect to ϕ is then given by

$$\frac{\mathrm{d}P_m^n(\cos\phi)}{\mathrm{d}\phi} = -(1 - z^2)^{1/2} \frac{\mathrm{d}P_m^n(z)}{\mathrm{d}z}, \tag{13.5.40b}$$

which can be seen to be everywhere finite. The function in Eq. (13.5.39b)

$$\csc\phi\, P_m^n(\cos\phi) = \frac{(-1)^m}{2^m\, m!} (1 - z^2)^{(1/2)(n-1)} \frac{\mathrm{d}^{m+n}(1 - z^2)^m}{\mathrm{d}z^{m+n}}, \tag{13.5.40c}$$

is finite at the poles if $n \geq 1$ and is multiplied by zero when $n = 0$. Thus all of the finiteness conditions are satisfied.

Eq. (13.5.38) defines two real positive values of Ω_{mn} for each value of m.† If m is such that km^4 is very much less than unity the frequency parameters given by Eq. (13.5.38) are quite insensitive to the shell thickness, indicating that the energy of stretching is very much greater than the energy of bending. Approximate values of the frequency parameters are given in this case by

$$\left.\begin{matrix}\Omega_{n1}\\ \Omega_{n2}\end{matrix}\right\} = \left\langle \tfrac{1}{2}\Big[(m + 2)(m - 1) + 3(1 + \nu) \mp \{[(m + 2)(m - 1) + 3(1 + \nu)]^2 - 4(1 - \nu^2)(m + 2)(m - 1)\}^{1/2}\Big]\right\rangle^{1/2}. \tag{13.5.41}$$

† The case $m = 0$ (pure radial vibrations) is an exception since one of the two roots is imaginary. We can show, however, that the corresponding solution is spurious and thus should not be considered. The vibration frequency is given by the simple expression [see Eq. (13.5.16)]

$$\Omega = \left[\frac{2(1 + \nu)}{1 + k}\right]^{1/2}.$$

Some values of the frequency parameters defined by Eq. (13.5.41) are given in the first two columns of values in Table 13.6 and are shown in Fig. 13.22. For much greater values of m or k the larger of the two roots, Ω_{n2}, is still insensitive to shell thickness. The values of the smaller root will approach

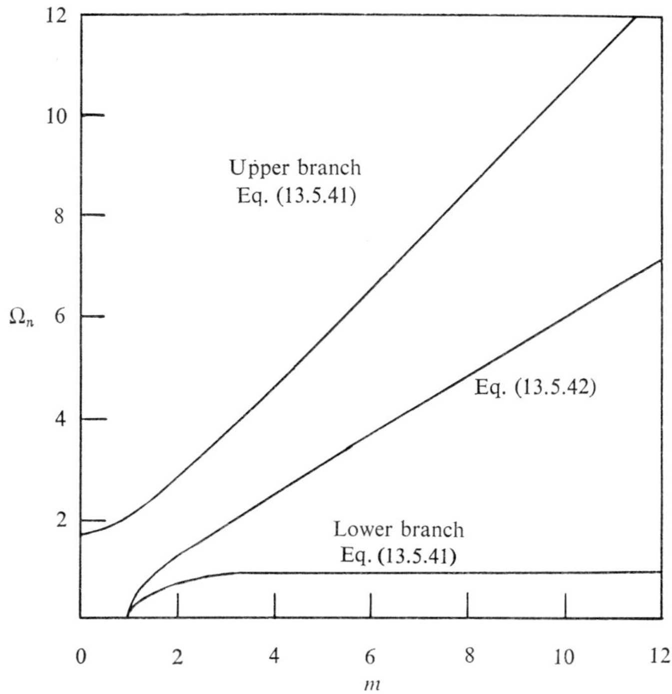

Fig. 13.22. Frequency variation for a thin closed spherical shell ($m \geqslant n$, $m^4 \ll 1/k$ $\nu = \frac{1}{3}$)

those of the larger root and thus will differ significantly from those given by Eq. (13.5.41). The behavior of the roots is illustrated in Fig. 13.23.† It is at first surprising that the parameter n, the number of circumferential waves of vibration, nowhere appears in Eq. (13.5.38) for the frequency parameters. This phenomenon is a consequence of the spherical symmetry of the shell which permits the superposition of vibration modes which are axisymmetric about different axes to yield nonsymmetric modes having the same frequency. The dependence of the frequency parameters upon n is implied, however, by the modes of vibration for, from Eq. (13.5.40a), $P_m^n(\cos \phi)$ is

† A. Kalnins, 'Effect of Bending on Vibrations of Spherical Shells,' op. cit.

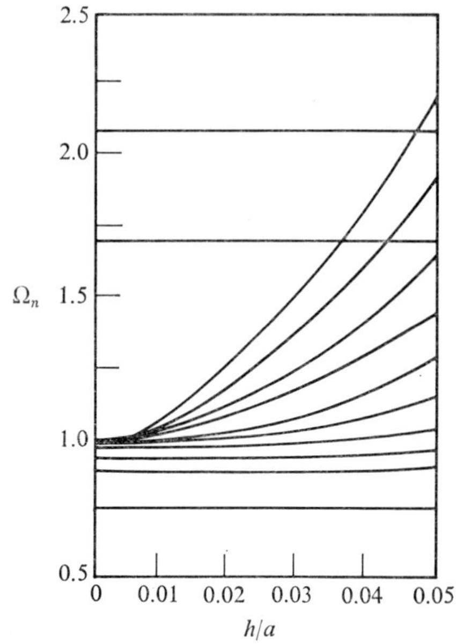

Fig. 13.23. Variation of natural vibration frequencies for a closed spherical shell

Table 13.6 Vibration frequencies of thin complete spherical shells ($\nu = 0.3$, $k \ll 1/m^4$)

m	Ω_{1n}	Ω_{2n}	Ω_{3n}
0	—	1.612	—
1	0.000	1.975	0.000
2	0.701	2.722	1.183
3	0.830	3.635	1.871
4	0.881	4.596	2.510
5	0.905	5.575	3.130
6	0.919	6.562	3.742
7	0.928	7.552	4.347
8	0.934	8.546	4.950
9	0.938	9.540	5.550
10	0.941	10.536	6.148
11	0.943	11.533	6.745
12	0.945	12.530	7.342
m very large	0.954	$m - \frac{1}{2}$	$0.592\,(m - \frac{1}{2})$

nonvanishing only if $m + n$ is less than or equal to $2m$ or if n is less than or equal to m. Thus for a given value of n only the vibration frequencies for those values of m greater than or equal to n are valid. It is interesting to note, then, that the vibration frequency for a given value of m will be the same for $m + 1$ distinct mode shapes. With $m = 2$ and $\Omega_n = 0.701$ the variation with ϕ of the deformations for $n = 0$, 1, and 2 is shown in Fig. 13.24 as an example. Other mode shapes may be obtained with the aid of tables of the associated Legendre functions or by direct calculation by means of tables of trigonometric functions and the function definitions.

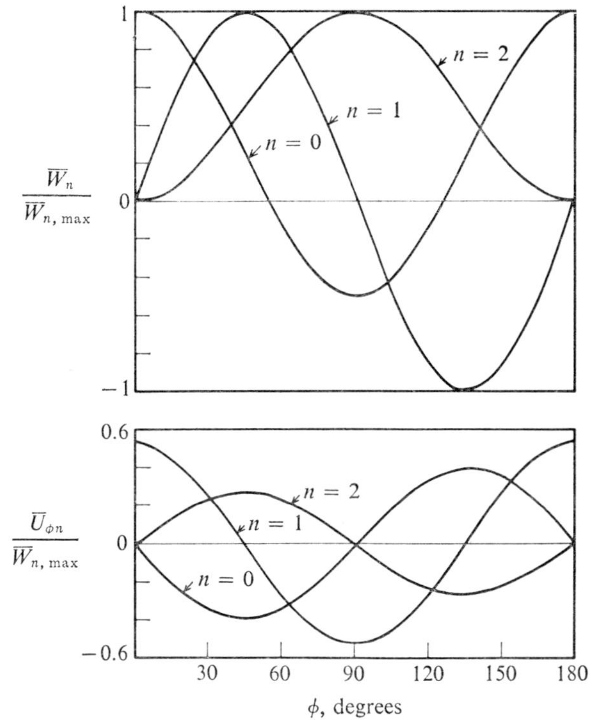

Fig. 13.24. Meridional variation of vibration mode patterns having the same frequency $(m = 2, \Omega_{mn} = 0.701)$

Another set of frequency parameters and associated mode shapes is obtained if we assume that the arbitrary constant A_{4n} is the only nonzero constant and that m_{4n} is an integer m. In this case we obtain from Eq. (13.5. 17d)

$$\Omega_{3n} = \left[\frac{1 - \nu}{2k_1}(m + 2)(m - 1)\right]^{1/2}. \tag{13.5.42}$$

The frequency parameters can be seen to be insensitive to shell thickness for all values of m and, again, are dependent on the circumferential wave number

n by the condition that m be less than or equal to n. Some values of Ω_{3n} are given in the third column of values in Table 13.6 and are shown in Fig. 13.22. The corresponding mode shapes can be readily obtained from Eqs. (13.5.18) as

$$\overline{W}_n = 0, \tag{13.5.43a}$$

$$\overline{U}_{\phi mn} = -n \csc\phi\, A_{mn} P_m^n(\cos\phi), \tag{13.5.43b}$$

$$\overline{U}_{\theta mn} = \frac{1}{m(m+1)} A_{mn} \frac{\mathrm{d}P_m^n(\cos\phi)}{\mathrm{d}\phi}, \tag{13.5.43c}$$

which is valid for the case of pure torsion as well.

Comparisons of the above frequencies and the corresponding frequencies obtained for a complete spherical shell of a completely isotropic material have been made.† These indicate that there is generally good agreement between the two sets of frequencies up to values of m which depend on the radius–thickness ratio of the shell. When a/h is 10, for example, the lowest branch of the corresponding frequency spectra agree up to m about 13, beyond which the isotropic shell has lower frequencies. The upper branches agree up to m of about 22. The frequencies given by Eq. (13.5.42) agree with those of isotropic elasticity theory up to the limit of $m = 25$ which was investigated. For larger values of a/h the pertinent values of m increase. The isotropic shell has other modes of vibration with higher frequencies which involve wall transverse shear and thickness deformations. These are neglected in the theory of transversely rigid shells.

13.6 Vibration characteristics of conical shells

The circular cylinder and the sphere exhaust the shell shapes for which exact solutions of free vibration problems can be found in terms of known, if not tabulated, functions. For other shells of revolution it is generally necessary to obtain approximate solutions by means of digital computer programs based on one of the many available methods of numerical analysis of shell vibration problems. Programs of this sort have been applied to the vibrations of conical shells, but primarily to obtain results for specific shells. Similarly in the one exact power series solution‡ available for axisymmetric vibrations

† A. H. Shah, C. V. Ramkrishnan, and S. K. Datta, 'Three-dimensional and Shell Theory Analysis of Elastic Waves in a Hollow Sphere; Part I: Analytical Results; Part II: Numerical Results,' *J. Appl. Mech.*, vol. 36, no. 3, September 1969, pp. 431–444.

‡ J. E. Goldberg, 'Axisymmetric Oscillations of Conical Shells,' *Proc. IXth Int. Cong. Appl. Mech.*, Brussels, 1956, pp. 333–343.

of conical shells, only a single configuration with a clamped small edge and a free large edge is investigated and the results used as a check on those of a numerical procedure.† These results, however, serve to indicate the phenomena which characterize the behavior of thin conical shells.

In Fig. 13.25 experimental and approximate theoretical results‡ are

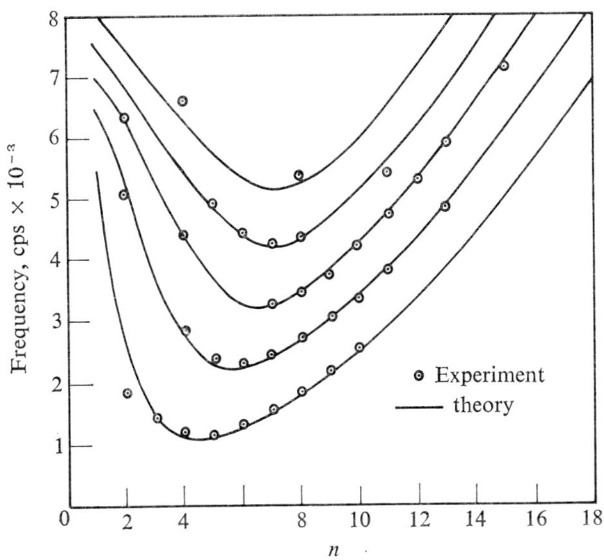

Fig. 13.25. Comparison of theoretical and experimental vibration frequencies for a conical shell

shown for the particular cone of Fig. 13.26. The edges of the cone have been assumed to be simply supported ($w = M_s = N_s = u_\theta = 0$). We see that the behavior of conical shells is similar to that of cylindrical shells in that the frequency decreases to a minimum with increasing number of circumferential waves and then increases. This phenomenon is similarly associated with a transfer of energy from stretching to bending energy and has been shown

† J. E. Goldberg, J. L. Bogdanoff, and L. Marcus, 'On the Calculation of Axi-symmetric Modes and Frequencies of Conical Shells,' *J. Acoust. Soc. Amer.*, vol. 32, 1960, pp. 738–742.

‡ P. Seide, 'On the Free Vibrations of Simply Supported Truncated Conical Shells,' *Proc. VII Israel Ann. Conf. Av. and Astro., Israel J. Tech.*, vol. 3, no. 1, 1965, pp. 50–61. See also V. I. Weingarten, 'Free Vibrations of Conical Shells,' *J. Eng. Mech. Div., Proc. ASCE*, vol. 91, no. EM 4, August 1965, pp. 70–87.

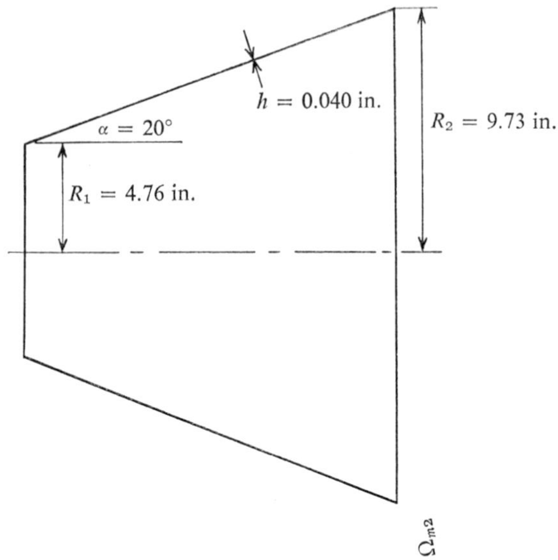

Fig. 13.26. Dimensions of cone of Fig. 13.25.

to occur for shells of positive or negative as well as zero Gaussian curvature.†

We may gain some insight into the behavior of conical shells by examining the variation of the first mode normal deflection shape with circumferential wave number (Fig. 13.27). We note that for the lower values of n the mode shapes are approximately sinusoidal with the maximum deflection occurring near the center of the generator. This circumstance suggests that the natural frequencies for this range of values of n may be similar to those of a cylindrical shell with a length equal to the slant length of the conical shell, a radius equal to the average radius of curvature of the conical shell, and the same wall thickness. Further investigation of the approximate theory used in the calculations‡ suggests that whereas the conical shell vibrates with n circumferential waves, the equivalent cylinder vibrates with $n/\cos\alpha$ circum-

† See A. L. Goldenveiser, 'Qualitative Analysis of Free Vibrations of an Elastic Thin Shell,' *Appl. Math. and Mech.* (*Trans. of Prik. Mat. i Mekh.*), vol. 30, no. 1, 1965, pp. 110–127; R. L. Carter, A. R. Robinson, and W. C. Schnobrich, 'Free Vibrations of Hyperboloidal Shells of Revolution,' *J. Eng. Mech. Div., Proc. ASCE,* vol. 95, no. EM 5, October 1969, pp. 1033–1052; D. S. Margolias and V. I. Weingarten, 'Free Vibrations of Pressure-loaded Paraboloidal Shells,' *AIAA J.*, vol. 9, no. 12, December 1971, pp. 2339–2345.

‡ P. Seide, 'On the Free Vibrations of Simply Supported Truncated Conical Shells,' op. cit.; also V. I. Weingarten, 'Free Vibrations of Conical Shells,' op. cit.

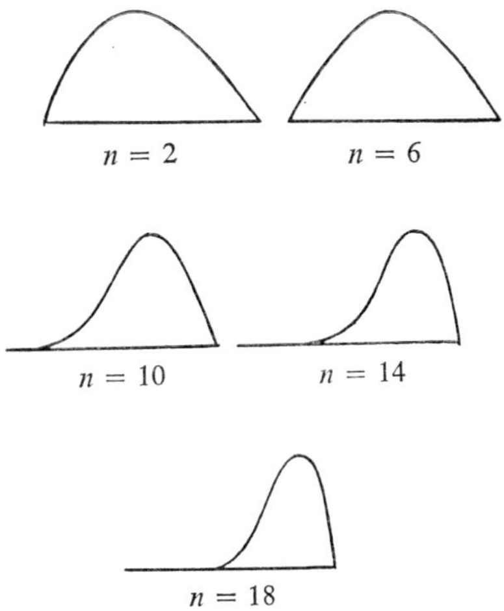

Fig. 13.27. Variation of first mode normal deflection shape with circumferential wave number

ferential waves. We may obtain an approximate expression for the frequencies for low values of n as (see Eq. (13.4.19))

$$k_c^2 = \left[\frac{2}{U(1+S)}\right]^4 \left[\left(\frac{m\pi}{2}\frac{1+S}{1-S}\right)^2 + \eta^2 - 1\right]^2 + \frac{\left[\frac{2}{U(1+S)}\right]^2 \left(\frac{m\pi}{2}\frac{1+S}{1-S}\right)^4}{\left[\left(\frac{m\pi}{2}\frac{1+S}{1-S}\right)^2 + \eta^2\right]^2}, \tag{13.6.1}$$

with

$$k_c = \left[\frac{\rho}{12(1-\nu^2)E}\right]^{1/2} \omega h \tan^2 \alpha, \tag{13.6.2a}$$

$$U = [12(1-\nu^2)]^{1/2} \frac{R_2 \cos \alpha}{h \sin^2 \alpha}, \tag{13.6.2b}$$

$$S = R_1/R_2, \tag{13.6.2c}$$

$$\eta = n/\cos \alpha, \tag{13.6.2d}$$

and m is the longitudinal mode number. A comparison of the results of Eq. (13.6.1) for m equal to unity with more exact values is shown in Fig. (13.28) for two values of the parameter U. We see that the agreement is good up to values of the circumferential wave number parameter η which are in the vicinity of the parameter for which the frequency parameter k_c is a minimum. For still larger values of η the agreement between the two sets of results becomes increasingly poor. Eq. (13.6.1) can be used for the second and higher longitudinal modes of vibration as well, but the agreement with the more exact results is not as uniformly good.

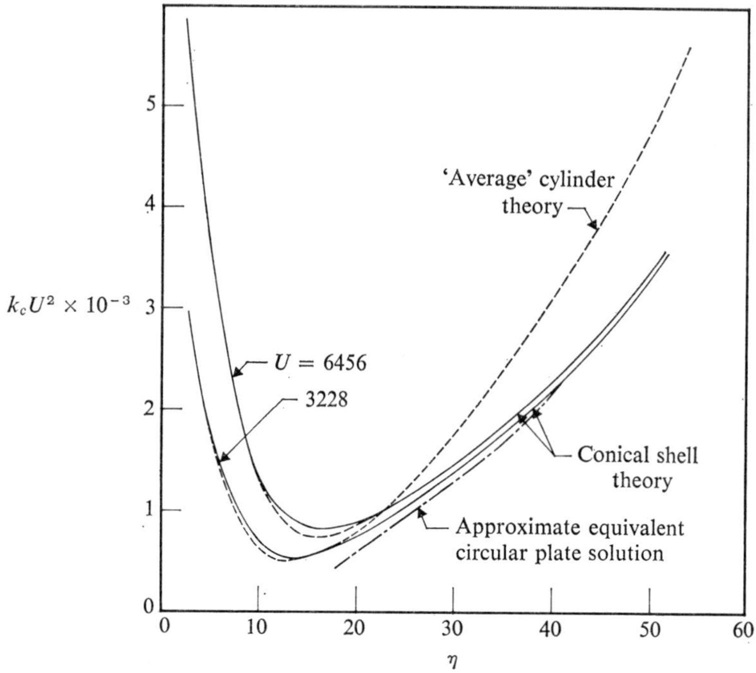

Fig. 13.28. Comparison of various methods of calculation of first mode frequency spectrum ($S = 0.438$)

The behavior of the conical shell for large values of η is explained by noting that these large values generally correspond to a vibration pattern with many circumferential waves. The portions of the cone between nodal generators can then be expected to approximate portions of flat plates. We hypothesize that stretching of the middle surface may be neglected so that the strain energy of deformation is entirely due to bending.† With this assump-

† We do not assume that the strains of the middle surface are identically zero as in the inextensional theory of shell vibrations but that the strain energy of stretching is negligible. Although inextensional theory yields reasonable results for cylindrical shells, the agreement with more accurate results for other shell shapes, including the conical shell, is poor.

tion we find that the behavior of a truncated simply supported conical shell vibrating in n circumferential waves is similar to that of a flat circular annulus plate with inner and outer radii equal to the meridional distances s_1 and s_2, respectively, to the small and large ends of the conical shell, the same thickness, and which vibrates in η circumferential waves. Some approximate results for flat plate vibration frequencies are also shown in Fig. 13.28 and can be seen in good agreement with the trends of the more exact results.

We may summarize the behavior of freely vibrating simply supported conical shells as being similar to cylindrical shells when the number of circumferential waves is small and to flat circular plates when the number of circumferential waves is large. The adequacy of these two approximations for a given value of n depends on the value of the geometry parameters U and S, the behavior being more cylinder-like as U and S increase and more plate-like as U and S decrease.

A somewhat more extensive investigation† has been made of the minimum frequency of vibration of a truncated simply supported conical shell by means of a Galerkin method (or, equivalently, the Rayleigh–Ritz method) solution of the dynamic counterpart of Vlasov's equations, with tangential and rotatory inertia neglected. The edge conditions are those of the previous discussion. In the calculations the value of η for conical shells with particular values of U and S was varied until a minimum value of the frequency parameter k_c was obtained. It was assumed that η was large enough that it could be assumed to be continuous rather than discrete. Thus for given values of

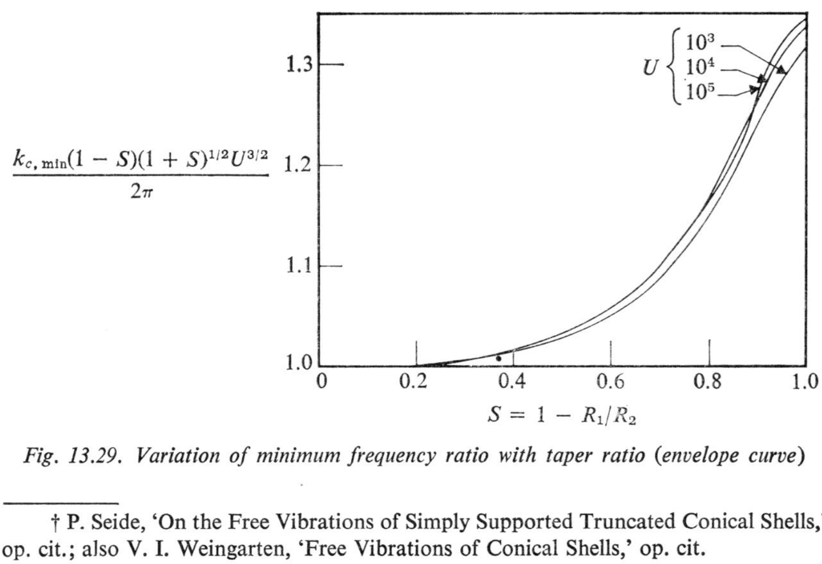

Fig. 13.29. Variation of minimum frequency ratio with taper ratio (envelope curve)

† P. Seide, 'On the Free Vibrations of Simply Supported Truncated Conical Shells,' op. cit.; also V. I. Weingarten, 'Free Vibrations of Conical Shells,' op. cit.

U and S the minimum frequency could be approximated by the minimum of the continuous curve of k_c versus η. The calculated results are shown in Fig. 13.29 where the parameter $[(1 - S)(1 - S)^{1/2}U^{3/2}k_c \min]/2\pi$ has been plotted as a function of the parameter S for various values of U. We see that in the range $U > 10^3$ there is little dependence of the modified frequency parameter on U. We may thus write an approximate expression for the minimum circular frequency as

$$\left(\frac{U\rho}{E}\right)^{1/2} \frac{R^2}{\cos\alpha} \frac{\omega_{\min}}{2\pi} = \frac{g(S)}{(1 - S)(1 + S)^{1/2}}, \tag{13.6.3}$$

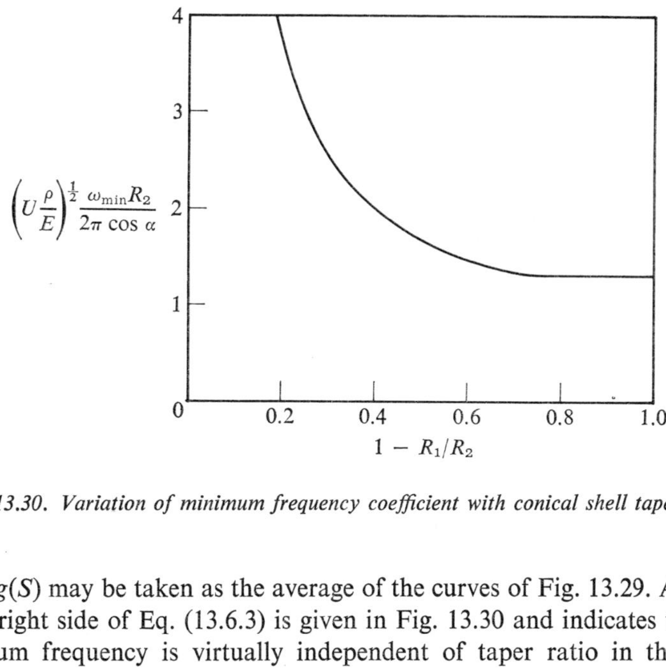

Fig. 13.30. Variation of minimum frequency coefficient with conical shell taper ratio

where $g(S)$ may be taken as the average of the curves of Fig. 13.29. A graph of the right side of Eq. (13.6.3) is given in Fig. 13.30 and indicates that the minimum frequency is virtually independent of taper ratio in the range $0 \lesssim S \lesssim 0.25$.

Results for torsional vibrations of conical shells are also available.† An extensive and quite complete survey of calculated mode shapes and frequencies of shells in the literature has recently been made by A. W. Leissa‡ and should be consulted for additional results.

† H. Garnet, M. A. Goldberg, and V. L. Salnero, 'Torsional Vibrations of Shells of Revolution,' *J. Appl. Mech.*, vol. 28, no. 4, December 1961, pp. 571–573.

‡ A. W. Leissa, 'Vibration of Shells,' *NASA* SP-288, 1973.

Supplementary references

[1] Herrmann, G., and Mirsky, I., 'Three-dimensional and Shell Theory Analysis of Axially Symmetric Motion of Cylinders,' *J. Appl. Mech.*, vol. 23, no. 4, December 1956, pp. 563–568.
[2] Baron, M. L., 'Circular-Symmetric Vibrations of Infinitely Long Cylindrical Shells with Equidistant Stiffeners,' *J. Appl. Mech.*, vol. 23, no. 2, June 1956, pp. 316–318.
[3] Baron, M. L., and Bleich, H. H., 'Tables for Frequencies and Modes of Free Vibration of Infinitely Long Thin Cylindrical Shells,' *J. Appl. Mech.*, vol. 21, no. 2, June 1954, pp. 178–184.
[4] Weingarten, V. I., 'Experimental Investigation of the Free Vibration of Multi-layered Cylindrical Shells,' *Experimental Mechanics*, vol. 4, no. 7, July 1964, pp. 200–205.
[5] Weingarten, V. I., 'Free Vibrations of Ring-Stiffened Conical Shells,' *AIAA J.*, vol. 3, no. 8, August 1965, pp. 1475–1481.
[6] Hoppmann, W. H., 'Some Characteristics of the Flexural Vibrations of Orthogonally Stiffened Cylindrical Panels,' *J. Acoust. Soc. Amer.*, vol. 30, no. 1, January 1958, pp. 70–82.
[7] Brogan, F., Forsberg, K., and Smith, S., 'Dynamic Behavior of a Cylinder with a Cutout,' *AIAA J.*, vol. 7, no. 5, May 1969, pp. 903–911.
[8] Forsberg, K., 'A Review of Analytical Methods used to determine the Modal Characteristics of Cylindrical Shells,' Lockheed Missiles and Space Co., Report 6-75-65-25, May 1965.
[9] Weingarten, V. I., and Gelman, A. P., 'Free Vibrations of Cantilevered Conical Shells,' *J. Eng. Mech. Div., ASCE*, vol. 93, no. EM 6, December 1967, pp. 127–138.
[10] Kurt, C. E., and Boyd, D. E., 'Free Vibrations of Noncircular Cylindrical Shell Segments,' *AIAA J.*, vol. 9, no. 2, February 1971, pp. 239–244.
[11] Bushnell, D., 'Stress, Buckling, and Vibration of Prismatic Shells,' *AIAA J.*, vol. 9, no. 10, October 1971, pp. 2004–2013.
[12] McDonald, D., 'A Problem in the Free Vibration of Stiffened Cylindrical Shells,' *AIAA J.*, vol. 8, no. 2, February 1970, pp. 252–258.
[13] Sewall, J. L., Thompson, W. M., Jr., and Pusey, C. G., 'An Experimental and Analytical Vibration Study of Elliptical Cylindrical Shells,' NASA TN D-6089, February 1971.
[14] Sewall, J. E., and Pusey, C. G., 'Vibration Study of Clamped-free Elliptical Cylindrical Shells,' *AIAA J.*, vol. 9, no. 6, June 1971, pp. 1004–1011.
[15] Culberson, L. D., and Boyd, D. E., 'Free Vibrations of Freely Supported Oval Cylinders,' *AIAA J.*, vol. 9, no. 8, August 1971, pp. 1474–1480.

14

Response of shells to dynamic loading

14.1 Solution of initial value problems

The orthogonality condition developed in Section 13.3 is the basis for a method of solution of dynamic loading problems called the method of modal analysis. In the method of modal analysis the displacements are expanded into infinite series of vibration mode functions. The orthogonality conditions enable us to explicitly determine the coefficients of the infinite series.

The simplest problem which can be solved by this technique is that of determining the subsequent motion of an unloaded shell when initial values of the displacements and velocities of the middle surface are given. In Section 13.2 we derived the result that free harmonic vibrations of the form

$$u_\alpha = \overline{U}_{\alpha n}(A_n \cos \omega_n t + B_n \sin \omega_n t), \tag{14.1.1a}$$

$$u = \overline{U}_{\beta n}(A_n \cos \omega_n t + B_n \sin \omega_n t), \tag{14.1.1b}$$

$$w = \overline{W}_n(A_n \cos \omega_n t + B_n \sin \omega_n t), \tag{14.1.1c}$$

will occur if the initial displacements and velocities are of the forms

$$u_\alpha\big|_{t=0} = A_n \overline{U}_{\alpha n}, \tag{14.1.2a}$$

$$u_\beta\big|_{t=0} = A_n \overline{U}_{\beta n}, \tag{14.1.2b}$$

$$w\big|_{t=0} = A_n \overline{W}_n, \tag{14.1.2c}$$

$$\left.\frac{\partial u_\alpha}{\partial t}\right|_{t=0} = \omega_n B_n \overline{U}_{\alpha n}, \tag{14.1.2d}$$

$$\left.\frac{\partial u_\beta}{\partial t}\right|_{t=0} = \omega_n B_n \overline{U}_{\beta n}, \tag{14.1.2e}$$

$$\left.\frac{\partial w}{\partial t}\right|_{t=0} = \omega_n B_n \overline{W}_n, \tag{14.1.2f}$$

In the case of arbitrary initial conditions, let the initial displacements and velocities be expressed as

$$u_{\alpha 0} = \sum_n A_n \bar{U}_{\alpha n}, \tag{14.1.3a}$$

$$u_{\beta 0} = \sum_n A_n \bar{U}_{\beta n}, \tag{14.1.3b}$$

$$w_0 = \sum_n A_n \bar{W}_n, \tag{14.1.3c}$$

$$\frac{\partial u_{\alpha 0}}{\partial t} = \sum_n \omega_n B_n \bar{U}_{\alpha n}, \tag{14.1.3d}$$

$$\frac{\partial u_{\beta 0}}{\partial t} = \sum_n \omega_n B_n \bar{U}_{\beta n}, \tag{14.1.3e}$$

$$\frac{\partial w_0}{\partial t} = \sum_n \omega_n B_n \bar{W}_n, \tag{14.1.3f}$$

where the harmonic mode functions satisfy edge conditions appropriate to the problem under consideration and the summation is over all the appropriate modal functions. We also have

$$\Theta_{\alpha 0} = \sum_n A_n \bar{\Theta}_{\alpha n}, \tag{14.1.3g}$$

$$\Theta_{\beta 0} = \sum_n A_n \bar{\Theta}_{\beta n}, \tag{14.1.3h}$$

and

$$\frac{\partial \Theta_{\alpha 0}}{\partial t} = \sum_n \omega_n B_n \bar{\Theta}_{\alpha n}, \tag{14.1.3i}$$

$$\frac{\partial \Theta_{\beta 0}}{\partial t} = \sum_n \omega_n B_n \bar{\Theta}_{\beta n}. \tag{14.1.3j}$$

We now multiply Eqs. (14.1.3a) and (14.1.3g) by $\bar{U}_{\alpha m}$ and $\bar{\Theta}_{\alpha m}$, Eqs. (14.1.3b) and (14.1.3h) by $\bar{U}_{\beta m}$ and $\bar{\Theta}_{\beta m}$, and Eq. (14.1.3c) by $\bar{W}_m$. On arranging the various equations in the manner suggested by the orthogonality condition and integrating over the middle surface of the shell, we obtain

$$\begin{aligned}\sum_n A_n \int_{\alpha_1}^{\alpha_2} \int_{\beta_1}^{\beta_2} \Bigg[& \left(1 + \frac{h^2}{12R_\alpha R_\beta}\right)(\bar{U}_{\alpha n}\bar{U}_{\alpha m} + \bar{U}_{\beta n}\bar{U}_{\beta m} + \bar{W}_n\bar{W}_m) \\ & - \frac{h^2}{12}\left(\frac{1}{R_\alpha} + \frac{1}{R_\beta}\right)(\bar{U}_{\alpha n}\bar{\Theta}_{\alpha m} + \bar{U}_{\beta n}\bar{\Theta}_{\beta m} + \bar{\Theta}_{\alpha n}\bar{U}_{\alpha m} + \bar{\Theta}_{\beta n}\bar{U}_{\beta m}) \\ & + \frac{h^2}{12}\left(1 + \frac{3}{20}\frac{h^2}{R_\alpha R_\beta}\right)(\bar{\Theta}_{\alpha n}\bar{\Theta}_{\alpha m} + \bar{\Theta}_{\beta n}\bar{\Theta}_{\beta m})\Bigg] g_\alpha g_\beta \, d\alpha \, d\beta\end{aligned}$$

$$= \int_{\alpha_1}^{\alpha_2} \int_{\beta_1}^{\beta_2} \left[\left(1 + \frac{h^2}{12R_\alpha R_\beta}\right)(u_{\alpha 0}\overline{U}_{\alpha m} + u_{\beta 0}\overline{U}_{\beta m} + w_0 \overline{W}_m) \right.$$
$$- \frac{h^2}{12}\left(\frac{1}{R_\alpha} + \frac{1}{R_\beta}\right)(u_{\alpha 0}\overline{\Theta}_{\alpha m} + u_{\beta 0}\overline{\Theta}_{\beta m} + \Theta_{\alpha 0}\overline{U}_{\alpha m} + \Theta_{\beta 0}\overline{U}_{\beta m})$$
$$\left. + \frac{h^2}{12}\left(1 + \frac{3}{20}\frac{h^2}{R_\alpha R_\beta}\right)(\Theta_{\alpha 0}\overline{\Theta}_{\alpha m} + \Theta_{\beta 0}\overline{\Theta}_{\beta m}) \right] g_\alpha g_\beta \, d\alpha \, d\beta. \tag{14.1.4}$$

From the orthogonality condition, we note that only the terms with n equal to m remain on the left side of Eq. (14.1.4) to yield

$$A_m = \frac{1}{L_m} \int_{\alpha_1}^{\alpha_2} \int_{\beta_1}^{\beta_2} \left[\left(1 + \frac{h^2}{12R_\alpha R_\beta}\right)(u_{\alpha 0}\overline{U}_{\alpha m} + u_{\beta 0}\overline{U}_{\beta m} + w_0 \overline{W}_m) \right.$$
$$- \frac{h^2}{12}\left(\frac{1}{R_\alpha} + \frac{1}{R_\beta}\right)(u_{\alpha 0}\overline{\Theta}_{\alpha m} + u_{\beta 0}\overline{\Theta}_{\beta m} + \Theta_{\alpha 0}\overline{U}_{\alpha m} + \Theta_{\alpha 0}\overline{U}_{\beta m})$$
$$\left. + \frac{h^2}{12}\left(1 + \frac{3}{20}\frac{h^2}{R_\alpha R_\beta}\right)(\Theta_{\alpha 0}\overline{\Theta}_{\alpha m} + \Theta_{\beta 0}\overline{\Theta}_{\beta m}) \right] g_\alpha g_\beta \, d\alpha \, d\beta. \tag{14.1.5}$$

An equation for $\omega_m B_m$ can be obtained similarly and is given by replacing the prescribed initial displacement and rotation distributions in Eq. (14.1.5) by the prescribed initial translational and rotational velocities. With the coefficients known it follows that the solution of the initial value problem is given by

$$u_\alpha(\alpha, \beta, t) = \sum_n U_{\alpha n}(A_n \cos \omega_n t + B_n \sin \omega_n t), \tag{14.1.6a}$$

$$u_\beta(\alpha, \beta, t) = \sum_n U_{\beta n}(A_n \cos \omega_n t + B_n \sin \omega_n t), \tag{14.1.6b}$$

$$w(\alpha, \beta, t) = \sum_n W_n(A_n \cos \omega_n t + B_n \sin \omega_n t). \tag{14.1.6c}$$

The equations for the coefficients are easily modified when the orthogonality condition given by Eq. (14.1.6) is replaced by Eq. (13.3.9) or Eq. (13.3.10).

14.2 Solution of problems involving time-dependent surface or edge loading

Problems involving motion of a shell subjected to given time dependent surface or edge forces may also be solved with the aid of the free vibration modes. To demonstrate the method of solution we start with the full principle of virtual work given by

$$\int_{\alpha_1}^{\alpha_2}\int_{\beta_1}^{\beta_2}[\delta W - (\hat{q}_\alpha \delta u_\alpha + \hat{q}_\beta \delta u_\beta + \hat{q}_\zeta\, \delta w - \hat{m}_\alpha\, \delta\Theta_\alpha - \hat{m}_\beta\, \delta\Theta_\beta)]g_\alpha g_\beta\, d\alpha\, d\beta$$
$$= \int_{\alpha_1}^{\alpha_2}\int_{\beta_1}^{\beta_2}(q_\alpha^*\, \delta u_\alpha + q_\beta^*\, \delta u_\beta + q_\zeta^*\, \delta w - m_\alpha^*\, \delta\Theta_\alpha - m_\beta^*\, \delta\Theta_\beta)g_\alpha g_\beta\, d\alpha\, d\beta$$
$$+ \int_{\beta_1}^{\beta_2}(N_\alpha^*\, \delta u_\alpha + N_{\alpha\beta}^*\, \delta u_\beta + Q_\alpha^*\, \delta w - M_\alpha^*\, \delta\Theta_\alpha - M_{\alpha\beta}^*\, \delta\Theta)g_\beta|_{\alpha=\alpha_1}^{\alpha=\alpha_2}\, d\beta$$
$$+ \int_{\alpha_1}^{\alpha_2}(N_{\beta\alpha}^*\, \delta u_\alpha + N_\beta^*\, \delta u_\beta + Q_\beta^*\, \delta w - M_{\beta\alpha}^*\, \delta\Theta_\alpha - M_\beta^*\, \delta\Theta_\beta)g\alpha|_{\beta=\beta_1}^{\beta=\beta_2}\, d\alpha. \tag{14.2.1}$$

Let the solution for the displacements be expressed as

$$u = \sum_n \overline{U}_n F_n(t), \tag{14.2.2a}$$

$$u = \sum_n \overline{U}_n F_n(t), \tag{14.2.2b}$$

$$w = \sum_n \overline{W}_n F_n(t), \tag{14.2.2c}$$

where $F_n(t)$ is an arbitrary function of time and where the harmonic mode functions satisfy edge conditions appropriate to the problem. In particular, any geometric homogeneous boundary conditions must be satisfied. If any of the edges are subjected to time dependent stress resultants, the harmonic mode functions with vanishing corresponding edge stress resultants should be chosen.† Now if we substitute Eqs. (14.2.2) into Eq. (14.2.1), making use of Eqs. (13.3.3), (13.3.4), and (13.3.7) we have

$$\sum_n \Big\{ h\Big(\frac{d^2F_n}{dt^2} + \omega_n^2 F_n\Big)L_n$$
$$- \Big[\int_{\alpha_1}^{\alpha_2}\int_{\beta_1}^{\beta_2}(q_\alpha^*\overline{U}_{\alpha n} - q_\beta^*\overline{U}_{\beta n} + q_\zeta^*\overline{W}_n - m_\alpha^*\overline{\Theta}_{\alpha n} - m_\beta^*\overline{\Theta}_{\beta n})g_\alpha g_\beta\, d\alpha\, d\beta$$
$$+ (N_\alpha^*\overline{U}_{\alpha n} + N_{\alpha\beta}^*\overline{U}_{\beta n} + Q_\alpha^*\overline{W}_n - M_\alpha^*\overline{\Theta}_{\alpha n} - M_{\alpha\beta}^*\overline{\Theta}_{\beta n})g_\beta|_{\alpha=\alpha_1}^{\alpha=\alpha_2}\, d\beta$$
$$+ (N_\beta^*\overline{U}_{\beta n} + N_{\beta\alpha}^*\overline{U}_{\alpha n} + Q_\beta^*\overline{W}_n - M_\beta^*\overline{\Theta}_{\beta n} - M_\beta^*\overline{\Theta}_{\alpha n})g_\alpha|_{\beta=\beta_1}^{\beta=\beta_2}\, d\alpha\Big]\Big\}$$
$$\times\ \delta F_n = 0. \tag{14.2.3}$$

Since the functions δF_n are independent, each coefficient of δF_n must independently vanish. The required equation for $F_n(t)$ may be written as

† This is a consequence of the requirement that the chosen functions in a Rayleigh-Ritz type of analysis be complete and capable of representing those arbitrary edge displacements which are not prescribed. The series need not converge to the (known) stress resultants.

$$\frac{d^2F_n}{dt^2} + \omega_n^2 F_n = G_n(t), \tag{14.2.4a}$$

where

$$G_n(t) = \frac{1}{\rho h L_n}\Bigg[\int_{\alpha_1}^{\alpha_2}\int_{\beta_1}^{\beta_2}(q_\alpha^*\overline{U}_{\alpha n} + q_\beta^*\overline{U}_{\beta n} + q_\zeta^*\overline{W}_n - m_\alpha^*\overline{\Theta}_{\alpha n} - m_\beta^*\overline{\Theta}_{\beta n})g_\alpha g_\beta \mathrm{d}\alpha\mathrm{d}\beta$$
$$+ \int_{\beta_1}^{\beta_2}(N_\alpha^*\overline{U}_{\alpha n} + N_{\alpha\beta}^*\overline{U}_{\beta n} + Q_\alpha^*\overline{W}_n - M_\alpha^*\overline{\Theta}_{\alpha n} - M_{\alpha\beta}^*\overline{\Theta}_{\beta n})g_\beta\Big|_{\alpha=\alpha_1}^{\alpha=\alpha_2}\mathrm{d}\beta$$
$$+ (N_\beta^*\overline{U}_{\beta n} + N_{\beta\alpha}^*\overline{U}_{\alpha n} - Q_\beta^*\overline{W}_n - M_\beta^*\overline{\Theta}_{\beta n} - M_{\beta\alpha}^*\overline{\Theta}_{\alpha n})g_\alpha\Big|_{\beta=\beta_1}^{\beta=\beta_2}\mathrm{d}\alpha\Bigg]. \tag{14.2.4b}$$

The solution of this equation subject to the conditions of zero initial displacement and zero initial velocity can readily be obtained by the method of variation of parameters as

$$F_n(t) = \frac{1}{\omega_n}\int_0^t G_n(\tau)\sin\omega_n(t-\tau)\,\mathrm{d}\tau, \tag{14.2.5}$$

which completes the solution to the problem. If initial conditions are prescribed the previous solution for the initial value problem can be superposed.

The free vibration mode shapes are so useful in dynamic problems of shell analysis that it is only natural to wonder if they might not be also useful in static analysis. In the case of static surface and edge force loading, all that is necessary to be modified in the above analysis is to assume that the functions $F_n(t)$ in Eqs. (14.2.2) are independent of time. Eq. (14.2.4a) then reduces to an explicit equation for the constants F_n as

$$F_n = \frac{1}{\rho h\omega_n^2 L_n}\Bigg[\int_{\alpha_1}^{\alpha_2}\int_{\beta_1}^{\beta_2}(q_\alpha^*\overline{U}_{\alpha n} + q_\beta^*\overline{U}_{\beta n} + q_\zeta^*\overline{W}_n$$
$$- m_\alpha^*\overline{\Theta}_{\alpha n} - m_\beta^*\overline{\Theta}_{\beta n})q_\alpha q_\beta\,\mathrm{d}\alpha\,\mathrm{d}\beta$$
$$+ \int_{\beta_1}^{\beta_2}(N_\alpha^*\overline{U}_{\alpha n} + \overline{N}_{\alpha\beta}^*\,\overline{U}_{\beta n} + Q_\alpha^*\overline{W}_n - M_\alpha^*\overline{\Theta}_{\alpha n} - M_{\alpha\beta}^*\overline{\Theta}_{\beta n})g_\beta\Big|_{\alpha=\alpha_1}^{\alpha=\alpha_2}\mathrm{d}\beta$$
$$+ \int_{\alpha_1}^{\alpha_2}(N_{\beta\alpha}^*\overline{U}_{\alpha n} + N_\beta^*\overline{U}_{\beta n} + Q_\beta^*\overline{W}_n - M_{\beta\alpha}^*\overline{\Theta}_{\alpha n} - M_\beta^*\overline{\Theta}_{\beta n})g\alpha\Big|_{\beta=\beta_1}^{\beta=\beta_2}\mathrm{d}\alpha\Bigg]. \tag{14.2.6}$$

14.3 Periodic surface and edge forces

If the forces acting on the shell are periodic, we would in general expect the motions of the shell to be roughly periodic in nature. It is possible, however, for periodic forces to lead to divergent behavior of the motions and to possible failure of the structure. A particular case of periodically varying forces indicates the importance of a knowledge of the circular frequencies of vibration of the shell. Suppose that the surface forces and edge stress resultants are such that $G_n(t)$ can be written as

$$G_n(t) = G_n \sin(\omega t + \alpha_0),$$

where G_n is a constant. Then $F_n(t)$ is given by

$$\begin{aligned} F_n(t) &= G_n \int_0^t \sin(\omega t + \alpha_0) \sin \omega_n(t - \tau)\, d\tau \\ &= \tfrac{1}{2} G_n \left\{ \frac{\sin[\omega + \omega_n)\tau - \omega_n t + \alpha_0]}{\omega + \omega_n} \right. \\ &\quad \left. - \frac{\sin[(\omega - \omega_n)\tau + \omega_n t + \alpha_0]}{\omega - \omega_n} \right\}_{\tau = 0}^{\tau = t} \\ &= \frac{G_n}{\omega^2 - \omega_n^2} [(\omega \sin \omega_n t - \omega_n \sin \omega t) \cos \alpha_0 \\ &\quad + \omega_n(\cos \omega t - \cos \omega t) \sin \alpha_0]. \end{aligned} \tag{14.3.2}$$

So long as ω is not equal to ω_n, $F_n(t)$ is bounded, and even periodic in t if the ratio of ω and ω_n is a rational number. If, however, ω is equal to ω_n both the numerator and denominator vanish and the expression is indeterminate. If we let ω differ from ω_n by a small quantity and take the limit of the expression as the quantity approaches zero we obtain

$$F_n(t) \to -\frac{G_n}{2} \left[t \cos(\omega_n t + \alpha_0) - \frac{\sin \omega_n t}{\omega_n} \cos \alpha_0 \right], \tag{14.3.3}$$

which can be seen to diverge in an oscillatory fashion. Such an unstable condition is called resonance and if not checked can lead to destruction of the structure. It should be remembered, however, that the validity of the theory is restricted to small deformations and that non-linear large deformation effects will limit the amplitude of the deformations in a resonance condition to a large but not necessarily infinite magnitude. It is also possible that the superposed motions due to initial conditions and external loading may be large, despite the smallness of the individual components, so that the avoidance of resonance may in itself not be sufficient to ensure the safety of a structure.

14.4 Solution of problems involving time-dependent edge displacements

When a shell is subjected to time-dependent edge displacements rather than time-dependent edge forces the solution of the response problem is similar but somewhat more difficult. Let us take for example the case when the edge displacements and slope normal to the edge are given as $u_\alpha^0(s, t)$, $u_\beta^0(s, t)$, $w^0(s, t)$, $\partial w^0(s, t)/(\partial n)$, where s is a measure of boundary location. We shall assume that the given values of edge deformation and their first derivative with respect to time vanish at time $t = 0$. We wish to determine the response of the shell subject to the initial conditions of zero displacement and velocity throughout the shell.

Let us define functions $u_\alpha^0(\alpha, \beta, t)$, $u_\beta^0(\alpha, \beta, t)$, $w^0(\alpha, \beta, t)$ which reduce to $u_\alpha^0(s, t)$, $u_\beta^0(s, t)$, $w^0(s, t)$ and $\partial w^0(s, t)/(\partial n)$ on the boundary. We shall take these functions to be the static solution for the shell subjected to the prescribed edge conditions at any instant of time. We next express the displacements as defined above and an auxiliary set of displacements as

$$u_\alpha(\alpha, \beta, t) = u_\alpha^0(\alpha, \beta, t) + u_\alpha'(\alpha, \beta, t), \tag{14.4.1a}$$

$$u_\beta(\alpha, \beta, t) = u_\beta^0(\alpha, \beta, t) + u_\beta'(\alpha, \beta, t), \tag{14.4.1b}$$

$$w(\alpha, \beta, t) = w^0(\alpha, \beta, t) + w'(\alpha, \beta, t), \tag{14.4.1c}$$

and substitute into Eqs. (13.1.4) and (13.1.5). The equations thus obtained can be interpreted as those for the deformations u_α', u_β', and w' of a shell subjected to the following equivalent forces and moments per unit middle surface area

$$q_{\alpha 0}^* = -h\left[\left(1 + \frac{h^2}{12R_\alpha R_\beta}\right)\frac{\partial^2 u_\alpha^0}{\partial t^2} - \frac{h^2}{12}\left(\frac{1}{R_\alpha} + \frac{1}{R_\beta}\right)\frac{\partial^2 \Theta_\alpha^0}{\partial t^2}\right], \tag{14.4.2a}$$

$$q_{\beta 0}^* = -h\left[\left(1 + \frac{h}{12R_\alpha R_\beta}\right)\frac{\partial^2 u_\alpha^0}{\partial t^2} - \frac{h^2}{12}\left(\frac{1}{R_\alpha} + \frac{1}{R_\beta}\right)\frac{\partial^2 \Theta_\beta^0}{\partial t^2}\right], \tag{14.4.2b}$$

$$q_{\zeta 0}^* = -h\left(1 + \frac{h^2}{12R_\alpha R_\beta}\right)\frac{\partial^2 w^0}{\partial t^2}, \tag{14.4.2c}$$

$$m_{\alpha 0}^* = -\frac{h^3}{12}\left[\left(\frac{1}{R_\alpha} + \frac{1}{R_\beta}\right)\frac{\partial^2 u_\alpha^0}{\partial t^2} - \left(1 + \frac{3}{20}\frac{h^2}{R_\alpha R_\beta}\right)\frac{\partial^2 \overline{\Theta}_\alpha^0}{\partial t^2}\right], \tag{14.4.2d}$$

$$m_{\beta 0}^* = -\frac{h^3}{12}\left(\left[\frac{1}{R_\alpha} + \frac{1}{R_\beta}\right)\frac{\partial^2 u_\beta^0}{\partial t^2} - \left(1 + \frac{3}{20}\frac{h^2}{R_\alpha R_\beta}\right)\frac{\partial^2 \Theta_\beta^0}{\partial t^2}\right], \tag{14.4.2e}$$

with the edge conditions

$$u_\alpha'(s, t) = u_\beta'(s, t) = w'(s, t) = \frac{\partial w'(s, t)}{\partial n} = 0.$$

But the deformations of the shell under these conditions are readily obtained from the work of the preceding section, with zero initial values of displacement and velocity, as

$$u'_\alpha(\alpha, \beta, t) = \sum_n \overline{U}_{\alpha n}(\alpha, \beta)F_n(t), \tag{14.4.4a}$$

$$u'_\beta(\alpha, \beta, t) = \sum_n \overline{U}_{\beta n}(\alpha, \beta)F_n(t), \tag{14.4.4b}$$

$$w'(\alpha, \beta, t) = \sum_n \overline{W}_n(\alpha, \beta)F_n(t), \tag{14.4.4c}$$

where the modal functions are those for free vibrations of the fully restrained shell, and

$$F_n(t) = \frac{1}{\omega_n}\int_0^t G_{n0}(\tau)\sin\omega_n(t-\tau)\,\mathrm{d}\tau, \tag{14.4.5a}$$

$$G_{n_0}(t) = \frac{1}{\rho h L_n}\iint (q^*_{\alpha 0}\overline{U}_{\alpha n} + q^*_{\beta 0}\overline{U}_{\beta n} + q^*_{\zeta 0}\overline{W}_n - m^*_{\alpha 0}\overline{\Theta}_{\alpha n} - m^*_{\beta 0}\overline{\Theta}_{\beta n})\,\mathrm{d}A. \tag{14.4.5b}$$

Thus the given problem is solved. There does not appear to be any method for expressing the static solution for the prescribed edge displacement conditions in terms of modal functions.

Problems involving mixed boundary conditions can be solved by a combination and slight modification of the techniques discussed above and in the preceding section.

14.5 Some examples of modal analysis

As an example of the use of the preceding analysis, let us consider the response of a cantilevered cylindrical shell of length L, the free end of which has been twisted through an angle (Fig. 14.1) and is suddenly released. The

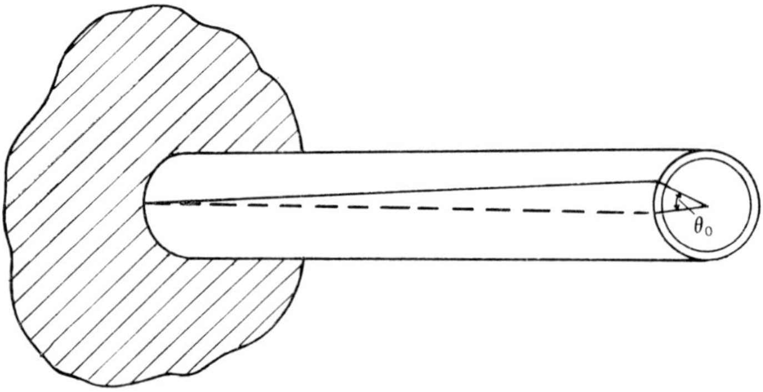

Fig. 14.1. Twisted cantilevered cylindrical shell

free torsional vibrations of a cantilevered cylinder are described by the mode shapes and frequencies.

$$\left(\frac{\rho}{G}\right)^{1/2} \omega_m L = \frac{2m+1}{2}\pi, \tag{14.5.1a}$$

$$\overline{U}_{\theta m} = \sin\frac{2m+1}{2}\pi\xi, \tag{14.5.1b}$$

$$\overline{U}_{xm} = \overline{W}_m = 0, \qquad m = 0, 1, 2, \ldots, \tag{14.5.1c}$$

with

$$\xi = x/L. \tag{14.5.1d}$$

The deflection shape prior to release can readily be found to be given by

$$u_{\theta_0} = \theta_0 L\xi, \tag{14.5.2a}$$

$$u_{x_0} = w_0 = 0, \tag{14.5.2b}$$

and the initial velocities are equal to zero. We then have from Eq. (14.1.5)

$$A_m = \frac{\displaystyle\int_0^1 \theta_0\,\xi\sin\frac{2m+1}{2}\pi\xi\,d\xi}{\displaystyle\int_0^1 \sin^2\frac{2m+1}{2}\pi\xi\,d\xi} = \frac{2(-1)^m\theta_0}{\left(\dfrac{2m+1}{2}\pi\right)^2}, \tag{14.5.3a}$$

$$B_m = 0, \tag{14.5.3b}$$

so that the response is given by

$$\begin{aligned} u_\theta(x,t) &= \frac{8\theta_0 L}{\pi^2}\sum_{m=0}^{\infty}\frac{(-1)^m}{(2m+1)^2}\sin\left(\frac{2m+1}{2}\frac{\pi x}{L}\right)\cos\left[\frac{2m+1}{2}\pi\left(\frac{G}{\rho}\right)^{1/2}\frac{t}{L}\right] \\ &= \frac{4\theta_0 L}{\pi^2}\sum_{m=0}^{\infty}\frac{(-1)^m}{(m+1)^2} \\ &\times\left\{\sin\left[\frac{2m+1}{2}\frac{\pi}{L}\left(x-\left(\frac{G}{\rho}\right)^{1/2}t\right)\right] + \sin\left[\frac{2m+1}{2}\frac{\pi}{L}\left(x-\left(\frac{G}{\rho}\right)^{1/2}t\right)\right]\right\}. \end{aligned} \tag{14.5.4}$$

The response can be expressed in closed form† as

$$u_\theta(x,t) = \frac{\theta_0 L}{2}\left\{H\left[\frac{x+(G/\rho)^{1/2}t}{L}\right] + H\left[\frac{x-(G/\rho)^{1/2}t}{L}\right]\right\}, \tag{14.5.5}$$

† R. V. Churchill, *Modern Operational Mathematics in Engineering*. First Edition, McGraw-Hill Book Co., Inc., pp. 221–222.

with

$$\begin{aligned} H(u) &= u && (0 < u < 1) \\ &= 2 - u && (1 < u < 2) \\ &= -H(-u), \end{aligned}$$

$$H(u + 4) = H(u) \text{ for all } u. \tag{14.5.6}$$

The graphical representation of $H(u)$ is shown in Fig. 14.2 and is seen to be a sawtooth function. Thus the response $u_\theta(x, t)$ is given as the superposition of two sawtooth waves, the first moving toward the root of the cantilever with

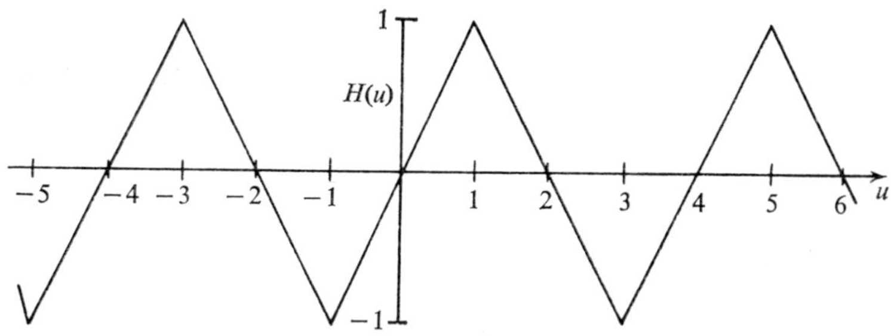

Fig. 14.2. Sawtooth function

wave speed $(G/\rho)^{1/2}$ while the other moves toward the tip with the same wave speed. The variation of the circumferential displacements u_θ with time is shown in Fig. 14.3. In this particular case the response is identical with that given by the theory of elasticity for a completely isotopic cylinder since there are no transverse displacements involved.

As an example of the solution of a problem involving transverse displacements, let us consider the problem of a finite simply supported cylindrical shell which is subjected to a suddenly applied pressure which is uniformly distributed over the region $\frac{1}{2}L - t_2 < x < \frac{1}{2}L + \epsilon_2$, $-\epsilon_1 < R\theta < \epsilon_1$ (Fig. 14.4). For simplicity we assume that the edge conditions at the edges of the shell ($x = 0, L$) are

$$u_\theta = w = M_x = N_x = 0. \tag{14.5.7}$$

We neglect tangential inertia and use Morley's equation as well. Then the pertinent free vibration radial deformation shapes and frequencies are given by Eqs. (13.4.5c) and (13.4.19) as

$$\overline{W}_{mn} = \sin\frac{m\pi x}{L}\cos n\theta, \tag{14.5.8a}$$

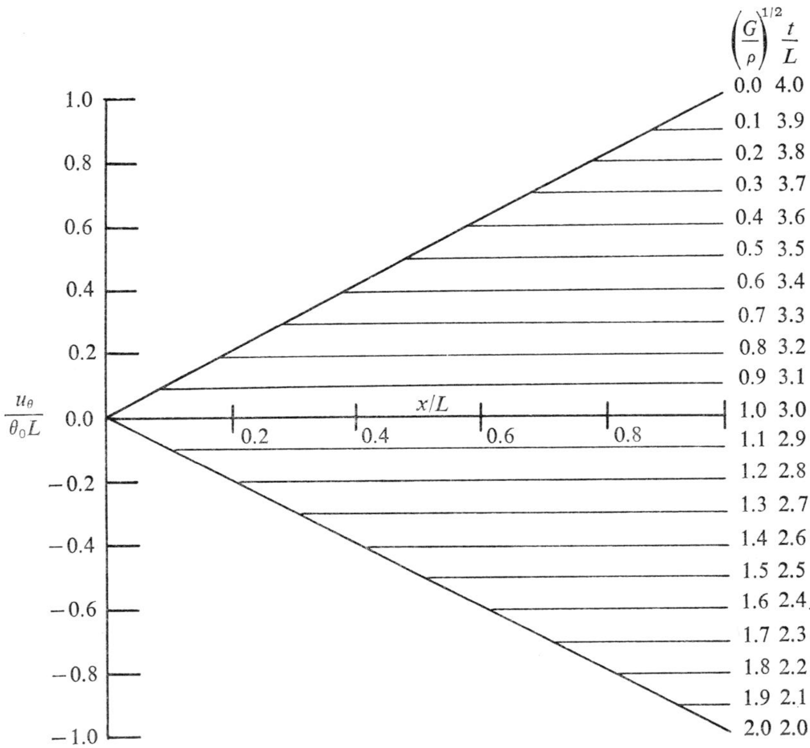

Fig. 14.3. Response of suddenly released twisted cantilevered cylinder

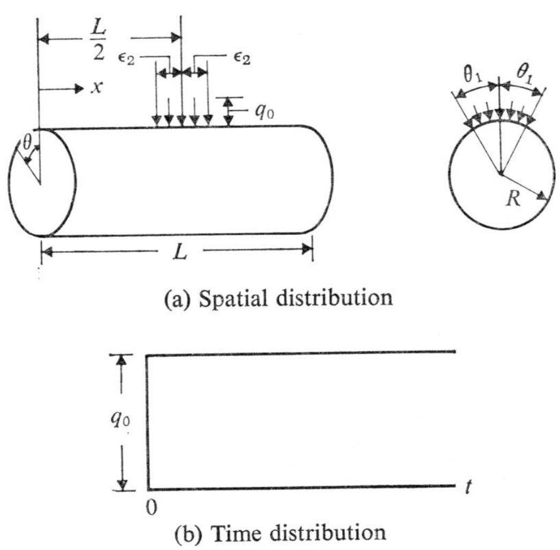

Fig. 14.4. Load distribution

$$\omega_{mn} = \left\{\frac{E/(1-\nu^2)}{\rho R^2}\left[(\lambda_m^2 + n^2 - 1)^2 k + \frac{(1-\nu^2)\lambda_m^4}{(\lambda_m^2 + n^2)^2}\right]\right\}^{1/2}$$
$$(n = 0, 1, 2, 3, \ldots) \quad (m = 1, 2, 3, \ldots) \qquad (14.5.8c)$$

From Eqs. (13.3.6) and (14.2.4b) we then have

$$G_{mn}(t) = G_{mn} = \int_{L/2-\varepsilon_2}^{L/2+\varepsilon_2}\int_{-\theta_1}^{\theta_1} \frac{2q_0}{\pi\rho h L}\sin\frac{m\pi x}{L}\cos n\theta \, \mathrm{d}x\, \mathrm{d}\theta$$

$$= \begin{cases} \dfrac{2q_0\theta_1}{\pi^2\rho h}\dfrac{(-1)^{(m-1)/2}}{m}\sin\dfrac{m\pi\epsilon_2}{L} & (n = 0, m = 1, 2, 3, \ldots) \\ \dfrac{\pi^2\rho h}{4q_0}\dfrac{(-1)^{(m-1)/2}}{mn}\sin\dfrac{m\pi\epsilon_2}{L}\sin n\theta_1 & (n = 1, 2, 3, \ldots;\ m = 1, 2, 3, \ldots) \\ 0 & (n = 0, 1, 2, 3, \ldots;\ m = 2, 4, 6, \ldots) \end{cases} \qquad (14.5.9)$$

so that Eq. (14.2.5) yields

$$F_{mn}(t) = \frac{1}{\omega_{mn}^2}(1 - \cos\omega_{mn}t)G_{mn}. \qquad (14.5.9b)$$

Thus the radial response of the shell is given by

$$w = \sum_{n=0,1,2,\ldots}^{\infty}\ \sum_{m=1,3,5,\ldots}^{\infty} \frac{G_{mn}}{\omega_{mn}^2}(1 - \cos\omega_{mn}t)\sin\frac{m\pi x}{L}\cos n\theta. \qquad (14.5.10)$$

The part of the infinite series which is independent of time is the static deflection due to the applied loading while the remainder of the series represents oscillatory motion about this "mean" position.

The solution as expressed above in series form is quite slowly convergent. In order to evaluate the radial deformation at the point $x = \frac{1}{2}L$, $\theta = 0$ for a particular case, values of n from 0 to 40 and of m from 1 to 41, or 861 terms of the series in all, had to be considered.† The results are shown in Fig. 14.5. Slowness of convergence of modal solutions occurs whenever we consider loading which is restricted to a small portion of the structure. In some cases it is possible to improve the convergence by mathematical transformations of the series‡ In cases involving impact of the shell it may be

† J. C. Yao, 'An Analytical and Experimental Study of Cylindrical Shells under Localised Impact Loads,' *Aeronaut. Q.*, vol. 17, Pt. 1, February 1966, pp. 72–82.

‡ D. Feit and M. C. Junger, 'High-frequency Response of an Elastic Spherical Shell,' *J. Appl. Mech.*, vol. 36, no. 4, December 1969, pp. 859–864.

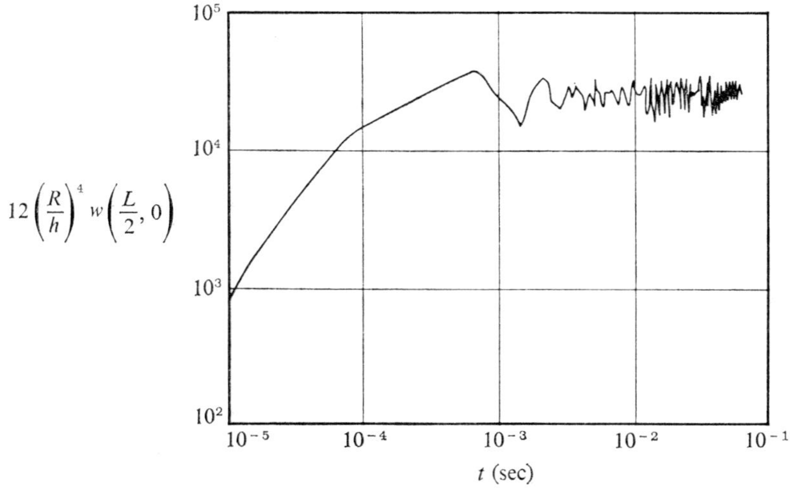

Fig. 14.5. Response of cylindrical shell to step-function load ($\epsilon_2/R = \theta_1 = \frac{1}{16}$, $L/R = 2$, $R/h = 100$)

advantageous to use the method of characteristics† to determine the response of the entire shell or to use asymptotic expansion‡ techniques to determine the behavior of the shell near the head of the induced stress wave. It should be pointed out, however, that such methods can be used only when the partial differential equations of motion are hyperbolic and that this is true for the theory of shells only if shearing deformations and/or rotatory inertia is included or if bending deformations are neglected (membrane theory).

14.6 Other methods of solution of shell response problems

Although the modal expansion technique always yields the solution of specific problems in principle, it is not necessarily the most convenient technique, especially if a great number of mode shapes and frequencies are required for convergence. In many applications it is better to solve the equations of motion approximately. The partial differential equations of motion are generally transformed into a set of coupled second-order ordinary differential equations for the displacements at discrete points. The mechanism for

† See, for example, W. R. Spillers, 'Wave Propagation in a Thin Cylindrical Shell,' *J. Appl. Mech.*, vol. 32, no. 2, June 1965, pp. 346–350; P. C. Chou, 'Analysis of Axisymmetrical Motions of Cylindrical Shells by the Method of Characteristics,' *AIAA J.*, vol. 6, no. 8, August 1968, pp. 1492–1497.

‡ H. M. Berkowitz, 'Longitudinal Impact of a Semi-infinite Elastic Cylindrical Shell,' *J. Appl. Mech.*, vol. 30, no. 3, September 1963, pp. 347–354.

effecting this transformation may be the method of finite differences, if feasible, or the finite element method. The resulting equations may then be integrated numerically with respect to time.

In one of the available numerical integration schemes, the Houbolt (implicit) Method of Numerical Integration,† the acceleration at time t is expressed in finite difference form in terms of the displacement at time t and at three previous times (backward differences). If the displacements at the three previous times are known, the displacements at time t can be obtained by solving all of the equations simultaneously as was done for the statically loaded structure. The new displacements are then used in conjunction with displacements at two previous times to find the displacements at the end of the next time interval and so on.

An explicit method of numerical integration may also be used. The acceleration at time t is expressed in terms of the displacements at the end of the next time interval, at time t, and at the previous time (central differences). If the displacements at time t and the previous time are known, the displacements at the end of the next time interval can be calculated. If rotatory inertia is neglected the equations of motion are then explicit equations for the required displacements. If rotatory inertia is included, however, the method is no longer explicit and the equations must again be solved simultaneously, although the coupling is not as severe as in the previous method.

Care must be taken, as in most numerical procedures, to use sufficiently small space and time increments. Investigations‡ indicate that the explicit method tends to be more efficient when the response varies rapidly. On the other hand, the Houbolt method is generally more flexible since it is always numerically stable.

Supplementary references

[1] Sagartz, M. J., and Forrestal, M. J., 'Transient Stresses at a Clamped Support of a Circular Cylindrical Shell,' *J. Appl. Mech.*, vol. 36, no. 2, June 1969, pp. 367–369.

[2] Longcope, D. B., and Forrestal, M. J., 'Bending Stress at a Clamped Support of an Impulsively Loaded Conical Shell,' *AIAA J.*, vol. 8, no. 9, September 1970, pp. 1715–1719.

[3] Hwang, C., and Pi, W. S., 'Random Acoustic Response of a Cylindrical Shell,' *AIAA J.*, vol. 7, no. 12, December 1969, pp. 2204–2210.

† J. C. Houbolt, 'A Recurrence Matrix Solution for the Dynamic Response of Elastic Aircraft,' *J. Aero. Sci.*, vol. 17, no. 9, September 1950, pp. 540–550.

‡ D. E. Johnson and R. Greif, 'Dynamic Response of a Cylindrical Shell: Two Numerical Methods,' *AIAA J.*, vol. 4, no. 3, March 1966, pp. 486–494.

[4] Schiffner, K., and Steele, C. R., 'Cylindrical Shell with an Axisymmetric Moving Load,' vol. 9, no. 1, January 1971, pp. 37–47.
[5] Longcope, D. B., and Forrestal, M. J., 'Transient Axisymmetric Bending Stress in an Infinite Shell at an Elastic Ring Stiffener,' *AIAA J.*, vol. 9, no. 2, February 1971, pp. 339–340.
[6] Fisher, H. D., 'Bending Stress in an Impulsively Loaded Cylindrical Shell of Exponentially Varying Thickness,' *AIAA J.*, vol. 10, no, 12, December 1972, pp. 1699–1700.
[7] Smith, T. A., 'Numerical Analysis of Rotationally Symmetric Shells under Transient Loadings,' *AIAA J.*, vol. 9, no. 4, April 1971, pp. 637–643.
[8] Kana, D. D., 'Response of a Cylindrical Shell to Random Acoustic Excitation,' *AIAA J.*, vol. 9, no. 3, March 1971, pp. 425–431.

Author index

S

T

U

V

W

Y

Z

Subject index

Subject index

D